The Changing Energy Mix

The Changing Energy Mix

A Systematic Comparison of Renewable and Nonrenewable Energy

Paul F. Meier

OXFORD
UNIVERSITY PRESS

OXFORD
UNIVERSITY PRESS

Oxford University Press is a department of the University of Oxford. It furthers the University's objective of excellence in research, scholarship, and education by publishing worldwide. Oxford is a registered trade mark of Oxford University Press in the UK and certain other countries.

Published in the United States of America by Oxford University Press
198 Madison Avenue, New York, NY 10016, United States of America.

Library of Congress Cataloging-in-Publication Data
Names: Meier, Paul F. author.
Title: The changing energy mix : a systematic comparison of renewable and nonrenewable energy / Paul F. Meier.
Description: New York, NY : Oxford University Press, [2020] |
Includes bibliographical references and index.
Identifiers: LCCN 2020005853 (print) | LCCN 2020005854 (ebook) |
ISBN 9780190098391 (hardback) | ISBN 9780190098407 (updf) |
ISBN 9780190098414 (epub) | ISBN 9780190098421 (online)
Subjects: LCSH: Power resources—United States. | Renewable energy sources—United States.
Classification: LCC HD9502.U52 M43 2020 (print) | LCC HD9502.U52 (ebook)
| DDC 333.793/20973—dc23
LC record available at https://lccn.loc.gov/2020005853
LC ebook record available at https://lccn.loc.gov/2020005854

1 3 5 7 9 8 6 4 2

Printed by Integrated Books International, United States of America

Contents

Introduction

I.A. Foreword

Energy! We use energy every day but it is easy to presume its availability and not appreciate what a wonderful thing it is. Electricity and natural gas are seamlessly delivered to our homes, and with a short drive we can refuel our vehicles with gasoline or diesel. Yet, if we could suddenly convey a person from the early 1800s, when wood and whale oil were common energy sources, into our time period, they would be truly amazed, just as we would be if we could glimpse into the next century.

The word "energy" is derived from the Greek word *ἐνέργεια* (*energeia*), and appeared for the first time in the work of Aristotle titled "Nicomachean Ethics,"[1] in the 4th century BC. Other than human attributes, such as vigor or vitality, *Webster's II New Riverside University Dictionary*[2] provides a definition of energy appropriate for this book as "usable heat or electrical power." That is, this definition of energy is the one that describes the use of energy to provide us with transportation fuel, residential heating and cooling, electrical energy, and commercial work. In this sense, energy is also commonly defined as the "ability to do work." Whatever we define as civilization, the use of energy to "do work" is very important and provides for the many conveniences that have driven the advancement of our current civilization.

Ever since early civilizations of mankind used wood to provide heating and cooking, energy has been an important part of our lives. And, as civilization has continued to advance, the percentage of people devoted to providing food decreased, allowing mankind to devote their time to other things, such as transportation, commerce, communication, construction, government, banking, arts and literature, science and engineering, and religion, just to name a few. This has led to mankind's greater dependence on energy, and we now live in a world where it is possible for people to routinely fly to foreign countries and live long distances from their place of work, and we can enjoy things like fresh lobster or avocados, even when they are harvested great distances from our homes. Without energy to provide air conditioning, would cities like Phoenix and Houston have the large populations they have today?

No doubt, the advancement of our civilization has resulted in many and extensive ways to harness and use energy, and these advances continue to shape our everyday life. Consider that if you lived in the 18th century it was commonplace to use whale

The Changing Energy Mix. Paul F. Meier, Oxford University Press (2020). © Oxford University Press.
DOI: 10.1093/oso/9780190098391.001.0001.

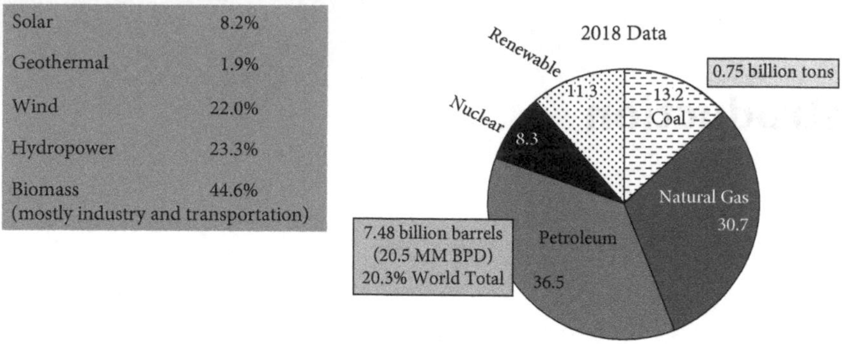

Figure I.1 Total Energy Consumption Was 17,476 MM boe/year[4]

oil to light street lamps and public buildings while coal was the primary means to generate steam for engines. And just as the drive for energy nearly brought about extinction of the whale, the current drive for energy will eventually consume all the coal, natural gas, and petroleum crude oil on our planet.

The use and availability of energy, currently an important issue, will likely become an acute issue as the world population increases. In 1700, there were only 0.6 billion people in the world, increasing to 0.9 billion in 1800, 1.65 billion in 1900, and almost doubling to 3.0 billion by 1960.[3] It doubled again to 6.1 billion by 2000 and stood at 7.4 billion in 2015; the world population is projected to be 8.5 billion by 2030. Considering the growth in the world population, there will be even greater pressure on the development of convenient and affordable energy in the future.

Where will this energy come from? Traditionally, the world has relied on fossil fuels including coal, petroleum, and natural gas. For various reasons, this energy mix is changing. Consider Figure I.1, which examines 2018 data for the United States taken from the Energy Information Administration (EIA).[4] The figure shows a percent breakdown of energy used from all sectors, including transportation, electricity, industry, and residential. It also shows total energy consumption in units of million barrels of oil equivalent per year, or MM boe/year. That is, the total energy consumption is represented in terms of the energy in an equivalent barrel of oil, even though not all the energy was used for transportation fuel. As the figure illustrates, 80.4% of our energy came from petroleum, natural gas, and coal with 11.3% renewables. However the next figure, Figure I.2, shows that 18 years earlier in 2000, those fossil fuels made up 85.9% of our energy and renewables were only 6.2%. Total energy consumption was relatively constant, increasing about 3% over this 18-year period.

Looking at a percent breakdown of renewables, hydroelectric and biomass made up 95% of all renewables in 2000. Although biomass can be used to generate electricity, this high use of biomass was primarily for industry and transportation fuels. However, in 2018, the amount of electricity generated from wind was almost equal to hydroelectric, and solar has moved from only 1% of renewables to over 8%. Also

Solar	1.0%
Geothermal	2.7%
Wind	0.9%
Hydropower	46.1%
Biomass (mostly industry and transportation)	49.3%

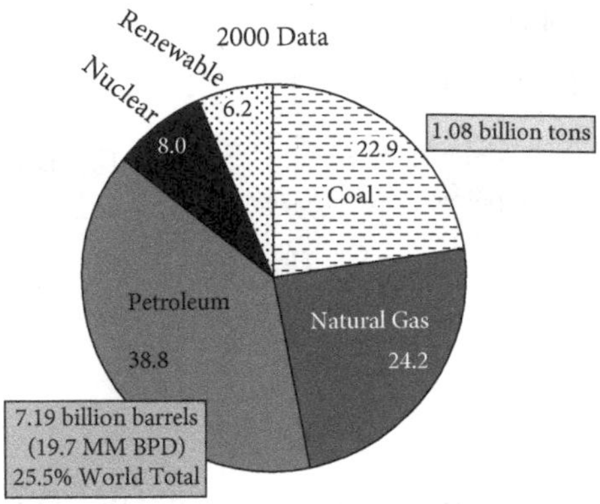

Figure I.2 Total Energy Consumption Was 16,999 MM boe/year[4]

Solar	8.4%
Geothermal	0.0%
Wind	18.2%
Hydropower	71.0%
Biomass	2.4%

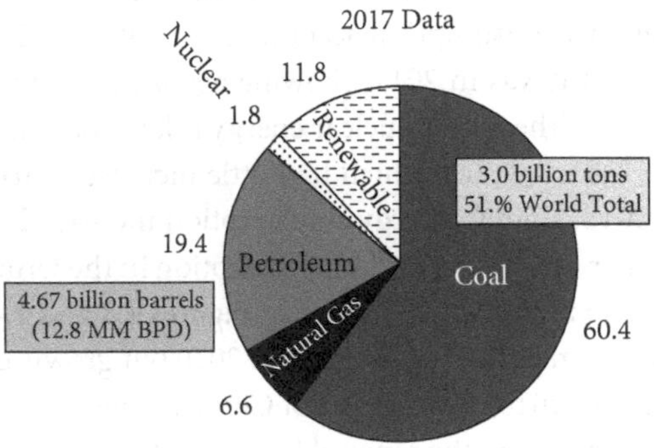

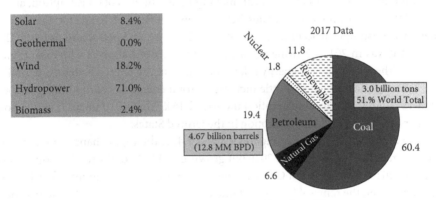

Figure I.3 Total Energy Consumption Was 21,413 MM boe/year[5,6]

significant is that coal use dropped from about 23% of the total energy mix in 2000 to only 13% in 2018, with natural gas mostly taking its place.

On the other side of the world is China, the largest country by population and one that has been undergoing rapid growth. From 2008 to 2014, I had a consulting job in China with a front-row seat to the rapid expansion of automobile use in Beijing. During this time period, the right to car ownership in Beijing was obtained by permit and, in 2010, more than 700,000 new cars were added to Beijing roads (~2,000/day), bringing the total to 4.7 million at year end! To help slow this down, the government limited 2011 permits to 240,000 new cars (~660/day). This staggering increase in automobile ownership shows you how fast the Chinese economy was expanding at the time, with a concomitant increase in energy consumption.

Figure I.3 shows data for China in 2017, assembled from two different sources.[5,6] As can be seen, they are highly dependent on coal, which makes up about 60% of their total energy consumption and accounts for about half of the entire world's coal

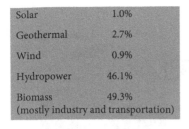

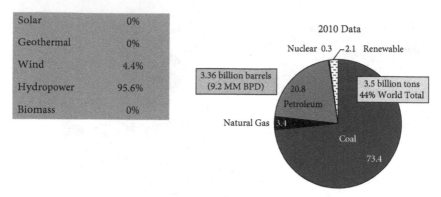

Solar	0%
Geothermal	0%
Wind	4.4%
Hydropower	95.6%
Biomass	0%

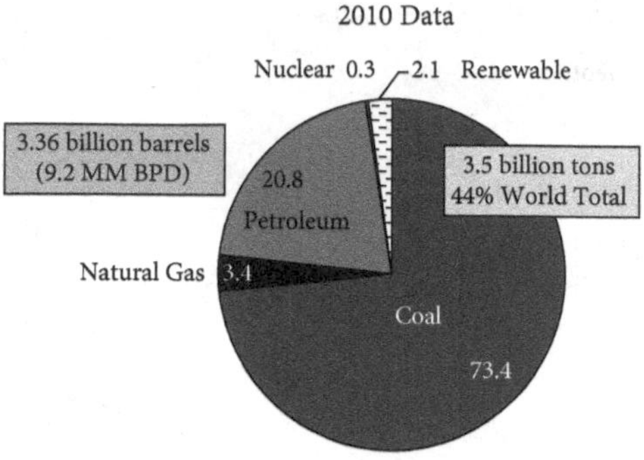

Figure I.4 Total Energy Consumption Was 14,655 MM boe/year[7]

consumption. Petroleum follows next, making up about 20% of consumption, about 62% of that used by the United States. Yet, as shown in Figure I.4, data for 2010[7] show that coal consumption accounted for 73% of all energy, and petroleum use was 72% of what it was in 2017, showing again the rapid transformation of China's automobile use. The use of nuclear energy is low compared to the United States. And, unlike the United States, which had little increase in total energy consumption from 2000 to 2018, China's energy consumption increased 46% from 2010 to 2017, and is now greater than total energy consumption in the United States.

Also surprising, comparing Figures I.3 and I.4, is the rapid change in renewable energy, making up only 2.1% in 2010 but growing to 11.8% of the total energy mix in 2017. In 2010, almost 96% of China's renewable energy came from hydroelectric. By 2017, however, the renewable mix contained significant amounts of electricity generation from wind and solar. And, although the percent of hydroelectric of all renewables decreased, there was still significant growth in hydroelectricity, most notably the addition of the Three Gorges Dam, one of the world's largest hydroelectric power plants, which generated 98.8 TWh (terawatts, or one trillion watts) of electricity in 2014. It is clear from these two figures that China is working hard to increase their use of renewable energy types.

So what is driving energy demand and the changing mix of energy to meet that demand? The increase in energy demand is relatively easy to understand. As mentioned, the world population is rapidly increasing. Figure I.5 shows 2015 data for 36 countries taken from the World Bank Organization.[8] The chart shows barrels of oil equivalent (boe) consumed per capita versus gross domestic product (GDP) per capita, in US dollars. Barrels of oil equivalent represents the amount of energy consumed, per person in each country, recognizing that this energy consumed includes heating, electricity, and other uses in addition to the traditional use of oil as a transportation fuel.

Gross domestic product per capita can be defined as the market value of all goods and services produced within a country in a given year. Alternatively, it can be

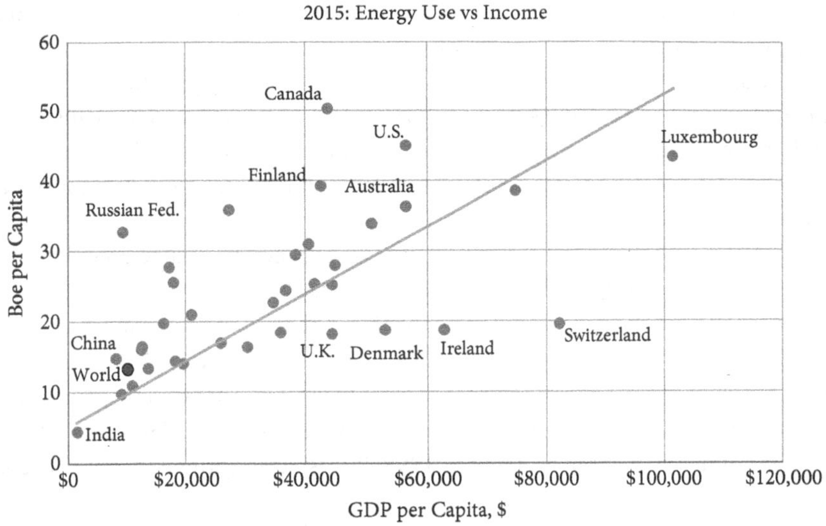

Figure I.5 Energy Use by Country versus GDP per Capita[8]

calculated as the sum of incomes for people living in a country for one year. Although there are some complications beyond the scope of this discussion, GDP per capita can be considered a good comparison of average salaries for each country.

Consider, for example, the United States with an average income of about $56,500 per year. The average energy consumption was 45 boe, meaning each person in the United States consumed an amount of energy equivalent to 45 barrels of oil. A statistical line drawn on this chart shows a general increasing trend. This is not surprising, as increasing income also means that people have more disposable income beyond basic living needs, such as housing, food, and transportation, and this money can be used for a better life style: bigger homes, cars and boats, home heating and cooling, and hopping on a plane to take a vacation. Although people in North America and western Europe may take these things for granted, they are not so common in developing countries. Thus, countries such as China and India, with average salaries less than $10,000 per year consume much less energy per person than the United States, Canada, and western Europe, with average salaries greater than $40,000 per year. The implication is that as developing countries increase their GDP, energy consumption will increase, and this is significant because China and India, both with populations exceeding one billion, are fast-growing economies.

The dark circle on the chart is the world average, a per capita income of $10,164 and energy consumption of 13.2 boe. The figure also shows that some countries lie above and below this line. Although a detailed country-by-country analysis has not been made, those countries above the line are, generally speaking, larger countries and therefore will have greater need of transportation fuels. In addition, in the case of the United States, Russia, and Canada, there is a greater need for energy used for heating. Countries below the line are smaller countries, suggesting more efficient use of energy, such as better use of public transportation.

According to a British Petroleum (BP) review of world energy, the consumption of all energy types increased by 2% from 2016 to 2017.[5] For renewables, while hydroelectric remained about the same, other renewables increased by 17% from 2016 to 2017! So the energy mix is changing, but the reason for this is less straightforward, because it is a result of many things.

One reason is that most of the world feels an urgency to reduce greenhouse gas (GHG) emissions. The Kyoto Protocol, adopted in late 1997,[9] reached a commitment to reduce GHG emissions, based on the belief that that global warming is occurring and that it is likely caused by GHGs, primarily carbon dioxide (CO_2). Originally, 37 countries and the European Union agreed to reduce GHG emission by 5% relative to their level in 1990, later reducing them by 18% from 2013 to 2020. However, although the United States signed the protocol in 1998, it was never ratified by the US Senate. As well, Canada withdrew in 2012.

In Europe, however, countries continue to develop aggressive goals for renewable energy. In 2009, the European Union (EU) Renewable Energy Directive set a binding target of 20% of energy consumption to come from renewables by 2020.[10] And Sweden had a target of 49% by 2020! There is also a target that 10% of transportation fuels will come from renewable sources by 2020. In the case of Denmark, they are building a fleet of fuel cell vehicles (FCVs) powered by hydrogen, which will be produced from water electrolysis. And in 2014, the EU reached a new agreement targeting 27% of energy consumption from renewables by 2030.

One factor driving the development of renewable fuels is the desire to diversify energy types, and therefore reduce imported energy. Also, the use of renewable energy can, in principle, lead to stable energy prices since there is no cost for the fuel and it is inexhaustible. Compare this to fossil fuels, for which the cost of the fuel can be quite unstable. For example, crude oil had a price of $61.73/bbl in 2009, which increased to $97.98/bbl in 2013 and then decreased to $48.66 in 2015.[11]

Another reason for the changing energy mix is that the cost for some renewable energy types has been decreasing. Stealing some information discussed later in the chapters for wind (Chapter 6) and solar (Chapter 7), the cost to build these renewable electric plants has decreased significantly. For an onshore wind farm, the capital cost in 2019 is 73% of what it was in 2015. For a solar photovoltaic (PV) electric plant, the capital cost reduction for the same years is 43%. These reductions have significantly closed the gap with traditional, fossil fuel, energy types in electricity generation.

There are also obstacles and disadvantages for moving toward renewable energy. Unlike the traditional energy types—including petroleum, coal, natural gas, and nuclear energy—for which the fuel source can be continuously fed to generate usable energy, renewable energy types are affected by weather. Obviously, wind turbines need wind, solar plants need sunshine, and hydroelectric plants need a continuous supply of flowing water. Also, renewable energy plants are typically smaller than those using traditional energy sources. And, for electricity generation, renewable energy will need access to the electric grid to transport the electricity to the consumer. Unlike coal, natural gas, and nuclear power plants, which can be built next to the existing electric grid, renewable electric plants are built where the conditions are favorable, be it wind, solar, or hydro, but these sites may be long distances from the existing grid. That means these plants will need new transmission lines to connect to the electric grid.

Finally, while it may not be the most important factor in Europe, things in the United States are normally driven by cost. Certainly, some people in the United States are willing to spend more for energy if it means developing domestic resources, thus avoiding the importation of energy from countries that are not necessarily our allies, or if it means developing renewable energy. But generally, cost is what drives the US economy. Because of the reduction in the cost to build, renewables, especially wind and solar energy, have made great advances in the United States since the turn of the century.

This book analyzes different renewable and nonrenewable energy types in a way that allows us to compare their advantages and disadvantages, and thereby better understand the changing energy mix in the United States and the world. This is done using a systematic approach, with twelve common criteria used for each energy type. These criteria include how much is available, the technology used to exploit it, how much of this energy is converted to useful work, the level of maturity and experience with a given energy type, and environmental issues, including the amount of CO_2 production. As much as possible, the criteria used to compare the positive and negative aspects of each energy type are done free of political and emotional bias. Although not a complete list, this book covers twelve sources of energy: petroleum crude oil, natural gas, coal, nuclear, hydroelectric, wind, solar, ethanol, biomass, geothermal, hydrogen, and the Fischer-Tropsch synthesis. Here ethanol and hydrogen are used as transportation fuels while biomass is for electricity generation. The Fischer-Tropsch synthesis is a technology that can be used to convert coal, natural gas, or biomass to transportation fuels.

This book focuses on the use of energy for transportation and electricity generation, recognizing that energy is also used for heating and cooling as well as various commercial processes.

Figure I.6, also taken from the EIA Monthly Energy Review,[4] shows 2017 data for how energy is used. For example, nuclear energy is used exclusively for electric power generation while natural gas is used in transportation, industrial, residential and commercial heating, and electric power. Coal is used primarily for electric power generation, as coal-fired heating is not as common as in years past. This same figure gives a snapshot of how much energy is used for a certain sector. For example, transportation and electric power generation consume about 65% of all the energy used. Some of the energy types traditionally used to generate electricity will see growth in their use for transportation as the technology for electric vehicles improves. Also, there are two technologies that allow coal, natural gas, and biomass

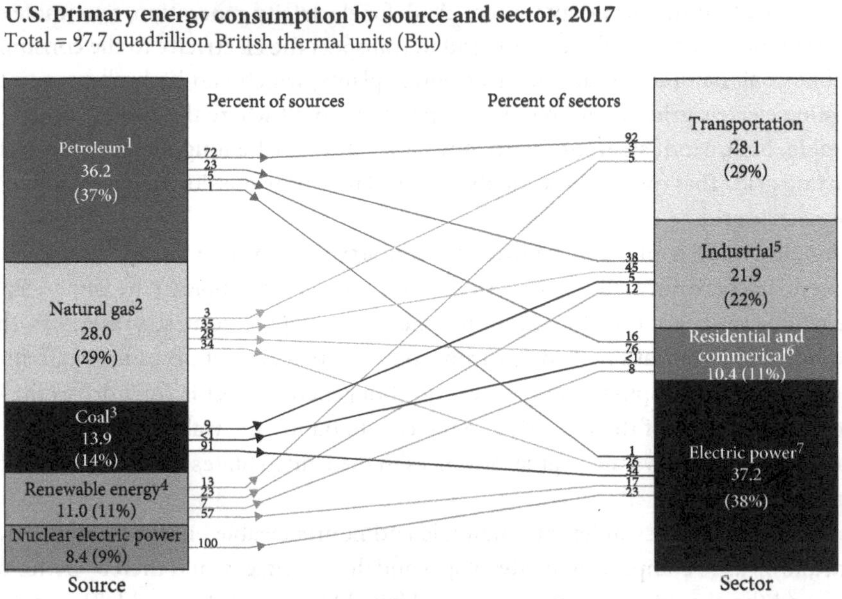

Figure I.6 US Primary Energy Consumption by Source and Sector, 2017 (Source: EIA[4])

to be transformed into transportation fuels. These are the hydrogen FCV and the Fischer-Tropsch synthesis, which are discussed in Chapters 11 and 12, respectively.

I.B. Common Criteria

A broad overview for the twelve different criteria used to evaluate each energy type is given in this section. As the previous section indicated, energy can be used for transportation, industrial, residential and commercial heating, and electric power generation. In spite of their different applications, it is important to be able to compare different types of energy on a consistent basis. In addition, it is likely that in the future, energy types will have expanding use in other areas. For example, if electric automobiles become common, then those energy types that are used to generate electricity will have an increasing role in the area of transportation.

I.B.1. Proven Reserves

For nonrenewable energy types, this section quantifies proven reserves, production rates, and consumption rates, and calculates the number of years before this energy type will be exhausted.

By definition, renewable energy is inexhaustible, so there is no time limit imposed on these types of energy. For example, wind and solar energy are inexhaustible energy types. Although more difficult to quantify than those for fossil fuels and nuclear energy, proven reserves are estimated for the different renewable energy types.

Given earlier discussions in the Foreword about increasing worldwide energy consumption, the numbers of years calculated using current proven reserves and current consumption rates may be a maximum value for nonrenewable energy, unless more sources of a given energy type are found or expected increases in energy consumption are wrong.

I.B.2. Overview of Technology

This section explains how each energy type is used. This includes a discussion of how the energy is found or harnessed and how the energy is refined or captured as a usable energy type.

I.B.3. Capital Cost, Operating Cost, Well-to-Wheels Levelized Cost of Electricity, and Well-to-Wheels Levelized Cost of Fuel

Cost is a major factor in the commercial introduction and success of any energy type, and in this book the "levelized cost of energy" is used to compare different energy technologies. The equations used to calculate the "levelized cost of energy" are meant to determine a price at which electricity or transportation fuel needs to be sold to break even. The calculations include the total cost to build and operate an energy plant, over an assumed lifetime, and also include the cost to deliver that energy to the consumer. In this way we can compare the cost of different technologies, such as electricity from coal versus solar, even though the plants have different lifetimes, capital costs, and plant capacities. This levelized cost includes the capital cost to build the plant, operating expenses, the cost to borrow money, the depreciation of the plant, the cost of fuel, and the cost to deliver the product to the consumer. The equations used, the details, and the specific calculations for each energy type are presented in Appendix A.

The levelized cost of energy is determined using what is generally called a "Well-to-Wheels" evaluation, which considers the cost for the different steps to take an energy resource out of the "Well" and deliver it to the "Wheels" of your vehicle. In the case of electricity, we could also call this "Well-to-Outlet" or "Mine-to-Outlet." Naturally, for renewable fuels such as wind, solar, and hydroelectric, there is no "Well," but this same terminology is used regardless.

For example, to make gasoline and diesel from petroleum crude oil, in addition to the construction and operation of the refinery, other steps in the "Well-to-Wheels" cost include the cost of oil drilling and production, transportation of the crude oil to the refinery, and transportation of the gasoline and diesel to the fueling station. For a natural gas electric plant, the "Well-to-Wheels" cost must include the natural gas drilling and production costs, natural gas processing, transportation to the electric plant (e.g., by pipeline), construction and operation of the power plant, and high-voltage transportation through the electric grid before coming to your home.

In the case of electric plants, the cost is defined as the levelized cost of electricity (LCOE) and, in the case of plants that produce transportation fuels, the cost is defined as the levelized cost of fuel (LCOF).

These equations need to capture those economic factors that control the break-even cost needed to recover initial and ongoing expenses for a given energy technology. One important factor is the capital cost, the initial cost to build the plant. For LCOE, the plant costs are generally reported in units of $/kilowatts, or $/kW and, for LCOF, the units are $/gallon of gasoline equivalent, or $/gge. The "gge" is the amount of fuel needed to equal the energy content of 1 gallon of gasoline.

To finance the capital cost to build the energy plant, companies may borrow the money or use existing capital. Even if the company has existing capital to build the plant, that money could, alternatively, have been invested at current interest rates to earn interest rather than investing in the energy plant. Therefore a term called "capital recovery factor" is included in the equations for the levelized cost of energy. The capital recovery factor includes the current interest rate and the useful life of the plant and is, essentially a method to see when that initial investment is recovered. Depreciation, the reduction in value for an asset over time, is also a factor included in the levelized cost. Table I.1 shows the useful plant life and depreciation time period allowed by the Internal Revenue Service (IRS) for each technology. As well, the levelized cost equations include a combined federal and state tax, set to 38%.

While capital cost is defined as the initial, one-time, cost to build a plant, there are also ongoing operating costs for the plant. Two types are included in the calculations, including fixed operation and maintenance (Fixed O&M) and variable operation and maintenance (Variable O&M). Fixed O&M are costs that occur regardless of the product flow from the plant. For example, if a coal electrical plant only operates at 50% of design, employee salaries, utilities, and insurance must still be paid. Variable O&M, on the other hand, is directly proportional to the plant output. As output increases, the cost of fuel, catalyst, and chemicals will also increase. For nonrenewable energy types, the variable O&M is highly dependent on the cost of fuel. If the price of oil, natural gas, or coal increases, the levelized cost must increase to recover this operating expense. To compare the different energy types on an equivalent basis, the best available fuel cost data from 2018 were used in the calculations.

The cost of product marketing and transportation is also included. Transportation is the cost to deliver the product to the fueling station, in the case of a transportation fuel, or the cost to deliver electricity to the customer.

Table I.1 Book Lives for Different Plants and the Number of Years Allowed by the IRS to Depreciate a Given Energy Type Plant.

Electric Plants	Book Life, years	US IRS Depreciation Schedule, years
Advanced Pulverized Coal	40	20
Advanced Pulverized Coal with CCS	40	20
Integrated Gasification Combined Cycle Coal	40	20
Integrated Gasification Combined Cycle Coal with CCS	40	20
Natural Gas Conventional Combined Cycle	30	20
Natural Gas Advanced Combined Cycle	30	20
Natural Gas Advanced Combined Cycle with CCS	30	20
Natural Gas Combustion Turbine	30	15
Advanced Nuclear	40	15
Wind—onshore	20	7
Wind—offshore	20	7
Solar Thermal	20	5
Solar Photo Voltaic	20	5
Geothermal—Binary	20	5
Biomass—BFB	40	7
Hydroelectric	30	20
Transportation Fuel Plants	**Book Life, years**	**US IRS Depreciation Schedule, years**
Refining	20	10
Fischer-Tropsch Coal to Liquids (CTL)	20	7
Fischer-Tropsch Gas to Liquids (GTL)	15	7
Fischer-Tropsch Biofuels to Liquids (BTL)	20	7
Dry Mill Ethanol	15	7
Cellulosic Ethanol	15	7
Hydrogen	20	7

a. Note: CCS stands for carbon capture and sequestration.

Another factor included in the levelized cost equations is the capacity factor, the yearly average of plant use as a fraction of capacity. For example, if a natural gas electric plant is designed to generate 1 gigawatt (GW), it would generate 8,760 gigawatt-hours (GWh) per year, as there are 8,760 hours in a year. However, plants rarely operate at 100% of design capacity. A typical natural gas plant operates at a capacity factor of 87%, so it would only generate 87% of the design rate, or 7,621 GWh. On the other hand, a typical onshore wind farm only operates at 34% of capacity, so a 1 GW wind farm would generate 2,978 GWh.

A final factor, used only in the case of electricity generation, is the heat rate, given in units of BTU/kWh. Here, BTU stands for British thermal unit, the amount of heat needed to raise the temperature of 1 pound of water by 1 degree Fahrenheit. The heat rate is the amount of energy used by the electric plant to generate one kilowatt-hour (kWh) of electricity. As such, it measures the efficiency of the electric plant to convert fuel into electricity, so the lower the heat rate the higher the efficiency.

Capital costs, fixed and variable operating costs, heat rates, and capacity factors were taken from various authors for the different energy types. While these authors may also report values for LCOE or LCOF, these values will be different than the LCOE and LCOF reported in this book because they may have used different calculation strategies as well as different interest and tax rates. For the calculations in this book, these were the same for all energy types in order to compare them on a common basis.

I.B.4. Cost of Energy

For nonrenewable energy types, fuel is needed to generate either transportation fuels (LCOF) or electricity (LCOE). Therefore, the cost of energy delivered to the commercial plant is an important economic factor. This cost is given in units of both mass ($/lb) and energy ($/MM BTU). Depending on the technology, the different energy costs include natural gas, bituminous coal, biomass wood (for electricity), crude oil, the cost of corn and corn stover (for ethanol production), and uranium. The technologies for wind, hydroelectric, geothermal, and solar do not have a feed energy cost.

Unfortunately, the cost of the fuel for different energy technologies is a moving target from year to year, and there has been much volatility in these costs over the last decade. For example, the cost of West Texas Intermediate (WTI), a typical benchmark crude oil, has shown a lot of volatility since 2005. Looking at 2-year intervals, the average annual price of WTI in 2005, 2007, 2009, 2011, 2013, 2015, and 2017 were $56.44, $72.26, $61.73, $94.88, $97.98, $48.66, and $50.88.[11] For 2018, the average price for WTI was $64.94, comparable to what it was in 2009 but far less than 2011 and 2013 prices. To compare the different energy types on an equivalent basis, the best available fuel cost data for 2018 were used. However, because of the volatility in fuel costs, sensitivity studies were also made to quantify the impact of fuel cost on LCOE and LCOF.

I.B.5. Capacity Factor

Capacity factor refers to the percent of the rated capacity used per year. Reasons why this number is less than 100% include:

a. Equipment failure or routine maintenance
b. Electricity not needed or price is too low
c. Variation in energy input

For example, refineries, coal plants, and nuclear plants have routine maintenance, equipment that is scheduled to be cleaned, inspected, and sometimes upgraded. That same equipment can also fail unexpectedly, which results in the need for maintenance. In some cases, equipment can be repaired without interruption in energy production but it is a rare refinery or electricity plant that operates 100% of any year.

For "b," plants that generate electricity may operate below capacity simply because demand is low or there is no economic incentive to operate at 100% capacity. If an electric plant is built with a capacity to meet peak demands in the summer, when air conditioning is in high use, that same capacity may not be needed in the winter.

For some renewable energy types, there is a variation in energy input. Clearly, for solar energy, there is no power generation at night. Likewise, the generation of electricity from wind energy varies greatly with wind speed. Similarly, hydroelectric energy may vary if the water flow through a dam is not constant. For some dams, such as Niagara Falls, water flow is controlled to create a more aesthetic waterfall for tourism during the day, but more power generation at night. For other dams, the variation in water flow is affected by seasonal climate changes.

I.B.6. Efficiency: Fraction of Energy Converted to Work

Energy efficiency is different from capacity. While capacity is affected by energy input, efficiency is affected by the ability of the process to convert energy to useful work. Unfortunately, energy is wasted to varying degrees for different types of energy generation and use.

To understand this a little better, a brief discussion of the first and second laws of thermodynamics is needed. For more detailed discussion, please refer to any thermodynamic textbook on the subject.[12,13,14]

The first law of thermodynamics is that while energy can assume many forms, the total quantity of energy in a "closed" system is constant. A "closed" system is one in which no external source of energy is added or subtracted. The second law of thermodynamics states that no apparatus or machine can operate in such a manner that it

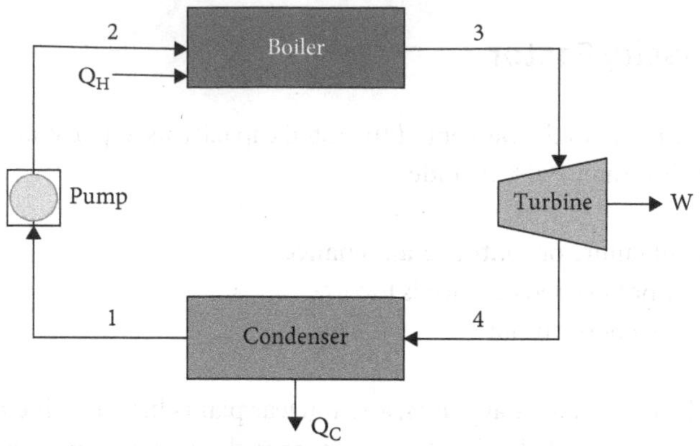

Figure I.7 Heat Engine Four Step Cycle

only converts heat into work. That is, some of the energy is not converted into work, or useful energy, and is wasted. And there is no apparatus or machine that can create energy. If this were possible, you could operate the engine without input of any energy type, a so-called perpetual motion machine.

The heat engine is reviewed here as a practical example.[15] The heat engine is a machine that produces work from heat in a cyclical process, such as a steam power plant that uses coal, natural gas, or nuclear energy. The process consists of four cyclical steps, including (1–2) pumping water to the boiler, (2–3) using energy to boil water into steam, (3–4) transferring steam to a turbine to generate electricity, (4–1) and taking steam from the turbine and condensing to water at a lower temperature. This is shown schematically in Figure I.7.

A French scientist, Sadi Carnot,[15] showed that that maximum efficiency of an engine can be defined by the following equation.

$$\eta = \frac{W}{Q_H} = 1 - \frac{T_C}{T_H} \tag{1}$$

Here:

W = net work taken from steam (for example, to make electricity)
Q_H = heat added to the system
T_C = temperature of the cold liquid (water)
T_H = temperature of the hot gas (steam)

Later, in the chapters for energy types that use a heat engine to produce electricity or vehicle power, you will see that heat engines only operate at efficiencies of 25% to 50%.

Although the Carnot cycle (heat engine) also applies to an internal combustion engine (ICE), it is worth noting here that the thermodynamic efficiency of an ICE can be improved using hybrid-electric vehicle (HEV) technology. The HEV uses a combination of the ICE with electric propulsion, thus achieving better use of the fuel. Basically the HEV recovers waste heat, heat not used to power the engine, by (a) generating electricity from braking, (b) reducing wasted energy by turning the ICE off, and (c) helping the ICE engine use a smaller engine. Because of the reduced fuel usage, there is also a reduction in air pollution emissions.

I.B.7. Energy Balance

While thermal efficiency is a measure of how well a process converts energy to useful work, an energy balance considers the percent of useful work derived from the initial energy source. In addition to the thermal efficiency energy losses from the process, be it a power plant or ICE, there are also energy losses from other parts of the entire "Well-to-Wheels" process. For example, in addition to the energy losses incurred using natural gas to generate electricity through combustion, the overall energy balance also incurs energy losses from natural gas drilling and production, natural gas processing, transportation to the electric plant (e.g., by pipeline), and high-voltage transportation through the electric grid before coming to your home. Likewise, before energy losses were incurred in the ICE of a gasoline vehicle, there were also energy losses from drilling and producing the crude oil, transporting it to a refinery, refining it, and then transporting the finished gasoline to a refueling station.

I.B.8. Maturity: Experience

Maturity is defined as the number of years of experience, including technical development, commercial operation, and general usage. This is important because when you make economic decisions on mature energy types, the numbers for capital and operating costs are based on many years of experience. On the other hand, a relatively new technology with little or no commercial experience will tend to have mostly "paper" evaluations for capital and operating costs. While these are based on thorough engineering evaluations, they still generally tend to be optimistic, presenting the best possible case for development. Of course, the more plants that are built, the better the economic numbers become.

I.B.9. Infrastructure

This section gives an overview of infrastructure needed to support an energy type. This includes the commercial plant to make the fuel or electricity as well as infrastructure used to deliver the fuel to the commercial plant and deliver the product to the user. For transportation fuels, delivery methods include pipelines, trucks, rail, and barges, while for electricity the delivery method is the electric grid. This is a very important issue for renewable energy types. While electric plants using fossil fuels or nuclear power can be built near existing parts of the electric grids, renewable energy plants—such as solar and wind—are built where the weather conditions are favorable for their generation. That means that for some new renewable energy plants, new infrastructure will be needed to transport the electricity to the electric grid.

I.B.10. Footprint and Energy Density

"Footprint" is a commonly used term in industrial plants to describe the area used by an energy type. Therefore we could, for example, compare the electrical plant footprint for a natural gas plant, coal plant, nuclear plant, solar plant, and a wind farm to produce the same amount of electric energy. Unfortunately, this is too simple an approach and does not take into account all of the land area involved in the entire "Well-to-Wheels" process of producing the energy. Coal, for example, incurs land usage to mine the coal, prepare the coal for electric plant, transport it to the plant, make the electricity, and then transport the electricity through the high voltage electric grid. Thus, as in the case of the levelized cost of electricity, the footprint must also take into account all of the land area used to take the fuel source from the "Well" to the "Outlet." Likewise, as in the case of the LCOF for petroleum crude oil, all of the land used to refine petroleum crude oil into fuel must be included, such as oil drilling and production, transportation to the refinery, the refinery, and transportation to the fueling station.

Similarly, energy density is calculated by considering all of the land area to take the fuel from "Well-to-Wheels." Although not specifically applicable to transportation fuels, energy densities for different energy types will be reported in units of gigawatt-hours per square mile per year, or GWh/(square mile-year). This will allow comparison of energy types on a common basis.

I.B.11. Environmental Issues

This section and the next consider environmental issues. Some examples of environmental issues are air pollution, water pollution, ground pollution, accidents, and nuclear waste. While the amount of CO_2 produced is quantified in the next section, other environmental issues are discussed in general terms. For example, we know that wildlife can be affected by wind turbines but it is difficult to quantify this in terms of numbers and specific species. Such comparisons can also tend to be subjective. For example, if you compare long-term nuclear waste to air pollution created by burning fossil fuels, it is subjective to specifically say which has worse long-term issues. Environmental issues for each energy type are listed and the potential problems are discussed.

I.B.12. CO_2 Production and the Cost of Capture and Sequestration

Although CO_2 is not normally considered a "pollutant," we consider its production because of growing concerns about the effect of GHGs on climate change. Therefore, this section quantifies the amount of CO_2 produced. The additional cost of capture and sequestration, when applicable, is also examined. For example, in electrical plants using coal or natural gas, or a refinery using crude oil, the amount of CO_2 produced and the additional cost of adding CO_2 capture and sequestration are quantified.

It may surprise you that renewable technologies such as wind, hydroelectric, geothermal, and solar also produce CO_2, primarily during the plant construction phase. Although the plants operate with carbon-free energy, fossil fuels are still used to build the plant and maintain it. Compared to fossil fuels it is, however, a small issue.

I.C. Units, Unit Prefixes, and Basic Definitions

There is probably nothing more frustrating, and nothing more important, than the understanding of units. Units, and the conversions between units, are essential to understand the variety of information reported in the field of energy. Ideally, the entire world would use the metric system, greatly simplifying the comparative understanding of different energy types. The metric system is more formally defined as SI units, or *système international d'unités* in French. In the metric system, you use meter for length, kilogram for mass, seconds for time, and Kelvin for temperature. The greatest advantage is the easy conversion between greater and lesser units by powers of the ten. For example, 1,000 grams is one kilogram and one kilometer is 1,000 meters.

Unfortunately some countries, notably the United States, still use the English system of units, making it necessary to provide some basic information for the purpose of understanding this book. English units come from England and evolved from the Roman system. Compared to the SI unit system, conversion between units is not as simple as the metric system. For example, one pound is 16 ounces and one mile is 5,280 feet.

Specific details for different types of units are given in Appendix B. However, to understand energy resources, consumption, and production for the United States and the world, it is essential to understand the terminology used for large numbers. If, for example, you want to describe the national debt for the United States, it is quite cumbersome to give it as $20,000,000,000,000 so we say, instead, $20 trillion.

In the electric power industry, people use the prefix "tera" in place of the word "trillion." Thus, one trillion watts is given as 1 terawatt, or one TW. On the other hand, in the transportation fuel industry, it is common to use scientific notation to describe large numbers. For example, one trillion BTUs of energy can also be represented as 1 trillion BTUs, or 1×10^{12} BTUs. The conversion between prefixes used in the electric power industry and scientific notation is shown here:

$$\text{Kilo (k)} = 1 \text{ thousand}, 1 \times 10^3 \tag{2}$$

$$\text{Mega (M)} = 1 \text{ million}, 1 \times 10^6 \tag{3}$$

$$\text{Giga (G)} = 1 \text{ billion}, 1 \times 10^9 \tag{4}$$

$$\text{Tera (T)} = 1 \text{ trillion}, 1 \times 10^{12} \tag{5}$$

$$\text{Peta (P)} = 1 \text{ quadrillion}, 1 \times 10^{15} \tag{6}$$

References

1. Aristotle, "Nicomachean Ethics," 1098b33, at Perseus, https://oll.libertyfund.org/titles/aristotle-the-nicomachean-ethics
2. *Webster's II New Riverside University Dictionary*, Boston, MA: Houghton Mifflin, 1984.
3. "World Population Growth," Max Roser and Esteban Ortiz-Ospina, April 2017, https://ourworldindata.org/world-population-growth
4. *Monthly Energy Review*, US Energy Information Administration (EIA), DOE/EIA-0035, February 2019, https://www.eia.gov/totalenergy/data/monthly/pdf/mer.pdf
5. "BP Statistical Review of World Energy," June 2018, (https://www.bp.com/content/dam/bp/business-sites/en/global/corporate/pdfs/energy-economics/statistical-
6. "China Renewable Energy Outlook 2018," file:///C:/Users/paulf/Documents/Book/References_Figures_Tables_Revised/Introduction/China-Renewable-Energy-Outlook-2018-Folder_ENG.pdf
7. "China," Country Analysis Briefs, Energy Information Administration, May 2011, https://www.eia.gov/beta/international/analysis.php?iso=CHN
8. World Bank Organization, 2015, http://data.worldbank.org/indicator/NY.GDP.PCAP.CD and http://data.worldbank.org/indicator/EG.USE.PCAP.KG.OE
9. "What is the Kyoto Protocol?" United Nations Climate Change, 1997, https://unfccc.int/process-and-meetings/the-kyoto-protocol/what-is-the-kyoto-protocol/what-is-the-kyoto-protocol
10. "Renewable Energy: Moving Towards a Low Carbon Economy," European Commission, March 2020, https://ec.europa.eu/energy/en/topics/renewable-energy
11. "Average Annual West Texas Intermediate (WTI) Crude Oil Price from 1976 to 2016 (in U.S. dollars per barrel), The Statistics Portal, 2017, http://www.statista.com/statistics/266659/west-texas-intermediate-oil-prices/
12. *Chemical Thermodynamics: Basic Theory and Methods*, 3rd Edition, I.M. Klotz and R.M. Rosenberg, 1972, W.A. Benjamin, Menlo Park, CA.
13. *Thermodynamics*, 2nd Edition, K.S. Pitzer and L. Brewer, McGraw Hill 1961, New York.
14. *Introduction to Chemical Engineering Thermodynamics*, 3rd Edition, J.M. Smith and H.C. van Ness, McGraw Hill, 1975, New York.
15. *Reflections on the Motive Power of Heat and on Machines Fitted to Develop That Power, Sadi Carnot* (R.H Thurston, editor and translator), Wiley & Son, 1890, New York.

1
Petroleum Crude Oil
A Nonrenewable Energy Type

1.1. Foreword

Petroleum is a complex mixture of hydrocarbons that resides in the earth in liquid, gaseous, or solid forms. The word "petroleum," or "rock oil," is derived from the Latin *petra* (rock) and *oleum* (oil).[1] This chapter focuses on crude oil found in geologic formation below the earth's surface, either under land or water. Although not covered in this book, there is also a type of oil called shale oil, extracted from a fine-grained sedimentary rock called oil shale. Generally speaking, all forms of petroleum are fossil fuels, meaning that they were formed by anaerobic decomposition of dead organisms, such as algae and plankton, in a process taking millions of years and, by one estimate, as long as 650 million years.[2] During this process of many million years, the dead organisms are subjected to the heat and pressure from the interior of the earth, and thus geothermal energy is used in the transformation.

Petroleum is a complex mixture of primarily hydrogen and carbon, but it also contains other elements such as nitrogen, sulfur, oxygen, vanadium, nickel, cadmium, zinc, arsenic, mercury, and selenium, and these elements can cause environmental problems if not handled appropriately. The boiling point range of the material in crude oil is large, with initial boiling points as low as 100°F (38°C) and final boiling points in excess of 1300°F (704°C). The hydrocarbons in crude oil have four general classifications, namely paraffins, olefins, naphthenes, and aromatics. A complete discussion of the organic chemistry nature of these different types is beyond the scope of this book, but a brief discussion is given here. Paraffins are carbon chain molecules completely saturated with hydrogen, meaning that it is not possible to add any more hydrogen. Olefins are unsaturated carbon chain molecules and are hydrogen deficient. In general, a paraffin has a carbon chain of C_nH_{2n+2} while that for olefins is C_nH_{2n}. That is, an olefin has one unsaturated carbon-carbon bond to which two additional hydrogen atoms could be added. Some olefins have more unsaturated carbon-carbon bonds and, for example, an olefin with two of these bonds is called a diolefin. Generally speaking, however, olefins and diolefins are rarely found in crude oil. They are made by many refining processes, as discussed later. To provide a visual image of the differences between paraffins and olefins, Figure 1.1 shows hexane and

The Changing Energy Mix. Paul F. Meier, Oxford University Press (2020). © Oxford University Press.
DOI: 10.1093/oso/9780190098391.001.0001.

hexene, 6-carbon compounds representing paraffins and olefins, respectively. In this figure, the vertices represent the carbon, to which hydrogen atoms are bonded.

Figure 1.1 Hexane and Hexene

Similarly, naphthenes and aromatics are saturated and unsaturated, but are cyclic compounds. Thus, while paraffins and olefins are linear chain molecules, with or without chains along the main carbon chain path, naphthenes and aromatics are compounds with a cyclic base of carbons. For example, a 6-carbon cyclic compound aromatic is called benzene if it is an unsaturated aromatic but is called cyclohexane if it is a fully saturated naphthene. These two compounds are shown in Figure 1.2 to provide examples of naphthenes and aromatics, respectively.

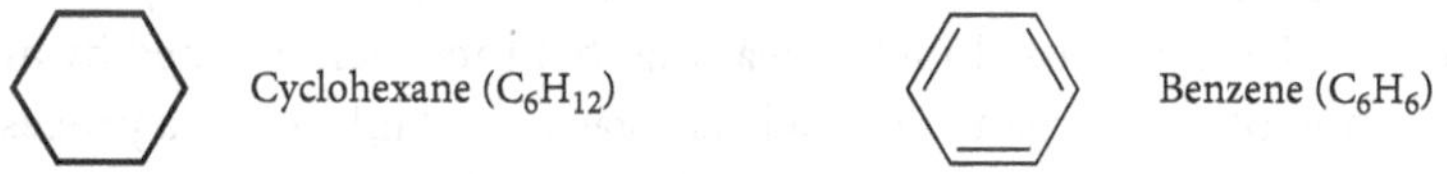

Figure 1.2 Cyclohexane and Benzene Structures

The composition and boiling point range of crude oil is important, and dictates its value and eventual transformation into products such as gasoline and diesel. A crude oil can contain carbon chains greater than C_{60}. Since gasoline typically has a carbon range of C_4 to C_{12} and diesel from C_{10} to C_{24}, the amount of gasoline and diesel from a given crude oil depends on its original hydrocarbon composition, thus dictating its eventual value to the refiner that purchases the crude oil.

The presence of the other compounds, especially the nitrogen and sulfur, is problematic, as combustion will eventually create nitrogen and sulfur oxides, molecules that need to be controlled for environmental reasons.

1.2. Proven Reserves

Making an estimate of the existing petroleum crude oil in the world is difficult, and the amount changes as new technology and new reserves are found. The industry uses two general terms to describe crude oil reserves, namely "proven reserves" and "unproven reserves."

As the name suggests, proven reserves are oil reservoirs that have a high confidence (at least 90%) of being recovered under existing economic and political conditions. In contrast, unproven reserves are oil reservoirs that may exist, based on geological data, but are not considered proven based on a variety of factors, such as economic and political viability. Naturally, new technology or an increase in crude oil prices could result in the reclassification of some previously unproven reserves as proven.

An *Oil & Gas Journal* estimate[3] in December 2018 gives a worldwide value of 1.67 trillion barrels for proven reserves. Similarly, a June, 2018 BP (British Petroleum) statistical review[4] gives a value of 1.70 trillion barrels and a 2018 US Energy Information Administration (EIA)[5] publication gives a value of 1.66 trillion barrels. Based on these three publications, the worldwide proven crude oil reserves are around 1.66–1.70 trillion barrels. Figure 1.3 shows the location of crude oil reserves, and the Middle East dominates the world with 47.6% of all reserves.

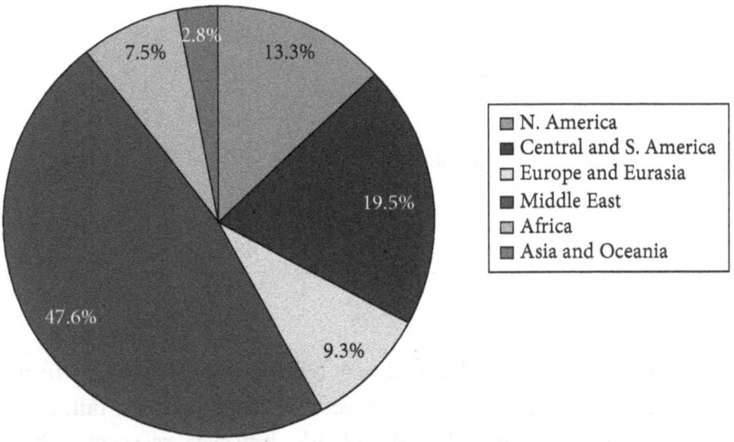

Figure 1.3 Worldwide Proven Crude Oil Reserves by Region[4]

Table 1.1 shows the top ten countries for crude oil reserves. These ten countries control 85.3% of the worldwide proven reserves based on a total of 1.7 trillion barrels.

While on the subject of crude oil reserves, there should be some discussion on the method called hydraulic fracturing, or fracking. This is a method that uses high-pressure liquids, generally a mixture of water and sand, to create cracks in deep rock formations so the oil can be more easily recovered. The drilling is done at depths exceeding three kilometers (~1.9 miles) and usually involves horizontal drilling. This oil is generally referred to as "tight oil." By using this method, the percent of oil recovered from a well can be greatly increased.

Countries that have performed fracking, or have at least considered it, include the United States, Canada, China, Argentina, and Mexico.[6] On the other hand, many European countries have banned fracking, as it can have negative impacts on air and water quality. With respect to air, one chemical released in the fracking process is methane. In terms of global warming potential, methane is 28 to 36 times greater than CO_2. Perhaps of greater importance is the issue of water quality, as millions of gallons of water are used in the fracking process. Roughly 20% to 40% of this water is returned to the ground, with the potential of contaminating the water with toxic metals. Crude oils contain many toxic metals such as cadmium, zinc, arsenic, mercury, selenium, vanadium, and nickel.

Table 1.1 Top Ten Crude Oil Reserves by Country, 2017 (billion barrels)[4]

Country	Billion Barrels
World	1,696.6
Venezuela	303.2
Saudi Arabia	266.2
Canada	168.9
Iran	157.2
Iraq	148.8
Russian Federation	106.2
Kuwait	101.5
United Arab Emirates	97.8
U.S.	50.0
Libya	48.4

Nevertheless, fracking has unleashed new oil reserves, particularly in the United States.[7] In 2010, the proven US reserves were 35 billion barrels but, in 2017, this number grew to 50 billion barrels![4] Worldwide, however, reserves only increased from 1,643 billion barrels in 2010 to 1,697 billion barrels in 2017; interesting enough, the difference is very close to the increased production in the United States.

The top ten countries in terms of crude oil production are shown in Table 1.2. These ten countries produce 69.2% of the world's daily production of crude oil. As the table illustrates, the United States, Saudi Arabia, and Russia lead the world. However, 7 years earlier in 2010, the United States was in third place and produced 7,549 thousand barrels per day (BPD), only 58% of the amount in 2017. Clearly, hydraulic fracking has had a big impact on US crude oil production.

On the consumption side, the BP statistical review[4] reports worldwide energy consumption for 2017 at 98.2 million barrels per day (MBPD). Data for the top ten countries are shown in Table 1.3. These ten countries account for about 60% of the world consumption, with the United States and China consuming 20.2% and 13.0%, respectively.

Given the earlier discussions in the Introduction, growing world population and growth in income, particularly in China and India, will likely lead to increased consumption in the future. However, ignoring this potential growth in oil consumption, the current estimate for proven reserves and crude oil consumption rate suggests about 47 years remaining for crude oil. Based on this assessment, crude oil is clearly a dwindling commodity, with a possibility of using all the remaining world supply around the year 2070. However, increasing crude oil reserves and increasing demand will change

Table 1.2 Top Ten Crude Oil Producing Countries, 2017 (thousand BPD)[4]

World	**92,649**
United States	13,057
Saudi Arabia	11,951
Russian Federation	11,257
Iran	4,982
Canada	4,831
Iraq	4,520
United Arab Emirates	3,935
China	3,846
Kuwait	3,025
Brazil	2,734

Table 1.3 Top Ten Crude Oil Consuming Countries, 2017 (thousand BPD)[4]

World	**98,186**
United States	19,880
China	12,799
India	4,690
Japan	3,988
Saudi Arabia	3,918
Russian Federation	3,224
Brazil	3,017
South Korea	2,796
Germany	2,447
Canada	2,428

this assessment up or down. One book[8] talks about oil development in three phases including (1) onshore, (2) offshore, and (3) deepwater. According to these authors, deepwater development is still in the early phases and, with improvement in this technology, the limit on the supply of oil and gas is only based on the ingenuity of people searching for and producing the oil. Of course, the price to produce these deepwater sources is important, and will have to be competitive with alternate energy sources.

1.3. Overview of Technology

Several steps are needed to get gasoline or diesel into your automobile, the entirety of the steps referred to as "Well-to-Wheels." These steps include (1) finding and producing crude oil, (2) transporting it to a refinery, (3) refining it, and (4) transporting it to gasoline stations. Producing crude oil involves several steps including locating the oil, drilling, and then several recovery steps called primary, secondary, and tertiary recovery.

The formation of crude oil, a process that started around 50–200 million years ago, involves the death of prehistoric animals and plants and then the action of the earth's heat and pressure. This process occurred in the absence of oxygen, and anaerobic microorganisms caused the decomposition into carbon-rich material. Crude oil is normally found in sedimentary rocks, rocks formed from the deposition of materials on the surface of the earth. These rocks, such as sand, silt, and clay, were formed from erosion of larger rocks, and then moved through water to pile up into sediments that underwent compacting to form hard rocks. Generally, crude oil pools are found in the faults and folds of the sedimentary rock layers.

Given this knowledge, geologists use the earth's geology to locate oil. Finding the crude oil is both a science and an art, but no method is guaranteed to find crude oil. The easiest method to locate crude oil is visual recognition of small amounts of oil and natural gas that seep through the surface of the earth. Another method is surface exploration, which uses different technologies to locate hydrocarbons, iron ore deposits, or just analyze the geography. Geologists can map the subsurface using magnetomometers to read the magnetic properties of the rock layers to indicate faults and folds.[9] However, the most sophisticated method of exploration is the use of seismic techniques, such that radio waves are shot at a possible crude oil site and these data are used to create a digital geological map to model the site. This method is essential in the discovery of deep underground reservoirs and offshore oil production. In 2015, offshore oil production accounted for 30% of the world oil production and 27% of the world natural gas production.[10] As of 2016, the water depth record for offshore production was 3,400 meters (11,156 feet), when a Maersk drill ship drilled at a site offshore Uruguay.[10]

Once a possible oil site is discovered, test drilling is used to confirm the crude oil location and potential reserves. After the potential crude oil site is found, examined by geological analysis, and confirmed with test drilling, it is time to drill an oil well. The basic technique used to extract oil is to drill a hole and then place steel pipe, or casing, into the hole. In the early days of oil production, the pipe was placed straight into the ground; however, today production uses slanted drilling methods, allowing oil producers to rotate the pipe and produce oil from different deposits without drilling a new hole.

Drilling in the ocean is basically the same, but a floating platform must be used to support the oil rig and its weight. This can be a submersible, semisubmersible floating oil rig, or a drillship. Generally, the submersible is used in shallow water, semisubmersibles in 200–5,000 feet of water, and drill ships in 200–10,000 feet.[9] The method used to stabilize the rig and compensate for water and vessel movements depends on the type of floating platform.

When oil is first produced, it is referred to as the *primary recovery stage*, and oil is removed from the well from natural forces such as natural water displacing oil downward into the well, expansion of natural gas at the top of the reservoir, expansion of gas initially dissolved in the crude oil, and gravity drainage resulting from the movement of oil within the reservoir.[11] Recovery during the primary stage is typically 5%–15%.[11]

As oil leaves the reservoir, the pressure will decrease, and at some point the pressure is too low to move the oil up the pipe. At this time, the *secondary recovery stage* begins. Basically, the secondary recovery stage uses fluid injection to increase reservoir pressure. Other methods include pumps, such as electrical submersible pumps (ESPs), water injection, natural gas reinjection, and gas lift. In gas lift, gases such as air and carbon dioxide (CO_2) are used. Typical recovery from both primary and secondary recovery is 30–50%.[11]

When secondary recovery is complete, the *tertiary recovery stage* may begin. The goal of tertiary recovery is to make the crude oil more mobile to increase recovery. One method is thermal recovery, such that the oil is heated to reduce its viscosity. This is sometimes called thermally enhanced oil recovery (TEOR) and is most commonly done with steam injection, but CO_2 can also be injected to reduce viscosity. Another method is gas injection, which employs gases such as nitrogen, natural gas, or CO_2 to push oil out of the oil well. Yet another method is to use chemical injection, such as surfactants, which reduce the surface tension of the oil and increase its mobility.[12]

Typical tertiary recovery is 5% to 15% of the reservoir.[11] Tertiary recovery methods are more expensive than primary and secondary, meaning that tertiary recovery begins when it is profitable to do so. Using the earlier percentages, oil produced from primary, secondary, and tertiary ranges from 35% to 65%. One 2011 reference[13] cites a worldwide average oil field recovery of about 35%, meaning that 65% of the oil already discovered remains in the reservoir. Increasing the recovery factor from 35% to 45% would result in 1 trillion barrels of additional oil. Achieving this goal will require better technology and cost reduction.

The cost to get crude oil from a well is a combination of the cost to both locate and produce the crude oil. As discussed, the cost to locate oil includes the cost to explore for the oil and perform test drilling as well as the cost for leases of the land or water that may contain the oil. This will also include the cost of retaining undeveloped property.

The cost of oil production includes the cost of primary, secondary, and tertiary recovery, which accounts for equipment, facilities, and manpower. Locating and production costs in 2016 for twelve major oil and natural gas producers are shown in Table 1.4.[14] Data in the table are listed in decreasing total cost. The total cost for

Table 1.4 Costs for Locating and Producing Crude Oil and Natural Gas, 2016 Data, \$/barrel of oil equivalent (5,618 ft^3 of natural gas equivalent to one barrel of oil)

Country	Locating Costs, \$/barrel	Production Costs, \$/barrel	Total, \$/barrel
United Kingdom	\$26.97	\$17.36	\$44.33
Brazil	\$25.55	\$9.45	\$34.99
Nigeria	\$20.18	\$8.81	\$28.99
Venezuela	\$19.68	\$7.94	\$27.62
Canada	\$15.09	\$11.56	\$26.64
US Shale	\$17.50	\$5.85	\$23.35
Norway	\$17.07	\$4.24	\$21.31
US Nonshale	\$15.84	\$5.15	\$20.99
Indonesia	\$12.83	\$6.87	\$19.71
Russia	\$16.23	\$2.98	\$19.21
Iraq	\$8.41	\$2.16	\$10.57
Iran	\$7.15	\$1.94	\$9.08
Saudi Arabia	\$5.99	\$3.00	\$8.98

producing oil in Middle East countries is 50% or less than the costs for other countries. On the other hand, the United Kingdom has very high costs because the production is mostly offshore and in deep water. Likewise, newer projects in Brazil and Nigeria focus on offshore production. Canada also has high costs, but this is not due to offshore drilling but rather production from oil sand deposits in Alberta, which also have high capital costs. For the United States, shale oil extracted using fracking costs over \$2/barrel more than normal drilling. If the total cost for the United Kingdom is representative of deepwater drilling, it is clear that the market price of crude oil has a great impact on whether or not a company will pursue production from deepwater sources.

Once the oil is produced, it is transported to a refinery to make transportation fuels and other saleble products. This transportation can occur by pipeline only, or pipeline plus shipping, or pipeline plus shipping plus pipeline, depending on the source of the crude oil. For example, oil produced onshore in the United States can be sent by pipeline to a refinery, while North Sea oil refined in the United States is transported from the bottom of the ocean by pipeline to a floating production, storage, and offloading (FPSO) vessel or floating storage and offloading (FSO) vessel and then transported by ship to the United States. If the refinery is on the Gulf Coast, the oil can be delivered from the ship but if the refinery is further inland, the oil is sent by pipeline. The processing performed by an FPSO will include water separation, gas treatment, and gas compression. An FSO does not perform any processing at the offshore crude oil production site.

These crude oil pipelines typically have diameters of 2 to 12 inches[15] but can be as large as 48 inches, such as the 800-mile long Trans-Alaska Pipeline System, or TAPS. Crude oil moves through the pipeline from pressure, and all pipelines will have pumping stations to increase pressure lost during its movement along the pipeline. In the United States, there are about 56,000 miles of crude oil pipelines.[16] This is more than the number of miles in the US interstate highway system which, as of 2015, had a total length of 46,876 miles.[17] For refinery products, there are about 90,000 miles of pipelines. The cost of transporting crude oil to refineries by pipeline is low, under $5/ barrel, compared to rail, which is generally $10–$15/barrel.[18]

By itself, crude oil has little value, because it is a complex mixture of hydrocarbon with a wide range of boiling points. With separation, a small amount would be in the boiling point ranges we identify with gasoline and diesel, but with poor quality for a transportation fuel. Therefore, the refinery is used to transform crude oil into high value and usable products, especially gasoline and diesel. This transformation must be done as efficiently as possible to maximize products and minimize costs, and also in a way that is environmentally sound.

A refinery can be classified into five basic process steps.[19] The first step is distillation, the separation of crude oil by boiling point into different groups such as gas, liquefied petroleum gas (LPG), straight run gasoline, distillate, heavy gas oil, and vacuum gas oil or residual oil. Gas is composed of C_1 (methane) and C_2 (ethane and ethylene) hydrocarbons, plus hydrogen (H_2). In the early days of refining, this gas was routinely burned, or flared. However, today it has many possible economic values such as fuel for the refinery, electricity generation, and even making polyethylene if the refinery is equipped with a chemical plant. LPG is primarily propane (C_3) with some butane (C_4). Straight run gasoline is composed of hydrocarbons with a boiling range of C_5–C_{12} (40°C–205°C) but this gasoline would not generally be suitable for motor vehicles because of low octane and high sulfur. Distillate will eventually become kerosene (C_{10}–C_{18}, 175°C–325°C), or fuel for jet engines and diesel fuel, C_{12}–C_{24} (250°C–350°C). Heavy gas oil has hydrocarbons ranging from C_{20}–C_{70} (343°C–565°C) and the residual oil boils above 565°C. Components of crude oil boiling above 343°C are not usable products and must undergo some conversion process to upgrade them. Table 1.5 shows crude oil assay examples of four crude oils including light crude oils such as West Texas Intermediate (WTI) and Brent, an intermediate oil such as Arab Medium, and a heavy crude oil such as Maya.[19] The amount of material boiling above 343°C increases as the crude oil gets heavier. WTI and Brent have 39.0 and 44.0% boiling above 343°C and this increases to 56.8% for Arab Medium and to 65.9% for Maya.

When the oil is heavier, there are fewer hydrocarbons initially in the desired product boiling ranges such as straight run gasoline and distillate. Another problem with crude oils is that they contain sulfur, cadmium, zinc, arsenic, mercury, selenium, vanadium, and nickel, elements that cause environmental problems and must be removed at the refinery. Not surprisingly, the high-quality crude oils are also the

Table 1.5 Simplified Assays for Four Crude Oils[20]

Crude Oil Feed	West Texas Intermediate	Brent	Arab Medium	Maya
Density, g/cm^3	0.821	0.833	0.884	0.921
Sulfur, wt%	0.34	0.40	2.85	3.30
Vanadium, ppm	1.6	6.0	12.0	52.0
Nickel, ppm	1.6	1.0	5.9	12.0
IBP–70°C (IBP–158°F), wt%	5.1	6.0	3.5	2.4
70°C–100°C (158°F–212°F), wt%	5.2	5.4	2.9	2.5
100°C–190°C (212°F–374°F), wt%	21.6	17.3	11.4	9.3
190°C–235°C (374°F–455°F), wt%	9.0	6.8	7.3	5.3
235°C–280°C (455°F–536°F), wt%	8.8	9.1	7.5	5.8
280°C–343°C (536°F–650°F), wt%	11.3	11.4	10.6	8.8
343°C–565°C (650°F–1049°F), wt%	26.6	29.3	31.4	24.4
565°C+ (1049°F+), wt%	12.4	14.7	25.4	41.5

Note: IBP stands for "initial boiling point," the temperature at which vapor is first formed from the crude oil. The first compounds to boil are the most volatile ones, such as methane, which boils at −161.5°C. Since this initial boiling point is difficult to precisely measure, it is reported as IBP.

ones that cost the most. Therefore, WTI and Brent cost more than Arab Medium and Maya.

The second basic step in the refining process involves a variety of processing steps that include cracking (to reduce molecular size), combination reactions to make the molecules larger, and reforming to rearrange molecules into more valuable components. This second step in the refining process is the one that receives the most attention, the most research and development, and the one that incurs material costs in the form of catalysts. The classical definition of a catalyst is something that increases the rate of a chemical reaction and may also change the product distribution (selectivity) as compared to the application of heat alone, while the catalyst itself is not consumed. In practice, catalysts are consumed, with some lasting days and others lasting years.

It is beyond the scope of this book to give a comprehensive review of each refining process, but a short overview is given here. Simply stated, the goal of a refinery is to rearrange the hydrogen and carbon in the starting molecules and make them into higher value product, primarily gasoline and diesel transportation fuels. Since gasoline typically boils from 40°C to 205°C and diesel from 250°C to 350°C, it is clear from Table 1.5 that crude oil does not initially contain these in large quantities. Therefore, it is necessary to reduce the molecular size of the higher boiling components in crude oil.

There are three primary processes that can be used. The first is coking, a thermal process using heat and pressure but no catalyst. A coker, operating at 480°C to 500°C, is used to convert heavy gas oil (343°C–565°C) and the residual oil (565°C+) into lower molecular weight hydrocarbons. The coker also produces petroleum coke, which can be used for fuel or the manufacture of electrodes.

The second process is fluid catalytic cracking (FCC), a catalytic process that cracks heavy gas oil and residual oil into high-value products such as gasoline and diesel. The process is called "fluid" because the catalyst is composed of very fine (20–120 micron) spherical particles that behave like a fluid. The "cracking" part of the name is used because the catalyst has acid sites that perform the catalytic cracking, breaking the carbon-carbon molecular bonds of the larger molecular weight compounds. The FCC is the main unit in most US refineries, with an estimated 45% of gasoline worldwide coming from the FCC unit.[21]

The third process is called catalytic hydrocracking, a process that utilizes hydrogen and a catalyst or catalysts, depending on the particular application. In comparison to FCC, hydrocracking uses a bifunctional catalyst that can perform both cracking and hydrogen addition. In this respect, it is better able to handle large polycyclic aromatic compounds than FCC.

In general, most US refineries use FCC, producing about 50% gasoline and 15% diesel for each barrel of crude oil, while Europe generally uses catalytic hydrocracking, producing about 25% gasoline and 25% diesel from that same barrel of oil. This is one of the reasons you see more gasoline-powered vehicles in the United States and more diesel vehicles in Europe.

The third basic step in the refining process involves the rearrangement of hydrocarbons to improve the octane content of the gasoline. There are three main processes for this including reforming, isomerization, and alkylation.

Catalytic reforming is one process to increase the octane of gasoline. The goal is to change the structure of gasoline, or naphtha, to molecules that have higher octane. The name of the feed to these processes is typically called naphtha and gasoline is referred to as the final product. Some of the reactions in a reforming process include dehydrogenation (the removal of hydrogen to change a naphthene to an aromatic) and dehydrocyclization (the removal of hydrogen and creation of a cyclic aromatic molecule from a paraffin).

Isomerization is another catalytic process to increase the gasoline octane. The chemical definition of isomerization is to rearrange the molecule's atoms into a different configuration. In petroleum refining, isomerization is specifically intended to increase the octane value of the gasoline. Some of the reactions in an isomerization process include paraffin isomerization to isoparaffins and isomerization of naphthenes to larger naphthenes and then to aromatics.

Like reforming and isomerization, alkylation is also used to increase the gasoline octane. However, while the previous two processes rearrange molecules already in the boiling range for gasoline, alkylation takes lower molecular weight olefins and paraffins to make a higher molecular weight isoparaffin that boils in the gasoline range and has high octane. A typical reaction for this is to alkylate isobutane with olefins, normally propylene and butene. The product of this reaction, or alkylate, is typically a C_7 or C_8 carbon chain molecule.

The fourth basic step in the refining process is blending, storage, and lights end recovery. As discussed, gasoline, jet fuel, heating oil, and diesel fuel are the main products of the refinery. As US consumers know, gasoline is normally sold with three different road octane ratings including 87 (regular unleaded), 89 (mid-grade), and 91 (premium). At high elevation, such as Denver and Salt Lake City, 85 octane gasoline is also sold because the reduced density of air requires less fuel and reduced compression in the engine's cylinder. The vapor pressure of gasoline is also controlled by blending normal butane into the gasoline. Butane helps increase the gasoline octane, but must be controlled between summer and winter.

By the way, if you have traveled outside the United States, you may have noticed higher octane numbers shown at fueling stations. For example, in Europe you may see octane numbers like 91 or 95 at their fueling stations. In the United States, road octane numbers are used to market gasoline, an average of research octane number (RON) and motor octane number (MON). Without going into the details, both are laboratory engine tests used to measure the knock resistance of gasoline during combustion. And the measured RON is always greater than the measured MON. In Europe, however, gasoline is marketed using RON only. As an example, if you buy 87 road octane gasoline in the United States, the measured RON and MON are typically 91 and 83, respectively. Therefore, purchasing 91 RON gasoline in Europe is equivalent to an 87 road octane gasoline in the United States.

The fifth basic step in the refining process is management of environmental issues. When burned, sulfur and nitrogen create nitrogen oxides and sulfur oxides. When these react with water in the atmosphere, they can create acids such as nitrous acid (HNO_2) nitric acid (HNO_3), sulfurous acid (H_2SO_3), and sulfuric acid (H_2SO_4), eventually leading to acid rain. This can be prevented by removal of both nitrogen and sulfur before gasoline and diesel are sold as fuels. There are two basic approaches to the removal of nitrogen and sulfur defined as pretreatment and posttreatment. Pretreatment means removal of sulfur and nitrogen before FCC. The two main processes for this include coking and high pressure hydrotreating. In coking, the sulfur is primarily converted to hydrogen sulfide (H_2S) while the nitrogen is converted to ammonia (NH_3) and nitrogen (N_2). The H_2S and NH_3 are treated at the refinery. The H_2S is converted to elemental sulfur (S), and the NH_3 is removed and sold for the production of fertilizer. In high-pressure hydrotreating, S is primarily converted to H_2S and N_2 is converted to NH_3. Likewise, these are treated at the refinery.

In posttreatment, the N_2 and sulfur are removed after FCC. For both gasoline and diesel, N_2 and sulfur can also be removed by hydrotreating, but this posttreatment hydrotreating process can be done at lower reactor pressure than the high pressure hydrotreating used in pretreatment. This type of hydrotreating converts N_2 and sulfur into the same products of H_2S and NH_3. In addition to hydrotreating, there is another process for gasoline, the S Zorb™ sulfur removal process,[22] a selective adsorption process that removes sulfur with less negative effect on octane than hydrotreating.

The removal of sulfur from gasoline is also important because of the effect sulfur has on the automobile catalytic converter. The purpose of the catalytic converter is to deal with three harmful compounds including volatile organic compounds (VOCs), carbon monoxide (CO), and nitrogen oxide. VOCs include propane and benzene. This catalytic converter oxidizes the VOCs to CO_2 and water, oxidizes CO to CO_2, and reduces nitrogen oxides to nitrogen gas (N_2). Sulfur will poison the catalytic converter, making it less efficient to perform these three reactions. In addition, any sulfur in gasoline will form sulfur dioxide (SO_2), which, as mentioned previously, can form sulfurous and sulfuric acid in the atmosphere. Up to 2017, the US regulation for sulfur in gasoline was 30 ppm, based on the Tier 2 part of the Clean Air Act, and this standard began in 2004. After January, 2017, the Tier 3 standard requires gasoline sulfur to meet an annual average of 10 ppm.

Another important aspect of the refinery environmental programs is wastewater management. Basically, harmful molecules must be controlled before refinery water is discharged from the refinery. These molecules include benzene, sulfur dioxide, and nitrogen oxide. In addition, harmful metals may also need to be removed, including selenium, arsenic, mercury, cadmium, zinc, nickel, and vanadium.

1.4. Capital Cost, Operating Cost, Well-to-Wheels Levelized Cost of Electricity, and Well-to-Wheels Levelized Cost of Fuel

The equations and methodology used to calculate levelized cost of fuel (LCOF), and specific input used for petroleum crude oil to make gasoline, are detailed in Appendix A. As a reference point, a refinery capacity of 200,000 BPD was chosen to make the calculations, a size fairly typical for US refineries. The units for LCOF are in \$/gallon of gasoline equivalent, or \$/gge.

Transportation energy derived from petroleum crude oil is created over several steps including (1) crude oil production, (2) transportation to the refinery, (3) refining, (4) transportation of products to market, (5) taxes (both federal and state), and (6) marketing cost at the filling station. The cost for these different steps will yield the "Well-to-Wheels" LCOF in \$/gge. As described in Appendix A, (5) taxes are included in the LCOF equation.

In general, some oil companies are fully integrated, meaning that they drill and produce crude oil and also refine the crude oil to make transportation fuels. On the other hand, some oil companies are exclusively refining companies, meaning that they must buy their crude oil at market prices. In addition, even in a fully integrated company, there are internal costs that have the refining part of the company paying market prices, or near market prices, for a barrel of crude oil. Therefore, from the refinery point of view, crude oil production costs and the cost of transporting the crude oil to the refinery are included in the purchase price of the crude oil feed. In other words, the (1) crude oil production and (2) transportation to the refinery are captured in the cost of crude oil.

The best way to establish the true cost of crude oil is from global crude oil prices. WTI is used as an economic benchmark crude oil for many analyses and is used here. Unfortunately, there has been a lot of volatility in crude oil prices since 2005. Looking at 2-year intervals, the average annual price of WTI in 2005, 2007, 2009, 2011, 2013, 2015, and 2017 were \$56.44, \$72.26, \$61.73, \$94.88, \$97.98, \$48.66, and \$50.88.[23] For 2018, the average price for WTI was \$64.94, comparable to what it was in 2009 but far less than 2011 and 2013 prices. The 2018 value is used here for the LCOF calculations but, because of the volatility in crude oil prices, a sensitivity study was made using different prices for the cost of WTI to the refinery to see the impact of the crude oil cost. Based on the previous data, a low of \$40/barrel (\$40/BBL) and a high of \$100/BBL were used for the sensitivity study.

For the refinery capital cost, a range of capital costs for new refineries and refinery expansions gave an average of \$40,000/BPD, with a range of \$20,000 to \$90,000/BPD for the sensitivity study. Details can be found in Table A.1.1 in Appendix A. Using a base case value of \$40,000/BPD means the capital cost is \$8 billion dollars.

Fixed and variable operation and maintenance (O&M) costs were set to \$5/BBL, with a range of \$3 to \$8/BBL used for the sensitivity study.

The two remaining items we need to calculate the LCOF include (4) transportation of products to market and (6) marketing cost at the filling station. For the sum of these two steps, an April 2019 update by the EIA[24] shows values of \$0.32/gallon for gasoline and \$0.60/gallon for diesel. Generally, the distribution cost for diesel is higher than gasoline—because it has a higher density, a gallon of diesel weighs more than a gallon of gasoline.

With these data, LCOF was calculated for the base case and sensitivity study cases, shown in Table 1.6. For the base case, the price of crude oil is \$64.94/BBL, the fixed and nonfuel variable O&M is \$5/BBL, and the capital cost is \$40,000/BPD installed capacity.

As the table illustrates, changing the operating costs from \$5/BBL to either \$3 or \$8/BBL had only a minor impact on total cost. More than doubling the capital cost from \$40,000/BPD to \$90,000/BPD resulted in a \$0.38/gge increase in total cost. The change in crude oil feed price had the greatest impact, and changing the crude oil price from the base case value of \$64.94/BBL to \$100/BBL resulted in an LCOF increase of \$0.82/gge. This means that at a price of \$100/BBL of crude oil, the refinery needs to sell gasoline at a price of \$3.07/gge just to break even.

Note that Table 1.6 also gives the LCOF in units of \$/megawatt-hours (\$/MWh) and \$/barrels of oil equivalent (\$/BOE). These units are used to put different energy types, be they transportation fuels or electric energy, on a common set of units for comparison in the Conclusions chapter.

Table 1.6 Levelized Cost of Fuel (LCOF) for Petroleum Crude Oil

	Levelized Capital Cost	Fixed and Nonfuel Variable O&M	Feed Cost	Transportation and Marketing	Total System Levelized Cost	Total System Levelized Cost	Total System Levelized Cost
Units	$/gge	$/gge	$/gge	$/gge	$/gge	$/MWh	$/BOE
Base Case	$0.30	$0.12	$1.52	$0.32	$2.25	$67.41	$114.58
Capex is $20,000/BBL	$0.15	$0.12	$1.52	$0.32	$2.10	$62.90	$106.92
Capex is $90,000/BBL	$0.68	$0.12	$1.52	$0.32	$2.63	$78.68	$133.75
Opex is $3/BBL	$0.30	$0.07	$1.52	$0.32	$2.21	$66.01	$112.21
Opex is $8/BBL	$0.30	$0.19	$1.52	$0.32	$2.32	$69.50	$118.14
Crude is $30/BBL	$0.30	$0.12	$0.93	$0.32	$1.67	$50.00	$84.99
Crude is $100/BBL	$0.30	$0.12	$2.33	$0.32	$3.07	$91.88	$156.18

Finally, it is worth noting that the consumer must pay an additional federal and state tax for gasoline and diesel. For gasoline, the federal tax is \$0.184/gallon and for diesel the federal tax is \$0.244/gallon. State taxes for gasoline range from a high of \$0.514 for Pennsylvania to a low of \$0.1225 for Alaska with an average state tax of \$0.2978/gallon.[25] Likewise state taxes for diesel range from a high of \$0.651 for Pennsylvania to a low of \$0.1275 for Alaska with an average state tax of \$0.2983/gallon. This means, on average, the combined state and federal tax for gasoline is \$0.4818/gallon while for diesel it is \$0.5423/gallon. While state and federal taxes do not directly affect LCOF and refinery profitability, it does show the additional cost the consumer must pay over the LCOF cost to the refiner.

1.5. Cost of Energy

Next, the cost of the crude oil in \$/lb and \$/MM BTU (\$/million British Thermal Units) will be calculated for comparison to other energy types.

The average cost for WTI for 2018 was earlier shown to be \$64.94/BBL and, in Table 1.5 above, the density of WTI is 0.821 g/cm^3. Also, Appendix B shows the energy content for a "typical" barrel of crude oil to be 5,800,000 BTU. With these values, we can calculate the mass for a barrel of WTI to be 287.8 lb and the BTU content to be 20,155 BTU/lb. These equations show these simple calculations.

$$lb\,/\,BBL = 350.51 * 0.821 = 287.8$$

$$BTU\,/\,lb = 5{,}800{,}000\,/\,287.8 = 20{,}155$$

From this we can calculate the cost of crude oil as \$0.23/lb in terms of mass and \$11.20/MM BTU in terms of energy content.

1.6. Capacity Factor

The capacity factor was set to 0.931, based on 2018 data from the EIA. The capacity factor refers to the percent of the rated capacity used each year. According to EIA,[26] this capacity factor has increased almost every year in the 9 years since 2010. In 2010, the capacity factor was 86.3%, increasing to 90.4% in 2014, and 93.1% in 2018.

1.7. Efficiency: Fraction of Energy Converted to Work

The thermal efficiency of an engine is defined as the percentage of the energy from combustion which is converted to mechanical work. In the case of an internal combustion engine (ICE), used by gasoline and diesel engines, the combustion of gases at high temperature and pressure produce mechanical work to move the piston or turbine blades. Basically, an ICE draws a mixture of air and fuel into the engine, and this mixture is compressed, ignited, expanded, and then released from the engine. It is also worth noting that an ICE is a heat engine such that thermal energy is converted into mechanical work and the reactants (fuel and air) and products (primarily CO_2, nitrogen, and water) serve as the working fluid.[27] This differs from steam engines, which use steam to drive the piston and not the combustion gases.

Generally, the ICE is a four-cycle engine. In the first cycle, the piston travels downward and creates a vacuum to draw in air and fuel, either gasoline or diesel, into the combustion cylinder. Next, the piston moves upward to compress the air-fuel mixture. In the third cycle, the compressed air-fuel mixture is ignited and the gas expansion from this burning drives the piston back down. Finally, in the fourth cycle, the piston travels back up and forces the exhaust gases out of the cylinder. Most of the exhaust gas is composed of N_2, water vapor, and CO_2. Other exhaust gases in smaller amounts include CO from incomplete combustion, light hydrocarbons that did not burn, and particulate matter that is mostly soot. Also, depending on the starting composition of the gasoline, you can have ppm levels of nitrogen oxides (NO and NO_2) and sulfur dioxide (SO_2).

Unfortunately, a high percentage of the heat from combustion is not used to power the car. One reference[28] gives a detailed description of how the energy is lost. Some of these losses include heat lost to the radiator and exhaust that goes out with the exhaust gases, engine friction, combustion inefficiency, things that use energy (such as power steering, the water pump, air conditioning, etc.), drivetrain losses, energy lost while the car is at idle, braking losses that occur through friction, wind resistance, and rolling resistance caused by the deformation of the tire as it rolls down the road.

Recalling equation (1) from the Introduction, the thermal efficiency is related to the cold liquid temperature and the temperature of the combusted gases. For gasoline and diesel, the auto-ignition temperatures are 495°F (257.2°C) and 600°F (315.6°C), respectively.[29] Converting these to the absolute temperatures of 530.4K and 588.7K and assuming a cold temperature of 25°C (298.15K) gives calculated efficiencies of 44% and 49% for gasoline and diesel, respectively.

$$\eta = \frac{W}{Q_H} = 1 - \frac{T_C}{T_H}$$

In practice, however, efficiencies are lower. One reference gives a thermal efficiency of 30% for a gasoline ICEs with a compression ratio of 8 to 1.[27] A problem with gasoline ICEs is "knocking," such that you have an explosion of the fuel-air mixture during compression before the intended spark ignition. Although this has been improved through better fuels with higher octane ratings, a diesel engine was invented by Rudolf Diesel to prevent the knock problem. In this engine, only the air is compressed and the diesel fuel is injected into the cylinder after compression to prevent the early combustion with gasoline ICEs. This results in higher thermal efficiencies.

Thermal efficiencies have improved over the last 55 years. The Japanese Society of Mechanical Engineers[30] shows the improvement of thermal efficiencies since 1960. The thermal efficiency for a gasoline ICE has improved from about 25% to a value of 35% in 2015 while the diesel ICE has improved from 28% to a 2015 value of 45%. Therefore, the current thermal efficiencies for gasoline and diesel ICEs are typically 35% and 45%, respectively. This means that gasoline and diesel ICEs have not reached their theoretical limit based on the auto-ignition temperatures.

1.8. Energy Balance

For a refinery, which produces energy in the form of petroleum products, the energy balance is defined by the following equation:

$$\text{Energy Balance} = \frac{\left[\begin{array}{c}\text{Energy from refined product (such as gasoline)}\\ \text{that powers vehicle}\end{array}\right]}{\left[\text{Energy from the crude oil}\right]}$$

For gasoline, there are five steps in the "Well-to-Wheels" analysis, including (1) petroleum crude oil production; (2) transport to and storage at the refinery; (3) refining to petroleum products; (4) transport, storage, and distribution of gasoline; and (5) the gasoline energy used to power the vehicle.

According to an Argonne National Laboratory (ANL) report,[31] Three percent of the energy in crude oil is lost in step (1), 1% is lost in step (2), 14%–15.5% is lost in step (3), and 0.5% is lost in step (4). These first four steps are often referred to as "Well-to-Tank," or WTT. Using 15% for the midrange of step (3), the refining step, gives an overall WTT energy recovery of 81.2%. That is, 81.2% of the original crude oil energy ends up in the tank of the vehicle. Note that the energy losses are calculated step-by-step. For example, since 3% of the energy is lost in step (1), the 1% loss in step (2) is based on the remaining 97% and not the original 100% of the energy.

It is worth noting that step (1), the energy lost during crude oil production, increases significantly as the quality of the crude oil decreases. In another ANL report,[32] conventional crude oil recovery had a 2% loss in energy during production, but the production of heavy and sour (i.e., high in sulfur) crude oil lost over 12%! It is also worth noting that refining energy efficiency has improved since the earlier-referenced 2003 ANL report.[31] In reports from 2010[33] and 2012,[32] the energy efficiency is reported as 90.6%. Updating the WTT energy recovery using a refinery efficiency of 90.6% gives a WTT energy recovery of 86.6%.

For step (5), the gasoline energy used to power the vehicle, the thermal efficiency of a gasoline ICE was given in the previous section as 35%. Putting these five steps together and using 100 units of energy as the amount of energy in the original petroleum crude oil, we can arrive at the overall energy balance:

Petroleum Crude Oil Energy Balance for Gasoline

100 (source) → 97.0 (produce crude oil) → 96.0 (transport to and store at refinery)
→ 87.0 (operate refinery to make gasoline) → 86.6 (transport, store, and distribute gasoline)
→ 30.3 (gasoline energy that powers vehicle)

Thus, for the overall "Well-to-Wheels" energy balance, only 30.3% of the original crude oil energy is used to power the gasoline vehicle.

1.9. Maturity: Experience

There are many years of experience in oil well drilling, automobile production, and crude oil refining. Of course, these businesses are interrelated and their individual growths are related.

Surprisingly, oil wells were drilled in China as early as AD 347, with depths of up to 800 feet using drill bits attached to bamboo poles. This oil was a byproduct that was burned to evaporate the brine and produce salt.[34] It is believed, however, that the first modern oil well was drilled in Asia in 1848, on the Aspheron Peninsula northeast of Baku, Azerbaijan.[35] In the United States, George Bissell and Edwin L. Drake produced oil near Titusville, Pennsylvania, on August 28, 1859.[36] This well produced 25 barrels a day in the first year.

In the same year, French engineer J.J. Etienne Lenoir made the first ICE powered by gasoline.[37] The history of the automobile began as early as 1769, when automobiles powered by steam were made.[38] However, the gasoline-powered ICE automobile era began in 1885 in Mannheim, Germany, when Karl Benz built his first automobile. Only a few years later, in 1889 Gottlieb Daimler and Wilhelm Maybach of Stuttgart, Germany, designed their version of an automobile.

In the United States, brothers Charles and Frank Duryea became the first automobile manufacturing company in 1893.[39] Henry Ford started building automobiles in 1896, founded the Ford Motor Company in 1903, and started the first conveyor belt assembly line in 1913, producing the Model T.[40] Later, Ransom E. Olds founded the Olds Motor Vehicle Company (later known as Oldsmobile) with a production line starting in 1902.

Naturally, the production of gasoline-powered automobiles created a demand for gasoline. Oil refineries did exist before the gasoline-powered ICE automobile, and the first oil refinery, using distillation to produce kerosene, was built in 1862.[37] With the construction of gasoline-powered automobiles, the demand for gasoline led to the development in 1913 of a process called thermal cracking, leading to the greatly increased production of gasoline and diesel. As discussed earlier in "Overview of

Technology," the modern US oil refinery is a very complex plant designed to maximize gasoline product.

To summarize timelines in the United States, the first oil well was drilled in 1859, the first gasoline-powered automobile was built in 1893, and the change of the refinery focus from kerosene to gasoline happened in 1913. Thus, in the United States, there are more than 100 years of experience in this industry.

1.10. Infrastructure

The petroleum industry is an important baseline for comparison with other energy types that could compete with the production of transportation fuels. The current infrastructure is immense and has made it very convenient to use gasoline and diesel-powered automobiles, thereby making it difficult for competitive energy types to compete and provide the same level of convenience.

As discussed, once crude oil is produced at the well, it must be transported to a refinery to upgrade the oil to gasoline, diesel, and other valuable products. The AOPL (Association of Oil Pipe Lines) 2009 annual report[41] shows that pipelines account for 70% of all petroleum transportation. According to this report, pipelines transported 70.2% of crude oil and products while water carriers accounted for 23.1%, motor carriers for 4.2%, and railroads 2.6%. The advantage of pipelines is significant. For example, replacing a pipeline transporting 150,000 BPD would require, for the same day, 750 tanker trucks or a 225-car railroad train.[42]

Also, there is an extensive network of both crude oil and petroleum product pipelines in the United States. For crude oil, there are about 55,000 miles of pipelines, typically 8 to 24 inches in diameter.[43] For petroleum products, there are about 95,000 miles, varying in size from 8 to 42 inches in diameter. Thus, all totaled, there are around 150,000 miles of crude oil and petroleum product pipelines. There are also natural gas pipelines, which are discussed in another chapter.

For crude oil, the pipelines deliver the oil for refineries to upgrade. According to a 2018 EIA report,[44] 135 refineries were operable, with none idle. These had a total operating capacity of 18.6 million BPD, 52% of that located along the Texas and Louisiana Gulf Coast.

Once the gasoline and diesel are produced, pipelines and other transportation methods are used to deliver the products to filling stations. According to the National Association of Convenience Stores (NACS), there were 121,998 convenience stores selling motor fuels in 2019, and these represent about 80% of the motor fuels purchased in the United States.[45] So, in 2019 there were more than 150,000 filling stations in the United States. Using these filling stations, according to a 2019 Information Handling Services (IHS) report,[46] were more than 278 million light vehicles with an average age of 11.8 years.

In summary, people who drive gasoline and diesel-powered automobiles have the convenience of an infrastructure that moves crude oil to refineries, moves refined products from refineries to filling stations, and provides the convenience of adding a liquid fuel to your automobile in just 2 to 3 minutes. In addition, even if gasoline and diesel produced from crude oil were replaced with another energy type, the turnover of automobiles will take about 10 years, meaning that the gasoline and diesel-based transportation industry is well entrenched and it will take decades to replace it with something different.

1.11. Footprint and Energy Density

The footprint for petroleum includes all the land area needed to bring gasoline and diesel to the fueling station including (1) area to drill and produce oil, (2) transport the oil to the refinery, (3) the refinery, and (4) transport the refined products of gasoline and diesel to the fueling station.

Using federal data for oil wells, according to the Bureau of Land Management (BLM) and Bureau of Ocean Energy Management (BOEM)[47] there were 11.1 million acres of onshore and 6.6 million acres of offshore federal land producing commercial volumes of oil in 2012. For these acres, onshore and offshore produced 339,700 BPD and 1,303,300 BPD, respectively for all of 2012. Stated differently, onshore oil production was 0.031 BPD/acre, offshore was 0.197 BPD/acre, and overall was 0.093 BPD/acre. Since there are 640 acres in a square mile, onshore oil production was 19.6 BPD/square mile, offshore was 126.4 BPD/square mile, and overall was 59.4 BPD/square mile. A "typical" barrel of crude oil contains 5,800,000 BTU. Converting BPD to units of energy and using the overall value of 59.4 BPD/square mile, the (1) energy density to drill and produce oil is 344.6 million BTU/(day-square mile) or 344.6×10^6 BTU/(day-square mile). Just as a refinery does not operate all of the time (capacity factor less than 100%), likewise not all oil wells operate continuously over the course of a year. However, the oil production data reported by the BLM and BOEM already take this into account, so no further correction was used.

For the second step in the overall "Well-to-Wheels" process, (2) transporting the oil to the refinery, three methods are commonly used. These include transporting crude oil via pipelines, oil tankers and barges, and trains. While trucks can be used to transport gasoline and diesel to fueling stations, they are not generally used for crude oil because most refineries process greater than 100,000 BPD. Pipelines have a negligible contribution to overall area and generally can coexist with farming and pasture land. Oil tankers transport oil to refineries that are based on coasts and therefore do not really impact land area. Trains move over tracks which can be used for other purposes. While it can be argued that some railroad tracks are used mainly for oil transportation and would not otherwise exist, the land area for railroad tracks is negligible compared to that used for drilling oil. Therefore, land area used for transporting the oil to the refinery will not be included in calculating the overall footprint.

For (3) the refinery land area, data for some typical US refineries were used to calculate a footprint. Table 1.7 shows that for the ten refineries examined, there is a

Table 1.7 Land Area for Some Typical US Refineries

Company	Refinery Location	Nameplate Capacity, BPD	Acreage	Square Miles	BPD/square mile
BP[48]	Carson, CA	265,000	630	0.98	270,408
BP[48]	Cherry Point, WA	230,000	2,400	3.75	61,333
BP[48]	Toledo, OH	160,000	585	0.91	175,824
ExxonMobil[49]	Torrance, CA	155,000	750	1.17	132,479
Chevron[50]	Richmond, CA	240,000	2,900	4.53	52,980
Chevron[51]	Pascagoula, LA	330,000	3,000	4.69	70,362
Phillips 66[52]	Wood River, IL	306,000	1,888	2.95	103,729
Phillips 66[52]	San Francisco, CA	80,000	1,100	1.72	46,512
Phillips 66[52]	Lake Charles, LA	239,000	690	1.08	221,296
Shell[53]	Deer Park, TX	340,000	1,500	2.34	145,299
Average		234,500	1,544	2.41	

rather wide range of land area on a BPD basis. Presumably, this is because different refineries have different equipment used in their approach to refining, different wastewater treatment, and different storage facilities, and they may have land they maintain for expansion. In addition, a refinery near a large metropolitan area may require more land area because of a need to have a greater barrier between the refinery and community. Indeed, the two refineries in the table showing the lowest BPD/square mile are the Chevron refinery in Richmond, CA, and the Phillips 66 refinery in San Francisco, CA.

As the table illustrates, the average refinery capacity is 234,500 BPD, which covers an average area of 2.41 square miles. Using these averages, we can calculate a value of 97,303 BPD/square mile for the energy density of a typical refinery. However, earlier an average capacity factor of 93.1% was established for a typical refinery. Since the data in the above table are the nameplate capacity for the unit, the average energy density must be corrected for the capacity factor, so the average value becomes 90,589 BPD/square mile instead. Converting BPD to units of energy and using 5,800,000 BTU as the typical energy content for a barrel of crude oil, the (3) energy density to refine oil is 52.5 billion BTU/(day-square mile) or 52.5×10^9 BTU/(day-square mile).

The land area needed to (4) transport the refined products of gasoline and diesel to the fueling station is negligible to the land area used to drill and produce oil and to refine oil. Refined products are generally transported by pipeline, train, or truck to the fueling station. Pipelines do not contribute very much to land area and generally can coexist with farming and pasture land, trains move over tracks which can be used for other purposes, and trucks move over roads that already exist.

Therefore, the footprint and energy density for petroleum is really based on the land area (1) to drill and produce oil and (3) the refinery. Although not directly applicable to petroleum crude oil and refining, the energy unit that will be used to compare energy densities is gigawatt-hour per square mile per year, or GWh/(square mile-year). The conversion from BTU to Watt-hour (Wh) is shown here:

$$1 \text{ BTU} = 0.293071 \text{ Wh}$$

Using this conversion and recognizing that a gigawatt (GW) is equivalent to one billion Watts (or 1×10^9), we can calculate the energy density to drill and produce oil as 36.8 GWh/(year-square mile) and the energy density to refine oil as about 56,000 GWh/(year-square mile).

Given that the energy density for refining is about 1,500 times that for drilling and production, it is clear that the land area used for oil and production dwarfs the land area needed for refining and essentially controls the overall energy density for the "Well-to-Wheels" process. An interesting part of this analysis is the future status for the land used in drilling and production. If the land can be remediated and returned to farming and pasture land, or other types of use, then the energy density for petroleum crude oil refining would be 1,500 times greater.

1.12. Environmental Issues

There have certainly been some catastrophic accidents in the petroleum industry. Some fairly recent ones include the British Petroleum Deepwater Horizon oil spill in April to July 2010 and the oil spill in Mayflower, Arkansas, on March 29, 2013. In the case of Deepwater Horizon, this was a problem at an oil drilling site in the Gulf of Mexico, resulting in an estimated oil release of 4.9 million barrels. In the case of Mayflower, Arkansas, the ExxonMobil Pegasus pipeline ruptured and spilled around 12,000 barrels of Canadian crude oil. From an environmental aspect, these spills have a large effect on water, fish, and birds, and sea life in the case of an ocean spill.

There are also some general environmental issues with crude oil and refining. Each crude oil is a complex mixture of many organic compounds primarily composed of carbon and hydrogen. Certainly some hydrocarbons are carcinogenic but, in general, the greatest danger of hydrocarbons is their flammability. Flammable liquids are defined as liquids that flash below 100°F; liquids that flash above this temperature are defined as combustible. When combusted, hydrocarbons form CO_2 and water (H_2O). These compounds are generally benign, although increasing amounts of CO_2 in the atmosphere have been attributed to the cause of the earth's warming.

In addition, crude oils contain small amounts of heteroatoms (meaning not carbon or hydrogen), the most significant of which are sulfur and nitrogen, as well as small amounts of cadmium, zinc, arsenic, mercury, selenium, vanadium, and nickel. Crude oils have a wide range of sulfur content. For example, the typical economic benchmark crude oils, WTI and Brent, have 0.34% and 0.37% sulfur. However, there are oils much heavier, such as Arab Heavy, Boscan, and Maya, which have 2.87%, 5.70%, and 3.33%, respectively. Most sulfur is captured at the refinery as elemental sulfur; otherwise, when sulfur burns in transportation fuels, it forms sulfur dioxide (SO_2), which will react with water and become sulfuric acid (H_2SO_4) or sulfurous acid (H_2SO_3) in the atmosphere, components of acid rain.

In general, the nitrogen content of crude oils is less than sulfur, about 0.1 to 0.9 wt%.[54] Similar to sulfur, most nitrogen compounds in crude oil are removed at the refinery. Otherwise, when nitrogen burns in transportation fuels, it forms nitrogen dioxide (NO_2), which will react with water and form nitric acid (HNO_3) in the atmosphere, another component of acid rain.

At the refinery, sulfur is mostly removed from the transportation fuels. That is because both gasoline and diesel must meet EPA standards for sulfur content. For gasoline, a Tier 2 sulfur regulation limited sulfur to an average of 120 ppm starting in 2004, and was reduced to 30 ppm by 2006. A Tier 3 regulation of 10 ppm took effect on January 1, 2017. For diesel, a so-called ultra-low-sulfur diesel (ULSD) of 15 ppm began in 2006 with refiners having to be compliant by 2010.

The main reason for sulfur removal is to protect the catalytic converter of the vehicle. Sulfur is a poison for the catalysts in the vehicle catalytic converter. This catalytic converter does three things including (1) taking the incomplete combustion product of CO to CO_2, (2) converting VOCs to CO_2 and water, and (3) converting nitrogen oxide to nitrogen and oxygen gas. This chemistry is shown in three equations. A fourth equation shows what happens to any sulfur that remains in the transportation fuel. As mentioned earlier, any sulfur and nitrogen oxides formed will result in acid rain. Thus, the removal of sulfur in gasoline and diesel prevents acid rain formation from sulfur, and protects the catalytic converter, which converts nitrogen oxide to nitrogen gas, thereby preventing acid rain formation from nitrogen.

$$CO + \tfrac{1}{2}\,O_2 \rightarrow CO_2 \tag{1}$$

$$C_xH_y + (x+y/4)\,O_2 \rightarrow x\,CO_2 + y/2\,H_2O \tag{2}$$

$$2NO \rightarrow N_2 + O_2 \tag{3}$$

$$S + O_2 \rightarrow SO_2 \tag{4}$$

At the refinery, most sulfur is converted to hydrogen sulfide (H_2S) and then sent to a sulfur recovery unit to be converted into elemental sulfur. The conversion to H_2S is performed by a variety of units generally called hydrotreaters and cokers. If this removal is done before the FCC unit, it is called pretreatment and, if after, is called posttreatment. Nitrogen is converted into ammonia during hydrotreating processes, and can be sold for the production of fertilizer. Typically, no refinery will capture 100% of the sulfur or nitrogen, but removals are well above 90%. As mentioned, the sulfur and nitrogen not captured and leaving the refinery in the transportation fuel are addressed in the vehicle catalytic converter.

Finally, the wastewater leaving a refinery must be treated to prevent heavy metals like cadmium, zinc, arsenic, mercury, selenium, vanadium, and nickel from entering our waterways. This is a fascinating topic, and involves many regulations and processes, such as ion exchange and adsorption. However, the details of these regulations and processes are beyond the scope of this book.

1.13. CO_2 Production and the Cost of Capture and Sequestration

For petroleum crude oil, the "Well-to-Wheels" life cycle analysis for the production of Greenhouse gas emissions includes five basic steps. Following a report by the National Energy Technology Laboratory (NETL),[55] these five steps are (1) raw material acquisition (RMA), (2) raw material transport (RMT), (3) liquid fuels production (LFP), (4) product transport and refueling (PTR), and (5) vehicle or aircraft operation use (USE). Looking in more detail at Greenhouse gases made for each of these steps, RMA is the extraction of crude oil from the ground, and any processing of the oil that may occur at the well site. Raw material transport is defined as the transportation of the crude oil from the well site to the refinery. Liquid fuels production is the refining process to make liquid fuels, with boundaries starting at the entrance to the refinery and ending at the entrance into either the pipeline or fuel storage. Product transport and refueling is the transportation of the fuel products from storage to the fueling station. Finally, USE involves Greenhouse gases made during combustion in a vehicle or aircraft. Table 1.8, and taken from this report, shows the Greenhouse gas emissions made for each step for gasoline, diesel, and jet fuel, as well as the percentage for each contribution. Compared to the report, the contributions have been

Table 1.8 Greenhouse Gas Emissions for Petroleum Transportation Fuels Sold or Distributed in the United States in Year 2005 (lb CO_2/MM BTU)

Life Cycle Stage	Conventional Gasoline		Conventional Diesel		Kerosene-Based Jet Fuel	
RMA	16.1	8%	14.6	7%	15.0	7%
RMT	3.1	1%	2.9	1%	2.9	1%
LFP	21.6	10%	20.9	10%	13.2	6%
PTR	2.4	1%	2.0	1%	2.2	1%
USE	168.9	80%	169.1	81%	171.3	84%
Total Well-to-Wheels	212.1	100%	209.4	100%	204.6	100%

converted from kg CO_2/MM BTU to lb CO_2/MM BTU, to be consistent with other chapters in this book.

As is evident from the table, combustion in the vehicle or aircraft makes up 80% or more than the Greenhouse gas emissions.

In general, crude oil contains 83 to 87 wt% carbon, 10 to 14 wt% hydrogen, and other minor components. As the carbon content increases, the crude oil has a higher density. And as it goes from a lower to higher density, the crude oil is described as "light," "medium," or "heavy." Clearly, as carbon content increases, the oil will, overall, make higher levels of CO_2 Greenhouse gases. A 1999 *Oil & Gas Journal* article[56] shows composite Greenhouse gas emissions for different crude oils. "Composite" means that the Greenhouse gas emissions are calculated using a composite of contributions from gasoline, diesel, and jet fuel. Table 1.9 shows emissions for five crude oils and how they increase as the oil goes from light to heavy. Comparing the lightest in the table, Canadian Light, with the heaviest, Venezuela Very Heavy, the emissions increase by 18%. This illustrates that crude oil choice will also affect Greenhouse gas emissions. Taking an average of these five crude oils gives a value of 242.6 lb CO_2/MM BTU.

Using the average value of 242.6 lb CO_2/MM BTU and the earlier definition that a barrel of crude oil contains 5,800,000 BTU of energy, we can convert the value to 1,407 lb CO_2/BBL. Of the five contributions to Greenhouse gas production, the raw material acquisition only accounts for about 7%–8% and is produced at the well site, so this CO_2 would be difficult to capture. Also, the CO_2 produced during raw material transport, product transport, and vehicle or aircraft combustion would be very difficult to capture, as CO_2 produced by vehicle or aircraft combustion would require capture from the vehicle exhaust. Therefore, the only likely place to capture CO_2 is at the refinery and, as Table 1.8 shows, this is about 10% of the overall CO_2 production.

While, at this time, no US refineries are currently doing CO_2 capture and sequestration (CCS), some model estimates have been made. In general, CCS is a three-step process including capture of the CO_2, transportation to the storage site (likely by pipelines), and storage by injection into geological formation or depleted oil and gas fields. According to one review,[57] the cost for electric power plants will range from

Table 1.9 Greenhouse Gas Emissions from Five Different Crude Oils (lb CO_2/MM BTU)

Crude Oil	Composite Greenhouse Emissions
Canadian Light	229.2
Brent North Sea	224.7
Saudi Light	238.8
Venezuela Heavy	250.7
Venezuela Very Heavy	269.5

\$40 to \$80 per ton of CO_2 including \$30–\$50 for capture, \$5–\$20 for onshore/offshore pipeline transport, and \$5–\$10 per ton for injection and storage.

However, another reference suggests that the cost will be higher for refineries than for power plants.[58] This is because the CO_2 is made by many different units and will require new piping and pumping to route it to a central location. Thus, there will have to be modifications to existing refinery units, and there is some uncertainty in the cost since the technology has not yet been built for a refinery. As an example, the burning (or regeneration) of carbon from the catalyst in the FCC unit is done with air, which produces a flue gas composed primarily of nitrogen, CO_2, and water. After water is condensed from the flue gas it contains around 80 volume% nitrogen. It is not practical to sequester the CO_2 with such large amounts of nitrogen, so the FCC regenerator will have to use pure oxygen instead of air, and this will require significant unit modification.

A 2010 paper[59] suggests the cost will be on the order of \$90–\$120 per ton of CO_2. Of the 1,407 lb CO_2 estimated production per barrel of crude oil, only about 10% can be captured at the refinery. In other words, about 141 lb or 0.07 tons CO_2 can be captured per barrel. Assuming a cost of \$100 per ton CO_2, this means the cost per barrel is about \$7/BBL. Since a barrel contains 42 gallons, this means a cost of about \$0.17/gge. Therefore, the addition of CCS at the refinery, capturing only about 10% of the CO_2 made from the crude oil, would add about \$0.17 to the LCOF, shown in Table 1.6.

The refinery production of CO_2 comes primarily from furnaces, boilers, FCC regeneration, flares, incineration, and power generation, and the amount of carbon that goes to these activities is refinery specific. One estimate was made using a refinery material balance reported by Gary et al.[19] From this, about 9.3 wt% went to things other than products transported from the refinery. Another estimate was made from a report stating that oil refineries pumped more than 250 million tons of CO_2 into the air in 2004.[60] Using EIA data,[26] the amount of crude oil processed in US refineries in 2017 was 16.9 million BPD. From the earlier reported density of WTI (0.821 g/cm^3) and assuming a carbon content of 87 wt%, we can calculate that this crude oil contained about 2,115,000 tons of carbon per day, or about 7,755,000 tons CO_2/day. This gives an annual amount of 2,830 million tons of CO_2 that would be produced if all carbon in crude oil sent to US refineries were burned at the refinery. Using the 250 million tons for the refinery alone gives a percentage of 9% for the carbon being burned at the refinery, similar to the values in Table 1.8. These two analyses suggest it is a reasonable estimate to assume that 10 wt% of the carbon in crude oil is burned at the refinery to produce CO_2.

Carbon dioxide production is also commonly listed in the units of lb CO_2/kWh and g CO_2/kWh. Of course, this chapter has focused on the use of petroleum to make transportation fuels through refining, but petroleum can also be used to generate electricity. This is not common however, as only 0.6% of the US production of electricity in 2018 came from petroleum.[61] Nevertheless, we can determine CO_2 production in terms of kWh by using a heat rate for electricity generation from petroleum. The EIA gives an average heat rate of 10,984 BTU/kWh for petroleum,[62] which includes distillate fuel oil, residual fuel oil, jet fuel, kerosene, petroleum coke, and waste oil. Using this heat rate, we can convert 242.6 lb CO_2/MM BTU to a kWh basis as 2.66 lb CO_2/kWh and 1,209 g CO_2/kWh.

In summary, the amount of CO_2 produced by a barrel of oil is 1,407 lb, equivalent to 242.6 lb CO_2/MM BTU, 2.66 lb CO_2/kWh and 1,209 g CO_2/kWh. Adding CCS at the refinery would only capture about 10% of this CO_2 and would add about \$0.17 to the LCOF.

1.14. Chapter Summary

A review of worldwide crude oil proven reserves shows an estimate of 1.7 trillion barrels which, at the current consumption rate of 98 MBPD, leads to a calculated value of 47 years remaining for crude oil. The number of years remaining could decrease if developing countries increase consumption, or could increase if proven reserves increase due to deepwater oil and fracking technologies. For these proven reserves, ten countries including Venezuela, Saudi Arabia, Canada, Iran, Iraq, Russia, Kuwait, the United Arab Emirates, the United States, and Libya control 85.3% of the worldwide proven reserves.

The "Well-to-Wheels" economic analysis gave an LCOF of \$2.25/gge, which can be converted to \$67.41/MWh or \$114.58/BOE. Doing CCS at the refinery would add an additional cost of \$0.17/gge, increasing the LCOF with CCS to \$2.42/gge. \$0.17/gge is equivalent to ~\$7/BBL of crude oil processed. In addition, the consumer must pay federal and state taxes for gasoline. The federal tax is \$0.184/gge, and the state taxes range from a high of \$0.514 for Pennsylvania to a low of \$0.1225 for Alaska, with an average state tax of \$0.2978/gge. Using the average state tax, the addition of federal and state taxes increases the total cost to \$2.73/gge.

The production of CO_2 was calculated to be 242.6 lb CO_2/MM BTU, equivalent to 2.66 lb CO_2/kWh and 1,209 g CO_2/kWh. Since about 90% of the CO_2 production takes place outside of the refinery, only about 10% can potentially be captured at the refinery.

The cost of the crude oil, on the basis of energy content, is \$11.20/MM BTU using a 2018 cost for WTI of \$64.94/BBL. On a mass basis, the cost of crude oil is \$0.23/lb.

On the average, refineries operate with a high capacity factor, 93.1% for 2018. Compared to the starting energy in the crude oil, 86.6% is available at the pump in the form of transportation fuel. Unfortunately, because of the low 35% thermal efficiency of a gasoline ICE, the overall energy balance is only 30.3%, meaning that 30.3% of the original crude oil energy is available to power the gasoline vehicle.

In terms of experience and maturity, the petroleum industry is quite mature, and oil wells and automobiles have been around for more than 100 years. Not surprisingly, the current infrastructure is immense and the use of gasoline or diesel powered vehicles is very convenient. To transport crude oil and petroleum products, we have around 150,000 miles of pipelines in the United States. To convert the crude oil into products, the United States currently has 135 refineries with an operating capacity of 18.3 MBPD. Once transportation products are made, there are more than 150,000 filling stations in the United States to fuel 278 million light vehicles.

The refinery footprint is quite small, about 2.2 square miles for a 200,000 BPD refinery. Based on this area, the energy density is an impressive 56,000 GWh/(year-square mile). However, the land area needed to produce oil for this 200,000 BPD refinery is much greater, decreasing the energy density to 36.8 GWh/(year-square mile). The land area footprint to produce oil to supply this 200,000 BPD refinery is around 3,370 square miles. Thus, the land needed for oil drilling and production dwarfs that of the refinery, effectively lowering the energy density by a factor of ~1,500.

Finally, most potential environmental issues for crude oil are resolved at the refinery. Most of the sulfur and a substantial amount of the nitrogen are captured there. Sulfur is now controlled to very low levels in both gasoline and diesel according to EPA regulations. Any remaining nitrogen in gasoline and diesel is converted to nitrogen gas in the vehicle catalyst converter. Also, any heavy metals originally in the crude oil are removed from the refinery wastewater streams before the water is returned to our waterways. Although CO_2 is not considered a pollutant at this time, only about 10% of the CO_2 produced from the crude oil can potentially be captured at the refinery.

References

1. *Webster's II New Riverside University Dictionary*, Boston, MA: Houghton Mifflin, 1984.
2. Paul Mann et al., "Tectonic Setting of the World's Giant Oil and Gas Fields," in Michel T. Halbouty (ed.) *Giant Oil and Gas Fields of the Decade, 1990–1999*, Tulsa, Okla: American Association of Petroleum Geologists, p. 50, accessed June 22, 2009.
3. "Worldwide Oil, Natural Gas Reserves Exhibit Marginal Increases," C. Xu and L. Bell, *Oil & Gas Journal*, December 3, 2018, https://www.ogj.com/articles/print/volume-116/issue-12/special-report-worldwide-report/worldwide-oil-natural-gas-reserves-exhibit-marginal-increases.html
4. "BP Statistical Review of World Energy," June 2018, https://www.bp.com/content/dam/bp/business-sites/en/global/corporate/pdfs/energy-economics/statistical-review/bp-stats-review-2018-full-report.pdf
5. "International Energy Statistics," U.S. Energy Information Administration, 2018 data, https://www.eia.gov/cfapps/ipdbproject/IEDIndex3.cfm?tid=5&pid=57&aid=6
6. "Countries With The Highest Fracking Potential," Shobhit Seth, September 16, 2014, https://www.investopedia.com/articles/investing/091614/countries-highest-fracking-potential.asp
7. "An Overview of Hydraulic Fracturing and Other Formation Stimulation Technologies for Shale Gas Production," Update 2015, Luca Gandossi and Ulrik Von Estorff, Publications Office of the European Union.
8. *Deepwater Petroleum—Exploration & Production—A Nontechnical Guide*, 2nd Edition, W.L. Leffler et al., Tulsa, OK: PennWell Corporation, 2011, p. 41 and 339.
9. *Oil and Gas Production in Nontechnical Language*, M.S. Raymond and W.L. Leffler, Tulsa OK: PennWell Corporation, 2006.

10. "Offshore Oil and Gas Production," Planete Energies, August 11, 2015, https://www.planete-energies.com/en/medias/close/offshore-oil-and-gas-production; "Maersk Drillship Spuds World's Deepest Well," Mike Schuler, April 1, 2016, https://gcaptain.com/maersk-venturer-begins-drilling-worlds-deepest-well/
11. "Enhanced Oil Recovery Using Carbon Dioxide in the European Energy System," E. Tzimas et al., *Institute for Energy*, Petten, The Netherlands, December 2005, http://www.claverton-energy.com/wordpress/wp-content/files/Tzimas_CO2_EOR_Report.pdf
12. "Enhanced Oil Recovery," U.S. Department of Energy, 2020, http://energy.gov/fe/science-innovation/oil-gas-research/enhanced-oil-recovery
13. "Increasing Recovery Factors: A Necessity," Alain Labastie, Journal of Petroleum Technology, 12–13, August 2011.
14. "Barrel Breakdown," *news graphic*, Wall Street Journal, April 15, 2016, http://graphics.wsj.com/oil-barrel-breakdown/
15. *Oil & Gas Pipelines in Nontechnical Language*, T.O. Miesner and W.L. Leffler, Tulsa, OK: PennWell Corporation, 2006.
16. "Table 1-10, U.S. Oil and Gas Pipeline Mileage," Bureau of Transportation Statistics, 2014, http://www.rita.dot.gov/bts/sites/rita.dot.gov.bts/files/publications/national_transportation_statistics/html/table_01_10.html
17. U.S. Department of Transportation, Federal Highway Administration, November 8, 2015, https://www.fhwa.dot.gov/interstate/faq.cfm#question3
18. "About Pipelines," Association of Oil Pipe Lines, 2014, http://www.aopl.org/page/pipeline-basics
19. *Petroleum Refining—Technology and Economics*, 5th Edition, James H. Gary, et al., Boca Raton, FL: CRC Press, 2007, p. 1.
20. "Evaluation of Crude Oil Quality," D. Stratiev et al., *Petroleum & Coal* 52(1) 35–43, February 1, 2010.
21. "Heavy Oil Upgrading," *Eloy Flores III, Technology Today*, 10–15, Spring, 2010.
22. "Removing Gasoline Sulfur," Paul F. Meier et al., *Hydrocarbon Engineering*, January 2001, 5 pages.
23. "Average Annual West Texas Intermediate (WTI) Crude Oil Price from 1976 to 2016 (in U.S. dollars per Barrel)," The Statistics Portal, 2017, http://www.statista.com/statistics/266659/west-texas-intermediate-oil-prices/
24. "Gasoline and Diesel Fuel Update," Energy Information Administration, April 1, 2019, http://www.eia.gov/petroleum/gasdiesel/
25. "State Motor Fuel Taxes: Notes Summary," American Petroleum Institute, July 1, 2016, http://www.api.org/~/media/Files/Statistics/State-Motor-Fuel-Excise-Tax-Update-July-2016.pdf
26. "Petroleum & Other Liquids: Refinery Utilization and Capacity," US Energy Information Administration, 2018 annual average, https://www.eia.gov/dnav/pet/pet_pnp_unc_dcu_nus_a.htm
27. "Hydrocarbon-Fueled Internal Combustion Engines: The Worst Form of Vehicle Propulsion ... Except for All the Other Forms," USC Viterbi School of Engineering, Spring 2002, http://ronney.usc.edu/whyicengines/WhyICEngines.pdf
28. "Fuel Economy: Where the Energy Goes," April 5, 2013, http://www.fueleconomy.gov/feg/atv.shtml
29. "Alternative Fuels Data Center—Fuel Properties Comparison," US Department of Energy, October 29, 2014, http://www.afdc.energy.gov/fuels/fuel_comparison_chart.pdf
30. "Thermal Efficiency of Engines," The Japan Society of Mechanical Engineers, http://www.jsme.or.jp/English/jsme%20roadmap/No-7.pdf
31. "Well-to-Wheels Energy and Emission Impacts of Vehicle/Fuel Systems: Development and Applications of the GREET Model," Michael Wang, Center for Transportation Research,

Argonne National Laboratory, California Air Resources Board, Sacramento, CA, April 14, 2003, https://greet.es.anl.gov/files/ea30hyon
32. "Well-to-Wheels Energy Use and Greenhouse Gas Emissions of Ethanol from Corn, Sugarcane and Cellulosic Biomass for US Use," Michael Wang et al., Environmental Research Letters 7 (2012), 13 pages, Argonne National Laboratory, December 13, 2012, http://www.ethanolrfa.org/wp-content/uploads/2015/09/Well-to-wheels-energy-use-and-greenhouse-gas-emissions-of-ethanol-from-corn-sugarcane-and-cellulosic-biomass-for-US-use.pdf
33. "Updated Estimation of Energy Efficiencies of U.S. Petroleum Refineries," Ignasi Palou-Rivera and Michael Wang, Center for Transportation Research, Argonne National Laboratory, July, 2010, 6 pages, http://www.transportation.anl.gov/pdfs/TA/635.PDF
34. "A Timeline of Highlights from the Histories of ASTM Committee D02 and the Petroleum Industry," George E. Totten, June 2014, http://www.astm.org/COMMIT/D02/to1899_index.html
35. "First Oklahoma Oil Was Produced in 1859," Muriel H. Wright, Chronicles of Oklahoma 4 (4), Oklahoma City, OK: Oklahoma Historical Society, December 1926.
36. *Trek of the Oil Finders*, Edgar Wesley Owen, Tulsa, Oklahoma, American Association of Petroleum Geologists, January 1, 1975, p. 12, https://www.amazon.com/Trek-Oil-Finders-Exploration-Petroleum/dp/B000YZG2IU
37. "How Oil Refining Transformed U.S. History and Way of Life," Katrina C. Arabe, January 17, 2003, http://news.thomasnet.com/IMT/2003/01/17/how_oil_refinin/
38. *World History of the Automobile*, Erik Eckermann, SAE Press, 2001, p. 14, https://www.sae.org/publications/books/content/r-272/
39. *Cars: Early and Vintage, 1886–1930*, G.N. Georgano, Nordbok, Sweden, 1985.
40. "Ford Motor Company History," 2004, http://www.fundinguniverse.com/company-histories/ford-motor-company-history/
41. "Intermodal Safety in the Transport of Oil," Diana Furchtgott-Roth and Kenneth P. Green, Fraser Institute, October 2013, https://www.fraserinstitute.org/sites/default/files/intermodal-safety-in-the-transport-of-oil-rev3.pdf
42. "Pipelines in the U.S.," American Association of Pipelines (AOPL), 2016, https://pipeline101.org/Why-Do-We-Need-Pipelines/Other-Means-Of-Transport
43. "How Pipelines Make the Oil Market Work—Their Networks, Operation and Regulation," A Memorandum Prepared for the Association of Oil Pipe Lines And the American Petroleum Institute's Pipeline Committee, Allegro Energy Group, December 2001, https://www.iatp.org/documents/how-pipelines-make-the-oil-market-work-their-networks-operation-and-regulation
44. "Number and Capacity of Petroleum Refineries," 2018 data, Energy Information Administration, https://www.eia.gov/dnav/pet/pet_pnp_cap1_dcu_nus_a.htm
45. "U.S. Convenience Store Count," NACS (National Association of Convenience Stores), 2020, https://www.convenience.org/Research/FactSheets/ScopeofIndustry/IndustryStoreCount
46. "IHS Markit: Average Age of Cars and Light Trucks in US Rises Again in 2019 to 11.8 Years," Green Car Congress, June 28, 2019, https://www.greencarcongress.com/2019/06/20190628-ihsmarkit.html
47. "U.S. Crude Oil and Natural Gas Production in Federal and Non-Federal Areas," Marc Humphries, Specialist in Energy Policy, April 10, 2014, Congressional Research Service, 7-5700, R42432, https://fas.org/sgp/crs/misc/R42432.pdf
48. BP Refinery Data, 2016, http://www.bp.com/sectiongenericarticle.do?categoryId=9050573&contentId=7085114
49. "Exxon Mobil's Outdated Equipment and Procedures Led to Torrance Explosion," May 3, 2017, https://www.latimes.com/business/la-fi-exxon-mobil-refinery-20170502-story.html

50. "Chevron Access Needed for Richmond Bay Trail Link," Genevieve Dubosca, Berkeley Daily Planet, March 27, 2007, access date June 4, 2009, http://www.berkeleydailyplanet.com/issue/2007-03-27/article/26643
51. Chevron Refinery Data, 2020, http://pascagoula.chevron.com/about
52. Phillips 66 Refinery Data, 2020, http://www.phillips66.com/EN/about/our-businesses/refining-marketing/refining/Pages/index.aspx
53. Shell Refinery Data, 2016, http://www.shell.us/aboutshell/projects-locations/deerpark.html
54. *The Chemistry and Technology of Petroleum*, Fourth Edition, James G. Speight, Boca Raton, FL: CRC Press, December 2010, p. 191.
55. "Development of Baseline Data and Analysis of Life Cycle Greenhouse Gas Emissions of Petroleum-Based Fuels," DOE/NETL-2009/1346, Timothy J. Skone and Kristin Gerdes, National Energy Technology Laboratory, November 26, 2008, https://www.netl.doe.gov/File%20Library/Research/Energy%20Analysis/Life%20Cycle%20Analysis/NETL-LCA-Petroleum-based-Fuels-Nov-2008.pdf
56. "Crude Oil Greenhouse Gas Life Cycle Analysis Helps Assign Values For CO 2 Emissions Trading," Tom McCann and Phil Magee, Oil & Gas Journal, February 22, 1999, http://www.ogj.com/articles/print/volume-97/issue-8/in-this-issue/general-interest/crude-oil-greenhouse-gas-life-cycle-analysis-helps-assign-values-for-co-2-emissions-trading.html
57. "CO_2 Capture & Storage," Energy Technology Systems Analysis Programme, October 2010, https://iea-etsap.org/E-TechDS/PDF/E14_CCS_oct2010_GS_gc_AD_gs.pdf
58. "Potential for CO_2 Capture and Storage in EU Refineries," Alan Reid, IEAGHG and IETS Workshop on CCS in Process Industries, Lisbon, March 10–11, 2015, https://ieaghg.org/docs/General_Docs/Lisbon%20presentations%20for%20website/14%20-%20A.%20Reid%20(Concawe).pdf
59. "CO2 Capture for Refineries, a Practical Approach," Jiri van Staelen et al., International Journal of Greenhouse Gas Control 4 (2), 316–320, February 2010, https://www.researchgate.net/publication/229195197_CO2_capture_for_refineries_a_practical_approach
60. "Oil Refineries Targeted for Global Warming Emissions Cuts," Earth Justice, August 28, 2007, http://earthjustice.org/news/press/2007/oil-refineries-targeted-for-global-warming-emissions-cuts
61. "What is U.S. Electricity Generation by Energy Source?," US Energy Information Administration, 2018, https://www.eia.gov/tools/faqs/faq.cfm?id=427&t=3
62. "Annual Energy Review 2011," Table A6, DOE/EIA-0384(2011), US Energy Information Administration, September 2012, https://www.eia.gov/totalenergy/data/annual/pdf/aer.pdf

2
Natural Gas
A Nonrenewable Energy Type

2.1. Foreword

Unlike petroleum, which is a complex mixture of hydrocarbons, natural gas is a rather simple mixture by comparison. This hydrocarbon mixture is mainly composed of methane, but has varying amounts of higher hydrocarbons, such as ethane, propane, and butane. In addition, it will also contain small amounts of hydrogen sulfide (H_2S), carbon dioxide (CO_2), and nitrogen. Like petroleum, it is a fossil fuel, formed by anaerobic decomposition of dead plants and organisms (diatoms) in a process taking millions of years. It can be found alone, with crude oil well production, and with coal mining.

And like petroleum, this transformation took place in the interior of the earth, under intense heat and pressure. Thus, geothermal energy was used in the formation of natural gas. A typical composition and range of components for natural gas are shown in Table 2.1.[1] As the table illustrates, methane and a small amount of ethane are the primary components for natural gas. Also, like petroleum crude oil, olefins are not found in natural gas. And as is the case with crude oil, the presence of heteroatoms such as nitrogen and sulfur will eventually create nitrogen and sulfur oxides, molecules that need to be controlled for environmental reasons.

2.2. Proven Reserves

Like crude oil, estimating proven reserves of natural gas in the world can be difficult, as new technology unleashes new supplies. One new technology that is increasing natural gas production is fracking, or hydraulic fracturing. Fracking is fracturing rock using the injection of high pressure fluids, typically water mixed with sand and other chemicals. This injection will make new channels in the rock, resulting in higher gas production. According to the International Business Times,[2] natural gas production in the U.S. increased 14% from 2005 to 2013 due to fracking. Production is still increasing and, according to a British Petroleum (BP) review of world energy,[3] worldwide natural gas production increased 25% from 2007 to 2017 while US production increased by 41%.

The effect of fracking also shows up in worldwide proven reserves, which increased from 5,776 trillion cubic feet (tcf) in 2007 to 6,832 tcf in 2017, an 18% increase. And,

The Changing Energy Mix. Paul F. Meier, Oxford University Press (2020). © Oxford University Press.
DOI: 10.1093/oso/9780190098391.001.0001.

Table 2.1 Typical Composition and Range for Natural Gas[1]

Component	Typical Analysis (mole %)	Range (mole %)
Methane (CH_4)	94.9	87.0–96.0
Ethane (C_2H_6)	2.5	1.8–5.1
Propane (C_3H_8)	0.2	0.1–1.5
Isobutane (C_4H_{10})	0.03	0.01–0.3
Normal Butane (C_4H_{10})	0.03	0.01–0.3
Isopentane (C_5H_{12})	0.01	trace–0.14
Normal Pentane (C_5H_{12})	0.01	trace–0.04
Hexanes plus (C_6H_{14}+)	0.01	trace–0.06
Nitrogen (N_2)	1.6	1.3–5.6
Carbon Dioxide (CO_2)	0.7	0.1–1.0
Oxygen (O_2)	0.02	0.01–0.1
Hydrogen trace (H_2)	trace	trace–0.02
Hydrogen Sulfide (H_2S)	7 ppm	7–10 ppm

in the United States, proven reserves increased from 228 tcf in 2007 to 309 tcf in 2017, a 36% increase.

In terms of proven reserves, the BP report gives worldwide reserves of 6,832 tcf in 2017. This is similar to the proven reserves reported by the Energy Information Administration (EIA), citing a value of 6,923 tcf in 2017.[4] The top ten countries in the world are shown in Table 2.2, accounting for 78.8% of the worldwide natural gas reserves. By themselves, Russia and Iran account for 18.1% and 17.2%, respectively.

The top ten natural gas producing countries are shown in Table 2.3. The United States and Russia are, by far, the top two natural gas producing countries in the world, producing 20.0% and 17.3% of the world total, respectively. These top ten countries account for 68.9% of the worldwide natural gas production, or about two-thirds.

Finally, the BP report shows that natural gas consumption in 2017 was 129,619 billion cubic feet.[3] Table 2.4 shows the top ten consuming countries, with the top ten countries accounting for 60.5% of consumption. Like production, the United States and Russia were the top two consuming countries, 20.1% and 11.6%, respectively.

Comparing consumption to production, we see that Japan, Germany, the United Kingdom are top ten consuming countries but are not in the top ten for production. The reverse of this is that Qatar, Norway, Australia, and Algeria are in the top ten producing countries but not in the top ten for consumption. Also, comparing the two tables shows that Russia produces more natural gas than they consume while the United States, Iran, and Saudi Arabia consume about what they produce.

Table 2.2 Top Ten Natural Gas Reserves Countries, 2017 (tcf)[3]

Total World	**6831.7**
Russia	1,234.9
Iran	1,173.0
Qatar	879.9
Turkmenistan	688.1
United States	308.5
Saudi Arabia	283.8
Venezuela	225.0
United Arab Emirates	209.7
China	193.5
Nigeria	183.7

Table 2.3 Top Ten Natural Gas Producing Countries, 2017 (Billion Cubic Feet per year)[3]

Total World	**129,971**
United States	25,939
Russia	22,445
Iran	7,907
Canada	6,226
Qatar	6,205
China	5,269
Norway	4,352
Australia	4,007
Saudi Arabia	3,935
Algeria	3,222

Natural gas consumption likely will continue to grow as more ways are found to use natural gas in the energy pool. In addition to electricity generation, there may be an increase in natural-gas powered vehicles, as well as use of the Fischer-Tropsch synthesis to convert natural gas into gasoline and diesel. Synthetic fuel formed from the Fischer-Tropsch reaction is discussed in a Chapter 12.

Table 2.4 Top Ten Natural Gas Consuming Countries, 2017 (Billion Cubic Feet per year)[3]

Total World	129,619
United States	26,114
Russia	15,000
China	8,491
Iran	7,573
Japan	4,135
Canada	4,088
Saudi Arabia	3,935
Germany	3,184
Mexico	3,094
United Kingdom	2,782

Using the current estimate for proven reserves of 6,832 tcf and the current consumption rate of 129.6 tcf per year gives a calculated value of 53 years remaining for natural gas. Therefore, like crude oil, natural gas is a dwindling commodity with a possibility of using up the world supply around 2070. Fracking could greatly extend natural gas supplies, while increasing use of natural-gas powered vehicles and the synthesis of liquid fuels from natural gas could result in a significant decrease in the estimated remaining years.

2.3. Overview of Technology

Natural gas has many uses, including electricity generation, home heating, cooking, transportation, fertilizers, and hydrogen generation. It is a flammable gas, consisting largely of methane and other light hydrocarbons, and occurs naturally underground. Some wells only produce natural gas while, in others, the natural gas is associated with crude oil.

Like crude oil, the formation of natural gas occurred over millions of years in the interior of the earth under intense heat and pressure. And, like crude oil, natural gas is formed from fossils of dead plants and creatures using the earth's geothermal energy. Therefore, the energy contained in natural gas comes partly from the sun that was used to grow the plants and partly from geothermal energy.

As mentioned, natural gas can be found alone or associated with crude oil or coal beds. Generally, natural gas can be found as "dry gas" or "wet gas."[5] Dry gas is a natural gas stream that consists almost entirely of methane, with a few percent of ethane and propane. Wet gas is a natural gas stream that consists of high percentage of methane (80%–90%) plus natural gas liquids (NGLs). Natural gas liquids include ethane, propane, butane, and natural gasoline as well as water, CO_2, O_2, nitrogen, and H_2S.

Before natural gas is used as a fuel, it must first be treated and processed. Compared to refining crude oil, treatment and processing is less complicated. If the natural gas is "dry gas," essentially methane and little heavy hydrocarbons, the gas can go to the pipeline with no further processing. If it is "wet gas," however, the heavy hydrocarbons and water need to be removed, otherwise they can condense into liquids in the pipeline. This will eventually cause blockage, especially for low spots in the pipeline.[5] Therefore, raw natural gas from a production well is normally treated, to remove liquid water and heavy hydrocarbons that are condensed into liquids, before being sent to a gas processing plant for further treatment.

At the gas processing plant, the initial purification step is the removal of acid gases, mostly H_2S and CO_2. If natural gas does not contain H_2S, it is called sweet gas. H_2S, as well as CO_2 and water, can cause corrosion in pipelines and other equipment.[5] In addition, depending on the length of exposure, H_2S is deadly to humans at a level of only 5 ppm. Therefore, these acid gases must be removed.

While there are many processes used for removing H_2S and CO_2, the most common is amine absorption. An amine is an organic compound in which the hydrogen atoms on ammonia (NH_3) are replaced with organic groups. In this process, natural gas is contacted with an amine solution, which will chemically absorb both CO_2 and H_2S and produce natural gas free of these acid gases. Then, the amine can be heated to strip out the CO_2 and H_2S, so it can be used again. The reaction is shown as follows, in which "R" refers to the organic hydrocarbon making up the amine. Some common amines used are MEA (monoethanolamine), MDEA (methyldiethanolamine), and DEA (diethanolamine).

$$2\ RNH_2 + H_2S \rightarrow (RNH_3)_2\ S$$

After removal, H_2S can be converted into elemental sulfur using the Claus process (see the next chemistry equation) or into sulfuric acid using the wet sulfuric acid process. At this time, the CO_2 can simply be released into the atmosphere.

$$2\ H_2S + O_2 \rightarrow 2S + 2\ H_2O$$

After removal of the acid gases, the next step is to remove water that was not removed at the production well. Two common methods of removal are glycol absorption and adsorption using solid material. Before discussing these two options, it is useful to understand the difference between absorption and adsorption. Absorption is a process such that a gas or liquid is dissolved by some liquid or solid absorbent. In other words, the volume of the absorbent is used to hold the absorbed gas or liquid material. On the other hand, adsorption is a process such that the surface of the adsorbent is used to hold the gas or liquid.

In glycol dehydration,[6] the glycol removes the water from the natural gas by absorption. A glycol is an alcohol containing two hydroxyl groups, or "OH" groups. This absorption occurs because the glycol has an affinity to absorb the water but not the natural gas. A common glycol used is triethylene glycol, or TEG. Its chemical formula is shown here:

$$OH\text{-}CH_2\text{-}CH_2\text{-}O\text{-}CH_2\text{-}CH_2\text{-}O\text{-}CH_2\text{-}CH_2\text{-}OH$$

For natural gas dehydration by adsorption, a solid desiccant material is used. Typical desiccants include activated alumina (Al_2O_3), or silica gel (SiO_2).[7] In the process, water in the natural gas is adsorbed on the surface of the desiccant particles. After the adsorption tower containing the desiccant becomes saturated with water, the desiccant can be regenerated using heated gas to remove the water.

After water removal, the dry gas enters a demethanizer tower in which the methane rises and larger hydrocarbons descend. The heavier hydrocarbons are collectively called NGLs. The methane leaving the demethanizer is ready to go to the product pipeline, to be distributed and sold. Since natural gas is odorless, ethyl- or methyl-mercaptan is added to detect leaks in the pipeline. If, instead, pipelines are not available or the natural gas is transported overseas, it is transported as liquefied natural gas (LNG). The LNG is methane that has been cooled to −260°F (−162°C) in an LNG plant, making methane a liquid; in liquid form it takes up about 1/600th of the volume, thus making it more economically feasible to load and transport in tankers across the ocean from producing countries to consuming countries.[8]

If desired, the NGL can be further separated. Briefly, this further processing includes a deethanizer (to recover ethane), depropanizer (to recover propane), and a debutanizer (to recover butane). The bottom of the debutanizer tower leaves C_5 (pentanes) hydrocarbons and larger. Also, a butane splitter can be used to separate isobutane and normal butane.

According to 2018 data from the American Gas Association,[9] in the United States there are more than 210 natural gas pipeline systems with about 301,000 miles of interstate and intrastate pipelines to deliver natural gas to various users in order to generate electricity, heat homes, or other industrial uses. Adding the around 18,000 miles of pipeline used for gathering and the 1.3 million miles of local distribution lines used to deliver the natural gas to the customer, there are over 1.6 million miles altogether. Figure 2.1, taken from the EIA,[10] gives an overall picture of natural gas from production to the final users.

The last part of the technology overview discusses the use of natural gas in electric power generation. The other end-user categories shown in Figure 2.1 do not need any special technology discussion, since use of natural gas in a residential or commercial furnace is pretty straightforward.

Three methods of electricity generation from natural gas are discussed here: (1) natural gas combustion, (2) natural gas combined cycle (NGCC), and (3) NGCC with carbon capture and sequestration (CSS).

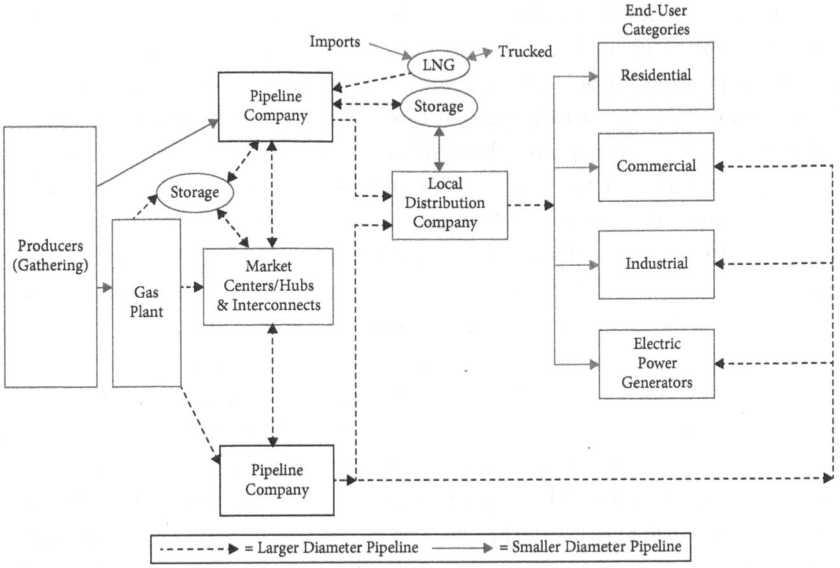

Figure 2.1 Natural Gas from Production to the Final Users (*Source*: EIA[10])

Thermal electricity generation is the process of making electricity from heat energy. The heat can be generated from many combustible materials such as coal, natural gas, nuclear fission, or biomass. Therefore, the following discussion will be revisited for other energy types that generate electricity from heat energy. In general, the production and delivery of electricity to your house involves ten steps. First, you need fuel to power the plant. Second, you need a furnace to burn the fuel and release the heat energy. Third, if your turbine uses steam, you have a boiler, such that cold water circulates through pipes and is turned into steam using the heat. This steam, at high pressure, powers the blades of a steam turbine in the fourth step. In this sense, the heat energy is converted into kinetic energy. After the steam leaves the turbine, the fifth step is the cooling of steam in the cooling tower for reuse. The turbine is connected via an axle to a generator, which uses the kinetic energy from the turbine to make electricity in the sixth step. Once this electricity is made, cables are used to transport the electricity to a transformer in the seventh step. The eighth step is the use of a transformer to step-up voltage from the power plant, to reduce transmission losses. Typically, a power plant will produce electricity at voltages from 2 to 30 kV (thousand volts) and the transformer will step-up this voltage to 115 to 765 kV. As an example of transmission losses, a 765 kV line carrying 1,000 MW of power will have losses of 0.5–1.1% over 100 miles but a 345 kV line with the same load will have losses of 4.2% over the same distance.[11] Historical data from the state of California shows annual average transmission losses around 5% to 7% over the time period of 2002 to 2008.[12] In the ninth step, metal towers are used to safely support overhead power cables and, finally, a step-down transformer is used to reduce voltage that comes into your home in the tenth step.

In the case of natural gas combustion, the turbine can be driven by steam or by the combustion gases. If the turbine is powered by steam, the previous discussion explains the production of electricity, such that the furnace and boiler get their energy and steam from the natural gas combustion. Alternatively, the natural gas combustion gases can directly power the turbine to generate electricity. That is, instead of making steam to drive the turbine, the combustion gases are used to drive the turbine. These combustion gases will have primarily nitrogen, CO_2 and steam, as well as smaller amounts of nitrogen oxides (NO_x), carbon monoxide (CO), methane (CH_4), sulfur dioxide (SO_2), volatile organic compounds (VOCs), and particulate matter (PM).[13] The VOCs are defined as organic chemicals with a high vapor pressure and include things such as benzene, ethylene glycol, formaldehyde, and toluene.

As mentioned, electricity generation from natural gas can also proceed by NGCC. For NGCC, there are two turbines—a gas turbine and steam turbine. The gas turbine operates as described previously, such that natural gas and air are compressed and ignited, causing expansion. This expansion creates pressure, which spins the turbine blades to create electricity. Next, the waste heat contained in the exhaust from the gas turbine is used to generate steam by using a heat recovery steam generator (HRSG), which then powers the steam turbine. Figure 2.2 shows a simplified view of this process, and demonstrates the use of waste heat to produce steam to power the steam turbine. A steam turbine will typically only use 33% to 35% of the thermal energy used to make steam, but because of the more efficient use of the thermal energy of natural gas combustion, the NGCC will have thermal efficiencies of 50% to 60%.[14]

Finally, we consider NGCC with CCS. There are basically two ways to capture and sequester CO_2 from the power plant flue gas. One way is to remove, or scrub, the CO_2 from the flue gas using an absorber unit containing an amine. The CO_2 reacts with the amine in the absorber and is then later separated by stripping. Typical amines include MEA and MDEA. The second way is to use oxygen firing, or oxy-firing, the combustion of the natural gas using pure oxygen instead of air. In this way, nitrogen is eliminated from the flue gas.[15]

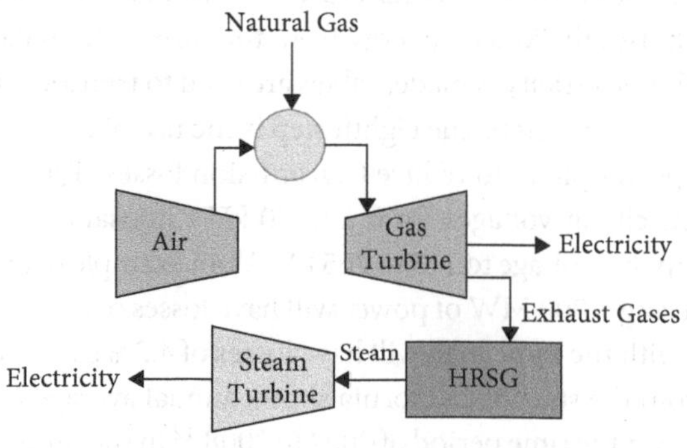

Figure 2.2 Schematic of Natural Gas Combined Cycle Plant[68]

An advantage of oxy-firing, over amine capture, is the higher removal efficiency. Amines typically have removal rates of 90%, while oxy-firing will have CO_2 removal of greater than 99%.[16] Disadvantages include the higher expense and energy demands of the air separation unit (ASU) and that it is a less mature technology than amine capture and it is more difficult to control the combustion temperature.[15]

According to the US EIA 2017 data there are 1,820 natural gas power plants in the United States with 5,878 generators and a nameplate capacity of 522.4 GW (gigawatts). Of these, about 56% of the capacity is for NGCC, 28% for natural gas with the turbine driven by combustion gases, and 15% with the turbine driven by steam.[17] At this time, there are no US natural gas plants with CCS.

2.4. Capital Cost, Operating Cost, Well-to-Wheels Levelized Cost of Electricity, and Well-to-Wheels Levelized Cost of Fuel

The equations and methodology used to calculate levelized cost of electricity (LCOE), and specific input used for natural gas to make electricity, are detailed in Appendix A. The units for LCOE are in \$/megawatt-hour, or \$/MWh. Data for the different cases were taken from two references.[18,19]

The costs of transmitting and distributing electricity were 0.9 cents/kWh (or \$9/MWh) and 2.6 cents/kWh (or \$26/MWh), set according to the EIA.[21] Here, transmission is defined as the high-voltage movement of electricity from the power station, and distribution is the cost of electrical lines that take power from a substation to the customer. The cost of transmission and distribution is essential to get a true "Well-to-Wheels" analysis of the total cost. For a natural gas generating plant, the "Well-to-Wheels" costs include (1) the cost of natural gas production and transportation to the power plant, (2) power plant generating costs, (3) cost of connecting the power plant to the electrical grid, (4) transmission and distribution costs, and (5) taxes (both federal and state).

The capacity factor was set according to the authors. Since this factor varies from 0.10 to 0.87, further explanation is needed. From the authors' data, the combined cycle units for natural gas operated with capacity factors from 0.80 to 0.87, but the combustion turbine capacity factors ranged from 0.10 to 0.30. The reason for this is that combined cycle natural gas units are used for primary electricity generation while combustion turbines are used primarily to meet "peak" electrical load. According to the EIA,[20] in 2017 natural gas combustion turbines had a nameplate capacity of 152.6 gigawatts (GW), 28% of the total 522.4 GW for all natural gas plants. However, these combustion turbines only contributed 8.6% of the electricity generation, operating at a capacity factor of about 0.08. Thus, natural gas combustion turbines do not fully utilize their nameplate capacity.

In addition, calculations used a current feedstock cost for natural gas of $3.17/MM BTU (million British thermal unit), based on Henry Hub spot price 2018 data.[22] This price captures the cost of natural gas production and transportation to the power plant. Results for the different LCOE cases are shown in Table 2.5.

Table 2.5 illustrates the large effect of capacity factor on the LCOE of combustion turbines versus combined cycle. The average cost of combined cycle is $73.93/MWh versus $140.87/MWh for combustion turbine. If the capacity factor is changed from the combustion turbine average of 0.23 to the combined cycle average of 0.85, the LCOE changes from $140.87/MWh to $92.53/MWh. Thus, the combustion turbine is more competitive with combined cycle if they have the same operating capacity factor.

Table 2.5 also allows us to directly compare EIA data such that combined cycle is used with, and without, CCS. The table shows that adding CCS increases the LCOE from $70.32 to $101.74, an increase of more than $31/MWh. The increase in LCOE is due primarily to capital costs, which increase from $10.74/MWh to $29.83/MWh with the addition of CCS, almost tripling the capital cost.

A sensitivity analysis was made for combined cycle generated electricity by varying the capital cost, the combined fixed and variable operating expenses (excluding the fuel cost), and the fuel cost. For the capital cost and operating expenses, the range of the reference data was used as a guide.

The base case used the operating conditions for the 2019 EIA conventional combined cycle case,[18] so the capital cost was $999/kW, the combined fixed and variable operation and maintenance (O&M) was $5.11/MWh excluding the cost of fuel, and the cost of natural gas was $3.17/MM BTU. For capital cost, the range for combined cycle without CCS was $700 to $1,300/kW and range for the combined fixed and variable operating costs, excluding the cost of fuel, was $2.80 to $7.40/MWh.

Like crude oil, the price for natural gas has undergone great change over the last 14 years. Table 2.6 shows historical Henry Hub natural gas prices,[22] indicating a range of $2.52 to $8.86/MM BTU over this time. Using these data as a guide, the sensitivity study considers natural gas prices of $2.50, $6.00, and $9.00/MM BTU.

Results for the sensitivity study for an advanced combined cycle natural gas plant are shown in Table 2.7.

As the table illustrates, changing the combined fixed and variable operating costs (excluding natural gas cost) had a minor impact on total cost. Increasing or decreasing the capital cost by $300/kW changed the LCOE by about $4/MWh. Clearly, the change in natural gas price had the greatest impact. The result of using the historical range of Henry Hub natural gas prices was a swing of about +/– $43/MWh. Since the 2018 cost for natural gas is near the bottom for the last 14 years, the calculated LCOE benefits greatly from this low price.

Table 2.5 Economic Analysis for Natural Gas–Fired Electric Plants

Energy Type	Capacity Factor	Capital Cost	Levelized Capital Cost	Fixed O&M	Variable O&M (including fuel)	Transmission Investment	Total System Levelized Cost (LCOE) to Generate	LCOE including Transmission and Distribution
		$/kW	$/MWh	$/MWh	$/MWh	$/MWh	$/MWh	$/MWh
Conventional CC[18]	0.87	$999	$13.52	$1.49	$24.5	$1.20	$40.73	$75.73
Conventional CC[19]	0.80	$1,000	$14.71	$0.82	$24.00	$1.20	$40.72	$75.72
Advanced CC[18]	0.87	$794	$10.74	$1.35	$22.0	$1.20	$35.32	$70.32
Average	0.85	$931	$12.99	$1.22	$23.52	$1.20	$38.93	$73.93
Advanced CC with CCS[18]	0.87	$2,205	$29.83	$4.52	$31.19	$1.20	$66.74	$101.74
Average	0.87	$2,205	$29.83	$4.52	$31.19	$1.20	$66.74	$101.74
Combustion Turbine[19]	0.10	$ 825	$93.78	$17.1	$34.3	$1.20	$146.42	$181.42
Combustion Turbine[18]	0.30	$1,126	$42.67	$1.4	$49.2	$3.50	$96.76	$131.76
Advanced Combustion Turbine[18]	0.30	$691	$26.18	$2.7	$42.1	$3.50	$74.44	$109.44
Average	0.23	$881	$54.21	$7.05	$41.88	$2.73	$105.87	$140.87

Table 2.6 Henry Hub Historical Natural Gas Prices, $/MM BTU[22]

Year	$/MM BTU (million BTU)
2005	8.69
2006	6.73
2007	6.97
2008	8.86
2009	3.94
2010	4.37
2011	4.00
2012	2.75
2013	3.73
2014	4.37
2015	2.62
2016	2.52
2017	2.99
2018	3.17

Table 2.7 Sensitivity Study Using the EIA Case for Natural Gas Combined Cycle[18]

	Natural Gas Price	Capital Cost	Fixed and Variable O&M (no fuel cost)	LCOE Including Transmission and Distribution
Units	$/MM BTU	$/kW	$/MWh	$/MWh
Base Case	$3.17	$999	$5.10	$75.73
Capex is $700/kW	$3.17	$700	$5.10	$71.69
Capex is $1,300/kW	$3.17	$1,300	$5.10	$79.81
Opex is $2.8/MWh	$3.17	$999	$2.80	$73.44
Opex is $7.4/MWh	$3.17	$999	$7.40	$78.04
NG price is $2.50/MM BTU	$2.50	$999	$5.10	$71.31
NG price is $6.00/MM BTU	$6.00	$999	$5.10	$94.41
NG price is $9.00/MM BTU	$9.00	$999	$5.10	$114.21

2.5. Cost of Energy

In this section, the cost of natural gas is converted from \$/MM BTU to \$/lb. The Henry Hub spot price[22] for 2018 was \$3.17/MM BTU. Using an energy content of 22,792 BTU/lb for natural gas, the energy cost of natural gas on a mass basis is \$0.07/lb.

As given in Chapter 1, the cost of crude oil for a similar 2018 time period was \$0.23/lb in terms of mass and \$11.20/MM BTU in terms of energy content. Thus, natural gas is less than one-third the cost of crude oil on both a mass and energy content basis.

2.6. Capacity Factor

The combined cycle units operated with capacity factors from 0.80 to 0.87, but combustion turbine units operated with factors from 0.10 to 0.30.[18,19] The reason for this difference was attributed to the use of combustion turbines primarily for meeting peak electrical loads. Because of this, combustion turbine plants have capacity factors well below their nameplate capacity.

The data taken from EIA[19] give a capacity factor of 0.87 for NGCC plants both with and without CCS, indicating no expected operating downtime due to the addition of CCS.

2.7. Efficiency: Fraction of Energy Converted to Work

Like petroleum crude oil, the thermal efficiency of a gas turbine is dictated by the maximum gas temperature. However, an easier way to calculate thermal efficiency, the fraction of energy converted to work, is to use the BTU content of one kilowatt-hour (kWh) of electricity, or 3,412 BTU and divide this by the heat rate. The heat rate is the amount of energy used, in this case natural gas, to make one kWh of electricity.

Table 2.8 shows the results of these calculations. As can be seen, combined cycle thermal efficiencies range from 50.9% to 54.2% with an average of 52.3%. Comparing the EIA case for advanced combined cycle with and without CCS shows a reduction in thermal efficiency from 54.2% to 45.3%. This is because some of the thermal energy of natural gas is now being used to capture and sequester CO_2. For a combustion turbine, the thermal efficiencies range from 34.7% to 38.3% with an average of 35.9%. Thus for a combined cycle plant, the use of waste heat to produce steam that powers a second turbine increases the overall efficiency by around 16%.

Table 2.8 Heat Rates and Thermal Efficiencies for the Cases Shown in Table 2.5

Energy Type	Heat Rate	Thermal Efficiency
	BTU/kWh	%
Conventional CC[18]	6,600	51.7
Conventional CC[19]	6,700	50.9
Advanced CC[18]	6,300	54.2
Average	6,533	52.3
Advanced CC with CCS[18]	7,525	45.3
Average	7,525	45.3
Combustion Turbine[19]	8,902	38.3
Combustion Turbine[18]	9,840	34.7
Advanced Combustion Turbine[18]	9,800	34.8
Average	9,514	35.9

2.8. Energy Balance

For natural gas, the energy balance is defined by this equation:

$$\text{Energy Balance} = \frac{[\text{Electrical energy from natural gas delivered to the customer}]}{[\text{Energy contained in the natural gas at the well site}]}$$

For the conversion of natural gas energy to electricity, there are several steps in the "Well-to-Wheels" analysis including (1) the energy used to extract and process the natural gas, (2) transport the natural gas to the electrical plant, (3) make the electricity, and (4) distribute the electricity through the grid to your home.

One NETL study[23] shows an analysis for what they call a "cradle-to-gate" use of natural gas. In this case, "cradle" is analogous to "Well." In their study, they show that natural gas that is not delivered to the power plant is vented as methane emissions, flared in environmental control equipment, or used to fuel process heaters, compressors, and other equipment. Their analysis shows that 0.7% is lost from extraction, 6.0% from processing, and 1.4% from transporting to the electrical plant. The total is 8.1%, meaning that 91.9% of the original natural gas energy arrives at the electrical plant.

Another study[24] shows 3.2% lost from extraction, 2.3% from processing, and 2.6% from transporting to the electrical plant, yielding the same 91.9% of the original natural gas energy arriving at the electrical plant but with different percentages in each step.

Table 2.8 in the previous section showed the thermal efficiencies for making electricity to be 52.3% for combined cycle, 45.3% for combined cycle with CCS, and 35.9% for combustion turbine.

It then remains to determine the energy needed to distribute the electricity through the grid to your home. As mentioned in the "Overview of Technology" section, transmission and distribution losses are affected by the voltage of the power line and the distance the electricity is transmitted. Historical data from the state of California shows annual average transmission losses around 5% to 7% over the time period of 2002 to 2008.[12] Similarly, EIA data for July 2018[25] shows that of the electricity generated, 7% is lost in transmission and distribution.

Of course, this 7% lost in transmission and distribution is the percent of the energy leaving the electrical plant, not the energy in the natural gas when it was extracted from the well. Using 100 units of energy as a reference, we can arrive at these equations:

Natural Gas Combined Cycle

$$100\text{ (well)} \rightarrow 99.3\text{ (extraction)} \rightarrow 93.3\text{ (processing)} \rightarrow 91.9\text{ (transporting)} \rightarrow 48.1\text{ (NGCC)} \rightarrow 44.7\text{ (transmission and distribution)}$$

Natural Gas Combined Cycle with Carbon Capture and Sequestration

$$100\text{ (well)} \rightarrow 99.3\text{ (extraction)} \rightarrow 93.3\text{ (processing)} \rightarrow 91.9\text{ (transporting)} \rightarrow 41.6\text{ (NGCC - CCS)} \rightarrow 38.7\text{ (transmission and distribution)}$$

Natural Gas Combustion Turbine

$$100\text{ (well)} \rightarrow 99.3\text{ (extraction)} \rightarrow 93.3\text{ (processing)} \rightarrow 91.9\text{ (transporting)} \rightarrow 33.0\text{ (CT)} \rightarrow 30.7\text{ (transmission and distribution)}$$

Summarizing, of the original energy in the natural gas at the well, 44.7% will end up as usable electricity for NGCC and 30.7% for combustion turbine. The addition of CCS will decrease the value for NGCC to 38.7%.

2.9. Maturity: Experience

Natural gas was used in China before 400 BC, when bamboo pipelines were used to carry it to an appropriate location to mix it with air to make fires.[26] The heat was used to boil salt water to extract the salt. Great Britain was probably the first country to commercialize the use of natural gas, when around 1785 natural gas produced from coal mining was used for streetlamps and light houses.[27]

In the United States, the first successful natural gas well was drilled in Fredonia, New York, in 1821, leading to the formation of Fredonia Gas Light Company, the first US gas distribution company.[28] Drilling for natural gas in the 1800s was first done with vertical wells, following the technology lead of drilling oil wells. In 1953 in New York State, a vertical well of 11,145 feet was drilled using cable tool technology.[29] In this technology, a cable made of rope or wire pulls a blunt bit up and down to smash the rock in a series of repeated blows. Rotary bits for drilling came into use in the 1830s. Originally powered by a mule walking in a circle, modern natural gas rigs use diesel or natural gas engines to power the drill.

The Bunsen burner was invented by Robert Bunsen in 1885, leading to the use of natural gas as a fuel for home cooking. As demand for natural gas increased, natural gas pipelines were built. In the United States, the development of a large pipeline network began in the 1950s. While prohibitive in cost before this time, the improvements in welding and pipe making during World War II allowed the economic construction of pipelines. With the construction of extensive pipeline networks, home heating followed.[30] According to 2018 data from the American Gas Association,[9] there are now over 1.6 million miles of natural gas pipelines including 1.3 million for local distribution, 301,000 for of interstate and intrastate for transmission, and 18,000 for gathering. And, if it is not cost-effective or practical to transport natural gas because of distance or lack of pipelines, it can be converted into LNG to be transported by sea or by cryogenic road tankers.

In addition to cooking and home heating, some new uses for natural gas have emerged such as powering vehicles, making synthetic fuels, and producing hydrogen. As of 2018, there were about 175,000 NGVs in the United States and about 23 million vehicles worldwide.[31] To fuel these NGVs, there are now 1,680 compressed natural gas (CNG) stations in the United States.[32] Natural gas can be used to make gasoline and diesel synthetically via the Fischer-Tropsch process. Invented in 1925 by Hans Fischer and Franz Tropsch, this process originally used coal, but now natural gas is used as well. This is discussed in more detail in Chapter 12. Natural gas can also be used to make hydrogen, and the production of hydrogen from natural gas is discussed in Chapter 11.

The focus of this chapter has been on the use of natural gas to make electricity. The electrical grid, used to accept power produced from various energy sources (coal, nuclear, natural gas, solar, wind, etc.), involves both transmission and distribution. Transmission and distribution includes a step-up transformer, transmission lines, and a substation step-down transformer. The first electric power distribution system was built in 1882 in New York City, overseen by none other than Thomas Edison. Between this time and the 1930s, the electrical grid system grew rapidly, roughly doubling every 6 years. Today, what we take for granted, involves more than 180,000 miles of high-voltage transmission lines connecting about 7,000 power plants.[33]

The first natural gas turbine used to generate electricity in the United States was built in 1937 by Sun Oil, although this was for private use at its Philadelphia chemical plant.[34] However, in 1945 and 1948, Switzerland built two power plants at Beznau for public use. The first combined cycle natural gas plant was built in Korneuburg, Austria, in 1960 and the first US plant was built in Oklahoma in 1963.[35]

In terms of CCS, the technology is still relatively new. In the United States, CCS was planned for two coal-generating power plants, one in Kempler County, Mississippi, and the Petra Nova plant near Houston, Texas. Both had planned startups in 2016.[36] While the Petra Nova plant did start up in 2017, the Kempler coal plant project was suspended and the plant switched to natural gas.[69] There is also an operating coal-fired plant with CCS near Estevan, Saskatchewan, which started up in fall of 2014.[37] At the time of this writing there were no natural gas power plants using CCS. A project was announced for Peterhead, Scotland, to be operational by 2018,[38] but this project was canceled.

To summarize timelines in the United States, natural gas drilling has been around since 1821, extensive pipeline networks have been around since the 1950s, and the first electrical grid system was built in 1882 followed by rapid development through the 1930s. The first US natural gas electrical plant was built in 1937, and the first combined cycle plant was built in 1963. For CCS, there is limited commercial experience using coal-fired electricity generation but no experience with natural gas. Summarizing, in the United states, there are more than 100 years of experience in the natural gas drilling industry and about 80 years of experience using natural gas to generate electricity.

2.10. Infrastructure

Like petroleum, there is already an immense infrastructure to transport and support the use of natural gas for cooking, heating, and electricity generation. The United States has the fifth largest amount of natural gas reserves in the world at 309 tcf and was the largest producer in the world in 2017, producing 25,939 billion cubic feet.[3] To move this natural gas, there are around 1.6 million miles of natural gas pipelines.[9] And, to make electricity, in 2017 the EIA listed 1,820 operating natural gas power plants with 5,878 generators and a nameplate capacity of 522,378 MW.[17] These units had a net generation of 1,296,415 thousand MWh (GWh).[39]

In summary, there is an established network of gas drilling, pipelines, electrical plants, and electrical transmission and distribution lines to support the use of natural gas. This network allows natural gas to be used for industrial and residential heating and cooking, as well as electricity generation. Although relatively new, there are a growing number of CNG stations to fuel a growing number of NGVs in the United States.[31]

2.11. Footprint and Energy Density

"Footprint" is a term commonly used in industrial plants to describe the area used by an energy type. For natural gas, the footprint will include the area to (1) extract natural gas from the ground, (2) process the natural gas in a gas processing plant,

(3) transport the natural gas to the natural gas electrical plant via pipeline, (4) make electricity at the natural gas electrical plant, and (5) transmit and distribute the electricity via the grid to commercial customers.

To determine the footprint to (1) extract the natural gas, the same reference is used as in Chapter 1 for petroleum crude oil. Using federal data for oil and gas wells, according to the Bureau of Land Management (BLM) and Bureau of Ocean Energy Management (BOEM),[40] there were 11.1 million acres of onshore and 6.6 million acres of offshore federal land producing commercial volumes of oil and gas in 2012. For these acres, onshore and offshore produced 2,919 and 1,351 billion cubic feet, respectively, for the year 2012. Since there are 640 acres in a square mile, this means that there were 17,344 square miles of onshore federal oil and gas production and 10,313 square miles of offshore federal oil and gas production. Dividing gas production by square miles gives 0.168 billion cubic feet per square mile per year for onshore production and 0.131 billion cubic feet per square mile per year for offshore production. The combined gas production for a square mile is 0.154 billion cubic feet per year. One cubic foot of natural gas contains 1,032 BTU. Therefore, we can calculate the energy content of this gas production as 159.3 billion BTU/(year-square mile).

The second footprint to consider for natural gas is (2) the land area needed for the gas processing plant. Anyone that has driven by a natural gas processing plant will notice that it does not appear to take up much land area, as it consists of several tall towers. However, the land needed for the plant will include a protected area, interfaces with the gas gathering equipment, and room for possible expansion. Therefore, to determine the land area needed for the plant, several commercial operations were examined. For the land area needed to transport the gas to the processing plant, there are two possible scenarios. First, the natural gas processing plant may be built next to the gas drilling site, meaning the two possibly occupy the same land area. Alternatively, if the natural gas is piped to the plant from long distance, pipeline transportation is needed. However, the area for pipelines is negligible and can coexist with farming and pasture land.

For the footprint of the natural gas processing plant, several references were examined. A DCP Midstream Partners facility was built southwest of Kersey, Colorado, with a capacity of 160 MMSCF/day on 160 acres.[41] In Ohio, a natural gas processing facility in Kensington processes 800 MMCF/day on a 117-acre site and a facility in Cadiz processes 325 MMSCF/day on 207 acres.[42] The East Toyah plant operated by EagleClaw Midstream Services processes 75 MMSCF/day on an 80-acre site in Reeves, Texas.[43] Pennant Midstream has one 200 MMSCF/day gas-processing facility under construction at a 90-acre site in Mahoning County, but will add two additional plants for a total of 600 MMcf/day.[44] The Williams Partners Opal Gas-Processing Plant has the capacity to process 1.5 billion cubic feet of natural gas daily and sites on 160-acres in Opal, Wyoming.[45] Energy Transfer has a plant in Karnes County, Texas, which can process 200 million cubic feet/day and sits on 325 acres of land. However, the

Table 2.9 Natural Gas Processing Plant Data

	Million ft^3/day	Acreage	Square miles	Billion ft^3/ (square mile-day)	Billion ft^3/ (square mile-year)
Kersey, CO	160	160	0.25	0.64	234
Kensington, OH	800	117	0.18	4.38	1,597
Cadiz, OH	325	207	0.32	1.01	367
East Toyah, TX	75	80	0.13	0.60	219
Mahoning County, OH	600	90	0.14	4.27	1,557
Opal, WY	1,500	160	0.25	6.00	2,190
Karnes, TX	200	325	0.51	0.39	144
Caymus, TX	1,000	330	0.52	1.94	708
Average	583	184	0.29	2.03	741

plant itself only takes up a small portion of the total acreage.[46] And Vaquero Midstream has its Caymus I in Pecos County, Texas. The first plant processes 200 MMSCF/day but, eventually there will be four more plants built for a total capacity of 1,000 MMSCF/day on 330 acres.[47] These data are summarized in Table 2.9, and, when applicable, the eventual future capacity was used for the total feed flow for natural gas. The average for these data is given, recognizing that this may be somewhat different than the average for all US natural gas processing plants.

From the average data, we can calculate an overall natural gas processing rate of 2.03 billion ft^3/(square mile-day), or 741.0 billion ft^3/(square mile-year). Comparing the average values for extracting natural gas with processing natural gas, we see that for extraction, it take one square mile to produce 0.154 billion cubic feet per year. However, for natural gas processing, one square mile can process around 741 billion cubic feet per year. Since 741 is almost 5,000 times larger than 0.154, we can conclude that the land area needed to process the natural gas is negligible compared to the area to extract it.

As was the case for natural gas extraction, the energy in the natural gas that was processed can also be calculated. Using the value of 1,032 BTU per cubic foot of natural gas, we calculate that the natural gas processing plant can process natural gas with the energy of 0.76 quadrillion BTU/(year-square mile) or 0.76×10^{15} BTU/(year-square mile).

The next use of land area is to (3) transport the natural gas to the natural gas electrical plant via pipeline. Like petroleum crude oil, the contribution of this land area is negligible compared to the overall footprint. Generally, pipelines have small diameters and can coexist with farming and pasture land.

The land area used to make electricity at the natural gas electrical plant, step (4) is the next part of the footprint. Although we do not expect the footprints for combustion turbine and combined cycle plants to be very much different, power plant footprints were only examined for NGCC plants. The data for 14 plants are summarized in Table 2.10 below.

The units of MW/acre can be converted to units of quadrillion BTU/(year-square mile) in order to provide a direct comparison to gas extraction and gas processing. A 1 MW plant operating at 100% capacity for 1 year will produce 8,760 MWh (1 MW * 24 hour/day * 365 days/year). Also, there are 640 acres per square mile and 3.412×10^6 BTU/MWh. From these value, we can determine that 33.7 MW/acre is equivalent to 0.645 quadrillion BTU/(year-square mile), assuming the plant operates at a

Table 2.10 Natural Gas Combined Cycle Electric Plants

Location	Company	Startup	MW	Acres	MW/Acre
Cape Canaveral, FL[48]	Florida Power & Light	April 2013	1,250	42	29.8
Colusa, CA[49]	E&L Westcoast, LLC	December 2010	660	31	21.3
Salem Harbour, MA[50]	Footprint Power	June 2017	674	20	33.7
Oregon, OH[51]	North American Project Development, LLC	2017	800	30	26.7
Shamokin Dam Borough, PA[52]	Panda Power and Sunbury Generation LP	July 2016	1,000	18	55.6
Pittsburg, CA[53]	Calpine	June 2002	880	20	44.0
Victorville, CA[53]	High Desert Power	April 2003	700	25	28.0
Los Medanos (Pittsburg), CA[53]	Calpine	July 2001	500	12	41.7
Blythe, CA[53]	Blythe Energy	July 2003	520	15	34.7
San Bernardino, CA[53]	Southern California Edison	December 2005	1,056	16.3	64.8
Otay Mesa, CA[53]	Otay Mesa	October 2009	510	15	34.0
Lebec, CA[53]	Calpine	May 2005	750	30	25.0
Moundsville, WV[54]	Moundsville Power	2018	549	37.5	14.6
Vineyard, UT[55]	PacifiCorp Energy	2007 (Lake Side 1) and 2014 (Lake Side 2)	1,203	65	18.5
Average			789	26.9	33.7

capacity factor of 100%. However, capacity factors are not 100% and the average value for NGCC plants was shown in Table 2.5 to be 85%. With this adjustment, the amount of energy produced for the power plant is 0.548 quadrillion BTU/(year-square mile), or 0.548×10^{15} BTU/(year-square mile).

As in the comparison of natural gas processing to gas extraction, the energy produced from the natural gas electric plant in 1 year per square mile is much larger than that for gas extraction, more than 3,000 times larger. Therefore, we can conclude that the land area needed to make electricity from natural gas is negligible compared to the area needed to extract it.

Now that the electricity has been made, the last part of the footprint is to (5) transmit and distribute the electricity to customers via the grid. However, the land area needed for step-up and step-down transformers is very small and can be ignored. Likewise, the metal towers and electrical lines take up very little land area and can co-exist with farming, pasture land, and roadways.

To summarize, the three main contributors to footprint and energy density are (1) gas extraction at 159.3 billion BTU/(year-square mile), (2) gas processing at 0.76 quadrillion BTU/(year-square mile), and (4) the combined cycle electrical plant at 0.548 quadrillion BTU/(year-square mile). Stated using scientific notation, these three values are 159.3×10^9, 0.76×10^{15}, and 0.548×10^{15}.

In order to compare energy densities for different technologies on a common basis of gigawatt-hours per square mile per year, or GWh/(square mile-year), energy was converted from BTU to Watt-hour, or Wh, using the following equation:

$$1 \text{ BTU} = 0.293071 \text{ Wh}$$

With this conversion and recognizing that a gigawatt (GW) is equivalent to one billion Watts (or 1×10^9), we can calculate the energy density to extract gas from the ground as 46.7 GWh/(year-square mile), the energy density to process gas as about 224,000 GWh/(year-square mile), and the energy density to make electricity at the combined cycle electrical plant as about 160,500 GWh/(year-square mile). The energy density to extract the gas from the ground is similar to that shown in Chapter 1 to drill and produce oil, which is 36.8 GWh/(year-square mile).

Calculating a footprint for a 1,000 MW combined cycle electrical plant, for example, yields a very small footprint of about 0.06 square miles. However, the land area needed to extract the gas from the ground and supply natural gas for this 1,000 MW plant is 188 square miles, which dwarfs the footprint of the electric plant.

According to the International Energy Agency,[56] the US electric power consumption per capita for 1 year was 12.8 MWh in 2016. Using the capacity factor of 0.85 shown in Table 2.5 for combined cycle plants, a simple calculation shows that 1 MW will power about 580 homes for one year, so a 1,000 MW plant will power nearly 600,000 homes!

2.12. Environmental Issues

Unlike petroleum, natural gas is a very simple mixture of light hydrocarbons, primarily methane. The light hydrocarbons present are paraffins, which are relatively inert, and there are no reactive type compounds such as olefins and aromatics.

At room temperature and atmospheric pressure, all the compounds in natural gas are gases; therefore, one danger is flammability. One way to understand this danger is the lower explosive limit (LEL) and upper explosive limit (UEL). For a fire or explosion to occur, you must have the following three conditions: fuel, oxygen in certain percentages, and an ignition source, such as a spark or flame. The amount of fuel and oxygen depends on the gas. The minimum concentration for the gas to have combustion in air is called the LEL. Below this percentage, the mixture is too lean in oxygen to burn. The maximum concentration for the gas to combust is called the UEL, above which the fuel mixture is too rich in oxygen to burn. Between the LEL and UEL, the gas is flammable. For methane, this range is fairly small, 5% to 15%.[57] By comparison, gasoline vapor has a range of 1.2% to 7.1% while hydrogen has a range of 4% to 75%. Thus, methane is more flammable than gasoline but less than hydrogen.

In 2016, there were some natural gas explosions. In the Salem Township, Pennsylvania, a natural gas explosion caused a fire, causing one injury and damage to nearby utilities. The explosion involved a 30-inch pipeline.[58] Also, in Kingsville, Texas, a natural gas pipeline caused an explosion on the King Ranch.[59]

The products of combustion are CO_2 and water (H_2O), and these are benign, although increasing amounts of CO_2 in the atmosphere have been attributed to global warming. However, CO_2 and water are not toxic to humans.

Although natural gas contains only ppm levels of H_2S, this is a very toxic gas. The American Conference of Governmental Industrial Hygienists (ACGIH®) has adopted limits of 1.0 ppm for the 8-hour time weighted average (TWA) and 5.0 ppm for the short term exposure limit (STEL).[60] Above these limits, there can be dizziness, irreversible damage, and even death. However, natural gas leaving the gas processing plant should have safe levels of H_2S.

Table 2.1 shows the nitrogen natural gas content varies from 1.3 to 5.6% and the pipeline requirement for nitrogen is typically 4%,[61] a limit set because nitrogen lowers the BTU content of the natural gas. Nevertheless, when the natural gas is burned at the power plant, it will produce NO_x. However, NO_x emissions from natural gas power plants are quite small compared to coal or products made from crude oil. One reference shows NO_x emissions in terms of pounds per billion BTU of energy input. For natural gas, this value is 92 versus 457 for coal and 448 for crude oil.[62]

Naturally, any oil or gas drilling will alter the land, and this can affect wildlife habitats and animal migration patterns and can cause soil erosion. And, if the natural gas is produced by hydraulic fracturing, there can be contamination of groundwater.

There are documented cases that show groundwater near oil and gas wells has been contaminated with fracking fluids, methane, and volatile organic compounds.[63]

One last potential environmental issue for natural gas is the effect of fugitive emissions on global warming, as methane has a global warming potential that is 28 to 36 times greater than CO_2. And, hydraulic fracking may further exacerbate this, since methane is one chemical released in the fracking process. Based on data from the National Oceanic and Atmospheric Administration (NOAA), one paper[64] found that methane emissions from the US natural gas industry accounted for only 1.2% of all 2016 global methane emissions. On the other hand, 12.4% of global methane emissions were attributed to oil and natural gas production. So, while oil and gas production does introduce significant amounts of methane, the natural gas industry has very low leakage rates and does not significantly contribute to global methane emissions.

2.13. CO_2 Production and the Cost of Capture and Sequestration

The amount of CO_2 produced from natural gas is a relatively straightforward. A small amount of the carbon in the natural gas is flared during extraction and processing, a small amount is used to provide energy for processing and pipeline transport, and most is burned at the electric plant to generate electricity. All of these actions result in the formation of CO_2, so the only methane that does not end up as CO_2 are those lost through emissions. Recalling an earlier reference,[23] for 1,088 kg of methane obtained from extraction, about 88 kg are lost before 1,000 kg are delivered to the electric power plant. These include 30.3 kg lost to flaring, 45.6 kg used to provide energy for processing and transport, and 12.5 kg lost to emissions. Therefore, if CCS is added to the electric power plant, 91.9% of the original 1,088 kg could be captured. Seven percent of CO_2 is made from flaring, processing, and pipeline transport, which cannot be easily captured at the power plant and another 1.1% is lost from emissions. Note that this amount lost to emissions amounts to 1.1%, similar to the 1.2% given in the previous section on fugitive emissions in the natural gas industry.

Carbon dioxide production is commonly listed in the units of lb CO_2/MM BTU, lb CO_2/kWh, and g CO_2/kWh. To calculate CO_2 production on a basis of lb CO_2/MM BTU (lb CO_2/million BTU), we use 1,032 BTU as an energy content for one cubic foot of natural gas. Also, there are 379.477 ft^3/lb-mole, one mole of natural gas will produce about one mole of CO_2, and the molecular weight of CO_2 is 44 lb/lb-mole. Thus for natural gas:

$$lb\ CO_2\ /\ MM\ BTU = \frac{44 * 1\times10^6}{1032 * 379.477} = 112.4\ lb\ CO_2\ /\ MM\ BTU$$

Next, the CO_2 production in terms of kWh can be determined using a heat rate for electricity generation. For NGCC, Table 2.8 shows an average heat rate of 6,533 BTU/

kWh. Using this heat rate, we can calculate the CO_2 produced on a kWh basis as 0.73 lb CO_2/kWh and 333 g CO_2/kWh.

At the time this book was written, there were no natural gas plants employing CCS. One project, the Peterhead Power Station in Aberdeenshire, Scotland, planned to capture up to 10 million metric tonnes of CO_2 and then transport it by pipeline 100 km offshore to be stored 2 km under the North Sea in a depleted gas reservoir.[65] Peterhead Power Station uses NGCC technology. Construction was to begin in 2016 with commissioning in 2019–20. Unfortunately, the United Kingdom canceled project funding in November 2015 and the future for this project is unknown.[66]

Therefore, as was the case for adding CCS to a refinery, we have to use economic model projections rather than actual commercial data. CCS is a three-step process including capture of the CO_2, transportation to the storage site by pipeline, and storage by injection into geological formation or depleted oil and gas field. The EIA data from Table 2.5 show that capital cost more than doubles with the addition of CCS, and there are increases in fixed and variable operating costs as well. The relevant excerpts from this table are shown here. Overall, the levelized cost of electricity increases by about $31/MWh.

Another report[67] shows a cost difference for building a new plant with and without CCS as $30/MWh. Therefore, these two economic studies agree and indicate that adding CCS to a combined cycle natural gas plant will cost $30/MWh.

For a 1,000 MW plant with a capacity factor of 85%, the cost of CCS will be about $600,000 per day, or about $225 MM per year. Also, since the 1,000 MW plant generates 20,400 MWh/day at a capacity factor of 85%, and since the CO_2 produced on a kWh basis is 0.73 lb CO_2/kWh, this means a 1,000 MW plant generates about 15 million lb CO_2/day or about 7,500 tons CO_2/day. On a mass basis, the cost of CCS ranges is about $82/ton.

In conclusion, the amount of CO_2 produced for a NGCC electrical plant is 112.4 lb CO_2/MM BTU, or 0.73 lb CO_2/kWh and 333 g CO_2/kWh. The cost of CCS is about $30/MWh, or about 3.0¢ per kWh.

2.14. Chapter Summary

A review of worldwide natural gas proven reserves shows an estimate of 6,832 tcf in 2017 which, at the 2017 consumption rate of 129.6 tcf per year leads to a calculated value of 53 years remaining for natural gas. While this would lead to a depletion in world supply around 2070, hydraulic fracking could greatly extend natural gas supplies while increasing use of NGVs and the synthesis of liquid fuels from natural gas via the Fischer-Tropsch synthesis could result in a significant decrease in the estimated remaining years.

For proven reserves, ten countries including Russia, Iran, Qatar, Turkmenistan, United States, Saudi Arabia, Venezuela, United Arab Emirates, China, and Nigeria

Table 2.5 Excerpts

Energy Type	Capacity Factor	Capital Cost	Levelized Capital Cost	Fixed O&M	Variable O&M (including fuel)	Transmission Investment	Total System Levelized Cost (LCOE) to Generate	LCOE including transmission and distribution
		$/kW	$/MWh	$/MWh	$/MWh	$/MWh	$/MWh	$/MWh
Advanced CC[18]	0.87	$794	$10.74	$1.35	$22.0	$1.20	$35.32	$70.32
Advanced CC with CCS[18]	0.87	$2,205	$29.83	$4.52	$31.19	$1.20	$66.74	$101.74

control 78.8 of the worldwide natural gas reserves. By themselves, Russia and Iran account for 18.1% and 17.2%, respectively.

The "Well-to-Wheels" economic analysis gives an average LCOE of $73.39/MWh for combined cycle and $140.87/MWh for combustion turbine. This large difference is due primarily to the low capacity factor for combustion turbine electrical generation, 0.23 versus 0.85 for combined cycle. If, instead, a 0.85 capacity factor is used for combustion turbine, the LCOE changes to $92.53/MWh, making it more competitive with combined cycle. Adding CCS at the electric plant increases the cost to generate electricity around $31/MWh, due primarily to capital costs, which almost triple.

For a 1,000 MW NGCC power plant operating at a capacity factor of 85%, the CO_2 generation is about 15 million lb CO_2/day or about 7,500 tons CO_2/day. In other units, this is equivalent to 112.4 lb CO_2/MM BTU, 0.73 lb CO_2/kWh, and 333 g CO_2/kWh. This compares to 2.66 lb CO_2/kWh for a refinery, so the natural gas only produces about 27% of that for a refinery, primarily due to the lower carbon content of methane relative to crude oil. The cost of adding CCS will be $30/MWh, or about 3.0¢ per kWh. On a mass basis, the cost of CCS ranges is $82/ton.

The cost of natural gas, on the basis of energy content, is $3.17/MM BTU using 2018 Henry Hub spot prices. This is less than one-third of the cost of crude oil, costing $11.20/MMBTU based on a 2018 cost for West Texas Intermediate of $64.94/BBL.

Combined cycle units operate with capacity factors from 0.80 to 0.87, but combustion turbine units operate with factors from 0.10 to 0.30. The reason for this difference is attributed to the use of combustion turbines primarily for meeting peak electrical loads, leading to electricity production well below their nameplate capacity. For combined cycle units, the overall energy balance was 44.7%, meaning that only 44.7% of the energy originally in the natural gas ends up as energy at the distribution point. Most of this energy is lost during electricity generation, as the thermal efficiency for combined cycle is 52.3%. The numbers are worse for combustion turbine, with a thermal efficiency of 35.9% and only 30.7% ending up as energy at the distribution point. And, if CCS is added to combined cycle, the thermal efficiency decreases from 52.3% to 45.3%, with only 38.7% of the energy reaching the distribution point.

In terms of experience and maturity, natural gas drilling has been around since 1821, pipeline networks since the 1950s, and the first US natural gas electrical plant was built in 1937. Thus, like the petroleum industry, it is quite mature with more than 100 years of natural gas drilling and about 80 years of using natural gas to generate electricity.

Also, like petroleum crude oil, the infrastructure is immense. In the United States, there are around 1.6 million miles of natural gas pipelines, 1,820 operating natural gas power plants (in 2017) with 5,878 generators, and more than 160,000 miles of transmission lines that exist in the electrical grid.

The energy density for natural gas electricity generation, 46.7 GWh/(year-square mile), is controlled by the land area needed to extract the natural gas from the ground. This corresponds to a footprint of 188 square miles needed to supply natural gas for a 1,000 MW plant, much larger than the footprint of only about 0.06 square miles for the 1,000 MW combined cycle electric plant. Thus, this 1,000 MW plant which will power nearly 700,000 homes uses only a small amount of area relative to the area needed for gas extraction. If the land used to extract the natural gas could simultaneously exist with farming or pasture land, or later be remediated to return to its original state, the overall footprint for natural gas would be quite small.

Finally, the main environmental issues for natural gas are H_2S and nitrogen. The H_2S leaving the plant will be at a level that is not dangerous to people. The small amount of nitrogen in the natural gas will produce NO_x, but this amount is about one-fifth of that made from coal or crude oil. Although CO_2 is not considered a pollutant at this time, around 92% of the CO_2 made during the natural gas Well-to-Wheels cycle can be captured at the electric power plant using CCS. Also, fugitive methane emissions from oil and gas production contribute 12.4% of all global methane emissions. Since methane has a global warming potential that is 28 to 36 times greater than CO_2, this is another potential environmental issue. However, after gas production, the natural gas industry is very efficient, with low leakage rates and accounting for only 1.2% of all global methane emissions.

References

1. North American Energy Standards Board, Natural Gas Spec Sheet, retrieved August 3, 2016, https://www.naesb.org/pdf2/wgq_bps100605w2.pdf
2. "Fracking, or Hydraulic Fracturing, Is Helping US to Produce More of Its Own Energy," David Kashi, International Business Times, December 11, 2013.
3. "BP Statistical Review of World Energy," June 2018, https://www.bp.com/content/dam/bp/business-sites/en/global/corporate/pdfs/energy-economics/statistical-
4. "Energy Information Administration (EIA) "Internal Energy Statistics," 2017, http://www.eia.gov/cfapps/ipdbproject/IEDIndex3.cfm?tid=3&pid=3&aid=6
5. *Oil and Gas Production in Nontechnical Language*, M.S. Raymond and W.L. Leffler, Tulsa, OK: PennWell Corporation, 2006.
6. *Gas Purification*, A. Kohl and F. Riesenfeld, Houston: Gulf Publishing Co., 1985, p. 953–957.
7. "Processing Natural Gas," September 25, 2013, NaturalGas.org
8. "Frequently Asked Questions about LNG," California Energy Commission, 2019, http://www.energy.ca.gov/lng/faq.html
9. "Annual Distribution and Transmission Miles of Pipeline," American Gas Association, November 28, 2018, https://www.aga.org/research/data/distribution-and-transmission-miles-of-pipeline/
10. "About U.S. Natural Gas Pipelines—Transporting Natural Gas," 2008, http://www.eia.gov/pub/oil_gas/natural_gas/analysis_publications/ngpipeline/transpath_fig.html
11. "American Electric Power, Transmission facts," p. 4, December 31, 2007, https://web.archive.org/web/20110604181007/https://www.aep.com/about/transmission/docs/transmission-facts.pdf
12. "A Review of Transmission Losses in Planning Studies," Lana Wong, California Energy Commission, Figure ES-1, August 2011, https://ww2.energy.ca.gov/2011publications/CEC-200-2011-009/CEC-200-2011-009.pdf
13. "Natural Gas Combustion," US Environmental Protection Agency, July 1998, https://www3.epa.gov/ttnchie1/ap42/ch01/final/c01s04.pdf

14. "Overview of Natural Gas," Natural Gas.org, September 20, 2013, http://naturalgas.org/overview/uses-electrical/
15. "Assessment of Natural Gas Combined Cycle Plants for Carbon Dioxide Capture and Storage in a Gas-Dominated Electricity Market," Eric Holden, Final Project Report, Contract Number 500-10-037, prepared for the California Energy Commission, December 2014.
16. "Carbon Capture Approaches for Natural Gas Combined Cycle Systems," National Energy Technology Laboratory (NETL), DOE/NETL-2011/1470, final report, revision 2, December 20, 2010, https://www.netl.doe.gov/projects/files/FY11_CarbonCaptureApproachesforNaturalGasCombinedCycleSystems_010111.pdf
17. "Count of Electric Power Industry Power Plants, by Sector, by Predominant Energy Sources within Plant, 2007 through 2017," Table 4.1, Energy Information Administration, http://www.eia.gov/electricity/annual/html/epa_04_01.html; "Existing Capacity by Energy Source, 2017 (Megawatts)," Table 4.3, Energy Information Administration; https://www.eia.gov/electricity/annual/html/epa_04_03.html
18. "Cost and Performance Characteristics of New Generating Technologies," Annual Energy Outlook 2019, January 2019, Energy Information Administration, Table 2, https://www.eia.gov/outlooks/aeo/
19. "Lazard's Levelized Cost of Energy Analysis—Version 12.0," November 2018, https://www.lazard.com/media/450784/lazards-levelized-cost-of-energy-version-120-vfinal.pdf
20. "Electric Power Monthly," Energy Information Administration, January 2019, https://www.eia.gov/electricity/monthly/current_month/epm.pdf
21. "Annual Energy Outlook 2015 with Projections to 2040," DOE/EIA-0383, Table A8, April 2015, https://www.eia.gov/outlooks/aeo/pdf/0383(2015).pdf
22. "Natural Gas: Henry Hub Natural Gas Spot Price," Energy Information Administration, 2018, https://www.eia.gov/dnav/ng/hist/rngwhhdm.htm
23. "Life Cycle Analysis of Natural Gas Extraction and Power Generation," T.J. Skone et al., National Energy Technology Laboratory (NETL), DOE Contract Number DE-FE0004001, Figure 4-3, May 29, 2014, https://www.energy.gov/sites/prod/files/2014/05/f16/Life%20Cycle%20GHG%20Perspective%20Report.pdf
24. "Real Energy Efficiency: Why Energy-Smart Consumers Heat with Natural Gas," American Oil & Gas, 2014, https://www.aogc.com/realenergyefficiency.aspx
25. "Monthly Energy Review," Energy Information Administration, March 2019, https://www.eia.gov/totalenergy/data/monthly/pdf/mer.pdf
26. "First Oil Wells," 2007, http://www.historylines.net/history/chinese/oil_well.html
27. "History," NaturalGas.org, September 20, 2013, http://naturalgas.org/overview/history/
28. "A Brief History of Natural Gas," American Public Gas Association, 2020, http://www.apga.org/apgamainsite/aboutus/facts/history-of-natural-gas
29. "A Brief Introduction to the History of Natural Gas Extraction," Jennifer Halpern, August 30, 2010, https://thesciencebeneaththesurface.wordpress.com/2010/08/30/a-brief-introduction-to-the-history-of-natural-gas-extraction/
30. "Historical Timeline: History of Alternative Energy and Fossil Fuels," ProCon.org, March 31, 2020, http://alternativeenergy.procon.org/view.timeline.php?timelineID=000015#
31. "Natural Gas Vehicles," Alternative Fuels Data Center, 2020, http://www.afdc.energy.gov/vehicles/natural_gas.html
32. "Stations," NGVAmerica, January 2016, https://www.ngvamerica.org/stations/
33. "United States Electricity Industry Primer," Office of Electricity Delivery and Energy Reliability U.S. Department of Energy DOE/OE-0017, July 2015, file:///C:/Users/paulf/Documents/Book/Chapters_for_OUP/New_Gen_Documents/united-states-electricity-industry-primer.pdf
34. "A Short History of the Evolving Uses of Natural Gas," Tim Miser, Power Engineering, February 13, 2015, https://www.power-eng.com/2015/02/13/a-short-history-of-the-evolving-uses-of-natural-gas/

35. “The History of the Industrial Gas Turbine (Part 1 The First Fifty Years 1940–1990),” Ronald J. Hunt, Paper 582, Version 2, January 14, 2011, http://www.idgte.org/IDGTE%20Paper%20582%20History%20of%20The%20Industrial%20Gas%20Turbine%20Part%201%20v2%20(revised%2014-Jan-11).pdf
36. “Chapter 4: Advancing Clean Electric Power Technologies,” Quadrennial Technology Review 2015 Technology Assessments, http://energy.gov/sites/prod/files/2015/12/f27/QTR2015-4D-Carbon-Dioxide-Capture-for-Natural-Gas-and-Industrial-Applications.pdf
37. “Boundary Dam Carbon Capture Project,” 2016 SaskPower, http://saskpowerccs.com/ccs-projects/boundary-dam-carbon-capture-project/
38. “Natural Gas Power Plant with CCS is a Positive Step for the Climate,” Michael Tubman, April 23, 2014, Center for Climate and Energy Solutions, http://www.c2es.org/blog/tubmanm/natural-gas-power-plant-ccs-positive-step-climate
39. “Electric Power Annual 2017,” October 2018, Revised December, 2018, Energy Information Administration, https://www.eia.gov/electricity/annual/pdf/epa.pdf
40. “U.S. Crude Oil and Natural Gas Production in Federal and Non-Federal Areas,” Marc Humphries, Specialist in Energy Policy, April 10, 2014, Congressional Research Service, 7-5700, R42432.
41. “Greenfield Gas Processing Facility in Colorado Supports Continued Development of Niobrara Shale,” DCP Midstream Partners, retrieved August 17, 2016, Black and Veatch Company, http://bv.com/Projects/DCP-midstream-greenfield-gas-processing-kersey-co
42. “Companies Investing Heavily in Ohio Natural Gas-Processing Plants, December 8, 2012, Ohio.com, http://www.ohio.com/news/local/companies-investing-heavily-in-ohio-natural-gas-processing-plants-1.356646
43. “EagleClaw Midstream Services, LLC Acquires East Toyah Natural Gas Gathering and Processing System in Reeves County, Texas,” December 9, 2014, http://www.eagleclaw-midstream.com/news/eagleclaw-midstream-services-llc-acquires-east-toyah-natural-gas-gathering-and-processing
44. “Marcellus, Utica Shales Make Northeast Focal Point of Growing U.S. Production,” Al Pickett, November 2013, American Oil & Gas Reporter, http://www.pdce.com/images/stories/pdc-documents/FeatureArticles/201311_AOGR_Final.pdf
45. “Williams Partners Provides Update Regarding April 23 Incident at Its Opal, Wyo., Gas-Processing Facility Developing Plans to Bring Facility Back in Service in Systematic, Timely Manner,” April 28, 2014, http://investor.williams.com/press-release/williams/williams-partners-provides-update-regarding-april-23-incident-its-opal-wyo-ga
46. “Cryogenic Plants Sprouting Up around South Texas,” by Christina Rowland, 2012, http://mysoutex.com/bookmark/19603173-Cryogenic-plants-sprouting-up-around-South-Texas
47. “Vaquero Commissions Natural Gas Processing Facility in Texas,” July 7, 2016, Hydrocarbons-technology.com, http://www.hydrocarbons-technology.com/news/newsvaquero-commissions-natural-gas-processing-facility-texas-us-4942754
48. “FPL Changes Space Coast Skyline to Add New, Clean Energy Center: Construction Nears for More Efficient and More Reliable Units,” Florida Power & Light, August 22, 2010, http://newsroom.fpl.com/news-releases?item=101533
49. “Colusa Generating Station (CGS), Electrical Power Plant Project,” 2016, California Energy Commission, http://www.energy.ca.gov/sitingcases/colusa/
50. “Salem Harbour Combined-Cycle Gas Turbine Power Plant, Massachusetts, United States of America,” power-technology.com, 2016, http://www.power-technology.com/projects/salem-harbour-combined-cycle-gas-turbine-power-plant-massachusetts/
51. “Natural-Gas-Fired Power Plant Planned on 30 Acres in Oregon,” The Blade, September 6, 2012, http://www.toledoblade.com/Energy/2012/09/06/Natural-gas-fired-power-plant-planned-on-1.html

52. "Panda Power to Jointly Develop 1,000 MW Combined Cycle Power Plant," February 18, 2015, Penn Energy, http://www.pennenergy.com/articles/pennenergy/2015/02/panda-power-to-jointly-develop-1-000-mw-combined-cycle-power-plant.html
53. "Update on Energy Commission's Review of California Power Projects," May 24, 2000, Proposed Power Projects—An Overview, http://www.energy.ca.gov/releases/2000_releases/2000-05-24_siting_update.html
54. "Moundsville Power Natural Gas Plant Still Set for 2018 Online Data in Marshall County," June 3, 2016, The Intelligencer, http://www.theintelligencer.net/news/top-headlines/2016/06/moundsville-power-natural-gas-plant-still-set-for-2018-online-date-in-marshall-county/
55. "Lake Side 1 and 2 Generation Facilities," PacifiCorp Energy, August 2014, http://www.pacificorp.com/content/dam/pacificorp/doc/Energy_Sources/EnergyGeneration_FactSheets/RMP_GFS_Lake_Side.pdf
56. "Electric Power Consumption (kWh per capita)," IEA Statistics for 2016, http://energyatlas.iea.org/#!/tellmap/-1118783123/1
57. "Lower and Upper Explosive Limits for Flammable Gases and Vapors (LEL/UEL)," Matheson Gas, 2001, https://www.mathesongas.com/pdfs/products/Lower-(LEL)-&-Upper-(UEL)-Explosive-Limits-.pdf
58. "Natural Gas Explosion Sparks Large Fire in Salem Township," WTAE.com, April 30, 2016, http://www.wtae.com/news/reports-gas-well-on-fire-in-salem-township/39279470
59. "Investigation Underway into Cause of Natural Gas Pipeline Explosion," July 18, 2016, KRGV.com, http://www.krgv.com/story/32472255/investigation-underway-into-cause-of-natural-gas-pipeline-explosion
60. "New TLV® Exposure Limit: Measuring Hydrogen Sulfide," June 2, 2011, Industrial Safety and Hygiene News, http://www.ishn.com/articles/91070-new-tlv-exposure-limit-measuring-hydrogen-sulfide
61. "New Technology Improves Nitrogen-Removal Economics," Oil & Gas Journal, April 23, 2001, http://www.ogj.com/articles/print/volume-99/issue-17/drilling-production/new-technology-improves-nitrogen-removal-economics.html
62. "Natural Gas and the Environment," NaturalGas.org, September 20, 2013, http://naturalgas.org/environment/naturalgas/
63. "Environmental Impacts of Natural Gas," Union of Concerned Scientists, June 19, 2014, http://www.ucsusa.org/clean_energy/our-energy-choices/coal-and-other-fossil-fuels/environmental-impacts-of-natural-gas.html#.V8ClqCX2ZMt
64. "Report Finds U.S. Natural Gas Methane Emissions Have Little Climate Change Impact," February 16, 2018, Seth Whitehead, Energy in Depth: Climate & Environment, https://eidclimate.org/report-finds-u-s-natural-gas-methane-emissions-little-climate-change-impact/
65. "The Peterhead Carbon Capture and Storage Project," March 25, 2015, https://www.spe-uk.org/aberdeen/wp-content/uploads/2014/01/Spence-SPE-Talk-25Mar15.pdf
66. "Peterhead Project Fact Sheet: Carbon Dioxide Capture and Storage Project," November 25, 2015, https://sequestration.mit.edu/tools/projects/peterhead.html
67. "Assessment of Natural Gas Combined Cycle Plants for CCS in a Gas-Dominated Electricity Market CB&I Report Summary and Discussion," April 16, 2015, http://www.energy.ca.gov/research/notices/2015-04-16_workshop/presentations/CBI_04_16_15.pdf
68. "An Overview of Combined Cycle Power Plant," Electrical Engineering Portal, August 25, 2012 http://electrical-engineering-portal.com/an-overview-of-combined-cycle-power-plant
69. "Carbon Capture Suffers a Huge Setback as Kemper Plant Suspends Work," Katie Fehrenbacher, Green Tech Media, June 29, 2017, https://www.greentechmedia.com/articles/read/carbon-capture-suffers-a-huge-setback-as-kemper-plant-suspends-work#gs.5eegq2

3

Coal

A Nonrenewable Energy Type

3.1. Foreword

Coal is one of the most important energy sources in the world. It has many uses, including heating, cooking, and electricity generation, and it can even be used to make synthetic fuels. Over a process of many million years, coal was formed from prehistoric plants in marshy environments as thermal and bacterial decomposition of the plants took place. Initially, a carbon-rich material called peat was formed. Then, over time, temperature, and high pressure, different ranks of coal were formed. There are generally four coal ranks considered, and increasing age corresponds to increasing carbon content. These four ranks are (1) lignite, with an age of 60 million years and a carbon content of 60%–75%, (2) subbituminous, with an age of 100 million years and carbon content of 75%–80%, (3) bituminous, with an age of 300 million years and carbon content of 80%–90%, and (4) anthracite, with an age of 350 million years and carbon content of 90%–95%.[1] Like petroleum crude oil and natural gas, coal is considered a fossil fuel since it originates from fossilized remains, in this case prehistoric plants.

The transformation of these prehistoric plants into peat and coal is commonly regarded as proceeding in two steps, called the biochemical and physicochemical stage of coalification.[2] During biochemical coalification organisms initiate and assist in the chemical decomposition of vegetal matter and its conversion into peat and coal and, during the subsequent physiochemical coalification, the transformation to different coal ranks (or types) occurs due to time, temperature, and pressure. The transition from peat to anthracite is characterized by a number of chemical changes including (1) disappearance of cellulose, (2) decreasing amounts of hydrogen and oxygen, (3) increasing amounts of carbon, and (4) decreasing proportion of volatile matter. For step (3), with the increase in carbon also comes the incorporation of this carbon into aromatic carbon (i.e., carbon bonded into benzene ring structures).[3]

Typical compositions are shown on a dry and ash-free basis. As Table 3.1 illustrates, as the coal rank increases toward the older, more mature forms such as bituminous and anthracite, the carbon and energy content increase. Elemental analysis shows that coal has complicated empirical formulas. As examples, one chemical formula for bituminous coal is $C_{137}H_{97}O_9NS$ while one chemical formula for anthracite is $C_{240}H_{90}O_4NS$.[4]

The Changing Energy Mix. Paul F. Meier, Oxford University Press (2020). © Oxford University Press.
DOI: 10.1093/oso/9780190098391.001.0001.

Table 3.1 Typical Composition for Peat and Coal Ranks[4,5]

	Peat	Lignite	Subbituminous	Bituminous	Anthracite
Age, million years		~60	~100	~300	~350
Carbon, wt%	50–60	60–75	75–80	80–90	90–95
Hydrogen, wt%	5–6	5–6	5–6	4–5	2–3
Oxygen, wt%	35–40	20–30	15–20	10–15	2–3
Sulfur, wt%		0.4–1.0	<2	0.7–4.0	0.6–0.8
Chlorine, ppm		120+/−20	120+/−20	340+/−40	340+/-40
Heating Content (BTU/lb)	4,000	4,000–8,300	8,500–13,000	11,000–15,000	13,000–15,000

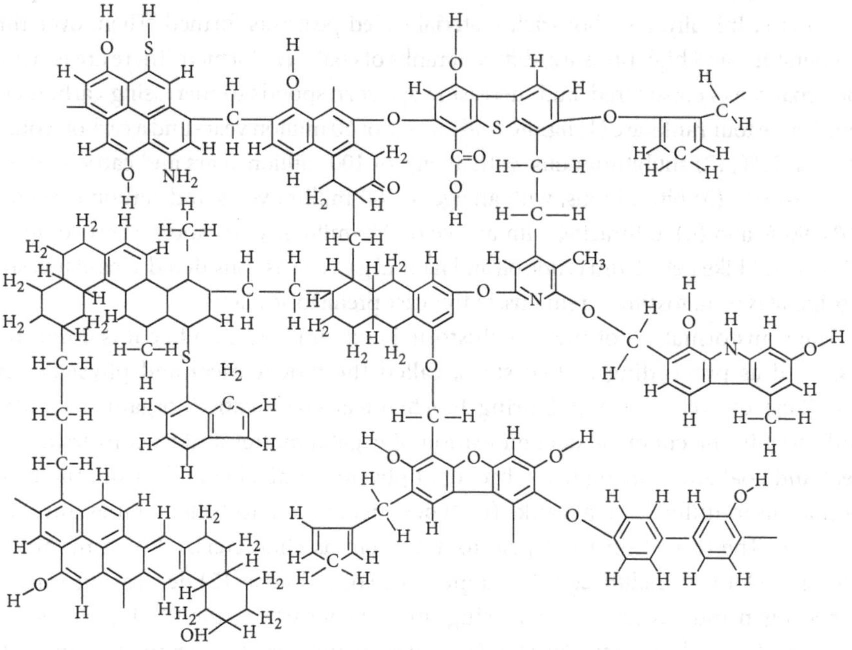

Figure 3.1 Typical Structure for Bituminous Coal (*Source*: EPRI[6])

A typical structure for bituminous coal is shown in Figure 3.1, such that the pentagons and hexagons represent five or six aromatic carbon structures.[6]

3.2. Proven Reserves

Estimating the amount of coal reserves in the world and the United States is difficult. The reason for this is shown in the three definitions the Energy Information

Administration (EIA) uses to classify coal reserves.[7] The first category, "Total Resources," is an estimate that also includes undiscovered coal, and is based on geologic data with different degrees of accuracy. For this category, total US resources are estimated to be around 3.9 trillion tons. Note here that "tons" refers to 2,000 lb. When describing coal reserves, the term "tonne" is sometimes used instead. One tonne is 1,000 kilograms, and has about 1.1 times the weight of one ton.

The second category is called "Demonstrated Reserve Base" and refers to coal that has been found or measured, and can be mined commercially. For this category, total US resources are estimated to be around 475 billion tons. Finally, the third category is called "Estimated Recoverable Reserves," and includes only coal that can be mined today. For this category, total US resources are estimated to be around 253 billion tons, or about 53% of the Demonstrated Reserve Base.

According to the EIA, recovery rates are different for underground and surface mining. For example, recovery rates for underground mines may be less than 40% while surface mines can have recovery rates of 90%. In underground mines there can be restrictions in recovery due to environmental and physical reasons. For example, in an underground mine, some of the coal may be left in place to prevent collapse, and geologic features may make the coal inaccessible.

Accordingly, the ranking of worldwide coal reserves is most accurately based on the third category, coal that can be mined today. Table 3.2 shows coal reserves for the top ten countries in the world, taken from a British Petroleum (BP) review of world energy.[8] The worldwide total reserves are shown as 1,140,904 million tons, so these ten countries account for 91% of the worldwide coal reserves. Note that the BP number for the United States differs from the 253,000 million tons reported earlier from the EIA.

Table 3.2 Top Ten Countries for Coal Reserves, 2017 (million tons)[8]

Million Tons	
Total World	1,140,904
United States	276,587
Russia	176,771
Australia	159,634
China	153,021
India	107,727
Germany	39,802
Ukraine	37,892
Poland	28,452
Kazakhstan	28,225
Indonesia	24,910

Table 3.3 Top Ten Coal Producing Countries, 2017 (million tons)[8]

Million Tons	
Total World	8,518
China	3,884
India	789
United States	774
Australia	531
Indonesia	508
Russia	453
South Africa	278
Germany	193
Poland	140
Kazakhstan	123

The top ten coal producing countries are shown in Table 3.3. As the table illustrates, China is by far the top coal producing country in the world. Worldwide coal production in 2017 was 8,518 million tons,[8] so these top ten countries account for 90% of the worldwide coal production. By itself, China accounted for 46% of the world coal production.

Finally, the EIA shows that the worldwide coal consumption rate in 2017 was 8,434 million tons.[8] Table 3.4 shows the top ten consuming countries. These ten countries account for 88% of the worldwide consumption. By itself, China accounted for 50% of the world coal consumption.

Comparing Tables 3.3 and 3.4, it can be seen that the United States and Poland consume about what they produce. Russia, South Africa, Australia, and Indonesia produce more than they consume and are net exporters. Japan, South Korea, and, to a lesser extent, Germany, consume more than they produce.

Before estimating the remaining years that coal will last at current consumption rates, it is worth discussing expectations for the future of coal. Table 3.5 shows coal consumption in the United States from 2008 to 2017. As shown, there has been a gradual decrease in coal consumption since 2008, and coal consumption in 2017 was 62% of what it was in 2008. Also, there was a significant decrease from 2014 to 2015, with 2015 consumption being ~85% of what it was in 2014.

The reason that US coal consumption has decreased is because many coal electric plants have been shut down or converted to another energy source, mainly natural gas. In terms of US electricity generation, natural gas passed coal in 2015. There are environmental advantages of natural gas over coal in electricity generation, which

Table 3.4 Top Ten Coal Consuming Countries, 2017 (million tons)[8]

Million Tons	
Total World	8,434
China	4,207
India	1,137
United States	693
Japan	241
Russia	203
South Korea	199
South Africa	160
Germany	348
Indonesia	107
Poland	137

Table 3.5 United States and China Coal Consumption, 2008–2017 (million tons)[8]

Year	United States	China
2008	1,118	3,577
2009	983	3,747
2010	1,040	3,887
2011	981	4,232
2012	867	4,285
2013	900	4,377
2014	898	4,344
2015	776	4,254
2016	710	4,199
2017	693	4,207

are discussed later in this chapter. Resolving coal electric plant environmental issues causes an increase in both capital and operating costs compared to electricity generated by natural gas.

Nevertheless, while consumption may be decreasing in the United States, Table 3.5 shows that coal consumption in China has increased significantly from 2008 to 2013, then decreasing slightly after 2013.

In addition, like natural gas, there could be increases in coal consumption using the Fischer-Tropsch synthesis to convert coal into synthetic diesel and, to a lesser extent, synthetic gasoline. Synthetic fuel formed from the Fischer-Tropsch reaction is discussed in Chapter 12.

Using the 2017 estimates for proven reserves of 1,140,904 million tons and consumption rate of 8,434 million tons per year gives a calculated value of 135 years remaining for coal. Therefore, like crude oil and natural gas, coal is a dwindling commodity with a possibility of using up the world supply by around the middle of the next century. Accessing the Demonstrated Reserve Base, coal that has been found or measured, could extend this time while additional environmental regulations could make coal less attractive from an economic standpoint.

3.3. Overview of Technology

According to one reference,[9] the first discovery of coal in what is now the United States was shown on a map of the Illinois River prepared by Louis Joliet and Father Jaques Marquette in 1673–1674. By 1736, several locations for coal were noted on a map of the Potomac River on the border of Maryland and West Virginia. Early settlers used coal to heat their homes, and in the early 1800s other uses for coal included lighting streets and using coal to evaporate water from salt brines to make salt. Of course, the advent of coal-burning locomotives greatly expanded coal use, with coal quickly replacing wood as the fuel of choice. In the latter part of the 1800s, another surge in coal demand was brought on by the iron and steel industry. Following this, yet another surge in coal demand came from electric power generation, when Thomas Edison built the first coal-fired electricity station in 1882.[9]

Today, coal is not used by locomotives and the use of coal to heat homes is rare. The primary uses for coal today are in electricity generation, steel production, and paper and cement manufacturing. In 2017, electrical power generation accounted for 93% of the total US coal consumption and coal was the fuel used to generate about 33% of our electricity.[10]

As mentioned, coal is also used to make iron and steel. While a detailed review of coal use in this industry is beyond the scope of this book, a brief description of the process is given here. First, coal is converted to coke in an oven. Next, raw iron is made by removing oxygen from iron oxide (or iron ore) using the coke for heat. In this step, carbon in the coke uses the oxygen in the iron oxide to make carbon dioxide (CO_2). In addition, the coke provides the furnace heat, which produces more CO_2. Finally, an alloy of iron and carbon are made to produce steel with coke carbon making up about 1% of the alloy. Therefore, the coke removes the oxygen from the iron oxide, provides heat for the furnace, and provides carbon for the steel.[11] In the production of cement and paper, coal is used to produce the heat needed for these processes.

Although not common, coal can also be used to make liquid products. These processes, categorized as liquefaction technologies, fall into two categories. These are direct coal liquefaction (DCL) and indirect coal liquefaction (ICL). In ICL the coal is gasified to carbon monoxide (CO) and hydrogen (H_2), a mixture referred to as synthesis gas, or syngas. Then, using the Fischer-Tropsch process, the syngas is converted to liquid hydrocarbons, mostly diesel and gasoline. This process is discussed in detail in Chapter 12; however, a simplified set of equations is shown here:

1. Gasification: Coal + Oxygen + Steam → Syngas (H_2 + CO)
2. Syngas Conversion: H_2 + CO → Hydrocarbons (Fischer-Tropsch synthesis)

At this time, there are no large scale commercial Fischer-Tropsch plants in the United States, but Germany used this process extensively during World War II and South Africa has used this process for many decades.

In contrast, DCL is a process to convert coal directly into liquid by adding hydrogen at high pressure and high temperature. In a typical rendition of this process, hydrogen is added to the coal to break down its organic structure into liquid hydrocarbons. Typical reaction conditions are temperatures of 750–850°F and pressures of 1,000–2,500 psia in the presence of a solvent. This solvent helps in the extraction of hydrocarbons from the coal and in hydrogen addition. The products of DCL are mainly aromatic compounds and some linear hydrocarbons, so further upgrading is needed to produce suitable liquid transportation fuels.[12] A simplified reaction equation is shown here:

$$\text{Coal} + \text{Hydrogen } (H_2) \rightarrow \text{Linear Hydrocarbons} + \text{Ring Hydrocarbons}$$

In general, DCL has high capital cost and energy consumption, and is therefore not competitive unless crude oil prices are high. In the 1990s, a decade of generally low crude oil prices, plants in Germany, Japan, Australia, and the United Kingdom all shut down. However, the Shenhua Group started up a DCL plant in 2008 in Inner Mongolia. This plant, with a capacity of 30 times any other DCL unit past or present, processes 6,000 tonnes/day of coal.[13]

There are two main methods of extracting coal, namely surface and underground mines. Surface mining is used when coal is deposited less than 200 feet below the surface, while underground mines are suitable for coal formations several hundred feet below the earth.[14] For surface mines, it is possible to recover more than 90% of the coal deposits, but recovery is less than 40% for underground mines. After the coal is mined, it can be sent to the customer by rail, barge, or truck. In the United States, rail is the primary method for moving coal long distances.[14] As an example, most of the coal mined in Wyoming is sent by rail to power plants in the eastern United States.

The United States has the largest coal reserves in the world, almost 25% of the world total. Coal is produced in twenty-four states, but, in 2017, 71% of the coal production was in five states:[15] Wyoming (316.5 million tons), West Virginia (92.8 million tons), Kentucky (41.8 million tons), Pennsylvania (49.1 million tons), and Illinois (48.2 million tons). Wyoming was by far the leader, producing 41% of all coal.

The remainder of the technology overview for coal focuses on electricity generation. In general, the new technology used to generate electricity from coal can be classified as "Clean Coal Technology." There is no single approach for clean coal technology but, rather, it is a variety and collection of different processes used to reduce the environmental impact of the use of coal in making electricity.

The four main categories for clean coal technology include (1) coal washing and preparation, (2) emission controls, (3) efficient coal conversion technologies, and (4) carbon capture and storage.

The first step, coal washing, is the method of upgrading mined coal to satisfy size and purity specifications for the electric power generating plant. This upgrade occurs after mining but before transport of the coal to market, and is achieved using solid-solid and solid-liquid separation processes that remove rock and water from the coal.[16] This washing removes organic and inorganic materials. Inorganic material includes shale, slate, and clay, impurities that reduce the heating value and result in ash during burning. One approach to washing involves grinding the coal into smaller particles and using a liquid with a density that floats the coal but allows impurities to sink (a form of gravity separation).

After producing electricity, emission control technologies are used to remove pollutants that would otherwise go into the air. For particulates, an electrostatic precipitator (ESP) can be used. The ESP electrically charges the ash particle in the flue gas stream, and these particles are attracted to parallel vertical collecting plates to remove the ash from the flue gas.[17]

It is also important to remove nitrogen oxides (NO_x) from the flue gas, as these oxides react with sunlight and oxygen in the air to make ground level ozone (O_3), a form of smog. Nitrogen oxides can also react with water in the atmosphere, creating acids such as nitrous acid (HNO_2) and nitric acid (HNO_3), components of acid rain. NO_x can be removed by selective noncatalytic reduction (SNCR) or selective catalytic reduction (SCR). Catalysts are used in many industrial processes. An oversimplified definition describes a catalyst as a material that increases the rate of reaction without itself being changed. Both of these technologies reduce the NO_x to nitrogen (N_2) and water.[17] Equations showing both of these processes are given here:

Selective Noncatalytic Reduction (SNCR)

- inject ammonia or urea with hot flue gas (760–1090°C) to react with NO_x

$$NH_2CONH_2\,(\text{urea}) + H_2O \rightarrow 2NH_3 + CO_2$$

$$4NO + 4NH_3 + O_2 \rightarrow 4N_2 + 6H_2O$$

Selective Catalytic Reduction (SCR)

- Catalyst Supports: TiO_2, CeO_2, and Al_2O_3
- Catalysts: V_2O_5, WO_2, zeolites, Pt, Rh, Pd
- Temperature: 350–450°C

$$4NO + 4NH_3 + O_2 \rightarrow 4N_2 + 6H_2O$$

$$6NO_2 + 8NH_3 \rightarrow 7N_2 + 12H_2O$$

A final flue gas pollutant that must be removed is sulfur dioxide (SO_2). If SO_2 is allowed to go into the atmosphere, it can react with water to make sulfurous (H_2SO_3) and sulfuric acid (H_2SO_4), major components of acid rain. There are many technologies for SO_2 removal, also called flue gas desulfurization (FGD), including wet scrubbing FGD, dry sorbent injection FGD, wet sulfuric acid (WSA) processes, and the Haldor Topsoe SNOX technnology.[17,18]

In the WSA process, SO_2 is converted into sulfuric acid, thereby turning the pollutant into a saleable product. Equations for this process are shown here:

$$\text{Oxidation: } SO_2 + \tfrac{1}{2}O_2 \rightarrow SO_3$$

$$\text{Hydration: } SO_3 + H_2O \rightarrow H_2SO_4(g)$$

$$\text{Condensation: } H_2SO_4(g) \rightarrow H_2SO_4(1)$$

The Haldor Topsoe SNOX process deserves special mention, as this process can be used to remove three flue gas pollutants including particulates, NO_x, and SO_2. Basically, it combines ESP with SCR of NO_x and the WSA process.

The third category for clean coal technology is the efficiency of the coal conversion process to generate electricity. Different methods of coal conversion, with different efficiencies, include (1) stoker-fired coal combustion, (2) pulverized coal combustion, (3) fluidized bed coal combustion, and (4) integrated gasification combined cycle (IGCC). A description taken from an EPA article[19] is given for each technology.

For simple coal combustion, the production of electricity from coal takes place in three steps. In the first step, coal is burned in the presence of air to produce heat, with by-products of CO_2 and a small amount of water. In the second step, the heat of combustion is used to produce steam in the boiler. This steam, at high pressure and temperature, is used to rotate the blades of a turbine. In the third step, the rotating turbine turns the generator to produce electricity. For the first three types of coal combustion listed earlier, the technologies differ only in the manner in which the coal combustion takes place.

Stoker-fired coal combustion was first used in the late 1800s. For this technology, coal is crushed into large lumps and burned on a grate. During the burning, air is heated which is subsequently used to boil water. No new coal stoker-fired plants are expected to be built, so this technology will become obsolete as existing units are retired.

Pulverized coal combustion uses a fine powder, allowing more efficient combustion compared to stoker-fired combustion. When the coal particle is large, as in stoker-fired combustion, some of the coal does not get good contact with air; however, for pulverized coal combustion, the particle sizes are generally less than 200 microns, providing more complete and efficient combustion of the coal due to better contact with the air. After combustion, the hot flue gas from the furnace is used to boil water in the boiler to produce the high pressure and high temperature steam, which is then sent to the turbine.

There are three types of pulverized coal combustion, namely subcritical, supercritical, and ultrasupercritical. The concept of a supercritical fluid is beyond the scope of this book but is discussed in numerous books on thermodynamics. Simply stated, a supercritical fluid is a material that, above a certain temperature and pressure, no longer has distinct phases of liquid and gas. In the case of pulverized coal combustion, if a boiler operating at supercritical conditions is used, it is no longer necessary to have a drum that separates steam from water. That means instead of boiling water to make steam which turns the turbine, the boiler operation is at such high pressure (3,200 psi or 221 bar and above) that the water is a supercritical fluid and no longer a liquid or gas.[20] Because less coal fuel is used than subcritical, you get higher efficiencies, meaning less coal is burned and less CO_2 is produced. A table comparing operating conditions and efficiencies, taken from two references, is shown in Table 3.6.

Fluidized bed coal combustion uses small coal particles (~1–10 mm) which are fluidized with air. A fluidized bed occurs when the upward gas flow, in this case air, causes the solid coal particles to be suspended and behave like a fluid. This type of combustion is done at lower temperature compared to pulverized coal combustion, as there is excellent heat transfer due to the mixing in the fluidized bed and the temperature is more uniform. Because of the lower temperature, there is less NO_x made than in pulverized coal combustion. Also, limestone can be mixed with the coal particles to absorb sulfur, and reduce SO_2 emissions; thus, SO_2 removal is done in situ rather than with flue gas desulfurization. Fluidized beds can handle larger particles compared to pulverized coal combustion. This allows for greater fuel flexibility, and fluidized beds can use all types of coal (from anthracite to lignite), petroleum coke (made from refinery coker units), and biomass. Since the bed is fluidized, it is possible to continuously add fresh coal.

There are two types of fluidized bed designs including operation at atmospheric and elevated pressure. Also, there are two types of fluidization used, a fixed bubbling fluidized bed and a circulating fluidized bed. A circulating fluidized bed will use bed materials, such as silica sand, to support the combustion coal.[22] An advantage of

Table 3.6 Pressures, Temperatures, and Efficiency Ranges for Different Pulverized Coal Power Plants[21,23]

Pulverized Coal Power Plant	Steam Pressure, MPa	Steam Temperature, °C	% Efficiency[21]	% Efficiency[23]
Subcritical	<22.1	Up to 565	33–39	34.3
Supercritical	22.1–25	540–580	38–42	38.5
Ultrasupercritical	>25	>580	>42	43.3

higher pressure is a smaller furnace used for combustion, because of an increase in contact time between the coal and air, and better transfer of heat. Advantages of a circulating bed versus a fixed bubbling fluidized bed include (1) operation at higher feed rates since the furnace operates at a higher gas velocity, (2) a lower combustion temperature which results in less NO_x production, (3) a greater ability to handle coal feed rates below design, and (4) higher thermodynamic efficiency. According to one reference,[23] circulating fluidized bed combustion has about the same thermodynamic efficiency as subcritical pulverized combustion, 34.8% versus 34.3%.

The last coal combustion technology discussed here is IGCC. Simply stated, in IGCC gasified coal is made into synthesis gas to power a gas turbine and heat from the gas turbine is used to run a steam turbine. Thus, in comparison to the three previously discussed technologies, IGCC uses two turbines to generate electricity. Looking at this in more detail, an IGCC electrical plant will use coal gasification to turn coal into synthesis gas, or syngas, a mixture of carbon monoxide (CO) and hydrogen (H_2). What follow are the primary equations in the production of syngas. Equation (1) is the primary equation, as coal is mostly carbon, and equation (3) represents the combustion of the small amount of hydrogen in the coal. Equation (4) is called the water-gas shift reaction, used to raise the content of hydrogen.

$$C + \tfrac{1}{2}O_2 \rightarrow CO \tag{1}$$

$$CO + \tfrac{1}{2}O_2 \rightarrow CO_2 \tag{2}$$

$$H_2 + \tfrac{1}{2}O_2 \rightarrow H_2O \tag{3}$$

$$CO + H_2O \rightarrow CO_2 + H_2 \tag{4}$$

This syngas is used as fuel to power the gas turbine, thus producing electricity. Next, the heat from the gas turbine is used to generate steam and power the steam turbine, thus producing additional electricity. For this process, oxygen is normally separated from the air, thus reducing the size of the reactor and removing what would be significant amounts of nitrogen in the synthesis gas. The cryogenic process to separate oxygen from air (an air separation unit or ASU) is old technology, but uses about 10% of the power output and represents ~15% of the plant capital cost.[24] Table 3.7, taken from two references, gives a summary of some existing IGCC plants. As shown, the gas turbine produces from 52% to 79% of the total power. Also, thermodynamic efficiencies range from 39% to 45%. A study by MIT suggests a typical efficiency for IGCC as 38.4%.[23] The expectation for IGCC is that the recovery of heat from the gas turbine, used to generate electricity with the steam turbine, would increase the overall efficiency to over 50%. The results show that while efficiencies equivalent to pulverized supercritical and ultrasupercritical are possible, they fall short of the 50% goal.

For the four types of coal combustion technologies described in this section, US EIA data show that there are currently 359 coal power plants in the United States with 789 generators and a nameplate capacity of 279.2 GW. Although not available for 2017, in 2015 the technology breakdown was 0.4% using stoker-fired coal

Table 3.7 Summary of IGCC Plant Data[25,26]

	Buggenum IGCC, Netherlands	Puertollano, Spain	Wabash River, Indiana, US	Tampa Electric, Florida, US	Nakoso, Japan	Negishi, Japan
Electrical Capacity, MW	253	300	262	250	250	342
Gas Turbine Power, MW	168	200	171	171	130	270
Steam Turbine Power, MW	85	100	91	79	120	72
% Gas Turbine Power	66	67	65	68	52	79
Efficiency	43%	45%	39%	41%	42%	

combustion, 93.5% using pulverized coal combustion, 2.5% using fluidized-bed coal combustion, and only 0.4% using IGCC.[28,32] Another 3.2% are classified as "other." Thus, pulverized coal combustion dominates the US coal electrical plant market.

A future consideration for electricity generation from coal is carbon capture and sequestration (CCS), such that CO_2 is captured and stored in the ground (old oil or gas fields), liquid storage in the ocean, or solid storage as a carbonate. The main problem with CO_2 capture from pulverized and fluidized bed coal combustion is the difficulty in capturing CO_2 from a flue gas that is mostly nitrogen. Since air is used as the combustion gas, the flue gas has a high concentration of nitrogen. One method to capture CO_2 is to substitute oxygen for air, thereby removing the nitrogen before combustion; this approach is referred to as oxy-fired combustion. In the case of IGCC, the nitrogen is already removed from the air before combustion by the air separation unit.

The removal of CO_2 from a coal electric power plant poses several challenges. As discussed in the next section on economics, it increases capital costs. In addition, it decreases the thermodynamic efficiencies. Table 3.8 shows that efficiency decreases around 7%–9%.[23]

The reason for this decrease in efficiency is that significant energy is used in the process to capture carbon. The increase in energy use is caused by H_2S removal, the energy for an air separation unit, and the power used to clean and capture CO_2.[29]

Another challenge is related to the sheer volume of CO_2 produced from US coal electric plants. If 60% of the CO_2 produced from US coal-based power generation were captured and compressed to a liquid for geologic sequestration, its volume would about equal the total US oil consumption of 20 million barrels per day.[23]

A final topic is distribution of the electricity. Once electricity is made, cables are used to transport the electricity to a transformer. The transformer is used to step-up the voltage from the power plant, to reduce transmission losses. Typically, a power plant will produce electricity at voltages from 2 to 30 kV and the transformer will

Table 3.8 Thermodynamic Efficiencies With and Without CCS[23]

Technology	Efficiency without CCS	Efficiency with CCS
Subcritical Pulverized Coal	34.3%	25.1%
Supercritical Pulverized Coal	38.5%	29.3%
Ultrasupercritical Pulverized Coal	43.3%	34.1%
Fluidized Bed	34.8%	25.5%
IGCC	38.4%	31.2%

step-up this voltage to 115 to 765 kV. As an example of transmission losses, a 765 kV line carrying 1,000 MW of power will have losses of 0.5%–1.1% over 100 miles but a 345 kV line with the same load will have losses of 4.2% over the same distance.[30] Historical data from the state of California shows annual average transmission losses around 5% to 7% over the time period of 2002 to 2008.[31] Next, metal towers are used to safely support overhead power cables and, finally, a step-down transformer is used to reduce voltage that comes into your home.

3.4. Capital Cost, Operating Cost, Well-to-Wheels Levelized Cost of Electricity, and Well-to-Wheels Levelized Cost of Fuel

The equations and methodology used to calculate levelized cost of electricity (LCOE), and specific input used for coal to make electricity, are detailed in Appendix A. The units for LCOE are in $/megawatt-hour, or $/MWh. While this section discusses the economics of using coal to generate electricity, it is recognized that coal has other uses in steel, cement, and paper industries, as well as the Fischer-Tropsch synthesis. The Fischer-Tropsch synthesis is discussed in Chapter 12.

For a coal generating plant, the "Well-to-Wheels" costs will include (1) the cost of coal production and transportation to the power plant, (2) power plant generating costs, (3) cost of connecting the power plant to the electrical grid, (4) transmission and distribution costs, and (5) taxes (both federal and state).

The cost of coal (based on Illinois No. 6 bituminous coal) was set to $2.06/MM BTU (British thermal unit) using a 12-month average of 2018 Energy Information Administration data.[40] The cost to the power plant also captures the cost of coal production and transportation.

For the power plant generating costs, we need capital costs, operating costs, and capacity factors. Finding current data for these is a little less straightforward than for other technologies. For example, in 2013, the EIA provided data for pulverized coal with and without CCS as well as data for IGCC with and without CCS.[33] However, for 2018 costs, the EIA only reported coal plant cases with 30% and 90% CCS.[34] Relative to 2013, cases were limited in 2018 because of the assumption that new coal plants without CCS cannot be built because of new plant emission standards.

In August 2015, the Environmental Protection Agency (EPA) released a final rule to limit greenhouse gas emission for new power plants.[35] This "Carbon Pollution Standard for New Power Plants" limits CO_2 emissions of no more than 1,000 lb of CO_2 per megawatt-hour (MWh) of electricity produced. Later in this chapter, in the section on CO_2 production, it is shown that CO_2 emission from coal plants are greater than 2 lb/kWh, or 2,000 lb/MWh. To meet this standard, new coal plants will require the use of CCS technology.

Therefore, in order to determine the LCOE for both pulverized coal and IGCC plants, EIA data from both 2013 and 2018 as well as Lazard data from 2014,[36] 2017,[37] and 2018[38] were used to set capital costs, operating costs, and capacity factors. For both technologies, cases were examined with and without CCS.

For transmission costs, the reference data provide the cost for new transmission lines to tie into existing substations or existing transmission lines. In other words, the generating station provides transmission lines and steps up the voltage to tie into the "electrical grid," a network that delivers the electricity to the customer. The grid has high-voltage transmission lines (sometimes distant) that carry power to demand centers, transformers to step down the voltage, and distribution lines that connect individual customers.

The costs of transmitting and distributing electricity were 0.9 cents/kWh (or $9/MWh) and 2.6 cents/kWh (or $26/MWh), set according to the EIA.[39] Here, transmission is defined as the high-voltage movement of electricity from the power station while distribution is the cost of electrical lines that take power from a substation to the customer. The cost of transmission and distribution is essential to get a true "Well-to-Wheels" analysis of the total cost. Results for LCOE are shown in Table 3.9 for the different cases.

Table 3.9 shows that IGCC is about $31/MWh more expensive than pulverized coal combustion, without CCS, and about $26/MWh more expensive with the addition of CCS. In the case without CCS, the difference is largely attributable to the air separation unit, which is about 15% of the plant capital cost. The table also shows that when CCS is added, variable costs including fuel increase about $11/MWh for pulverized coal combustion and $4/MWh for IGCC. The increase in fuel costs are due to the significant energy used to capture carbon, namely the energy used for H_2S removal, the air separation unit, and the cleaning and capturing of CO_2. In the case of pulverized coal combustion, the LCOE more than doubles with the addition of CCS while IGCC increases about 82%. Since an air separation unit is needed to capture CO_2, this must be added to pulverized coal combustion but IGCC already has this equipment. This is why the addition of CCS changes the LCOE less for IGCC than pulverized coal combustion.

A sensitivity analysis was made for both pulverized coal combustion and the IGCC technology. The sensitivity study was made by varying the capital cost, the combined fixed and variable operating expenses (excluding the fuel cost), and the fuel cost. To determine the range for capital cost and operating expenses, all of the available reference data were used as a guide.

Table 3.9 Economic Analysis for Coal-Fired Electric Plants

Energy Type	Capacity Factor	Capital Cost	Levelized Capital Cost	Fixed O&M	Variable O&M (Including Fuel)	Transmission Investment	Total System Levelized Cost (LCOE) to Generate	LCOE Including Transmission and Distribution
		$/kW	$/MWh	$/MWh	$/MWh	$/MWh	$/MWh	$/MWh
Pulverized coal[33]	0.85	$3,246	$41.84	$5.08	$22.60	$1.20	$70.71	$105.71
Pulverized coal[36,38]	0.93	$3,000	$35.34	$4.91	$20.03	$1.20	$61.48	$96.48
Average	0.89	$3,123	$38.59	$4.99	$21.31	$1.20	$66.09	$101.09
Pulverized coal with 30% CCS[34]	0.85	$5,169	$66.62	$9.69	$27.40	$1.20	$104.90	$139.90
Pulverized coal with 90% CCS[34]	0.85	$5,716	$73.67	$11.25	$33.89	$1.20	$120.01	$155.01
Pulverized with 90% CCS[33]	0.85	$5,227	$67.37	$10.82	$34.23	$1.20	$113.61	$148.61
Pulverized with 90% CCS[36,38]	0.93	$8,400	$98.95	$9.82	$29.72	$1.20	$139.69	$174.69
Average	0.88	$6,448	$80.00	$10.63	$32.61	$1.20	$124.44	$159.44
IGCC[33]	0.85	$4,400	$56.71	$8.36	$25.14	$1.20	$91.41	$126.41
IGCC[37]	0.75	$ 4,175	$60.99	$11.11	$32.60	$1.20	$105.90	$140.90
IGCC[36]	0.75	$4,000	$58.43	$9.47	$25.13	$1.20	$94.23	$129.23
Average	0.78	$4,192	$58.71	$9.65	$27.62	$1.20	$97.18	$132.18
IGCC with 90% CCS[33]	0.85	$6,599	$85.05	$9.78	$30.49	$1.20	$126.53	$161.53
IGCC with 90% CCS[37]	0.75	$8,350	$121.97	$11.11	$32.60	$1.20	$168.88	$201.88
IGCC with 90% CCS[36]	0.75	$8,000	$116.86	$11.11	$30.17	$1.20	$159.34	$194.34
Average	0.78	$7,650	$107.96	$10.67	$31.09	$1.20	$151.58	$185.92

Note: The pulverized coal data from Lazard for 2014 and 2018 (references 36 and 38) were identical.

For capital cost, the cases with and without CCS were used to establish the maximum and minimum, since the "Carbon Pollution Standard for New Power Plants" may force new plants to be built with CCS. For pulverized coal, these were \$3,000 and \$8,400/kW, respectively while for IGCC these were \$4,000 and \$8,350/kW, respectively. Excluding the cost of fuel in the variable operating cost, the combined fixed and variable operating costs minimum and maximum were \$6.9/MWh and \$21.1/MWh for pulverized coal and \$15.6/MWh and \$19.6/MWh for IGCC.

Unlike crude oil and natural gas, the price for coal has been relatively stable over the last decade. Table 3.10 shows historical prices for Illinois No. 6 bituminous coal, which range from \$2.06 to \$2.39/MM BTU. Using a BTU content of 12,712 BTU/lb for the bituminous coal, these prices are equivalent to a range of \$52.4 to \$60.8 per ton. Although the range for the last decade is not large, the price of coal was as low as \$1.20/MM BTU in 2000. In order to better understand the possible impact of coal prices on LCOE, the range was broadened from a low of \$1.50/MM BTU to a high of \$2.50/MM BTU.

The base cases were set using EIA data for pulverized coal and IGCC.[33] The results are shown in Table 3.11.

As the table illustrates, changing the combined fixed and variable operating costs (excluding coal cost) had a minor impact on LCOE. Likewise, changing the cost of coal from \$1.50 to \$2.50/MM BTU had only a minor impact on LCOE. However, increasing the capital cost to as high as \$8,400/kW increased the LCOE \$66/MWh for the pulverized coal case. Likewise, increasing the capital cost to as high as \$8,350/kW increased the LCOE for the IGCC case by \$51/MWh. This is not surprising, as the increases in capital cost essentially reflect the cost of adding CCS, so the impact on LCOE mirrors that shown earlier in Table 3.9 for the addition of CCS.

Table 3.10 Illinois No. 6 Bituminous Coal Historical Prices, \$/MM BTU[40]

Year	\$/MM BTU
2009	2.21
2010	2.27
2011	2.39
2012	2.38
2013	2.34
2014	2.37
2015	2.22
2016	2.11
2017	2.06
2018	2.06

Table 3.11 Sensitivity Study Using the EIA Case for Integrated Gas Combined Cycle

	Coal Price	Capital Cost	Fixed and Variable O&M (no fuel cost)	LCOE Including Transmission and Distribution
Units	$/MM BTU	$/kW	$/MWh	$/MWh
Pulverized Coal				
Base Case	$2.06	$3,246	$9.55	$105.71
Capex is $3,000/kW	$2.06	$3,000	$9.55	$102.54
Capex is $8,400/kW	$2.06	$8,400	$9.55	$172.14
Opex is $6.9/MWh	$2.06	$3,246	$6.90	$103.06
Opex is $21.1/MWh	$2.06	$3,246	$21.10	$117.26
Coal price is $1.50/MM BTU	$1.50	$3,246	$9.55	$100.79
Coal price is $2.50/MM BTU	$2.50	$3,246	$9.55	$109.59
IGCC				
Base Case	$2.06	$4,400	$8.36	$126.41
Capex is $4,000/kW	$2.06	$4,000	$15.58	$121.26
Capex is $8,350/kW	$2.06	$8,350	$15.58	$177.32
Opex is $15.6/MWh	$2.06	$4,400	$15.60	$126.43
Opex is $19.6/MWh	$2.06	$4,400	$19.60	$130.43
Coal price is $1.50/MM BTU	$1.50	$4,400	$15.58	$121.54
Coal price is $2.50/MM BTU	$2.50	$4,400	$15.58	$130.24

3.5. Cost of Energy

The cost of coal in units of $/lb, $/ton and $/MM BTU are presented in this section. The cost of coal (based on Illinois No. 6 bituminous coal) was set to $2.06/MM BTU using a 12-month average of 2018 EIA data.[40] Using an energy content of 12,712 BTU/lb for bituminous coal, the energy cost of coal on a mass basis is $0.026/lb or $52/ton.

As given in chapter 1, the cost of crude oil for a similar 2018 time period was $0.23/lb in terms of mass and $11.20/MM BTU in terms of energy content. Thus, coal is about one-ninth the cost of crude oil on a mass basis and about one-fifth on an energy content basis. Chapter 2 shows the cost of natural gas to be $3.17/MM BTU on an energy basis and $0.07/lb on a mass basis. Thus, coal is also cheaper than natural gas on both an energy and mass basis.

3.6. Capacity Factor

Capacity factors are shown in Table 3.9. For the two studies shown for pulverized coal, the capacity factors were 0.85 and 0.93, respectively. These factors were projected to be the same if CCS is added. Thus, these plants are very reliable, operating about 90% of the time.

In the case of IGCC, the capacity factors are somewhat lower, 0.85 and 0.75 for the two studies shown. The lower capacity factor for IGCC versus pulverized coal combustion is likely due to the additional maintenance needed for an extra turbine and an air separation unit. As in the case of pulverized coal, the addition of CCS is not expected to hurt time on stream. Thus, IGCC plants are also very reliable, operating about 80% of the time.

3.7. Efficiency: Fraction of Energy Converted to Work

Table 3.6 shows the thermal efficiencies of steam turbines for pulverized coal power plants. For subcritical operation, one study gives a range of 33% to 39%[21] and another gives an efficiency of 34.3%.[23] Increasing the turbine steam or gas inlet temperature will increase efficiency. For supercritical pulverized coal combustion, with steam temperatures ranging from 540–580°C, the same two studies referenced for subcritical operation show an efficiency range of 38%–42%[21] and an efficiency of 38.5%.[23] And, for ultrasupercritical pulverized coal combustion, with steam temperature in excess of 580°C, efficiencies are shown to be greater than 42%[21] and 43.3%.[23]

For IGCC, efficiencies were shown in Table 3.7 to range from 39 to 43% while another study gives a value of 38.4%.[23] And, for fluidized bed coal combustion, the efficiency is shown in Table 3.8 to be 34.8%. Considering only the values from Table 3.8,[23] subcritical pulverized coal combustion and fluidized bed coal combustion have the lowest efficiencies, about 34%–35%. With supercritical and ultrasupercritical pulverized coal combustion, efficiencies increase to 38.5% and 43.3%, respectively. IGCC, with an efficiency of 38.4%, is equivalent to supercritical pulverized coal combustion. For convenience, Table 3.8 is shown again.

Table 3.8 Thermodynamic Efficiencies With and Without CCS[23]

Technology	Efficiency without CCS	Efficiency with CCS
Subcritical Pulverized Coal	34.3%	25.1%
Supercritical Pulverized Coal	38.5%	29.3%
Ultrasupercritical Pulverized Coal	43.3%	34.1%
Fluidized Bed	34.8%	25.5%
IGCC	38.4%	31.2%

Table 3.12 Heat Rates and Thermal Efficiencies for the Cases Shown in Table 3.9

Energy Type	Heat Rate	Thermal Efficiency
	BTU/kWh	%
Supercritical Pulverized Coal[33]	8,800	38.8
Supercritical Pulverized Coal[36]	8,750	39.0
Average		38.9
Supercritical Pulverized with CCS[33]	12,000	28.4
Supercritical Pulverized with CCS[36]	12,000	28.4
Average		28.4
IGCC[33]	8,700	39.2
IGCC[36]	8,800	38.8
Average		39.0
IGCC with CCS[33]	10,700	31.9
IGCC with CCS[36]	10,520	32.4
Average		32.2

Also, as the table shows, efficiencies drop dramatically when CCS is added, about 7% to 9%. The reason for this is that significant energy is used in the process to capture carbon. The increase in energy use is caused by H_2S removal, the energy for an air separation unit, and the power used to clean and capture CO_2.[29]

Efficiencies can also be calculated from the "heat rate." The heat rate is how much energy is used to make useful work. In the case of a coal power plant, the coal is the source of energy and the useful work is the electrical power made. To calculate the thermal efficiency, the work needed to make one kilowatt-hour (kWh) of electricity, or 3,412 BTU, is divided by the heat rate. The heat rate is the amount of energy used to make one kWh of electricity.

Table 3.12 shows the results of these calculations. The values shown in Table 3.12, comparing supercritical pulverized coal combustion and IGCC, with and without CCS, are similar to those values shown in Table 3.8.

3.8. Energy Balance

For coal combustion to electricity, the energy balance is defined by this equation:

$$\text{Energy Balance} = \frac{[\text{Electrical energy from coal delivered to the customer}]}{[\text{Energy contained in the coal gas at the mine site}]}$$

For the conversion of coal energy to electricity, there are several steps in the "Well-to-Wheels" analysis including (1) the energy used to extract and process the coal, (2) transport the coal to the electrical plant, (3) make the electricity, and then (4) distribute the electricity through the grid to your home.

For the first step of extracting and processing the coal, there are three basic steps. The first is extraction, which includes blasting, ventilation, drilling, digging, and dewatering. The second step is materials handling, which includes the use of diesel fuel and electric energy. The third step is beneficiation and processing, which includes crushing, grinding, and separation. Beneficiation is any process that improves the value of the coal by removing worthless material. A 2007 DOE study[41] provides energy consumption for the coal mining industry. In this study, 1,309 million tons of coal were mined in 2007 with a recovery of 1,073 million tons, a recovery ratio of 82%. For the tons mined, energy consumption is given as 370,628 BTU/ton, meaning that the US coal mining industry consumed 485.3×10^{12} BTU/year in 2007. At an energy consumption of 370,628 BTU per ton, the energy consumed per pound is 185.3 BTU. As the energy content for bituminous coal was given earlier as 12,712 BTU/lb, this means that 1.5% of the coal energy was consumed to extract and process the coal.

To transport the coal, some estimates will be made. According to the EIA,[42] 69% of coal shipments were delivered by rail, 13% by water, 11% by truck, and 7% by conveyor belts and tramways. Because rail was used the most, rail transportation is used for the estimates. Freight trains can transport anywhere from 1,500 to 6,000 tons for distances around 100 to 500 miles.[43] Based on this, we assume a typical load of 4,000 tons being transported 300 miles. Next, another study[44] shows that one gallon of diesel fuel can move one ton of freight by rail 450 miles. Finally, as shown in Appendix B, that one gallon of diesel contains 138,690 BTU, so we can write the following equation:

$$BTU\,/\,ton = \left[\frac{gallon}{450\ tons - miles}\right] * 300\ miles * 138{,}690\ BTU\,/\,gallon$$
$$= 92{,}460\ BTU\,/\,ton = 46.2\ BTU\,/\,lb$$

As the energy content for bituminous coal is assumed to be 12,712 BTU/lb, this means that 0.4% of the coal energy was consumed to transport the coal.

Examining only supercritical pulverized coal combustion and IGCC, Table 3.8 shows the thermal efficiencies for making electricity to be 38.5% and 29.3% for supercritical coal without and with CCS and 38.4% and 31.2% for IGCC without and with CCS.

It then remains to determine the energy needed to distribute the electricity through the grid to your home. As mentioned in the "Overview of Technology" section, transmission and distribution losses are affected by the voltage of the power line and the distance the electricity is transmitted. Historical data from the state of California shows annual average transmission losses around 5 to 7% over the time period of 2002 to 2008.[45] Similarly, EIA data for 2018[46] show that of the electricity generated, 7% is lost in transmission and distribution. Of course, this 7% lost in transmission and distribution is the percent of the energy lost leaving the electrical plant, not energy from the mine where the coal was extracted.

Putting these results together, after deducting the energy used to extract, process, and transport the coal, 98.1% of the original coal energy at the mine arrives at the electrical plant. The thermal efficiency for supercritical pulverized coal combustion are 38.5% and 29.3% and for IGCC are 38.4% and 31.2% for IGCC without and with CCS. Finally, 7% of the energy leaving the electrical plant is lost in transmission and distribution. Using 100 units of energy as a reference, we can arrive at the following equations.

Supercritical Pulverized Coal Combustion (SPC)

100 (mine) → 98.5 (extraction) → 98.1 (transporting) → 37.8 (SPC)
→ 35.1 (transmission and distribution)

Supercritical Pulverized Coal Combustion with Carbon Capture and Sequestration (SPC-CCS)

100 (mine) → 98.5 (extraction) → 98.1 (transporting) → 28.7 (SPC-CCS)
→ 26.7 (transmission and distribution)

Integrated Gasification Combined Cycle (IGCC)

100 (mine) → 98.5 (extraction) → 98.1 (transporting) → 37.7 (IGCC)
→ 35.0 (transmission and distribution)

Integrated Gasification Combined Cycle (IGCC) with Carbon Capture and Sequestration (IGCC-CCS)

100 (mine) → 98.5 (extraction) → 98.1 (transporting) → 30.6 (IGCC-CCS)
→ 28.5 (transmission and distribution)

Summarizing, this means that of the original energy in the coal at the mine, 35.1% will end up as useable electricity for pulverized coal combustion and 35.0% for IGCC. The addition of CCS will decrease these values to 26.7% and 28.5%, respectively.

3.9. Maturity: Experience

Coal has been in use for a long time. There is archeological evidence in China indicating surface mining and household use around 3490 BC.[47] And, the Romans used coal in the late 2nd century AD, for heating public baths and homes of wealthy people.[48] The first known use of coal in North America was by the Hopi Indians,[49] who in the

1300s used it for cooking, heating, and baking pottery, and by the Aztecs, who used it for fuel and making ornaments during their 14th- to 16th-century existence.[50]

As mentioned earlier in this chapter, the first discovery of coal was shown on a map of the Illinois River prepared by Louis Joliet and Father Jaques Marquette in 1673–1674,[9] and by 1736 several locations for coal were noted on a map of the Potomac River on the border of Maryland and West Virginia. Early settlers used coal to heat their homes and in the early 1800s, other uses for coal included street lights and using coal to evaporate water from salt brines to make salt. The first known commercial coal mine was in the Manakin, Virginia, area in 1748, and the first surface mine was near Danville, Virginia, in 1866.[51] The first electric coal plant, the Pearl Street Power Station, was built in 1882 in New York City by Thomas Edison.[9] Thus, coal mining has been around for nearly 270 years, surface mining for about 150 years, and coal-fired power generation for about 135 years.

Looking at new methods of coal combustion, commercial supercritical and ultra-supercritical pulverized coal combustion plants have been operating since the 1960s and 1990s, respectively.[21] The first use of IGCC was in the mid-1980s in a 110 MW demonstration plant near Barstow, California, and the first full-scale commercial IGCC plants started operating in the 1990s.[52] In addition, the Polk IGCC Power Station near Mulberry, Florida, capable of 313 MW, started operation in 1996 and the Wabash River IGCC plant, capable of 260 MW, started operation in 1995. The first commercially successful fluidized bed combustion plant, although only around 10 MW, started operation in 1979.[53] Thus, supercritical and ultrasupercritical pulverized coal combustion have been around 55 and 25 years of experience, respectively, and full-scale IGCC has been around for more than 20 years.

From the first electrical plant in 1882 and the 1930s, the electrical grid system grew rapidly, roughly doubling every 6 years. Today, what we take for granted, involves more than 180,000 miles of high-voltage transmission lines connecting about 7,000 power plants.[54]

Like natural gas electricity generation, the use of CCS in coal combustion is new. The world's first CCS plant is the Boundary Dam CCS project near Estevan, Saskatchewan. The plant, with a capacity of 115 MW, started operation in the fall of 2014.[55] The captured CO_2 is sold and transported by pipeline to nearby oil fields in southern Saskatchewan to be used for enhanced oil recovery (EOR). In the United States, the Petra Nova CCS project started operation in January, 2017.[56] The plant uses a 240 MW equivalent slipstream of the flue gas, with CO_2 concentrations less than 15%, from the 640 MW pulverized coal plant, capturing about 90% of the CO_2. Amine scrubbing is used to capture the CO_2, which is then transported through a pipeline to an operating oil field for use in EOR.

To summarize timelines in the United States, coal mining has been around nearly 270 years and surface mining about 150 years. Coal-fired power generation has been used for about 135 years and Supercritical pulverized coal combustion has about 55 years of experience while IGCC has about 20 years of experience. Experience with CCS is very new, with no commercial operation on power plants greater than 500

MW. Thus, there are around 270 years of coal mining experience and 135 years of using coal to generate electricity.

3.10. Infrastructure

There is an immense infrastructure to transport coal. For rail, by which 69% of coal is delivered, a 2012 report indicates that large railroads operated on 95,264 miles of track.[57] For trucks, by which 11% of coal is delivered, the interstate highway system currently has 46,876 miles.[58]

To make electricity, the EIA listed 359 operating coal combustion power plants with 789 generators in 2017.[59] Although they no longer report specific plant types, 2015 data from the same source show 93.5% using pulverized coal technology, 2.5% using fluidized bed technology, 0.4% using IGCC, 0.4% using stoker technology, and 3.2% listed as other. Clearly, pulverized coal dominates the coal-to-electricity industry. These units had a nameplate capacity of 279,221.3 MW[59] with a net generation of 1,205,835 thousand MWh.[60] Once made, this electricity can be transmitted around the United States into the more than 180,000 miles of high-voltage transmission lines that exist in our grid.[54]

3.11. Footprint and Energy Density

Footprint is a commonly used term in industrial plants to describe the area used by an energy type. For coal, the footprint will include the area to (1) mine coal, (2) upgrade the coal before transport, (3) transport the coal to the coal plant via rail, (4) make electricity at the coal electrical plant, and (5) transmit and distribute the electricity via the grid to commercial customers.

Since a US Geological survey (USGS) conducted in the mid-1970s, no US government agency has reported acres of coal disturbed by coal mining. Accordingly, an approach similar to that from another source[61] will be used to take old USGS data and combine it with more current data to estimate the land area affected by coal mining.

To determine the area to (1) mine coal, we first recognize that there are two basic approaches to coal mining, namely underground and surface mining. Also, for underground mining, there are basically two methods. For the more traditional method, sometimes called "room-and-pillar mining," the rooms are excavated and pillars of coal are left in place between the rooms to support the mine roof. In contrast, longwall mining involves complete extraction of the coal and the roof in the mined area is allowed to collapse.[62] With these definitions, the "room-and-pillar mining" method does not really occupy surface area, as the area above the mine can conceivably be used for pasture land, farming, or other uses. On the other hand, a longwall mine is area used, and lost, unless it undergoes remediation.

In order to estimate the area used for longwall mining, we first make the following premises for the calculation. Although dated, a 1995 DOE report[62] gives an average mining height of 81 inches for US coal mines in 1993. The report defines mining height as the vertical distance of the cut through the coal seam. For the conversion of volume to mass, the density of the coal is also needed. One reference gives density ranges of 0.5–1.3 g/cm^3 for lignite, 1.15–1.5 g/cm^3 for bituminous, and 1.29–1.65 g/cm^3 for anthracite coal.[63] Since bituminous coal has been used for previous calculations in this chapter, and is far more common in use than lignite and anthracite, this coal type is used for the calculation of area. Taking the mid-range given earlier, a density of 1.3 g/cm^3, equivalent to 81.2 lb/ft^3, is used for an average. Recognizing that one acre of land area is equivalent to 43,560 ft^2, the following equation estimates that there are around 12,000 tons of bituminous coal in an acre of land when the coal thickness is 81 inches.

$$tons\,/\,acre = \frac{\left[\frac{81}{12}\right] ft * 43{,}560\ ft^2\,/\,acre * 81.2\ lb\,/\,ft^3}{2000\ lb\,/\,ton} = 11{,}938\ tons\,/\,acre$$

The Energy Information Administration annual coal report[64] shows that, for 2017, underground coal mining resulted in 273,129 thousand tons of coal, of which 170,018 thousand tons (59%) was from longwall mining. Using the previous equation, the 170,018 thousand tons of longwall mining is estimated to use 14,242 acres while the remaining 103,111 thousand tons from "room-and-pillar mining" is assumed to not disturb any land area. Stated differently, the amount of land disturbed for all underground mining was 14,242 acres (or 22.3 square miles) for 273,129 thousand tons, or 19,178 tons/acre. Using an energy content of 12,712 BTU/lb for bituminous coal, we calculate the energy density for underground mining to be 0.31 quadrillion BTU/(year-square mile) or 0.31×10^{15} BTU/(year-square mile).

$$BTU\,/\,(year - square\ \ mile) = \frac{273{,}129 \times 10^3\ tons\,/\,year * 2000\ lb\,/\,ton * 12{,}712\ BTU\,/\,lb}{22.3\ square\ \ miles}$$
$$= 0.31 \times 10^{15}$$

The next step is to estimate the area used for surface mining. The EIA annual coal report[64] shows that, for 2017, surface coal mining resulted in 501,480 thousand tons of coal. Since specific data on acres used in surface mining are not generally reported, one way to estimate the amount of land disturbed by surface mining is to look at new acres permitted. From 1977 to 2011, a total of 4,494,004 acres were permitted.[65] During this same time period, a total of 22,792 million tons were produced from surface mining.[66] Simple division shows that the production is, on average, 5,072 tons/acre for surface mining. Not surprisingly, this is considerably less than the 19,178 tons/acre for underground mining. Since, for 2017, surface mining produced 501,480 thousand tons of coal, using the value of 5,072 tons/acre for surface mining leads to a calculated area of 98,874 acres (or 154.5 square miles) used for surface mining in 2017.

With an energy content of 12,712 BTU/lb for bituminous coal, we calculate the energy density for surface mining to be to be 0.083 quadrillion BTU/(year-square mile) or 0.083×10^{15} BTU/(year-square mile). Therefore underground mining has an energy density about four times larger than surface mining.

$$BTU\,/\,(year - square\ mile) = \frac{501{,}480 \times 10^{3}\ tons\,/\,year * 2000\ lb\,/\,ton * 12{,}712\ BTU\,/\,lb}{154.5\ square\ \ miles}$$
$$= 0.083 \times 10^{15}$$

Combining underground and surface mining, for 2017 a total of 774,609 thousand tons of coal were produced over a land area of 176.8 square miles. This leads to an energy density of 0.11 quadrillion BTU/(year-square mile) or 0.11×10^{15} BTU/(year-square mile). It is recognized that some of the land used for coal mining, both surface and longwall mining, are recovered through remediation. However, given the approximations made to determine the land areas above and the difficulty in assessing the status of remediated land area recovered from coal mining, no additional adjustments were made.

The next contributions to the footprint of coal are to (2) upgrade the coal before transport and (3) to transport the coal to the coal plant via rail. For coal upgrading, the coal is washed and this is generally done before transport of the coal to market; however, the land area needed for this is small compared to area used for mining and is ignored here. In addition, transport of the coal to market occurs on existing railways, roads, and waterways. While one could argue that the railways would not exist, or that some rail would be eliminated if they were not used to transport coal, it is assumed here that the land area for transporting coal to market would exist in the absence of coal transport. Therefore, for the steps of (1) mining coal, (2) upgrading the coal before transport, and (3) transporting the coal to the coal plant via rail or other modes, the total land area for these three steps is assumed to be 176.8 square miles producing (in 2017) a total of 774,609 million tons of coal. For this, the energy density is 0.11 quadrillion BTU/(year-square mile) or 0.11×10^{15} BTU/(year-square mile).

The next contribution to the coal footprint is to (4) make electricity at the coal electrical plant. Determining the land area for coal plants is somewhat difficult, as most new fossil fuel power plant construction is for natural gas. Also, the land area cited may include wetlands and other protected area as well as future expansion, and some coal power plant may coexist with natural gas and nuclear plants. One reference[67] shows land area requirements to include the generating station, ash disposal area, an evaporation pond, access roads, railroad access, transmission lines, a source of limestone, and even coal mines that may be adjacent to the power plant. This reference examined four Arizona power plants and showed an average of 0.1 MW/acre. However, excluding the land area for the coal mine raises this to 0.3 MW/acre.

Land areas for ten pulverized coal combustion plants are shown in Table 3.13. Only supercritical and subcritical plants are shown. For the IGCC Polk and Wabash River plants, which are relatively new and small, land areas are likely skewed by coexisting with other plants and plans for future expansion. As the table illustrates, values range from 0.2 to 5.3 MW/acre with an average of 1.7 MW/acre. Finally, another reference suggests a generic need of 200 to 250 acres for a 1,000 MW plant,[78] or 4 to 5 MW/acre.

The average of 1.7 MW/acre from Table 3.13 will be used to calculate the energy density for the electric coal plant. To convert this to units of quadrillion BTU/(year-square mile), we recognize that a 1 MW plant operating at 100% capacity for 1 year will produce 8,760 MWh (1 MW * 24 hours/day * 365 days/year). Also, there are 640 acres per square mile and 3.412×10^6 BTU/MWh. From these values, we can determine that 1.7 MW/acre is equivalent to 0.033 quadrillion BTU/(year-square mile), assuming the plant operates at a capacity factor of 100%.

However, we know from Table 3.9 that the capacity factors at not 100% and the average value for pulverized coal combustion plants was earlier shown to be 89%. With

Table 3.13 Land Areas for Pulverized Coal Combustion Plants

Location	Company	Type	Startup	MW	Acres	MW/ Acre
Fort Martin, WV[68]	Longview Power, LLC	SC PC	2011	600	200	3.0
West Olive, MI[69]	J.H. Campbell	SubC PC	1962, 1967, 1980	1,450	1,000	1.5
Shippingport, PA[70]	Bruce Mansfield	SC PC	1976, 1977, 1980	2,490	473	5.3
Conesville, OH[71]	AEP Generation Resources Inc	SC PC	1973	450	200	2.3
Navajo, AZ[72]	Salt River Project	SC PC	1974, 1975, 1976	2,250	1,786	1.3
East Bend, KY[73]	Duke Energy	SubC PC	1981	772	1,800	0.4
Sherburne County, MN[74]	Northern States Power	SubC PC	1976, 1977, 1987	3,200	1,700	1.9
Crystal River, FL(a)[75]	Duke Energy	SubC PC	1969, 1982, 1984	3,155	4,700	0.7
W.A. Parish, TX (b)[76]	NRG Texas Power	SubC PC	1977, 1978, 1980, 1982	3,565	4,650	0.8
St. Marys, KS[77]	Jeffrey Energy Center	SubC PC	1978, 1980, 1983	2,400	10,500	0.2
Average				2,033.2	2,700.9	1.7

a. Includes 860 MW nuclear plant

b. Includes 4 natural gas plants with ~840 MW capacity

this adjustment, the energy produced for the power plant is 0.029 quadrillion BTU/(year-square mile).

The final step in the coal footprint is to (5) transmit and distribute electricity to customers via the grid. However, the land area needed for step-up and step-down transformers is very small and can be ignored. Likewise, the metal towers and electrical lines take up very little land area and can coexist with farming, pasture land, and roadways.

To summarize, the two main contributors to land area are (1) coal mining at 0.11 quadrillion BTU/(year-square mile) and (4) pulverized coal combustion plants at 0.029 quadrillion BTU/(year-square mile). Stated using scientific notation, these two values are 0.11×10^{15} and 0.029×10^{15}.

In order to compare energy densities for different technologies on a common basis of gigawatt-hours per square mile per year, or GWh/(square mile-year), energy was converted from BTU to Watt-hour, or Wh, using this equation:

$$1 \text{ BTU} = 0.293071 \text{ Wh}$$

Using this conversion and recognizing that a gigawatt (GW) is equivalent to 1 billion Watts (or 1×10^9), we calculate the energy density to produce coal from mining as 32,500 GWh/(year-square mile) and the energy density to make electricity at the pulverized coal electrical plant as 8,500 GWh/(year-square mile).

The footprint for a 1,000 MW pulverized coal electrical plant is small, only about 1.0 square mile. Adding the land area needed to supply coal for this 1,000 MW plant only increases the footprint by 0.3 square miles, for a total footprint of 1.3 square miles. Thus, while the coal electric plant footprint is larger than that for a natural gas plant, the land area needed to supply coal to the plant is much smaller than the 188 square miles needed to provide natural gas for the same size 1,000 MW plant. These numbers indicate it takes less land area to extract energy from mining for coal versus natural gas, but the electrical power plant for coal uses more land than natural gas.

According to the International Energy Agency,[79] the US electric power consumption per capita for one year was 12.8 MWh in 2016. Using the capacity factor of 0.89 shown in Table 3.9 for pulverized combustion power plants, a simple calculation shows that 1 MW will power about 600 homes for one year, so a 1,000 MW plant will power nearly 600,000 homes!

3.12. Environmental Issues

Like petroleum, coal is a complex mixture of hydrocarbons, but with higher carbon and much lower hydrogen content. And, at room temperature coal exists as a solid, compared to petroleum as a liquid and gaseous natural gas. As a solid, coal is not flammable like natural gas. However, coal dust can and does cause explosions and fires. The conditions for coal dust combustion are complicated and involve oxygen, coal dust particle size, and temperature. Depending on the concentration of the coal dust, one reference shows coal igniting around 550°C.[80] However, ignition can also be caused by heat sources, open flames, mechanical sparks, and static electricity. While coal mine safety has certainly improved, explosions are still possible. For example, in November 2016, an explosion occurred near Chongqing, China, where 33 miners were killed.[82] And in August 2016 a West Virginia coal miner was killed in an explosion.[83]

In addition, coal has volatile compounds, such as methane, other light hydrocarbons, hydrogen, and carbon monoxide, which will help precipitate coal combustion. And, when coal combustion occurs, sulfur and nitrogen compounds in the coal will produce SO_2, sulfur trioxide, and NO_x. As well, incomplete combustion of coal will produce carbon monoxide, which reduces the respiratory system's ability to carry oxygen, as carbon monoxide has a much greater affinity to combine with the blood hemoglobin than oxygen. For carbon monoxide, the Occupational Safety and Health Administration (OSHA) give a short term exposure limit (STEL) of 200 ppm over a 15-minute period and a permissible exposure limit (PEL) of 50 ppm averaged over an 8-hour period.[81]

Coal also contains things other than carbon and hydrogen, such as sulfur, nitrogen, and chlorine. In addition, there are an astounding number of trace metals that have been found in bituminous and subbituminous coal, with concentrations ranging from less than 1 ppm to greater than 50 ppm. These can include barium (Ba), manganese (Mn), strontium (Sr), titanium (Ti), zinc (Zn), arsenic (As), cerium (Ce), chromium (Cr), copper (Cu), lead (Pb), lithium (Li), nickel (Ni), rubidium (Rb), vanadium (V), zirconium (Zr), antimony (Sb), beryllium (Be), cadmium (Cd), cobalt (Co), gallium (Ga), germanium (Ge), lanthanum (La), molybdenum (Mo), niobium (Nb), scandium (Sc), selenium (Se), thorium (Th), uranium (U), mercury (Hg), silver (Ag), and tantalum (Ta).[84] These metals can affect humans and wildlife. For example, arsenic and chromium are human carcinogens, mercury and lead can result in brain damage and mental retardation, and selenium is known to poison fish.

The combustion of coal to generate electricity also leaves ash, defined as fly ash and bottom ash. Fly ash is made during combustion and captured by electrostatic precipitators before it leave plant as flue gas, while bottom ash is ash that falls to the bottom of the steam boiler.[84] Coal ash can be disposed in landfills, discharged into waterways under the plant water discharge permit, or recycled into products like concrete or wallboard.[85] Relative to their initial composition in the coal, some of the heavy metals listed earlier will be enriched in both the fly ash and bottom ash. Clearly, proper disposal of these ashes is an important environmental issue.

As mentioned, sulfur and nitrogen will combust to form SO_2, sulfur trioxide, and NO_x and, if allowed to leave the plant, will react with water in the air to form acids such as H_2SO_4, H_2SO_3, HNO_2, and HNO_3, components of acid rain. According to the EIA,[86] emissions of $SO_2$and NO_x from the electric power sector in 2012 declined to their lowest level since the passage of the Clean Air Act Amendments of 1990, due primarily to an increasing number of coal-fired units retrofitted with FGD, or scrubbers, to coal plants switching to lower sulfur coal, and to SCRs and SNCR. Also, in 2005, the

EPA developed the Clean Air Interstate Rule, a cap-and-trade program intended to further reduce SO_2 and NO_x emissions in the eastern half of the United States. A cap-and-trade program sets upper limits for atmospheric pollutants but allows businesses to buy credit from other businesses that do not use their full allowance.

There are also EPA rules for wastewater discharges from coal combustion power plant set in 1974 and amended in 1977, 1978, 1980, 1982, and 2015.[87] This topic is beyond the scope of this book, but these rules are incorporated into the National Pollutant Discharge Elimination System, or NPDES.

As in the case of natural gas, hydrogen sulfide (H_2S) is also a danger to miners. The American Conference of Governmental Industrial Hygienists (ACGIH) has adopted limits of 1.0 ppm for the 8-hour time weighted average (TWA) and 5.0 ppm for the STEL.[88] Above these limits, there can be dizziness, irreversible damage, or even death. Potential H_2S poisoning can be detected by the use of monitors.

The products of combustion are CO_2 and water (H_2O), and these are benign, although increasing amounts of CO_2 in the atmosphere have been attributed to the cause of the earth's warming. However, CO_2 and water are not toxic to humans.

Finally, like oil and natural gas drilling, surface coal mining requires the clearing of trees, plants, and topsoil and therefore destroys landscape and forests that are wildlife habitats.

3.13. CO_2 Production and the Cost of Capture and Sequestration

For coal, the "Well-to-Wheels" life cycle analysis for the production of greenhouse gas emissions includes four basic steps. Following a report by the National Energy Technology Laboratory (NETL),[89] these four steps are (1) raw material acquisition, (2) raw material transport, (3) the power plant, and (4) transmission and distribution. Looking in more detail at greenhouse gases made for each of these steps, raw material acquisition is bituminous coal mining and processing, raw material transport is the diesel-powered rail transportation of coal, the power plant is the construction, operation, and decommissioning of a supercritical pulverized coal plant, and transmission and distribution include losses from when power is placed on the grid up to when the power is pulled from the grid.

This NETL report gives greenhouse gas emissions in units of equivalent g CO_2 emissions produced per delivered kWh. Although CO_2 was the primary greenhouse gas, the analysis also included the greenhouse gases N_2O (nitrous oxide) and CH_4 (methane). In units of g CO_2/kWh, 71.6 were produced by raw material acquisition, 4.8 by raw material transport, 863.8 by the power plant, and 3.3 by transmission and distribution, giving a total of 943.5 g CO_2/kWh. This can also be reported as 2.08 lb CO_2/kWh. Clearly, most of the greenhouse gases are produced by the power plant, about 92%, so adding CCS at the power plant will capture most of the greenhouse gases produced.

In order to convert these reported results to units of lb CO_2/MM BTU, a heat rate of 8,800 BTU/kWh was used. With this heat rate, the amount of CO_2 equivalent produced was 17.9 lb CO_2/MM BTU for raw material acquisition, 1.2 lb CO_2/MM BTU for raw material transport, 216.4 lb CO_2/MM BTU from the power plant, and 0.8 lb CO_2/MM BTU from transmission and distribution, giving a total of 236.4 lb CO_2/ MM BTU.

At the time this book was written, there were at least two coal-fired power plants equipped with CCS. In October 2014, SaskPower officially opened the world's first commercial-scale plant with CCS at their Boundary Dam plant in Saskatchewan.[90] The 110 MW plant uses solvent-based processors to strip CO_2 from the flue gas and then sends it to a nearby oil field to be used for EOR.

Also, in January 2017,[56] the Petra Nova project, near Houston, started operation. For this project, a slipstream, equivalent to 240 MW, is taken from a pulverized coal combustion 640 MW plant, amine scrubbing is used to capture 90% of the CO_2, and the CO_2 enriched stream is then transported by pipeline to an operating oil field for use in EOR. According to one source,[91] the cost of capturing this 240 MW slipstream is 1 billion dollars. Converting this to $/kW gives a capital cost of $4,167/kW. In what follows, the average values for pulverized coal have been extracted from Table 3.9 and show a capital cost projection difference of $3,325/kW, in reasonable agreement with the W.A. Parish project. The addition of CCS increases the LCOE by about $58/MWh (about 5.8¢ per kWh).

For a 1,000 MW plant with a capacity factor of 89%, the plant will generate 21,360 MWh/day. If CCS costs $58/MWh, the cost of CCS for the pulverized coal plant will be $1,238,880 per day, or about $452 MM per year. As the CO_2 produced on a kWh basis is 2.08 lb CO_2/kWh, this means a 1,000 MW plant generates 44,428,800 lb CO_2/ day or about 22,214 tons CO_2/day.

In conclusion, the amount of CO_2 produced for a pulverized coal-fired combustion plant is 236.4 lb CO_2/MM BTU, or 2.08 lb CO_2/kWh and 943.5 g CO_2/kWh. The cost of CCS will be $58/MWh, or 5.8¢ per kWh. On a mass basis, CCS will cost about $56/ ton.

3.14. Chapter Summary

Before summarizing this chapter, it is interesting to examine recent trends in coal use in the United States and the world. As mentioned earlier, in terms of US electricity generation, natural gas passed coal in 2015. And, in 2017, electrical power generation accounted for 93% of the total US coal consumption. But, in 1960 only 44% of coal was used to generate electricity, rising to 81% in 1980 and 91% in 2000.[92] So, coal is no longer used that much to heat homes and business and it is losing its market share in the electric world, primarily to natural gas.

In the United States, the last significantly sized coal plant was built in 2015, while new natural gas electric plants are still common.[27] There are some coal plants

Table 3.9 Excerpt

Energy Type	Capacity Factor	Capital Cost	Levelized Capital Cost	Fixed O&M	Variable O&M (including fuel)	Transmission Investment	Total System Levelized Cost (LCOE) to Generate	LCOE Including Transmission and Distribution
		$/kW	$/MWh	$/MWh	$/MWh	$/MWh	$/MWh	$/MWh
Average Pulverized Coal	0.89	3,123	38.59	4.99	21.31	1.20	66.09	101.09
Average Pulverized Coal with CCS	0.88	6,448	80.00	10.63	32.61	1.20	124.44	159.44

proposed, however. At the University of Alaska, Fairbanks, a 17 MW plant is scheduled to start in 2019.[93] This plant is intended to provide both heating and electricity for the campus. On a larger scale, the Two Elk Generation Station in Wyoming was proposed for a capacity of 320 MW. However, its construction has been halted and its future is uncertain as the developer pleaded guilty to stealing government funding and is facing jail time.[94] Another large plant, the Power4Georgians 850 MW plant, is still proceeding with plans but while the regulatory approvals have been received, it is still not under construction.[94]

Worldwide, generation of electricity from coal doubled to 2,000 GW since 2000 with explosive growth in China and India.[95] Another 236 GW is under construction and another 336 GW is planned while 227 GW were closed due to retirements in Europe and the United States A list of the top ten countries in terms of operating plants and those proposed or under construction is shown in Table 3.14. China and India are responsible for 51% of the new plant growth.

So, in the United States and Europe, the use of coal to generate electricity is decreasing while in other parts of the world its use is growing, although there was only a net increase of 20 GW in 2018.

A review of worldwide coal reserves shows an estimate of 1,140,904 million tons which, at the 2017 consumption rate of 8,434 million tons per year gives a calculated value of 135 years remaining for coal. Of course, if the Demonstrated Reserve Base can be mined commercially, the years remaining for coal would almost double, as the Estimated Recoverable Reserves (includes only coal that can be mined today) is about 53% of the Demonstrated Reserve Base (coal that has been found or measured and can be mined commercially). For proven reserves, ten countries—the United States, Russia, Australia, China, India, Germany, Ukraine, Poland, Kazakhstan, and Indonesia—control 91% of the worldwide coal reserves.

Table 3.14 Operating and Planned Coal Electric Plants

Country	Operating (GW)	Share	Country	Proposed or Under Construction (GW)	Share
China	973	48%	China	199	35%
United States	261	13%	India	94	16%
India	221	11%	Vietnam	42	7%
Russia	48	2%	Turkey	37	7%
Germany	48	2%	Indonesia	25	4%
Japan	46	2%	Bangladesh	21	4%
South Africa	42	2%	Japan	15	3%
South Korea	37	2%	South Africa	14	2%
Poland	30	1%	Egypt	13	2%
Indonesia	29	1%	Philippines	13	2%

The "Well-to-Wheels" economic analysis gave an average LCOE, including transmission and distribution, of \$101.09/MWh for pulverized coal combustion and \$132.18/MWh for IGCC. This difference of about of \$31/MWh is largely attributable to the air separation unit, which can be 15% of the plant capital cost. Adding CCS increases the LCOE by about \$58/MWh for the pulverized coal combustion plant and increases LCOE by about \$54/MWh for the IGCC. In the case of pulverized coal combustion, the addition of CCS increases LCOE by 58% versus 41% for IGCC. Since an air separation unit is needed to capture CO_2, this must be added to pulverized coal combustion but IGCC already has this equipment. This is part of the reason the addition of CCS changes the LCOE less for IGCC than pulverized coal combustion.

For a 1,000 MW supercritical pulverized coal combustion power plant with a capacity factor of 89%, the plant will make 44,428,800 lb CO_2/day or about 22,214 tons CO_2/day. This is much more than the 1,000 MW natural gas combined cycle plant, generating about 7,500 tons CO_2/day. On an energy basis, the coal plant generates 2.08 lb/kWh versus 0.78 lb/kWh for natural gas, about 2 ½ times as much. Compared to natural gas, coal produces more CO_2 because it is mostly carbon, so most of the energy made from coal combustion will result in CO_2 formation.

Using a 12-month average of 2018 EIA data, the cost of coal, on the basis of energy content, is \$2.06/MM BTU This is about one-fifth of the cost of crude oil, costing \$11.20/MM BTU based on a 2018 cost for West Texas Intermediate of ~\$65/barrel.

Pulverized coal combustion capacity factors were, on average, 0.89 and IGCC capacity factors were 0.80. The lower factor for IGCC may be due to increased maintenance issues due to having two turbines and an air separation unit. Adding CCS is not expected to affect the capacity factors for either technology.

For pulverized coal combustion and IGCC plants, the overall energy balances were 35.1% and 35.0%, respectively, meaning that about 35% of the energy originally in the coal ends up as energy at the distribution point for both technologies. Most of this energy is lost during electricity generation, as the thermal efficiencies for pulverized coal combustion and IGCC are 38.5% and 38.4%, respectively. If CCS is added to pulverized coal combustion and IGCC, the thermal efficiencies decrease from 38.5% to 29.3% and 38.4% to 31.2%, respectively. In the case of CCS, only 26.7% of the original energy reaches the distribution point for pulverized coal combustion and 28.5% for IGCC.

In terms of experience and maturity, coal mining has been around nearly 270 years and surface mining for about 150 years. Coal-fired power generation has been used for about 135 years, with supercritical pulverized coal combustion having about 55 years of experience and IGCC having about 20 years of experience. Thus, like the petroleum industry, the industry of using coal to make electricity is very mature.

Also, like petroleum crude oil, the infrastructure is immense. To transport coal, there are 95,264 miles of track and 46,876 miles of interstate highway. To make the electricity, there were 359 operating coal combustion power plants in 2017, and more than 180,000 miles of high-voltage transmission lines that exist in the electrical grid.

For a 1,000 MW pulverized coal combustion plant, the footprint is only 1.0 square mile. Adding the coal mining area needed to supply coal to this plant adds only 0.3

square miles, giving a total footprint of 1.3 square miles. This takes much less land area than the same plant size for petroleum crude oil and natural gas, which use 238 and 188 square miles, respectively. Part of this difference is that a lot of coal is extracted from underground mining, thus having less impact on land surface area than crude oil production.

However, looking at only the footprint for the pulverized coal electrical plant, a 1,000 MW plant will cover about 1.0 square mile versus only 0.06 square miles for a natural gas combined cycle plant. This indicates that it takes less land to get energy for coal but more land to convert it to an amount of electricity equivalent to natural gas. Presumably, the differences in the size of the electrical plants are due to the additional processes needed for coal, including coal washing, NO_x capture, and SO_2 capture.

Finally, the main environmental issues for coal are mine explosions, CO poisoning, ash disposal, capture of sulfur and NO_x from the flue gas, wastewater disposal, and the destruction of land and wildlife habitats. Although CO_2 is not considered a pollutant at this time, most than 90% of the CO_2 made during the coal Well-to-Wheels cycle can be captured at the electric power plant using CSS.

References

1. "Coal," *Energy and Fuels in Society*, Ljubisa R. Radovic, University Penn State University, 1992,
2. "The Coalification Process," *Coal-Bearing Depositional Systems*, Claus F.K. Diessel, Springer Berlin Heidelberg, 1992, p. 41–85.
3. "Greening Coal—New Generation Low Emission Coal Technology, Part 1 The Formation of Coal & Thermal Electricity Generation," Dale Simmons, Teacher Earth Science Education Programme, 2016.
4. "Coal Characteristics, Basic Facts File #8," B.H. Bowen and M.W. Irwin, The Energy Center at Discovery Park, Purdue University, October 2008.
5. "The Chemistry and Technology of Coal," 3rd Edition, James G. Speight, Figure 3.1, 2013.
6. "Understanding the Chemistry and Physics of Coal Structure (A Review)," D.G. Levine et al., Proceedings of the National Academy of Sciences 79, 3365–3370, Figure 2, May 1982.
7. "U.S. Coal Reserves," US Energy Information Administration, release date November 2, 2018, http://www.eia.gov/coal/reserves/
8. "BP Statistical Review of World Energy," June 2018, (https://www.bp.com/content/dam/bp/business-sites/en/global/corporate/pdfs/energy-economics/statistical-
9. "History of Coal Use," National Energy Technology Laboratory (US Department of Energy), 2004, http://energybc.ca/cache/historyofenergyuse/www.netl.doe.gov/KeyIssues/historyofcoaluse.html
10. "Annual Coal Report," US Energy Information Administration, November 2, 2018, https://www.eia.gov/coal/annual/
11. "Can We Make Steel without Coal?," T. Jonescan, April 24, 2013, http://coalaction.org.nz/carbon-emissions/can-we-make-steel-without-coal
12. "Liquid Fuels: Direct Liquefaction Process," National Energy Technology Laboratory, 2009, http://www.netl.doe.gov/research/Coal/energy-systems/gasification/gasifipedia/direct-liquefaction
13. "Shenhua's DCL Project: Technical Innovation and Latest Developments," Shu Geping, Chief Engineer, China Shenhua Coal to Liquid and Chemical Co., Ltd., Cornerstone, The

Official Journal of the World Coal Industry, October 11, 2013, http://cornerstonemag.net/shenhuas-dcl-project-technical-innovation-and-latest-developments/
14. "Fast Facts About Coal," Rocky Mountain Coal Mining Institute, 2006, http://www.rmcmi.org/education/coal-facts#.XoTdQ0BFw2w
15. "Which States Produce the Most Coal?," US Energy Information Administration, December 4, 2018, https://www.eia.gov/tools/faqs/faq.php?id=69&t=2
16. "Coal Production Demands: Chapter 4—Coal Preparation," Virginia Center for Coal & Energy Research, December, 2008, https://energy.vt.edu/content/dam/energy_vt_edu/vccer-publications/Coal_Production_Demands.pdf
17. "Advanced Emissions Control Technologies for Coal-Fired Power Plants," A.L. Moretti and C.S. Jones, Babcock & Wilcox, Technical Paper BR-1886, Power-Gen Asia, October 3–5, 2012, Bangkok, Thailand.
18. "WSA & SNOXTM Technology for the Production of Sulfuric Acid in Power Plants," Patrick Polk, January 17, 2013.
19. "Available and Emerging Technologies for Reducing Greenhouse Gas Emissions from Coal-Fired Electric Generating Units," Sector Policies and Programs Division, Office of Air Quality Planning and Standards, US Environmental Protection Agency, Research Triangle Park, North Carolina, October 2010.
20. "The Benefits of Supercritical Boilers," National Boiler Service, December 8, 2014, http://www.nationalboiler.com/blog/boiler-repair/the-benefits-of-supercritical-boilers/
21. "Performance and Risks of Advanced Pulverized-Coal Plants," H. Nalbandian, Energeia 20 (1), 1–6, 2009.
22. "Chapter 1 Coal-Fired Power Generation and Circulating Fluidized Bed Combustion (CFBC)," Global CCS Institute, January, 2012, https://hub.globalccsinstitute.com/publications/scoping-study-oxy-cfb-technology-alternative-carbon-capture-option-australian-black-and-brown-coals/chapter-1-coal-fired-power-generation-and-circulating-fluidized-bed-combustion-cfbc
23. "The Future of Coal—Options for a Carbon-Constrained World," MIT Study on the Future of Coal, Massachusetts Institute of Technology, 2007, ISBN 978-0-615-14092-6.
24. "High Efficiency Electric Power Generation; The Environmental Role," János Beér, Massachusetts Institute of Technology, Progress in Energy and Combustion Science 33 (2), 107–134, April 2007.
25. "Energy Production in Selected Integrated Gas-Steam IGCC Systems Powered by Gas from Coal Gasification Processes," A. Leśniak and M. Bieniecki, Central Mining Institute (GIG), Katowice, Poland, CHEMIK 68 (12), 1074–1085, 2014.
26. "Power Plants with Coal Gasification," BINE Informationdienst, September 2006, ISSN 0937-8367
27. "Count of Electric Power Industry Power Plants, by Sector, by Predominant Energy Sources within Plant, 2007 through 2017," Table 4.1, Energy Information Administration, http://www.eia.gov/electricity/annual/html/epa_04_01.html; "Existing Capacity by Energy Source, 2017 (Megawatts)," Table 4.3, Energy Information Administration, https://www.eia.gov/electricity/annual/html/epa_04_03.html
28. "Electricity: Form EIA-860 Detailed Data," US Energy Information Administration, 2015 data, https://www.eia.gov/electricity/data/eia860/
29. "IGCC Efficiency / Performance," National Energy Technology Laboratory (NETL), Department of Energy, https://www.netl.doe.gov/research/coal/energy-systems/gasification/gasifipedia/igcc-efficiency
30. "American Electric Power, Transmission facts," p. 4, December 31, 2007, https://web.archive.org/web/20110604181007/https://www.aep.com/about/transmission/docs/transmission-facts.pd

31. "A Review of Transmission Losses in Planning Studies," Lana Wong, California Energy Commission, Figure ES-1, August 2011.
32. "Electricity: Form EIA-860 Detailed Data," US Energy Information Administration, 2018 data, September 13, 2018, https://www.eia.gov/electricity/data/eia860/
33. "Updated Capital Cost Estimates for Utility Scale Electricity Generating Plants," U.S. Energy Information Administration, April 2013; and "Levelized Cost and Levelized Avoided Cost of New Generation Resources in the Annual Energy Outlook 2015," June, 2015, Table 1.
34. "Cost and Performance Characteristics of New Generating Technologies," Annual Energy Outlook 2019, Table 2, Energy Information Administration, January 2019, https://www.eia.gov/outlooks/aeo/
35. "Regulating Power Sector Carbon Emissions," Center for Climate and Energy Solutions, 2018, https://www.c2es.org/content/regulating-power-sector-carbon-emissions/
36. "Lazard's Levelized Cost of Energy Analysis—Version 8," September 2014, https://www.lazard.com/media/1777/levelized_cost_of_energy_-_version_80.pdf
37. "Lazard's Levelized Cost of Energy Analysis—Version 11.0," November 2017, https://www.lazard.com/media/450337/lazard-levelized-cost-of-energy-version-110.pdf
38. "Lazard's Levelized Cost of Energy Analysis—Version 12.0," November 2018, https://www.lazard.com/media/450784/lazards-levelized-cost-of-energy-version-120-vfinal.pdf
39. "Annual Energy Outlook 2015 with Projections to 2040," DOE/EIA-0383, Table A8, April 2015.
40. "Table 9.9 Cost of Fossil-Fuel Receipts at Electric Generating Plants," US Energy Information Administration / Monthly Energy Review, p. 163, March 2019, http://www.eia.gov/totalenergy/data/monthly/pdf/sec9_13.pdf
41. "Mining Industry Energy Bandwidth Study," June 2007, US DOE, Industrial Technologies Program.
42. "Railroads and Coal—Association of American Railroads," July 2016, https://www.aar.org/BackgroundPapers/Railroads%20and%20Coal.pdf
43. "Transportation of Coal," Navira Younas, June 25, 2015, http://www.slideshare.net/NainaRajput1/transportation-of-coal-ppt
44. "Railroad Transportation Energy Efficiency," Chris Barkan, University of Illinois at Urbana-Champaign, 2007, http://www.istc.illinois.edu/about/seminarpresentations/20091118.pdf
45. "A Review of Transmission Losses in Planning Studies," Lana Wong, California Energy Commission, Figure ES-1, August 2011.
46. "Monthly Energy Review," US Energy Information Administration, March 2019, https://www.eia.gov/totalenergy/data/monthly/pdf/mer.pdf
47. "Use of Coal in the Bronze Age in China," J. Dodson et al., The Holocene (5), 525–530, March 3, 2014.
48. "Provenance of Coals from Roman Sites in England and Wales," A.H.V. Smith, 1997, Britannia 28, 297–324 (322–324).
49. "Fossil Energy Study Guide," US Department of Energy, https://www.energy.gov/sites/prod/files/Elem_Coal_Studyguide.pdf
50. *Coal: A Human History*, B. Freese, London: Penguin Books, 2004, p. 137.
51. "Timeline of Coal in the United States," American Coal Foundation, 2005, http://www.paesta.psu.edu/sites/default/files/timeline_of_coal_in_the_united_states.pdf
52. "Pioneering Gasification Plants," Energy.gov, Office of Fossil Energy, 2005, http://energy.gov/fe/science-innovation/clean-coal-research/gasification/pioneering-gasification-plants
53. "Fluidized Bed Technology—An R&D Success Story," Energy.gov, Office of Fossil Energy, 2005, http://www.energy.gov/fe/science-innovation/clean-coal-research/advanced-combustion-technologies/fluidized-bed-technolog-0

54. "United States Electricity Industry Primer," Office of Electricity Delivery and Energy Reliability U.S. Department of Energy DOE/OE-0017, July 2015, https://www.energy.gov/sites/prod/files/2015/12/f28/united-states-electricity-industry-primer.pdf
55. "Boundary Dam Carbon Capture Project," Saskpower CCS, 2014, https://www.saskpower.com/Our-Power-Future/Infrastructure-Projects/Carbon-Capture-and-Storage/Boundary-Dam-Carbon-Capture-Project
56. "Petra Nova CCUS Project in USA," Noriaki Shimokata, JX Nippon Oil & Gas Exploration Corporation, June 8, 2018, https://d2oc0ihd6a5bt.cloudfront.net/wp-content/uploads/sites/837/2018/06/Noriaki-Shimokata-Petra-Nova-CCUS-Project-in-USA.pdf
57. "Class I Railroad Statistics," Association of American Railroads, May 3, 2016, https://www.aar.org/Documents/Railroad-Statistics.pdf
58. "Highway History," Federal Highway Administration, US Department of Transportation, December 18, 2018, https://www.fhwa.dot.gov/interstate/faq.cfm#question3
59. "Count of Electric Power Industry Power Plants, by Sector, by Predominant Energy Sources within Plant, 2007 through 2017," Table 4.1, Energy Information Administration, http://www.eia.gov/electricity/annual/html/epa_04_01.html; "Existing Capacity by Energy Source, 2017 (Megawatts)," Table 4.3, Energy Information Administration, https://www.eia.gov/electricity/annual/html/epa_04_03.html
60. "Net Generation by Energy Source: Total (All Sectors)," 2007–February 2017, Table 1.1, Energy Information Administration, https://www.eia.gov/electricity/monthly/epm_table_grapher.cfm?t=epmt_1_01
61. "The Footprint of Coal," Global Energy Monitor, December 25, 2019, https://www.gem.wiki/The_footprint_of_coal
62. "Longwall Mining," DOE/EIA-TR-0588, Distribution Category UC-950, Energy Information Administration, Office of Coal, Nuclear, Electric and Alternate Fuels, March 1995.
63. "Coal, Its Properties, Analysis, Classification, Geology, Extraction, Uses, and Distribution," E.S. Moore, School of Mines, Pennsylvania State College, New York: John Wiley & Sons, Inc., 1922.
64. "Annual Coal Report 2017," Table 3, US Energy Information Administration, November 2018.
65. "OSMRE 2012 Annual Report," US Department of the Interior, Office of Surface Mining Reclamation and Enforcement (OSMRE).
66. "Annual Energy Review 2011," DOE/EIA-0384, September 2012, www.eia.gov/aer
67. "Land Requirements for the Solar and Coal Options," M.J. Pasqualetti and B.A. Miller, The Geographical Journal 150 (2), July 1984, Table I.
68. "Groundbreaking Ceremony Held for Longview Power Plant, First Reserve," May 30, 2007, http://www.firstreserve.com/news-articles/groundbreaking-ceremony-held-for-longview-power-plant
69. "J.H. Campbell Generating Complex," Michigan Public Power Agency, 2020, http://www.mpower.org/Projects/ID/8/James-H-Campbell-Unit-No-3
70. "Bruce Mansfield Plant, First Energy Generation," August 2016, https://www.firstenergycorp.com/content/dam/corporate/generationmap/files/Bruce%20Mansfield%20Plant%20Facts.pdf
71. "Retrofit Potential for Indiana Coal-Fired Power Plants, Including Oxy-Fuel," E.J. Miklaszewski et al., Purdue University, April 12, 2011, https://www.purdue.edu/discoverypark/energy/assets/pdfs/cctr/CCTR-Miklaszewski-041211.pdf
72. "Navajo Generating Station," Navajo Reservation, Page, Arizona, August 3, 2003, https://online.platts.com/PPS/P=m&s =1029337384756.1478827&e= 1096495472414.22400233 08785804128/ ?artnum=2F0ed0SY40727 J0h9232q1_1
73. "Duke Energy Works to Lessen Impact of Power Plant on Environment, Educate Community," July 5, 2013, http://www.kyforward.com/duke-energy-takes-on-projects-to-lessen-impact-of-power-plant-on-the-environment/

74. "Survey of Flue Gas Desulfurization Systems: Sherburne County Generating Plant," US EPA, EPA-600/7-79-199d, August 1979, https://nepis.epa.gov/Exe/ZyNET.exe/9101QCIP.TXT?ZyActionD=ZyDocument&Client= EPA&Index=1976+Thru+1980&Docs=&Query=&Time=& EndTime=&SearchMethod= 1&Toc
75. "Progress Energy Florida (Duke): Crystal River Generation Complex, History and Issues of Value," 2012, http://citrus.granicus.com/MetaViewer.php?view_id=2&clip_id=1666&meta_id=301164
76. "W. A. Parish Electric Generating Station," August, 2015, http://www.texas-flyer.com/Fly-In-EngineOut/powerplant.htm
77. "Jeffrey Energy Center Garners User Group's Plant of the Year Award," R. Peltier, April 3, 2015, https://online.platts.com/ PPS/P=m&s=1029337384756.1478827 &e=109649521325 3.1080337080681645232/?artnum=Z2004WYA07z 21m1335b2e3_1
78. "Comparison of Fuels Used for Electric Generation in the U.S.," Natural Gas Supply Association, 2016, http://www.ngsa.org/analyses-studies/beck-data-rev/
79. "Electric Power Consumption (kWh per capita)," IEA Statistics for 2016, http://energyatlas.iea.org/#!/tellmap/-1118783123/1
80. "Overview of Dust Explosibility Characteristics," K.L. Cashdollar, Journal of Loss Prevention in the Process Industries 13(3–5), 183–199, May 2000.
81. "Carbon Monoxide: Health Information Summary," ARD-EHP-20 2007, Environmental Fact Sheet, New Hampshire Department of Environmental Services.
82. "Gas Explosion Kills 33 Chinese Miners," Serenitie Wang and Serena Dong, CNN, November 1, 2016, http://www.cnn.com/2016/10/31/asia/china-chongqing-mine-explosion/
83. "West Virginia Coal Miner Killed in Explosion," N. Spencer, August 9, 2016, https://www.wsws.org/en/articles/2016/08/09/wvir-a09.html
84. "Trace Element Emissions from Coal," Herminé Nalbandian, IEA Clean Coal Centre, September 2012, CCC/203 ISBN 978-92-9029-523-5, https://www.usea.org/sites/default/files/092012_Trace%20element%20emissions%20from%20coal_ccc203.pdf
85. "Coal Ash Basics," Environmental Protection Agency, 2014, https://www.epa.gov/coalash/coal-ash-basics#02
86. "Power Plant Emissions of Sulfur Dioxide and Nitrogen Oxides Continue to Decline in 2012," February 27, 2013, http://www.eia.gov/todayinenergy/detail.php?id=10151
87. "Steam Electric Power Generating Effluent Guidelines," 2015, https://www.epa.gov/eg/steam-electric-power-generating-effluent-guidelines
88. "New TLV® Exposure Limit: Measuring Hydrogen Sulfide," June 2, 2011, Industrial Safety and Hygiene News, http://www.ishn.com/articles/91070-new-tlv-exposure-limit-measuring-hydrogen-sulfide
89. "Life Cycle Analysis: Supercritical Pulverized Coal (SCPC) Power Plant," September 30, 2010, Timothy Skone and Robert James, DOE/NETL-403-110609, National Energy Technology Laboratory, https://www.netl.doe.gov/File%20Library/Research/Energy%20Analysis/Life%20Cycle%20Analysis/SCPC-LCA-Final-Report---Report---9-30-10---Final---Rev-1.pdf
90. "Saskpower Unveils World's First Carbon Capture Coal Plant," October 2014, http://www.saskpowerccs.com/newsandmedia/news-archive/saskpower-unveils-worlds-first-carbon-capture-coal-plant/
91. "Petra Nova W.A. Parish Fact Sheet: Carbon Dioxide Capture and Storage Project," MIT Carbon Capture and Sequestration Technologies program, September 2016, https://sequestration.mit.edu/tools/projects/wa_parish.html
92. "Table 6.2—Coal Consumption by Sector (thousand short tons)," Energy Information Administration, March 2020, http://www.eia.gov/totalenergy/data/monthly/pdf/sec6_4.pdf

93. "Combined Heat and Power Plant," University of Alaska Fairbanks, 2020, https://www.uaf.edu/heatandpower/scope.php
94. "Will the U.S. Ever Build Another Big Coal Plant?," Benjamin Storrow, August 21, 2017, Scientific American, https://www.scientificamerican.com/article/will-the-u-s-ever-build-another-big-coal-plant/
95. "Global Coal Power," Carbon Brief, March 25, 2019, https://www.carbonbrief.org/mapped-worlds-coal-power-plants

4
Nuclear
A Nonrenewable Energy Type

4.1. Foreword

Unlike natural gas and petroleum crude oil, which are used in transportation, heating, and electricity generation, nuclear fuel is essentially used only to generate electricity. To do this, nuclear fission is used to create heat, boil water to make steam, and power a turbine to make electricity. There are several nuclear atoms that are fissile, including Uranium-233, Uranium-235, Plutonium-238, Plutonium-239, Plutonium-241, Neptunium-237, and Curium-244. A fissile material is one that can undergo nuclear fission, such that the large atom will split into two or more smaller atoms to release energy. Thorium-232 is not fissile but can be converted to Thorium-233 by neutron capture and subsequent beta decay.

Because uranium is relatively abundant in nature, it is a common choice for nuclear power plants. Uranium-235 (U-235) is the only naturally occurring material that can sustain a chain reaction and fuel a nuclear reactor, but only makes up 0.7% of uranium ore; Uranium-238 makes up the rest.[1] Plutonium is also a possible fuel for power plants, but is only available in trace quantities in nature. Plutonium can be made from Uranium-238 in the power plant nuclear reactor, such that U-238 captures a neutron, changing it into U-239, which then undergoes two "beta" decays to Plutonium-239. Beta decay is when a neutron is changed into a proton or a proton into a neutron; in the case of the transformation of uranium to plutonium, two neutrons are changed into protons, as Plutonium-239 has two more protons than Uranium-239.

Plutonium-239 can also be made in "breeder reactors," a reactor defined as one that makes more fissile material than it consumes.[2] Interestingly, the decommissioning of excess nuclear weapons by the United States and Russia led to large stocks of plutonium, which now provides plutonium fuel for nuclear reactors.[1] Like plutonium, neptunium is available in only trace quantities in nature. And, while minute amounts of curium probably exist in natural deposits of uranium, the presence of natural curium has never been detected.[3]

Thorium can also be used in a nuclear power plant. Naturally available, thorium was discovered in 1828 by the Swedish chemist Jons Jakob Berzelius, who named it after Thor, the Norse god of thunder.[4] Although thorium (Th-232) is not fissile, it can be changed into U-233, which is.[4] There have been several attempts in the United States to use thorium, such as a nuclear plant in Indian Point, New York City, which began operation in 1962, and the last serious attempt in 1979 at Fort St. Vrain in

The Changing Energy Mix. Paul F. Meier, Oxford University Press (2020). © Oxford University Press.
DOI: 10.1093/oso/9780190098391.001.0001.

Colorado, a plant that closed 10 years later after equipment failures, leaks, and fuel failures.[5]

Thorium continues to be discussed as a replacement, or at least a supplement, to uranium fuel. For example, one discussion cites some of the advantages and concerns.[6] Thorium is three to four times more available than uranium and, in countries like India where thorium resources vastly exceed uranium resources, there is great interest and development taking place. Nuclear weapons proliferation is always a concern, and if thorium is used as a new fuel, it creates U-233 in the process, which can be used to make a nuclear weapon. From an economic standpoint, the use of thorium commercially has not progressed to the point such that a direct economic comparison can be made to uranium. Therefore, the US private sector is unlikely to proceed without government investment. While thorium may be developed in India, and perhaps China, it is unlikely to happen in the United States anytime soon.

On the basis of the previous discussion, this chapter focuses on the use of U-235 as the fuel used in the nuclear reactor power plant.

4.2. Proven Reserves

Uranium is thought to have been formed in supernovas about 6.6 billion years before the formation of earth.[7] It is relatively abundant and found in various ores in the earth's crust and even in seawater. However, not all of this uranium can be easily recovered, so recovery from the ocean or granites, such as such as quartz and mica, are only possible if prices are sufficient to support mining costs.[8] Likewise, as for other metals, it must be sufficiently concentrated to be economically recoverable. Therefore, defining proven reserves has a lot to do with mining costs and current market prices, and uranium reserves are, therefore, calculated as tonnes recoverable up to a certain cost.

Following the definitions set forth in a report by the Nuclear Energy Agency and International Atomic Energy Agency,[9] reasonably assured resources, or RARs, and inferred resources, refer to uranium deposits that have been discovered by direct measurements. For RARs, there is high confidence in estimates while inferred resources are not defined with the same degree of confidence. Identified resources are defined as the sum of RARs and inferred resources that are recoverable for a cost at or below $130/kg of uranium. "Identified resources" for the top ten countries are shown in Table 4.1, taken from Table 1.2a of the earlier referenced report.

At this cost, the worldwide total for uranium in 2017 is 6,142,200 tonnes and the top ten countries account for 88% of the worldwide uranium reserves. By themselves, Australia and Kazakhstan account for 30% and 14%, respectively, of these reserves.

If the recovery cost is allowed to increase to $260/kg of uranium, the total reserves increase to 7,988,600 tonnes, an increase of 30%. The top ten countries at this recovery cost are also shown in Table 4.1 and, for the most part, the order of these countries is unchanged. Notably, the United States is not in the top ten in either scenario.

Table 4.1 Top Ten Countries for Uranium Reserves based on Identified Resources, 2017 (tonnes)[9]

Country	<$130/kgU	Country	<$260/kgU
World Total	6,142,200	World Total	7,988,600
Australia	1,818,300	Australia	2,054,800
Kazakhstan	842,200	Kazakhstan	904,500
Canada	514,400	Canada	846,400
Russia	485,600	Russia	656,900
Namibia	442,100	Namibia	541,700
South Africa	322,400	South Africa	449,300
China	290,400	Niger	425,600
Niger	280,000	China	290,400
Brazil	276,800	Brazil	276,800
Uzbekistan	139,200	Ukraine	219,100

At $130/kg, the United States ranks 15th with 47,200 tonnes and at $260/kg it ranks 16th with 100,800 tonnes.

So, at a price of $130/kg, the world has about 6 million tonnes, increasing to 8 million tonnes at $260/kg. But there is a lot more uranium on earth, and one report estimates the uranium in the ocean at 4.5 billion tonnes, sufficient to power the world's electric plants for 13,000 years![10] In seawater, it exists as water soluble uranyl (UO_2^{+2}) at very low concentration, about 3 parts per billion (ppb). Accordingly, its production costs are much higher, estimated to be $400 to $1,000/kg. Therefore, the limit on worldwide uranium resources is controlled by the market price for uranium and, without this limit, potential reserves are almost inexhaustible.

The top ten uranium-producing countries are shown in Table 4.2. Kazakhstan and Canada are, by far, the top two uranium-producing countries in the world, producing 39% and 22%, respectively. In 2017, the worldwide uranium production was 59,342 tonnes, and these top ten countries accounted for 98% of the worldwide uranium production.

Worldwide uranium consumption for 2017 was 62,825 tonnes.[9] Table 4.3 shows the top ten consuming countries, which accounted for 85% of the worldwide consumption. Notably, Kazakhstan is a net supplier to much of the world, as the earlier referenced report shows that this country has no nuclear power plants. By themselves, the United States and France consume 34% and 13%, respectively, of the worldwide consumption. The United States gets around 95% of its uranium from other countries, and France, which has almost no production, imports virtually all of its uranium needs. Canada produces about seven times what they use.

Table 4.2 Top Ten Uranium Producing Countries, 2017 (tonnes)[9]

Country	2017
World Total	59,342
Kazakhstan	23,400
Canada	13,130
Australia	5,800
Namibia	4,000
Niger	3,485
Russia	2,900
Uzbekistan	2,400
China	1,700
United States	960
Ukraine	615

Table 4.3 Top Ten Uranium Consuming Countries, 2017 (tonnes)[9]

Country	U, tonnes
World Total	62,825
United States	21,070
France	8,000
China	6,700
Russia	4,800
South Korea	3,400
Ukraine	2,480
Japan	2,315
Canada	1,830
Germany	1,400
Belgium	1,305

Predicting the future for uranium-powered nuclear power plants is difficult. The Fukushima nuclear power plant accident has brought about policy changes in many European countries. While growth is expected in East Asia and non–European Union (EU) countries in Europe, North America and the EU may either have increases or decreases in nuclear power plants. The Nuclear Agency Report[9] indicates that, by the year 2035, the low-demand scenario has uranium requirements decreasing to 53,010 tonnes, a 16% reduction, while the high-demand scenario has uranium requirements increasing to 90,820 tonnes, a 45% increase.

Using the 2017 estimate for reserves of 6,142,200 tonnes and the 2017 consumption rate of 62,825 tonnes per year leads to a calculated value of 98 years remaining for uranium. If the low and high demand growth scenarios discussed previously come to fruition, the years remaining would be 116 and 68 years, respectively. Of course, increasing demand and prices would result in finding uranium from other sources other than mining, as discussed earlier.

4.3. Overview of Technology

In 1789, a German chemist named Martin Heinrich Klaproth was the first to discover uranium, while studying the mineral pitchblende,[11] a form of the mineral uraninite. He named it Uranium after the planet Uranus. Uraninite is a mineral containing the uranium oxides UO_2 and U_3O_8, and is the primary source of uranium.

Goldschmidt[12] gave a historical overview of uranium in 1989 at the International Symposium in London. Excerpts from this overview are discussed here. While uranium was discovered by Klaproth, the French chemist Eugène Peligot in 1841 was the first to prepare it, using a thermal reaction of tetrachloride with potassium. In 1870, Russian chemist Dimitri Mendeleev established that uranium is the last and heaviest element naturally present on earth. This is, by the way, the same Dmitri Mendeleev credited with inventing the periodic table of elements, with table rows having increasing atomic number and columns exhibiting similar atomic structures. Before being used for its radioactive properties, uranium was actually used to produce yellow glass and ceramic glazes because of its color. What is interesting is that uraninite was used as a source for radium before it was used to produce uranium. Since all uraninite minerals contain small amounts of radium, formed from radioactive decay of uranium, Marie Curie used the uraninite as a source for her research on radium in 1898. The demand for radium began to grow as early as 1904, when researchers started using it for cancer therapy. The use of radium for its radioactivity, as well as polonium, which is also found in uraninite, eventually led to the discovery of the use of uranium to produce energy through nuclear fission.

In 1903, Pierre Curie measured energy released from the decay of radium, and this was later confirmed by Ernest Rutherford and his student Frederick Soddy, who were the first to learn that the release of energy was due to an element changing into another by radioactive decay, a process known as transmutation. Thus began the knowledge of the awesome energy possible from nuclear fission and its potential in generating cheap electrical power.

Other than for the purpose of generating electricity, uranium has other purposes. It is, of course, well known that uranium is used to make atomic weapons. However, the military also uses uranium for nonradioactive purposes. Since it is dense, hard, and flammable, it can be used to make ammunition capable of penetrating armored vehicles.[13] It is also used to power submarines.

Outside of the military, radio-isotopes of uranium are used in medicine for diagnosis and research, in food processing to sterilize fresh products, and in industrial X-rays for safety and quality inspection.[14] Also, in space probes where solar energy is not available for power generation, radio-isotopes of uranium can provide heat and electricity. Finally, on a small scale, uranium is used to make transuranium elements, defined as chemical elements with atomic numbers greater than 92, the atomic number of uranium. These transuranium elements include unstable elements such as neptunium (Np), plutonium (Pu), and americium (Am).

Perhaps the first time nuclear energy was used to generate electricity was in 1951 when an experimental breeder reactor in Idaho was used to illuminate four light bulbs.[15] However, the purpose of this reactor was to validate the breeder reactor concept. In June, 1954 at Obninsk, Russia, a nuclear power plant with a net output of 5 MW was connected to the power grid, so this is the first nuclear power plant that made electricity for commercial use. In August, 1956, England's Calder Hall 1 power plant produced a more substantial output of 50 MW, which also supplied electricity to the grid.

A technical description of using U-235 to have a controlled nuclear reaction that produces electricity can be found in many references, but the brief description here follows that from the World Nuclear Association.[16] The basic reaction to release energy, such that U-235 absorbs a neutron and splits into two or more smaller atoms and free neutrons, is shown in equation (1), where the smaller atoms are barium and krypton. The lower number represents the number of protons and the upper number is the sum of protons and neutrons.

$$^{235}_{92}U + ^{1}_{0}n \rightarrow ^{141}_{56}Ba + ^{92}_{36}Kr + 3\,^{1}_{0}n \qquad (1)$$

In a nuclear power plant, the neutrons made from equation (1) are used to create more fission, creating a nuclear chain reaction. Of course, the chain reaction must be controlled to avoid having the reactor becoming a nuclear bomb, so moderators are used to slow down neutrons and change the amount that result in continued fission reactions. Moderators include materials such as water or graphite. Figure 4.1 gives a conceptual picture of this chain reaction.

In addition to the reaction for U-235, U-238 changes into Plutonium-238 (Pu-238), and the fission of Pu-238 provides about one-third of the total reactor energy output.

The nuclear fission of a single atom of U-235 releases 202.5 MeV, or 83.14 TJ/kg (terajoule or 10^{12} J).[17] This is equivalent to 35.74 Giga BTU/lb (35.74×10^9 BTU/lb). Chapter 3 shows that bituminous coal has an energy content of 12,712 BTU/lb so, on an equivalent mass basis, U-235 has about 2.8 million times more energy than coal combustion!

The nuclear reaction takes place in a steel pressurized vessel, surrounded by liquid water. The reactor pressure is sufficient to keep the water as liquid, even at operating

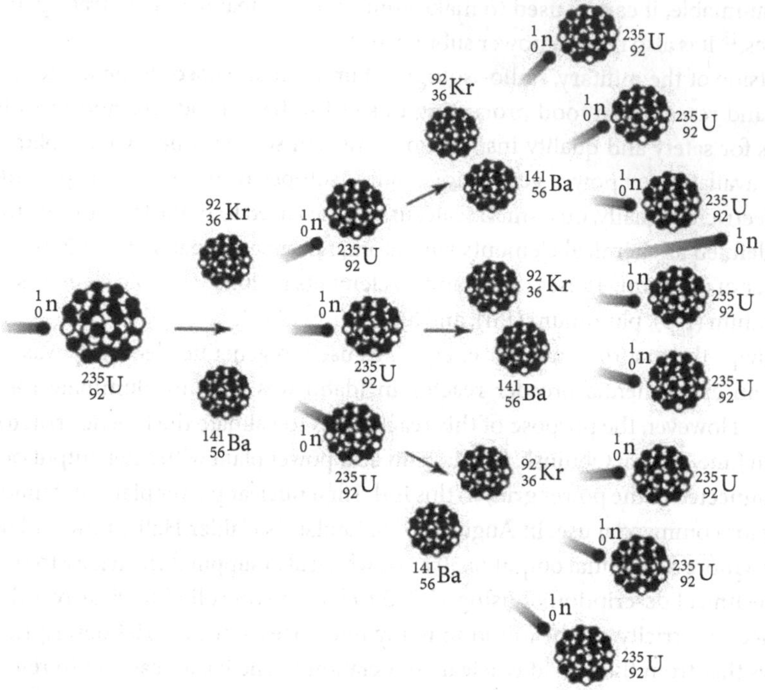

Figure 4.1 Nuclear Chain Reaction

temperatures over 320°C.[16] Steam used to drive the turbine to produce electricity is either above the reactor core or in a separate vessel.

There are two main types of nuclear reactors including the pressurized water reactor (PWR) and the boiling water reactor (BWR). The PWR is more common, and in this design the water flows through the reactor as a moderator but is isolated from the turbine and contained in a pressurized primary loop. This loop produces steam in a secondary loop, powering the turbine. This design prevents radioactive contamination of the turbine and condenser.

For the BWR, one water loop is used for moderator and the steam source to power the turbine. In contrast to the PWR, the BWR will therefore have the disadvantage that a fuel leak will make the water radioactive, and thus contaminate the turbine and condenser.

The next part of the overview discusses the percent needed for U-235 as a fuel and how enrichment is done to raise this level above that in uranium obtained from mining. Natural uranium is composed of 99.2745 wt% U-238, 0.720 wt% U-235, and 0.0055 wt% U-234, but, depending on the source, the content of U-235 can vary by as much as 0.1 wt%.[18] The Department of Energy (DOE) has adopted an official value of 0.711 wt% for the U-235 content in natural uranium.[3] However, to make the uranium suitable for a nuclear power plant, the content of U-235 must be enriched to 3%–5%.[19]

A review of enrichment of natural uranium is given by Manojlović[20] and is summarized here. Figure 4.2 shows a mass balance for U-235, such that "M_f" is the mass flow of uranium before enrichment with a U-235 concentration of "x_f." After enrichment, the mass flow of enriched product is "M_p" with an enriched concentration of "x_p" while the mass flow of waste, or tails, uranium is "M_t" with a uranium concentration of "x_t."

From this figure, we can write the mass flow of uranium as equation (2). The total mass balance is:

$$M_f = M_p + M_t \tag{2}$$

The mass flow for U-235 is simply the mass flow times the stream concentration of U-235, as shown in equation (3):

$$M_f * x_f = M_p * x_p + M_t * x_t \tag{3}$$

The value for "M_t" can be eliminated from equation (3) using equation (2) to give equation (4):

$$M_f * x_f = M_p * x_p + (M_f - M_p) * x_t = M_p * x_p + M_f * x_t - M_p * x_t \tag{4}$$

Simple rearrangement gives equation (5):

$$M_f = M_p \left[\frac{(x_p - x_t)}{(x_f - x_t)} \right] \tag{5}$$

Following an example from Manojlović, if it is desired to make 50 tons of uranium with an enriched concentration of U-235 at 4.3%, with the tails concentration at 0.24% and the starting concentration of U-235 in the natural uranium at 0.711%, then it will take a feed rate of 430 tons of the natural uranium. Stated differently, the feed requirements are 8.6 times the product requirements for this level of enrichment.

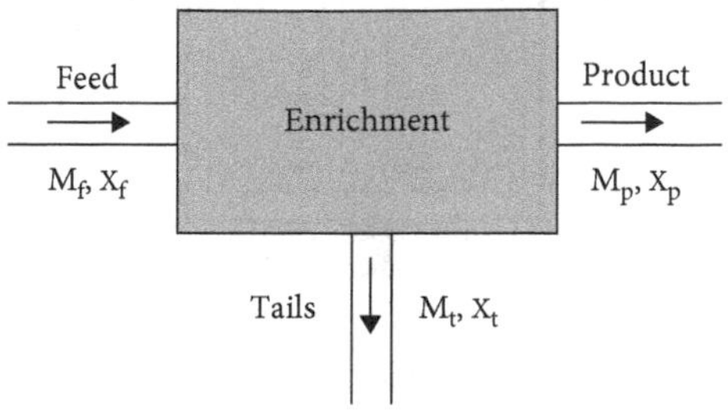

Figure 4.2 Mass Balance for U-235

The three main methods to obtain the enriched uranium are all based on gas separation and, since uranium has the very high boiling point of 4,131°C, uranium hexafluoride (UF_6) gas is used in the separation methods.[20] Each of these three methods is briefly described.

The first enrichment method is "Gaseous Diffusion." First, the natural uranium is converted to UF_6 gas. The diffusion process requires pumping the UF_6 through a large number of porous barriers and is very energy intensive.[21] Because the UF_6 containing the U-235 is lighter than that with U-238, the UF_6 with the U-235 will diffuse through the barriers at a higher rate. As the difference in mass between U-235 and U-238 is very small, the enrichment process will require thousands of diffusions through these barriers to enrich the uranium to the 3% to 5% level. To give you an idea of the scale involved, when the Oak Ridge plant in Tennessee was built in the early 1940s, it was the largest industrial building in the world.

The second enrichment method uses "Gas Centrifuges" and is also the most commonly used technology.[21] In this method, the UF_6 gas is fed into the centrifuge at rotational speeds approaching the speed of sound. Because of the heavier mass for the U-238 compared to U-235, these molecules concentrate to the outside of the centrifuge cylinder. Through heating, a convection cycle is started such that the gas is circulated to drive the depleted uranium to the outer wall and out through the top while the enriched gas is driven to the bottom, and then out a central tube. These centrifuges are quite large, typically 4 to 5 meters tall, and some as tall as 12 meters.[20] This process is shown in Figure 4.3.

As in the case of "Gaseous Diffusion," many thousand steps are needed to enrich the uranium to the 3% to 5% level. An important advantage for this process is that it uses 40 to 50 times less energy than "Gaseous Diffusion."

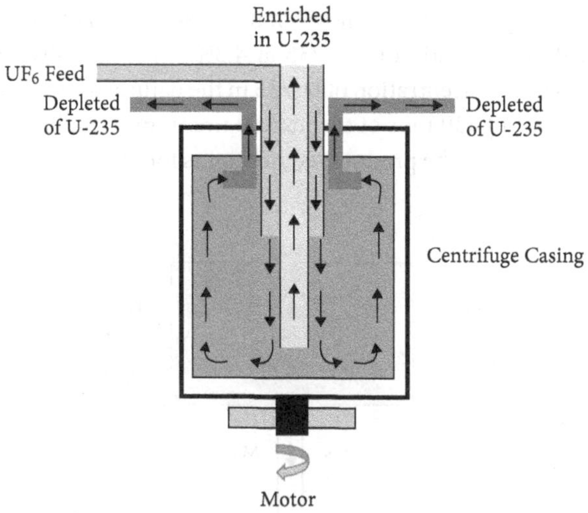

Figure 4.3 Cross-Sectional Diagram of One Gas Centrifuge

The third method is called "Jet Nozzle/Aerodynamic Separation.[21]" For this method, the enrichment process is similar to "Gaseous Diffusion." UF_6 gas is pressurized in helium or hydrogen gas to increase the overall gas velocity, and this mixture is sent through small pipes that separate the stream on the inside of the pipe, the enriched stream, from the stream on the outside of the pipe, the depleted stream. Because of the difficulty in making the separation nozzles and the energy requirements for compressing the gas mixture, this process is not very economical.

Based on the previous discussion, it is not surprising that the primary method for the enrichment of uranium is the "Gas Centrifuge" method, listed as the only technology commercially used in 2015.[22]

According to the World Nuclear Association,[23] every tonne (1,000 kg) of natural uranium produced and enriched for use in a nuclear reactor yields about 130 kg of enriched U-235 fuel, thus producing about 870 kg of uranium tails. These tails are also known as depleted uranium (DU), and DU is stored as either UF_6 or converted back to U_3O_8. Every year over 50,000 tonnes of DU are added to the world DU stockpile of 1.6 million tonnes.

The last discussion concerns the spent nuclear fuel rods. As mentioned, the U-235 isotope is enriched to a level of 3% to 5% but, after 18 to 36 months in the reactor, used fuel rods are removed from the reactor. This used fuel will typically have about 1% U-235, 0.6% fissile plutonium (Pu), and around 95% U-238.[24] At these levels, the fuel is still hot and radioactive. Typically, the fuel rods will be removed underwater before transfer to dry, ventilated concrete containers.[25]

At this time, the used nuclear fuel is stored at the nuclear power plant using steel-lined, concrete pools filled with water.[26] The water acts as a natural barrier for radiation, as well as keeping the fuel cool. About 2,000 tonnes of used fuel rods are produced each year.

This on-site storage was not meant to be permanent and, according to legislation passed in 1987, the federal government began planning a permanent underground storage site at Nevada's Yucca Mountain, about 100 miles northwest of Las Vegas.[27] However, after two decades of research and federal spending of about $6.7 billion, the plan to use this site to hold about 64,000 tonnes of nuclear waste in tunnels 600 feet below the mountain surface has been stopped for the foreseeable future.

4.4. Capital Cost, Operating Cost, Well-to-Wheels Levelized Cost of Electricity, and Well-to-Wheels Levelized Cost of Fuel

The equations and methodology used to calculate levelized cost of electricity (LCOE), and specific input used for nuclear fuel to make electricity, are detailed in Appendix A.

The units for LCOE are in $/megawatt-hour, or $/MWh and data for nuclear power plants were taken from two sources.[28,29] Although the Energy Information Administration (EIA) reference[28] only specifies the reactor design as "advanced nuclear," an earlier EIA reference[30] further identifies it as a PWR. The data of Lazard[29] do not specify reactor type. Regardless, Figure 1 from Mott MacDonald[31] shows that the capital cost for a PWR and BWR are the same and Figure 2 from the same report shows that the LCOE for the two technologies are very nearly the same. Therefore, the various costs reviewed in this section are applied equally to either reactor design.

For a nuclear power plant, the "Well-to-Wheels" costs include (1) the cost of uranium mining and production, preparation for the nuclear power plant through conversion to UF_6, enrichment, and manufacture of the fuel elements; (2) power plant generating costs; (3) transmission and distribution costs; and (4) taxes (both federal and state). The costs must also include (5) the cost of storing nuclear waste.

A price is needed for the uranium ore. According to one source,[33] uranium does not trade on an open market like other commodities but, rather, buyers and sellers negotiate private contracts. Spot prices are, however, given by independent consultants such as Ux Consulting and TradeTech. The 12-month average spot price for uranium in 2018 was \$24.59/lb, or \$54.21/kg, for U_3O_8.

Some more work has to be done to get the cost of nuclear fuel for the power plant, however. To obtain usable nuclear fuel, the U_3O_8 has to be converted to UF_6, enriched using a gas centrifuge, and fabricated into fuel elements. According to the World Nuclear Association,[34] the cost to obtain 1 kg of uranium as reactor fuel included 46% for the uranium ore, 6% to convert it to UF_6, 32% to enrich it, and 16% to fabricate the fuel elements.

Assuming a U-235 enrichment level of 4.3%, a starting concentration of 0.711% U-235 in the natural uranium, and a tails concentration of 0.24%, the feed requirements are 8.6 times the product requirements. Therefore, at a price of \$54.21/kg, the different preparation costs are \$466/kg to purchase the U_3O_8, \$61/kg to convert it to UF_6, \$324/kg to enrich it to 4.3% U-235, and \$162/kg to fabricate the fuel elements. This gives a total cost of \$1,013/kg or \$459/lb for the uranium reactor fuel.

The nuclear fission of a single atom of U-235 releases 202.5 MeV, equivalent to 35.74 Giga BTU/lb (35.74×10^9 BTU/lb). At an enrichment level of 4.3%, the energy release on a total uranium mass basis is 1,536,820,000 BTU/lb. Therefore, at a cost of \$459/lb for the reactor fuel, this is equivalent to \$0.30/MM BTU.

Even at the low enrichment level of 4.3%, it is interesting to compare the energy content to that of fossil fuels. In Chapters 1, 2, and 3, it was reported that the energy content of a barrel of West Texas Intermediate is 20,155 BTU/lb, natural gas contains 22,792 BTU/lb, and bituminous coal contains 12,712 BTU/lb. This shows the awesome potential of nuclear fuel, with ~76,000 times more energy than crude oil, ~67,000 times more energy than natural gas, and ~120,000 times more energy than coal.

The capital costs, capacity factors, operating and maintenance (O&M) costs, and transmission costs were taken from the authors. For transmission costs, they provide the capital cost for new transmission lines to tie into existing substations or existing transmission lines. In other words, the generating station provides transmission lines and steps up the voltage to tie into the "electrical grid," a network that delivers the electricity to the customer. The grid has high-voltage transmission lines (sometimes distant) that carry power to demand centers, transformers to step down the voltage, and distribution lines that connect individual customers.

The costs of transmitting and distributing electricity were 0.9 cents/kWh (or $9/MWh) and 2.6 cents/kWh (or $26/MWh), set according to the EIA.[32] Here, transmission is defined as the high-voltage movement of electricity from the power station, while distribution is the cost of electrical lines that take power from a substation to the customer.

In the case of a natural gas or coal electric plant, it is easy to measure the fuel consumption, as there must be continuous flow of the fuel. For a nuclear power plant, however, the fuel elements are placed in the reactor, so there is not a continuous measurement of volume or mass flow as there is in the case of natural gas and coal. According to the World Nuclear Association,[24] these fuel elements are in the reactor several years, and about one-third of the fuel elements are replaced every 18 to 24 months to maintain efficient reactor performance.

For the cost for nuclear waste disposal, one report available was published by the Organisation for Economic Co-Operation and Develop (OECD), of which the United States is a member.[35] The report discusses the economics of both direct disposal and partial recycling. Direct disposal is a process such that the fuel is used once and is then regarded as waste for disposal, and partial recycling is a process such that the spent fuel is reprocessed to recover unused uranium and plutonium, which are recycled back into reactors. According to Figure ES.1 in the report, the cost of direct disposal is about $1/MWh while the cost of direct disposal with partial recycling is about $7/MWh. In another study, Berry and Tolley[36] report direct disposal costs at $1.09/MWh and recycling costs of $2.40/MWh. The choice between direct disposal and recycling is an environmental issue, as well as the choice of temporary on-site storage versus permanent disposal. In addition, in the United States there is no permanent long-term storage site as the Yucca Mountain project, started in 1987, was canceled in 2009. This is discussed in more detail in the section "Environmental Issues." For the purpose of completing the economic analysis, a cost of $1/MWh is assumed for nuclear waste storage, based on direct disposal.

Table 4.4 shows the LCOE for the cost of generating electricity, the additional cost of transmission and distribution, and the additional cost of direct disposal. As can be seen, the cost of direct disposal has a small impact on the overall LCOE. Also, the EIA case is similar to the low-cost case for Lazard. An obvious advantage of nuclear-generated electricity is that there is no carbon dioxide (CO_2) produced from the power plant. Thus, if carbon capture and sequestration (CCS) become mandatory in the future, this improves the economic comparison of nuclear plants to natural gas and coal plants.

A sensitivity analysis was made by varying the capital cost, the combined fixed and variable operating expenses (excluding the fuel cost), and the fuel cost. The capital cost considered lower and upper values of $6,034/kW and $12,250/kW, while the combined fixed and variable operating cost, excluding the cost of fuel, considered

Table 4.4 Economic Analysis for Nuclear-Powered Electric Plants

Energy Type	Capacity Factor	Capital Cost	Levelized Capital Cost	Fixed O&M	Variable O&M (including fuel)	Transmission Investment	Total System Levelized Cost (LCOE) to Generate	LCOE Including Transmission and Distribution	LCOE Including Direct Disposal
EIA[28]	0.90	$6,034	$70.94	$13.10	$5.51	$1.10	$90.65	$125.65	$126.65
Lazard, low cost case[29]	0.90	$6,500	$76.42	$14.59	$3.89	$1.10	$95.99	$130.99	$131.99
Lazard, high cost case[29]	0.90	$12,250	$144.01	$17.12	$3.89	$1.10	$166.12	$201.12	$202.12
Average	0.90	$8,261	$97.12	$14.94	$4.43	$1.10	$117.59	$152.59	$153.59

lower and upper values of \$8/MWh and a high of \$24/MWh. For the fuel costs, lower and upper values were selected based on historical spot prices for uranium, shown in Table 4.5.[33] As the table illustrates, the price of uranium ore has been steadily decreasing, losing nearly half its value since 2012. Basically, the sensitivity study used the lower and upper values of Table 4.5, which are \$21.66 and \$48.40/lb. In terms of energy, these are \$0.26/MM BTU and \$0.59/MM BTU.

All other data for the calculations used the EIA case.[28] The results are shown in Table 4.6.

As the table illustrates, changing the combined fixed and variable operating costs (excluding nuclear fuel cost) by 50% had a small impact on LCOE. Likewise, doubling

Table 4.5 Historical Spot Prices for Uranium, \$/lb of U_3O_8[33]

Year	\$/lb U_3O_8
2012	48.40
2013	38.17
2014	33.21
2015	36.55
2016	25.64
2017	21.66
2018	24.59

Table 4.6 Sensitivity Study Using the EIA Nuclear Case[28]

	Nuclear Fuel Preparation Cost	Capital Cost	Fixed and Variable O&M (no fuel cost)	Total System Levelized Cost (LCOE) to Generate	Total LCOE Cost to Generate, Transmit, and Distribute
Units	\$/MM BTU	\$/kW	\$/MWh	\$/MWh	\$/MWh
Base Case	\$0.30	\$6,034	\$15.47	\$90.65	\$125.65
Capex is \$6,034/kW	\$0.30	\$6,034	\$15.47	\$90.65	\$125.65
Capex is \$12,250/kW	\$0.30	\$12,250	\$15.47	\$163.72	\$198.72
Opex is \$8/MWh	\$0.30	\$6,034	\$8.00	\$83.18	\$118.18
Opex is \$24/MWh	\$0.30	\$6,034	\$24.00	\$99.18	\$134.18
Nuclear price is \$0.26/MM BTU	\$0.26	\$6,034	\$15.47	\$90.23	\$125.23
Nuclear price is \$0.59/MM BTU	\$0.59	\$6,034	\$15.47	\$93.68	\$128.68

the cost of fuel made a very small change in LCOE. However, doubling the capital cost increased LCOE by \$73/MWh, an increase of 58%.

4.5. Cost of Energy

The cost of uranium in units of \$/lb and \$/MM BTU are presented in this section. Of course, this cost must reflect the cost of the fuel used by the nuclear power plant, not the cost of the uraninite ore, and therefore includes the cost of the ore, conversion to UF_6, enrichment to a usable level of U-235, and the fabrication of the fuel elements. This total cost was previously shown to be \$1,013/kg or \$459/lb for an enrichment level of 4.3% U-235.

The energy release from nuclear fission of U-235 is 35.74×10^9 BTU/lb and, at an enrichment level of 4.3%, the energy release is 1,536,820,000 BTU/lb. Therefore, we can also calculate the cost of the fuel on an energy basis to be \$0.30/MM BTU. Compared to other traditional energy types, uranium fuel is very inexpensive compared to crude oil at \$11.20/MM BTU, natural gas at \$3.17/MM BTU, and coal at \$2.06/MM BTU.

4.6. Capacity Factor

Capacity factors are shown in Table 4.4, taken from two references.[28,29] For both of these studies, the capacity factors were reported as 0.90. Thus, nuclear-power electric plants are very reliable, operating about 90% of the time.

4.7. Efficiency: Fraction of Energy Converted to Work

Thermodynamic efficiencies can be calculated from the "heat rate," the amount of energy used to make useful work. In the case of a nuclear power plant, the nuclear fuel is the source of energy and the useful work is the electrical power made. To calculate the thermal efficiency, the work needed to make one kilowatt-hour (kWh) of electricity, or 3,412 BTU, is divided by the heat rate.

The heat rate reported by the EIA study is 10,461 BTU/kWh[28] and that reported by the Lazard study is 10,450 BTU/kWh.[29] Using these heat rates gives thermodynamics efficiencies of 32.6% and 32.7%, respectively. In addition, EIA average data for 10 years of US nuclear plant operation are shown in Table 4.7.[37] As can be seen, the thermal efficiencies are identical to within a few tenth of a percent for both of these studies.

Table 4.7 EIA Data for Average Operating Heat Rates and Thermal Efficiencies for 2008 through 2017[37]

Year	Heat Rate, BTU/kWh	Thermal Efficiency, %
2008	10,452	32.6
2009	10,459	32.6
2010	10,452	32.6
2011	10,464	32.6
2012	10,479	32.6
2013	10,449	32.7
2014	10,459	32.6
2015	10,458	32.6
2016	10,459	32.6
2017	10,459	32.6

4.8. Energy Balance

For nuclear-fueled power generation, the energy balance is defined by this equation:

$$\text{Energy Balance} = \frac{[\text{Electrical energy from nuclear fuel delivered to customer}]}{[\text{Energy contained in the nuclear fuel at the mine site}]}$$

For the conversion of nuclear fuel energy to electricity, there are several steps in the "Well-to-Wheels" analysis including the energy used to (1) mine and produce the uraninite ore, (2) convert the uranium to UF_6, (3) enrich the uranium through the use of centrifuges, (4) manufacture the fuel elements, (5) transport the nuclear fuel to the power plant, (6) make the electricity, (7) manage the nuclear waste, and (8) distribute the electricity through the grid to your home.

A brief review is made of what is involved in each step before looking at the energy consumption. Issues for each step can be found in many sources, such as the one published by van Leeuwen and Smith.[38] For example, in mining the uranium ore, if the ore has 0.1% U_3O_8 versus 1%, ten times the amount of ore mass will have to be processed. Also, uranium can be mined underground or in open pit mines. After mining the uranium, it must next be milled. Milling is an extraction process, such that the uranium ore is ground to a fine powder and then treated, usually with sulfuric acid. As mentioned, there are several methods for enrichment and each of these will have different levels of energy consumption.

It is clear, then, that the amount of energy required to make uranium suitable for the nuclear power plant can vary somewhat. To do the energy balance for this section, data from the World Nuclear Organization are used.[39] For their analysis, they assume a 1,000 MW power plant with a 40-year life, an 86% capacity factor, centrifuge enrichment to 2.3% U-235 with 0.25% in the tails, and a 33% thermal efficiency. The uranium ore is assumed to contain 0.26% uranium. Energy use over the 40-year life is shown in Table 4.8, which also includes construction and decommissioning of the plant as well as waste management. Their data are given in PetaJoules, equal to one quadrillion Joules or 1×10^{15} Joules. The World Nuclear Organization[24] also indicates that 27 tonnes of fuel are required each year for a 1,000 MW reactor.

For the 1,000 MW plant with a capacity factor of 86%, the electricity production is 7.5 terawatt-hours/year, or 7.5 TWh/y. A terawatt-hour is a trillion watts-hours, or 1×10^{12} watts-hours. Thus, over a 40-year life the plant will produce 300 TWh. Assuming the plant has a thermal efficiency of 33% and the fact that 1 TWh is equal to 3.6 PetaJoules, the output energy over these 40 years is 3,273 PetaJoules. Comparing this to the total energy used in Table 4.8, we see that the energy use is only 1.6% of the energy produced by the power plant. The percent for each step is also shown in the table.

Of course, the energy used to perform the activities in Table 4.8 does not come from the energy in the uranium, but from other energy sources. Nevertheless, this energy is deducted from the starting energy in the uranium to show its overall effect on the energy balance.

It then remains to determine the energy needed to distribute the electricity through the grid to your home in order to construct an overall energy balance.

Transmission and distribution losses are affected by the voltage of the power line and the distance the electricity is transmitted. Historical data from the state of California shows annual average transmission losses around 5 to 7% over the time period of 2002 to 2008.[40] Similarly, EIA data for July, 2018[37] show of the electricity generated, 7% is lost in transmission and distribution. Therefore, a value of 7% will be

Table 4.8 Energy Use of a Nuclear Power Plant over a 40-Year Life[39]

Step	PetaJoules	% of Output
Mining and milling	2.5	0.08
Conversion to UF_6	9.2	0.28
Enrichment	2.6	0.08
Fuel fabrication	5.8	0.18
Build, operate, & decommission plant	30.7	0.94
Waste management	1.5	0.05
Total input energy	52.3	1.60
Total output energy	3,273	

used for the amount of electricity lost in transmission and distribution after leaving the nuclear electric plant.

Using 100 units of energy as a reference, we can arrive at the following equations. Summarizing, this means that of the original energy in the uranium at the mine, 30.2% will end up as useable electricity. By far, the greatest loss in the energy is the nuclear plant energy efficiency of 33%.

Nuclear Power Energy Balance

100 → 99.92 (mining and milling) → 99.64 (conversion to UF_6) → 99.56 (enrichment) → 99.38 (fuel fabrication) → 98.44 (construction and decommissioning the plant) → 32.49 (power generation) → 32.44 (waste management) → 30.17 (transmission and distribution)

4.9. Maturity: Experience

Uranium was first discovered in 1789 by a German chemist, Martin Heinrich Klaproth.[11] Further, the French chemist Eugène Peligot was the first to prepare it, in 1841, using a thermal reaction of tetrachloride with potassium.[14] Uranium minerals (pitchblende) were observed by miners as early as 1565 in the Saxony Ore Mountains.[42] In Jáchymov, located in the Ore Mountains, systematic mining of uranium began as early as the 19th century.[43] In this time period, the uranium compounds were used for coloring glass and porcelain glazing, as well as making photographs. However, in the 20th century, uranium ore mining was used for the post–World War II arms race, and more than 7,000 tonnes were extracted from these mines following World War II.

Uranium mining in the United States can be divided into four periods.[44] From 1905 to 1925, the focus of uranium mining was to obtain radium, and this led to extensive mining on the Colorado Plateau. From 1925 to 1947, vanadium was needed to increase the tensile strength and elasticity of steel. The US government formed the Atomic Energy Commission (AEC) in 1946, and from 1947 to 1970, the need for uranium was spurred by the development of nuclear arms. In fact, during the 1940s and 1950s, uranium was mined on the Navajo Reservation to support the atomic weapons program,[45] unfortunately leading to the death of 133 Navajo miners. The fourth period began in 1970. As the AEC already had ample reserves of uranium, production waned until a second boom in mining occurred when nuclear power plants came into existence in the mid-1970s.

The first commercial nuclear power plant, albeit it small with an output of 5 MW, started operation in June 1954 in Obninsk, Russia. In August 1956, the Calder Hall 1 in England was the first substantial commercial power plant, producing 50 MW.[15]

The first US commercial power plant was built in 1957 in Shippingport, Pennsylvania,[46] at a cost of $72,500,00, which included some government funding. This was also the first PWR. The Dresden Nuclear power plant, which began operation in 1960, located in Grundy County, Illinois, was the first commercial plant built without government funding. This was also the first BWR built in the United States.[46]

Since the first coal-fired electrical plant was built in 1882 and through the 1930s, the electrical grid system grew rapidly, roughly doubling every 6 years. Today, what we take for granted, involves more than 180,000 miles of high-voltage transmission lines connecting about 7,000 power plants.[47]

To summarize, there has been uranium mining in the United States since 1905 and since 1947 the mining was specifically for uranium. The first nuclear plant was built in 1957, and was also the first PWR. The first BWR was subsequently built in 1960. The first electrical grid system was built in 1882 followed by rapid development through 1930s. Thus, there are over 100 years of experience in uranium mining, about 60 years of experience in nuclear power electric plants, and more than 135 years of experience in electric grids.

4.10. Infrastructure

As shown earlier in Table 4.2, the United States had a uranium production of 960 tonnes in 2017, much less than the consumption listed in Table 4.3 of 21,070 tonnes. In fact, uranium mining today is done by only a few companies on a small scale.[48] Current production facilities are located in Wyoming, Nebraska, Louisiana, and Texas. Thus, most of the infrastructure for uranium mining is outside the United States, and more than 95% is imported.

If uranium ore or the finished fuel rods are transported by rail or truck, there are 95,264 miles of track and 46,876 miles of Interstate highway in the United States.

The EIA listed 99 operating reactors on 60 plant sites making electricity in 2016.[49] This includes 34 BWRs and 65 PWRs. For the BWRs, 13 plants have 1 reactor, 9 have 2 reactors, and 1 has 3 reactors. For PWRs, 13 plants have 1 reactor, 23 have 2 reactors, and 2 have 3 reactors. In 2017, EIA data listed a total capacity of 104,792.4 MW[50] with a net generation of 804,950 thousand MWh.[41,51] Once made, this electricity can be transmitted around the United States using the more than 180,000 miles of high-voltage transmission lines in the grid.[47]

In the United States, the last nuclear power plant was built in 2016, the Watts Bar plant operated by the Tennessee Valley Authority.[88] However, this was the first plant built since 1996, when its sister reactor was built, also bearing the Watts Bar name. Two new reactors are planned for Georgia, Vogtle 3 and 4, each rated for 1,250 MW and scheduled to start operation in 2021 and 2022, respectively.[89] Worldwide, the World Nuclear Association shows 51 reactors under construction, scheduled to start operation from 2019 to 2026.[89]

4.11. Footprint and Energy Density

Footprint is a term commonly used in industrial plants to describe the area used by an energy type. For nuclear energy we must consider all the land area used to produce the energy from "Well-to-Wheels." This includes land used (1) for mining, (2) for enrichment, (3) for manufacture of the fuel elements, (4) to generate electricity, (5) for nuclear waste storage, and (6) and to transmit and distribute the electricity via the grid to commercial customers.

For the land area used for (1) mining, we first consider the different types of uranium mining. According to the World Nuclear Association[52] there are three types of uranium mining—open pit, underground, and in situ leaching (ISL) mining. Open pit mining is used when the uranium ore is near the surface, and the ore can be extracted by cutting open a large pit and removing the ore, as well as waste rock. When the uranium ore is deeper and not near the surface, underground mining is used, making shafts and tunnels to extract the ore. In the case of open pit or underground mining, considered conventional mining, the recovered ore will next be sent to a mill, where the ore is crushed and ground to extract the ore and then leached with sulfuric acid to dissolve the uranium oxide from the ore. However, in the case of ISL mining, the uranium ore is in porous materials in groundwater, such as gravel or sand, and can be extracted by dissolving the uranium and then pumping it out. This is only used when the water aquifer is confined, and will not contaminate potable water systems. Leaching solutions used to extract the uranium ore are usually acidic solutions of sulfuric acid or carbonate solutions such as sodium bicarbonate, ammonium carbonate, or dissolved CO_2.

According to the World Nuclear Association,[52] in 2017, 53% of the world's uranium came from ten mines in four countries including Canada, Kazakhstan, Australia, and Niger. Normalizing this 53%, 38% was mined by ISL, 49% by underground, and 13% by open pit. However, in the United States, there are currently no active open pit uranium-mining operations.[53] Uranium is mined at six ISL plants in Nebraska (Crow Butte) and Utah (Smith Ranch-Highland, Willow Creek, Lost Creek, Nichols Ranch, and Ross).[54] Uranium is mined by both ISL and underground mining at White Mesa, Utah, which is also the site of the only operating mill. Two other uranium mills are authorized for operation, but the Shootaring Canyon mill in Utah and the Sweetwater mill in Wyoming are on standby. Since 85% of the uranium mined in the United States came from ISL, for the purpose of uranium mining footprint, this section focuses on uranium obtained through ISL.

For ISL, the uranium is recovered from the ore by dissolving it and pumping the uranium-containing solution to the surface. In this respect, there is little disturbance to the land used, and very little land area is needed. Nevertheless, the land around the ISL-capturing area would be restricted from farming and ranching. A few examples will be considered to get an idea of the land area involved. One example is the Christensen Ranch ISL mine in the Powder River Basin in Wyoming,[55] where production figures from January 2010 show that their Unit 7, which encompasses 30 to 40 acres, was expected to yield about 1 million pounds of yellowcake over a 3-year period. Using 40 acres for the land area gives a value of 12.5 tons/acre or 4.2 tons/[year-acre]. Table 2.13 in another reference[56] shows land requirements for the production of yellowcake as 3.2–6.3 tons/[year-acre] for open-pit mining and 4.5–9.0 tons/[year-acre] for ISL, an average of 6.75 tons/[year-acre] for ISL.

A good example of the difference between land allocated for a project and that actually used to mine uranium can be seen in the Uranium One project in Converse County, Wyoming.[57] The Ludeman Project is proposed for nearly 20,000 acres but the area proposed for the ISL is estimated to be 763 acres. The company estimates that 6.3 million pounds of uranium will be produced over a 12-year life, giving a yield of 4.1 tons/acre, or 0.34 tons/[year-acre]. Finally, using information from two references,[48,58] a production rate per acre can be estimated for the Smith Ranch-Highland Project in Wyoming, one of the largest ISL-producing uranium mines in the country. From 2002 to 2016, about 10,800 tons of uranium were produced over an estimated land area of 600 acres.[58] This gives an average yield of 18.1 tons/acre or 1.2 tons/[year-acre] over this 15-year period. Thus, for the four examples examined here, the uranium production yield for ISL mining ranges from a low of 0.34 tons/[year-acre] to a high of 6.75 tons/[year-acre]. From these four cases, an average of 3.1 tons/[year-acre] was calculated for uranium yield from ISL mining.

Next we convert the ISL mining uranium production of 3.1 tons/[year-acre] to units of energy. For the calculation, it is assumed the U-235 content of the ore is 0.711%. Since the energy release for U-235 fission is 35.74 G BTU/lb (35.74×10^9 BTU/lb), this amounts to 254,111,400 BTU/lb at this content. As shown in the next equation, the energy density is 1.0 quadrillion BTU/(year-square mile).

$$\frac{BTU}{year - square\ mile} = 3.1\frac{tons}{year - acre} * 2000\frac{lb}{ton}640\frac{acre}{square\ mile} * 35.74\times 10^9\frac{BTU}{lb} * \frac{0.711}{100} = 1.0\times 10^{15}$$

For the land area associated with (2) uranium enrichment, Urenco Company operates the only uranium enrichment facility in the US, supplying about one-fourth of the 21,070 tonnes of uranium fuel needed for the US nuclear power plants.[60] The plant has an enrichment capacity of 4,700 tonnes.[61] This $4 billion plant occupies 640 acres, or one square mile, and uses centrifuge technology to enrich the uranium.[59] Since 4,700 metric tonnes is equivalent to 5,170 tons, this means the plant can process about 8 tons/acre per year. Assuming a U-235 enrichment level of 4.3%, the energy density is calculated to be 15.7 quadrillion BTU/(year-square mile).

$$\frac{BTU}{year - square\ mile} = 8.0\frac{tons}{year - acre} * 2000\frac{lb}{ton}640\frac{acre}{square\ mile} * 35.74\times 10^9\frac{BTU}{lb} * \frac{4.3}{100} = 15.7\times 10^{15}$$

For step (3), the manufacture of uranium fuel elements, the World Nuclear Association[62] lists two of the largest US manufacturers as Areva and Westinghouse, capable of producing fuel elements containing uranium mass of 1,200 and 1,500 tonnes/year, respectively. The Areva plant,[63] located in Richland, WA, makes elements that produce about 5% of the total US electricity. About 200,000 fuel rods are made each year in a building with an area of 400,000 square feet, or 9.2 acres. The Westinghouse plant,[64] located in Columbia, SC, makes elements that produce about 10% of the total US electricity. This building is sized at 550,000 square feet, or 12.6 acres. At these plants, the UF_6 is converted into UO_2, and then into fuel rods and assemblies. Using the World Nuclear Association production values, and converting from tonnes to tons, we see that the Areva and Westinghouse facilities produce 1,320 and 1,650 tons in 9.2 and 12.6 acres, respectively, leading to area yields of 143 and 131 tons/[year-acres]. Compared to ISL mining at 3.1 tons/[year-acre] and uranium enrichment at 8 tons/[year-acre], the area yield for making the fuel elements is much larger, and therefore this land area is insignificant compared to mining and enrichment.

It is worth noting that the mass of uranium converted into fuel elements occurs after enrichment, and therefore the masses just given represent enriched uranium. Recalling an earlier discussion in the "Overview of Technology" section, an enriched concentration of U-235 at 4.3% will have feed requirements that are 8.6 times the product requirements for this level of enrichment. That means these 1,320 and 1,650 tons processed by Areva and Westinghouse represent uranium weight before enrichment of about 11,350 and 14,200 tons. Data in Table 4.3 show that the US consumption for 2017 was 21,070 tonnes, or about 23,200 tons. Thus, these two plants likely account for all of the uranium fuel rod assemblies used in the United States.

Another part of the nuclear energy footprint that involves the land area is (4) generating electricity at the nuclear power plant. Land areas for ten nuclear power plants are shown in Table 4.9. As can be seen from the table, the areas vary from a low of 0.12 MW/acre to a high of 3.55 MW/acre with an average of 1.17 MW/acre.

Units of MW/acre can be converted to the units of quadrillion BTU/(year-square mile) to compare directly to land area used for mining, enrichment, and manufacture of the fuel elements. For this, the average from Table 4.9, 1.17 MW/acre, is used. A 1 MW plant operating at 100% capacity for 1 year will produce 8,760 MWh (1 MW * 24 hour/day * 365 days/year). Also, there are 640 acres per square mile and 3.412×10^6 BTU/MWh. From these values, we can determine that 1.17 MW/acre is equivalent to 0.022 quadrillion BTU/(year-square mile), assuming the plant operates at a capacity factor of 100%.

However, we know from Table 4.4 that capacity factors are not 100% and the average value for nuclear power plants is 90%. With this adjustment, the energy produced for the power plant is 0.020 quadrillion BTU/(year-square mile).

The next step in the nuclear energy plant footprint is (5) to store nuclear waste. Given earlier discussions in the "Overview of Technology" section, nuclear waste is currently stored on-site at the nuclear power plant; therefore, the area for nuclear waste storage is included in the area for the nuclear power plant. Even if the waste were transferred to a permanent location, such as the Yucca Mountain site in Nevada, the area involved would be very small compared to the land area used for mining, enrichment, fuel element manufacture, and electricity generation.

Now that the electricity has been made and stored, step (6) is to transmit and distribute it to customers via the grid. However, the land area needed for step-up and

Table 4.9 Land Areas for Nuclear Power Plants

Location	Company	Type	Startup	MW	Acres	MW/acre
Grand Gulf, MS[65]	System Energy Resources Inc.	BWR	1985	1,207	2,300	0.52
Palo Verde, AZ[66]	Arizona Public Service Company	PWR	1985, 1986, 1987	3,937	4,000	0.98
Matagorda County, TX[67]	STP Nuclear Operating Company	PWR	1983	2,760	12,220	0.23
Perry, OH[68]	FirstEnergy Nuclear Operating Co.	BWR	1987	1,268	1,100	1.15
Seabrook, NH[69]	NextEra Energy Seabrook LLC	PWR	1990	1,250	900	1.39
Berwick, PA[70]	TalenEnergy Susquehanna LLC	BWR	1983, 1985	2,630	1,200	2.19
Rhea, TN[71]	Tennessee Valley Authority	PWR	1996, 2016	2,300	1,700	1.35
Fulton, MO[72]	Union Electric Company	PWR	1984	1,190	7,200	0.17
Burlington, KS[73]	Kansas Gas and Electric Company	PWR	1985	1,160	9,818	0.12
Peach Bottom Township, PA[74]	Exelon Nuclear	BWR	1974	2,200	620	3.55
Average				1,990	4,106	1.17

step-down transformers is very small and can be ignored. Likewise, the metal towers and electrical lines take up very little land area and can coexist with farming, pasture land, and roadways.

To summarize, the main contributors to land area are (1) ISL mining at 1.0 quadrillion BTU/(year-square mile), (2) enrichment at 15.7 quadrillion BTU/(year-square mile), and (4) nuclear electricity generation at 0.020 quadrillion BTU/(year-square mile). Stated using scientific notation, these three values are 1.0×10^{15}, 15.7×10^{15}, and 0.020×10^{15}.

In order to compare energy densities for different technologies on a common basis of gigawatt-hours per square mile per year, or GWh/(square mile-year), energy was converted from BTU to Watt-hour, or Wh, using this equation:

$$1 \text{ BTU} = 0.293071 \text{ Wh}$$

Using this conversion and recognizing that a gigawatt (GW) is equivalent to 1 billion Watts (or 1×10^9), the energy density to mine the uranium is 293,000 GWh/

(year-square mile), the energy density for uranium enrichment is 4,601,000 GWh/(year-square mile), and the energy density to make electricity at the uranium plant is 5,900 GWh/(year-square mile).

For a uranium plant energy density of 5,900 GWh/(year-square mile), a footprint of 1.5 square miles can be calculated for a 1,000 MW uranium power plant. The land area needed to supply uranium to this 1,000 MW plant is insignificant, only 0.03 square miles as well as the land area for enrichment, only 0.002 square miles. Thus, the footprint of the nuclear power plant controls the land area needed to produce electricity. The land area for the electric plant is similar to that for a 1,000 MW pulverized coal combustion at 1.0 square mile but much greater than that for a natural gas combined cycle plant at 0.06 square miles. Presumably part of the greater land area for nuclear energy compared to natural gas includes onsite nuclear waste storage as well as land dedicated to an emergency planning zone (EPZ) around the plant.

According to the International Energy Agency,[75] the US electric power consumption per capita for 1 year was 12.8 MWh in 2016. Using the capacity factor of 0.90 shown in Table 4.4 for combined cycle plants, a simple calculation shows that 1 MW will power about 620 homes for one year, so a 1,000 MW plant will power about 600,000 homes!

4.12. Environmental Issues

Of the environmental issues related to nuclear energy, the two most important are nuclear accidents and spent fuel rods. With respect to nuclear accidents, there have been three major ones including Three Mile Island in March 1979, Chernobyl in April 1986, and Fukushima in March 2011.

The Three Mile Island accident was caused by a malfunction in the cooling of the Unit 2 reactor, causing the reactor to be destroyed.[76] As a result of the accident, some radioactive gas was released but did not affect local residents. Significantly, there were no injuries or adverse health effects from the Three Mile Island accident. The plant, which is located near Harrisburg, Pennsylvania, is still in operation.

Far more notable was the Chernobyl accident in 1986, which was caused by a flawed reactor design and poorly trained personnel.[77] During a turbine test, automatic shutdown procedures were disabled and before operators were able to shut down the reactor, it was in an unstable mode. As a result, a dramatic power surge occurred in the reactor leading to a rapid increase in steam and pressure. This caused an explosion, with an estimated 5% of the reactor core being released into the atmosphere. Two people died during the accident and 28 people later died from acute radiation syndrome (ARS). This disaster remains the only one in the nuclear power industry where fatalities occurred due to radiation. Chernobyl is no longer operating and, in fact, has actually become a tourist destination.

The third disaster occurred at the Fukushima Daiichi reactors. A 15-meter (~50 feet) tsunami, caused by an earthquake, disabled power and cooling of three reactors, and the reactor cores melted over a 3-day period.[78] There have been no deaths due to radiation sickness, but 100,000 people were moved from their homes to help prevent this. The reactors are no longer operating and, unfortunately, radiation levels are still so high that it is not possible to remove the melted fuel rods. In fact, some fuel rods melted through the containment vessels and their location is unknown.

"Spent" uranium may no longer be efficient in producing electricity, but the spent fuel rods still contain high-level radioactive waste.[79] This is because during fission to produce electricity, uranium atoms also capture neutrons to form heavier elements, notably plutonium. Elements heavier than uranium, also called transuranic elements, take longer to decay and account for most of the radioactive waste after 1,000 years. This high-level waste is hazardous and produces fatal radiation doses.

For the first 100 years or so, radioactivity is dominated by the fission products, notably Strontium-90 and Cesium-137.[80] After this the radioactivity is controlled by the transuranic compounds: plutonium, americium, neptunium, and curium. To give an example of the longevity of this problem, Plutonium-239 has a half-life of 24,000 years!

In the United States, the long-term storage solution for nuclear waste has not been resolved. And it is expensive, as the cost of a long-term solution is currently $38 billion and rising.[81] That is because the US government has already spent $15 billion on Nevada's Yucca Mountain and another $23 billion is an estimate of the damages the US government will have to pay to nuclear power facilities, who have been paying a fee of $0.001 per kWh since 1983 to pay for the collection of nuclear waste, which was originally to begin in 1998. Instead, nuclear fuel plants continue to store spent fuel on site, where they pay the cost for fuel storage and security. Although the Yucca Mountain project was canceled in 2009, the country of Finland is going forward with the Onkalo repository project. It is estimated to cost about 3.5 billion Euros, and will store nuclear waste in tunnels 1,400 feet underground.[82] Their plan, to happen sometime in the next decade, is to place 3,000 sealed copper canisters each about 17 feet long and containing two tons of plutonium and other products of nuclear fission, into these tunnels. Eventually, there will be as much as 20 miles of tunnels, which will be packed with clay and eventually abandoned. Finish officials say there should be no risk of contamination to future generations.

In addition to these major environmental issues, there are some other minor issues with nuclear power plants.[83] One is related to heat removal and, if the 60% to 70% of the thermal heat rejected from the plant is cooled by use of lakes and rivers, it can negatively affect aquatic life in these water bodies. For example, studies have shown that the heat rejected into water bodies can cause significant population drops in some fish species.

4.13. CO_2 Production and the Cost of Capture and Sequestration

Since a nuclear power plant does not use fossil fuels to generate heat, no CO_2 is directly produced when electricity is made. However, when you consider a life cycle analysis of the plant, some CO_2 is still produced. One paper[84] classifies these other processes as "upstream," "operational," and "downstream." As these parts of the process use fossil fuels, a nuclear power plant still produces some CO_2 emissions.

Upstream processes refer to facility construction, as well as supplying the materials of construction and these are evaluated to determine the CO_2 emissions made over the entire lifetime of the plant. Operational processes include uranium mining, milling, conversion, enrichment, fuel rod fabrication, transportation, and maintenance. Finally, downstream processes include plant decommissioning and waste disposal. For this study,[84] the median CO_2 emissions calculated were 22 g CO_2/kWh for a PWR and 11 g CO_2/kWh for a BWR.

Another paper[85] breaks different parts of the process into percentages. According to the paper, 38% of the emissions are due to mining, milling, conversion, and enrichment. Another 12% comes from operation, 17% through the use of backup generators using fossil fuels during plant downtime, 14% for fuel processing and waste, and 18% for decommissioning. They cite a mean value of 66 gCO_2 /kWh.

Finally, two other papers cite median CO_2 emissions values of 12 g CO_2/kWh[86] and 29 g CO_2/kWh.[87] Taking a rough average of all five values, the CO_2 production for nuclear energy is 30 g CO_2/kWh, or 0.066 lb CO_2/kWh.

Carbon dioxide production is commonly listed in the units of lb CO_2/MM BTU, lb CO_2/kWh, and g CO_2/kWh. To calculate CO_2 production on a basis of lb CO_2/MM BTU (lb CO_2/million BTU), a heat rate of 10,450 BTU/kWh will be used. The simple conversion is shown in this equation:

$$lb\ CO_2\ /\ MM\ BTU = \frac{0.066 * 1\times 10^6}{10{,}450} = 6.3\ lb\ CO_2\ /\ MM\ BTU$$

For a 1,000 MW plant with a capacity factor of 90%, the plant will generate 21,600 MWh/day. As the CO_2 produced on a kWh basis is 0.066 lb CO_2/kWh, this means the life cycle production of CO_2 is equivalent to 1,425,600 lb CO_2/day or about 713 tons CO_2/day. This is only 3% of the CO_2 produced by a pulverized coal combustion plant (22,214 tons CO_2/day) and 10% of the CO_2 produced by a natural gas combined cycle plant (7,500 tons CO_2/day).

As this CO_2 is not directly produced by the power plant, but rather by other processes, including construction, mining, milling, conversion enrichment, fuel rod fabrication, and waste disposal, it is not practical to capture it. Therefore, it is unlikely that the small amount of CO_2 produced as part of the nuclear power plant process will be removed by capture and sequestration (CCS)

4.14. Chapter Summary

A review of worldwide uranium reserves shows an estimate of 6,142,200 metric tonnes for 2017, which at the consumption rate of 62,825 metric tonnes per year gives a calculated value of 98 years remaining for uranium. Low and high consumption growth scenarios could change the remaining years to 116 and 68 years, respectively. The reserves are based on a uranium recovery cost of $130/kg of uranium or less. If the recovery cost is raised to $260/kg, the worldwide reserves increase to 7,988,600 tonnes. Moreover, if the recovery cost is raised to as high as $1,000/kg, there is substantial uranium in the ocean that can be tapped, estimated to be as much as 4.5 billion tonnes.

For proven reserves, ten countries including the Australia, Kazakhstan, Canada, Russia, Namibia, South Africa, China, Niger, Brazil, and Uzbekistan control 88% of the worldwide uranium reserves. Two countries, Australia and Kazakhstan, dominate worldwide reserves with 30% and 14%, respectively of the world's reserves.

The "Well-to-Wheels" economic analysis gave an average levelized cost of electricity (LCOE) of $117.59 /MWh to generate electricity from uranium, increasing to $153.59/MWh when transmission, distribution, and waste disposal are included. Direct waste disposal only adds $1/MWh.

For a 1,000 MW nuclear power plant with a capacity factor of 90%, a life cycle analysis for the plant shows that, on a kWh basis, the plant will produce 30 g CO_2/kWh or 0.066 lb CO_2/kWh. Although no CO_2 is made during electricity generation, upstream, operational, and downstream processes, including construction, mining, milling, conversion enrichment, fuel rod fabrication, and waste disposal, do produce CO_2. Capturing this CO_2 is not practical.

Based on a 12-month average of 2018 uranium spot price data, the cost of uranium, on the basis of mass is $1,013/kg or $459/lb. This price assumes a U-235 enrichment level of 4.3%. On the basis of energy, the cost is $0.30/MM BTU, very inexpensive in comparison to crude oil at $11.20/MM BTU, natural gas at $3.17/MM BTU, and coal at $2.06/MM BTU.

For a uranium-power nuclear power plant, average capacity factors of 0.90 were reported.

The overall energy balance for the nuclear power plant was 30.2%, meaning that about 30% of the energy originally in the uranium ends up as energy at the distribution point as electricity. Most of this energy is lost during electricity generation, as the thermal efficiency for a nuclear power plant is only 33%.

In terms of experience and maturity, uranium mining has been around since 1905 and the first nuclear power plant, a PWR plant, was built in 1957. The first BWR plant was built in 1960. Thus, while not as old as coal and petroleum, the industry of using uranium to make electricity is very mature.

The infrastructure needed for uranium nuclear plants includes transportation, generation, and electricity transmission. To transport finished uranium fuel rods, there are 95,264 miles of track and 46,876 miles of interstate highway. To make the electricity, there were 99 operating reactors in 2016, located on 60 plant sites. In 2017, these nuclear power plants had a total capacity of 105 GW with a net generation of about 805,000 GWh, accounting for 20% of the total electricity generated in the United States. Once made, this electricity can be transmitted around the United

States using the more than 180,000 miles of high-voltage transmission lines in the electric grid.

Assuming a nuclear power plant with a size of 1,000 MW, the footprint is 1.5 square miles. The land area needed to support this plant is dominated by the electric power plant, as the land area for enrichment and uranium mining are insignificant in comparison.

Finally, the main environmental issues for nuclear power plants are major disasters, such as occurred at Three Mile Island in March 1979, Chernobyl in April 1986, and Fukushima in March 2011. In addition, long-term waste disposal is still a very important unresolved issue in the United States.

References

1. "Plutonium as an Energy Source," Arjun Makhijani, Institute for Energy and Environmental Research, February 1997, http://www.ieer.org/ensec/no-1/puuse.html
2. "Fast Breeder Reactors," A.E. Waltar and A.B. Reynolds, London: Pergamon Press, 1981, p. 853.
3. "The Elements," C.R. Hammond, Handbook of Chemistry and Physics, 81st Edition, CRC Press, 2000,
4. "Thorium," World Nuclear Organization, September 2015, http://www.world-nuclear.org/information-library/current-and-future-generation/thorium.aspx
5. "Thorium: The Wonder Fuel That Wasn't," Robert Alvarez, Bulletin of the Atomic Scientists, May 1, 2014, http://thebulletin.org/thorium-wonder-fuel-wasnt7156
6. "Thorium vs. Uranium," Brian Schimmoller, Power Engineering, July 17, 2014, http://www.power-eng.com/articles/print/volume-118/issue-7/departments/nuclear-reactions/thorium-vs-uranium.html
7. "What Is Uranium? How Does It Work?," World Nuclear Organization, March 2014, http://www.world-nuclear.org/information-library/nuclear-fuel-cycle/introduction/what-is-uranium-how-does-it-work.aspx
8. "Supply of Uranium," World Nuclear Organization, July 2016, http://www.world-nuclear.org/information-library/nuclear-fuel-cycle/uranium-resources/supply-of-uranium.aspx
9. "Uranium 2018: Resources, Production and Demand," A Joint Report by the Nuclear Energy Agency and the International Atomic Energy Agency, NEA No. 7413, 2018, https://www.oecd-nea.org/ndd/pubs/2016/7301-uranium-2016.pdf
10. "Review of Cost Estimates for Uranium Recovery from Seawater," Harry Lindner and Erich Schneider, Energy Economics 49, 9–22, May 2015.
11. "Chemische Untersuchung des Uranits, einer neuentdeckten metallische Substanz," M.H. Klaproth, Chemische Annalen für die Freunde der Naturlehre (2), 387–403, 1789.
12. "Uranium's Scientific History 1789–1939," B. Goldschmidt, 14th International Symposium, London, September 1989, http://ist-socrates.berkeley.edu/~rochlin/ushist.html
13. "The Pentagon said it wouldn't use depleted uranium rounds against ISIS. Months later, it did — thousands of times," The Washington Post, Thomas Gibbons-Neff, February 16, 2017, https://www.washingtonpost.com/news/checkpoint/wp/2017/02/16/the-pentagon-said-it-wouldnt-use-depleted-uranium-rounds-against-isis-months-later-it-did-5265-times/
14. "Uranium and Its Applications," Zeus Resources Ltd., 2020, http://www.zeusresources.com/uranium-and-its-applications.asp

15. "Nuclear Power Plants, World-Wide," European Nuclear Society, 2019, https://www.euro-nuclear.org/info/encyclopedia/n/nuclear-power-plant-world-wide.htm
16. "How a Nuclear Reactor Makes Electricity," World Nuclear Association, 2020, http://world-nuclear.org/nuclear-basics/how-does-a-nuclear-reactor-make-electricity.aspx
17. "4.7 Nuclear Fission and Fusion, and Neutron Interactions," National Physics Laboratory, Kaye & Laby Tables of Physical & Chemical Constants, July 30, 2015, http://www.kayelaby.npl.co.uk/atomic_and_nuclear_physics/4_7/4_7_1.html
18. "Reactor Fuel Isotopics and Code Validation for Nuclear Applications," ORNL/TM-2014/464, M.W. Francis et al., Oakridge National Laboratory, September 2014, http://web.ornl.gov/sci/nsed/rnsd/staff/Publications/WeberPubs/PTS%2052057_ReactorFuelIsotopics_Approved%20for%20Release.pdf
19. "Nuclear Fuel Processes," Nuclear Energy Institute, 2020, http://www.nei.org/Knowledge-Center/Nuclear-Fuel-Processes
20. "Uranium Enrichment Methods," Stanko Manojlović, Ljubljana, Slovenia, May 13, 2010, http://mafija.fmf.uni-lj.si/seminar/files/2009_2010/SeminarUraniumEnrichment.pdf
21. "Uranium Enrichment: Just Plain Facts to Fuel an Informed Debate on Nuclear Proliferation and Nuclear Power," A. Makhijani et al., Institute for Energy and Environmental Research, October 15, 2004, http://www.helencaldicott.com/uranium.pdf
22. "Uranium Enrichment," World Nuclear Association, November 2016, http://www.world-nuclear.org/information-library/nuclear-fuel-cycle/conversion-enrichment-and-fabrication/uranium-enrichment.aspx
23. "Uranium and Depleted Uranium," World Nuclear Association, September 2016, http://www.world-nuclear.org/information-library/nuclear-fuel-cycle/uranium-resources/uranium-and-depleted-uranium.aspx
24. "Nuclear Fuel Cycle," World Nuclear Association, June 2016, http://www.world-nuclear.org/information-library/nuclear-fuel-cycle/introduction/nuclear-fuel-cycle-overview.aspx
25. "What Are Nuclear Wastes and How Are They Managed?," World Nuclear Association, 2020, http://www.world-nuclear.org/nuclear-basics/what-are-nuclear-wastes.aspx
26. "At-Reactor Storage of Used Nuclear Fuel: Safe Temporary Storage in Water-Filled Pools," Nuclear Energy Institute, 2020, https://www.nei.org/Issues-Policy/Used-Nuclear-Fuel-Management/At-Reactor-Storage-of-Used-Nuclear-Fuel
27. "Nuclear Waste: Do We Know What to Do with It?," Amanda Onion, May 17, 2010, http://abcnews.go.com/Technology/story?id=98555&page=1
28. "Cost and Performance Characteristics of New Generating Technologies," Annual Energy Outlook 2019, Table 2, Energy Information Administration, January 2019, https://www.eia.gov/outlooks/aeo/
29. "Lazard's Levelized Cost of Energy Analysis—Version 12.0," November 2018, https://www.lazard.com/media/450784/lazards-levelized-cost-of-energy-version-120-vfinal.pdf
30. "Updated Capital Cost Estimates for Utility Scale Electricity Generating Plants," Energy Information Administration, April 2013.
31. "Costs of Low-Carbon Generation Technologies," Mott MacDonald, Committee on Climate Change, May 2011, https://www.theccc.org.uk/archive/aws/Renewables%20Review/MML%20final%20report%20for%20CCC%209%20may%202011.pdf
32. "Annual Energy Outlook 2015 with Projections to 2040," DOE/EIA-0383, Table A8, April 2015.
33. "Uranium Price," Cameco, 2018, https://www.cameco.com/invest/markets/uranium-price
34. "The Economics of Nuclear Power," World Nuclear Association, July 2016, http://www.world-nuclear.org/information-library/economic-aspects/economics-of-nuclear-power.aspx

35. "The Economics of the Back End of the Nuclear Fuel Cycle," Organisation for Economic Co-Operation and Develop (OECD), NEA No. 7061, October 23, 2013, https://www.oecd-nea.org/ndd/pubs/2013/7061-ebenfc.pdf
36. "Nuclear Reprocessing: Technological, Economic, and Social Problems," Final paper for BPRO 29000: Energy and Energy Policy, R. Stephen Berry and George S. Tolley, October 21, 2015, http://franke.uchicago.edu/bigproblems/BPRO29000-2015/Team25-NuclearReprocessing.pdf
37. "Electric Power Annual 2017," Energy Information Administration, October 2018, Revised December 2018, https://www.eia.gov/electricity/annual/pdf/epa.pdf
38. "Nuclear Power: The Energy Balance, Chapter 2, From Ore to Electricity," J.W.S. van Leeuwen and P. Smith, 6th revision, August 5, 2005, http://jayhanson.us/_Energy/NuclearPower.pdf
39. "Energy Balances and CO2 Implications," World Nuclear Organization, March 2014, http://www.world-nuclear.org/information-library/energy-and-the-environment/energy-balances-and-co2-implications.aspx
40. "A Review of Transmission Losses in Planning Studies," Lana Wong, California Energy Commission, Figure ES-1, August 2011.
41. "Monthly Energy Review," Energy Information Administration, March 2019, https://www.eia.gov/totalenergy/data/monthly/pdf/mer.pdf
42. "Uranium Ore Deposits," Franz J. Dahlkamp, Springer Berlin Heidelberg, December 1, 2010.
43. "Fame of the Ore Mountains Mining," Wirtschafts Forderung Erzgebirge, 2013, http://www.muzeum-most.cz/montanregion/files/Pr-vodce.EN.pdf
44. "Abandoned Uranium Mines and the Navajo Nation: Section 1: Mining History and Mine Site Information," Environmental Protection Agency, August, 2007, https://www.epa.gov/sites/production/files/2017-01/documents/navajo_nation_aum_screening_assess_report_atlas_geospatial_data-2007-08.pdf
45. "Uranium Mining and Activities, Past and Present," Arizona Game and Fish Department and Commission, May 2007, http://www.grandcanyontrust.org/sites/default/files/gc_agfUraniumUpdate.pdf
46. "Nuclear Power," Paul Breeze, Academic Press, October 14, 2016, https://books.google.com/books?id= PwlmDAAAQBAJ&pg=PA35&lpg=PA35 &dq=first+commercial+ boiling+water+reactor+in+usa& source=bl&ots=uEWL5-mhnv&sig=knQGX_3nODCd r8KDm_A9tVKhudU&hl=en&sa=X&ved= 0ahUKEwjmtI3VlIvRAhVJyVQKHfkeBfo-Q6AEIczAS#v= onepage&q=first%20commercial %20boiling%20water%20reactor%20in %20usa&f=false
47. "United States Electricity Industry Primer," Office of Electricity Delivery and Energy Reliability U.S. Department of Energy DOE/OE-0017, July 2015, https://www.energy.gov/sites/prod/files/2015/12/f28/united-states-electricity-industry-primer.pdf
48. "US Uranium Mining and Exploration: US Nuclear Fuel Cycle Appendix 1," November 2016, http://www.world-nuclear.org/information-library/country-profiles/countries-t-z/appendices/us-nuclear-fuel-cycle-appendix-1-us-uranium-mining.aspx
49. "Spent Nuclear Fuel: Table 2. Nuclear Power Plant Data," Energy Information Administration, February 3, 2016, https://www.eia.gov/nuclear/spent_fuel/ussnftab2.php
50. "Count of Electric Power Industry Power Plants, by Sector, by Predominant Energy Sources within Plant, 2007 through 2017," Table 4.1, Energy Information Administration, http://www.eia.gov/electricity/annual/html/epa_04_01.html; "Existing Capacity by Energy Source, 2017 (Megawatts)," Table 4.3, Energy Information Administration, https://www.eia.gov/electricity/annual/html/epa_04_03.html

51. "Net Generation by Energy Source: Total (All Sectors) 2007–February 2017," Table 1.1, Energy Information Administration, https://www.eia.gov/electricity/monthly/epm_table_grapher.cfm?t=epmt_1_01
52. "Uranium Mining Overview," World Nuclear Association, February 2019, http://www.world-nuclear.org/information-library/nuclear-fuel-cycle/mining-of-uranium/uranium-mining-overview.aspx
53. "Conventional Mining and Milling of Uranium Ore," Uranium Producers of America, 2014, http://www.theupa.org/uranium_technology/conventional_mining/
54. "Domestic Uranium Production Report—Annual," 2017 data, Energy Information Administration, May 22, 2018, https://www.eia.gov/uranium/production/annual/
55. "Uranium Mining—Wyoming's Main Uranium Players," Dustin Bleizeffer, Casper Star Tribune, September 30, 2010, http://trib.com/business/energy/wyoming-s-main-uranium-players/article_b747f7e1-1253-5717-8aa7-cced984a5589.html
56. "Land Use and Energy," Office of Scientific Information, K.E. Robeck et al., Report Number ANL/AA-19, Argonne National Lab., IL (USA), July 1, 1980, https://www.osti.gov/scitech/servlets/purl/6300166
57. "Uranium One Looks to Establish Big Presence in Converse Co.," Douglas Budget & Glenrock Independent, Dave Robatcek, March 6–7, 2013, http://www.douglas-budget.com/content/tncms/assets/v3/eedition/f/a1/fa14cfbb-9930-5cc1-a401-27613c3c0a06/51378353193b9.pdf.pdf
58. "The Smith Ranch Project: A 1990s in situ Uranium Mine," M.D. Freeman and D.E. Stover, The Uranium Institute, 24th Annual International Symposium 1999, http://mdcampbell.com/SmithRanchProjectUraniumMine.pdf
59. "Urenco Spins Uranium into Fuel at High-Tech Desert Facility," June 8, 2012, Kevin Robinson-Avila, Albuquerque Journal, http://www.bizjournals.com/albuquerque/news/2012/06/08/urenco-spins-uranium-into-fuel-at.html
60. "Desert Facility Spins Uranium," May 11, 2015, Kevin Robinson-Avila, Albuquerque Journal, https://www.abqjournal.com/582632/spins.html
61. "Uranium Enrichment," World Nuclear Association, November 2016, http://www.world-nuclear.org/information-library/nuclear-fuel-cycle/conversion-enrichment-and-fabrication/uranium-enrichment.aspx
62. "Nuclear Fuel Fabrication," February 2017, World Nuclear Association, http://www.world-nuclear.org/information-library/nuclear-fuel-cycle/conversion-enrichment-and-fabrication/fuel-fabrication.aspx
63. "Richland's Areva Makes Nuclear Fuel," May 31, 2015, Tri-City Herald, http://www.tri-cityherald.com/news/local/article32226900.html
64. "Columbia Plant Overview," Marc Rosser, June 14, 2012, https://www.admin.sc.gov/files/nac/Westinghouse_Plant_Overview_061412_SCNucAdvComm.pdf
65. "Economic Benefits of Grand Gulf Nuclear Station," December 2006, http://www.entergy-nuclear.com/content/resource_library/ESP/Economic_Benefits_Grand_Gulf.pdf
66. "Palo Verde Nuclear Generating Station," November 1, 2015, Robert Peltier, Electric Power, http://www.powermag.com/palo-verde-nuclear-generating-station-wintersburg-arizona/
67. "The South Texas Project Electric Generating Station (Colorado River Basin)," Texas Water Development Board, 2020, http://www.twdb.texas.gov/surfacewater/rivers/reservoirs/stpegs/index.asp
68. "Ohio Nuclear Profile 2010," U.S. Energy Information Administration, 2010, https://www.eia.gov/nuclear/state/ohio/
69. "Seabrook Factsheet," NextEra Energy, 2008, https://www.nexteraenergyresources.com/pdf_redesign/seabrook.pdf
70. "The Susquehanna: A Division of Talen Energy," 2012, https://susquehannanuclear.com/

71. "Watts Bar Nuclear Plant," Tennessee Valley Authority, 2020, https://www.tva.gov/Energy/Our-Power-System/Nuclear/Watts-Bar-Nuclear-Plant
72. "Missouri Nuclear Profile 2010," Energy Information Administration, April 26, 2012, https://www.eia.gov/nuclear/state/missouri/
73. "Kansas Nuclear Profile 2010," Energy Information Administration, April 26, 2012, https://www.eia.gov/nuclear/state/kansas/
74. "Peach Bottom Atomic Power Station Fact Sheet," 2010, https://www.pema.pa.gov/Preparedness/Power-Plant-Safety/Resources/Documents/Peach-Bottom-Atomic-Power-Station-Fact-Sheet.pdf
75. "Electric Power Consumption (kWh per capita)," IEA Statistics for 2016, http://energyatlas.iea.org/#!/tellmap/-1118783123/1
76. "Three Mile Island Accident," March 2001, World Nuclear Association, http://www.world-nuclear.org/information-library/safety-and-security/safety-of-plants/three-mile-island-accident.aspx
77. "Chernobyl Accident 1986," World Nuclear Association, November 2016, http://www.world-nuclear.org/information-library/safety-and-security/safety-of-plants/chernobyl-accident.aspx
78. "Fukushima Accident," January 2017, World Nuclear Association, http://www.world-nuclear.org/information-library/safety-and-security/safety-of-plants/fukushima-accident.aspx
79. "Backgrounder on Radioactive Waste," US Nuclear Regulatory Commission, July 23, 2019, https://www.nrc.gov/reading-rm/doc-collections/fact-sheets/radwaste.html
80. "Spent Fuel from Nuclear Power Reactors," H. Feiveson et al., June 2011, International Panel of Fissile Materials, http://fissilematerials.org/library/ipfm-spent-fuel-overview-june-2011.pdf
81. "The $38 billion nuclear waste fiasco," by Darius Dixon, November 30, 2013, http://www.politico.com/story/2013/11/nuclear-waste-fiasco-100450
82. "On Nuclear Waste, Finland Shows U.S. How It Can Be Done," New York Times, Henry Fountain, June 9, 2017, https://www.nytimes.com/2017/06/09/science/nuclear-reactor-waste-finland.html?rref=collection%2Ftimestopic%2FNuclear%20Waste&action= click&contentCollection=timestopics&region=stream&module=stream_unit&version=latest&contentPlacement=2&pgtype=collection
83. "Impact of Nuclear Power Plants," Misam Jaffer, Standford University, March 26, 2011, http://large.stanford.edu/courses/2011/ph241/jaffer2/
84. "Life Cycle Greenhouse Gas Emissions of Nuclear Electricity Generation," Ethan S. Warner and Garvin A. Heath, Journal of Industrial Ecology16, S73–S92, May 5, 2012, https://papers.ssrn.com/sol3/papers.cfm?abstract_id=2051332
85. "Nuclear Energy: Assessing the Emissions," Kurt Kleiner, Nature, September 24, 2008, http://www.nature.com/climate/2008/0810/full/climate.2008.99.html
86. "Technology-Specific Cost and Performance Parameters," Steffen Schlömer, Annex III, 2014, https://www.ipcc.ch/pdf/assessment-report/ar5/wg3/ipcc_wg3_ar5_annex-iii.pdf#page=5
87. "Greenhouse Gas Emissions Avoided through Use of Nuclear Energy," World Nuclear Association, 2018, http://www.world-nuclear.org/nuclear-basics/greenhouse-gas-emissions-avoided.aspx
88. "It's the First New U.S. Nuclear Reactor in Decades. And Climate Change Has Made That a Very Big Deal," Chris Mooney, Washington Post, June 17, 2016, https://www.washingtonpost.com/news/energy-environment/wp/2016/06/17/the-u-s-is-powering-up-its-first-new-nuclear-reactor-in-decades/?utm_term=.25d37716a29d
89. "Plans for New Reactors Worldwide," World Nuclear Association, April 2019, http://www.world-nuclear.org/information-library/current-and-future-generation/plans-for-new-reactors-worldwide.aspx

5
Hydroelectric
A Renewable Energy Type

5.1. Foreword

Many years before the development of hydroelectric power, the movement of water was used to do work. Water moves in brooks, streams, and rivers providing kinetic energy, sometimes overpowering. Anyone that has seen roads wash away after a hard rain or has been thrown from a canoe or raft into white water can attest to the power of water. As far back as the 3rd century BC, there is evidence to suggest that the Nanyang area in China used water-powered bellows in iron production.[1] The Chinese may also have been the first to use a water wheel to grind grain and irrigate fields.[2] When the Greeks conquered Persia around 400 BC, the Persians were using water wheels, and the Greeks copied this to grind wheat and cut wood. Of course, the Romans are famous for the many aqueducts built to provide water for drinking, bathing, and watering crops as far back as the first century BC.[3]

Harnessing the power of water was done using waterwheels, allowing the transfer of the water's kinetic energy to the mechanical energy of grinding wheat, cutting wood, polishing stone, or many other uses. The use of waterwheels in Europe became so prevalent that in England alone, "The Doomsday Survey" recorded 5,624 waterwheels in use in 1086![2] There were also many mills to grind wheat into flour in the United States, and the ruins of what was once the world's largest flour mill can be seen in the Mill City Museum in Minneapolis. Starting in the 1870s, Gold Medal flour was made there by harnessing water power from St. Anthony Falls on the Mississippi River.

The concept of using water power, or hydropower, to generate electricity began in the mid-1700s, when Bernard Forest de Bélidor (a French military engineer) published "L'architecture hydraulique," describing vertical and horizontal hydraulic machines. Combining this knowledge with the advent of the generator, a water turbine was used in 1880 to provide arc lighting to a theater and storefront in Grand Rapids, Michigan.[4] Similarly, in 1881 a water-powered generator was used to provide street lighting in Niagara Falls, New York. Shortly after, the world's first hydroelectric power plant began operation in 1882 in Appleton, Wisconsin.[5]

As discussed in more detail later, the production of electricity from hydropower is similar to coal- and natural gas-fired power plants in that a source of energy is used to rotate a turbine, which then turns the electric generator shaft and produces electricity. In the case of coal and natural gas, steam is produced, and synthesis gas if it

The Changing Energy Mix. Paul F. Meier, Oxford University Press (2020). © Oxford University Press.
DOI: 10.1093/oso/9780190098391.001.0001.

is a combined cycle plant. In the case of hydroelectric power, the moving and perhaps falling water rotates the turbine to generate electricity. In this chapter this type of power generation is referred to as "conventional hydroelectric."

However, not all hydroelectric power in the United States, or the world for that matter, comes from conventional hydroelectric. Some of the power is generated from "pumped storage hydroelectric" facilities. For this type of power generation, water is pumped from a lower to an upper reservoir in order to store energy in the form of gravitational potential energy to be used later. In this respect, the system is operating as a battery to store energy for future use. During periods of low electrical demand, usually nights or weekends, the upper water reservoir is refilled using lower-cost electricity from the power grid to pump water from the lower resevoir.[6] Then, when demand is high, or at "peak" demand, this water is released through turbines in the same way as a conventional hydroelectric plant to make electricity. Pumped storage operates at a net energy loss, but it makes economic sense because of the higher value per kWh during peak usage compared to the cost to pump the water during periods of low value kWh. In the United States and Europe, pumped storage has been used since the 1920s, and there are forty pumped storage facilities operating in the United States.[6]

Looking at 2017 Energy Information Administration (EIA) data,[7] there were 1,498 hydroelectric facilities in the United States, of which forty are pumped storage. The total capacity was 101,238 MW, of which 79,595 MW was conventional hydroelectric and 21,643 MW was pumped storage. There are some very large dams, such as Grand Coulee in Washington with a capacity of 6,809 MW and Robert Moses Niagara in New York with a capacity of 2,675 MW. In fact, the eleven largest dams in the United States account for a capacity of 27,219 MW, 34% of the total capacity but less than 1% of the plants. Normally, conventional hydroelectric dams are typically small, with an average plant capacity of 55 MW.

Most of these hydroelectric plants generate their electricity from a dam with a reservoir behind it. The general approach is to build a dam on a river with water stored behind the dam in a reservoir, and then release the water with a significant drop in elevation. When water is released it moves a turbine which then turns the generator, thereby producing electricity.

So it sounds simple. Build a dam, use water flow to move a generator and make electricity, then send it to the grid to be used in people's homes or business. No fossil fuel use, no CO_2 emissions, no nuclear waste, and completely renewable. But wait, while hydroelectric is very "green" from an energy standpoint, it is certainly not without controversy. Dams have been known to displace people from their homes, even affecting entire cities. Sometimes, these dams affect historical sites and even ancient gravesites. In addition, dams can cause massive environmental damage, affecting fish populations and migration, affecting plants and animals living in or next to the river, and even causing landslides and earthquakes. Oh well, it seems no energy type is completely free from some negative impacts.

5.2. Proven Reserves

Unlike petroleum crude oil, natural gas, coal, and nuclear energy, the amount of proven reserves for hydroelectric power is subjective and difficult to assess. Certainly, there are existing dams not generating hydroelectric power and various waterways that have the potential of creating hydroelectric power, either by building a dam or using existing water flow. While having this potential is essential for new development, there are also technical, environmental, and economic considerations for the project to proceed.

A 2012 study by the Department of Energy (DOE), performed by Oak Ridge National Laboratory (ORNL), reports that in contrast to the roughly 2,500 dams that provide 79 gigawatts (GW) of conventional and 22 GW of pumped storage hydropower, the United States has more than 80,000 nonpowered dams (NPDs) that are not currently producing electricity.[8] These dams were built to provide things such as water supplies and inland navigable water. Therefore if these dams were further exploited to generate electricity, the cost of construction and possible environmental impacts of the dam and its construction have already happened, and construction time would also be reduced. According to this study, 54,391 dams were analyzed out of the more than 80,000 NPDs in the United States. The other dams were excluded because of unresolved data issues. The conclusion of the study was that these NPDs have the potential to add as much as 12 GW, or 12,000 MW, of new capacity. As the current US hydroelectric capacity is about 101 GW, this means it would increase hydroelectric power by 12%.

Another study, also performed by ORNL and sponsored by DOE, evaluated the potential of new stream-reach development (NSD) from the more than 3 million US streams.[9] This study indicates that the undeveloped NSD capacity is 84.7 GW or 460 TWh/year assuming a capacity factor of 62%. A TWh, or terawatt-hour, is 1×10^{12} watt-hours. Some of these waterways are protected by federal legislation that limits development in national parks, scenic rivers, and wilderness areas. If these areas are excluded from the study, the NSD capacity is about 65.5 GW, or 356 TWh/year assuming the same capacity factor. As the current US capacity is about 101 GW, this means complete development of NSDs would add 84% more hydroelectric capacity, while adding the more realistic development which does not affect federally protected land would add 65%. Although capacity factors are discussed later, it is worth noting here that the NSD streams had factors ranging from 53% to 71%, much higher than conventional hydroelectric power with capacity factors around 30%. The reason for this is that the NSDs are assumed to use "Run-of-River" power, and would not be subjected to the various flow control issues used in hydroelectric dams.

In summary, taking the value of 12 GW for NPDs and using the more conservative value of 65.5 GW for NSDs, the United States has the potential of adding 77.5 GW of new hydroelectric power. Using the same 62% capacity factor suggests this 77.5 GW of power could produce 420 TWh/year!

A 2016 report published by the World Energy Council[10] states that while there is no clear consensus for new hydropower development throughout the world, estimates indicate there are about 10,000 TWh/year of worldwide hydroelectric potential. Using data from this report, the top ten countries in terms of undeveloped hydroelectric power are shown in Table 5.1. It is important to note that these numbers, and hence rankings, are based on reporting from individual countries and do not necessarily indicate whether the development of these resources is economically possible or even feasible under current conditions and political leadership. Based on a 10,000 TWh/year estimate for worldwide potential, the top ten countries control about 59% of the unutilized potential.

Another study, published in 2017, performs a detailed evaluation for hydroelectric potential in different countries, based on the slope and discharge of each river in the world.[11] The theoretical potential is 52,000 TWh/year, cumulated from 11.8 million locations. The top ten countries are also shown in Table 5.1. Six of the countries are common to both lists, although the ordering is different. These top ten countries account for 49% of the worldwide potential.

Table 5.1 Top Ten Countries by Unutilized Hydroelectric Potential, 2015 Data[10–12]

Country	Undeveloped Potential (TWh/year)[10]	Country	Undeveloped Potential (TWh/year)[11]	Country	Exploitable, TWh/yr[12]
World	10,000	World	52,000	World	16,000
Russia	1,510	China	7,168	Russia	2,401
China	1,014	Brazil	3,630	China	2,329
Canada	805	Russia	3,503	Canada	1,610
India	540	Canada	3,064	Democratic Republic of the Congo	1,064
Brazil	436	U.S.	2,564	U.S.	1,009
Indonesia	388	Colombia	1,641	Congo	579
Peru	369	Peru	1,145	Colombia	518
Democratic Republic of the Congo	307	Papua New Guinea	1,087	Indonesia	477
Tajikistan	299	Bolivia	816	India	387
United States	279	Nepal	789	Papua New Guinea	355
10 Country Total	5,947	Total	25,407	Total	10,729

Yet another study, published in 2015, discusses four categories of worldwide potential including (1) gross potential, calculated from water runoff with elevation data; (2) technical potential, estimated from monthly stream flow and elevation differences and consideration of design capacity; (3) economic potential, which takes into account the generation cost; and (4) exploitable potential, derived from the economic potential but adjusted by environmental restrictions.[12] The worldwide (1) gross potential is 128,000 TWh/year and for (4) the exploitable potential is 16,000 TWh/year. The exploitable potential for the top ten countries is also given in Table 5.1, and these ten countries account for 67% of the worldwide potential. Once again, some countries are common to the other lists while some countries are new.

Ignoring the gross potential of 128,000 TWh/year in the previous paragraph, the worldwide totals shown in Table 5.1 range from 10,000 to 52,000 TWh/year. According to the International Hydropower Association,[13] in 2017 the total worldwide generation of electricity was about 25,000 TWh/year, so even if we use the most conservative value from Table 5.1 of 10,000 TWh/year, this hydropower potential is equivalent to 40% of the total worldwide generation of electricity from all energy types.

Focusing on the United States, the ORNL study discussed earlier[9] showed a total US potential of 420 TWh/year, while the three studies in Table 5.1 show 279,[10] 2,564,[11] and 1,009.[12] In 2017, the total US electricity generation, from all energy types, was 4,058 TWh/year, so even if we use the lowest estimate of 279 TWh/year for potential hydropower, this represents 7% of all electricity generation.

A 2018 hydropower status report[13] can be used to rank the top ten countries based on current capacity as well as electricity generated. Table 5.2 shows these data, and the rankings are based on capacity instead of generation, as capacity factors may vary from year to year, and from country to country. Pumped storage capacity is also shown in the table.

These top ten countries control 68% of the worldwide hydropower capacity, and China alone controls 27%, with the United States a distant 8%. Based on electricity generation, the top ten countries accounted for 71%, with China generating 29% of the worldwide total. Interestingly, the capacity factors vary widely, and Brazil and Canada produced more electricity than the United States due to higher capacity factors. However, these capacity factors are based on the combined conventional and pumped storage capacities. Since pumped storage plants are a net consumer of electricity, the US average capacity factor suffers relative to Brazil and Canada, which have very little pumped storage capacity. Likewise Japan, which has more pumped storage capacity than conventional has a low average capacity factor.

The same 2018 hydropower status report[13] can also be used to rank the top ten countries based on pumped storage current capacity. Table 5.3 shows these data. Comparing Tables 5.2 and 5.3, it is interesting that some countries are leaders in pumped storage, even though their worldwide ranking on total hydropower capacity is lower. Notable, there are five countries from Europe, and two others could be

Table 5.2 Top Ten Countries by Capacity, 2017 Data[13]

	Total Hydropower Capacity Including Pumped Storage	Pumped Storage	Generation	Capacity Factor
	MW	MW	TWh	%
World	1,266,955	153,041	4,185.0	37.7
China	341,190	28,490	1,194.5	39.9
United States	102,867	22,809	322.4	35.8
Brazil	100,273	30	401.1	45.6
Canada	80,985	177	403.4	56.8
Japan	49,905	27,637	95.2	21.8
India	49,382	4,786	135.5	31.3
Russia	48,450	1,385	178.9	42.1
Norway	31,837	1,392	143.0	51.3
Turkey	27,273	—	59.2	24.7
France	25,517	6,985	53.2	23.8

Table 5.3 Top Ten Countries by Pumped Storage Capacity, 2017 Data[13]

	Pumped Storage Capacity, MW
World	153,041
China	28,490
Japan	27,637
United States	22,809
Italy	7,555
France	6,985
Germany	6,806
Austria	5,212
India	4,786
South Korea	4,700
Spain	3,329

Table 5.4 Top Ten Countries by Consumption, 2015 Data[10]

Country	Net Hydropower Generation (TWh)	Consumption of Hydroelectricity (TWh)
World	3,969	3,946
China	1,126	1,126
United States	250	254
Brazil	382	361
Canada	376	383
India	129	124
Russia	160	170
Japan	91	97
Norway	139	138
Venezuela	80	76
Sweden	74	75
Total	2,807	2,803

included if we extended the table a little further. The UK capacity for pumped storage is 2,744 MW, and for Portugal it is 2,613 MW.

Altogether, these ten countries account for 77% of the worldwide pumped storage capacity, with China, Japan, and the United States accounting for 19%, 18%, and 15%, respectively. Even more interesting is that Japan has more pumped storage capacity than conventional hydropower capacity, 55% of their total. Japan is a world leader in using pumped storage to store power from nuclear plants during low demand as well as balancing the variable generation of renewable energy types.

Although 2017 data were not available for consumption, 2015 data comparing generation and consumption are shown in Table 5.4.[10] Not surprisingly, consumption and generation match closely, as it is not likely there is significant transfer of hydroelectric electricity between countries.

5.3. Overview of Technology

Conceptually, the generation of electricity from hydropower is simple. The kinetic motion and energy of moving water turns a turbine which, in turn, turns the generator rotor to generate electricity.

Hydroelectric plants can be generally classified by three schemes including (1) run-of-river, (2) storage dam, and (3) pumped storage.[15] A run-of-river scheme generates electricity by using a flowing river, taking a portion of the river to generate electricity.

This is done where rivers have a flow significant enough to justify construction of the hydroelectric plant, and may not require the construction of a dam. When a dam is constructed, it is a small dam used to create the necessary water flow to the penstocks, gates that regulate the flow of water. Because the dam is small, or perhaps not even present, run-of-river schemes can be strongly affected by seasonal variations in the rain and snow that feed the river.

A storage dam scheme uses water impoundment upstream of a dam to make a reservoir to store water. This has at least two advantages. By building a dam, a large vertical drop in water elevation can be created over a much shorter distance than the original flow of the river. We discuss shortly how the dam height relates to the amount of electricity generated. The dam also stores water behind it in the reservoir, thus having much greater control over water flow than the run-of-river scheme and mitigating possible seasonable variations in water flow upstream of the dam.

The pumped storage scheme pumps water from a lower reservoir or river to a higher reservoir during period of low electricity need, usually at night, and then uses the water during periods of peak demand.

Looking at the storage dam scheme in more detail, the basic idea is to build a dam on a large river that has a large elevation drop over short distances. Once the dam is built, gravity causes the water to flow through the penstocks inside the dam, such that a turbine is turned by the flowing water. A shaft connects the turbine to the generator, thus producing electricity. A good visual way to understand how a dam is constructed can be seen in the PBS American Experience show "Grand Coulee Dam,"[16] first aired in April 2017. To build the dam, water in the river is diverted through a tunnel, created by drilling holes and using explosives to create the tunnel. Diverting the river occurs in the summer when river levels are typically the lowest during the year. Next, a small dam, or cofferdam, is built upstream to force the river through the diversion tunnel. After this, pumps are used to remove any water downstream of the cofferdam before the dam is constructed. This area is cleared of rock and rubble, the foundation is laid, and the dam is built. While the dam is built, the things needed to generate electricity are also included, such as penstocks, turbines, and generators. A general schematic of the finished dam is shown in Figure 5.1.[15] After dam construction is finished, the diversion tunnel is closed, which allows the reservoir above the dam to fill.

When there is too much water, the finished dam must also have the ability to allow this excess water to bypass the hydroelectric power generation, and this is accomplished by using a spillway.

The penstock shown in Figure 5.1 directs water from the reservoir to the turbines. It is the height of the water and the amount of water flowing through the penstock that dictates the amount of hydroelectric power generated. The sluice gate controls water flow into the penstock.

There are two types of turbines used, namely a "reaction" turbine and an "impulse" turbine. The rotor of the reaction turbine is fully immersed in water, while an impulse turbine operates in air and is driven by jets of water. Turbine selection is based on the pressure head, which is directly related to the water height. The pressure heads are

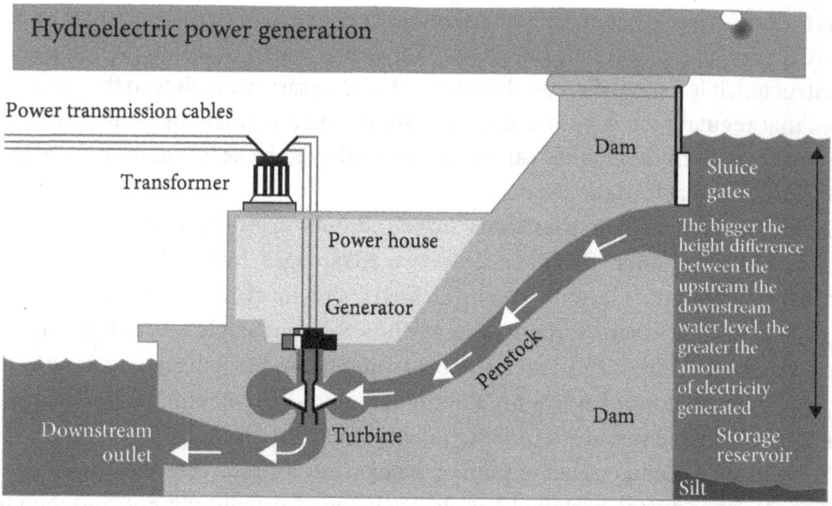

Figure 5.1 Hydroelectric Power Generation (Source: Environment Canada[15])

generally placed into three categories, namely low head (<10 m), medium head (10 to 50 m), and high head (>100 m).[11]

An impulse turbine is generally suitable for high head, low flow applications.[17] There are three main types of impulse turbines, namely Pelton Wheel, Turgo Wheel, and Cross-Flow. A Pelton turbine has a wheel with split buckets, and a jet or jets of water that strike the buckets to turn the wheel. The Turgo Wheel is a variation of the Pelton Wheel, the main difference being the design of the buckets. The Pelton turbine has two curved slices to make up the bucket while the Turgo has half of the slice, and the water jet strikes the bucket at a 20 to 25° angle. The Crossflow turbine has a rotor shaped like a drum, and the water strikes the blades twice, once when water flows from the outside to inside and one when the water flows from inside back out.

The reaction turbine power comes from a combination of pressure head and moving water, and is used for sites with lower head and higher flows compared with the impulse turbines.[17] These turbines use the flowing water to propel the blades in a way similar to a ship propeller, but in the opposite direction. The two main reaction turbines are Kaplan and Francis. For the Kaplan turbine, the angle of the blades is adjustable, allowing for a wider range of operation. In the Francis turbine, the water flows inward toward the blades at a radial direction, causing the turbine to spin, and then the water flows axially when it leaves the blades.

The theoretical power from a storage dam can be calculated from equation (1).[15]

$$P(MW) = \left[\frac{\eta^* \rho^* g^* Q^* H}{10^6} \right] \tag{1}$$

Where:

- P is the power produced at the transformer, MW
- η is the overall efficiency of power plant
- ρ is the density of water, 1,000 kg/m^3
- g is the acceleration of gravity, 9.81 m/s^2
- Q is the volume flow rate passing through the turbine, m^3/s
- H is the net pressure head, m

Assuming a plant efficiency of 87%, less than 100% because of losses due to friction in the penstocks as well as other mechanical effects in the system, and using the constants for water density and gravity, the equation can be simplified, as shown, with power now given in kW instead of MW:

$$P(kW) = 8.5 * Q * H \tag{2}$$

Although this is an oversimplification, estimates of the theoretical dam output can be made based on the dam height and design peak water flow. For example, we can make a simple comparison between Grand Coulee and Hoover Dam. For their designs, the dam heights are 168 and 222 m, peak water flows are 4,300 and 1,060 m^3/s, and design capacities are 6,809 and 2,078 MW, for Grand Coulee and Hoover Dams, respectively.[18,19] Inserting the values for dam height and water flow into equation (2), we can calculate 6,140 and 2,000 MW for Grand Coulee and Hoover, respectively, in reasonable agreement with the design capacities.

When it comes to hydroelectric dams, there are some really large plants. According to the Energy Information Administration (EIA),[20] nine of the ten largest power plants in the world are hydroelectric. A list of these nine dams is shown in Table 5.5.

To put the enormous size of these plants in perspective, let's consider just how many homes they can power. We can calculate the number of homes powered by 1 MW using the per capita electric power consumption in the United States. According to the International Energy Agency,[23] the US electric power consumption per capita for 1 year was 12.8 MWh in 2016. From 2014 to 2018, the 5-year average capacity factor for US hydroelectric plants was about 40%.[24] Using this as a capacity factor, a simple calculation shows that 1 MW will power about 270 homes for one year, so a 1 GW plant will power about 270,000 homes. Although the Three Gorges plant is in China, it would power nearly 6 million homes based on the US per capita consumption; however, since per capita electrical consumption is less in China, this plant provides power to more people than that. And, the Grand Coulee dam will provide power to nearly 2 million people!

It is also clear from Table 5.5 that China has been very aggressive in hydroelectricity development, with four dams having finished construction since 2009.

A run-of-river hydroelectric plant uses the natural elevation gradient of a flowing body of water to generate electricity.[25] Water is diverted from the main channel

Table 5.5 Top Nine Hydroelectric Plants in the World[21,22]

Hydroelectric Plant	Size, GW	Construction Start	Construction Finish
Three Gorges, China	22.5	1993	2012
Itaipu, Brazil and Paraguay	14.0	1975	1982
Xiluodu, China	13.86		2013
Guri, Venezuela	10.235	1963	1978 and 1986
Tucuruí, Brazil	8.37	1975 (phase 1) and 1998 (phase 2)	1984 and 2010
Xiangjiaba, China	6.4		2012
Grand Coulee, United States	6.809	1933	1941
Longtan, China	6.426	2007	2009
Krasnoyarsk, Russia	6.0	1956	1972

through a series of pipes, or penstocks, to move the power plant turbines before returning to the river downstream. As mentioned earlier, a disadvantage of this type of plant is the seasonal fluctuation in water, meaning the amount of electricity generated will vary. According to one source,[26] run-of-river projects were developed primarily for navigation and hydroelectric generation, not water storage. Water levels behind run-of-river dams vary only 3 to 5 feet in normal operations. Although normally smaller than storage dam hydroelectric plants, not all run-of-river plants are small. Examining nine run-of-river dams along the Columbia and Snake Rivers, you have Bonneville (1.19 GW, 197 feet[27]), Chief Joseph (2.620 GW, 236 feet[28]), Ice Harbor (693 MW, 213 feet[29]), John Day (2.485 GW, 184 feet[30]), Little Goose (932 MW, 253 feet[31]), Lower Granite (932 MW, 181 feet[32]), Lower Monumental (930 MW, 152 feet[33]), McNary (1.133 GW, 183 feet[34]), and The Dalles (2.038 GW, 260 feet[35]). So, while certainly not the size of the massive Grand Coulee Dam (6.809 GW, 550 feet[21]) on the Columbia River, these run-of-river projects generate significant amounts of electricity.

Pumped storage is a hydroelectric plant that pumps water from a lower reservoir or river to a higher reservoir during period of low electricity need, usually at night, and then uses the water during periods of peak demand.

Originally, in the 1960s and 1970s, pumped storage projects were built in conjunction with large nuclear power plants but, today, they are being considered to complement intermittent renewable energy generation, such as wind and solar energy.[36] As shown in Figure 5.2, water is pumped from the lower to upper reservoir during the period of low demand, and then later released to generate electricity during the period of high electrical demand. In this sense, the upper reservoir acts as a battery, storing potential energy which is later released as kinetic energy (flowing water) to

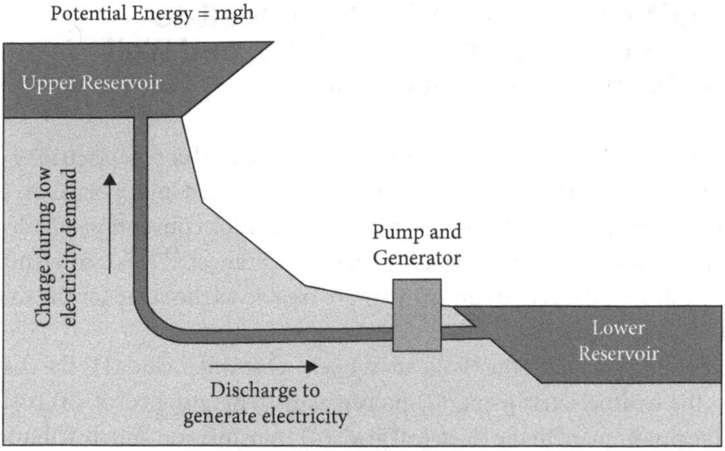

Figure 5.2 Schematic of Pumped Storage Facility[37]

make electricity. The amount of electricity that can be generated is proportional to the height of the stored water and the volume of water stored. These two things control the volumetric water flow rate, and thus the amount of kinetic energy sent to the turbines to make electricity.

Unfortunately, no mechanical system is 100% efficient, and the overall energy cycle ranges between 70% and 80%. In other words, the energy recovered by generating electricity is about 70% to 80% of that used to raise the water to the upper reservoir as well as the energy spent to make electricity. According to one paper,[36] the largest loss, about 8%, comes from pumps used to raise the water plus the turbines used to generate electricity.

A great example of a pumped storage system is in Bath County, Virginia, with a capacity of 3 GW and which began operation in December 1985.[39] At the time of this book's writing, it was the largest pumped storage power station in the world. However, when the 3.6 GW Fengning plant in China starts operation in 2021, it will exceed the size of the Bath County plant.[40]

A 2018 national hydropower dataset published by ORNL gives a breakdown for these three different types of hydroelectric power.[38,41] According to this report, there are 80 GW of conventional hydroelectric and 22 GW of pumped storage in the United States, and these plants represent 7% of all US electrical generation.[7] However, hydroelectricity is much more important in Washington, California, and Oregon, three states that control about half of the total US capacity. Although the report did not have the operational mode for all the plants, it did analyze 75% of the plants, which represented 64 GW, or about 80% of the installed capacity. Of these plants, 43.3 GW were storage dams and 20.7 GW were run-of-river, or 67.6% and 32.4%, respectively. For the storage dam plants, the average capacity factor was 40.8% while that for the run-of-river was 48.0%. The overall average capacity factor, which also included the plants for which the operational mode was not identified, was 44.8%,

5.4. Capital Cost, Operating Cost, Well-to-Wheels Levelized Cost of Electricity, and Well-to-Wheels Levelized Cost of Fuel

The equations and methodology used to calculate levelized cost of electricity (LCOE), and specific input used for hydroelectric power, are detailed in Appendix A. The units for LCOE are in $/megawatt-hour, or $/MWh. Data for conventional hydroelectric and pumped storage plants were taken from two references.[45,46] For conventional hydroelectric, storage dams and run-of-river are treated as the same for the calculation of LCOE.

For hydroelectric plants, the "Well-to-Wheels" cost will include (1) the cost of constructing the hydroelectric plant, (2) power plant generating costs, (3) cost of connecting the power plant to the electrical grid, (4) transmission and distribution costs, and (5) taxes (both federal and state). Unlike nuclear, natural gas, and coal, there are no fuel costs for a hydroelectric plant so there are no drilling or mining costs.

The capital, operating and maintenance (O&M), and transmission costs were taken from the authors. For storage dam / run-of-river, the Energy Information Administration (EIA)[45] reports a capital cost of $2,948/kW, fixed O&M costs of $40,85/kW-yr, and variable O&M costs of $1.36/MWh. The capacity factor is listed in a separate reference as 37.5%,[46] which is an average of annual US data for the years 2013 to 2016.

For pumped storage, the EIA[46] reports a capital cost of $5,288/kW, fixed O&M costs of $18.00/kW-yr, no variable O&M costs, and a capacity factor of 11.8%.

These capital costs include the capital cost for new transmission lines to tie into existing electrical substations or existing transmission lines. Therefore, the generating station supplies the transmission lines and steps up the voltage to tie into the "electrical grid," a network that delivers electricity to the customer. The grid has high-voltage transmission lines (sometimes distant) that carry power to demand centers, transformers to step down the voltage, and distribution lines that connect individual customers.

The costs of transmitting and distributing electricity were 0.9 cents/kWh (or $9/MWh) and 2.6 cents/kWh (or $26/MWh).[47] Here, transmission is defined as the high-voltage movement of electricity from the power station while distribution is the cost of electrical lines that take power from a substation to the customer. The cost of transmission and distribution is essential to get a true "Well-to-Wheels" analysis of the total cost.

Table 5.6 shows the LCOE for the cost of generating electricity and the additional cost of transmission and distribution. An obvious advantage of hydroelectric power is that there is no carbon dioxide (CO_2) produced from the hydroelectric plant operation. If carbon capture and sequestration (CCS) becomes mandatory in the future, it will enhance the comparison of hydroelectric to natural gas and coal. Table 5.6 also shows the importance of capacity factor on levelized capital cost. Even if the capital cost for pumped storage was the same as that for storage dam / run-of-river, the much lower capacity factor results in an LCOE that is still more than twice as much.

Table 5.6 Economic Analysis for Hydroelectric Plants

Energy Type	Capacity Factor	Capital Cost	Levelized Capital Cost; with MACRS Depreciation	Fixed O&M	Variable O&M (including fuel)	Transmission Investment	Total System Levelized Cost (LCOE) to Generate	LCOE Including Transmission and Distribution
		$/kW	$/MWh	$/MWh	$/MWh	$/MWh	$/MWh	$/MWh
Storage / Run-of-River	0.375	2,948	92.53	12.44	1.36	2.00	108.32	143.32
Pumped Storage	0.118	5,288	527.46	17.41	0.00	2.00	546.87	581.87

A sensitivity analysis was made for electricity generated from storage dam / run-of-river and pumped storage by varying the capital cost, fixed O&M, and capacity factor.

A report by the International Renewable Energy Agency (IRENA) shows capital costs for large hydroelectric plants (>20 MW) with a range of $1,050/kW to $7,650/kW while small hydroelectric plants have a range of $1,300/kW to $8,000/kW.[48] Given these data, the sensitivity analysis used a range of $1,000/kW to $8,000/kW for the capital cost.

For O&M costs, the IRENA report indicates it is quoted as a percent of capital cost, with a typical value of 2% to 2.5%. However, for the EIA base case,[45] the fixed O&M is actually 1.4% of capital cost. A report by the International Finance Corporation[15] suggests that 1.5% of capital costs is more typical of projects they have analyzed. Also, for 19 projects analyzed from countries around the world, fixed O&M ranged from $2.3/kW to $63.9/kW. They attribute this large spread as primarily due to labor costs, which vary widely among different countries. Based on this information, the sensitivity study used a low of $2.3/kW and high of $63.9/kW for the fixed O&M costs.

For capacity factors, a 2017 hydroelectric report prepared by ORNL indicates that for the US hydroelectric fleet, capacity factors as low as 25% and as high as 75% are not exceptional,[42] so these are used as the lower and upper range of the sensitivity analysis. Otherwise, the other data used in the sensitivity study are the same as used to prepare Table 5.6.

For the capital cost of a pumped storage plant, Table 5.6 shows a value of $5,288/kW. However, this value given by the EIA is much higher than those reported by the US Army Corp of Engineers and the DOE. In the Corp of Engineers study, the capital cost range is given as $617–$1,752/kW,[36] while in the DOE study, the capital cost range is given as $600–$1,800/kW.[49] Given these data, the sensitivity analysis uses $600/kW as a minimum cost and $5,288/kW as a maximum.

Likewise, for fixed O&M costs, the cost used to prepare Table 5.6 is $18.00/(kW-yr).[45] In contrast, the Corp of Engineer study presents a range of $2.43–$22.23/MWh[36] while the DOE study presents a range of $2.51–$22.98/(kW-yr).[49] Given these data, the sensitivity analysis uses $2.5/kW-yr as a minimum and $23.0/kW-yr as a maximum.

Since the capacity factor is the percent of rated capacity used per year, this definition is not as easy to apply for a pumped storage plant, since these plants are really used as batteries to store energy from another type of electric plant, be it a fossil fuel plant or one that uses renewable energy. While the capacity factor for a pumped storage plant can certainly be affected by equipment failure or routine maintenance, the plant operation really has more to do with how many hours of low demand electricity are available to use from the other plant.

Nevertheless, capacity factor still has a significant impact on the LCOE, so it is included in the sensitivity analysis. The EIA data used to prepare Table 5.7 used a capacity factor of 11.8%.[45] However, a study of several US pumped storage plants showed capacity factors ranging from around 10% to 40%.[44] Given these data, the sensitivity analysis used a minimum of 10% and a maximum of 40%. Results of the sensitivity analysis are shown in Table 5.7.

As the table illustrates, for both storage dam / run-of-river and pumped storage, changes in capital cost had much greater impact than changes to the fixed variable operating costs. Essentially, the LCOE increased proportionately to the increase in capital cost.

The table also illustrates the profound impact of capacity factor on LCOE. For storage dam / run-of-river, decreasing the base case value from 37.5% to 25% resulted in a~$50/MWh increase in LCOE; increasing the base case value to 75% resulted in a similar decrease in LCOE. For pumped storage, increasing the base case capacity factor from 11.8% to 40% resulted in a decrease of around $380/MWh. Thus, for both types of hydroelectric, the percent of plant capacity has a large impact on the LCOE. Unfortunately, improving the capacity factor may be out of the control of the plant, as

Table 5.7 Sensitivity Study for Storage / Run-of-River and Pumped Storage

	Capacity Factor	Capital Cost	Fixed O&M	Total System Levelized Cost (LCOE) to Generate	Total LCOE Cost to Generate, Transmit, and Distribute
Units		**$/kW**	**$/(kW-yr)**	**$/MWh**	**$/MWh**
Base Case—Storage / Run-of-River	0.375	$2,948	$40.85	$108.32	$143.32
Capex is $1,000/kW	0.375	$1,000	$40.85	$47.18	$82.18
Capex is $8,000/kW	0.375	$8,000	$40.85	$266.89	$301.89
Opex is $2.3/kW-yr	0.375	$2,948	$2.30	$96.59	$131.59
Opex is $63.9/kW-yr	0.375	$2,948	$63.90	$115.34	$150.34
Capacity Factor is 0.25	0.25	$2,948	$40.85	$160.81	$195.81
Capacity Factor is 0.75	0.75	$2,948	$40.85	$55.84	$90.84
Base Case—Pumped Storage	0.118	$5,288	$18.00	$546.87	$581.87
Capex is $600/kW	0.118	$600	$18.00	$79.26	$114.26
Capex is $3,000/kW	0.118	$3,000	$18.00	$318.65	$353.65
Capex is $5,288/kW	0.118	$5,288	$18.00	$546.87	$581.87
Opex is $2.5/kW-yr	0.118	$5,288	$2.50	$531.88	$566.88
Opex is $23.0/kW-yr	0.118	$5,288	$23.00	$551.71	$586.71
Capacity Factor is 0.10	0.10	$5,288	$18.00	$644.95	$679.95
Capacity Factor is 0.40	0.40	$5,288	$18.00	$162.74	$197.74

the capacity factor can vary year to year because of hydrologic conditions, competing use for water, and environmental and regulatory restrictions.

5.5. Cost of Energy

Unlike natural gas, coal, and nuclear electricity generation, hydroelectric power generation does not use fuel, generating the electricity from the kinetic energy of moving water. Thus hydroelectricity is not affected by changes, sometimes unpredictable, in fuel cost.

5.6. Capacity Factor

Average annual capacity factors for US storage dam / run-of-river hydroelectric plants are shown in Table 5.8.[24] In Table 5.5, the EIA gives a capacity factor of 37.5%, which is exactly the average for the 4 years 2013 to 2016. These capacity factors vary from year to year because of hydrologic conditions, competing use for water, environmental and regulatory restrictions, and plant power outages. They also reflect significant plant-to-plant variability. In fact, within the US hydroelectric fleet, capacity factors as low as 25% and as high as 75% are not unusual.[42] The lowest annual value in Table 5.7, 35.8% in 2015, was caused by drought conditions in much of the West. As Washington, California, and Oregon control about half of the total US capacity, this will certainly bring down the US annual average.

Figure 5.3 shows monthly data for 2017 and 2018, taken from an EIA report.[24] As the figure illustrates, the highest capacity factors are in the months of April and May, when rivers and creeks swell from rain and snow melt, and lowest in October, after the long, dry summer.

For pumped storage dams, the capacity factor is not so straightforward to understand. After all, a pumped storage dam is really just a large battery, which stores energy from another electric plant during periods of low economic value. And that can

Table 5.8 Average Annual Capacity Factors for US Storage Dam / Run-of-River Plants[24]

Year	Capacity Factor
2013	38.9%
2014	37.3%
2015	35.8%
2016	38.2%
2017	43.1%
2018	42.8%

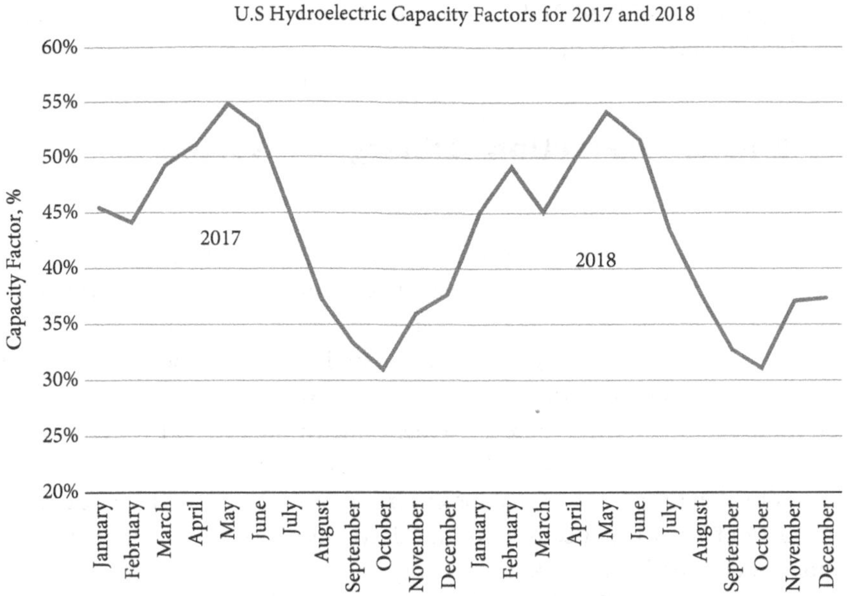

Figure 5.3 Average Monthly Capacity Factors for US Storage Dam/ Run-of-River Plants[24]

Table 5.9 Capacity Factors for Pumped Storage Plants[44]

Plant Name (Location)	Capacity, MW	Daily Generation, MWh	Capacity Factor
Taum Sauk (Missouri)	408	2,700	27.6%
Northfield Mountain (Massachusetts)	1,000	8,500	35.4%
Ludington (Michigan)	2,076	20,760	41.7%
Blenheim-Gilboar (New York)	1,000	4,700	19.6%

be any type of electric plant. Japan, for example, stores power from nuclear plants during low demand. Pumped storage can also be used to store energy from renewable energy types with intermittent generation, such as wind and solar. Overall, the net generation of a pumped storage plant is negative, since the energy recovered by generating electricity is about 70% to 80% of that used to raise the water to the upper reservoir as well as the energy spent to generate electricity.

Nevertheless, we can calculate a capacity factor using gross electricity generation from the pumped storage plant. According to the EIA,[43] in 2011 pumped storage plants produced 23 thousand gigawatt-hours (GWh) from 22 GW of capacity. This gives a capacity factor of 11.9%, meaning the average daily operation was only about 3 hours. Some specific data from four plants in the United States are shown in Table 5.9.[44] For these plants, the capacity factor ranged from 19.6% to 41.7%, much better than the US average of 11.9%. Like conventional hydroelectric plants, there is a lot of variability from plant to plant.

5.7. Efficiency: Fraction of Energy Converted to Work

For coal, natural gas, and nuclear power plants, the thermal efficiency of the turbine used to generate electricity can be calculated using the BTU content of one kilowatt-hour (kWh) of electricity, or 3,412 BTU, and dividing this by the heat rate. The heat rate is the amount of energy used to make one kWh of electricity. However, for storage dam / run-of-river hydroelectric power, as well as other renewables such as solar, wind, and geothermal, there is no fuel being consumed, and therefore energy used to make the electricity is not frequently measured. This is because renewable energy wasted in the process of making electricity has no fuel cost. If some of the kinetic power of water is not converted into electricity, there was no alternate use, or economic value, for the water's kinetic energy.

However, there are still inefficiencies in the process of making electricity from renewable fuels because no mechanical device is perfect.

To calculate thermal efficiencies for renewable energy, the EIA[50] proposed an "Incident Energy Approach." In this approach, the amount of electricity made can be compared to the incident energy that was input to the mechanical device. In the case of hydroelectric power, the incident energy is the kinetic energy of the water passing through the penstocks. In Table F1 of this paper, they report an efficiency of 90% for hydroelectric power.

Similarly, a European group, the Union of the Electricity Industry, listed hydroelectric efficiencies of 90% to 95%.[51] They cite hydroelectric power generation as the most efficient method to generate electric power, as there are no inefficient thermodynamic or chemical processes, and no heat losses. It is, however, impossible to achieve 100% efficiency, because using 100% of the water's kinetic energy means the water flow would have to stop.

A US Department of the Interior paper[52] cites an efficiency of 90% for the hydroelectric turbine, with losses mainly due to friction in the penstock.

Considering these three sources, the thermal efficiency is set to 90%. That is, of the kinetic energy entering the hydroelectric plant penstock, 90% will result in the formation of electricity.

Pumped storage is different from storage dam / run-of-river in that the incident energy is electricity that comes from another power plant. There is an alternate economic use for that electricity, but the decision to use this electrical energy to pump water to a higher reservoir and store this energy is based on the electricity being more valuable later, when demand is higher and electricity can be sold at a higher cost. The round trip energy efficiency of a pumped storage plant was given earlier as 70% to 80%,[36] such that most of the lost energy comes from the pumps used to raise the water and the turbines used to generate electricity.

5.8. Energy Balance

For the hydroelectric energy balance, only storage dam / run-of-river are considered, since pumped storage uses energy made from other power plants. For this, the energy balance is defined by the following equation:

$$\text{Energy Balance} = \frac{\left[\text{Electrical energy from hydroelectric power delivered to customer}\right]}{\left[\text{Kinetic energy contained in the water flowing into the hydroelectric plant}\right]}$$

This energy balance is greatly simplified compared to coal, natural gas, and nuclear. That is because in the "Well-to-Wheels" analysis, there are no mining or extraction costs, and no fuel transportation costs of fuel to the hydroelectric plant. Therefore, the energy balance only includes (1) making the electricity and (2) distributing the electricity through the grid to your home.

For the first step, the previous section gave the thermal efficiency for making electricity as 90%.

It then remains to determine the energy needed to distribute the electricity through the grid to your home in order to construct the final energy balance.

Transmission and distribution losses are affected by the voltage of the power line and the distance the electricity is transmitted. Historical data from the state of California shows annual average transmission losses around 5% to 7% over the time period of 2002 to 2008.[53] Similarly, EIA data for July 2016[54] show that of the electricity generated, 7% is lost in transmission and distribution. Therefore, a value of 7% is used for the amount of electricity lost in transmission and distribution after leaving the hydroelectric plant. Of course, this 7% lost in transmission and distribution is the percent of the energy leaving the electrical plant, not the kinetic energy of the water entering the hydroelectric plant.

Using 100 units of energy as a reference, we can arrive at the following equations. Summarizing, for the original kinetic energy of flowing water entering the hydroelectric plant, 83.7% will end up as usable electricity. This is much better than coal, natural gas, and nuclear range, with energy balances ranging from around 25% to 45%.
Storage Dam / Run-of-River Hydroelectric

$$100\ (\text{entering plant}) \rightarrow 90\ (\text{generation}) \rightarrow 83.7\ (\text{transmission and distribution})$$

5.9. Maturity: Experience

Water has been used as far back as the 3rd century BC to do work when the Chinese used water-powered bellows in iron production[1] and perhaps were the first to use a water wheel to grind grain and irrigate fields.[2] The concept of using water power, or hydropower, to generate electricity began in the mid-1700s when Bernard Forest de Bélidor (a French military engineer) published "L'architecture hydraulique," describing vertical and horizontal hydraulic machines. Combining this knowledge with the advent of the generator, a water turbine was used in 1880 to provide arc lighting to a theater and storefront in Grand Rapids, Michigan.[4] Similarly, in 1881 a water-powered generator was used to provide street lighting in Niagara Falls, New York. Shortly after, on September 30, 1882, the world's first hydroelectric power plant began operation along the Fox River, in Appleton, Wisconsin.[79] The plant produced only 12.5 kW, enough for paper manufacturer H.F. Rogers to run his plant and power his home.

Following the 1882 Appleton, Wisconsin, plant until the initial planning for the Niagara Falls hydroelectric plant in 1889, there was large growth in hydroelectricity. In 1886, an Electrical World article[55] stated that "there were 40–50 electric light plants depending largely or wholly upon water power." These plants were only as large as 6 MW. Niagara Falls first generated power on August 26, 1895, and by 1899 the United States had almost 500 hydroelectric plants.

In 1849, an engineer named James Francis developed the Francis Turbine, which is considered the first modern water turbine.[56] Today, it is still the most widely used water turbine.

The first use of pumped storage was in the 1890s in Italy and Switzerland, and the first use in the United States was in 1930, built by Connecticut Electric and Power Company along the Housatonic River near New Milford, Connecticut.[57]

Beginning when the first coal-fired electrical plant was built in 1882 and through the 1930s, the electrical grid system grew rapidly, roughly doubling every 6 years. Today, what we take for granted, involves more than 180,000 miles of high-voltage transmission lines connecting about 7,000 power plants.[58] Also, since hydroelectric power output can be quickly increased or decreased, and the power grid must always meet customer demands, hydroelectric plants help stabilize the grid. This will be especially important when less flexible sources of renewable energy, such as wind, solar, and geothermal, are integrated into the grid system.

To summarize, the first hydroelectric plant was built in 1882 and the modern water turbine has been around since 1849. The first use of pumped storage was in 1890, and starting in 1930 in the United States. The first electrical grid system was built in 1882 followed by rapid development through 1930s. Thus, there are about 135 years of experience in hydroelectric plants and almost 170 years of experience in hydroelectric turbines. Pumped storage technology has almost 130 years of experience and there are more than 135 years of experience in electric grids.

5.10. Infrastructure

As shown earlier in Table 5.2, the United States ranks number two in the world in total hydropower capacity, with 102,867 MW. However, due to low capacity factors compared to Brazil and Canada, in 2017 the United States ranked fourth in terms of hydroelectric power generation, when it produced 322.4 TWh.[13] About half of the total US capacity is in Washington, California, and Oregon.

In 2017 the EIA listed 1,498 operating hydroelectric power plants in the United States of which 40 are pumped storage.[7,14] Although somewhat different than the data from the International Hydropower Association given in the previous paragraph, 2017 data from the EIA shows the total US hydroelectric capacity as 101,238 MW, of which 79,595 MW come from conventional hydroelectric and 21,643 MW come from pumped storage. For the conventional hydroelectric plants, 67.6% were storage dams and 32.4% were run-of-river. The net electricity generation in 2017 was 293,838 MWh, with 300,333 MWh from conventional hydroelectric and −6,495 MWh from pumped storage. Pumped storage plants store energy for later use, and are net consumers of electrical energy.

Once made, this electricity can be transmitted around the United States using the more than 180,000 miles of high-voltage transmission lines in the grid.[58]

5.11. Footprint and Energy Density

Footprint is a term commonly used in industrial plants to describe the area used by an energy type. For hydroelectric power, the footprint includes the area to (1) make electricity at the hydroelectric plant and (2) transmit and distribute the electricity via the grid to commercial customers.

For the footprint and energy density analysis, run-of-river and storage dam plants are considered, but not pumped storage. There are at least two ways to look at the land area impacted by these hydroelectric plants. The first is to consider only the land impacted by the dam itself and the power station and the second is to also include the land impacted by the reservoir.

For the first, it was difficult to find much data for the land associated with just the dam and power station. One reference states that Hoover Dam covers 220 acres.[59] Given the design capacity of 2,078 MW cited earlier, this would translate into 9.4 MW/acre. Also, for a run-of-river plant that does not impound water, another reference estimates about 2.5 acres for a 10 MW facility, or 4 MW/acre.[60] In lieu of other data, we can compare these two plant facility values to those for natural gas (Chapter 2), coal (Chapter 3), and nuclear (Chapter 4). For these, the average plants sizes are 33.7, 1.7, and 1.2 MW/acre. This suggests that 1 MW of capacity requires less land than coal and nuclear but more land than natural gas. Given the environmental and containment issues of coal and nuclear, this seems reasonable. Therefore, the land associated with just the dam and power station will be set to 7 MW/acre, a rough average of the limited data available.

Calculating the footprint and energy density for a run-of-river plant can be difficult, as there are at least three types of plants including some that do not store water, some that use storage of water behind small dams, and some that divert part of the river. For the footprint, five run-of-river dams in Washington State were examined

Table 5.10 Land Areas for US Run-of-River Dams

Name	Location	Water Storage	Startup	MW	Acres	MW/acre
Chief Joseph Dam[61]	Washington	Rufus Woods Lake	1958	2,620	8,384	0.31
Little Goose Dam[62]	Washington	Lake Bryan	1970	932	10,025	0.09
Lower Granite Dam[63]	Washington	Lower Granite Lake	1975	810	8,900	0.09
Lower Monumental Dam[64]	Washington	Lake Herbert G. West	1969	810	6,590	0.12
McNary Dam[65]	Washington	Lake Waulla	1957	1,127	20,700	0.05
Average				1,260	10,920	0.13

where there was water impoundment, with the data shown in Table 5.10. The data ranged from 0.05 to 0.31 MW/acre with an average of 0.13 MW/acre.

For storage dams the impounded water land area can be quite large and, therefore, the MW produced per land acre is much smaller than run-of-river. Table 5.11 shows data for ten US plants, with a range of 0.0004 to 0.0830 MW/acre and an average of 0.0149 MW/acre.

Similarly, for other storage dam hydroelectric plants outside the United States, the data ranged from 0.0032 to 0.0420 MW/acre and an average of 0.0175.

It is clear from these data that the land area for water impoundment, whether run-of-river or storage dam, greatly exceeds the land area for the hydroelectric plant. Summarizing the discussion thus far, to (1) make electricity at the hydroelectric plant, the land associated with just the dam and power station will have an energy density of 7 MW/acre, the run-of-river energy density is about 0.13 MW/acre while the storage dam energy density is about 0.015 MW/acre.

Before continuing with the calculations, it is worth noting that land area used for hydroelectric power generation is different than land used for uranium mining, oil and gas exploration and production, and coal mining. That is, the reservoir that supplies the water for hydroelectricity generation is not used exclusively for generation and has other purposes. For example, large water reservoirs control water sent downriver during rainy and snow melting seasons to prevent flooding, provide water during dry seasons to water crops and provide water for homes and businesses, and provide a site for recreational activities such as boating and fishing. Therefore, depending on how the reservoirs impact local environment or affect public interests, the energy density calculated in the absence of the reservoir may be a more reasonable assessment of the footprint for a hydroelectric plant.

We can convert the units of MW/acre to the units of quadrillion BTU/(year-square mile). A 1 MW plant operating at 100% capacity for one year will produce 8,760 MWh

Table 5.11 Land Areas for US Storage Dam Plants

Name	Location	Water Storage	Startup	MW	Acres	MW/acre
Grand Coulee Dam[66]	Washington	Lake Roosevelt	1941	6,809	82,000	0.0830
Hoover Dam[67]	Nevada	Lake Mead	1936	2,078	158,720	0.0131
Fort Randall Dam[68]	South Dakota	Lake Francis	1954	320	102,000	0.0031
Toledo Bend[68]	Louisiana and Texas	Toledo Bend Lake	1963	92	205,000	0.0004
Shasta Dam[68]	California	Lake Shasta	1945	710	29,740	0.0239
Libby Dam[68]	Montana	Lake Koocanusa	1972	600	46,700	0.0128
Fort Peck Dam[68]	Montana	Fort Peck Lake	1943	185	245,000	0.0008
Garrison Dam[68]	North Dakota	Lake Sakakawea	1953	583	307,000	0.0019
Oahe Dam[68]	South Dakota	Lake Oahe	1962	768	374,000	0.0021
Glen Canyon Dam[68]	Arizona	Lake Powell	1966	1,296	161,000	0.0080
Average				1,344	171,116	0.0149

(1 MW * 24 hour/day * 365 days/year). There are 640 acres per square mile and 3.412×10^6 BTU/MWh. Also, the capacity factor for both storage dam and run-of-river hydroelectric plants was reported earlier to be 37.5%.

From these values, the 7 MW/acre for the land associated with just the dam and power station can be converted to 0.050 quadrillion BTU/(year-square mile) (0.050×10^{15}). Likewise, the run-of-river energy density of 0.13 MW/acre can be converted to 0.933 trillion BTU/(year-square mile), or 0.933×10^{12} and the energy density of 0.015 MW/acre for the storage dam can be converted to 0.108 trillion BTU/(year-square mile), or 0.108×10^{12}. Therefore, if the reservoir land area is not included in the energy density assessment, the energy density is on the order of fifty to five hundred times larger!

Now that the electricity has been made, the next step is to (2) transmit and distribute it to customers via the grid. However, the land area needed for step-up and step-down transformers is very small and can be ignored. Likewise, the metal towers and electrical lines take up very little land area and can coexist with farming, pasture land, and roadways. This means that the energy densities for storage dam and run-of-river hydroelectric plants are essentially based on the land area for the dam, power station, and the water reservoir.

Table 5.12 Land Areas for Non-US Storage Dam Plants[69]

Reservoir	Country	MW	Acres	MW/acre
Itaipu	Brazil-Paraguay	14,000	333,592	0.0420
Guri	Venezuela	10,300	1,052,667	0.0098
Tucurui	Brazil	8,400	600,465	0.0140
Sayano Shushenskaya	Russia	6,400	153,452	0.0417
Robert-Bourossa	Canada	7,722	695,601	0.0111
Yacyreta	Argentina/Paraguay	2,700	425,021	0.0064
Cahora Bassa	Mozambique	2,075	657,299	0.0032
Karakaya	Turkey	1,800	73,637	0.0244
Sao Simao	Brazil	1,635	166,549	0.0098
Marimbondo	Brazil	1,400	108,232	0.0129
Average		5,643	426,651	0.0175

In order to compare energy densities for different technologies on a common basis of gigawatt-hours per square mile per year, or GWh/(square mile-year), energy was converted from BTU to Watt-hour, or Wh, using this equation:

$$1 \text{ BTU} = 0.293071 \text{ Wh}$$

Using this conversion and recognizing that a gigawatt (GW) is equivalent to one billion Watts (or 1×10^9), we can calculate the energy density for the land associated with just the dam and power station as 14,700 GWh/(year-square mile), the energy density for the run-of-river as 273 GWh/(year-square mile), and the energy density for the storage dam as 32 GWh/(year-square mile).

Without including the land area of the water reservoir, the footprint for a 1,000 MW power station is 0.60 square miles. Expanding this area to include the water reservoir, the footprint for a 1,000 MW storage dam facility is about 280 square miles and that for a Run-of-River facility is 32 square miles. Clearly, the land area associated with the impounded water reservoir has a great effect on the assessment of footprint and energy density.

According to the International Energy Agency,[23] the US electric power consumption per capita for one year was 12.8 MWh in 2016. Using the capacity factor of 0.375 shown in Table 5.6, a simple calculation shows that 1 MW will power about 250 homes for one year, so a 1,000 MW plant will power about 250,000 homes!

5.12. Environmental Issues

Hydroelectric power is an inexpensive way to generate electricity, there are no fuel costs (compared to nuclear, natural gas, and coal), and it is a renewable energy source. There are some drawbacks, however, to hydroelectric power.

One problem is that damming rivers can destroy or disrupt wildlife.[5] The dam can affect the ability of fish, such as salmon, to swim upstream and spawn. Fish ladders can be built to help this, but it may still change migration patterns and decrease fish populations. In the case of Grand Coulee Dam with a height of around 500 feet, this is a difficult climb to make even if fish ladders are available.

The reservoir behind the dam can also create low levels of dissolved oxygen, which is harmful to the fish. This happens because water released from deep in a reservoir behind a high dam is usually cooler in summer and warmer in winter than river water, while water from outlets near the top of a reservoir will tend to be warmer than river water all year round.[70] This warming or cooling affects the amount of dissolved oxygen.

Another environmental issue is that reservoirs make methane, which, like CO_2 (discussed in the next section), is a greenhouse gas. Methane is produced by microbes living in now stagnant pools of water with oxygen-starved environments. These microbes eat organic carbon from plants, or biomass, for energy, but then emit methane.[71] It is believed that more than 20% of all man-made methane comes from the surface of reservoirs. According to the Environmental Protection Agency (EPA),[72] methane is estimated to have a global warming potential (GWP) of 28 to 36 times that of CO_2. While methane in the atmosphere has a life of about a decade, compared to thousands of years for CO_2, it absorbs more solar energy than CO_2 and, will produce ozone, which is also a greenhouse gas. These factors result in methane having a higher GWP.

In addition to these environmental issues, the building of a dam can affect, and displace, communities. When the Three Gorges Dam was built on China's Yangtze River, it is estimated that 1.3 million people were displaced and thousands of villages were flooded. One such village was Fuling, described in Peter Hessler's book *River Town*.[73] It is believed the dam caused droughts upstream of the Yangtze River and led to landslides and earthquakes.[74] The reservoir also submerged factories, mines, and waste sites, which caused pollution of the river. In the case of the Grand Coulee Dam, the reservoir disturbed burial grounds and ancient villages on the Colville Indian Reservation.[16]

5.13. CO_2 Production and the Cost of Capture and Sequestration

Clearly, since a hydroelectric power plant does not use any fuel to generate heat, no CO_2 is directly produced when electricity is made. However, over the life cycle of a plant, some CO_2 is produced from construction and, to a lesser extent, operation. In one reference, Hondo[75] gives a total of 11.3 g CO_2/kWh from a hydroelectric plant with over 80% being due to construction. In another study, Tremblay et al.[76] give a range of 4 to 18 g CO_2/kWh for a hydroelectric plant.

In addition, reservoir flooding results in biomass decomposition to methane, which is also a greenhouse gas. For a run-of-river hydroelectric plant, Steinhurst et al.[77] report a much broader range of CO_2 equivalent production at 0.5 to 152 g CO_2eq/kWh. Here, "eq" means that the effect of other greenhouse gases are taken into account, in this case methane, to report the GWP in equivalent amounts of CO_2. Some research suggests that methane accounts for 79% of the equivalent CO_2 produced from reservoirs.[78]

Given these different reports, we consider two different calculations for CO_2 production, namely one without and one with the equivalent effect of methane. Given the previous data, the CO_2 production without methane is set to 11 g CO_2/kWh and the production with methane is set to 75 g CO_2eq/kWh, basically taking the midpoint of the range given earlier.

Carbon dioxide production is commonly listed in the units of lb CO_2/MM BTU. However, hydroelectric power generates electricity without the combustion of a fuel. Because of this, there is no heat rate to convert from CO_2 produced on a mass basis and that produced on an energy basis. The conversion from g CO_2/kWh to lb CO_2/kWh is a simple conversion. For CO_2 production without methane, this value is 0.024 lb CO_2/kWh while the CO_2eq production with methane included is 0.165 lb CO_2eq/kWh.

For a 1,000 MW plant with a capacity factor of 37.5%, the plant will generate 9,000 MWh/day. For CO_2 produced on a kWh basis without methane included, the life cycle production of CO_2 is equivalent to 216,000 lb CO_2/day or about 108 tons CO_2/day. If instead, we include methane as another contributor to greenhouse gases, the life cycle production of CO_2 is equivalent to 1,485,000 lb CO_2/day or 743 tons/day. Considering CO_2 and methane together, the CO_2 equivalent production is roughly equivalent to a nuclear power plant (713 tons CO_2/day), 3% of the CO_2 produced by a pulverized coal combustion plant (22,214 tons CO_2/day) and 10% of the CO_2 produced by a natural gas combined cycle plant (7,500 tons CO_2/day).

As this CO_2 is not directly produced by the power plant, but rather by other processes, including construction and biomass decomposition and formation of methane, it is not practical to capture any of this CO_2 or methane. Therefore, it is unlikely that the greenhouse gases produced from the hydroelectric plant and reservoir would be removed by CCS.

5.14. Chapter Summary

Since it is renewable, unlike petroleum crude oil, coal, natural gas, and nuclear, there is no clock running to a day when there will not be any more hydroelectric power. And, there is potential of adding new hydroelectric power, using either the more than

80,000 NPDs that are not currently producing electricity or new stream development. According to some analyses, the United States has the potential of adding 77.5 GW of new hydroelectric power producing from 279 to 420 TWh/year! On the world stage, three different reports suggest the potential ranges from 10,000 to 52,000 TWh/year. Using the conservative generation value of 10,000 TWh/year, the top ten countries control about 59.5% of the unutilized potential. These countries are, in order of potential, Russia, China, Canada, India, Brazil, Indonesia, Peru, Democratic Republic of the Congo, Tajikistan, and the United States.

The "Well-to-Wheels" economic analysis gave an average LCOE, including transmission and distribution, of $143.32/MWh for storage dam / run-of-river plants and $581.87/MWh for a pumped storage plant. The low capacity factor for pumped storage, 11.8% versus 37.5% for storage dam / run-of-river, as well as much higher capital cost, account for the sizable difference in LCOE.

For a 1,000 MW hydroelectric power plant with a capacity factor of 37.5%, a life cycle analysis for the plant shows that, on a kWh basis, the plant will produce 75 g CO_2eq/kWh or 0.165 lb CO_2eq/kWh. The term "CO_2eq" is used because along with CO_2 made during plant construction, the methane formed by biomass decomposition in the reservoir is also included to calculate the combined greenhouse gas effect. Capturing the CO_2 formed during plant construction and capturing the methane made by biomass decomposition in the reservoir using CCS is not practical.

For a storage dam / run-of-river power plant, the average US capacity factor is 37.5%. These number vary year-to-year because of hydrologic conditions, competing use for water, environmental and regulatory restrictions, plant power outages, and plant-to-plant variability. On average, capacity factors over the last 6 years have varied from a low of 35.8% to a high of 43.1%. Within the US hydroelectric fleet, individual plant capacity factors vary from as low as 25% to as high as 75%. The US average capacity factor for pumped storage is even lower, only 11.8% on average, although some individual plants exceed 40%. As pumped storage plants are really just big batteries that store energy from other plants, the capacity factor is strongly affected by the low and high value periods for electricity sales, and therefore how many hours each day it makes sense to store, rather than sell, electricity.

For a storage dam / run-of-river power plant, the overall energy balance was 83.7%, meaning that about 84% of the kinetic energy originally in the flowing water ends up as energy at the distribution point as electricity. The reason this number is so high, compared to other technologies, is due to the high efficiency of the various turbines typically 90%.

In terms of experience and maturity, hydroelectric plants have been around since 1882, the modern water turbine has been around since 1849, and pumped storage plants have been in existence since 1890, and in the United States since 1930.

Also, there is a lot of infrastructure for hydroelectric power and for electrical transmission. To make the electricity, as of 2017 there were 1,498 operating hydroelectric power plant in the United States, of which 40 were pumped storage. Of the 1,458 conventional hydroelectric plants, the capacity was 79,595 MW, while the 40 pumped storage plants had a capacity of 21,643 MW. For the conventional hydroelectric plants, 67.6% were storage dam and 32.4% were run-of-river plants. The net electricity generation in 2017 was 293,838 GWh, representing 7% of all US electrical generation. Conventional hydroelectric plants generated 300,333 MWh while pumped storage generated -6,495 MWh. Once made, this electricity can be transmitted around the United States using the more than 180,000 miles of high-voltage transmission lines in the grid.

Assuming a hydroelectric plant with the size of 1,000 MW, the footprint for just the dam and power station is 0.60 square miles. Expanding this area to include the water reservoir, the footprint for a 1,000 MW storage dam facility is about 280 square miles and that for a run-of-river facility is about 32 square miles. Clearly, the land area associated with the impounded water reservoir has a great effect on the assessment of footprint and energy density.

The energy density for the land associated with just the dam and power station is 14,700 GWh/(year-square mile). However, if the reservoir land area is included, the energy densities decrease to 273 GWh/(year-square mile) for the run-of-river and to 32 GWh/(year-square mile) for the storage dam.

For a capacity factor of 37.5%, a 1 GW hydroelectric plant will power about 250,000 homes.

Finally, the main environmental issues for hydroelectric plants are the destruction or disruption of wildlife caused by the dam, low levels of dissolved oxygen which are harmful to the fish due to water temperature gradients in the reservoir, and the production of methane, a greenhouse gas with 28 to 36 times the effect of CO_2. The methane is produced by microbes decomposing biomass in the stagnant pools of water in the reservoir. In addition, the creation of new water reservoirs can result in the displacement of people, as well as pollution caused by submerged factories, mines, and waste sites.

References

1. "Iron and Steel in Ancient China," Donald B. Wagner, The Netherlands: E.J. Brill Publishing, 1996,
2. Xeneca Power Development Inc., 2015, http://www.xeneca.com/history/
3. "Electricity & Alternative Energy," Alberta Culture and Tourism, 2020, http://www.history.alberta.ca/energyheritage/energy/hydro-power/hydro-power-in-ancient-times.aspx
4. "History of Hydropower," Energy.gov, 2020, https://energy.gov/eere/water/history-hydropower
5. "Hydropower," National Geographic, May 13, 2019, http://www.nationalgeographic.com/environment/global-warming/hydropower/
6. "Pumped Hydroelectric Storage," Energy Storage Association, 2020, http://energystorage.org/energy-storage/technologies/pumped-hydroelectric-storage
7. "Count of Electric Power Industry Power Plants, by Sector, by Predominant Energy Sources within Plant, 2007 through 2017," Table 4.1, Energy Information Administration, http://www.eia.gov/electricity/annual/html/epa_04_01.html; "Existing Capacity by Energy Source, 2017 (Megawatts)," Table 4.3, Energy Information Administration, https://www.eia.gov/electricity/annual/html/epa_04_03.html

8. "An Assessment of Energy Potential at Non-Powered Dams in the United States," US Department of Energy, Budget Activity Number ED 19 07 04 2, prepared by Oak Ridge National Laboratory, Boualem Hadjerioua et al., April 2012, https://energy.gov/sites/prod/files/2013/12/f5/npd_report_0.pdf
9. "New Stream-Reach Development: A Comprehensive Assessment of Hydropower Energy Potential in the United States," US Department of Energy, Wind and Water Power Technologies Office, Budget Activity Number ED 19 07 04 2, prepared by Oak Ridge National Laboratory, Shih-Chieh Kao et al., April 2014, https://nhaap.ornl.gov/sites/default/files/ORNL_NSD_FY14_Final_Report.pdf
10. "World Energy Resources: Hydropower," World Energy Council, 2016, https://www.worldenergy.org/wp-content/uploads/2017/03/WEResources_Hydropower_2016.pdf
11. "Systematic High-Resolution Assessment of Global Hydropower Potential," Olivier A. C. Hoes et al., PLOS One, February 8, 2017, https://journals.plos.org/plosone/article?id=10.1371/journal.pone.0171844
12. "A Comprehensive View of Global Potential for Hydro-Generated Electricity," Energy & Environmental Science, July 2015, Y. Zhou et al., https://www.researchgate.net/publication/279911353_A_Comprehensive_View_of_Global_Potential_for_Hydro-generated_Electricity
13. "2018 Hydropower Status Report," International Hydropower Association, https://www.hydropower.org/sites/default/files/publications-docs/2018_hydropower_status_report_0.pdf
14. "Net Generation by Energy Source: Total (All Sectors)," 2007–February 2017, Table 1.1, Energy Information Administration, https://www.eia.gov/electricity/monthly/epm_table_grapher.cfm?t=epmt_1_01
15. "Hydroelectric Power: A Guide for Developers and Investors," International Finance Corporation, February, 2015, http://www.ifc.org/wps/wcm/connect/topics_ext_content/ifc_external_corporate_site/ifc+sustainability/learning+and+adapting/knowledge+products/publications/hydroelectric_power_a_guide_for_developers_and_investors
16. "Grand Coulee Dam," Public Broadcasting System American Experience, April 18, 2017, http://www.pbs.org/wgbh/americanexperience/films/coulee/
17. "Types of Hydropower Turbines," Office of Energy Efficiency & Renewable Energy, 2020, https://energy.gov/eere/water/types-hydropower-turbines
18. "Grand Coulee Dam Statistics and Facts," US Department of the Interior Bureau of Reclamation, 2019, https://www.usbr.gov/pn/grandcoulee/pubs/factsheet.pdf
19. "How Much Dam Energy Can We Get?," December 20, 2011, https://dothemath.ucsd.edu/2011/12/how-much-dam-energy-can-we-get/
20. "The World's Nine Largest Operating Power Plants Are Hydroelectric Facilities," October 18, 2016, https://www.eia.gov/todayinenergy/detail.php?id=28392
21. "The 10 Biggest Hydroelectric Power Plants in the World," Power Technology, October 28, 2013, http://www.power-technology.com/features/feature-the-10-biggest-hydroelectric-power-plants-in-the-world/
22. "The World's 10 Largest Hydroelectric Dams," OilPrice.com, September 10, 2014, http://oilprice.com/Latest-Energy-News/World-News/The-Worlds-10-Largest-Hydroelectric-Dams.html
23. "Electric Power Consumption (kWh per capita)," IEA Statistics for 2016, http://energyatlas.iea.org/#!/tellmap/-1118783123/1
24. "Electric Power Monthly," Energy Information Administration, January 2019, https://www.eia.gov/electricity/monthly/current_month/epm.pdf
25. "Analysis of Reservoir-Based Hydroelectric versus Run-of-River Hydroelectric Energy Production," Cassie Modal et al., LRES Capstone, December 9, 2014, http://landresources.montana.edu/archives/capstone/2014_Capstone_Hydro.pdf

26. "The Columbia River System Inside Story," 2nd Edition, Federal Columbia River Power System, Bonneville Power Administration, US Bureau of Reclamation, and US Army Corps of Engineers, April 2001, https://www.bpa.gov/power/pg/columbia_river_inside_story.pdf
27. "The Bonneville Lock and Dam Fact Sheet," US Army Corps of Engineers, March 1, 2015, The Bonneville Lock and Dam Fact Sheet.
28. "Fundamentals of Renewable Energy: Chief Joseph Dam Hydroelectric Power Plant," March 1, 2015,https://www.nws.usace.army.mil/Missions/Civil-Works/Locks-and-Dams/Chief-Joseph-Dam/.
29. "Ice Harbor Lock and Dam," US Army Corps of Engineers, March 1, 2015, www/gpp.
30. "John Day Dam—Hydroelectric Project Information," Columbia Basin Research, March 1, 2015, https://usace.contentdm.oclc.org/digital/collection/p16021coll11/id/2893.
31. "Little Goose Lock and Dam," US Army Corps of Engineers, March 1, 2015, https://www.nww.usace.army.mil/Locations/District-Locks-and-Dams/Little-Goose-Lock-and-Dam/.
32. "Lower Granite Lock and Dam," US Army Corps of Engineers, March 1, 2015, https://web.archive.org/web/20110517163500/http://www.nww.usace.army.mil/dpn/dpn_project.asp?project_id=91.
33. "Lower Monumental Lock and Dam," US Army Corps of Engineers, March 1, 2015, https://web.archive.org/web/20051126051648/http://www.nww.usace.army.mil/dpn/dpn_project.asp?project_id=92.
34. "McNary Lock and Dam," US Army Corps of Engineers, March 1, 2015, https://www.nww.usace.army.mil/Locations/District-Locks-and-Dams/McNary-Lock-and-Dam/.
35. "The Dalles Dam—Hydroelectric Project Information," Columbia Basin Research, March 1, 2015, https://usace.contentdm.oclc.org/digital/collection/p16021coll11/id/2892.
36. "Technical Analysis of Pumped Storage and Integration with Wind Power in the Pacific Northwest," Prepared for US Army Corps of Engineers, Northwest Division, Hydroelectric Design Center, August 2009, http://www.hydro.org/wp-content/uploads/2011/07/PS-Wind-Integration-Final-Report-without-Exhibits-MWH-3.pdf
37. "Hydroelectric Pumped Storage," 2012, http://www.lowcarbonfutures.org/sites/default/files/PHS_final_0.pdf
38. "2014 Hydropower Market Report," US Department of Energy, Wind and Water Power Technologies Office, April 2015, https://www.energy.gov/sites/prod/files/2015/04/f22/2014%20Hydropower%20Market%20Report_20150424.pdf
39. "Bath County Pumped Storage Station," 2020, https://www.dominionenergy.com/company/making-energy/renewable-generation/water/bath-county-pumped-storage-station
40. "New Chinese Pumped-Storage Hydro Plant to Be "World's Largest" When Complete in 2021," Elizabeth Ingram, Hydroworld, September 5, 2017, https://www.hydroworld.com/articles/2017/09/new-chinese-pumped-storage-hydro-plant-to-be-world-s-largest-when-complete-in-2021.html
41. "National Hydropower Plant Dataset, Version 1 (Update FY18Q2)," Oak Ridge National Laboratory (ORNL), 2018, https://hydrosource.ornl.gov/national-hydropower-plant-dataset-version-1-update-fy18q2
42. "2017 Hydropower Market Report April 2018," Office of Energy Efficiency and Renewable Energy, Department of Energy, prepared by Oak Ridge National Laboratory, https://www.ornl.gov/content/2017-hydropower-market-report
43. "Pumped Storage Provides Grid Reliability Even with Net Generation Loss," US Energy Information Administration, July 8, 2013, https://www.eia.gov/todayinenergy/detail.php?id=11991
44. "The Cost of Pumped Hydroelectric Storage," Oscar Galvan-Lopez, December 11, 2014, http://large.stanford.edu/courses/2014/ph240/galvan-lopez2/

45. "Cost and Performance Characteristics of New Generating Technologies," Annual Energy Outlook 2019, Table 2, Energy Information Administration, January 2019, https://www.eia.gov/outlooks/aeo/
46. "Updated Capital Cost Estimates for Utility Scale Electricity Generating Plants," April 2013, US Energy Information Administration; and "Levelized Cost and Levelized Avoided Cost of New Generation Resources in the Annual Energy Outlook 2015," Table 1, June 2015.
47. "Annual Energy Outlook 2015 with Projections to 2040," DOE/EIA-0383, Table A8, April 2015.
48. "Renewable Energy Technologies: Cost Analysis Series: Hydropower: International Renewable Energy Agency (IRENA)," Volume 1: Power Sector Issue 3/5, June 2012, https://www.irena.org/documentdownloads/publications/re_technologies_cost_analysis-hydropower.pdf
49. "Pumped Storage and Potential Hydropower from Conduits," Report to Congress, US Department of Energy, February 2015, https://energy.gov/sites/prod/files/2015/06/f22/pumped-storage-potential-hydropower-from-conduits-final.pdf
50. "Alternatives for Estimating Energy Consumption," Appendix F, Annual Energy Review 2010, https://www.eia.gov/totalenergy/data/annual/pdf/sec17.pdf
51. "Efficiency in Electricity Generation," EURELECTRIC "Preservation of Resources" Working Group's "Upstream" Sub-Group in collaboration with VGB, July 2003, EEGJulyrevisedFINAL1-2003-030-0548-2-.pdf
52. "Reclamation: Managing Water in the West, Hydroelectric Power," US Department of the Interior, Bureau of Reclamation Power Resources Office, July 2005, https://www.usbr.gov/power/edu/pamphlet.pdf
53. "A Review of Transmission Losses in Planning Studies," Lana Wong, California Energy Commission, Figure ES-1, August 2011.
54. "Monthly Energy Review," US Energy Information Administration, March 2019, https://www.eia.gov/totalenergy/data/monthly/pdf/mer.pdf
55. "Some Early History of Hydroelectric Power," Hydro Review, June 1988, http://www.hydroworld.com/content/dam/hydro/Document/earlyhistoryrev2.pdf
56. "Francis and His Turbine," Hydro Review, February 1989, http://www.hydroworld.com/content/dam/hydroworld/downloads/Q%26Aturbine2.pdf
57. "A Ten-Mile Storage Battery," Popular Science, July 1930, p. 60.
58. "United States Electricity Industry Primer," Office of Electricity Delivery and Energy Reliability U.S. Department of Energy DOE/OE-0017, July, 2015, https://www.energy.gov/sites/prod/files/2015/12/f28/united-states-electricity-industry-primer.pdf
59. "Knowing Vegas: Are There Really Bodies in the Cement of Hoover Dam?," Chris Kudialis, Las Vegas Review Journal, October 8, 2015, https://www.reviewjournal.com/uncategorized/knowing-vegas-are-there-really-bodies-in-the-cement-of-hoover-dam/
60. "Renewable Electricity Futures Study," Volume 2: "Renewable Electricity Generation and Storage Technologies," Chad Augustine et al., National Renewable Energy Laboratory (NREL), June, 2012, http://www.nrel.gov/docs/fy12osti/52409-2.pdf
61. "Chief Joseph Dam," 2015, https://en.wikipedia.org/wiki/Chief_Joseph_Dam
62. "Little Goose Dam," July 2010, https://en.wikipedia.org/wiki/Little_Goose_Dam
63. "Lower Granite Dam," July 2010, https://en.wikipedia.org/wiki/Lower_Granite_Dam
64. "Lower Monumental Dam," July 2010, https://en.wikipedia.org/wiki/Lower_Monumental_Dam
65. "McNary Dam," July 2010, https://en.wikipedia.org/wiki/McNary_Dam
66. "Grand Coulee Dam Facts," 2020, http://www.gcdvisitor.com/grand-coulee-dam-facts/
67. "Hoover Dam," March 2015, https://www.usbr.gov/lc/hooverdam/faqs/lakefaqs.html

68. "The 10 Largest Reservoirs in the United States," August 26, 2016, Tata and Howard, http://tataandhoward.com/2016/08/10-largest-reservoirs-united-states/
69. "The Water Footprint of Electricity from Hydropower," M.M. Mikonnen and A.Y. Hoekstra, Value of Water Research Report Series No. 51, University of Twente, Enschede, The Netherlands, June 2011, Report51-WaterFootprintHydropower_1.pdf
70. "Dams and Water Quality," International Rivers, 1996, https://www.internationalrivers.org/dams-and-water-quality
71. "Hydropower May Be Huge Source of Methane Emissions," Climate Central, October 2014, http://www.climatecentral.org/news/hydropower-as-major-methane-emitter-18246
72. "Greenhouse Gas Emissions, Understanding Global Warming Potentials," US Environmental Protection Agency, February, 2017, https://www.epa.gov/ghgemissions/understanding-global-warming-potentials
73. "River Town: Two Years on the Yangtze," Peter Hessler, Harper Collins, January 2001.
74. "Dangerous Waters—The World's Most Controversial Hydropower Dam Projects," Power Technology, June, 2014, http://www.power-technology.com/features/featuredangerous-waters---the-worlds-most-controversial-hydropower-dam-projects-4290847/
75. "Life Cycle GHG Emission Analysis of Power Generation Systems: Japanese Case," H. Hondo, Energy 30 (2005) 2042–2056.
76. "The Issue of Greenhouse Gases from Hydroelectric Reservoirs: From Boreal to Tropical Regions," A. Tremblay et al., University of Québec in Montréal, February, 2014, http://www.un.org/esa/sustdev/sdissues/energy/op/hydro_tremblaypaper.pdf
77. "Hydropower Greenhouse Gas Emissions State of the Research," February 14, 2012, W. Steinhurst et al., Hydropower-GHG-Emissions-Feb.-14-2012.pdf
78. "Reservoirs Are a Major Source of Global Greenhouse Gases, Scientists Say," Chris Mooney, Washington Post, September 28, 2016, https://www.washingtonpost.com/news/energy-environment/wp/2016/09/28/scientists-just-found-yet-another-way-that-humans-are-creating-greenhouse-gases/?utm_term=.222134afcbaa
79. "1882: First Hydroelectric Plant Opens," National Geographic Society, December 16, 2013, https://www.nationalgeographic.org/thisday/sep30/first-hydroelectric-plant-opens/

6
Wind Energy
A Renewable Energy Type

6.1. Foreword

In the context of this chapter, wind energy is the conversion of the kinetic energy of the wind into electricity. It is recognized, however, that wind energy can also be used directly for mechanical power, such as propelling ships through sailing, windmills, and other wind devices. Wind is a weather variable caused by regional differences in atmospheric pressure and temperature, causing air movement.

For electrical power generation, wind energy can be harnessed in onshore and offshore wind farms, which are connected to the electric power grid. On a small scale, it is also possible to have small wind turbines for individual power generation. As discussed later, a "wind farm" is a term describing a cluster of wind turbines arranged in various configurations in a given geographical area, the cumulative total of which gives the total capacity of the wind farm. These "farms" can be onshore, in mountainous or flat and possibly remote areas, or offshore, which has less visual environmental impact but with higher construction costs due to location and tying into the electrical power grid through underwater cables. Since a typical wind turbine is 2 MW, a wind farm should consist of several hundred individual wind turbines if the goal is to approach the size of a typical coal or natural gas power plant. A complicating factor in assessing the cost of the electricity generated is that the land area used for the wind farm can also be employed for other purposes, such as agricultural and ranching.

Generally speaking, wind turbines are similar in design, consisting of a vertical tower holding a horizontally disposed wind turbine with three blades.

Wind is a very fast growing renewable energy technology. According to the Global Wind Energy Council,[1] at the end of 2016 the worldwide wind power capacity was 487.3 GW (gigawatts), increasing by 51.8 GW to 539.1 GW in 2017. And, in 2018, the worldwide capacity increased another 51.3 GW to 590.4 GW.[2] That means installed wind capacity increased by 11% from 2016 to 2017 and 10% from 2017 to 2018, a 21% increase over just 2 years! And there were 341,320 wind turbines worldwide at the end of 2016 for these 487.3 GW.[3] Although a value for the number of wind turbines at the end of 2018 was not available, a 21% extrapolation gives a value over 400,000 turbines. And this number does not include small scale turbines. According to the 2018 Renewables Global Status Report,[4] there were approximately one million small scale turbines in 2015 with a capacity of about 1 GW, an average of one kW (kilowatt) per

The Changing Energy Mix. Paul F. Meier, Oxford University Press (2020). © Oxford University Press.
DOI: 10.1093/oso/9780190098391.001.0001.

turbine. These small turbines are used for a variety of applications, typically residential homes in remote locations.

For 2017, China, the United States, and Germany led the world in new installations with 19.7 GW, 7.0 GW, and 6.6 GW, respectively.

Using the number of turbines and nameplate capacity for 2016 gives an average turbine nameplate capacity of about 1.4 MW. While 1.5–2 MW are typical ratings for wind turbines, there are also commercial onshore wind turbines with ratings over 5 MW and offshore turbines over 10 MW. These massive offshore wind turbines have large rotor diameters. For example, three turbines with 8 MW ratings include MHI Vestas V164 with a rotor diameter of 164 meters, the Adwen AD-180 with a rotor diameter of 180 meters, and the Siemens SWT-8.0-154 with a rotor diameter of 154 meters. Amazingly, these diameters are longer than a football field, with a length of 120 yards, or 109 meters, including the end zones! And the GE Haliade-X offshore turbine, with a 12 MW rating, has a rotor diameter of 220 m, more than two football fields.

The Renewables 2018 Global Status Report[4] reported that, in 2017, wind power was responsible for generating about 1,430 terawatt-hours (TWh), or about 5.6% of the total electricity used in the world. Using a worldwide capacity of 539.1 GW, the average capacity factor is 30.3%.

6.2. Proven Reserves

Unlike petroleum crude oil, natural gas, coal, and nuclear energy, the amount of proven, or potential, reserves for wind energy are difficult to assess. This is because it involves many factors, such as turbine rating, turbine height, what land and ocean areas are available for use, and the wind speed available at a given height, just to name a few factors.

One estimate for the United States was made by the National Renewable Energy Laboratory (NREL) employing a model that used geographic information system (GIS) data.[5] For onshore wind power, they assumed an 80 m (~262 feet) turbine height (measured at the hub) and typical utility-scale wind turbine power curves. In their analysis, they eliminated land unlikely to be developed, such as urban areas, federally protected lands, and recreational water reservoirs. For this, they projected a potential capacity of 10,955 GW on 845,931 square miles of land, which would generate 32,784,004 GWh at a capacity factor of 34%. The capacity factor, which is discussed in more detail later, is the percent of rated capacity used per year. The total area for the US 50 states is 3,796,742 square miles, so the potential capacity of 10,955 GW would use 22.3% of the total US land area. Naturally, turbines are spaced to provide optimum performance, so the total land area also includes the land area between turbines.

Another report from NREL[6] shows a map of average wind speeds at a turbine height of 80 m for the 48 contiguous states, shown in Figure 6.1. While it is probably

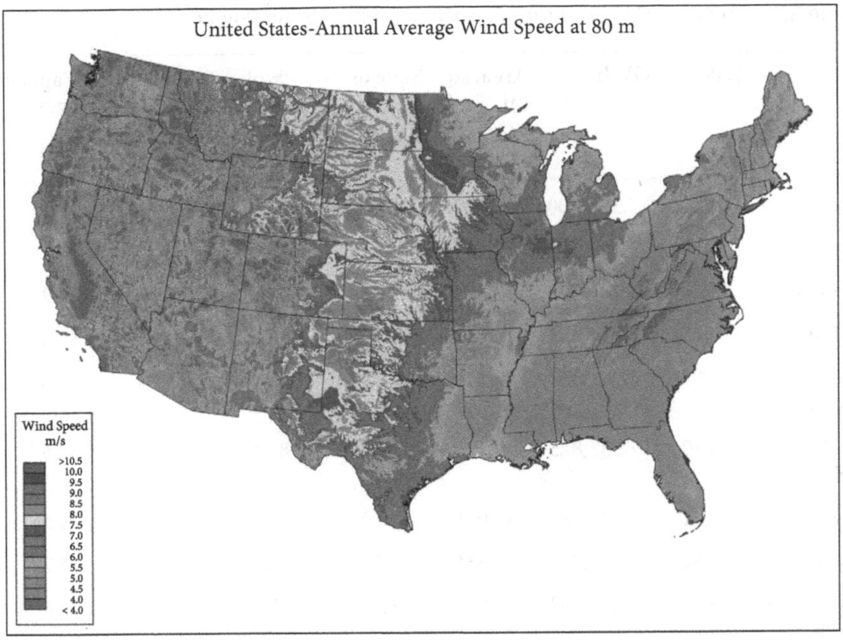

Figure 6.1 Average Wind Speeds at a Turbine Height of 80 m (Source: NREL[6])

not a surprise to anyone who has lived in or driven through Kansas, Oklahoma, or the Texas panhandle, average wind speeds are highest in the western and central Great Plains. This part of the country experiences average wind speeds of 8.5–9.0 m/s, equivalent to 19–20 mph.

The NREL report also shows the potential state-by-state, and the top twelve states are shown in Table 6.1. The table shows the potential installed capacity and electricity generation. It also shows the area used to meet this potential compared to the total area of the state, the percent this represents of the total US potential generation, and the capacity factor for each state. Altogether, these twelve states represent 88% of the entire US potential, with Texas alone representing 17%. For the five states of Kansas, Nebraska, North Dakota, South Dakota, and Iowa, the percent of land area exceeds 75%, indicating that these state would virtually become one big wind farm to reach their potential. Twelve states were shown, rather than ten, because Alaska, New Mexico, and Minnesota had almost the same capacity and electricity generation.

In the same report, they also examined the potential for offshore wind power, with the criteria that wind speeds should be greater than 6.4 m/s (~14 miles/hour), turbine heights would be 90 m (~295 feet), and the locations would be no further than 50 nautical miles from shore.[5] As in the case of onshore wind, some areas were eliminated included shipping lanes, marine sanctuaries, and a variety of other areas deemed unlikely to be developed. For offshore, they projected a potential capacity of

Table 6.1 Top Twelve US States for Onshore Potential Wind Power[4]

State	GW	GWh	Area, sq mi	State or US Area, square mile	% of State Area	% of Potential US Generation	Capacity Factor, %
Texas	1,902	5,552,400	146,837	268,596	54.7	16.9	33.3
Kansas	952	3,101,576	73,542	82,278	89.4	9.5	37.2
Montana	944	2,746,272	72,896	147,040	49.6	8.4	33.2
Nebraska	918	3,011,253	70,888	77,348	91.6	9.2	37.4
South Dakota	882	2,901,858	68,140	77,116	88.4	8.9	37.6
North Dakota	770	2,537,825	59,475	70,698	84.1	7.7	37.6
Iowa	571	1,723,588	44,071	56,273	78.3	5.3	34.5
Wyoming	552	1,653,857	42,631	97,813	43.6	5.0	34.2
Oklahoma	517	1,521,652	39,909	69,899	57.1	4.6	33.6
Alaska	493	1,373,433	38,096	665,384	5.7	4.2	31.8
New Mexico	492	1,399,157	37,999	121,590	31.3	4.3	32.5
Minnesota	489	1,428,525	37,782	86,936	43.5	4.4	33.3
Twelve State Total	9,482	28,951,396	732,268	1,820,971		88.3	34.9
US Total	10,955	32,784,004	845,931	3,796,742	22.3		34.2

4,223 GW on 326,142 square miles of water, which would generate 16,975,802 GWh at a capacity factor of 45.9%.

Looking at this potential state-by-state, the top ten states are shown in Table 6.2. The only state common to both lists is Texas, as the states shown for onshore wind power are, for the most part, land locked. Altogether, these ten states represent 81% of the entire US offshore potential.

Combining onshore and offshore potential gives a total of 15,178 GW on 1,172,073 square miles of land, with a generation of 49,759,806 GWh. To put this number in perspective, the total installed US capacity for electrical generation from all sources was 1,203 GW in 2017, and net generation was 4,058,258 GWh![7,8] And, the numbers get even better if the height of the wind turbine can be increased, as wind speeds generally increase with height. Using one wind speed calculator,[9] increasing the turbine height from 50 to 80, 100, and 150 m resulted in wind speed increases of 8%, 11%, and 18%, respectively.

Table 6.2 Top Ten US States for Offshore Potential Wind Power[4]

State	GW	GWh	Area, square miles	Capacity Factor
Hawaii	737	2,836,735	56,907	43.9
California	655	2,662,580	50,567	46.4
Michigan	423	1,739,801	32,631	47.0
Louisiana	341	1,200,699	26,302	40.2
North Carolina	306	1,269,627	23,631	47.4
Texas	271	1,101,063	20,961	46.4
Oregon	225	962,727	17,375	48.8
Massachusetts	184	799,344	14,214	49.6
Maine	147	631,960	11,384	49.1
New York	146	614,280	11,280	48.0
Ten State Total	3,435	13,818,816	265,253	45.9
US Total	4,223	16,975,802	326,142	45.9

Another NREL report[6] showed that increasing the wind turbine height from 80 to 100 m increased the onshore potential wind capacity from 10,955 GW to 12,000 GW. So the potential will increase as new generations of wind turbines become available.

According to the US Office of Energy Efficiency & Renewable Energy,[10] some of the newer wind turbines are 100 to 140 m tall, with the height measured at the center of the rotor, up to 1½ times the height of the Statue of Liberty! Also, the capacity factor improves because taller turbines are not obstructed by trees and buildings and the blades are longer than 60 m, allowing more efficient generation of electricity.

There have also been several studies of the worldwide wind energy potential. A study by the National Academy of Sciences[11] calculated a worldwide potential of 479,452 GW. This study was based on 2.5 MW turbines for onshore and 3.6 MW offshore at a hub height of 100 m. The same study gives the US potential as 38,813 GW, more than double the 15,178 GW from the NREL report discussed earlier.

A paper published by the World Wind Energy Association[12] used data from individual countries to determine a much smaller value of 94,500 GW. Unfortunately, this total has some issues, as the potential reported for all of Europe is 37,500 GW while the "rest of the world" is given as 10,400 GW. The "rest of the world" includes everything else except for the United States and Russia, so this potential seems too low considering all the land area in the "rest of the world" compared to Europe. The US potential was given as 11,000 GW, in good agreement with the NREL report.

Finally, a paper published by Stanford University[13] shows a worldwide potential of 72,000 GW. In this study, average wind speed was 6.9 m/s (~15 miles/hour), the turbine rating was 1.5 MW, and the height was 80 m.

Therefore, these three studies show worldwide potential ranging from 72,000 GW to 479,452 GW. To put these numbers in perspective, one source gives the total worldwide installed capacity for all electrical generation as 6,473 GW in 2016.[14] Thus, using just the low end number shows that the worldwide potential for new wind capacity is more than 11 times the current global installed capacity from all sources!

Naturally, these very optimistic projections ignore some complicating factors, such as regional political instability, whether the land is actually available, connecting wind power to the electric grid, global capital cost differences, labor costs, environmental resistance, and the like. Nevertheless, it does show the enormous worldwide potential for wind energy.

Table 6.3 shows the top ten countries for wind power potential, estimated by combining data from two references.[11,12]

These top ten countries control 82% of the 94,500 GW potential with Russia dominating the list at 38%. Given their large area and the fact average wind speeds tend to be stronger in the middle latitudes (roughly 25° to 65° north or south), it is not surprising that Russia, Canada, and the United States lead the world in wind potential.

In terms of current installed capacity, a Global Wind Energy Council report was used to rank the top ten countries using 2017 data.[1] The data are shown in Table 6.4, and 2016 data are also shown for comparison. The top ten countries have 84.7% of the worldwide capacity, and China and the United States dominate installed capacity

Table 6.3 Top Ten Countries by Potential Wind Power, 2014 Report[11,12]

Country	Wind Power Potential (GW)
Russia	36,000
Canada	12,375
United States	11,000
China	5,500
India	4,100
France	1,766
Sweden	1,695
Finland	1,766
United Kingdom	1,480
Germany	1,349
Total	77,031
World Total	94,500

Table 6.4 Wind Energy Installed Capacity (MW)[1]

Country	2016 (MW)	New in 2017	2017 (MW)	% of World Total
China	168,732	19,660	188,392	34.9
United States	82,060	7,017	89,077	16.5
Germany	50,019	6,581	56,132	10.4
India	28,700	4,148	32,848	6.1
Spain	23,075	96	23,170	4.3
United Kingdom	14,602	4,270	18,872	3.5
France	12,065	1,694	13,759	2.6
Brazil	10,741	2,022	12,763	2.4
Canada	11,898	341	12,239	2.3
Italy	9,227	252	9,479	1.8
Total Top Ten	411,119	46,081	456,731	84.7
World Total	487,279	52,492	539,123	100.0

with 34.9% and 16.5%, respectively. In fact, the 19,660 MW installed by China in 2017 exceeded the total installed capacity for five of the top ten countries!

To examine the rapid growth in wind power for China and the United States, an International Renewable Energy Agency (IRENA) report shows data for the last 10 years, reproduced in Table 6.5.[15] Note that these numbers differ somewhat from those shown in Table 6.4. As the table illustrates, there has been remarkable growth in the last 10 years for installed wind capacity. In China, installed capacity in 2018 is more than ten times larger than in 2009. For the United States, installed capacity increased almost three times from 2009 to 2018 and, for the world, it increased almost four times. This rapid growth is a result of reduced costs for wind energy as well as improvements in the technology, subjects discussed later in this chapter.

The Global Wind Energy Council report also shows the countries that have offshore wind capacity.[1] As of 2017, only 17 countries had any offshore wind installations, and the top twelve are shown in Table 6.6. The table was extended from top ten to top twelve to show the position of the United States. In 2017, the top ten countries accounted for 99.4% of all offshore capacity, and Europe accounted for 83.9% of the worldwide total, 15,785 MW. In addition to the seven European countries shown in the table, Ireland, Spain, Norway, and France also have some offshore wind farms. The United Kingdom and Germany are the world leaders, accounting for 36.3% and 28.5%, respectively. The 30 MW capacity in the United States is the Block Island Wind Farm off of the coast of Rhode Island.

Table 6.5 Wind Energy Installed Capacity (MW) for China, United States, and World by Year[15]

Year	China	United States	World
2009	17,599	34,296	150,096
2010	29,633	39,135	180,854
2011	46,355	45,676	219,984
2012	61,597	59,075	266,866
2013	76,731	59,973	299,941
2014	96,819	64,232	349,185
2015	131,048	72,573	416,225
2016	148,517	81,386	467,052
2017	164,392	87,543	514,622
2018	184,696	94,295	563,726

Table 6.6 Offshore Wind Energy Installed Capacity (MW)[1]

	2016 Total	New 2017	2017 Total
United Kingdom	5,156	1,680	6,836
Germany	4,108	1,247	5,355
China	1,627	1,164	2,788
Denmark	1,271	0	1,271
Netherlands	1,118	0	1,118
Belgium	712	165	877
Sweden	202	0	202
Vietnam	99	0	99
Finland	32	60	92
Japan	60	5	65
South Korea	35	3	38
United States	30	0	30
Top Ten Total	14,385	4,321	18,703
Total	14,483	4,334	18,814

6.3. Overview of Technology

An excellent review on the history of wind power is given by D.M. Dodge,[16] some of which is summarized here. The earliest known use of wind power is the sailboat, used as long ago as 4,000 BC by the Phoenicians and Egyptians. The first windmills, used to grind grain and pump water, were built in China, perhaps as much as 2,000 years ago. Documents found in Persia around AD 500–900 show the use of a vertical shaft with sails that rotated in a horizontal plane, spinning like a merry-go-round.

However, when Europe started using windmills, the shafts were disposed horizontally, with the sails rotating in the vertical plane, spinning like a pinwheel. The use of these large windmills declined with the advent of steam power.

Windmills were first used to generate electricity in Cleveland, Ohio, in 1888, by Charles F. Brush. The Brush windmill operated for 20 years, but only had 12 kilowatts (kW) of capacity. Further developments were made in 1891 by Poul La Cour, with a total capacity of 25 kW, but the use of fossil fuels was cheaper, and wind power electricity generation came to a stop.

Larger, utility-scale, turbine development occurred in Russia in 1931 with the introduction of the 100 kW Balaclava generator. Similar developments occurred in the United States, Denmark, France, Germany, and the United Kingdom from 1935 to 1970. The largest of these in the United States was the 1.25 megawatt (MW) turbine installed in Vermont in 1941. Government-sponsored wind energy R&D started in earnest in 1973 during the period of the Middle East oil embargo. When this crisis was over, federal funding dried up.

From 1973 to 1986, the wind turbine market evolved. Early in this time period, turbines sizes increased from 1–25 kW to 50–600 kW, and were used primarily for agriculture. Then, in the 1990s, California started constructing wind farms with turbine sizes of 20–350 kW. Eventually, after 1990, market development shifted to Europe and Asia, and turbines increased in size to the 1–5 MW turbines common today.

Moving now from history to theory, the amount of power in the wind depends on the air volume, velocity, and density. Using the well-known equation for kinetic energy, the wind energy is defined as:

$$\text{K.E.}\ (\text{Kinetic Energy}) = \tfrac{1}{2}\, m\, v^2$$, such that "m" is the air mass and "v" is the velocity.

This equation can also be framed in terms of the air density:

$$\text{K.E.} = \tfrac{1}{2}(A\, v\, t\rho)\, v^2 = \tfrac{1}{2} A\, t\rho v^3$$

Here "A" is the area covered by the turbine blades, "v" is the velocity, "t" is time, and ρ is the air density. Therefore, "A v t" is the volume of air passing through the turbine blade area "A" and "A v t ρ" is the mass of air passing through a given area "A."

Power is defined as energy per unit time, so the equation for power is:

$$P\text{ (Power)} = K.E./t = ½\, A\rho\, v^3$$

There is a limit on the efficiency of extracting power from the wind, known as the "Betz Limit." A brief overview is given here but more details can be found in reference 17. The power generated by a wind turbine is the difference between the power in and power out of the turbine. This is shown in equation form here:

$$P_{extracted} = P_{in} - P_{out} = ½\, m\,(v^2_{in} - v^2_{out})$$

In order for the wind turbine rotor to generate electricity, it must rotate, which means that the wind both enters and leaves the turbine rotor. It is a little abstract, but there must be air leaving the turbine in order for fresh air to come in. Therefore, air does exit the turbine but at a lower velocity, and is impossible to capture 100% of the original energy in the wind.

It can be shown that the wind velocity at the turbine is the geometric mean of the velocity into and out of the turbine, or

$$v_{turbine} = ½\,(v_{in} + v_{out})$$

Using this equation leads to the following equation for the power extracted from the wind.

$$P_{extracted} = ¼\,\rho\, A_{turbine}\,(v^2_{in} - v^2_{out})\,(v_{in} + v_{out})$$

This power can be compared to the total power in the wind, shown in the next equation.

$$P_{wind} = ½\,\rho\, A_{turbine}\, v^3_{in}$$

High wind speeds yield more power because wind power is proportional to the cube of wind speed.

Dividing $P_{extracted}$ by P_{wind} gives

$$\varepsilon_{Betz} = \frac{P_{extracted}}{P_{wind}} = \frac{\frac{1}{4}\rho\, A_{turbine}\,(v^2_{in} - v^2_{out})(v_{in} + v_{out})}{\frac{1}{2}\rho\, A_{turbine}\, v^3_{in}} = \frac{1}{2}\left[1 - \left(\frac{v_{out}}{v_{in}}\right)^2\right]\left[1 + \left(\frac{v_{out}}{v_{in}}\right)\right]$$

This equation shows that the Betz efficiency depends on the wind speed entering and leaving the turbine. By simple substitution, it can be shown that the maximum efficiency occurs when v_{out}/v_{in} is 1/3; that is, the maximum efficiency is when the gas velocity out of the turbine is 1/3 of the gas velocity into the turbine. At this ratio of velocity out and in, the efficiency is 59.3%.

The Betz limit is the maximum mechanical efficiency possible but, in practice, there are other factors reducing efficiency such as rotor blade friction and losses during the conversion of mechanical energy to electricity in the generator. Thus, the best designed real turbines operate at efficiencies of 44%, or about 75% of the Betz efficiency.

In addition, there are daily and seasonal variations in wind speed velocity so that overall capacity factors range from 20% to 40%. This is discussed in more detail later in the chapter.

Figure 6.2 shows a typical wind turbine and its parts.[18] Basically, there are four main parts to a wind turbine: the base, the tower, the nacelle, and the blades. The blades turn to capture the kinetic energy of the wind, which spins the generator in the nacelle. The tower supports the blades and nacelle and also houses the electrical wires that connect to the power collection system. The base supports the entire structure.

There are two basic types of wind turbines including Horizontal-Axis Wind Turbines (HAWTs) and Vertical-Axis Wind Turbines (VAWTs).[19] Blade length is the biggest factor in amount of electricity generated (recall the earlier equations showing the relationship of turbine blade area to power). The second biggest factor is the height of the turbine, and, recalling an earlier discussion, increasing turbine height from 50 to 150 m can increase wind speed by 18%. The largest turbines today have generating capacities of 5 to 12 MW and some of the tallest HAWTs have heights exceeding 100 m (330 feet), as tall as a 33-story building!

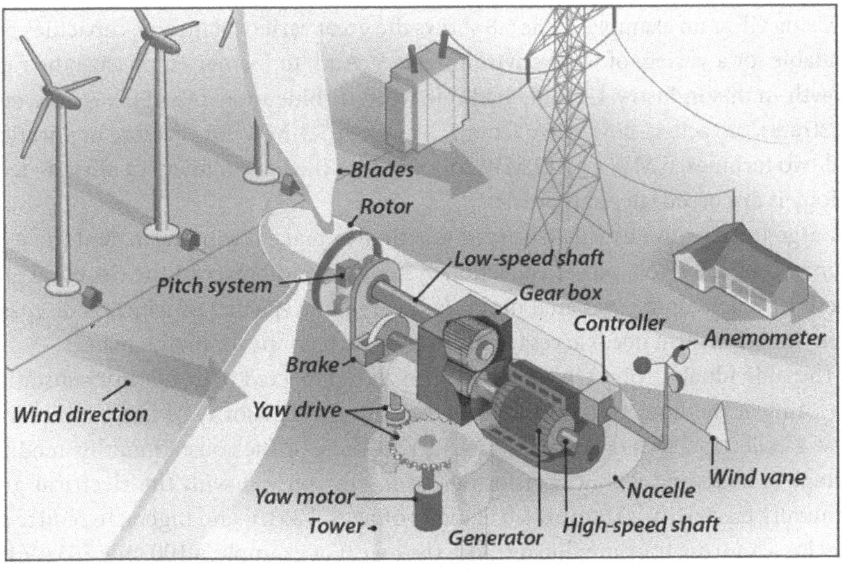

Figure 6.2 Typical Parts of a Wind Turbine (Source: Office of Energy Efficiency & Renewable Energy[18])

The HAWTs have blades similar to airplane propellers, and these commonly have three blades. This is also the most common type of turbine used commercially. VAWTs look like egg beaters in appearance, but few are used today because their performance is not as good as HAWTs.

For the HAWTs, their advantages include a tall tower base giving access to stronger wind and higher efficiency than VAWTs because their blades mostly move perpendicularly to the wind.[20] The HAWTs harness the most energy when the blades are perpendicular to the wind, and this is controlled by a yaw system that moves the turbine based on wind direction. Some of their disadvantages include the need for a massive tower to support the heavy blades, gearbox, and generator; their heights make them visible from long distances, considered by some as visual pollution; they need a mechanism to turn the blades toward the wind; and they need a braking device in high winds to stop the turbine, to prevent damage or even destruction. For the VAWTs, some advantages are that they do not need a mechanism to turn into wind, can be located nearer the ground for easier maintenance, will start generating at lower wind speeds, and can be built at locations where taller structures are prohibited. Their disadvantages include a decreased mechanical efficiency compared to HAWTs (because of more drag in rotating the blades) and their blades are closer to the ground where wind speeds are reduced.

In terms of company market shares, two US Department of Energy reports[21,22] were compiled to produce Table 6.7. These data show that General Electric (GE), Vestas, and Siemens control most of the market. Of the 7,017 MW installed in 2017, 29% used GE turbines, 35% used Vestas turbines, and 23% used Siemens turbines. This table also demonstrates what was shown earlier in Table 6.5, that US installed capacity has grown rapidly over the last decade.

Using GE as an example, Table 6.8 shows the great variety of turbine capacities now available for a variety of wind environments.[23] And, to further emphasize the rapid growth in this industry, GE only had one wind turbine model in 2002. As the table illustrates, capacities now range from 1.7 MW to 5.3 MW for onshore applications and two turbines, 6 MW and 12 MW, for offshore. The development of offshore technology is discussed later in more detail.

Large turbines are usually clustered together to make a wind farm, and the wind farm is connected to a power collection system to send power to the electrical grid. These turbines can be scattered over a large area, the spacing of which is discussed later. The wind farm needs access roads to each turbine site for maintenance.

The individual turbines in the wind farm are connected to a collector substation, operating at medium voltage. In North America, the majority of large wind farms have a collection system voltage of 34.5 kV.[24] Then, at the substation, this medium voltage is increased with a transformer so it can connect with the electrical grid. Generally, electricity is transmitted at high voltages, 120 kV and higher, to reduce energy losses during transmission due to resistance. For example, a 100 mile 765 kV line carrying 1,000 MW of power can have losses of 0.5% to 1.1% while a 345kV line carrying the same load across the same distance has losses of 4.2%.[25] Generally speaking,

Table 6.7 Annual US Turbine Installation Capacity (MW) by Manufacturer[21,22]

Manufacturer	2006	2007	2008	2009	2010	2011	2012	2013	2014	2015	2017
GE Wind	1,146	2,342	3,585	3,995	2,543	2,006	5,016	984	2,912	3,468	2,066
Vestas	439	948	1,120	1,489	221	1,969	1,818	4	584	2,870	2,481
Siemens	573	863	791	1,162	828	1,233	2,638	87	1,241	1,219	1,625
Acciona	0	0	410	204	99	0	195	0	0	465	0
Gamesa	74	494	616	600	566	154	1,341	0	23	402	0
Nordex	0	3	0	63	20	288	275	0	90	138	806
Sany	0	0	0	0	0	10	2	8	0	20	0
Goldwind	0	0	0	5	0	5	155	0	0	8	6
Mitsubishi	128	356	516	814	350	320	420	0	0	0	0
Suzlon	92	198	738	702	413	334	187	0	0	0	0
Other	2	50	587	973	180	502	1,086	4	2	2	31
TOTAL	2,457	5,253	8,362	10,005	5,216	6,820	13,131	1,087	4,854	8,598	7,017

Table 6.8 Types of Onshore and Offshore Turbines Available from GE[23]

Turbine Model	MW Rating	Hub Height, m	Rotor Diameter, m	Blade Length, m
2 MW-116 (onshore)	2	80, 90, 94	116	56.9
2 MW-127 (onshore)	2	89	127	62.2
3 MW-(onshore)	3	85, 110, 131, 134, 155	130,137	65, 68.5
Cypress 5.3 MW (onshore)	5.3		158	
1.7-100 (onshore)	1.7	80, 96	100	48.7
1.7-103 (onshore)	1.7	80, 96	103	50.2
1.85-87 (onshore)	1.85	80	87	42
1.85-82.5 (onshore)	1.85	121.25	82.5	40.3
2.75-120 (onshore)	2.75	85, 110, 120, 139	120	60
3.2-103 (onshore)	3.2	75, 85, 98	103	50.2
Haliade 150-6 (offshore)	6	100	150	73.5
Haliade-X (offshore)	12	260	220	107

the ten states in the United States with the greatest onshore wind energy potential—Texas, Kansas, Montana, Nebraska, South Dakota, North Dakota, Iowa, Wyoming, Oklahoma, and Alaska—are not close to large urban areas. Certainly Texas is a largely populated state, but wind generation in the panhandle needs to travel about 400 miles to reach Dallas and 600 miles to reach Houston. Therefore, while it is not impossible to transmit electricity long distances, the future transmission from sparsely populated states to large urban areas will depend on available transmission lines and the economics of accepting transmission losses compared to electricity produced closer to the urban areas.

Wind energy and wind farms have been developing rapidly, and the United States has some of the largest wind farms in the world. As of 2018, the top eight US onshore wind farms are shown in Table 6.9.[26–28] Five of these eight are located in four counties near Abilene, Texas, including Roscoe, Horse Hollow, Capricorn Ridge, Sweetwater, and Buffalo Gap. These five wind farms account for 2,143 turbines and a capacity of 3,288. Figure 6.1, shown earlier, indicates these counties have high average wind speeds.

And, wind farms are not exclusively onshore, as offshore wind farms are undergoing rapid development as well. Similar to an onshore farm, an offshore wind farm has wind turbines, inter-turbine cables to collect the electricity, possibly two substations (offshore and onshore), transmission cables to shore, and connection to the grid.[29] Compared to onshore, turbine ratings are generally much larger, 5 to 12 MW. The offshore substation is used to reduce electrical losses by increasing voltage before being exported to shore. This can be eliminated for small projects, projects close to shore (10 miles or less), or when the grid connection voltage is under 36 kV. Needless to say, relative to onshore, capital costs will be higher for offshore because of the complexity of turbine connection, the need for expensive turbines to withstand high wind and corrosive conditions of the sea, and the expense of underwater transmission cables to shore.

From 2016 to 2017, a total of 4,334 MW of new offshore wind power was installed worldwide, bringing the total to 18,814 MW.[1] Table 6.4 shows that the total new capacity installed in 2017 for all wind, onshore and offshore, was 52,492 MW for a total of 539,123 MW, so in 2017, 48,158 MW of new onshore capacity was installed for a total of 520,309 MW. Clearly, new offshore installations lag well behind onshore wind, and only 17 countries in the world have any offshore wind installations. Europe accounted for 83.9% of the 2017 worldwide total for offshore wind, spread among 11 countries, with the United Kingdom and Germany accounting for 36.3% and 28.5% of the world total.

The sole US offshore wind farm is the 30 MW Block Island Wind Farm located off of the coast of Rhode Island and operated by Deepwater Wind.[30] This wind farm became operational on December 12, 2016, and is located three miles southeast of Block Island, Rhode Island.[31] The farm has five turbines, each rated at 6 MW, and is connected to the New England grid by a new “sea2shore submarine transmission cable system.” Deepwater Wind is planning two larger offshore projects along the Atlantic Coast.

Table 6.9 Eight of the Largest Onshore Wind Farms in the United States[26–28]

Name	Owner	Location	Startup Year	Capacity, MW	Area, acres	Number and Types of Turbines
Alta Wind Energy Centre	Terra-Gen Power	Tehachapi, CA (turbines at altitude of 3,000 to 6,000 ft)	2011	1,548	9,000	586 from GE and Vestas; sizes 1.5, 1.7, 2.85, and 3.0 MW
Shepherds Flat	Caithness Energy	Near Arlington, OR	2012	845	19,200	338 GE 2.5 MW on three wind farms
Roscoe	E.ON Climate and Renewables	Nolan County, TX (near Abilene)	2008, 2009	781.5	100,000	627 (406 Mitsubishi 1 MW, 55 Siemens 2.3 MW, and 166 GE 1.5 MW)
Fowler Ridge Wind Farm	BP Alternate Energy and Dominion Resources	Benton County, Indiana (near Indianapolis)	2008, 2009	750	50,000	455 (182 Vestas, 1.65 MW, 40 Clipper 2.5 MW, and 233 GE 1.5 MW
Horse Hollow Wind Energy Centre	NextEra Energy Resources	Taylor and Nolan Counties, TX (near Abilene)	2009	735.5	47,000	421 (291 GE 1.5 MW and 130 Siemens 2.3 MW split into 3 parts)
Capricorn Ridge	NextEra Energy Resources	Sterling and Coke Counties, TX (near Abilene)	2007	662.5	52,645	407 (342 GE 1.5 MW and 65 Siemens 2.3 MW)
Sweetwater Windpower	Duke Energy and Infigen Energy	Nolan County, TX (near Abilene)	2003	585.3		392 GE, MHI, and Siemens
Buffalo Gap Wind Farm	AES Wind Generation	Nolan and Taylor Counties, TX (near Abilene)	2007, 2008	523.3	15,000	296 (67 Vestas, 1.8 MW, 155 GE, 1.5 MW, and 74 Siemens, 2.3 MW)

Returning to onshore wind farms, the area of land needed has some interesting aspects, and a good overall discussion is provided in an NREL technical report.[32] This report classifies two types of wind farm land use including (1) direct impact (land disturbed due to infrastructure development) and (2) total area (land associated with the complete wind plant project). A schematic of direct impact and total area land use is shown in Figure 6.3 (taken from the aforementioned report).

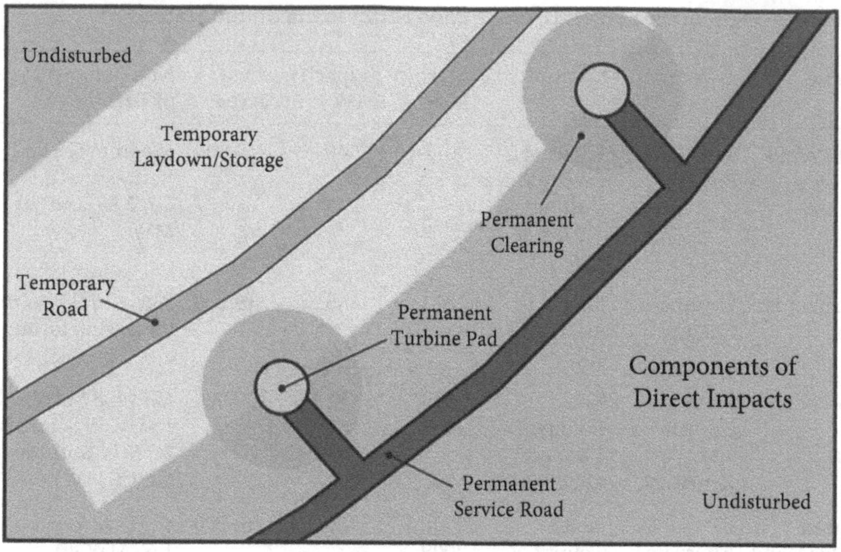

Figure 6.3 Illustration of the Two Types of Wind Plant Land Use: Total Area and Direct Impact Area (including permanent and temporary) (*Source*: NREL[32])

As the figure illustrates, things that directly occupy land in the wind farm include the wind turbine pad, service roads, and storage. Also, if the land was not previously cleared, development may include removal of trees. Land use defined in the figure as "temporary" are things done during construction that have a temporary impact such as construction roads, storage, and construction material that was placed on the land. After construction is complete, these areas will return to their original state, but the amount of time required is estimated to be 2 to 3 years for grasslands.

However, the total footprint of the wind farm really includes all the area containing the wind farm. That is not to say that the land cannot be used for other purposes, such as agriculture and ranching, but the turbines themselves must be sufficiently spaced to avoid wind interference. This is similar to the interference that can occur in sailing, where boats can cast a wind shadow (or "dirty wind") on other boats, blocking their wind and reducing the efficiency of the sail.

Using the terminology of one reference,[32] the arrangement of wind turbines can be classified into four general categories including (1) Single String, (2) Multiple Strings, (3) Parallel Strings, and (4) Clusters. As the terminology suggests, a Single String is a long string of turbine, Multiple Strings are strings of turbines that are not uniform, Parallel Strings are strings of turbines parallel to each other, and Clusters are wind farms with no observable turbine strings. Ideally, a wind farm would have Parallel Strings with grid spacing. Visual examples of these four configurations are shown in Figure 6.4.

In this NREL report, land-use data were reported for 172 projects representing 26,462 MW of capacity. Table 1 of this report shows that the direct impact area is 0.4 hectare/MW (1.0 acres/MW) and the total area is 34.5 hectare/MW (85.3 acres/MW).

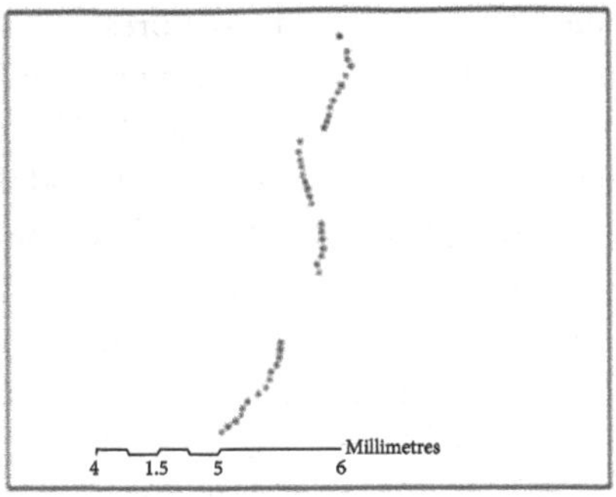

Figure 3.1. Single string configuration (Waymart, Pennsylvania)

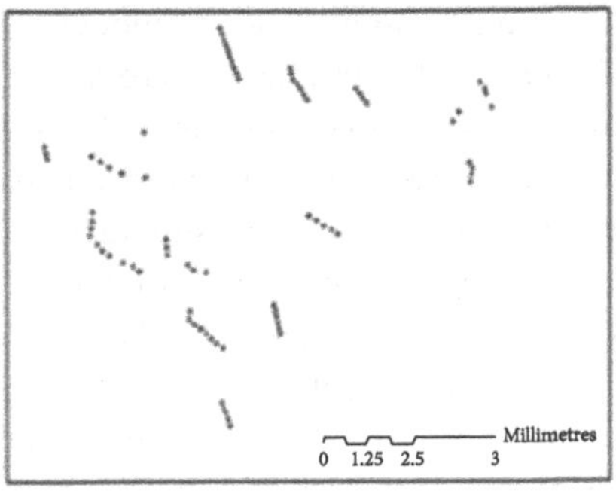

Figure 3.2. Multiple strings configuration (Wyoming Wind Energy Center, Wyoming)

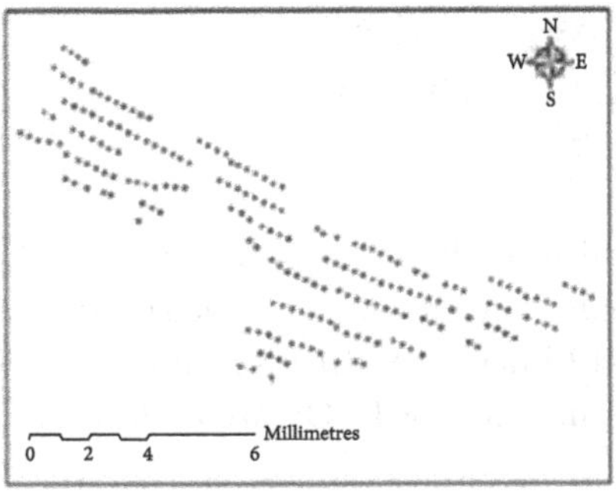

Figure 3.3. Parallel string configuration (Roscoe Wind Project, Texas)

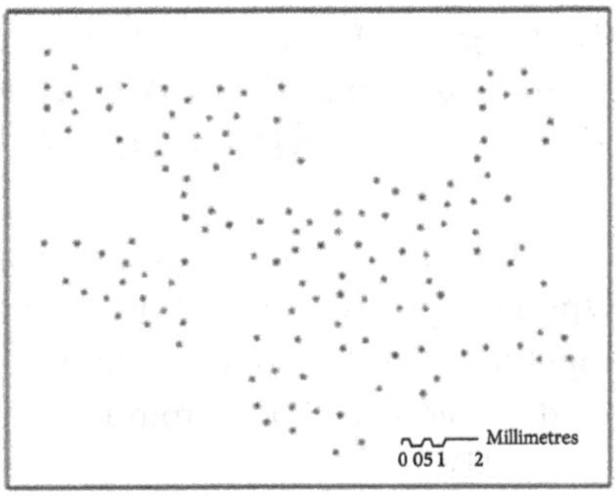

Figure 3.4. Cluster configuration (Spring Creek Wind Farm, Illinois)

Figure 6.4 Wind Farm Configurations (*Source*: NREL[32])

For the different types of configurations, the hectares/MW used were 34.7 for Parallel Strings, 27.7 for Multiple Strings, 30.3 for Single Strings, and 39.8 for Clusters. The authors state that the cluster configuration has greater total area per MW compared to the other configurations because of irregular turbine placement.

An average turbine spacing can be calculated from the previous numbers. As one hectare is the square of 100 meters, or 10,000 m^2, this means that the total area of 34.5 hectare/MW is 345,000 m^2/MW. If then, the turbines used were rated at 1 MW each, the area would be 345,000 m^2 for 1 MW and, taking the square root gives 587 m. If, instead, we use a turbine rated at 2 MW, the area would be 690,000 m^2 and the square root is 831 m. Table 6.8 shows blade lengths of about 60 m for a 2 MW turbine, giving a rotor diameter of 120 m, so a spacing of 831 m represents a spacing of about 7 rotor diameters (7 D).

The spacing of "7 D" comes from an average of existing wind farms and is not necessarily an optimal spacing. If the wind farm is laid out as a grid, similar to the parallel string configuration shown earlier, the common rule of thumb is for 3 to 5 rotor diameters in the cross-wind direction and 6 to 8 in the main wind direction.[33]

Looking at some economic modeling studies, one study used a net present value (NPV) model to optimize the cross-wind and main wind distances in a grid configuration, coming up with values of "3 D" in the cross-wind direction and "8 D" in the main wind direction.[34] Another study used an economic cost model and found that the optimal turbine spacing should be considerably higher, around "15 D," compared to the current convention of around "7 D.[35]" Taking the results together, and planning for a grid configuration that does not necessarily rely on one dominant wind direction, a spacing of around "7 D" is a reasonable value to calculate turbine spacing. Looking at Table 6.8 with onshore wind farm blade lengths varying from 40.3 to 68.5 m (rotor diameters 80.6 to 137 m), a "7 D" spacing for these turbines would range from 560 m (~1,850 feet) to 960 m (3,150 feet).

6.4. Capital Cost, Operating Cost, Well-to-Wheels Levelized Cost of Electricity, and Well-to-Wheels Levelized Cost of Fuel

The equations and methodology used to calculate levelized cost of electricity (LCOE), and specific input used for onshore and offshore wind farms, are detailed in Appendix A. Input for the calculations came from 2019 Energy Information Administration (EIA) data[36] and 2018 data from Lazard.[37] The units for LCOE are in $/megawatt-hour, or $/MWh.

For a wind farm, the "Well-to-Wheels" costs include (1) the cost of constructing the wind farm, (2) power plant generating costs, (3) cost of connecting the power plant to the electrical grid, (4) transmission and distribution costs, and (5) taxes (both federal and state). Unlike nuclear, natural gas, and coal, there are no fuel costs for a wind farm so there are no drilling or mining costs.

For the 2019 EIA data, the capital cost for an onshore wind farm was $1,624/kW,[36] similar to the $1,350/kW reported in 2018 data from Lazard.[37] For offshore wind, the EIA reported a value of $6,542/kW, more than double that of $3,025/kW reported by Lazard. Nevertheless, both studies show substantially higher capital cost than onshore wind.

Interesting, in 2015 the EIA reported an onshore capital cost of $2,213/kW and an offshore capital cost of $6,230/kW.[38] Thus, while the offshore capital cost increased by about 5%, the capital cost for onshore wind decreased by a sizable 27%! The current EIA value for onshore wind capital cost, $1,624/kW is still larger than that for natural gas combined cycle, $931/kW (see Chapter 2), but this 27% reduction does make wind more competitive, especially since there are no fuel costs.

The capital costs include the capital cost for new transmission lines to tie into existing electrical substations or existing transmission lines. Therefore, the generating station supplies the transmission lines and steps up the voltage to tie into the "electrical grid," a network that delivers electricity to the customer. The grid has high voltage transmission lines (sometimes distant) that carry power to demand centers, transformers to step down the voltage, and distribution lines that connect individual customers.

The costs of transmitting and distributing electricity were 0.9 cents/kWh (or $9/MWh) and 2.6 cents/kWh (or $26/MWh), taken from the EIA.[39] Here, transmission is defined as the high voltage movement of electricity from the power station while distribution is the cost of electrical lines that take power from a substation to the customer. The cost of transmission and distribution is essential to get a true "Well-to-Wheels" analysis of the total cost.

Table 6.10 shows the LCOE for the cost of generating electricity and the additional cost of transmission and distribution. An obvious advantage of wind power is that there is little carbon dioxide (CO_2) produced. Thus, if carbon capture and sequestration (CCS) become mandatory in the future, this will enhance any advantage wind power has over natural gas and coal.

Not surprisingly, the capital cost, operating costs, and transmission investment are much larger for an offshore wind farm. An offshore wind farm is attractive because wind speeds are generally higher and more stable. However, laying the foundation for the turbine pads is more difficult, and the structure must accommodate heavy storms and waves. In addition, it is expensive to connect the farm to the electric grid as the power cables must be installed under the ocean; this additional cost is reflected in the higher transmission investment, shown previously. Overall, the LCOE for offshore wind, including transmission and distribution, is more than double that for onshore wind.

A sensitivity analysis was made for both onshore and offshore electricity generation by varying the capital cost, the fixed operating and maintenance (O&M)

Table 6.10 Economic Analysis for Onshore and Offshore Wind Farms

Energy Type	Capacity Factor	Capital Cost	Levelized Capital Cost; with MACRS Depreciation	Fixed O&M	Transmission Investment	Total System Levelized Cost (LCOE) to Generate	LCOE Including Transmission and Distribution
		$/kW	$/MWh	$/MWh	$/MWh	$/MWh	$/MWh
Onshore							
EIA[36]	0.41	$1,624	$48.17	$13.48	$3.10	$64.75	$99.75
Lazard[37]	0.465	$1,350	$35.31	$7.92	$3.10	$46.32	$81.32
Average	0.44	$1,487	$41.74	$10.70	$3.10	$55.54	$90.54
Offshore							
EIA[36]	0.45	$6,542	$176.80	$20.33	$5.80	$202.93	$237.93
Lazard[37]	0.50	$3,025	$73.57	$21.69	$5.80	$101.06	$136.06
Average	0.48	$4,784	$125.19	$21.01	$5.80	$151.99	$186.99

expenses, and the capacity factor. Annual technology cost data from the NREL[40] were used to set the ranges for these different variables. The base cases for both the onshore and offshore wind farms used the EIA data.[36]

The capacity factor for onshore wind farms had a wide range, 11% to 48%. The NREL report used ten different areas of wind speed across the United States, so the wide range for capacity factor reflects wind speed differences. For offshore wind farms, the capacity factor had a smaller range, from 31% to 55%. As wind turbine heights increase, these capacity factors will improve since increasing turbine heights will expose the turbine to higher average wind velocities.

For an onshore wind farm, the capital cost range was $1,495 to $1,713/kW and the fixed O&M cost range was $8.63 to $14.20/MWh ($31 to $51/kW-yr). For an offshore wind farm, the capital cost range was $3,439 to $6,458/kW and the fixed O&M cost range was $26.89 to $40.33/MWh ($106 to $159/kW-yr). The results of the sensitivity study are shown in Table 6.11.

For onshore wind farms, the range used for capacity factor resulted in an LCOE difference of than $177/MWh, much higher than that shown by varying capital cost or fixed O&M. For offshore wind farms, which had a smaller range for capacity factor, the difference in LCOE was still a sizable $112/MWh. This shows the importance of plant utilization and, not surprisingly, if an onshore wind farm only operates at 11% of capacity versus 48%, this has a strong effect on the levelized cost of electricity (LCOE). Unlike onshore wind, the capital cost for offshore wind had a larger range and, thus, larger impact. Roughly doubling the capital cost increased the LCOE by about $82/MWh.

6.5. Cost of Energy

Unlike natural gas, coal, and nuclear electricity generation, wind farms do not use fuel, generating the electricity from the kinetic energy of wind. Thus, unlike these other technologies, electricity generated from wind is not affected by changes, sometimes unpredictable, in fuel cost.

6.6. Capacity Factor

For nonrenewable energy types, including natural gas, coal, and nuclear, capacity factors are less than 100% due to equipment failure, routine maintenance, or plant turndown because electricity is not needed or the price is too low. These types of plants typically have high capacity factors, ranging from 80% to 90%.

For renewable energy types, it is primarily the variation in energy input that keeps plant operation below their rated capacity. In the case of hydroelectric plants, discussed in Chapter 5, flowing water is the energy input and this varies year-to-year and month-to-month because of hydrologic conditions, competing use for water, and environmental and regulatory restrictions. Typical capacity factors range from

Table 6.11 Sensitivity Study for Onshore and Offshore Wind Farms

	Capacity Factor	Capital Cost	Fixed and Variable O&M (no fuel cost)	Total System Levelized Cost (LCOE) to Generate	Total LCOE Cost to Generate, Transmit, and Distribute
		$/kW	$/MWh	$/MWh	$/MWh
Base Case—Onshore Wind Farm[36]	0.41	$1,624	$13.48	$64.75	$99.75
Capex is $1,495/kW	0.41	$1,495	$13.48	$60.92	$95.92
Capex is $1,713/kW	0.41	$1,713	$13.48	$67.39	$102.39
Fixed O&M is $31/kW-yr	0.41	$1,624	$8.63	$59.90	$94.90
Fixed O&M is $51/kW-yr	0.41	$1,624	$14.20	$65.47	$100.47
Capacity Factor is 11%	0.11	$1,624	$50.25	$232.89	$267.89
Capacity Factor is 48%	0.48	$1,624	$11.52	$55.76	$90.76
Base Case—Offshore Wind Farm[36]	0.45	$6,542	$20.33	$202.93	$237.93
Capex is $3,439/kW	0.45	$3,439	$20.33	$119.07	$154.07
Capex is $6,458/kW	0.45	$6,458	$20.33	$200.66	$235.66
Fixed O&M is $106/kW-yr	0.45	$6,542	$26.89	$209.49	$244.49
Fixed O&M is $159/kW-yr	0.45	$6,542	$40.33	$222.93	$257.93
Capacity Factor is 31%	0.31	$6,542	$29.51	$291.95	$326.95
Capacity Factor is 51%	0.51	$6,542	$17.94	$179.73	$214.73

35%–40%, much less than that for nonrenewable electric plants, keeping hydroelectric plants well below their rated capacity.

For wind turbines, the energy extracted from the wind is limited to a maximum of 59.3%, the Betz efficiency discussed earlier in the "Overview of Technology" section. In practice, however, efficiencies are more typically 40 to 45%, and lower as the wind velocity increases. Recasting the equation for power extracted from the wind to include this efficiency leads to this equation:

$$P_{extracted} = (\pi/2)*r^2*v^3*\rho*\eta$$

Where: "r" is the rotor radius in meters (approximately the length of the turbine blade), "v" is the wind velocity in meters/second, "ρ" is the density of air in kg/m^3, and "η" is the turbine efficiency. A typical air density is 1.2 kg/m^3.

A typical manufacturer's performance curve may be found for the GE 2.75-120 onshore wind turbine, with the turbine parameters shown earlier in Table 6.8.[41] This turbine is rated for 2.75 MW with a rotor diameter of 120 m and blade length of 60 m. A generic performance curve for 1 2.5 MW wind turbine is also shown in Figure 6.5. The figure gives the rated power as a function of wind speed and the efficiency (C_p).

As the figure illustrates, the turbine does not generate any power until a wind speed of 3 m/s (6.7 mph), and increases to the rated output of 2.5 MW at 12 m/s (26.8 mph). The peak efficiency is almost 50% from 6 to 8 m/s (13.4 to 17.9 mph). At 7 m/s, and an efficiency of 50%, the previous equation for power yields 1.16 MW for a radius of 60 m, in reasonable agreement with the power curve. Thus, even though the turbine is operating at peak efficiency, the power output is well below its 2.5 MW design. The curve and the equation show that 2.5 MW is achieved at a speed of 12 m/s but the efficiency is only about 23%. Also, above 12 m/s, the power output does not exceed 2.5 MW. At higher wind speeds, the turbines are designed to limit the power to this maximum with no further increase in power.[42] Therefore, unless the wind speed is consistently above 12 m/s (26.8 mph), the turbine does not generate its maximum power.

Referring to Figure 6.1, shown earlier in the chapter, average wind speeds across the United States at a turbine height of 80 m vary from about 4 m/s (8.9 mph) to 8 m/s (17.9 mph). Applying the power curve of Figure 6.5 to these different wind speeds, the power outputs at 4 and 8 m/s would be 0.2 MW and 1.7 MW, respectively. Thus, at speeds of 4 m/s, the capacity factor would be a dismal 9% while the 1.7 MW output at 8 m/s would give a very respectable capacity factor of 69%. However, Figure 6.1 also shows that there are a limited number of geographic locations that have this superior wind velocity. Further complicating wind turbine performance is the variation in

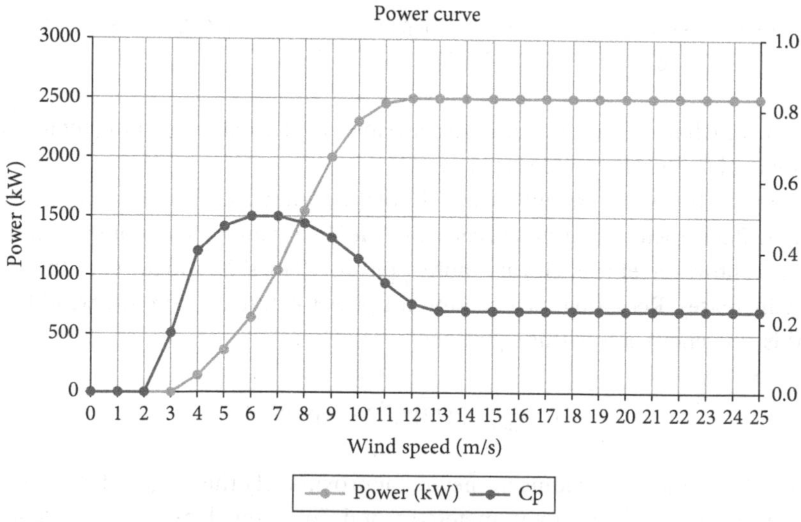

Figure 6.5 Performance Curve for Generic 2.5 MW Onshore Wind Turbine

Table 6.12 US Capacity Factors as a Function of Year[43]

Year	Capacity Factor
2013	32.4%
2014	34.0%
2015	32.2%
2016	34.5%
2017	34.6%
2018	37.4%

wind speed month-to-month and day-to-day. Wind speeds are typically high in the spring but tend to diminish in the summer months.

Another factor affecting capacity factor is that the wind speed can be either too little or even too much. As Figure 6.5 shows, the "cut-in speed," the speed at which the turbine first starts to rotate and generate power, is 3 m/s, equivalent to 6.7 mph. Likewise, there is a "cut-out speed," at which the wind force could cause damage to the turbine rotors, so a braking system is used to stop the rotation. This speed is typically about 25 meters per second, or 55.9 mph.

Given all these factors, it is not surprising that the average capacity factor shown in Table 6.10 for onshore wind farms is 44% while that for offshore wind farms is 48%. At a wind speed of 7 m/s, our generic 2.5 MW wind turbine would generate, on average 1.02 MW for onshore operation and 1.12 MW for offshore operation.

Capacity factors have been improving. Table 6.12 shows capacity factors for the last 6 years, increasing from 32.4% in 2013 to 37.4% in 2018.[43] In fact, capacity factors increased every year except for 2015, when there were continent-wide decreases in wind velocity.[44] The improvement is generally attributed to taller wind turbine hub heights and building new wind farms on sites with higher average wind speeds.[22] Wind speed increases with increasing altitude. On the other hand, curtailment of wind farms, defined as transmission and grid inflexibility issues, have a negative effect on capacity factors.

6.7. Efficiency: Fraction of Energy Converted to Work

For coal, natural gas, and nuclear energies, the thermal efficiency of the turbine used to generate electricity can be calculated using the BTU content of 1 kilowatt-hour (kWh) of electricity, or 3,412 BTU, and dividing this by the heat rate. The heat rate is the amount of energy used to make one kWh of electricity. However, for a wind turbine, as well as other renewables such as solar, hydroelectric, and geothermal, there is no fuel being consumed, and therefore energy used to make the electricity is not frequently measured. This is because renewable energy that is wasted in the process of making electricity has no fuel cost. If some of the kinetic power of the wind is not converted into electricity, there was no alternate use, or economic value, for the wind's kinetic energy.

Nevertheless, there are still inefficiencies in the process of making electricity from renewable fuels because no mechanical device is perfect. For a wind turbine, the mechanical efficiency is generally defined as the electrical power produced by the turbine divided by the wind power that went into the turbine.

To calculate efficiencies for renewable energy, the EIA[45] proposes an "Incident Energy Approach." In this approach, the amount of electricity made can be compared to the incident energy that is input to the mechanical device. In the case of wind, the incident energy is the kinetic energy of the wind entering the turbine blades. From Table F1 in this report, they report an efficiency of 26% for wind power. They further note that this average efficiency is based on the wind speed recommended by the manufacturer to achieve its nameplate rated output power. The rated output power typically occurs somewhere between wind speeds of 12 and 17 meters per second (or 27–38 mph), at which point the electrical generator reaches the power output limit.[46] At higher wind speeds, the turbine is designed to limit the power to this maximum level and there is no further rise in the output power.

A European group, the Union of the Electricity Industry, reports a greater wind turbine efficiency of up to 35%,[47] based on 40-meter blade diameters and 8 meters per second (18 mph) wind speed.

Earlier, we noted the maximum Betz efficiency is 59.3% but real turbines are designed to operate at efficiencies of 44%, clearly higher than the efficiencies cited earlier. However, at the maximum power output, Figure 6.5 shows the efficiency at about 23%, similar to the incident energy value of 26%. Unfortunately, the mechanical efficiency is a strong function of wind speed, and above a certain wind speed the efficiency decreases while the power output remains constant.

For calculations in the next section, the EIA value of 26% is used to represent a typical wind turbine mechanical efficiency.

6.8. Energy Balance

For a wind turbine, the energy balance is defined by this equation:

$$\text{Energy Balance} = \frac{[\text{Electrical energy from wind turbine delivered to customer}]}{[\text{Kinetic energy contained in the wind flowing into the wind turbine}]}$$

This energy balance is greatly simplified compared to coal, natural gas, and nuclear. That is because in the "Well-to-Wheels" analysis, there are no mining or extraction costs, and no fuel transportation costs of fuel to the wind turbine. Therefore, the energy balance only includes (1) making the electricity and (2) distributing the electricity through the grid to your home.

For the first step, the previous section set the wind turbine mechanical efficiency for making electricity at 26%.

It then remains to determine the energy needed to distribute the electricity through the grid to your home in order to construct the final energy balance.

Transmission and distribution losses are affected by the voltage of the power line and the distance the electricity is transmitted. Historical data from the state of California shows annual average transmission losses around 5 to 7% over the time period of 2002 to 2008.[48] Similarly, EIA data for July, 2016[49] show that of the electricity generated, 7% is lost in transmission and distribution. Therefore, a value of 7% is used for the amount of electricity lost in transmission and distribution after leaving the wind farm. Of course, this 7% lost in transmission and distribution is the percent of the energy leaving the electrical plant, not the kinetic energy of the wind entering the wind turbine.

Using 100 units of energy as a reference, we can arrive at the equation that follows. Summarizing, this means that of the original kinetic energy in the wind entering the wind turbine, 24.2% will end up as usable electricity. If the mechanical efficiency could be increased to 40%, this number would increase to 37.2%.

Wind Turbine

100 (entering plant) → 26 (generation) → 24.2 (transmission and distribution)

6.9. Maturity: Experience

As mentioned in the "Overview of Technology," the use of wind power dates back to the sailboat, as long ago as 4,000 BC.[16] And windmills were used to grind grain and pump water in China as much as 2,000 years ago.

Windmills were first used to generate electricity in Cleveland, Ohio, in 1888 by Charles F. Brush, but this was a small application, with the power rated at only 12 kilowatts (kW). Larger electrical plants followed in 1891 by Poul La Cour with a rating of 25 kW, but the use of fossil fuels was cheaper, and wind power electricity generation came to a stop.

In the 20th century, large utility scale turbine development occurred in 1931 Russia with the introduction of the 100 kW Balaclava generator. Similar developments followed in the United States, Denmark, France, Germany, and the United Kingdom from 1935 to 1970. The largest of these in the United States was the 1.25 megawatt (MW) turbine installed in Vermont in 1941. Government-sponsored wind energy R&D started in earnest in 1973 during the period of the Middle East oil embargo but, when this crisis was over, federal funding dried up.

From 1973 to 1986, the wind turbine market evolved to utility-size turbines from 50 to 600 kW. In the 1990s, California started constructing wind farms with turbine sizes of 20–350 kW. Eventually, after 1990, market development shifted to Europe and Asia, and turbines increased in size to the 1–5 MW turbines common today.

A review showing the timeline for the history of wind power gives dates for some important achievements.[50] In 1887, the first windmill for electricity production was built by Professor James Blyth of Anderson's College, Glasgow, Scotland, followed in 1888 by the 12 kW wind turbine built by Charles F. Brush. The first vertical axis wind turbine was built in 1931 by Frenchman George Darrieus and the first modern horizontal wind turbine was built the same year in Yalta, generating 100 kW. The first megawatt wind turbine, with 75-foot blades, was built in Vermont in 1941, and connected to the power grid.

The world's first onshore wind farm, with 20 turbines in total, was built in 1980 in New Hampshire. As shown earlier in Table 6.9, there are now some significantly sized onshore wind farms including Alta Wind Energy Centre (2011, 1,548 MW), Shepherds Flat (2012, 845 MW), Roscoe (2009, 781.5 MW), Fowler Ridge (2009, 750 MW), Horse Hollow Wind Energy Centre (2009, 735.5 MW), Capricorn Ridge (2007, 662.5 MW), Sweetwater Windpower (2003, 585.3 MW), and Buffalo Gap (2008, 523.3 MW).[26–28]

The world's first offshore wind farm was built in Vindeby, Denmark, in 1991, with eleven 450 kW turbines. And, at the time of this writing, the largest offshore wind farm in the world was the Walney Extension Wind Farm, with 659 MW and 87 turbines, and covering an area of 105 square km (41 square miles).[51] Next, and equally impressive, is the London Array,[52] located 20 km off England's southeast coast near Kent. The farm has 175, 3.6 MW turbines, for a total of 630 MW, and covers an area of 90 square km (35 square miles).

Since the first coal-fired electrical plant was built in 1882 and through the 1930s, the electrical grid system has grown rapidly, roughly doubling every 6 years. Today, what we take for granted, involves more than 180,000 miles of high-voltage transmission lines connecting about 7,000 power plants.[53]

To summarize, the first wind turbine to generate electricity dates back to 1887, with the modern horizontal wind turbine dating from 1931. The first megawatt turbine was built in 1941. The first onshore wind farm started in 1980 and the first offshore wind farm started operation in 1991. And, the first electrical grid system was built in 1882 followed by rapid development through 1930s. Thus, there are about 130 years of turbine development and about 78 years for megawatt-sized turbines. Onshore wind farms have about 39 years of experience and offshore wind farms have about 28 years of experience. Finally, there are more than 135 years of experience with electric grids.

6.10. Infrastructure

As shown earlier in Table 6.4, in 2017 the United States was ranked second in the world in total wind capacity with 89,077 MW, 16.5% of the world's 539,123 MW. According to data from the third quarter of 2017, there were more than 52,000 wind turbines operating in the United States,[54] an average of about 1.7 MW per turbine. And, in 2018, this same source listed 97,223 MW generated from 56,600 turbines, also an average of 1.7 MW per turbine. Similarly, a Windpower Engineering & Development report[55] listed 58,893 wind turbines spread over 1,604 wind farms, an average of 51 MW per farm. Of course, a "farm" can be anything from a single wind turbine to several thousand and these data lead to an average of 37 turbines per farm.

For 2017, a CNBC special report[56] reported that worldwide there are now 341,000 turbines. Based on a worldwide installed capacity total of 539,123 MW, this gives an average of about 1.6 MW per turbine.

In 2017, the United States generated 254,303 GWh (gigawatt-hours), 6.3% of the total of 4,058,258 GWh from all sectors.[57] Worldwide, this was second to the 305,700 GWh generated by China.[58] Iowa, Kansas, and Oklahoma produced more than 30% of their electricity from wind with 36.9%, 36.5%, and 32.0%, respectively while Texas led the nation on electricity produced at 67,061 GWh. Fourteen states generated more than 10% of their electricity from wind: Maine, Vermont, Iowa, Kansas, Minnesota, Nebraska, North Dakota, South Dakota, Oklahoma, Texas, Colorado, Idaho, New Mexico, and Oregon, all land-locked states with the exception of Oregon, Maine, and Texas.

Although China has about double the US installed capacity, 188,392 MW versus 89,077 MW, it only produced 305,700 GWh from this capacity versus the 254,303 GWh in the United States. This means China had a capacity factor of only about 18% versus 33% for the United States. Part of the greater use of capacity in the United States is due to strong wind resources, but another reason is improvement to the electric grid in terms of new or upgraded transmission lines to bring electricity to more densely populated parts of the United States.

Certainly, a key to wind energy, as well as other renewable types of energy, is having access to high-voltage transmission lines. Naturally, wind energy developers want to build a wind farm in the windiest locations, but these locations may not be located near high-voltage transmission lines in the grid. A common problem is defined as "grid curtailment," such that the output of the wind farm is curtailed because of either problems connecting to the grid or grid connection agreements that limit electricity sent to the grid. The NREL published a study in 2017 which showed that a small amount of new transmission lines added to the electrical grid could reduce this curtailment.[59] They focused their study on western states and found that without upgrading transmission lines, 15.5% of the wind power would be curtailed; however, with just four new transmission lines, this curtailment could be cut in half to 7.8%. The four new transmission lines, costing an estimated $10.1 billion, would bring access to 10.5 GW of power. These new transmission lines would carry electricity from wind-heavy states such as Wyoming, Montana, and New Mexico to Idaho, Nevada, and Arizona.

Another expansion of transmission lines was approved in 2011 by the Midcontinent Independent System Operator,[60] involving 17 new projects that encompass all or part of ten states from Indiana and Missouri on the east to the eastern edge of Montana.

And, in 2016 the Department of Energy (DOE) approved a project by Clean Line Energy Partners to build a 705-mile power line to carry 4 GW of power from the wind-rich Oklahoma panhandle to Arkansas and Tennessee.[61] This project, also referred to as the "Windcatcher Project" and expected to come into service in 2020, would have a new 600 kV transmission line costing $2.5 billion. Unfortunately, this partnership was terminated in March 2018, so, for now, it does not appear this transmission line will be built.[62]

Interestingly, there can be opposition to new transmission lines for various reasons. Sometimes, it is the expense of building these lines, as well as opposition by people who would be living in the path of these lines. Fossil fuel plants may resist new transmission lines, as the importing of alternate energy such as wind energy can undercut their business. And, in New Mexico, there was opposition to a new 515-mile transmission line, a $2 billion project that would send around 3,000 MW of stranded renewable energy west toward Arizona.[63] The federal government approved this new transmission line in January 2015. However, this project, which would take wind from New Mexico and solar and geothermal from New Mexico and Arizona, was first proposed in 2009. It received resistance from New Mexico politicians, who said it would create national security problems at the Army's White Sands Missile Range. The Bureau of Land Management resolved these concerns, and will bury some of the transmission lines when they go past the missile range.

Therefore, while some wind farms can use the more than 180,000 miles of high-voltage transmission lines in the grid,[53] there is also a need for new transmission line infrastructure to deliver electricity where the electrical grid is not available, or not sufficient. At the present, this new infrastructure is slowly being built.

6.11. Footprint and Energy Density

Footprint is a term commonly used to describe the land area needed to deliver energy for a certain energy type. In the case of wind power, the footprint includes the area to (1) make electricity at the wind farm and (2) transmit and distribute the electricity via the grid to commercial customers.

For the footprint and energy density to (1) make electricity at the wind farm, land area is based on total area used for the wind farm. That is, while direct impact land use only includes land disturbed by infrastructure, the total area includes all of the land containing the wind farm. Of course, most of this land can also be used for other things, such as agriculture or ranching, but the total area is needed so that the turbines are sufficiently spaced to avoid wind interference.

Data from twelve onshore wind farms were chosen to calculate an average MW per acre of land, shown in Table 6.13. The data ranged from 0.008 to 0.222 MW/acre, with an average of 0.021 MW/acre.

The 2009 NREL report discussed earlier,[32] which included 172 locations, gives an average area requirement of 85.3 acres/MW, or 0.012 MW/acre. The larger number

Table 6.13 Land Areas for Onshore Wind Farms

Name	Location	MW	Acres	Mw/ acre	Turbines
Alta Wind Energy Center[27,64]	California	1,548	9,000	0.172	566 (100 GE 1.5 MW SLE and 466 Vestas V 90-3 MW)
Shepherds Flat[26,27]	Oregon	845	19,200	0.044	338 2.5 MW GE2.5XL
Roscoe[26,27]	Texas	781.5	100,000	0.008	627 (406 Mitsubishi 1 MW; 55 Siemens 2.3 MW; 166 GE 1.5 MW)
Horse Hollow[26,27]	Texas	735.5	47,000	0.016	421 (291 GE 1.5 MW and 130 Siemens 2.3 MW)
Cedar Creek[65]	Colorado	300	32,000	0.009	274 (221 1 MW Mitsubishi and 53 1.5 MW GE)
Fantanele-Cogealac[26]	Romania	600	2,700	0.222	240 GE 2.5 MW XL
Fowler Ridge[26,27]	Indiana	750	50,000	0.015	420 (182 Vestas V82 1.65 MW; 40 Clipper C-96 2.5 MW; 133 GE 1.5 MW; 65 Siemens SWT 2.3 MW)
Whitelee[66]	Scotland	539	19,768	0.027	215 (140 2.3 MW Siemens, 69 3 MW ECO-100, 6 1.67 MW ECO-74)
Hopkins Ridge[67]	Washington	157	11,000	0.014	87 1.8 MW Vestas
Biglow Canyon[68]	Oregon	450	25,000	0.018	225 Vestas 2.5 MW
Capricorn Ridge[26]	Texas	662.5	52,645	0.013	407 (342 GE 1.5 MW; 65 Siemens 2.3 MW)
Buffalo Gap[26]	Texas	523.3	15,000	0.035	296 (67 Vestas V-80 1.8 MW; 155 GE 1.5 MW; 74 Siemens 2.3 MW)
Average		658	31,943	0.021	

for the data in Table 6.13 is heavily skewed by the Alta Wind Energy Center and the Fantanele-Cogealac wind farm; without these two cases, the average for Table 6.13 is 0.015 MW/acre. This NREL report also gives an average area requirement for just the direct impact area, which is 1 MW/acre. The direct impact area includes the turbine pads, roads, substations, and transmission equipment.

Table 6.13 also shows the number and types of turbines used at the different wind farms. Converting acres to square meters and assuming an equally spaced square grid of wind turbines, the turbine spacing can be calculated for each case. These are shown in Table 6.14, with the spacings ranging from a low of 213 m to a high of 803 m.

Table 6.14 Turbine-to-Turbine Spacing

Name	Location	MW	Acres	Turbine Spacing, m
Alta Wind Energy Center	California	1,548	9,000	254
Shepherds Flat	Oregon	845	19,200	479
Roscoe	Texas	781.5	100,000	803
Horse Hollow	Texas	735.5	47,000	672
Cedar Creek	Colorado	300	32,000	687
Fantanele-Cogealac	Romania	600	2,700	213
Fowler Ridge	Indiana	750	50,000	694
Whitelee	Scotland	539	19,768	610
Hopkins Ridge	Washington	157	11,000	715
Biglow Canyon	Oregon	450	25,000	671
Capricorn Ridge	Texas	662.5	52,645	724
Buffalo Gap	Texas	523.3	15,000	453
Average		658	31,943	581

Earlier, in the "Overview of Technology" section, a grid configuration that does not rely on one dominant wind direction had a spacing of "7 D" for turbine spacing, with "D" defined as the rotor diameter. Table 6.8 showed data for various GE wind turbines, and wind turbines with ratings of 2 MW to 3 MW had blade lengths of 50 to 70 m. With a blade length of 60 meters, the diameter would be 120 meters, suggesting a turbine spacing of 840 meters. Table 6.14 suggests that tighter spacing is being used, and the average of 581 m is a spacing of about "5 D." Of course, turbine spacing is also affected by turbine height. As well, the wind farms shown in Table 6.13 may not have optimal spacing or the total area may include plans for future expansion.

Similarly, twelve offshore wind farms were chosen to calculate an average MW per acre of land, shown in Table 6.15. Although not yet operational, Vineyard Wind off the coast of Massachusetts was included in the table. Construction was scheduled to begin in 2019 but the project has been delayed beyond 2020. The data ranged from 0.014 to 0.057 MW/acre, with an average of 0.025 MW/acre. Although there are wide ranges for both onshore and offshore wind farms, the average value for offshore was about 20% higher than onshore. In general, turbine spacing for the offshore wind farms is greater, due to the larger rotor diameters, but the MW/acre average is still higher. The likely reason is that offshore wind farms use higher rated turbines, leading to more power for the same area.

Table 6.15 Land Areas for Offshore Wind Farms

Name	Location	MW	Acres	MW/acre	Turbines	Turbine Spacing, m
London Array[26,52]	United Kingdom	630	22,239	0.028	175 Siemens 3.6 MW	717
Gemini[69]	Netherlands	600	17,297	0.035	150 Siemens SWT-4 MW	683
Greater Gabbard[70]	United Kingdom	500	36,324	0.014	140 SWT3.6 MW	1,025
Anholt[71]	Denmark	400	21,745	0.018	111 SWT 3.6 MW	890
BARD[72]	Germany	400	14,579	0.027	80 Bard 5 MW	859
Walney[51]	United Kingdom	659	25,946	0.025	87 (47 MHI Vestas and 40 Siemens Gamesa)	1,099
Thorntonbank[73]	Belgium	325	5,683	0.057	54 (6 Repower 5 MW and 48 Repower 6.15 MW)	653
Sheringham Shoal[74]	United Kingdom	317	8,649	0.037	88 Siemens 3.6 MW	631
Thanet[75]	United Kingdom	300	8,649	0.035	100 Vestas V90 3 MW	592
Lincs[76]	United Kingdom	270	8,649	0.031	75 Siemens 3.6 MW	683
Horns Rev II[77]	Denmark	209	8,154	0.026	91 Siemens SWT 2.3 MW	602
Vineyard Wind (a) [78]	United States (MA)	800	40,000 (b)	0.020	84 MHI Vestas 9.5 MW	1,388
Average		451	18,160	0.025		818

a. Although not yet operational, construction is scheduled to begin in 2019

b. Total land area leased is 160,000 acres, with a plan to have an eventual capacity of 3,200 MW. The area for the 800 MW wind farm was scaled accordingly.

Summarizing the discussion thus far, an energy density of 0.021 MW/acre is used for onshore wind farms and a value of 0.025 MW/acre is used for offshore wind farms.

The units of MW/acre can be converted to units of trillion BTU/(year-square mile). A 1 MW plant operating at 100% capacity for 1 year will produce 8,760 MWh (1 MW * 24 hour/day * 365 days/year). Also, there are 640 acres per square mile and 3.412 ×

10^6 BTU/MWh. For the capacity factors, Table 6.10 shows the capacity factor used for onshore wind farms as 44%, and for offshore wind farms as 48%.

From these values, we can calculate that 0.021 MW/acre for the onshore wind farm is equivalent to an energy density of 0.177 trillion BTU/(year-square mile), or 0.177×10^{12}. For the offshore wind farm, the calculated energy density is 0.230 trillion BTU/(year-square mile), or 0.230×10^{12}.

Now that the electricity has been made, the next step is to (2) transmit and distribute it to customers via the grid. However, the land area needed for step-up and step-down transformers is very small and can be ignored. Likewise, the metal towers and electrical lines take up very little land area and can coexist with farming, pasture land, and roadways. This means that the energy density for both onshore and offshore wind farms is essentially the land area for the wind farms.

In order to compare energy densities for different technologies on a common basis of gigawatt-hours per square mile per year, or GWh/(square mile-year), energy was converted from BTU to Watt-hour, or Wh, using the equation here:

$$1 \text{ BTU} = 0.293071 \text{ Wh}$$

Using this conversion and recognizing that a gigawatt (GW) is equivalent to one billion Watts (or 1×10^9), the calculated energy density for an onshore wind farm is 51.8 GWh/(year-square mile) and the energy density for an offshore wind farm is 67.3 GWh/(year-square mile).

As discussed earlier, an interesting aspect for an onshore wind farm is whether to calculate the footprint and energy density based on the total area or the direct impact area. The total area is needed to provide the proper turbine spacing and avoid wind interference among turbines; however, the area not directly impacted can still be used for agriculture and ranching. The average area requirement for just the direct impact area was cited earlier as 1 MW/acre, where direct impact area includes the turbine pads, roads, substations, and transmission equipment. Using instead an energy density of 1 MW/acre based on the direct impact area and the same onshore capacity factor of 44% used earlier, the direct onshore energy density is of 8.42 trillion BTU/(year-square mile), or 8.42×10^{12}. Converting this to a basis of gigawatt-hours per square mile per year yields 2,466.7 GWh/(square mile-year).

Assuming a 1,000 MW wind farm, the footprint for the onshore wind farm is 169 square miles and for the offshore wind farm is 130 square miles. For the onshore wind farm based solely on the direct impact area, the footprint is 3.6 square miles. Another way to look at this is to say that that about 2% of the wind farm is direct impact area.

According to the International Energy Agency,[79] the US electric power consumption per capita for 1 year was 12.8 MWh in 2016. Using the capacity factor of 44% for an onshore wind farm, a simple calculation shows that 1 MW will power about 300 homes for 1 year, so a 1 GW plant will power about 300,000 homes! For an offshore wind farm with a capacity factor of 48%, a 1 GW plant will power about 330,000 homes.

6.12. Environmental Issues

Before discussing environmental issues for wind turbines and farms, it is useful to present their positive benefits to the environment, such as avoiding air pollution that can occur from sulfur and nitrogen oxides from fossil fuels, reducing greenhouse gas emissions, and not using water to generate electricity. Also, once land is allocated to build a wind farm, there is no additional land use to provide fuel, as is the case with petroleum, coal, natural gas, and uranium, where drilling and exploration or mining are needed.

However, wind farms are not without some environmental issues, including land use, habitat disturbance, impact on birds and bats, soil erosion, noise, and even visual impact issues. These six different issues are discussed in a Penn State University 2010 report.[80] With respect to land use, some sites are just not ideal for a wind farm, such as housing developments, airport approaches, and radar installations. A wind farm can also disrupt animal habitats. For example, some studies have shown that birds and animals avoid nesting on hunting grounds near wind farms.

When birds and bats do not avoid the areas near a wind farm, bird death can result from the turbines. Unfortunately, the land geography that makes a site a good wind farm location may also be attractive to the birds and bats. In the early 1980s, it was discovered that a large number of raptors, some endangered, were killed at the Altamont Pass, California, wind farm. However, a 2014 report by the American Wind Wildlife Institute[81] indicates that the newer and larger wind turbines have reduced raptor deaths. They cite several factors for this, including that the larger turbines have fewer rotations per minute (thus reducing raptor collision rates), there are fewer perching sites on the modern turbines using tubular supports, and there are simply fewer turbines needed to produce the same total megawatts of power needed for the wind farm.

According to a 2008 Department of Energy report,[82] bird death caused by wind energy is small compared to other human causes. Using their data, for every 10,000 bird deaths, only 0.75 were caused by wind turbines versus 5,500 from building and windows, 1,000 from cats, 1,000 from other animals, 800 from electrical transmission lines, 700 from vehicles, 700 from pesticides, and 250 from communication towers. Nevertheless, a 2015 Institute for Energy Research report[83] cites a study saying that every year, 573,000 birds (including 83,000 raptors) and 888,000 bats are killed by wind turbines. Limited research suggests that bats are attracted to wind turbines for several reasons, including blade movement and sound.

Soil erosion can also be an issue, especially if the wind turbines are placed on a slope or if the natural soil is removed in the wind farm development.[80] In some cases, rain can create gullies through this erosion. Of course, this problem is easily prevented or mitigated if the wind farm design and construction are done in a way to minimize erosion.

Yet another issue is the visual impact of wind turbines, although this can be said for any electrical plant or refinery. In the case of wind turbines, they are visible from great distances, which can raise concerns with local communities and land owners. And the general public does not want to see a wind farm built in a scenic vista or state and national park.

Noise is another issue, but newer turbines are less noisy than older ones. The noise, created by rotor blade movement, typically ranges from 35 to 40 decibels (dB), roughly equivalent to the sound made by a kitchen refrigerator. This is, however, much less than things such as a vacuum cleaner (70 dB), freight train (70 dB), or jet engine (120 dB).

In addition to the six different issues discussed in the Penn State University report, another issue concerning wind farms is that they can actually change air temperature in the vicinity of the wind farm. According to a *Scientific American* article,[84] turbine blades chop up incoming wind that mixes different layers of the atmosphere having different temperatures. In an area near Palm Springs, California, temperature measurements found that the wind farm made it warmer at night and cooler during the day. This could be a good thing if it protects crops from frost, such as an orange farm.

One last topic associated with the environmental impact of a wind farm is the need for new transmission lines. Unlike traditional electric plants that use fuel such as natural gas, coal, or nuclear, renewable energy electric plants are built where the conditions are favorable, and this may require a new transmission line to integrate into the electric grid. And these new transmission lines will probably be several hundred miles long. For example, even though the project has been canceled, the Clean Line Energy Partner line that was to bring wind energy from the Oklahoma panhandle to Arkansas and Tennessee was going to be 705 miles long. Also, a TransWest Express 725-mile, 600 kV DC transmission line is scheduled to start service in 2023, to bring wind energy from Wyoming to Utah, Nevada, and California.[85]

According to the Environmental Protection Agency (EPA), new transmission lines can disturb forests and wetlands as new power lines and access roads are built in formerly undeveloped areas.[86] Access roads are needed for routine maintenance, such as trimming trees and other plants. As well, new transmission lines can affect bird migration routes and other ecosystems.

Interestingly, high-voltage transmission lines can also have negative effects on humans, animals, and plant life because of the electromagnetic field they create.[87] For humans, short-term health issues include headaches, fatigue, anxiety, and even insomnia while long-term issues include risk to the DNA, cancer, and even miscarriages. For plants, a study found that increasing the power line voltage from 100 kV to 230 kV, thereby increasing the electromagnetic field, reduced growth and production.

6.13. CO_2 Production and the Cost of Capture and Sequestration

Since a wind farm does not use any fuel to generate heat, no CO_2 is directly produced when electricity is made. However, CO_2 is produced from construction and operation. Analyzing this CO_2 production is referred to as a "lifecycle greenhouse gas emissions analysis," because it examines the production of CO_2 over the life of the plant including construction, use, and even dismantling. In one reference, Hondo[88] gives a total of 29.5 g CO_2/kWh from a wind farm with 21.2 g (72%) due to construction and 8.3 g (28%) from operation. Of the 21.2 g CO_2/kWh produced from construction, Hondo breaks this down as 7.4 for the turbine foundations, 1.4 for the blades, 5.9 for the nacelle, 3.4 for the turbine tower, and 3.0 for the turbine. This produced CO_2 comes mainly from the production of steel and concrete.

An NREL 2013 report examined forty-nine different references and adjusted the reported values to a consistent set of methods and assumptions.[89] From this study, they report average values of 30 g CO_2/kWh for onshore wind farms and 40 g CO_2/kWh for offshore wind farms.

Some studies show wide ranges of CO_2 production per kWh. For example, a 2015 report[90] reports an average of 34.1 g CO_2/kWh with a low estimate of 0.4 and a high of 364.8 g CO_2/kWh. To explain the wide range, the authors cite factors such as input, technology, location, sizing, and capacity as well the use of different calculation methods. Another 2015 report[91] gives an average value of 15 g CO_2/kWh for onshore wind with a range of 3 to 45 g CO_2/kWh. The same report gives an average value of 12 g CO_2/kWh for offshore wind with a range of 7 to 23 g CO_2/kWh. Finally, a 2011 report[92] reports an average value of 26 g CO_2/kWh with a low of 6 and a high of 124.

Summarizing these five different studies leads to an average value of 27 g CO_2/kWh. Unfortunately, for the two reports that differentiate between onshore and offshore wind farms, one report shows onshore CO_2 emissions being higher while the other report shows offshore being higher. Logically, we would expect the steel and concrete needed for an offshore wind farm to be more substantial, thus consuming more CO_2 from construction but, on the other hand, an offshore wind farm uses higher MW rated turbines, which could reduce total steel and concrete needed. Given the wide ranges of CO_2 emissions calculated by the different reports, no differentiation is made for onshore versus offshore, and an average value of 27 g CO_2/kWh is applied to both.

Carbon dioxide production is commonly listed in the units of lb CO_2/MM BTU. However, wind farms generate electricity without the combustion of a fuel. Because of this, there is no heat rate to convert from CO_2 produced on a mass basis and that produced on an energy basis. The conversion from g CO_2/kWh to lb CO_2/kWh is a simple conversion, and this value is 0.060 lb CO_2/kWh.

For a 1,000 MW onshore wind farm with a capacity factor of 44%, the plant will generate 10,560 MWh/day. For CO_2 produced on a kWh basis, the life cycle production of CO_2 is equivalent to 628,585 lb CO_2/day or about 314 tons CO_2/day. For the offshore wind farm with a capacity factor of 48%, the plant will generate 11,520 MWh/day and 685,729 lb CO_2/day, or about 343 tons CO_2/day.

Compared to other technologies, the CO_2 production for an onshore wind farm is about 44% of a nuclear power plant (713 tons CO_2/day), 1.4% of a pulverized coal combustion plant (22,214 tons CO_2/day), and 4% of a natural gas combined cycle plant (7,500 tons CO_2/day).

As this CO_2 is not directly produced by the wind farm, but rather by other processes such as construction and operational maintenance, it is not practical to capture any of this CO_2. Therefore, it is not likely that CO_2 produced from the construction and operation of a wind farm would be removed by CCS.

6.14. Chapter Summary

Frankly, the advent of wind energy in the United States, as well as the world, has been staggering. For the United States, data from the DOE[93] since the turn of the century are shown in Table 6.16, illustrating the rapid growth in wind-generated electricity. Since 2000, when installed capacity was a mere 2,539 MW, this has now grown to 84,143 MW in just 18 years, an increase of 3300%! This growth has been especially rapid since 2004, with the biggest single-year increase from 2011 to 2012. For hydroelectric power, the total installed capacity for 2017 was 102,867 MW. Given the growth for wind power, it is likely that wind will exceed hydroelectric power before 2025.

Table 6.16 Installed Capacity (MW) by year[93]

Year	Installed Capacity, MW
2000	2,539
2001	4,232
2002	4,687
2003	6,350
2004	6,723
2005	9,147
2006	11,575
2007	16,907
2008	25,410
2009	34,863
2010	40,267
2011	46,916
2012	60,005
2013	61,108
2014	65,877
2015	74,472
2016	82,171
2017	84,143

Since it is renewable, unlike petroleum crude oil, coal, natural gas, and nuclear, there is no clock running to a day when there will no longer be wind power. And, there is potential of adding new wind-power electrical plants, with a potential of as much as 11,000 GW in the United States. The potential is even higher in Russia and Canada, with potential wind power of 36,000 and 12,375 GW, respectively. In all, there is a potential of 94,500 GW of wind energy in the world. To put this number in perspective, if the 94,500 GW operated at a capacity factor of 44%, it would generate 364,241 terawatt-hours. In 2017, the total worldwide electricity generation was 25,551 terawatt-hours,[94] so this means that full utilization of the worldwide wind energy potential would make enough electricity to meet fourteen times the needs of the world in 2017! Naturally, there are many hurdles to jump over to make this happen, including improvements to the electrical grid and economical, geographical, technical, and environmental hurdles, just to name a few.

The "Well-to-Wheels" economic analysis gave an average LCOE, including transmission and distribution, of $90.54/MWh for an onshore wind farm and $186.99/MWh for an offshore wind farm. Not surprisingly, the capital cost, operating costs, and transmission investment are much larger for an offshore wind farm. Capital costs are decreasing rapidly, especially for onshore wind farms. In 2015, the EIA reported an onshore capital cost of $2,213/kW, but this decreased to $1,624/kW in 2019. For the same time period, however, there was little change in the capital cost for offshore wind farms.

Although a wind farm produces some amounts of CO_2 from construction and operation, capturing this CO_2 is not practical.

For a 1,000 MW onshore wind farm with a capacity factor of 44%, a life cycle analysis for the plant shows that, on a kWh basis, the plant will produce 27 g CO_2eq/kWh or 0.060 lb CO_2/kWh. For CO_2 produced on a kWh basis, the life cycle production of CO_2 is equivalent to 628,585 lb CO_2/day or about 314 tons CO_2/day. For an offshore wind farm with a capacity factor of 48%, the plant will generate 11,520 MWh/day and 685,729 lb CO_2/day, or about 343 tons CO_2/day.

For a 1,000 MW onshore wind farm with a capacity factor of 44%, the footprint for the plant is 169 square miles. If we based the footprint on only the direct impact area, defined as turbine pads, roads, substations, and transmission equipment, the footprint is only 3.6 square miles. For the offshore wind farm with a capacity factor of 48%, this footprint decreases to 130 square miles. The capacity factors for onshore and offshore wind farms are expected to increase as wind turbine heights increase and turbine technology improves.

The overall energy balance for a typical wind farm was calculated to be 24.2%, meaning that about 24% of the kinetic energy originally in the wind ends up as energy at the distribution point as electricity. This low number is primarily due to the efficiency of the wind turbine, which has a mechanical efficiency of 26%. Since there are no fuel costs, the mechanical efficiency of the turbine is not as important as fossil fuel-based electricity generation but, improving efficiency would have the benefits of reducing the number of wind turbines needed to generate the same amount of electricity.

In terms of experience and maturity, wind turbines for the purpose of generating electricity have been around since 1887 with the modern horizontal wind turbine dating from 1931. The first onshore wind farm started operation in 1980 and the first offshore wind farm in 1991.

Also, there is a growing amount of infrastructure for wind-generated electricity, as well as new electrical transmission lines to bring this electricity from remote locations to consumers. In 2017, the United States had a total wind capacity of 89,077 MW, second in the world to China's 188,392 MW. For the same year, the United States generated 254,303 GWh, 6.3% of the total of 4,058,259 GWh from all sectors. Worldwide, this generation from wind was second to the 305,700 GWh generated by China.

Looking at individual states, fourteen states generated more than 10% of their electricity from wind: Maine, Vermont, Iowa, Kansas, Minnesota, Nebraska, North Dakota, South Dakota, Oklahoma, Texas, Colorado, Idaho, New Mexico, and Oregon, all land-locked states with the exception of Oregon, Maine, and Texas. Iowa, Kansas, and Oklahoma produced more than 30% of their electricity from wind with 36.9%, 36.5%, and 32.0%, respectively, while Texas led the nation on electricity produced at 67,061 GWh.

Worldwide, in 2017 there were more than 341,000 wind turbines. Based on a worldwide installed total of 539,123 MW, this gives an average of about 1.6 MW per turbine. For the United States, there were more than 52,000 wind turbines in 2017, with an average of about 1.7 MW per turbine.

The energy densities for the onshore and offshore wind farms were determined to be 51.8 GWh/(year-square mile) and 67.3 GWh/(year-square mile), respectively. For the onshore wind farm based on the direct impact area, the energy density is 2,466.7 GWh/(square mile-year).

According to the International Energy Agency, the US electric power consumption per capita for 1 year was 12.8 MWh in 2016. Using the capacity factor of 44% for an onshore wind farm, a simple calculation shows that 1 MW will power about 300 homes for 1 year, so a 1 GW plant will power about 300,000 homes! For an offshore wind farm with a capacity factor of 48%, a 1 GW plant will power about 330,000 homes.

Finally, the main environmental issues for wind farms are land use, habitat disturbance, impact on birds and bats, soil erosion, noise, and visual impact. And, for the long-distance transmission lines that are needed to bring wind electricity to urban areas, there are additional issues such as disturbance of forest and wetlands, and the negative effect of electromagnetic fields created from the transmission lines on humans, animals, and plant life.

References

1. "Annual Market Update 2017," Global Wind Report, Global Wind Energy Council, April 2018, https://www.researchgate.net/publication/324966225_GLOBAL_WIND_REPORT_-_Annual_Market_Update_2017

2. "GWEC Reports 51.3 GW of New Wind Capacity in 2018," Jannah Mason, Composites World, March 1, 2019, https://www.compositesworld.com/news/gwec-reports-513-gw-of-new-wind-capacity-in-2018
3. "Wind in Numbers," Global Wind Energy Council, 2017, http://www.gwec.net/global-figures/wind-in-numbers/
4. "Renewables 2018: Global Status Report," Renewable Energy Policy Network for the 21st Century, 2018, http://www.ren21.net/wp-content/uploads/2018/06/17-8652_GSR2018_FullReport_web_final_.pdf
5. "U.S. Renewable Energy Technical Potentials: A GIS-Based Analysis," A. Lopez et al., National Renewable Energy Laboratory, Technical Report NREL/TP-6A20-51946, July 2012, http://www.nrel.gov/docs/fy12osti/51946.pdf
6. "New Wind Energy Resource Potential Estimates for the United States," D. Elliott et al., National Renewable Energy Laboratory, 91st Annual Meeting of the American Meteorological Society, Seattle, Washington, January 27, 2011, http://www.nrel.gov/docs/fy11osti/50439.pdf
7. "Count of Electric Power Industry Power Plants, by Sector, by Predominant Energy Sources within Plant, 2007 through 2017," Table 4.1, Energy Information Administration, http://www.eia.gov/electricity/annual/html/epa_04_01.html; "Existing Capacity by Energy Source, 2017 (Megawatts)," Table 4.3, Energy Information Administration, https://www.eia.gov/electricity/annual/html/epa_04_03.html
8. "Electric Power Annual 2017," Energy Information Administration, October 2018, revised December 2018, https://www.eia.gov/electricity/annual/pdf/epa.pdf
9. "Wind Speed Calculator," Danish Wind Industry Association, May 2003, http://xn--drmstrre-64ad.dk/wp-content/wind/miller/windpower%20web/en/tour/wres/calculat.htm
10. "Unlocking Our Nation's Wind Potential," May 19, 2015, Office of Energy Efficiency & Renewable Energy, https://energy.gov/eere/articles/unlocking-our-nation-s-wind-potential
11. "Global Potential for Wind-Generated Electricity," Xi Lua et al., Proceedings of the National Academy of Sciences of the United States of America (PNAS) 106 (27), 10933–10938, July 7, 2009, http://www.pnas.org/content/106/27/10933.full.pdf
12. "World Wind Resource Assessment Report," Jami Hossain, World Wind Energy Association, WWEA Technical Paper Series (TP-01-14), December 2014, WWEA_WWRAR_Dec2014_2.pdf
13. "Evaluation of Global Wind Power," C.L. Archer and M.Z. Jacobson, Department of Civil and Environmental Engineering, Stanford University, Journal of Geophysical Research 110, D12110, June 30, 2005, https://web.stanford.edu/group/efmh/winds/2004jd005462.pdf
14. "Q3 2017: Global Power Markets at a Glance," Power Technology, December 12, 2017, https://www.power-technology.com/comment/q3-2017-global-power-markets-glance/
15. "Renewable Capacity Statistics 2019," International Renewable Energy Agency, https://www.irena.org/publications/2019/Mar/Capacity-Statistics-2019
16. "Illustrated History of Wind Power Development," D.M. Dodge, http://www.telosnet.com/wind/
17. Energy Devices—The wind (Lecture 4), Eric R. Switzer, October 24, 2009, http://kicp.uchicago.edu/~switzer/compton/lecture4_notes.pdf
18. "The Inside of a Wind Turbine," Office of Energy Efficiency & Renewable Energy, 2020, http://energy.gov/eere/wind/inside-wind-turbine-0
19. "Types of Wind Turbines," Energy Information Administration, December 4, 2019, https://www.eia.gov/Energyexplained/index.cfm?page=wind_types_of_turbines
20. "Types of Wind Turbines," Centurion Energy, 2020, https://education.seattlepi.com/horizontal-vs-vertical-wind-turbines-3500.html

21. "2015 Wind Technologies Market Report," R. Wiser and M. Bolinger, US Department of Energy, Energy Efficiency and Renewable Energy, August 2016, https://energy.gov/sites/prod/files/2016/08/f33/2015-Wind-Technologies-Market-Report-08162016.pdf
22. "2017 Wind Technologies Market Report," US Department of Energy, Energy Efficiency and Renewable Energy, 2017, https://www.energy.gov/sites/prod/files/2018/08/f54/2017_wind_technologies_market_report_8.15.18.v2.pdf
23. "Wind Turbines Overview," General Electric, 2018, https://www.gerenewableenergy.com/wind-energy/turbines.html
24. "Wind Power Plant Collector System Design Considerations," IEEE PES Wind Plant Collector System Design Working Group, E.H. Camm et al., October 28, 2009, http://power.eecs.utk.edu/pubs/Fangxing_li_ieeepes2009_4.pdf
25. "American Electric Power, Transmission facts," p. 4, December 31, 2007, https://web.archive.org/web/20110604181007/https://www.aep.com/about/transmission/docs/transmission-facts.pd
26. "Top 10 Biggest Wind Farms," Power Technology, September 30, 2013, http://www.power-technology.com/features/feature-biggest-wind-farms-in-the-world-texas/
27. "Top 10 Onshore Wind Farms," Tom Wadlow, Energy Digital, January 5, 2017, https://www.energydigital.com/top-10/top-10-onshore-wind-farms
28. "The 11 Biggest Wind Farms and Wind Power Constructions That Reduce Carbon Footprint," Kashyap Vyas, February 15, 2018, Interesting Engineering, https://interestingengineering.com/the-11-biggest-wind-farms-and-wind-power-constructions-that-reduce-carbon-footprint
29. "Offshore Electrical System," Wind Energy: The Facts, 2020, https://www.wind-energy-the-facts.org/electrical-system.html
30. "First Offshore Wind Farm in the United States Begins Construction," August 14, 2015, https://www.eia.gov/todayinenergy/detail.php?id=22512
31. "America's First Offshore Wind Farm Powers Up," December 12, 2016, http://dwwind.com/press/americas-first-offshore-wind-farm-powers/
32. Land-Use Requirements of Modern Wind Power Plants in the United States, Paul Denholm et al., Technical Report NREL/TP-6A2-45834, August 2009, http://www.nrel.gov/docs/fy09osti/45834.pdf
33. "Layout of Wind Projects," Energypedia, June 30, 2015, https://energypedia.info/wiki/Layout_of_Wind_Projects
34. "Optimal Layout for Wind Turbine Farms," K. Attias and S.P. Ladany, World Renewable Energy Congress 2011, May 8–13, 2011, Optimal_Layout_for_a_Wind_Farm.pdf
35. "Optimal Turbine Spacing in Fully Developed Wind-Farm Boundary Layers," J. Meyers and C. Meneveau, Wind Energy 15, 305–317, 2012, Optimal_Turbine_Spacing.pdf
36. "Cost and Performance Characteristics of New Generating Technologies," Annual Energy Outlook 2019, Table 2, Energy Information Administration, January 2019, https://www.eia.gov/outlooks/aeo/
37. "Lazard's Levelized Cost of Energy Analysis—Version 12.0," November 2018, https://www.lazard.com/media/450784/lazards-levelized-cost-of-energy-version-120-vfinal.pdf
38. "Updated Capital Cost Estimates for Utility Scale Electricity Generating Plants," April 2013, Energy Information Administration, https://www.eia.gov/outlooks/capitalcost/pdf/updated_capcost.pdf; "Levelized Cost and Levelized Avoided Cost of New Generation Resources in the Annual Energy Outlook 2015," June, 2015, Table 1, http://large.stanford.edu/courses/2015/ph240/allen2/docs/electricity_generation.pdf
39. "Annual Energy Outlook 2015 with Projections to 2040," DOE/EIA-0383, Table A8, April 2015, https://www.eia.gov/outlooks/aeo/pdf/0383(2015).pdf
40. "2018 Annual Technology Baseline Cost and Performance Summary," National Renewable Energy Laboratory, 2018, https://atb.nrel.gov/

41. "General Electric GE 2.75–120," January 27, 2018, https://en.wind-turbine-models.com/turbines/983-general-electric-ge-2.75-120#powercurve
42. "How to Calculate Power Output of Wind," January 26, 2010, Wind Power Program, https://www.windpowerengineering.com/calculate-wind-power-output/
43. "Table 6.7.B. Capacity Factors for Utility Scale Generators Not Primarily Using Fossil Fuels," January 2013–February 2019, Electric Power Monthly, https://www.eia.gov/electricity/monthly/epm_table_grapher.php?t=epmt_6_07_b
44. "Wind Power Could Blow Past Hydro's Capacity Factor by 2020," Jan Dell and Matthew Klippenstein, February 8, 2017, Green Tech Media, https://www.greentechmedia.com/articles/read/wind-power-could-blow-past-hydros-capacity-factor-by-2020
45. "Alternatives for Estimating Energy Consumption," Appendix F, Annual Energy Review 2010, https://www.eia.gov/totalenergy/data/annual/pdf/sec17.pdf
46. "How to Calculate Power Output of Wind," Wind Power Program, January 26, 2010, https://www.windpowerengineering.com/calculate-wind-power-output/
47. "Efficiency in Electricity Generation," EURELECTRIC "Preservation of Resources" Working Group's "Upstream" Sub-Group in Collaboration with VGB, July 2003, EEGJulyrevisedFINAL1-2003-030-0548-2-.pdf
48. "A Review of Transmission Losses in Planning Studies," Lana Wong, California Energy Commission, Figure ES-1, August 2011.
49. "Monthly Energy Review," Energy Information Administration, July 2016, https://www.eia.gov/totalenergy/data/monthly/pdf/mer.pdf
50. "Timeline: The History of Wind Power," Niki Nixon, The Guardian, October 17, 2008, https://www.theguardian.com/environment/2008/oct/17/wind-power-renewable-energy
51. "The Largest Offshore Wind Farm on the Planet Opens," Anmar Frangoul, September 6, 2018, CNBC, https://www.cnbc.com/2018/09/06/the-largest-offshore-wind-farm-on-the-planet-opens.html
52. "World's Largest Wind Farm, London Array, Brought Fully Online," James Holloway, New Atlas, April 10, 2013, http://newatlas.com/worlds-largest-wind-farm-london-array-revs-up-to-full-output/27006/
53. "United States Electricity Industry Primer," Office of Electricity Delivery and Energy Reliability U.S. Department of Energy DOE/OE-0017, July 2015, https://www.energy.gov/sites/prod/files/2015/12/f28/united-states-electricity-industry-primer.pdf
54. "Wind Energy Facts at a Glance," American Wind Energy Association, 2018, https://www.awea.org/wind-energy-facts-at-a-glance
55. "Wind Projects Map," Windpower Engineering & Development, 2020, https://www.windpowerengineering.com/wind-project-map
56. "There Are Over 341,000 Wind Turbines on the Planet: Here's How Much of a Difference They're Actually Making," Anmar Frangoul, CNBC Special Report, September 8, 2017, https://www.cnbc.com/2017/09/08/there-are-over-341000-wind-turbines-on-the-planet-why-they-matter.html
57. "Electric Power Annual 2017," Energy Information Administration, October 2018, revised December, 2018, https://www.eia.gov/electricity/annual/pdf/epa.pdf
58. "Annual Market Update 2017," Global Wind Report, Global Wind Energy Council, https://www.tuulivoimayhdistys.fi/filebank/1191-GWEC_Global_Wind_Report_April_2018.pdf
59. "New Transmission Lines Required to Avoid Curtailment," David Weston, Wind Power, January11,2017,http://www.windpowermonthly.com/article/1420468/new-transmission-lines-required-avoid-curtailment
60. "Midwest Wind Farms Follow in the Wake of New Transmission Lines," Midwest Energy News, Karen Uhlenhuth, January 23, 2017, http://midwestenergynews.com/2017/01/23/midwest-wind-farms-follow-in-the-wake-of-new-transmission-lines/

61. "US Government Approves 705-mile Clean Power Transmission Line," Jonathan Crawford and Jim Polson, Bloomberg, March 29, 2016, http://www.renewableenergyworld.com/articles/2016/03/us-government-approves-705-mile-clean-power-transmission-line.html
62. "DOE, Clean Line End Partnership Over Wind Transmission Project," Electric Light & Power/ POWERGRID International, March 26, 2018, https://www.elp.com/articles/2018/03/doe-ends-partnership-with-clean-line-wind-transmission-project.html
63. "Transmission Line That Could Bring Wind and Solar Power to Millions in West Gets Go-Ahead," Ari Phillips, Think Progress, January 26, 2015, https://thinkprogress.org/transmission-line-that-could-bring-wind-and-solar-power-to-millions-in-west-gets-go-ahead-f5eebc480b43/
64. "Alta Wind Energy Center Onshore Wind Turbine Power Plant," CleanEnergy, January 2013, http://www.infrastructureusa.org/wp-content/uploads/2013/01/Alta-Wind-Energy-Center-.pdf
65. "Cedar Creek Wind Farm, Colorado, United States of America," Power Technology, 2020, http://www.power-technology.com/projects/cedarcreek/
66. "Whitelee Wind Farm, Scotland, United Kingdom," Power Technology, 2019, http://www.power-technology.com/projects/whiteleewindfarm/
67. "Hopkins Ridge Wind Facility," Puget Sound Energy, 2020, https://www.pse.com/pages/facilities/hopkins-ridge
68. "Citizen's Oversight Projects," August 4, 2008, http://www.copswiki.org/Common/WindFarmResearch
69. "Gemini Offshore Wind Farm," 4C Offshore, 2019, http://www.4coffshore.com/windfarms/gemini-netherlands-nl18.html
70. "Greater Gabbard Offshore Wind Project, United Kingdom," Power Technology, 2020, http://www.power-technology.com/projects/greatergabbardoffsho
71. "Anholt Offshore Wind Farm," Dong Energy, 2018, https://orsted.com/-/media/WWW/Docs/Corp/COM/Our-business/Wind-power/Wind-farm-project-summary/Anholt_UK_2018.ashx?la=en&hash=7707175E111C65A4AEFC353291091EE4
72. "BARD Offshore I Wind Farm: A Case Study," July 16, 2008, http://coastalenergyandenvironment.web.unc.edu/2018/07/16/bard-offshore-i-wind-farm-a-case-study/
73. "Thornton Bank Offshore Wind Farm, North Sea, Belgium," Power Technology, 2020, http://www.power-technology.com/projects/thornton-bank-offshore-wind-farm/
74. "Sheringham Shoal Offshore Wind Farm," 2010, https://ramboll.com/projects/re/sheringham_shoal_wind_farm
75. "Thanet Offshore Wind Farm—The World's Largest Operational Offshore Wind Farm, Kent, United Kingdom," 2020, https://www.power-technology.com/projects/thanetwindfarm/
76. "Lincs Offshore Wind Farm," 4C Offshore, 2019, http://www.4coffshore.com/windfarms/lincs-united-kingdom-uk13.html
77. "Horns Rev II Offshore Wind Farm," 4C Offshore, 2019, http://www.4coffshore.com/windfarms/horns-rev-2-denmark-dk10.html
78. "With Vineyard Wind, the U.S. Finally Goes Big on Offshore Wind Power," Jean Kumagai, January 1, 2019, https://spectrum.ieee.org/energy/renewables/with-vineyard-wind-the-us-finally-goes-big-on-offshore-wind-power
79. "Electric Power Consumption (kWh per capita)," IEA Statistics for 2016, http://energyatlas.iea.org/#!/tellmap/-1118783123/1
80. "The Significant Potential of Wind Energy in America: A Transformative Force in Struggling U.S. Rural Economies?," Michael Patullo, Penn State University, June, 2010, Patullo The Significant Potential of Wind Energy in America.pdf
81. "Wind Turbine Interactions with Wildlife and Their Habitats: A Summary of Research Results and Priority Questions," America Wind Wildlife Institute, January

2014, https://awwi.org/wp-content/uploads/2014/05/AWWI-Wind-Wildlife-Interactions-Factsheet-05-27-14.pdf
82. "20% Wind Energy by 2030: Increasing Wind Energy's Contribution to U.S. Electricity Supply," US Department of Energy, Energy Efficiency and Renewable Energy, May 2008, Report DOE/GO-102008-2567, 20_percent_wind_2.pdf
83. "License to Kill: Wind and Solar Decimate Birds and Bats," April 29, 2015, Institute for Energy Research, http://instituteforenergyresearch.org/analysis/license-to-kill-wind-and-solar-decimate-birds-and-bats/
84. "How Wind Turbines Affect Your (Very) Local Weather: Wind Farms Can Change Surface Air Temperatures in Their Vicinity," David Biello, Scientific American, October 4, 2010, https://www.scientificamerican.com/article/how-wind-turbines-affect-temperature/
85. "New 725-Mile Transmission Line Slated to Start in 2020," Amy Fischbach, T&D World, August 8, 2018, https://www.tdworld.com/overhead-transmission/new-725-mile-transmission-line-slated-start-2020
86. "Electricity Delivery and Its Environmental Impacts," US Environmental Protection Agency, January 19, 2017, https://www.epa.gov/energy/electricity-delivery-and-its-environmental-impacts
87. "Effects of High Voltage Transmission Lines on Humans and Plants," February 17, 2012, https://www.efis.psc.mo.gov/mpsc/commoncomponents/viewdocument.asp?DocId=935874247
88. "Life Cycle GHG Emission Analysis of Power Generation Systems: Japanese Case," H. Hondo, Energy 30, 2042–2056, 2005.
89. "Life Cycle Greenhouse Gas Emissions from Electricity Generation," National Renewable Energy Laboratory, January 2013, http://www.nrel.gov/docs/fy13osti/57187.pdf
90. "Lifecycle Greenhouse Gas Emissions from Solar and Wind Energy: A Critical Meta-Survey," Denise-Marie Ordway and Leighton Walter Kille, November 28, 2015, https://journalistsresource.org/studies/environment/energy/lifecycle-greenhouse-gas-emissions-solar-wind-energy
91. "Life Cycle Costs and Carbon Emissions of Wind Power: Executive Summary," R Camilla Thomson and Gareth P Harrison, University of Edinburgh, Climate Change Report, 2015, http://www.climatexchange.org.uk/files/4014/3324/3180/Executive_Summary_-_Life_Cycle_Costs_and_Carbon_Emissions_of_Wind_Power.pdf
92. "Comparison of Lifecycle Greenhouse Gas Emissions of Various Electricity Generation Sources," World Nuclear Association Report, July 2011, http://www.world-nuclear.org/uploadedFiles/org/WNA/Publications/Working_Group_Reports/comparison_of_lifecycle.pdf
93. "Installed Wind Capacity," US Department of Energy, Energy Efficiency & Renewable Energy, 2018, https://apps2.eere.energy.gov/wind/windexchange/wind_installed_capacity.asp
94. "Electricity," British Petroleum, BP Global, 2018, http://www.bp.com/en/global/corporate/energy-economics/statistical-review-of-world-energy/electricity.html

7
Solar

A Renewable Energy Type

7.1. Foreword

Solar energy is, by definition, the radiant energy emitted by the sun. And, in many ways, we owe much of our different types of energy to the sun. Fossil fuel energy, discussed in Chapters 1–3, is certainly the result of solar energy, as well as millions of years of heat and pressure in the interior of the earth. Since fossil fuels are formed from the anaerobic decomposition of dead organisms, such as algae, prehistoric plants and animals, and plankton, there would be no organisms to decompose if it were not for their growth from solar energy.

And, the hydroelectric power discussed in Chapter 5 also owes its energy, in part, to the sun. Rivers and streams flow downhill but the water for them comes from winter snow, glaciers, and melting; melting is caused by solar energy. As well, rivers and streams grow from rain. The rain occurs because of the sun heating the atmosphere, causing ground moisture to evaporate, rise into the atmosphere, form clouds, and eventually rain.

Likewise, wind energy discussed in Chapter 6, is caused by solar energy heating the earth and causing air movement. And some energy types discussed in later chapters, ethanol (Chapter 8), biomass (Chapter 9), hydrogen (Chapter 11), and the Fischer-Tropsch synthesis (Chapter 12), use fuels that were generated by the sun. Ethanol and biomass use biofuels while hydrogen and the Fischer-Tropsch synthesis rely on fossil fuels and are therefore also the indirect result of solar energy.

The only two energy types discussed in this book that do not rely on solar energy are nuclear and geothermal energy. Uranium is thought to have been formed from supernovas around 6.6 billion years ago, with a supernova being defined as a star reaching the end of its life that has an explosion ejecting most of its mass. This is certainly not something we want from our sun and, in fact, the sun is thought to not have enough mass to explode as a supernova. Geothermal energy comes from heat within the earth's core, and the source of this heat is from when the planet was formed, frictional heating from material movement, and heat from radioactive decay.

This chapter focuses on the direct use of solar energy to make electricity. There are basically two methods of generating electricity from solar energy including photovoltaic (PV) solar energy and solar thermal energy. In the case of PV solar energy, sunlight is directly converted into electricity using panels made of semiconductor cells while solar thermal energy uses solar energy to heat a fluid (such a gas, oil, or molten

The Changing Energy Mix. Paul F. Meier, Oxford University Press (2020). © Oxford University Press.
DOI: 10.1093/oso/9780190098391.001.0001.

salt) to high temperature. This fluid is then used to heat water, produce steam, and drive a turbine to make electricity in the same manner as natural gas, coal, and nuclear plants.[1] Both of these are discussed in more detail later.

Data from the International Renewable Energy Agency (IRENA)[2] for 2018 show that PV solar energy continues to experience rapid growth, with global additions of 94 GW to give a worldwide total of 480.4 GW, 24% growth relative to 2017. And, compared to a decade earlier in 2009, when the worldwide capacity was only 22.6 GW, capacity has increased more than 20 times! Germany was the world leader up until 2014, then with a capacity of 37.9 GW, but was overtaken in 2015 by China with 43.5 GW. Since then, the solar PV capacity in China has grown to 175.0 GW in 2018, well ahead of second place Japan at 55.5 GW and the United States in third place with 49.7 GW. Considering that 10 years earlier in 2009 China, Japan, and the United States had 0.4, 2.6, and 1.6 GW, respectively, it is clear to see that solar PV is experiencing rapid growth. In the United States, the growth is driven, in part, by the Federal Investment Tax Credit (ITC), which will expire in 2021. Nevertheless, as shown later, the capital cost of solar PV technology has seen a substantial decrease in cost, which has helped fuel this rapid growth.

For solar thermal energy, also known as concentrating solar power (CSP), growth has been much slower and, in some countries, has stopped. In 2018 the worldwide capacity was 5.47 GW, growing from 0.77 GW 10 years earlier. Spain and the United States lead the world with 2.30 and 1.76 GW, respectively, in 2018 but Spain and the United States have not seen any growth since 2013 and 2015, respectively. The last CSP plant built in the U.S was in 2015, the 110 MW Crescent Dunes facility in Nevada. The only countries that added CSP facilities between 2017 and 2018 were Morocco (0.35 GW), South Africa (0.1 GW), Saudi Arabia (0.05 GW), and Mexico (0.014 GW).

Although China led the world in solar PV capacity, these facilities had low capacity factors because of insufficient capacity to hook up to the grid. With these transmission issues, solar PV generated only 67.9 TWh in China for 2016, a capacity factor of 10.0% (based on a capacity of 77.79 GW).

One problem in tracking data for solar energy, and especially solar PV, is that there are many small-scale PV installations in the United States. According to the Energy Information Administration (EIA),[3] a "small-scale solar PV" installation is defined as having a capacity of less than 1 megawatt (MW), and these installations are normally at the site where the customer consumes the electricity. These types of installations may not even be metered and, for a residential system, a typical size is 5 kilowatts (kW). By contrast, a "Utility-Scale solar PV" facility is one greater than 1 MW, and which likely feeds its electricity into the grid.

The tables show US data taken from Electric Power Monthly.[4] Table 7.1 shows that capacities for small-scale PV range from 39 to 46% of the PV total, a substantial amount of the total PV capacity. The data in Tables 7.1 and 7.2 also allow us to calculate capacity factors for different types of solar energy applications as well as the aggregate total. The aggregate totals shown in Table 7.3 are 18 to 21%, better than the 10.0% for China. Tables 7.1 and 7.2 also show that electricity generation from solar

Table 7.1 US Capacity by Year, MW

MW	Utility-Scale PV	Small-Scale PV	PV Total	Solar Thermal	Solar Total
2014	8,656.6	7,326.60	15,983.2	1647.9	17,631.1
2015	11,905.40	9,778.50	21,683.9	1757.9	23,441.8
2016	20,192.9	12,765.1	32,958.0	1757.9	34,715.9
2017	25,209.0	16,147.8	41,356.8	1757.9	43,114.7
2018	30,170.5	19,521.5	49,692.0	1757.9	51,449.9

Table 7.2 US Electricity Generation by Year, Thousand MWh

Thousand MWh	Utility-Scale PV	Small-Scale PV	PV Total	Solar Thermal	Solar Total
2014	15,250	11,233	26,483	2,441	28,924
2015	21,666	14,139	35,805	3,227	39,032
2016	32,670	18,812	51,482	3,384	54,866
2017	50,017	23,990	74,007	3,269	77,276
2018	63,012	29,543	92,555	3,592	96,147

Table 7.3 Capacity Factors, %

	Utility-Scale PV	Small-Scale PV	PV Total	Solar Thermal	Solar Total
2014	20.1	17.5	18.9	16.9	18.7
2015	20.8	16.5	18.8	21.0	19.0
2016	18.5	16.8	17.8	22.0	18.0
2017	22.6	17.0	20.4	21.2	20.5
2018	23.8	17.3	21.3	23.3	21.3

thermal energy has been stagnant from 2015 to 2018, and thus losing percent market share as solar PV has grown. In terms of capacity, the market share has decreased from 9.3% of the solar total in 2014 to 3.4% in 2018. Likewise, in terms of generation, solar thermal decreased from 8.4% in 2014 to 3.7% in 2018.

7.2. Proven Reserves

Unlike petroleum crude oil, natural gas, coal, and nuclear energy, the amount of proven, or potential, reserves for solar energy are difficult to assess. There are a great number of factors to consider, such as how much solar energy reaches the surface of the earth, what is the efficiency of converting solar energy to electricity, what land space is available for the installation of solar energy generation, and so on.

Potentials for different countries can be estimated using a broad mathematical approach. According to a Department of Energy (DOE) report,[5] the average solar flux striking the outer atmosphere of the earth is 342.5 W/m^2 (watts/meter^2). The radius of the earth is 6,378 km (~4,000 miles) and the surface area of a sphere is $4\pi r^2$. We can use equation (1) to calculate the solar power striking the earth.

$$P = (342.5\,W/m^2) * 4\pi * (6{,}378\,km)^2 * (10^6\,m^2/km^2) * (10^{-12}\,TW/W) = 175{,}081\,TW \quad (1)$$

Thus, there are about 175,000 TW (terawatt = 10^{12} watt) of solar radiation from the sun that strike earth. However, en route to the earth's surface, about 30% is reflected back to space and about 19% is absorbed by the atmosphere and clouds. With these adjustments, we use equation (2) to calculate that about 89,300 TW reach the earth and are absorbed by land and oceans, or about half of the original incoming solar energy.

$$P = (342.5\,W/m^2) * (1 - 0.30 - 0.19) * 4\pi * (6{,}378\,km)^2 * (10^6\,m^2/km^2) * (10^{-12}\,TW/W) = 89{,}291\,TW \quad (2)$$

Of course, unless we install solar panels on the surface of the ocean, not all of the earth's area is available. Also, some of earth's area is in what is termed as "frigid zones," defined as north of the Arctic Circle and south of the Antarctic Circle. The percentage of earth that is ocean is 70.8% and the percentage of the land defined as frigid zones is 3.5%. Deducting this area and assuming a PV cell efficiency of 40%, the potential for generating electricity for PV solar is 10,065 TW, calculated in equation (3). Assuming an efficiency of 30% for solar thermal, the potential for solar thermal electricity is 7,550 TW, shown in equation (4).

$$P = 89{,}300 * (1 - 0.708) * (1 - 0.035) * PV\ efficiency = 89{,}300 * (1 - 0.708) * (1 - 0.035) * 0.40 = 10{,}065\,TW \quad (3)$$

$$P = 89{,}300 * (1 - 0.708) * (1 - 0.035) * solar\ thermal\ efficiency = 89{,}300 * (1 - 0.708) * (1 - 0.035) * 0.30 = 7{,}550\,TW \quad (4)$$

According to the Global Energy Yearbook, 2018,[6] the world consumed 22,015 TWh in 2017. If we assume a capacity factor of only 10% for solar energy, a simple calculation shows that 10,000 TW will generate 8,760,000 TWh in 1 year. This means that the total world energy consumption in 2018 was only 0.25% of the potential electricity generation from solar energy based on equation (3)!

A study[7] that took into account various physical limits found a potential of 16,300 TW for solar PV and 11,600 TW for solar thermal, even higher than the DOE estimates.

These two estimates include all of the earth's land area, excluding the "frigid zones," so they are very optimistic indeed. Also, there are other challenges, such as the variation in the amount of solar energy reaching any given area and the fact that northern and southern latitudes have limited sunlight in their winters. Nevertheless, there is staggering potential for harvesting energy and making electricity from the sun.

Now that we have considered solar energy from a worldwide perspective, some rough estimates can be made for potential reserves of individual countries. Recall that the average solar flux striking the outer atmosphere of the earth is 342.5 W/m^2. If we adjust this by the 30% reflected back to space and the 19% absorbed by the atmosphere, the adjusted average solar flux striking earth is 174.7 W/m^2. Of course, this is an average and, on any given day, the amount of solar flux will depend on latitude and the month of the year. Figure 7.1, taken from Solargis,[8] shows the Global Horizontal Irradiance reaching the surface of the earth. As a guide to interpreting this figure, the average solar flux of 174.7 W/m^2 translates into 4.2 kWh/m^2 for one day and 1,530 kWh/m^2 for 1 year, placing it in the light-colored part of the figure, a northern latitude of around 45°. Global Horizontal Irradiance is defined as the radiation received from above by a surface horizontal to the ground, and is one convention is measuring solar radiation. The data shown in Figure 7.1 represent long-term annual and daily averages. Not surprisingly, the averages increase as you get closer to equator and decrease as you go further north in the Northern Hemisphere and further south in the Southern Hemisphere. Therefore, while 174.7 W/m^2 is the worldwide average, the average for a given country depends a lot on latitude.

It is also clear that the solar energy potential for any given country will depend on land area. Of course, it is impossible to cover all land area with solar panels but those

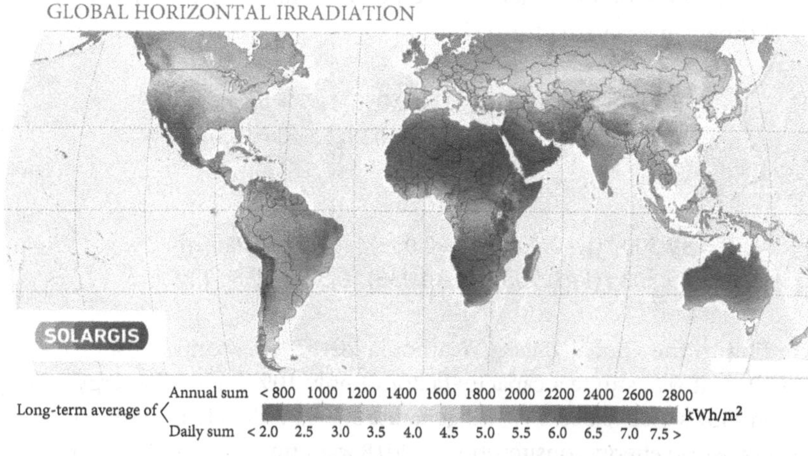

Figure 7.1 Long-Term Average Annual and Daily Irradiation Data (*Source*: Solargis[8])

countries with the most area certainly have an advantage in exploiting solar energy. To make a rough estimate of potential for some countries, the amount of solar radiation reaching a given area (insolation) was examined for the ten largest countries in the world by land area. Data taken from one reference[9] are shown in Table 7.4. This table illustrates many things. As expected, countries in the Northern Hemisphere have greater insolation in the summer months while the opposite is true for countries in the Southern Hemisphere. Also, in general, as latitude decreases, the average annual insolation increases. Thus, for example, while Russia is the country with the greatest land mass, its average annual insolation value of 1,486 kWh/m^2/year is only 72% of India's average insolation value of 2,076 kWh/m^2/year.

Using the data in Table 7.4, the solar flux and power can be calculated for each country, shown in Table 7.5.

This table illustrates several things. First, not all countries have the worldwide average solar flux of 174.7 W/m^2. Notably, Russia, Canada, and Argentina are below the average, as the countries are at higher latitudes. And, India and Algeria are significantly above the average, as they are in the lower latitudes. When you consider the potential power, however, Russia is still number one as their land area more than compensates for the lower solar flux. But Canada, which is the second largest country by land mass, is leapfrogged by three countries, the United States, China, and Brazil when you look at potential power. This means that in terms of solar energy potential, the top ten countries are ranked in the order of Russia, United States, China, Brazil, Canada, Australia, India, Argentina, Kazakhstan, and Algeria. Naturally, this ranking is based just on potential and does not consider technology, social and economic factors, the electrical grid status in each country, and even whether the population density warrants the installation of significant solar energy. Nevertheless, the world's ten largest countries by land area have great potential for solar energy.

Next, we examine the current world rankings in terms of installed capacity. In terms of capacity, data from IRENA[2] were used to rank the top ten countries for both solar PV and solar thermal, also known as CSP. In order to show a 3-year trend, data are reported for 2016, 2017, and 2018 in Tables 7.6 and 7.7.

There are some interesting things to note from the data in Table 7.6. First, when we consider the solar energy potential of Table 7.5, countries with large land area and potential for solar energy, such as Russia, Canada, and Brazil, are not even listed in the top ten in terms of installed solar PV capacity. Further, small countries by land area, such as Germany, Japan, Italy, France, and the United Kingdom have significant amounts of installed capacity. In fact, until 2015, Germany had the largest installed capacity in the world! Since then, however, several countries have had meteoric growth in installed capacity. Notably, from 2016 to 2017 and 2017 to 2018, China increased its capacity by 68% and 34%, respectively. Over the same period, Japan increased by 17% and 13%, respectively while India increased by 87% and 52%, respectively. And, the United States increased by 25% and 20% over these same years. As a result, Germany has moved from first in the world in 2014 to fourth in 2018. Considering only the data

Table 7.4 Daily Average Insolation by Month in $kWh/m^2/day$ for the Ten Largest Countries in the World by Land Area[9]

	Russia	Canada	US	China	Brazil	Australia	India	Argentina	Kazakhstan	Algeria
January	1.52	1.55	3.41	3.41	6.22	7.14	5.84	6.52	2.27	3.91
February	3.18	2.55	4.02	4.02	5.97	6.18	6.34	5.45	3.24	4.49
March	4.70	4.90	5.12	5.12	5.28	5.24	6.92	4.04	4.23	5.81
April	5.40	5.35	5.70	5.70	4.52	3.95	6.79	2.71	5.72	5.68
May	6.20	5.80	5.72	5.72	3.62	2.89	6.12	1.84	7.11	6.31
June	6.50	6.20	6.21	6.21	3.43	2.56	4.88	1.63	7.36	7.11
July	5.80	6.30	6.43	6.43	3.52	2.78	4.14	1.72	7.45	7.18
August	5.04	5.20	6.12	6.12	4.48	3.54	4.01	2.25	7.02	6.92
September	4.19	4.02	4.87	4.87	5.53	4.53	5.25	3.45	6.32	5.92
October	3.10	2.90	4.88	4.88	5.87	5.10	6.12	4.73	4.51	4.81
November	1.90	1.70	3.53	3.53	5.55	6.47	6.08	6.18	2.40	4.08
December	1.32	1.10	3.16	3.16	5.67	6.90	5.76	6.54	1.98	3.32
Average	4.07	3.96	4.93	4.93	4.97	4.77	5.69	3.92	4.97	5.46
Area, km^2	17,098,242	9,984,670	9,826,675	9,596,961	8,514,177	7,741,220	3,287,263	2,780,400	2,724,900	2,381,741
$kWh/m^2/year$	1,486	1,447	1,800	1,800	1,815	1,742	2,076	1,431	1,813	1,994
Latitude, °	60° N	45.4° N	38.9° N	35° N	15.8° S	35.3° S	21° N	37.2° S	48° N	28° N

Table 7.5 Solar Flux and Power for the Ten Largest Countries in the World by Land Area

Country	Area, km^2	Solar Flux, W/m^2	Power, TW	Power Ranking
Russia	17,098,242	169.6	2,900	1
Canada	9,984,670	165.2	1,649	5
United States	9,826,675	205.5	2,019	2
China	9,596,961	205.5	1,972	3
Brazil	8,514,177	207.2	1,764	4
Australia	7,741,220	198.9	1,540	6
India	3,287,263	237.0	779	7
Argentina	2,780,400	163.4	454	8
Kazakhstan	2,724,900	207.0	564	9
Algeria	2,381,741	227.6	542	10

Table 7.6 Top Ten Countries for Solar PV by Capacity (MW), 2016–2018 Data[2]

Country	2016	Country	2017	Country	2018
World	292,440	World	386,107	World	480,356
China	77,788	China	130,802	China	175,018
Japan	42,040	Japan	49,040	Japan	55,500
Germany	40,714	Germany	42,337	United States	49,692
United States	33,100	United States	41,273	Germany	45,930
Italy	19,283	Italy	19,682	India	26,869
United Kingdom	11,912	India	17,644	Italy	20,120
India	9,418	United Kingdom	12,776	United Kingdom	13,108
France	7,702	France	8,610	Australia	9,763
Australia	4,718	Australia	5,988	France	9,483
Spain	4,716	Korea Republic	5,835	Korea Republic	7,862

for 2018, these ten countries control 86% of the worldwide installed solar PV capacity, with China controlling about 36%.

For CSP, the story is quite different. There has been little additional capacity added over the last 3 years. Worldwide, the capacity increased only 103 MW from 2016 to 2017 and, from 2017 to 2018, the capacity increased only 514 MW. Also, there is much less capacity, 5,469 MW in 2018, compared to 480,356 MW for solar PV. Spain and the United States, the world leaders, did not add any capacity over the least 3 years.

Table 7.7 Top Ten Countries for Concentrating Solar Power (CSP) by Capacity (MW), 2016–2018 Data[2]

Country	2016	Country	2017	Country	2018
World	4,853	World	4,956	World	5,469
Spain	2,304	Spain	2,304	Spain	2,304
United States	1,758	United States	1,758	United States	1,758
India	229	South Africa	300	Morocco	530
South Africa	200	India	229	South Africa	400
Morocco	180	Morocco	180	India	229
United Arab Emirates	100	United Arab Emirates	100	United Arab Emirates	100
Algeria	25	Algeria	25	Saudi Arabia	50
Egypt	20	Egypt	20	Algeria	25
China	14	China	14	Egypt	20
Italy	6	Italy	6	Mexico	14
Israel	6	Israel	6	China	14

Considering only the 2018 data, these eleven countries control 99.5%, virtually all of the CSP. Eleven countries, instead of ten, were shown since Mexico and China had the same installed capacity. Spain and the United States control 42% and 32%, respectively.

Finally, Tables 7.8 and 7.9 show IRENA data[2] for the top ten countries for solar energy generation. Unfortunately, data for generation lags behind that for capacity, so 2016 is the most recent year shown in the IRENA data.

Considering data for 2016, Table 7.8 shows that the top ten countries control 84.3% of the worldwide solar PV generation. No country is truly dominant, with China, Japan, the United States, and Germany controlling 21.4%, 16.0%, 14.7, and 12.0% of the worldwide generation. Using 2016 data from Tables 7.6 and 7.8, we calculate the worldwide capacity factor as 12.4%. Of these top ten countries, capacity factors ranged from a low of 10.0% for China and the United Kingdom to a high of 19.5% for Spain. The US capacity factor for 2016 was 16.1%.

Considering data for 2016, Table 7.9 shows that for CSP, the top ten countries account for 99.9% of the worldwide generation. By far, Spain and the United States dominate CSP electricity generation, with 50.5% and 33.5%, respectively, a combined total of 84.0%. Capacity factors are better than for solar PV. Using data from Tables 7.7 and 7.9, the worldwide capacity factor was 26.0% compared to 12.4% for solar PV. Of these top ten countries, capacity factors ranged from a low of 13.6% for Egypt to a high of 61.4% for Algeria. The US capacity factor for 2016 was 24.0%.

Table 7.8 Top Ten Countries for Solar PV Generation (GWh), 2014–2016 Data[2]

Country	2014	Country	2015	Country	2016
World	181,670	World	240,126	World	317,673
Germany	36,056	China	38,978	China	67,865
China	23,751	Germany	38,726	Japan	50,952
Japan	22,952	Japan	34,802	United States	46,633
Italy	22,306	United States	32,091	Germany	38,098
United States	21,915	Italy	22,942	Italy	22,104
Spain	8,218	Spain	8,266	United Kingdom	10,421
France	5,913	United Kingdom	7,546	India	9,430
United Kingdom	4,054	France	7,262	France	8,160
Australia	4,007	India	5,311	Spain	8,070
Greece	3,792	Australia	5,019	Australia	6,205

Table 7.9 Top Ten Countries for Concentrating Solar Power (CSP) Generation (GWh), 2014–2016 Data[2]

Country	2014	Country	2015	Country	2016
World	9,042	World	10,258	World	11,037
Spain	5,455	Spain	5,593	Spain	5,579
United States	2,688	United States	3,544	United States	3,701
India	360	India	360	South Africa	498
United Arab Emirates	243	South Africa	263	Morocco	439
Algeria	197	United Arab Emirates	243	India	360
Morocco	39	Algeria	148	United Arab Emirates	261
Egypt	16	Morocco	43	Algeria	134
Italy	13	Egypt	23	Egypt	24
Israel	11	Italy	13	Italy	13
China	7	Israel	11	Israel	11

7.3. Overview of Technology

To be sure, there are a variety of technologies that fall under the category of "Solar Energy." One reference[10] lists five specific technologies including (1) solar PV systems, (2) solar hot water, (3) solar electricity, (4) passive solar heating and daylighting, and (5) solar process heating and cooling.

Solar hot water is pretty simple and easy to understand. A flat-plate collector, a flat rectangular box with a transparent cover, is mounted on the roof and installed to face the sun. Tubes in this box carry some liquid, such as water or antifreeze, which is solar heated. This heated liquid is then sent to a storage tank. If the heated liquid is water, the water can be stored in a hot water heater. If it is a liquid other than water, it can be sent through heating coils to heat water in the hot water heater.

In the case of passive solar heating and daylighting, a building is constructed with large windows on the wall receiving the most sunlight, usually the south side in the Northern Hemisphere. In addition, materials that absorb and store solar heat can be selected for floors and walls. Similarly, daylighting simply uses sunlight to brighten the interior of the building, which can include windows placed near, or on, the roof.

The concept of solar process heating and cooling is used when large buildings need ventilated air to maintain air quality. During cold weather, heating incoming air takes energy, so solar energy is used to preheat the air. During the hot weather, solar energy can be used to cool buildings using evaporative cooling. Solar energy that heated the air is used to evaporate water, thus taking heat from the air and lowering its temperature.

These three technologies can be useful in reducing industrial and residential operating costs, but are not the focus of this chapter, which is to generate electricity from solar energy. Therefore, the technology overview focuses on the two main technologies used to generate electricity, solar PV and solar electricity, also known as solar thermal or CSP.

The first of these two, solar PV, is simply the production of electricity from sunlight. Unfortunately, the simplicity ends here and you can easily become overwhelmed by the physics and chemistry needed to fully understand PV science. While a fully comprehensive treatise of photovoltaics is beyond the scope is this book, it is useful to understand the basic chemistry and physics involved.

The PV effect is defined as the generation of electric current when a material is exposed to light, and is both a physical and chemical phenomenon. French scientist Edmond Becquerel is credited with discovering this effect in 1839,[11] when he placed two silver electrodes in a solution and generated electricity when the system was exposed to sunlight. And it was none other than Albert Einstein who published a paper to theoretically explain this phenomenon in 1905, for which he received a Nobel Prize in 1921. The American physicist Robert Millikan, also a Nobel Prize winner, provided experimental proof of this PV effect in 1916.

The generation of electricity occurs when photon particles strike a PV cell. Without delving too deeply into this, light behaves as both a wave and particle, known as wave-particle duality, and these photon particles carry energy. When photons strike the PV cell, they may be reflected, pass through it, or be absorbed by the semiconductor material in the cell,[12] but it is only the absorbed photons that generate the electricity. A semiconductor is defined as a chemical atom or compound that has limited ability to conduct electrical current. Stated differently, it is a material that is neither a good insulator nor a good conductor.

Generally, a solar cell is composed of two different types of semiconductors, a p-type and an n-type, joined together to create a p-n junction.[13] The p-type semiconductor has four "valence" electrons, electrons that can form chemical bonds with neighboring atoms. To promote electron movement and generate electricity, these p-type semiconductors are "doped" with another element or compound to create an electron hole. That is, an electron hole is created so that the p-type semiconductor is electron deficient and will want an electron. Typical doping agents for p-type semiconductors are boron (B) and gallium (Ga). If you examine a periodic table, you see that boron and gallium belong to Group IIIA. So, a p-type semiconductor acts as an acceptor of the negatively charged electron, and because these holes are positively charged, the "p" stands for positive.

The solar cell generally also has an n-type semiconductor, such that the semiconductor is "doped" with another element or compound to create excess electrons, making it a donor of electrons. Here "n" stands for negative. Typical doping agents for an n-type semiconductor are phosphorous (P) and arsenic (As), which belong to Group VA in the period table. In this way then, when the photon is absorbed by the solar cell, electrons are made free to move through these holes, and create electrical current.

So what makes a good semiconductor and what are good materials for semiconductors? If you took high school or perhaps college physics, you know that insulators are poor conductors of electricity (like plastics, rubber, paper, and Styrofoam), and conductors are excellent conductors of electricity (such as copper, iron, and silver). Semiconductors lie in between, and silicon (Si) and germanium (Ge), which reside in period table Group IVA are typical examples. A useful property to define a good semiconductor is the "bandgap." Although somewhat of a simplification, electrons reside in orbitals and the valence electrons used to make the chemical bond occupy the outermost orbitals, or valence band. To have electrons move freely through the material (to be conducted), we need to add energy to move the electron to the "conduction band," such that it is now free from the chemical bond and can move through the material. The conduction band consists of the next energy level of vacant electron orbitals. The difference in energy between the valence and conduction bands is defined as the "bandgap," and for PV solar energy, 1.5 eV (electrons volts) is considered to be the optimum energy bandgap.[14]

An IRENA report discusses three generations of photovoltaics materials.[15] The first generation, which is fully commercial, uses wafer-based crystalline silicon (c-Si) technology. Crystalline indicates the solid has an ordered structure, and thus the material melts at one temperature while amorphous indicates the structure is not ordered, and thus does not have a single, sharp melting temperature. For c-Si technology, two types dominate the market including monocrystalline (mono c-Si) and polycrystalline (poly c-Si). In the case of mono c-Si, a single ingot of silicon is made by the Czochralski process, resulting in a material with an almost perfect crystalline structure.[16] This ingot, which has excellent electron flow and high efficiency, is sliced into thin wafers to make the solar cell. The downside of this process is that it is more expensive than making poly c-Si. To make poly c-Si, molten silicon is poured into a cast, resulting in a material with a crystalline structure not as pure as mono c-Si. The process is less expensive, but the efficiency is also less than mono c-Si. Since the mono c-Si has higher efficiency than poly c-Si, less surface area is needed to generate the same amount of electricity.

The second generation, also commercially available, includes thin film amorphous silicon (a-Si), cadmium-telluride (CdTe), copper-indium-selenide (CIS), and copper-indium-gallium-diselenide (CIGS). One method used to make a thin film semiconductor is chemical vapor deposition, such that the material is vaporized and then allowed to deposit on some surface, such as glass. Although thin films generally are less efficient than c-Si technology, the CdTe thin film semiconductors have efficiencies of around 17%, and are thus competitive with c-Si technology. For other thin films less efficient than c-Si semiconductors, more surface area is needed to generate the same amount of electricity. However, the manufacturing costs are less than making the crystalline panels. The materials c-Si, a-Si, CdTe, CIS, and CIGS have energy bandgaps in the range of 1.1 to 1.7 eV,[14] and are thus near the optimum energy bandgap.

The third generation, with only a small market at this time, includes concentrating photovoltaic (CPV) and organic photovoltaic (OPV). A CPV system uses mirrors to focus sunlight onto highly efficient and expensive solar cells, although crystalline silicon (c-Si) cells can also be used. This technology should not be confused with CSP, which also uses mirrors but takes this focused solar energy to make steam and drive a turbine; this technology is discussed later in the section. An OPV system uses organic or polymer materials for the solar cell. While these systems are generally not expensive, they are also not very efficient, with typical commercial systems having efficiencies of 4%–5%. Since these organic systems can be applied to plastic sheets, and are lightweight and flexible, they can be used for mobile applications such as battery chargers, mobile phones, and PCs. With these efficiencies, it is not likely that OPV systems will be used for utility scale solar PV power plants.

A 2019 report[17] indicates the market shares in 2017 were 4.5% thin film, 32.2% mono c-Si, and 60.8% poly c-Si. Thus, the Si-wafer technology accounted for 93% of the total market, with poly c-Si dominating the market. For the thin film market, the 4.5% were distributed as 51.1% Cd-Te, 6.7% a-Si, and 42.2% CIGS.

This same report also discusses efficiency, defined as the percent of solar energy striking the solar cell that is converted into usable electricity. According to the report, the record efficiency measured under laboratory conditions is 26.7% for mono c-Si and 22.3% for poly c-Si.[17] For thin films, the highest measured laboratory efficiencies are 21.0% for CdTe and 22.9% for CIGS.

Commercial efficiencies are lower, however. Product datasheets for two of the major crystalline silicon manufacturers, Jinko Solar[18] and Trina Solar,[19] show that polycrystalline efficiencies range from 16.2% to 17.5% while those for monocrystalline range from 16.8% to 19.6%. For this comparison, the Trina Allmax and Tallmax panels and the Jinko Eagle panels were used. A solar panel, or module, is an assembly of solar cells, typically 60 or 72 for a panel. For an example of thin film semiconductors, the First Solar Series 6 panel datasheet for a CdTe panel shows an efficiency range of 17.0%–18.0% efficiency.[20]

Table 7.10, taken from one reference,[21] reports the range and average cost per watt for the poly c-Si, mono c-Si, and thin film CdTe panels. As well, the table shows the area required for one kW of power. As expected because of the higher efficiency, the mono c-Si requires the least area to generate one kW (kilowatt), but is also more expensive. On the other hand, the poly c-Si requires somewhat more area but is less expensive than the mono c-Si. The CdTe thin film lies in between for both area required and cost.

It is worth noting that the cost per watt shown in Table 7.10 is a moving target, as the prices of panels continue to decrease. For example, the average cost for a Jinko Solar panel has decreased from \$0.48/W in 2013 to \$0.34/W in 2017.[22] Also, the cost data in Table 7.10 are for the panel only; total installation costs also include labor, land acquisition, structural components, and electrical components. An important electrical components is an inverter, which converts the direct current (DC) output of the solar panel into alternating current (AC) that can be used directly or fed into the electric grid.

When a solar panel power rating is listed, for example as 300 watts (W), this is the power the panel can produce at standard test conditions. Whether it is mono c-Si, poly c-Si, or a thin film, the power rating already takes into account the efficiency. Therefore, if a material has a higher efficiency, it will produce more power for the same surface area. Mono c-Si and poly c-Si panels currently dominate the commercial market. It is easy to differentiate between these two different panels based on their appearance. Mono c-Si, with high purity silicon, is black and the corners of the individual cells are missing because of the shape of the wafer made from the Czochralski manufacturing process. On the other hand, poly c-Si panels are dark blue in color, and the individual

Table 7.10 Area Requirements and Cost Per Watt (2018 data) for Commercial Solar Semiconductors[21]

	Mono c-Si	Poly c-Si	CdTe Thin Film
Area required for 1 kW, m^2	5.1–5.5	5.8–6.2	5.6–5.9
\$/watt (panel cost)	\$0.35 average (\$0.31–\$0.55)	\$0.27 average (\$0.24–\$0.41)	\$0.30 average (\$0.27–\$0.42)

cells are typically squares. While not intending to sponsor any particular manufacturer, data for some typical solar panels from two leading manufacturers, Jinko Solar[18] and Trina Solar,[19] are shown in Table 7.11. These panels are available in 5W increments over the range shown, with efficiency increasing with higher wattage.

Some things are evident from this table. First, for the same sized panel, the average watts per square foot for the monocrystalline silicon solar panels are higher than those for the polycrystalline ones. As the surface area for panels with either 60 or 72 cells are the same, the higher wattage is due to the higher efficiency. Therefore, for constant area, you can generate more watts with the more expensive monocrystalline solar panels. Regardless, power ratings vary over a small range, from 15.6 to 17.7 watts per square foot and 4.6 to 5.2 watts per cell.

The second technology for generating electricity from solar energy is CSP, also referred to as solar thermal. Basically, CSP uses mirrors to concentrate sunlight and heat some fluid to produce steam. This steam, once made, will power a turbine to generate electricity. In this respect, CSP generates electricity in the same way as conventional technologies such as natural gas, coal, and nuclear power. In fact, with CSP technology, it is possible to use fossil fuels as back-up fuel for those times when the solar power is not available.

CSP technologies can be divided into two groups, based on whether the solar collectors concentrate the sunlight along a focal line or a single focal point.[23] The focal line technologies include the (1) parabolic trough and (2) the linear Fresnel. The single focal point technologies include the (1) solar dish system and (2) solar tower plants. A simple depiction of these four technologies is shown in Figure 7.2, taken from one reference.[24]

For the Parabolic Trough Collector, parabolic mirrors concentrate sunlight into a central receiver tube, positioned along the focal line. These mirrors can be as long as 100 meters (~330 feet) and a tracking system is used to point the mirrors toward the sun as the sun moves from east to west.[23] A heat transfer fluid flowing through the receiver tube is heated by the concentrated sunlight. After heating, this hot fluid is used to boil water to make steam and power the steam turbine, which in turn powers the generator to make electricity. Heat transfer fluids include oils and molten salts. Molten salts are an inexpensive way to store energy, are designed to be compatible with the operating temperature of the steam turbine, and are not flammable. An example of a molten salt used to store solar energy is a blend of 60 wt% sodium nitrate and 40 wt% potassium nitrate.[25]

The second focal line technology is called the Linear Fresnel Collector, named after French physicist Augustin Jean Fresnel who introduced a new lens design that revolutionized lighthouse optics.[26] This lens takes light that normally would scatter and focuses it into a single beam. For CSP use, the Linear Fresnel Collector employs long flat or slight curved, mirrors to concentrate sunlight for a receiver tube fixed in space above the mirrors.[23] Each line of mirrors has its own tracking system to ensure sunlight is focused on this fixed receiver tube. Unlike the Parabolic Trough Collector, the Linear Fresnel Collector is distorted by astigmatism, a defect in the lens caused by deviation

Table 7.11 Commercial Solar Panel Data for Jinko Solar[18] and Trina Solar[19]

	Trina Solar	Trina Solar	Trina Solar	Trina Solar	Jinko Solar	Jinko Solar	Jinko Solar	Jinko Solar
Product	Allmax	Allmax	Tallmax	Tallmax	Eagle 60P	Eagle 60M	Eagle 72P	Eagle 72M
Type (c-Si)	Poly	Mono	Poly	Mono	Poly	Mono	Poly	Mono
Cells	60	60	72	72	60	60	72	72
Cell Size, inches	6 × 6	6 × 6	6 × 6	6 × 6	6 × 6	6 × 6	6 × 6	6 × 6
Orientation	6 × 10	6 × 10	6 × 12	6 × 12	6 × 10	6 × 10	6 × 12	6 × 12
Dimensions, inches	65.0 × 39.1 × 1.38	65.0 × 39.1 × 1.38	77.2 × 39.1 × 1.57	77.2 × 39.1 × 1.57	65×39.1 ×1.37	65×39.1 ×1.37	77.0×39.1 ×1.57	77.0×39.1 ×1.57
Watts	265–285	275–315	320–340	340–380	265–285	300–320	320–340	360–380
Efficiency, %	16.2–17.4	16.8–19.2	16.5–17.5	17.5–19.5	16.2–17.4	18.3–19.6	16.5–17.5	18.6–19.6
Average Watts/ Square Foot	15.6	16.7	15.7	17.2	15.6	17.6	15.8	17.7
Average Watts/Cell	4.6	4.9	4.6	5.0	4.6	5.2	4.6	5.1

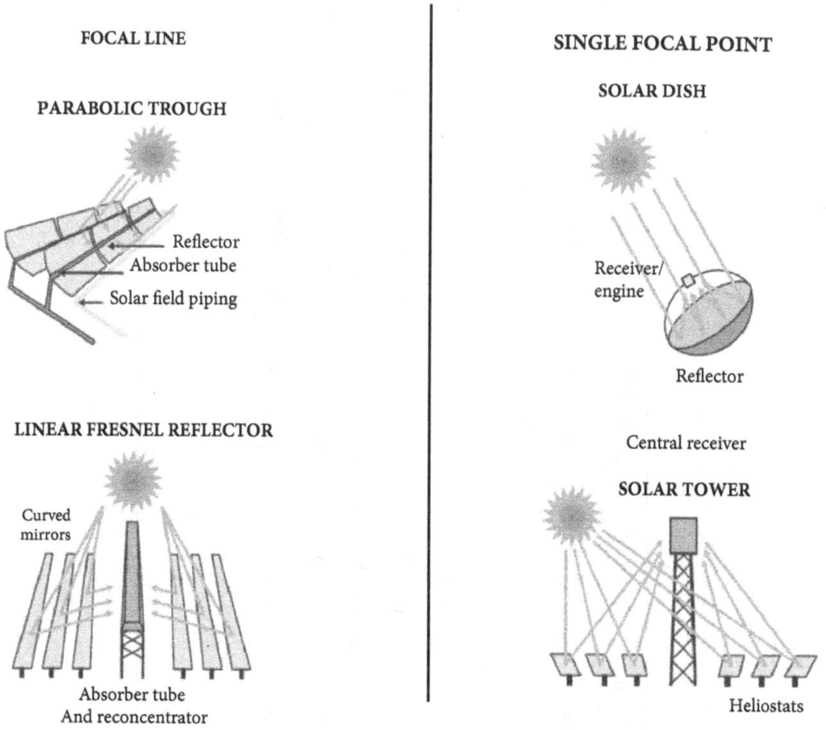

Figure 7.2 Schematic of focal line and single point technologies (left side parabolic and linear Fresnel; right side solar dish and solar tower) (*Source*: IEA (2014). Technology Roadmap: Solar Thermal Electricity. All Rights reserved[24])

from a spherical curvature. As a result, the ability to collect sunlight is less than parabolic trough collectors (see Collector Concentration in Table 7.12). Like the Parabolic Trough Collector, the heat transfer fluids can be oils, molten salts, and even water.

An advantage of linear Fresnel CSP systems compared to parabolic trough systems is that they can use less expensive flat glass mirrors. In addition, the metal support structure is not as heavy as the Parabolic Trough Collector, and so uses less steel and concrete.

There are also "single focal point" technologies, such as the Solar Tower power plant. This technology uses a field of mirrors to focus sunlight onto a receiver mounted high on a central tower, where the sunlight is captured and converted into heat.[23] The mirrors in this field are computer-controlled to track the sun. As shown in Table 7.12, the sunlight is more concentrated with the Solar Tower compared to Parabolic Trough and Linear Fresnel systems, more than >1,000 suns compared to 60–80 suns, so a higher temperature can be achieved. The receiver can use oil or molten salt as the heat transfer fluid. Like the Parabolic Trough and Linear Fresnel systems, the heat collected by this fluid is then used to generate steam and then electricity. Using an oil limits the operating temperature to around 390°C versus 550–650°C for molten salts.

Table 7.12 Comparison of Different CSP Technologies[23]

Technology	Parabolic Trough	Linear Fresnel	Solar Tower	Dish
Maturity of Technology	Commercially proven	Pilot projects	Commercially proven	Pilot projects
Operating Temperature, °C	350–550	390	250–565	550–750
Plant Peak Efficiency, %	14–20	18	23–35	30
Annual Solar-to-Electricity Efficiency, %	11–16	13	7.20	12–25
Annual Capacity Factor, %	25–28	22–24	55	25–28
Collector Concentration	70–80 suns	>60 suns	>1,000 suns	>1,300 suns

Thus, the thermodynamic efficiency will be greater when molten salts are used (see equation (7) in Section 7 on efficiency).

The other "single focal point" technology is the parabolic dish shown in Figure 7.2, also known as the Stirling Dish technology since it is used in conjunction with the Stirling engine. This technology uses a dish similar in shape to a satellite dish, where the sunlight is reflected onto a single focal point.[23] The receiver is placed at the focal point, and absorbs concentrated solar energy to heat gas inside the Stirling engine. As Table 7.12 indicates, the dish is very good at concentrating sunlight, giving greater than 1,300 suns and resulting in high operating temperatures. The Stirling engine is a sealed system filled with a working gas, typically hydrogen or helium, that is alternately heated and cooled.[27] It is called a working gas because it is continually recycled inside and is not consumed. The Stirling engine works by gas compression and expansion, and the electrical generator uses this mechanical energy to make electricity. Because the generator is located in each dish, heat losses are less than other technologies, resulting in higher efficiencies.

Commercial plants in the United States using these four technologies are shown in Table 7.13. The table shows that the total installed CSP capacity is 1,757.8 MW, with 1,731 MW operating. This agrees with the total installed US capacity shown earlier in Table 7.1. For the Linear Fresnel and Stirling Dish technologies, these plants were very small, likely demonstration plants, and are no longer operating. Of the 1,731 MW operating, the Parabolic Trough technology dominates the market with 1,239 MW, 72% of the market, and Solar Tower with 492 MW, 28% of the market. This table also shows that only four states have operating plants, including Arizona, California, Florida, and Nevada, and California dominates the market with 70.6% of the operating capacity.

Table 7.13 Concentrating Solar Power Projects in the United States[28]

Name	Owner	Location	Technology	Capacity, MW	Status	Start Year	Area, acres	Cost, $MM
Maricopa Solar Project	Tessera Solar	Peoria, AZ	Stirling Dish	1.5	Not operating	2010	15	
Saguaro Power	Arizona Public Service	Red Rock, AZ	Parabolic Trough	1.0	Not operating	2006	16	6
Solana Generating Station	Abengoa	Phoenix, AZ	Parabolic Trough	250	Operating	2013	1928	$2,000
Genesis Solar	Genesis Solar	Blythe, CA	Parabolic Trough	250	Operating	2014	1950	
Ivanpah Solar	NRG Energy, Bright Source Energy, Google	Primm, CA	Tower	377	Operating	2014	3500	$2,200
Kimberlina Solar	Ausra	Bakersfield, CA	Linear Fresnel	5	Not operating	2008	12	
Mojave Solar	Mojave Solar	Harper Dry Lake, CA	Parabolic Trough	250	Operating	2014	1765	$1,600
Sierra Sun Tower	eSolar	Lancaster, CA	Tower	5	Operating	2009		
SEGS I (a)	Cogentrix	Daggett, CA	Parabolic Trough	13.8	Not operating	1984	a	
SEGS II (a)	Cogentrix	Daggett, CA	Parabolic Trough	30	Operating	1985	a	
SEGS III (a)	NexEra	Kramer Junction, CA	Parabolic Trough	30	Operating	1985	a	
SEGS IV (a)	NextEra	Kramer Junction, CA	Parabolic Trough	30	Operating	1989	a	
SEGS V (a)	NextEra	Kramer Junction, CA	Parabolic Trough	30	Operating	1989	a	
SEGS VI (a)	NextEra	Kramer Junction, CA	Parabolic Trough	30	Operating	1989	a	
SEGS VII (a)	NextEra	Kramer Junction, CA	Parabolic Trough	30	Operating	1989	a	
SEGS VIII (a)	NextEra	Harper Dry Lake, CA	Parabolic Trough	80	Operating	1989	a	
SEGS IX (a)	NextEra	Harper Dry Lake, CA	Parabolic Trough	80	Operating	1990	a	

Colorado Integrated	Xcel Energy	Palisade, CO	Parabolic Trough	2.0	Not operating	2010	6	$4.5
Martin Next Generation	Florida Power & Light	Indiantown, FL	Parabolic Trough	75	Operating	2010	500	$476.3
Holaniku	Keahole Solar	Keahole Point, HI	Parabolic Trough	2.0	Not operating	2009	3	
Crescent Dunes	Tonopah Solar Energy	Tonopah, NV	Tower	110	Operating	2015	1600	
Nevada Solar One	Acciona Energia	Boulder City, NV	Parabolic Trough	72	Operating	2007	400	$266
Stillwater GeoSolar Hybrid Plant	Enel Green Power	Fallon, NV	Parabolic Trough	2.0	Operating	2015	21	
Tooele Army Depot	Tooele Army Depot	Tooele, UT	Stirling Dish	1.5	Not Operating	2013	17	

a. SEGS = Solar Electric Generating Station with 1,600 acres total[29]

7.4. Capital Cost, Operating Cost, Well-to-Wheels Levelized Cost of Electricity, and Well-to-Wheels Levelized Cost of Fuel

The equations and methodology used to calculate levelized cost of electricity (LCOE), and specific input used for solar energy to make electricity, are detailed in Appendix A. The units for LCOE are in $/megawatt-hour, or $/MWh.

Data for both solar PV and CSP were taken from a 2019 EIA report[30] and 2018 data from Lazard.[31] For comparison, the LCOE was also calculated using earlier data from the same sources, EIA data from 2013[32] and 2014 data from Lazard.[33] This was done to show how both LCOE and capital costs have decreased over the last 5 to 6 years.

A 2019 report from IRENA[34] shows capital cost data over the last 9 years for both solar PV and CSP. These data are global weighted averages for utility-scale plants. In the case of CSP, eight hours of thermal energy storage (TES) were built into the plants. CSP uses focused solar energy to heat fluids that are used to make steam and drive a turbine to make electricity. By including TES, these heated fluids are stored for later use to generate electricity after the solar conditions are no longer suitable to generate electricity. Data from this report are shown in Table 7.14. It is clear from these data that capital costs have decreased dramatically for solar PV over the last 9 years, and the capital cost in 2018 was 26% of what it was in 2010. For CSP, the capital cost in 2018 was 59% of what it was in 2010, so there has been a reduction albeit less dramatic than for solar PV. It is also clear from Table 7.14 that CSP is more capital intensive than solar PV, between two to five times that of solar PV over this 9-year period.

For a solar energy plant, the "Well-to-Wheels" costs include (1) the cost of constructing the solar plant, (2) power plant generating costs, (3) cost of connecting the power plant to the electrical grid, (4) transmission and distribution costs, and (5) taxes (both federal and state). Unlike nuclear, natural gas, and coal, there are no fuel costs for a solar plant so there are no drilling or mining costs.

Table 7.14 Global Weighted Average Capital Costs for Solar PV and CSP[34]

	Solar PV	CSP
2018	$1,210	$5,204
2017	$1,389	$7,196
2016	$1,609	$7,602
2015	$1,825	$7,232
2014	$2,323	$5,414
2013	$2,569	$6,307
2012	$2,933	$8,039
2011	$3,891	$10,403
2010	$4,621	$8,829

For the solar PV capital cost, the 2019 EIA data[30] report a value of $1,783/kW while the 2018 data from Lazard[31] report a value of $1,100/kW. In 2013, the EIA reported a capital cost of $4,183/kW while 2014 data from Lazard reported a lower value of $1,625/kW. The average of these two cases for each time period gives a capital cost similar to the IRENA global weight averages in Table 7.14. For CSP, the 2019 EIA data show a capital cost of $4,291/kW versus $6,925/kW for the 2018 data of Lazard. Compared to the 2013 EIA and 2014 Lazard data, these capital costs have been reduced by around 15%.

These capital costs include the capital cost for new transmission lines to tie into existing electrical substations or existing transmission lines. Therefore, the generating station supplies the transmission lines and steps up the voltage to tie into the "electrical grid," a network that delivers electricity to the customer. The grid has high-voltage transmission lines (sometimes distant) that carry power to demand centers, transformers to step down the voltage, and distribution lines that connect individual customers.

The costs of transmitting and distributing electricity were 0.9 cents/kWh (or $9/MWh) and 2.6 cents/kWh (or $26/MWh), taken from the EIA.[35] Here, transmission is defined as the high-voltage movement of electricity from the power station while distribution is the cost of electrical lines that take power from a substation to the customer. The cost of transmission and distribution is essential to get a true "Well-to-Wheels" analysis of the total cost.

Table 7.15 shows the LCOE for the cost of generating electricity and the additional cost of transmission and distribution. The EIA report[30] states the capital cost for the CSP plant is based on a size of 100 MW and, although not specified, is presumed to be based on the Parabolic Trough technology, for which there are more data than other technologies. Indeed, an average of the capital cost data shown in Table 7.13 for Parabolic Troughs gives an average of $5,339/kW, similar to the value of $5,608/kW shown in Table 7.15.

The solar PV plant is based on a size of 150 MW and, although not stated, is presumed to be based on polycrystalline (poly c-Si), since this is currently the most common form of solar PV.

An obvious advantage of solar power is that there is little carbon dioxide (CO_2) produced. Thus, if carbon capture and sequestration (CCS) become mandatory in the future, this will enhance any advantage solar power has over natural gas and coal.

An advantage of the CSP technology compared to solar PV is that it can utilize TES. Generally speaking, there are no large scale solar PV plants using TES,[36] so there is a capacity factor advantage for CSP versus solar PV. While this helps lower the calculated LCOE, the advantage in capacity factor is not great enough to overcome the differences in capital cost, fixed O&M, and transmission investment, so the LCOE to generate electricity for solar PV is lower than for CSP, $113.55/MWh versus $239.26/MWh (or 14¢/kWh versus 24¢/kWh).

Comparing the LCOE calculated with data from 2018 versus data from the 2013–2014 timeframe, the LCOE for solar PV decreased by 39% and that for CSP decreased by 25% over this 5-year period. In the case of solar PV, the decrease in LCOE was due almost entirely to the change in capital cost while, for CSP, the reduction in both capital cost and capacity factor had an impact on LCOE.

Table 7.15 Economic analysis for solar PV and solar CSP

Energy Type	Capacity Factor	Capital Cost	Levelized Capital Cost; with MACRS depreciation	Fixed and Variable O&M	Transmission Investment	Total System Levelized Cost (LCOE) to Generate	LCOE including transmission and distribution
Units		$/kW	$/MWh	$/MWh	$/MWh	$/MWh	$/MWh
Solar PV							
EIA[30]	0.25	$1,783	$84.78	$10.26	$4.1	$99.14	$134.14
Lazard[31]	0.27	$1,100	$49.35	$4.52	$4.1	$57.97	$92.97
Average	0.26	$1,442	$67.07	$7.39	$4.1	$78.55	$113.55
EIA[32]	0.25	$4,183	$198.91	$12.67	$4.1	$215.68	$250.68
Lazard[33]	0.26	$1,625	$75.76	$7.39	$4.1	$87.24	$122.24
Average	0.25	$2,904	$137.33	$10.03	$4.1	$151.46	$186.46
Solar CSP							
EIA[30]	0.29	$4,291	$175.90	$28.67	$6.0	$210.57	$245.57
Lazard[31]	0.48	$6,925	$173.31	$18.63	$6.0	$197.94	$232.94
Average	0.38	$5,608	$174.61	$23.65	$6.0	$204.26	$239.26
EIA[32]	0.20	$5,067	$301.18	$38.39	$6.0	$345.57	$380.57
Lazard[33]	0.52	$8,400	$192.04	$21.40	$6.0	$219.44	$254.44
Average	0.36	$6,734	$246.61	$29.90	$6.0	$282.51	$317.51

A sensitivity analysis was made for both solar electricity generation technologies by varying the capital cost, the fixed operating and maintenance (O&M) expenses, and the capacity factor. The base cases for both solar PV and CSP used the EIA data.[30]

For solar PV capital cost, the 2019 IRENA data shown in Table 7.14 demonstrate that, over the last 5 years, the global weighted average capital cost for utility scale solar PV plants has dropped dramatically.[34] Capital costs decreased from $2,323/kW in 2014 to $1,210/kW in 2018. Based on these data, the capital cost for solar PV is examined from a low of $1,200/kW to a high of $2,400/kW. For operating costs (O&M), the data of Lazard[31] showed a range of about 25%, so this was used to establish the low and high. Since the EIA base case O&M was about $10/MWh, the low was set to $7.5/MWh and the high to $12.5/MWh. Also, Lazard shows the capacity factor varying from 0.21 to 0.32, so these were used as the low and high relative to the base case value of 0.25.

For CSP, the 2018 data from Lazard[31] were used to set the ranges for all three variables. Capital cost was varied from a low of $3,850/kW to a high of $10,000/kW, O&M

was varied from a low of $16.7/MWh to a high of $38.4/MWh, and the capacity factors of 0.43 and 0.52 were used in the sensitivity analysis. Both of these values were higher than the base case capacity factor of 0.29. For CSP technology, there is some relationship between capital cost and capacity factor. Thermal energy storage (TES), where heated fluids are stored for later use, can be built into the system to provide electricity generation after the solar conditions are no longer suitable for heating the fluids. Thus, if the system is built with 10-hour versus 4-hour storage, it will have a higher capacity factor but the capital costs will also be higher.

The results of the sensitivity study are shown in Table 7.16.

For solar PV, the range used for capacity factor resulted in an LCOE difference of about $40/MWh, while the capital cost range gave a difference of about $57/MWh and the operating cost range gave a difference of only $5/MWh. Therefore, both capital costs and capacity factors have a significant impact on the solar PV LCOE. Like onshore wind energy, discussed in Chapter 6, the reduction in capital cost continues to push the LCOE lower, making it more competitive with electricity generated from natural gas, the current low cost provider. This is discussed in more detail in the book conclusions.

Table 7.16 Sensitivity Study for Solar PV and Solar CSP

	Capacity Factor	Capital Cost	Fixed and Variable O&M (no fuel cost)	Total System Levelized Cost (LCOE) to Generate	Total LCOE Cost to Generate, Transmit, and Distribute
		$/kW	$/MWh	$/MWh	$/MWh
Solar PV Base Case[30]	0.25	$1,783	$10.26	$99.14	$134.14
Capex is $1,200/kW	0.25	$1,200	$10.26	$71.42	$106.42
Capex is $2,400/kW	0.25	$2,400	$10.26	$128.48	$163.48
Opex is $7.5/MWh	0.25	$1,783	$7.50	$96.39	$131.39
Opex is $12.5/MWh	0.25	$1,783	$12.50	$101.39	$136.39
Capacity Factor is 0.21	0.21	$1,783	$12.21	$117.24	$152.24
Capacity Factor is 0.32	0.32	$1,783	$8.01	$78.35	$113.35
Solar CSP Base Case[30]	0.29	$4,291	$28.67	$210.57	$245.57
Capex is $3,850/kW	0.29	$3,850	$28.67	$192.50	$227.50
Capex is $10,000/kW	0.29	$10,000	$28.67	$444.60	$479.60
Opex is $16.7/MWh	0.29	$4,291	$22.04	$203.94	$238.94
Opex is $38.4/MWh	0.29	$4,291	$36.06	$217.96	$252.96
Capacity Factor is 0.43	0.43	$4,291	$19.34	$143.97	$178.97
Capacity Factor is 0.52	0.52	$4,291	$15.99	$120.09	$155.09

For solar CSP, the range used for capacity factor gave an LCOE difference of about \$90/MWh, while the range for capital cost gave difference of about \$250/MWh and the operating costs range gave a difference of about \$14/MWh. Thus, like solar PV, the capacity factor and operating costs have the greater impact on LCOE. The effect of operating costs is minor by comparison.

It is also interesting that in spite of the higher capital cost for CSP versus solar PV, CSP becomes quite competitive with solar PV when the capacity factor is 0.52, the maximum used for the sensitivity study. Thus, if the investor can overcome the much higher initial capital cost of building a CSP system, the higher capacity factor, aided by the addition of TES, makes the CSP technology competitive with solar PV.

7.5. Cost of Energy

Unlike natural gas, coal, and nuclear electricity generation, solar PV and solar CSP do not use fuel, generating the electricity solely from solar power. Thus, unlike these other technologies, electricity generated from solar power is not affected by changes, sometimes unpredictable, in fuel cost.

7.6. Capacity Factor

Before discussing capacity factor, it is useful to review again the differences between solar PV and CSP. The PV effect is defined as the generation of electric current when a material is exposed to light. That is, the solar cell absorbs light which promotes the electron to a "conduction band," where it is now free to move through the material and generate electricity. Figure 7.3 shows those wavelengths that are absorbed by a crystalline silicon cell.[37] The wavelengths shown cover part of infrared light (greater than 700 nm (nanometer)), all of visible light (400 to 700 nm), and part of ultraviolet light (10 to 400 nm). Thus, solar PV systems can still generate electricity on a cloudy day, albeit less efficiently, because visible, infrared, and ultraviolet light penetrates through clouds.

Conversely, CSP systems need direct sunlight. CSP uses mirrors to concentrate sunlight and heat fluids to produce steam, which then generates electricity by powering a turbine. In this respect, solar PV has an advantage for capacity factor, since it can generate electricity when it is cloudy.

On the other hand, because solar CSP heats fluids to produce steam, some of the heated fluids can be stored for later use, referred to as TES. With TES, the hours when CSP can generate electricity is extended by using this stored heat during intermittent sunlight and nighttime.

To summarize, the capacity factor for solar PV is closely related to the amount of solar insolation, as well as the composition of the solar panel, while CSP is related to both solar insolation and the amount of built-in TES.

Table 7.17 shows solar PV and CSP capacity factor data for 9 years, taken from an IRENA report.[34] For solar PV, the global weighted average capacity factor for 2018

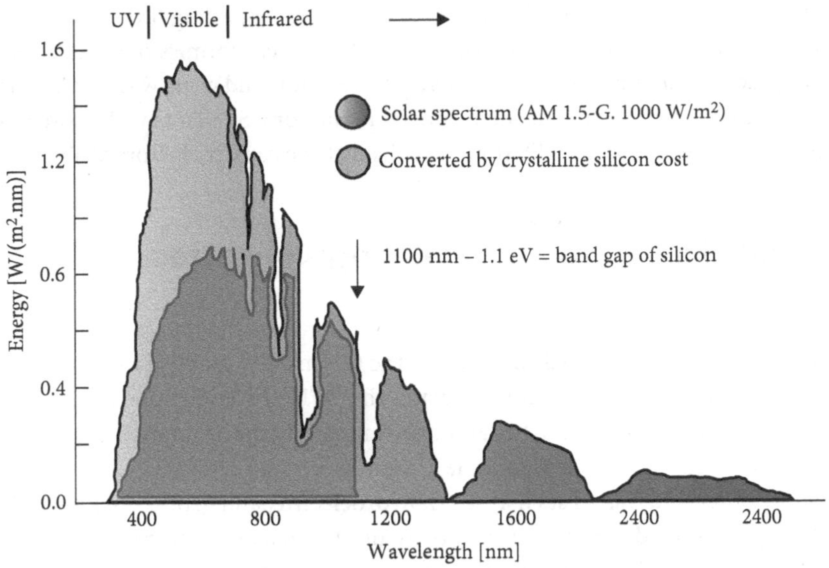

Figure 7.3 Energy Converted by a Crystalline Silicon Cell (Source: Daystar[37])

Table 7.17 Global Weighted Average Capacity Factors for Solar PV and CSP[34]

	Solar PV	CSP
2018	0.18	0.45
2017	0.18	0.39
2016	0.17	0.36
2015	0.17	0.40
2014	0.17	0.29
2013	0.16	0.31
2012	0.15	0.27
2011	0.14	0.35
2010	0.14	0.30

was 0.18, slightly improved from 0.14 in 2010. It is less than the value of 0.25 reported in the EIA report[30] which was used in section 4 to calculated the LCOE.

For CSP, the global weighted capacity factor for 2018 was 0.45, considerably better than the value of 0.30 reported in 2010. This improvement can be attributed to the increased in built-in TES storage. In 2010, the commercial projects used to determine the average either had no or 4–8 hours built-in TES while, in 2018, the commercial projects had either 4–8 hours or more than 8 hours of built-in TES. The EIA report[30] used to calculate LCOE gives a value of 0.29, suggesting little or no built-in TES. According to the IRENA report, new CSP projects commissioned in 2018 had capacity factors ranging from 0.31 to 0.64, where the low end value was for CSP plants with 4–8 hours of TES while the high end value was for CSP plants with more than 8 hours. CSP projects with no TES storage typically have capacity factors of about 0.25.

7.7. Efficiency: Fraction of Energy Converted to Work

For coal, natural gas, and nuclear energies, the thermal efficiency of the turbine used to generate electricity was calculated using the BTU content of one kilowatt-hour (kWh) of electricity, or 3,412 BTU, and dividing this by the heat rate. The heat rate is the amount of energy used to make one kWh of electricity. However, for solar power, as well as other renewables such as wind, hydroelectric, and geothermal, there is no fuel being consumed, and therefore energy used to make the electricity is not frequently measured. This is because renewable energy that is wasted in the process of making electricity has no fuel cost. If some of the solar energy is not converted into electricity, there was no alternate use, or economic value, for the energy.

However, there are still inefficiencies in the process of making electricity from solar energy, and these affect the capital cost of making the desired power output, as well as the land area needed to produce that power.

For solar PV technology, there are several factors that affect efficiency. However, the main one is that not all solar radiation is absorbed and converted into electricity by the solar cells. Some photons of energy have too low energy to be absorbed while others have energy too high. One method to calculate the efficiency of a solar PV system is shown in equation (5), such that the efficiency (η) is determined from the maximum power output from the panel (P_{max}), divided by the surface of the panel ($Area_{panel}$), and the solar flux (E_{sun}).[38]

$$\eta = \frac{P_{max}}{E_{sun} * Area_{panel}} \tag{5}$$

For example, the average solar flux for the United States is shown as 205.5 W/m^2 in Table 7.5. However, not every hour of sunlight may generate electricity because of the angle at which solar panels are pointed, and fixed solar panels are set to an angle that absorbs sunlight at midday when the sun is at its brightest.[39] Thus, solar panels distributors talk about "peak sun hours," defined as an hour of sunlight that gives PV power of 1,000 W/m^2. Given the average US solar flux of 205.5 W/m^2 and using 1,000 W/m^2 as the solar radiation that strikes the panel and generates electricity during these "peak sun hours," we can calculate about five hours of "peak sun hours" for the US average. Using data for Trina Solar from the first column of Table 7.11, the panel has dimensions of 65 by 39.1 inches, or 1.65 meters by 1 meter, an area of 1.65 m^2.

Assuming an average of 275 W for the maximum power output from the panel, equation (5) can be used to calculate an average efficiency of 0.167, or 16.7%. An IRENA report[15] shows that, for the most common types of commercial solar panels, the efficiency for mono c-Si panels varies from 15%–19% while that for poly c-Si varies from 13%–15%. Therefore, overall commercial solar efficiencies range from 13%–19%, with an average of 16%. That means that, of the solar radiation striking the panel, only about 16% is converted to electricity.

Another way to look at the previous example is that if the panel produces 275 W over a panel area of 1.65 m^2, the output is 167 W/m^2 versus the peak solar radiation of 1,000 W/m^2. Using five hours of peak sun, the 275 W panel will produce 1,375 Watt-hours over these five hours, or 1.375 kWh per day. In one month, the panel will generate about 41 kWh.

For CSP, the efficiency is the amount of solar radiation that is converted into mechanical work to operate the steam turbine and generate electricity. There are two components to this efficiency including the efficiency of converting solar radiation into heat using the receiver and the efficiency of using this heat to make electricity. Generally, the efficiency can be described by this simple equation:

$$\eta = \eta_{receiver} * \eta_{Carnot} \tag{6}$$

For the Carnot efficiency, the Introduction reviewed the maximum efficiency of an engine, as described by French scientist Sadi Carnot. In this equation, shown again here, "W" is the work taken from the steam to make electricity, "Q_H" is the heat added to the system, "T_C" is cold temperature of the liquid used to store the heat, and "T_H" is the hot temperature of this liquid.

$$\eta_{Carnot} = \frac{W}{Q_H} = 1 - \frac{T_C}{T_H} \tag{7}$$

Recalling the earlier discussion in Overview of Technology, when oils are used for CSP, they limit the operating temperature to around 390°C but molten salts can increase this to 550–650°C. Thus, the thermodynamic efficiency of the engine is greater when molten salts are used versus oils.

The receiver efficiency is described by equation (8), such that efficiency is simply the amount of solar radiation absorbed versus incoming, plus energy lost by the solar receiver system.

$$\eta_{Receiver} = \frac{E_{absorbed} - E_{lost}}{E_{sun}} \tag{8}$$

As examples, one report[40] gives individual data for both the steam turbine efficiency of equation (7) and the receiver efficiency of equation (8). For the Nevada Solar One plant, the report shows a peak efficiency of 66% for the receiver and a peak efficiency of 35% for the steam turbine, giving an overall peak efficiency of 23%. The average efficiency for Solar One was 12%. For the Andasol I plant in Spain, the receiver had a peak efficiency of 70% and an average of 50% while the steam turbine had a peak efficiency of 40% with an average of 30%. This results in overall efficiencies of 28% and 15% for the peak and average efficiencies, respectively.

An extensive theoretical review of the receiver efficiency for each of the four CSP technologies, Parabolic Trough Collector, Linear Fresnel Collector, Solar Tower, and Stirling Dish technology is beyond the scope of this book. Data taken from an IRENA report,[23] shown earlier in Table 7.12, cite annual solar-to-electricity efficiencies ranging from 11%–16% for Parabolic Trough Collectors, 13% for Linear Fresnel Collectors, 7–20% for Solar Towers, and 12%–25% for Stirling Dish technology.

As Parabolic Trough Collectors are, by far, the most common commercially used type of CSP, the efficiency is set to the midpoint of the previous range, or about 14%.

In summary, for the fraction of solar energy converted to work, the solar PV efficiency is set to 16%, based on c-Si panels while the efficiency is set to 14% for CSP, based on the Parabolic Trough Collectors.

7.8. Energy Balance

For solar energy, the energy balance is defined by this equation:

$$\text{Energy Balance} = \frac{\left[\begin{array}{c}\text{Electrical energy from the solar plant delivered}\\ \text{to customer}\end{array}\right]}{\left[\begin{array}{c}\text{Energy contained in the solar radiation coming to the}\\ \text{solar photovoltaic or concentrating solar plant}\end{array}\right]}$$

This energy balance is greatly simplified compared to coal, natural gas, and nuclear. That is because in the "Well-to-Wheels" analysis, there are no mining or extraction costs, and no fuel transportation costs of fuel to solar plant. Therefore, the energy balance only includes (1) making the electricity and (2) distributing the electricity through the grid to your home.

For the first step, the previous section set the efficiency for making electricity from a solar PV plant using crystalline silicon (c-Si) panels to 16% while the efficiency for making electricity from a concentrating solar plant (CSP) using Parabolic Trough Collectors was set to 14%.

It then remains to determine the energy needed to distribute the electricity through the grid to your home in order to construct the final energy balance.

Transmission and distribution losses are affected by the voltage of the power line and the distance the electricity is transmitted. Historical data from the state of California shows annual average transmission losses around 5 to 7% over the time period of 2002 to 2008.[41] Similarly, EIA data for July 2016[42] show that of the electricity generated, 7% is lost in transmission and distribution. Therefore, a value of 7% is used for the amount of electricity lost in transmission and distribution after leaving the solar generating plant. Of course, this 7% lost in transmission and distribution is the percent of the energy leaving the electrical plant, not the kinetic energy of the solar radiation entering the plant.

Using 100 units of energy as a reference, we can arrive at the equations that follow. Summarizing, of the energy contained in the solar radiation, only 14.9% ends up as usable electricity for solar PV and 13.0% for CSP. Naturally, for a renewable resource such as solar energy, there is an inexhaustible supply of solar radiation, so efficiency is only important as to how it affects capital cost and land area.

Solar PV (c-Si)

100 (entering plant) → 16 (generation) → 14.9 (transmission and distribution)

Solar CSP (Parabolic Trough Collectors)

100 (entering plant) → 14 (generation) → 13.0 (transmission and distribution)

7.9. Maturity: Experience

A good review on the history of solar energy is given in a DOE report[43] and some of that is summarized here. The utility of solar power dates back as early as the 7th century B.C., when magnifying glasses were first used to concentrate solar radiation and make fire. There is even a legend that, in the 2nd century B.C., the Greek scientist Archimedes used bronze shields to focus sunlight to set fire to Roman Empire wooden ships. And, in 1767, Swiss scientist Horace de Saussure was credited with building the first solar collector in the world using glass boxes to trap solar heat.

More specific to this chapter, the solar PV effect, defined as the generation of electric current when a material is exposed to light, was discovered by French scientist Edmond Becquerel in 1839.[11] In 1905, Albert Einstein published a paper on the photoelectric effect and in 1916, American physicist Robert Millikan provided experimental proof of this PV effect, for which both won Nobel prizes. In 1918, Polish scientist Jan Czochralski developed a method to grow single crystal silicon material, a process used to make mono c-Si. In 1954, scientists at Bell Laboratory patented a way of making electricity from solar radiation using silicon solar cells, thus bringing PV technology to the United States, although initially, these cells had efficiencies of only 4%. In order to promote the development of solar energy research, the DOE started the Solar Energy Research Institute in 1977, now more broadly called the National Renewable Energy Laboratory, or NREL. Construction of a solar PV manufacturing facility in Camarillo, CA began in 1979, built by ARCO (Atlantic Richfield

Company).[44] In 1982, ARCO completed construction of a 1-MW solar PV power station plant in Hisperia, California.

For CSP, conceptual ideas date back to 1878, when William Adams, a deputy registrar for the English Crown in Bombay, built a reflector with mirrors arranged in a semicircle, that could track the sun's movement by manually moving it around a semicircular track.[45] Seventy-two mirrors were used to concentrate solar radiation onto a boiler, generating temperatures of 1200°F (649°C), and the steam generated was used to run an engine. Essentially, this was the Solar Tower CSP plant of today. Likewise, John Ericsson invented the concept for Parabolic Trough Collectors in 1870, when he built a single linear boiler pipe, placed in the focus of the trough, and connected to a steam engine. Interesting enough, this same John Ericsson designed the Civil War battleship called The Monitor, used by the Union forces.

The first operational CSP power plant was built in Sant'llario, Italy, in 1968 by Professor Giovanni Francia,[46] using the Solar Tower concept. And, in 1982, the DOE, along with a consortium of industrial companies, started operation of Solar One, a 10 MW plant based on the Solar Tower concept. Two years later, in 1984, the first Parabolic Trough Collector system started operation in Daggett, CA.

Since the first coal-fired electrical plant was built in 1882 and through the 1930s, the electrical grid system has grown rapidly, roughly doubling every 6 years. Today, what we take for granted, involves more than 180,000 miles of high-voltage transmission lines connecting about 7,000 power plants.[47]

To summarize, the process to make solar PV cells dates back to 1918 and the first solar PV power plant dates back to 1982. The concept of CSP with the Solar Tower concept dates to 1878 and the first CSP power plant dates back to 1968. And, the first electrical grid system was built in 1882 followed by rapid development through 1930s. Thus, there are around 100 years of solar cell research and about 37 years of solar PV commercial plant operation. There are about 140 years of work on collecting solar power with mirrors and about 50 years of commercial CSP plant operation. Finally, there are more than 135 years of experience in electric grids.

7.10. Infrastructure

As shown in Table 7.1, the total US solar PV capacities in 2017 and 2018 were 41,357 and 49,692 MW, respectively, and the CSP capacity in both 2017 and 2018 was 1,758 MW. However, the data shown in Table 7.1 for small-scale PV are estimates, as tracking these solar PV data is difficult since most of these are residential applications with capacities less than 1 MW. Table 7.1 shows that these small-scale PV units are 39% of the total solar PV for both 2017 and 2018.

Through November 2018, the EIA[48] lists more than 2,500 utility-scale operating solar PV units plants, most of which are 5 MW or smaller. However, if we count only those sites that have capacities greater than 50 MW, solar PV has 82 locations with a total of 163 units and a nameplate capacity of 14,074MW.[49] Since the total utility-scale solar PV capacity for 2018 was 30,171 MW, that means those facilities greater than 50 MW made up about 47% of the capacity.

In 2018 there were 18 CSP plants on 8 locations in the United States with a nameplate capacity of 1,758 MW,[49] a capacity that has not changed since 2015. Combining solar PV and CSP, the total solar capacity for 2018 was 51,450 MW.

For electricity generation in 2018, utility-scale solar PV generated 66,604 MWh and small-scale PV generated 29,543 MWh, for a solar PV total of 92,555 MWh. CSP generated 3,592 MWh, so the total for solar generation was 96,147 MWh. As total US electricity generation in 2018 was 4,207,353 MWh, solar made up 2.3% of this total.[50,51]

Of the total US electricity generation, California was, by far, the largest producer of all types of solar electricity with 45.2% of the total capacity in 2017.[52] The next closest states were Arizona, North Carolina, and Nevada with 8.8, 6.9, and 5.9%, respectively.

One issue for building new utility-scale solar power plants is access to high-voltage transmission lines.[53] The lack of transmission lines serving areas suitable for solar electricity generation is a real barrier toward building these plants. The Solar Energy Industries Association (SEIA) recommends that transmission planning should be done on a regional level, rather than by a state-to-state basis, and that this planning should meet the timeframe of the developers of solar electricity plants.

Interestingly, there can be opposition to new transmission lines for various reasons. Sometimes, it is the expense of building these lines, as well as opposition by people who would be living in the path of these lines. Perhaps, not surprising, is that fossil fuel plant owners may resist new transmission lines, as the importing of alternate energy such as solar energy can undercut their business. And, in New Mexico, there was opposition to a new 515 mile transmission line, a $2 billion project that would send around 3,000 MW of stranded renewable energy west towards Arizona.[54] The federal government approved this new transmission line in January 2015. However, this transmission project, which would take wind from New Mexico and solar and geothermal from New Mexico and Arizona, was first proposed in 2009. It received resistance from New Mexico politicians, who said it would create national security problems at the Army's White Sands Missile Range. The Bureau of Land Management resolved these concerns, and will bury some of the transmission lines when they go through the missile range.

Therefore, while some solar electric plants, either solar PV or CSP, can use the more than 180,000 miles of high-voltage transmission lines that exist in the grid,[47] there is also a need for new transmission line infrastructure to deliver electricity generated from solar power where the electrical grid is not available.

7.11. Footprint and Energy Density

Footprint is a commonly used term to describe the land area needed to deliver energy for a certain energy type. For solar power, the footprint includes the area to (1) make electricity at the solar power plant and (2) transmit and distribute the electricity via the grid to commercial customers.

To determine an average footprint to (1) make electricity at the solar power plant for both solar PV and CSP, utility-scale plants were chosen but no small, demonstration-sized plants. The land area of a solar PV plant includes the solar panels, a system to mount the panels, inverters, transformers, and connection to the grid. Inverters are used to convert DC electricity generated by the PV module into AC electricity. The transformer increases the AC electricity from the inverter to the high voltage needed to connect to the grid. In the case of a CSP plant, mirrors and a receiver are needed to convert solar energy to heat and the steam turbine and generator are used to convert this heat to electricity. Like solar PV, transformers and grid connection are also needed.

Eleven PV plants from around the world were chosen to calculate an average MW per acre of land, and these data are shown in Table 7.18. The data ranged from 0.08 to 0.23 MW/acre, with an average of 0.14 MW/acre. Compared to the data for onshore and offshore wind farms, shown in Tables 6.13 and 6.15 of Chapter 6, the MW/acre

Table 7.18 Land Areas for Photovoltaic Plants

Name	Location	Number of Panels, million	Startup	MW	Acres	MW/acre
Topaz Solar Farms[55]	California	8	March 2015	550	4,700	0.12
Solar Star Projects[56]	California	1.72	March 2015	579	3,200	0.18
Agua Caliente[57]	Arizona	5.2	2011	290	2,400	0.12
Puertollano Solar[58]	Spain	0.35	July 2015	70	432	0.16
Waldpolenz Solar Park[59]	Germany	0.65	2008	52	544	0.10
Arnedo Solar Plant[60]	Spain	0.17	2008	34	173	0.20
Moura PV Power Station[60]	Portugal	0.38	2008	62	618	0.10
Sarnia PV Power Plant[61]	Canada	1.30	December 2009	80	950	0.08
Olmedilla PV Power Plant[61]	Spain	0.16	2008	60	266	0.23
Solarpark Strasskirchen[61]	Germany	0.23	2009	54	334	0.16
Solarpark Lieberose[61]	Germany	0.7	2009	53	402	0.13
Average				171	1,274	0.14

is larger on the average, 0.14 MW/acre for solar PV versus around 0.02 MW/acre for both onshore and offshore wind farms. However, if only the direct impact area of the onshore wind farm is included, defined as turbine pads, roads, substations, and transmission equipment, the average is 1 MW/acre.

For CSP, fifteen plants from around the world were chosen to calculate an average MW per acre of land, and these data are shown in Table 7.19. The data ranged from 0.07 to 0.22 MW/acre, with an average of 0.13 MW/acre. Therefore, even though the solar PV and CSP are quite different in their technological approach, it is interesting that the range and average MW/acre are quite similar.

The averages shown in Tables 7.18 and 7.19 are similar to averages given in an NREL report[64] showing data only for plants in the United States. For solar PV plants greater than 20 MW, data for 32 plants yielded an average land area use of 0.13 MW/

Table 7.19 Land Areas for Concentrating Solar Power Plants

Name	Location	Technology	Startup	MW	acres	MW/acre
Solano Generating Station[28]	Arizona	Parabolic Trough	2013	250	1,928	0.13
Genesis Solar[28]	California	Parabolic Trough	2014	250	1,950	0.13
Ivanpah Solar[28]	California	Tower	2014	377	3,500	0.11
Mojave Solar[28]	California	Parabolic Trough	2014	250	1,765	0.14
SEGS I-IX[28]	California	Parabolic Trough	1984 to 1990	353.8	1,600	0.22
Martin Next Generation[28]	Florida	Parabolic Trough	2010	75	500	0.15
Crescent Dunes[28]	Nevada	Tower	2015	110	1,600	0.07
Nevada Solar One[28]	Nevada	Parabolic Trough	2007	72	400	0.18
Shams I[62]	UAE	Parabolic Trough	2010	100	640	0.16
Andasol-I[63]	Spain	Parabolic Trough	November 2008	50	494	0.10
Arcosol 50[63]	Spain	Parabolic Trough	December 2011	50	568	0.09
Astexol II[63]	Spain	Parabolic Trough	2012	50	395	0.13
Casablanca[63]	Spain	Parabolic Trough	2013	50	494	0.10
La Florida[63]	Spain	Parabolic Trough	June 2010	50	494	0.10
Moron[63]	Spain	Parabolic Trough	May 2012	50	395	0.13
Average				143	1,115	0.13

acre. And, for 25 CSP plants, the average total land area use was 0.10 MW/acre. The NREL report also differentiates between the total area of the plant and the direct impact area, defined as land occupied by solar panels, access roads, substations, service buildings, and other infrastructure. For the direct impact area, the solar PV plant data average was 0.14 MW/acre and the CSP plant average was 0.13 MW/acre. Taking a ratio of direct impact area to total shows that solar PV plant components typically occupy 93% of the total area and CSP plant components typically occupy 77% of the total area. This is quite a contrast to an onshore wind farm, where the direct impact area is only about 2% of the total area (Chapter 6). While wind farms can coexist with ranching and agriculture, a solar plant, either PV or CSP, cannot be used for other purposes.

Summarizing the discussion thus far, an energy density of 0.14 MW/acre is used for solar PV plants and a value of 0.13 MW/acre is used for CSP plants.

We can convert units of MW/acre to the units of quadrillion BTU/(year-square mile). A 1 MW plant operating at 100% capacity for 1 year will produce 8,760 MWh (1 MW * 24 hour/day * 365 days/year). Also, there are 640 acres per square mile and 3.412×10^6 BTU/MWh. For the capacity factors, Table 7.15 shows the average capacity factor for solar PV as 0.26, or 26%, and for CSP plants as 0.38, or 38%.

From these values, we can calculate that 0.14 MW/acre for a solar PV plant is equivalent to an energy density of 0.696 trillion BTU/(year-square mile), or 0.696×10^{12}. For a CSP plant, we calculate the energy density is 0.945 trillion BTU/(year-square mile), or 0.945×10^{12}.

Now that the electricity has been made, the next step is to (2) transmit and distribute it to customers via the grid. However, the land area needed for step-up and step-down transformers is very small and can be ignored. Likewise, the metal towers and electrical lines take up very little land area and can coexist with farming, pasture land, and roadways. This means that the energy density for both solar PV and CSP is essentially the land area for the power plants.

In order to compare energy densities for different technologies on a common basis of gigawatt-hours per square mile per year, or GWh/(square mile-year), we convert from BTU to Watt-hour, or Wh, using this equation:

$$1 \text{ BTU} = 0.293071 \text{ Wh}$$

Using this conversion and recognizing that a gigawatt (GW) is equivalent to one billion Watts (or 1×10^9), the calculated energy density for a solar PV plant is 204.0 GWh/(year-square mile) and the energy density for a CSP plant is 276.9 GWh/(year-square mile).

Assuming a 1,000 MW solar plant, the footprint for the solar PV plant is 42.9 square miles and for CSP plant is 31.6 square miles. Compared to coal and nuclear power, the footprints are much larger but smaller than natural gas when the land area used to extract the natural gas from the ground is included. Both footprints are smaller than those for onshore and offshore wind farms.

According to the International Energy Agency,[65] the US electric power consumption per capita for 1 year was 12.8 MWh in 2016. Using the capacity factor of 0.26 for a solar PV plant and 0.38 for a CSP plant, simple calculations show that 1 MW will power about 180 homes per year for a solar PV plant and 260 homes per year for a CSP plant. Therefore, a 1 GW plant will power about 180,000 and 260,000 homes, respectively, for solar PV and CSP plants!

7.12. Environmental Issues

Certainly, the construction of any electrical generation plant will have some effect on the local environment, such as effects to the land (soil erosion), the local environment (flora and fauna), water use, and the visual appearance of the plant. In addition, there are some risks for the employees involved in the plant construction and daily operation. Solar PV plants introduce some new problems during the solar panel manufacturing process, as well as the end-of-life, because some materials in the solar panels are toxic. Concentrating solar power (CSP) also introduces some new problems, because of the intense heat from concentrated solar energy.

From the standpoint of land disturbance and use impact, any solar energy plant will need substantial land area, which is likely to interfere with existing land uses such as grazing, military use, and minerals production.[66] Clearing land for the solar plant can affect local flora and fauna, including loss of habitats. And, unlike wind farms, there is less chance for solar plants to share land with agriculture and ranching.[67] On the other hand, small-scale solar PV panels on homes and commercial buildings have a minimal effect compared to a utility-scale generation plant.

An environmental study performed in northwestern China[68] showed that CSP collectors lowered the soil temperature between 0.5 to 4°C (0.9 to 7.2°F) in spring and summer and increased temperatures by 0.5 to 4°C during the winter. This was attributed to changes in the air flow as well as shading. In situations where a CSP plant shares land with a farm, these changes could be significant in decisions about growing crops.

CSP plants use conventional steam turbines to generate electricity, so water is used in cooling. Unfortunately, many of the best places in the United States with potential for solar energy are arid, desert settings, where water demands are already strained. However, per acre, CSP plants consume only 1/3 of water use to grow alfalfa and cotton, and only ½ of water used for a typical golf course.[69] Current estimates are that CSP plants use about 202 million gallons of water per year.

It has also been discovered that CSP plants threaten birds.[70] Because CSP mirrors concentrate solar radiation, creating high temperatures, some birds get burned by this concentrated radiation. At three CSP sites in the California desert, 233 birds were recovered, where they had been fatally burned to death. In fact, the term "streamer" is used to describe birds whose feathers ignite in mid-air. It is thought that the mirrors appear to look like a lake, which attracts birds and insects. This high intensity light could also cause problems for airline travel, if the light is reflected into the airplane pathway.[66]

For workers at CSP plants, special training is required to deal with intense reflected light and temperature, including the use of special sunglasses as well as heat-insulating uniforms. As well, CSP plants use heat transfer materials, such as oils and molten salts which are hazardous materials, and thus require training and good maintenance practices.[71]

A big issue for solar PV is the environmental impact of the solar panels, both during the manufacturing process and their disposal. The manufacture of PV panels includes many hazardous materials, which are used to clean and purify the semiconductor surface. These include hydrochloric acid, sulfuric acid, nitric acid, hydrogen fluoride, 1,1,1-trichloroethane, and acetone.[67] Also, during the manufacture of polycrystalline silicon panels, workers are exposed to silicon dust, which can cause silicosis in the respiratory system. And, when thin-film PV cells are made, these contain more toxic materials than polycrystalline silicon, such as gallium arsenide, copper-indium-gallium-diselenide, and cadmium-telluride.

A big issue for solar PV panels is their disposal. If not properly disposed, there is the potential for environmental contamination upon decommissioning. A study by Environmental Progress (EP) warns that toxic waste from solar panels is now a global environmental threat.[72] According to this Berkeley-based group, solar panels make 300 times more toxic waste per energy unit than nuclear power plants. In developing countries, such as India and China, scavengers will burn the waste to salvage copper wire, the burning of which releases carcinogenic gas fumes. To put some perspective on the scope of solar PV panels, a 2016 IRENA report[73] states that by the end of 2016, global PV waste is expected to be in the range of 43,500–250,000 metric tonnes, representing 1 to 6% of the mass of currently-installed panels (4 million metric tonnes). Assuming an average panel life of 30 years, solar panel waste could be a very significant problem by 2030. At present, only the European Union has adopted waste regulations for solar PV panels, requiring all PV panel producers to finance the cost of collecting and recycling these PV panels.

Clearly, the best long-term solution is to reduce, reuse, or recycle these panels. Trends in research and development show that panels are expected to eventually require less materials to manufacture them.[73] Panel reuse is also a possibility, such that repaired PV panels could be resold at a reduced price. The best long-term solution is to recover materials from these panels and recycle their use. Currently, glass, aluminum, and copper in c-Si panels can be recovered at yields greater than 85%.

Of course, there are major environmental benefits from the use of solar energy. An NREL-sponsored 2007 report[74] projected annual reductions in CO_2, NO_x, and SO_2 emissions by 2015 and 2030 for the use of solar PV. These projected reductions assumed that solar PV would replace fossil fuel-base generation on a one-to-one basis, with natural gas being replaced at 75% and coal at 25%. In the case of CO_2 emissions, annual emissions were projected to be reduced by 5 to 11 million tons by 2015 and 69 to 100 million tons by 2030. In the case of NO_x emissions, reductions were 5,000 to 11,000 tons by 2015 and 68,000 to 99,000 tons by 2030. And, for SO_2 emissions, reductions were 9,000 to 21,000 tons by 2015 and 126,000 to 184,000 tons by 2030

7.13. CO_2 Production and the Cost of Capture and Sequestration

Since a solar plant does not use any fuel to generate heat, no CO_2 is directly produced when electricity is made. However, during the life cycle of a solar PV or concentration solar power (CSP) plant, there are emissions from other stages of the plant including manufacturing, materials transportation, installation, maintenance, and decommissioning and dismantlement. And, while overall each of these stages contribute small amounts of CO_2 compared to a petroleum, natural gas, or coal plant, they are still significant.

A fact sheet published by the National Renewable Energy Laboratory (NREL)[75] separate these contributions into three life cycle stages including (1) Upstream, (2) Operational Processes, and (3) Downstream processes. For a CSP plant, upstream includes raw materials extraction, materials production, module manufacture, plant component manufacture, and plant construction. Operational processes include power generation, plant operation, maintenance, and auxiliary natural gas combustion. Downstream processes include plant decommissioning and disposal. For these three stages, upstream contributes 60%–70% of the CO_2, operational processes 21%–26%, and downstream processes 5%–20%. This NREL fact sheet reports CO_2 production for parabolic trough plants (using 19 estimates from 7 references) and power tower plants (using 17 estimates from 6 references). First, however, the data were "harmonized," meaning the reported values were adjusted to a common irradiance of 2,400 kWh per square meter per year (a value typical for the US Southwest), a solar-to-electric efficiency of 15% for parabolic trough and 20% for power tower, and an operating lifetime of 30 years. For parabolic trough CSP plants, the numbers varied from 15–55 CO_2eq/kWh and for power tower CSP plants, the numbers varied from 10–40 CO_2eq/kWh. "CO_2eq" corresponds to an equivalent global warming potential (GWP) meaning the calculated value also includes emissions from other greenhouse gases (GHG), such as methane (CH_4). Overall, the report suggests a mean value of 20 g CO_2eq/kWh for both CSP technologies.

A different report,[76] published in 2012, gives a mean value of 26 g CO_2eq/kWh for parabolic trough plants with range of 12 to 240 and 38 g CO_2eq/kWh for power tower plants with range of 11 to 200. Given the broad range of values reported, especially in the second study, a value of 20 g CO_2eq/kWh is used for both parabolic trough and power tower CSP technologies.

Another NREL fact sheet[77] gives CO_2eq production values for solar PV using the same three stages. And, like solar CSP, the life cycle stages are the same with upstream contributing 60%–70%, operational processes 21%–26%, and downstream processes

5%–20% of the CO_2eq emissions. According to the fact sheet, reported data for crystalline silicon and thin film PV systems that have been "harmonized" varied from 25 to 215 g CO_2eq/kWh with a mean of about 40 g CO_2eq/kWh. There were not enough data for thin film PV systems to differentiate any CO_2 production differences compared to crystalline silicon PV. Another study[78] gives a mean of 49.9 g CO_2eq/kWh for the life cycle GHG emissions of solar PV, with a range of 1 to 218 g CO_2eq/kWh. This study suggests the large ranges in the reported emissions are due to various factors, such as resource inputs, technology, location, sizing and capacity and longevity, as well different calculation methods used by source studies. For example, if coal is used to generate energy in the manufacturing process, this would lead to a higher emission value compared to manufacturing processes using other fuels.

A World Nuclear Association report,[79] published in 2011, reports a range of 13 to 731 g CO_2eq/kWh with a mean of 85 g CO_2eq/kWh for solar PV. Here, they cite the wide range as being due to rapid advancement of the solar PV panels over the last decade, such that manufacturing processes have become more efficient. Finally, another 2011 report[80] gives separate values for three types of solar PV technology including 29 g CO_2eq/kWh for monocrystalline silicon, 28 for polycrystalline silicon, and 18 for CdTe thin films. Given the wide range for these four reports, and to use the two NREL fact sheets for consistency in comparing solar PV and CSP, a value of 40 g CO_2eq/kWh is used to represent all types of solar PV technologies.

In summary, a value of 20 g CO_2eq/kWh (0.044 lb CO_2eq/kWh) is used to represent emissions for both parabolic trough and power tower CSP technologies while a value of 40 g CO_2eq/kWh (0.088 lb CO_2eq/kWh) is used to represent all types of solar PV technologies. While these data indicate the life cycle global warming emissions for solar PV are twice those of CSP plants, they are still quite small compared to the global warming emissions for coal (943 g CO_2eq/kwh) and natural gas (333 g CO_2eq/kwh.).

Carbon dioxide production is commonly listed in the units of lb CO_2/MM BTU. However, solar plants generate electricity without the combustion of a fuel. Because of this, there is no heat rate to convert from CO_2 produced on a mass basis and that produced on an energy basis. The conversion from g CO_2/kWh to lb CO_2/kWh is a simple conversion, and these values are 0.088 lb CO_2/kWh for solar PV and 0.044 lb CO_2/kWh for CSP.

For a 1,000 MW solar PV plant with a capacity factor of 26%, the plant will generate 6,240 MWh/day. For CO_2 produced on a kWh basis, the life cycle production of CO_2 is equivalent to 550,277 lb CO_2/day or about 275 tons CO_2/day. For a 1,000 MW CSP plant with a capacity factor of 38%, the plant will generate 9,120 MWh/day, equivalent to 402,125 lb CO_2/day or about 201 tons. Compared to other technologies, solar PV and CSP are about 33% of a nuclear power plant (713 tons CO_2/day), 1% of the CO_2 produced by a pulverized coal combustion plant (22,214 tons CO_2/day) and 3% of the CO_2 produced by a natural gas combined cycle plant (7,500 tons CO_2/day). Also, solar PV and CSP produce about 70% of the CO_2 produced from onshore and offshore wind farms.

As this CO_2 is not directly produced by the solar plant, but rather by upstream, operational, and downstream processes, it is not practical to capture any of this CO_2. Therefore, it is not likely that CO_2 produced solar PV or CSP plants would be removed by CCS.

7.14. Chapter Summary

Like wind energy in Chapter 6, the growth of solar energy has been rapid. Expanding on the data shown in Table 7.1 for the United States, a comparison is shown in Table 7.20 for utility-scale solar PV and CSP plants. For the 11 years of data shown, the capacity for utility-scale solar PV plants has increased by a factor of 426, compared to a factor of 4 for CSP. Interestingly, CSP dominated solar PV in 2008 and 2009, but remained relatively stagnant until 2013 and 2014, when a number of new plants came on line. Since then, no new CSP plants have started operation.

Part of the slowdown in building new CSP plants can be attributed to the LCOE. The "Well-to-Wheels" economic analysis, including transmission and distribution, gave an LCOE of $113.55/MWh to generate electricity for solar PV plants compared to $239.26/MWh to generate electricity for a CSP plant. Also, the average capital cost for solar PV is significantly lower than CSP, $1,442/kW compared to $5,608/kW for CSP. Thus, both LCOE and capital costs are higher.

In addition, solar PV technology has some technology advantages. While it is true that solar PV-generated electricity has to be converted from DC to AC before being placed on the grid (not true for CSP), PV panels work in both direct and diffuse light,

Table 7.20 US Capacity by Year, MW

MW	Utility-Scale PV	Solar Thermal
2008	70.8	432.3
2009	145.5	439.8
2010	393.4	439.8
2011	1,052.0	439.8
2012	2,694.1	439.8
2013	5,336.1	1,244.6
2014	8,656.6	1,647.9
2015	11,905.4	1,757.9
2016	20,192.9	1,757.9
2017	25,209.0	1,757.9
2018	30,170.5	1,757.9

even when not pointed directly at the sun.[81] Because of this, they are suitable for non-utility scale applications, and can be placed on rooftops where solar tracking is not required. As CSP relies on direct sunlight, it is most suitable for very sunny locations in the United States, like the American southwest.

Although solar fluxes vary with different latitudes, Table 7.5 illustrates the tremendous global solar power potential for even high latitude countries solely based on their land area. As solar PV panel prices decrease and technological advances improve efficiency, there is likely to be continued growth in the use of PV panels on rooftops, as well as new utility-scale plants.

A life cycle analysis showed that the CSP technologies of parabolic trough and power tower made 20 g CO_2/kWh, or 0.044 lb CO_2/kWh. For a 1,000 MW CSP plant with a capacity factor of 38%, the plant will generate 9,120 MWh/day, equivalent to 402,125 lb CO_2/day or about 201 tons. The life cycle analysis for solar PV shows a CO2 production of 40 g CO_2/kWh, or 0.088 lb CO_2/kWh. For a 1,000 MW solar PV plant with a capacity factor of 26%, the plant will generate 6,240 MWh/day, equivalent to 550,277 lb CO_2/day or about 275 tons CO_2/day. As this CO_2 is not directly produced by the solar plant, it is not practical to capture any of this CO_2. Therefore, it is not likely that CO_2 produced solar PV or CSP plants would be removed by CCS.

For a 1,000 MW solar PV plant with a capacity factor of 26%, the footprint is 42.9 square miles. For the same CSP plant with a capacity factor of 38%, the footprint is 31.6 square miles. For solar PV, the direct impact area is typically 93% of the total land area; for CSP, the direct impact area is typically 77% of the total area. Therefore, unlike wind farms where the direct impact area is only about 2% of the total area, and activities such as ranching and agriculture can coexist, land used for solar utility plants are dedicated solely to plant operation.

For a solar PV plant, the overall energy balance was 14.9% compared to 13.0% for CSP. This means that, respectively, 14.9% and 13.0% of the solar energy ends up as energy at the distribution point in the form of electricity. The reason these numbers are so low is due to the low efficiency of the PV panels and the CSP technology. Since there are no fuel costs, the mechanical efficiency of the turbine (in the case of CSP) is not as important as fossil fuel-based electricity generation but, improving efficiency would have the benefits of reducing the land area footprint or the number of PV panels or mirrors needed to generate the same amount of electricity. Laboratory efficiencies as high as about 27% have been reported for crystalline silicon panels and about 23% for thin films, so research is still ongoing to increase the efficiency of commercial PV panels.

In terms of experience and maturity, solar PV cells have been around since 1918 with the first solar PV power plant starting operation in 1982. Concentrating solar power with the Solar Tower concept dates to 1878 and the first CSP power plant started operation in 1968. And, the first electrical grid system was built in 1882 followed by rapid development through 1930s.

The energy densities for a solar PV plant and CSP plant were determined to be 204.0 GWh/(year-square mile) and 276.9 GWh/(year-square mile), respectively.

According to the International Energy Agency,[65] the US electric power consumption per capita for 1 year was 12.8 MWh in 2016. Using the capacity factor of 0.26 for a solar PV plant and 0.38 for a CSP plant, simple calculations show that a 1 GW plant will power about 180,000 and 260,000 homes, respectively, for solar PV and CSP plants!

Finally, some minor environmental issues for solar plants are habitat disturbance (soil erosion, flora, and fauna), water use, and visual appearance. There are new risks for employees at CSP plants, requiring special training for working around reflected light and high temperature. CSP plants have also been shown to be hazardous to birds because of the high temperature from concentrating solar energy with mirrors. For solar PV, the main environmental issues are solar panel manufacture and disposal, because of the hazardous materials involved. To help reduce this issue, work is being done to reduce the amount of panels needed, reuse them, or recycle the panel materials.

References

1. "The Two Types of Solar Energy, Photovoltaic and Thermal," Planete Energies, February 4, 2015,http://www.planete-energies.com/en/medias/close/two-types-solar-energy-photovoltaic-and-thermal
2. "International Renewable Energy Agency (IRENA)," Renewable Electricity Capacity and Generation Statistics, 2018, http://resourceirena.irena.org/gateway/dashboard/?topic=4&subTopic=54
3. "EIA Electricity Data Now Include Estimated Small-Scale Solar PV Capacity and Generation," Energy Information Administration, December 2, 2015, https://www.eia.gov/todayinenergy/detail.php?id=23972
4. "Electric Power Monthly with Data for January 2019," Energy Information Administration, March 2019, https://www.eia.gov/electricity/monthly/pdf/epm.pdf
5. "Solar FAQs," J. Tsao et al., Department of Energy, Office of Basic Energy Science, April 20, 2006, http://www.sandia.gov/~jytsao/Solar%20FAQs.pdf
6. "Electricity Domestic Consumption," Global Energy Statistical Yearbook 2018, https://yearbook.enerdata.net/electricity-domestic-consumption-data-by-region.html
7. "Solar Energy for Fuels, Physical Limits of Solar Energy Conversion in the Earth System," A. Kleidon et al., *Topics in Current Chemistry* 371, 1–22, May 24, 2015, https://link.springer.com/chapter/10.1007/128_2015_637
8. "World Solar Resource Maps," © 2019 The World Bank, Source Global Solar Atlas 2.0, Solargis, http://solargis.com/products/maps-and-gis-data/free/download/world
9. "Solar Insolation of 10 Largest Countries in the World," Aditya Greens, April 24, 2015, http://www.adityagreens.com/blog/the-solar-insolation-levels-of-10-largest-countries-area-wise-in-the-world
10. "What Type of Energy Is Solar," Renewable Energy World, 2019, https://www.renewableenergyworld.com/types-of-renewable-energy/what-is-solar-energy/#gref
11. "The History of Solar," US Department of Energy, Energy Efficiency and Renewable Energy, 2002, https://www1.eere.energy.gov/solar/pdfs/solar_timeline.pdf
12. "Photovoltaics and Electricity," Energy Information Administration, October 7, 2019, https://www.eia.gov/EnergyExplained/index.cfm?page=solar_photovoltaics
13. "Photovoltaic Effect," Energy Education, August 26, 2015, http://energyeducation.ca/encyclopedia/Photovoltaic_effect

14. "Inorganic Photovoltaic Cells," R.W. Miles et al., Materials Today 10 (11), 20–27, November 2007, http://ac.els-cdn.com/S1369702107702754/1-s2.0-S1369702107702754-main.pdf?_tid=1abaa5c4-68ce-11e7.9748-00000aacb360&acdnat=1500062188_88494f9da10989fc09be046461dc5d1f
15. "Renewable Energy Technologies: Cost Analysis Series: Solar Photovoltaics," Volume 1: Power Sector, Issue 4/5, International Renewable Energy Agency (IRENA), June 2012, https://www.irena.org/DocumentDownloads/Publications/RE_Technologies_Cost_Analysis-SOLAR_PV.pdf
16. "What Is the Difference between Monocrystalline, Polycrystalline, and Thin Film Solar Panels?," Adam Loucks, Sun Electronics, April 10, 2013, https://sunelec.com/blog/what-is-the-difference-between-monocrystalline-polycrystalline-and-thin-film-solar-panels/
17. "Photovoltaics Report," Fraunhofer Institute for Solar Energy Systems, ISE, March 14, 2019, https://www.ise.fraunhofer.de/content/dam/ise/de/documents/publications/studies/Photovoltaics-Report.pdf
18. "Jinko Solar Datasheets," 2019, https://www.jinkosolar.com/download_356.html
19. "Trina Solar Datasheets," 2019, https://www.trinasolar.com/us/resources/downloads
20. "First Solar Series 6™ Datasheet," Next Generation Thin Film Solar Technology, January 2020, http://www.firstsolar.com/-/media/First-Solar/Technical-Documents/Series-6-Datasheets/Series-6-Datasheet.ashx
21. "Monocrystalline Solar Prices Fall," Sandra Enkhardt, PV Magazine, June 27, 2018, https://pv-magazine-usa.com/2018/06/27/monocrystalline-solar-prices-fall/
22. "Q4 2017/Q1 2018 Solar Industry Update," David Feldman et al., National Renewable Energy Laboratory, p. 67, May 2018, https://www.nrel.gov/docs/fy18osti/71493.pdf
23. "Concentrating Solar Power," Renewable Energy Technologies: Cost Analysis Series, Volume 1: Power Sector, Issue 2/5, International Renewable Energy Agency (IRENA), June 2012, https://www.irena.org/DocumentDownloads/Publications/RE_Technologies_Cost_Analysis-CSP.pdf
24. "Technology Roadmap: Solar Thermal Electricity", 2014 edition, Figure 3, International Energy Agency, http://www.solarconcentra.org/wp-content/uploads/2017/06/Technology-Roadmap-Solar-Thermal-Electricity-2014-edition-IEA-1.pdf
25. "Domestic Material Content in Molten-Salt Concentrating Solar Power Plants," C. Turchi et al., Technical Report, National Renewable Energy Laboratory (NREL), August 2015, https://www.nrel.gov/docs/fy15osti/64429.pdf
26. "The Fresnel Lens," Cape Hatteras, National Park Service, April 14, 2015, https://www.nps.gov/caha/learn/historyculture/fresnellens.htm
27. "A Compendium of Solar Dish/Stirling Technology," W.B. Stine and R.B. Diver, Sandia National Laboratory Report 93-7026, January 1994, http://www.dtic.mil/dtic/tr/fulltext/u2/a353041.pdf
28. "Concentrating Solar Power Projects in the United States," National Renewable Energy Laboratory (NREL), 2020, https://www.nrel.gov/csp/solarpaces/by_country_detail.cfm/country=US
29. "The 8 Most Exciting Solar Projects in the U.S.," Fast Company, September 16, 2010, https://www.fastcompany.com/1689125/8-most-exciting-solar-projects-us-updated
30. "Cost and Performance Characteristics of New Generating Technologies," Annual Energy Outlook 2019, January 2019, Energy Information Administration, Table 2, https://www.eia.gov/outlooks/aeo/
31. "Lazard's Levelized Cost of Energy Analysis—Version 12.0," November 2018, https://www.lazard.com/media/450784/lazards-levelized-cost-of-energy-version-120-vfinal.pdf
32. "Updated Capital Cost Estimates for Utility Scale Electricity Generating Plants," April 2013, Energy Information Administration, and "Levelized Cost and Levelized Avoided Cost of New Generation Resources in the Annual Energy Outlook 2015," Table 1, June 2015.

33. "Lazard's Levelized Cost of Energy Analysis—Version 8," September 2014, https://www.lazard.com/media/1777/levelized_cost_of_energy_-_version_80.pdf
34. "Renewable Power Generation Costs in 2018," International Renewable Energy Agency, 2019, https://www.irena.org/publications/2019/May/Renewable-power-generation-costs-in-2018
35. "Annual Energy Outlook 2015 with Projections to 2040," DOE/EIA-0383, Table A8, April 2015, https://www.eia.gov/outlooks/aeo/pdf/0383(2015).pdf
36. "Solar Intermittency: Australia's Clean Energy Challenge: 5.2 Storage and Capacity Factor," Saad Sayeef et al., June 2012, https://publications.csiro.au/rpr/download?pid=csiro:EP121914&dsid=DS1
37. "Test Equipment for Photovoltaic Systems," Daystar, 2020, http://www.zianet.com/daystar/Spectrum.html
38. "Critical Factors Affecting Efficiency of Solar Cells," F. Dincer and M. Emin Meral, Smart Grid and Renewable Energy 1, 47–50, May 2010http://citeseerx.ist.psu.edu/viewdoc/download?doi=10.1.1.619.2476&rep=rep1&type=pdf
39. "How to Use Peak Sun Hours to Maximize Solar Efficiency," Camsolar, January 27, 2017, http://www.gocamsolar.com/how-to-use-peak-sun-hours-to-maximize-solar-efficiency/
40. "Advanced CSP Teaching Materials, Chapter 5, Parabolic Trough Technology," M. Günther et al., Enermena, Deutsches Zentrum fur Luft-und Raumfahrt, 2011, http://edge.rit.edu/edge/P15484/public/Detailed%20Design%20Documents/Solar%20Trough%20Preliminary%20analysis%20references/Parabolic%20Trough%20Technology.pdf
41. "A Review of Transmission Losses in Planning Studies," Lana Wong, California Energy Commission, Figure ES-1, August 2011.
42. "Monthly Energy Review," Energy Information Administration, June 2019, https://www.eia.gov/totalenergy/data/monthly/pdf/mer.pdf
43. "The History of Solar," Department of Energy (DOE), Energy Efficiency and Renewable Energy, 2003, https://www1.eere.energy.gov/solar/pdfs/solar_timeline.pdf
44. "History of Solar Energy in California," Go Solar California, 2006, http://www.gosolarcalifornia.ca.gov/about/gosolar/california.php
45. "History of Solar Energy—Revisiting Solar Power's Past," Charles Smith, Solar Energy, 2015, http://solarenergy.com/info-history
46. "History of Concentrated Solar Power (CSP)," Aalborg CSP, 2015, https://www.aalborgcsp.com/business-areas/solar-district-heating/csp-parabolic-troughs/history-of-csp/
47. "United States Electricity Industry Primer," Office of Electricity Delivery and Energy Reliability U.S. Department of Energy DOE/OE-0017, July 2015, https://www.energy.gov/sites/prod/files/2015/12/f28/united-states-electricity-industry-primer.pdf
48. "Most U.S. Utility-Scale Solar Photovoltaic Power Plants Are 5 Megawatts or Smaller," Energy Information Administration, February 7, 2019, https://www.eia.gov/todayinenergy/detail.php?id=38272
49. "Electricity: Form EIA-860 detailed data," Energy Information Administration, 2018 data, https://www.eia.gov/electricity/data/eia860/
50. "Existing Capacity by Energy Source, 2017 (Megawatts)," Table 4.3, Energy Information Administration, https://www.eia.gov/electricity/annual/html/epa_04_03.html
51. "Net Generation by Energy Source: Total (All Sectors)," 2007–February 2017, Table 1.1, Energy Information Administration, https://www.eia.gov/electricity/monthly/epm_table_grapher.cfm?t=epmt_1_01
52. "Electric Power Annual 2017," Energy Information Administration, October 2018, Revised December, 2018, https://www.eia.gov/electricity/annual/pdf/epa.pdf
53. "Transmission," Solar Energy Industries Association, 2020, http://www.seia.org/policy/power-plant-development/transmission

54. "Transmission Line That Could Bring Wind and Solar Power to Millions in West Gets Go-Ahead," Ari Phillips, Think Progress, January 26, 2015, https://thinkprogress.org/transmission-line-that-could-bring-wind-and-solar-power-to-millions-in-west-gets-go-ahead-f5eebc480b43/
55. "Topaz Solar Farms," BHE Renewables Fact Sheet, February 2019, https://www.bherenewables.com/include/pdf/fact_sheet_topaz.pdf
56. "Solar Star Projects," BHE Renewables Fact Sheet, 2016, https://us.sunpower.com/sites/default/files/cs-solar-star-projects-fact-sheet.pdf
57. "World's Largest Operational Solar PV Project, Agua Caliente, Achieves 250 Megawatts of Grid-Connected Power," First Solar, September 10, 2012.
58. "70MW Puertollano Solar Photovoltaic Power Plant," IJ Global, July 21, 2015, https://ijglobal.com/data/project/16547/70mw-puertollano-solar-photovoltaic-power-plant
59. "Waldpolenz Solar Park," Revolvy, 2016, https://www.revolvy.com/main/index.php?s=Waldpolenz%20Solar%20Park
60. "Top 5 Solar Farms in the World," Chloe Lewis, November 7, 2016, http://www.energydigital.com/top-10/top-5-solar-farms-world
61. "Solar Resources," The Future of Energy, December 20, 2014, http://energyfuture.wikidot.com/printer--friendly//solar-resources
62. "World's Largest Concentrated Solar Power Plant Opens in the UAE," Darren Quick, New Atlas, March 18, 2013, http://newatlas.com/shams-1-worlds-largest-concentrated-solar-power-plant/26707/
63. "Concentrating Solar Power Projects in Spain," National Renewable Energy Laboratory (NREL), 2020, https://www.nrel.gov/csp/solarpaces/by_country_detail.cfm/country=ES
64. "Land-Use Requirements for Solar Power Plants in the United States, S. Ong et al., National Renewable Energy Laboratory (NREL), Technical Report NREL/TP-6A20-56290, June 2013, https://www.nrel.gov/docs/fy13osti/56290.pdf
65. "Electric Power Consumption (kWh per capita)," IEA Statistics for 2016, http://energyatlas.iea.org/#!/tellmap/-1118783123/1
66. "Solar Energy Development Environmental Considerations," Solar Energy Development Programmatic Environmental Impact Statement, 2020, http://solareis.anl.gov/guide/environment/
67. "Environmental Impacts of Solar Power," Union of Concerned Scientists, March 5, 2013, http://www.ucsusa.org/clean_energy/our-energy-choices/renewable-energy/environmental-impacts-solar-power.html#.WekmDiVK1jo
68. "Environmental Impacts of Large-Scale CSP Plants in Northwestern China," Zhiyong Wu et al., Environmental Science: Processes & Impacts, Issue 10, 2014, http://pubs.rsc.org/-/content/articlelanding/2014/em/c4em00235k#!divAbstract
69. "Concentrating Solar Power and Water Issues in the U.S. Southwest," N. Bracken et al., Joint Institute for Strategic Energy Analysis, Technical Report, NREL/TP-6A50-61376, March 2015, https://www.nrel.gov/docs/fy15osti/61376.pdf
70. "Solar Farms Threaten Birds," John Upton, Scientific American, August 27, 2014, https://www.scientificamerican.com/article/solar-farms-threaten-birds/
71. "Environmental Impacts from the Solar Energy Technologies," Theocharis Tsoutsosa et al., Energy Policy 33, 289–296, 2005, http://citeseerx.ist.psu.edu/viewdoc/download?doi=10.1.1.405.7333&rep=rep1&type=pdf
72. "A Clean Energy's Dirty Little Secret," Julie Kelly, June 28, 2017, National Review, http://www.nationalreview.com/article/449026/solar-panel-waste-environmental-threat-clean-energy
73. "End-of-Life Management: Solar Photovoltaic Panels, Stephanie Weckend et al., International Renewable Energy Agency (IRENA), June 2016, http://www.irena.org/

DocumentDownloads/Publications/IRENA_IEAPVPS_End-of-Life_Solar_PV_Panels_2016.pdf

74. "Energy, Economic, and Environmental Benefits of the Solar America Initiative," S. Grover, ECONorthwest Portland, Oregon, Subcontract Report, NREL/SR-640-41998, August 2007, https://www.nrel.gov/docs/fy07osti/41998.pdf
75. "Life Cycle Greenhouse Gas Emissions from Concentrating Solar Power," National Renewable Energy Laboratory (NREL) Fact Sheet, November 2012, https://www.nrel.gov/docs/fy13osti/56416.pdf
76. "Life Cycle Greenhouse Gas Emissions of Trough and Tower Concentrating Solar Power Electricity Generation: Systematic Review and Harmonization," J. J. Burkhardt et al., Journal of Industrial Ecology 16 (s1) S93–S109, April 9, 2012, http://onlinelibrary.wiley.com/doi/10.1111/j.1530-9290.2012.00474.x/full
77. "Life Cycle Greenhouse Gas Emissions from Solar Photovoltaics," National Renewable Energy Laboratory (NREL) Fact Sheet, November 2012, https://www.nrel.gov/docs/fy13osti/56487.pdf
78. "Assessing the Life Cycle Greenhouse Gas Emissions from Solar PV and Wind Energy," D. Nugent and B. Sovacool, Energy Policy 65, 229–244, February 2014, http://www.sciencedirect.com/science/article/pii/S0301421513010719
79. "Comparison of Lifecycle Greenhouse Gas Emissions of Various Electricity Generation Sources," World Nuclear Association Report, July 2011, http://www.world-nuclear.org/uploadedFiles/org/WNA/Publications/Working_Group_Reports/comparison_of_lifecycle.pdf
80. "Life Cycle Inventories and Life Cycle Assessments of Photovoltaic Systems," International Energy Agency (IEA), Photovoltaic Systems Power Programme, Report IEA-PVPS T12-02:2011, October 2012, http://www.seas.columbia.edu/clca/Task12_LCI_LCA_10_21_Final_Report.pdf
81. "A Review of Concentrated Solar Power in 2014," Tom Lombardo, Engineering.com, January 4, 2015, http://www.engineering.com/DesignerEdge/DesignerEdgeArticles/ArticleID/9286/A-Review-of-Concentrated-Solar-Power-in-2014.aspx

8

Ethanol

A Renewable Energy Type

8.1. Foreword

This chapter focuses on ethanol, or bioethanol, such that biomass is used as an energy source to make ethanol as a liquid fuel. Chapter 9 is devoted to the use of biomass to generate electricity. For both of these chapters, biomass is defined as any organic material that can be used as a fuel.

In general, biomass can come from both plants and animals and is therefore a renewable source of energy. In the case of plants, energy is stored from the sun via photosynthesis, such that plants make food from carbon dioxide and water. Biomass can take on many forms, such as wood, agricultural crops, agricultural waste, and solid waste. In addition, vegetable oils, animal fats, recycled greases, and soybean oil can be used to make biodiesel.

Ethanol is made from fermentation of sugar and used as a fuel for vehicles. Many different agricultural crops can be used to provide this sugar, and some of the more important ones include corn, sorghum, barley, switchgrass, sugarcane, sugar beets, potato skins, and rice. To keep the discussion manageable, the discussion focuses mainly on ethanol from corn, but with some consideration of sugarcane and switchgrass. Switchgrass is a tall North American grass, and has some yield advantages over corn, to be discussed later. Nevertheless, according to the Energy Information Administration (EIA),[1] most of the ethanol used as a fuel in the United States is distilled from corn.

Even the definition of "corn" deserves some explanation. As discussed in the "Overview of Technology," corn kernels can be transformed into ethanol using dry milling and wet milling. However, corn stover can also be used to make ethanol, in which case the production is referred to as cellulosic ethanol. Cellulosic ethanol is ethanol made from the fibrous part of the plant instead of the plant's seed or fruit. In the case of corn, the fruit is the corn kernel and the cellulosic part is the corn stover. The corn stover includes the leaves, stalks, and cobs left from the corn harvest.

Ethanol is a chemical compound with the simple formula of C_2H_5OH and is also called ethyl alcohol. It has a density of 0.7893 g/cm^3 at 20°C (or 68°F, about room temperature) and a boiling point of 78°C or 172°F. Also, it has a research octane number (RON) of 108.6 and motor octane number (MON) of 89.7, yielding a road octane number of 99.15. Road octane is the simple average of RON and MON. The United States sells gasoline using the road octane number designation, usually 87, 89, and 91

The Changing Energy Mix. Paul F. Meier, Oxford University Press (2020). © Oxford University Press.
DOI: 10.1093/oso/9780190098391.001.0001.

while Europe and Asia use the RON designation for sales. Since gasoline typically has densities ranging from 0.71%–0.77 g/cm^3 and a boiling range of about 35°C (95°F) to about 200°C (395°F), ethanol is a good candidate for blending with gasoline to raise the overall octane value.

However, while ethanol has a higher octane value than your typical gasoline, the energy content is lower. As shown in Appendix B, equations 28 and 31, the energy content of ethanol and a typical gasoline are 75,700 BTU and 124,238 BTU per gallon, respectively. Therefore, the energy content of ethanol is only 60% of that for gasoline. Because of the lower energy content, vehicles typically go 3%–4% fewer miles per gallon on E10 (10% ethanol with gasoline) and 4%–5% fewer on E15 (15% ethanol with gasoline).[2]

In addition, there are some limits on ethanol solubility with gasoline. Ethanol is naturally attracted to water, or hydrophilic, and therefore is miscible with water at any concentration. The danger is phase separation, such that ethanol separates from the gasoline into the water phase. At small amounts of water, phase separation does not occur. For example, with 5,500 ppm water, a 10% blend of ethanol with gasoline (or E10) does not separate until the ambient temperature is -18 to -27°C (-0.4 to 16.6°F).[3] The temperature range at which separation occurs is related to the composition of the gasoline. However, if water continues to be absorbed into an ethanol-gasoline blend, phase separation will occur at the saturation point. For an E10 blend at 15°C (59°F), this saturation point is 0.5% water.[4]

Of course, there are advantages using ethanol, such as cleaner combustion than gasoline, it is renewable, it is carbon neutral (meaning the CO_2 produced by combustion is the same amount the crops absorbed during photosynthesis), and it reduces some of the need to import foreign oil.

As ethanol is not made at the petroleum refinery (Chapter 1) but rather an ethanol production facility, ethanol blending normally occurs at the fuel terminal. A fuel terminal, also called a tank farm, is used to store and blend refinery products and can be located adjacent to a refinery or at a separate location, receiving the products via pipeline, tanker, or barge. Ethanol is transported from the production facility to the fuel terminal, where the blending occurs to make E10, E15, or E85.[5] E85 is a flexible fuel containing 51%–83% ethanol and can only be used in flexible fuel vehicles (FFVs). An FFV has an internal combustion engine capable of running on more than one type of fuel. Although a detailed discussion of the differences between an FFV and normal automobile are beyond the scope of this chapter, some of the differences include building the fuel tank to withstand ethanol corrosion, as well as fuel hoses, gaskets, and seals appropriate for high ethanol fuels.

Regardless of some of the disadvantages related to ethanol as a fuel, ethanol production has grown steadily in the United States and the world over the last 12 years. Table 8.1 shows data for the last 12 years in millions of gallons.[6,7] Globally, ethanol production has more than doubled from 2007 to 2018, increasing by a factor of 2.2, while in the United States production has also more than doubled from 2007 to 2018, going up by a factor of 2.4.

Table 8.1 Annual Ethanol Production Growth over the Last 12 Years (millions of gallons)[6,7]

Region	United States	Brazil	European Union	China	Canada	Rest of World	World
2007	6,521	5,019	570	486	211	315	13,123
2008	9,309	6,472	734	502	238	389	17,644
2009	10,938	6,578	1,040	542	291	914	20,303
2010	13,298	6,922	1,209	542	357	985	23,311
2011	13,948	5,573	1,168	555	462	698	22,404
2012	13,300	5,577	1,179	555	449	752	21,812
2013	13,300	6,267	1,371	696	523	1,272	23,429
2014	14,313	6,190	1,445	635	510	1,490	24,583
2015	14,807	7,093	1,387	813	436	1,147	25,682
2016	15,329	7,295	1,377	845	436	1,301	26,583
2017	15,800	7,060	1,415	875	450	1,450	27,050
2018	16,061	7,920	1,430	1,050	480	1,629	28,570

8.2. Proven Reserves

Unlike petroleum crude oil, natural gas, coal, and nuclear energy, the amount of proven, or potential, reserves for ethanol from biomass is difficult to assess. There are a great number of factors to consider, such as how much land can be devoted to growing biomass solely for ethanol as a fuel versus using that land to grow food, what agricultural improvements can be made to increase the yield per acre for crops, what types of biomass are technically and economically feasible to make ethanol, and how much ethanol can be absorbed by the gasoline/diesel market.

To make an estimate of the potential biomass that could be used to produce ethanol in the United States, data from a 2016 Department of Energy (DOE) report[8] were used. This report suggests that one billion or more tons of biomass could be available in the United States by 2030 and as much as one and a half billion by 2040, depending on crop yields, climate change impacts, harvest issues, and systems integration of production, harvest, and conversion. The report makes many projections based on the price of biomass per ton, and based on annual yield improvements for crops. Table 8.2 shows results with biomass priced at $60/ton. The base case scenario assumes a 1% annual yield improvement in energy crops while the high-yield case assumes 3% growth. Energy crops are defined as plants cultivated solely for the purpose of producing non-food energy.

Table 8.2 Currently Used and Potential Biomass for the United States Available at $60 per Dry Ton (million dry tons) per Year[8]

Feedstock, Million Dry Tons	2017	2022	2030	2040
Currently Used Resources				
Forestry Resources	154	154	154	154
Agricultural Resources	144	144	144	144
Waste Resources	68	68	68	68
Total Currently Used	365	365	365	365
Potential: Base-Case Scenario				
Forestry Resources (all timberland)	103	109	97	97
Forestry Resources (no federal timberland)	84	88	77	80
Agricultural Residues	104	123	149	176
Energy Crops		78	239	411
Waste Resources	137	139	140	142
Total (Base Case Scenario)	343	449	625	826
Total (Currently Used + Potential)	709	814	991	1,192
Potential: High-Yield Scenario				
Forestry Resources (all timberland)	95	99	87	76
Forestry Resources (no federal timberland)	78	81	71	66
Agricultural Residues	105	135	174	200
Energy Crops		110	380	736
Waste Resources	137	139	140	142
Total (High-Yield Scenario)	337	483	782	1,154
Total (Currently Used + Potential)	702	848	1,147	1,520

In this table, "Forestry Resources" are defined as whole-tree biomass and residues; "Agricultural Resources" include corn, hay, soybeans, wheat, cotton, sorghum, barley, oats, and rice; "Agricultural Residues" are defined as corn stover, cereal straws, sorghum stubble rice hulls, wheat dust, and sugarcane trash (bagasse); "Energy Crops" are defined as sorghum, energy cane (perennial tropical grass), eucalyptus, miscanthus (perennial bamboo-like grasses), pine trees, switchgrass (perennial native grass), and willow; and "Waste Resources" are defined as secondary and tertiary wastes from processing agricultural and forestry products, such agricultural secondary wastes, municipal solid waste, forestry and wood wastes, and animal fats.

In order to convert the data of Table 8.2 from tons of biomass to gallons of ethanol for these different types of biomass, two classifications are used to determine an ethanol yield per dry ton, namely ethanol made from starch crops and ethanol made from cellulosic material. Although this is discussed in more detail later, ethanol (or bioethanol) is made by fermenting sugar. For a starch crop, such as corn, the starch first undergoes saccharification to sugar followed by fermentation to ethanol. For cellulosic material, such as corn stover, enzymes are used to break down the complex cellulose into sugars which are then fermented to ethanol.

For the conversion of corn grain to ethanol, 5 years of data from the US DOE were examined, shown in Table 8.3.[9] Corn production for the last 5 years has been around 14 to 15 billion bushels, with 36%–38% used to make ethanol. Over these 5 years, ethanol production has increased from 2.77 to 2.87 gallons per bushel. According to the 2016 DOE report,[8] a bushel of shelled corn weighs 56 lb and has a water content of 15.5%. Accounting for the water content, 2.87 gallons per bushel is equivalent to 121 gallons of ethanol per dry ton of corn.

The same 2016 DOE report[8] gives a yield of 85 gallons per dry ton of cellulosic material, based on data from three commercial biorefineries using different feedstocks including yard and wood waste, corn stover, and agricultural waste.

To convert the data of Table 8.2 to million gallons of ethanol, a value of 120 gallons of ethanol per dry ton of corn grain was applied to the categories "Agricultural Resources" and "Energy Crops, and a value of 85 gallons of ethanol per dry ton of cellulosic materials was applied to the categories of "Forestry Resources," "Agricultural Residues," and "Waste Resources." For the year 2040, the sum of current plus the potential base case scenario is 120,745 million gallons and the sum of current plus the high-yield potential scenario is 160,000 million gallons. For the year 2017, the sum of current plus potential is about 65,000 million gallons, with the current production at about 36,000 million gallons of ethanol based on the 365 million dry tons. However, Table 8.1 shows that the ethanol produced in the United States for 2017 was

Table 8.3 Annual US Corn Production and Corn Used for Ethanol Production (billion bushels)[9]

Date	Corn Production (billion bushels)	Corn used for Ethanol (billion bushels)	Percent Corn used for Ethanol	Ethanol Produced (million gallons)	Gallons Ethanol per Bushel
2014	14.22	5.17	36.4%	14,313	2.77
2015	13.60	5.22	38.4%	14,807	2.84
2016	15.14	5.43	35.9%	15,329	2.82
2017	14.60	5.60	38.4%	15,800	2.82
2018	14.62	5.60	38.3%	16,061	2.87

15,800 million gallons, about 44% of the 36,000 million gallons calculated based on current biomass production. This difference is because the 365 million dry tons of biomass are not all used to make ethanol, as biomass is also used to generate electricity and heat. Table 2.7 in the DOE report[8] shows that in 2014, 39% of biomass went to ethanol, 59% to electricity and heat, and 2% to biobased chemicals. Therefore, the projections for ethanol production in 2040 will compete with the alternate uses of biomass, primarily in electricity generation and heating.

As the energy content of ethanol is 75,700 BTU/gallon and typical gasoline is 124,238 BTU/gallon, it takes 1.64 gallons of ethanol to equal the energy content of a gallon of gasoline. Using something else to replace the energy content of gasoline, in this case ethanol, is referred to as a gasoline gallon equivalent (gge). Using the value of 1.64 gallons of ethanol per gallon of gasoline, the 65,000 million gallons of ethanol for 2017 is equivalent in energy to about 40,000 million gge. For 2040, the current plus potential base case scenario is equivalent in energy to about 74,000 million gge and the current plus high-yield potential scenario is equivalent to about 98,000 million gge.

According to the EIA, the United States consumed 142,860 million gallons of gasoline in 2018.[10] Comparing these to the ethanol yields from the previous paragraph on a gge basis, the 40,000 gge for 2017 represents 28% of the 2018 consumption while the base case and high-yield scenario projections for 2040 represent 52% and 69%, respectively of the 2018 gasoline consumption. So, a significant percentage of gasoline could be replaced with ethanol if the scenarios in Table 8.2 become true. Of course, these percentages could increase if the average yield on a gallon per ton basis increases, but could also decrease if gasoline consumption increases by 2040. This exercise does show, however, that if ethanol is an economically viable alternative to gasoline, there are potential biomass resources in the United States to replace a substantial amount of the gasoline consumed.

It is worth noting that the biomass shown in Table 8.2 can also be used for electricity generation (to be discussed in Chapter 9) and heating. Whether these types of biomass are used to make ethanol, electricity, or provide heating depends on many factors, such as geographical logistics, demand, and economic viability.

According to the Union of Concerned Scientists, the DOE data shown in Table 8.2 are optimistic, as there are concerns about using food crops (such as corn, sugar, and vegetable oil) as a primary source for transportation fuels.[11] Increased use of these fuels has resulted in rising food prices and price volatility, as well as accelerated expansion of agriculture in the tropics. Limiting the analysis to nonfood sources of biomass, they suggest that by 2030, about 680 million tons of biomass could be made each year. This is less than the base case scenario of 991 million tons and the high-yield scenario 1,147 million tons shown in Table 8.2. Even so, the US potential to make ethanol from biomass is still substantial, even if food sources are eliminated from the projections.

Before considering the global potential for biomass, it is useful to recast the data in Table 8.2 to units of Exajoules per year, or EJ/year. An Exajoule is one quintillion Joules, or 1×10^{18} Joules. Since research reports that examine the global biomass potential use units of EJ/year, this will allow comparison of the data for the United States given in Table 8.2 to the global potential. To make the conversion from dry tons per year to EJ/year, the yield for corn grain is set to 120 gallons of ethanol per ton and the yield for cellulosic material is set to 85 gallons of ethanol per ton. Also, the energy content of ethanol is 75,700 BTU/gallon and 1 BTU (British thermal unit) is equivalent to 1,055 Joules. Lastly, for comparison to global biomass potential, it is useful to recognize that the data for Table 8.2 can be organized into three general categories, namely "Surplus forest growth," "Energy crops," and "Agricultural residues and waste." In Table 8.2, "Forestry Resources" are categorized as "Surplus forest growth," "Agricultural resources" are categorized are "Energy crops," and "Agricultural residues," and "Waste resources" are categorized as "Agricultural residues and waste." With these definitions, the base case and high-yield scenarios for the United States in 2040 are 9.6 and 12.8 EJ/year, respectively.

For global biomass, there are quite a few reports available that project the potential biomass energy for the year 2050. The ten different reports examined showed a very wide range of potential biomass energy in the year 2050, 26 to 1,471 EJ/year. Some of these reports only consider the category of "Energy crops" while others also consider "Surplus forest growth" and "Agricultural residues and waste." Needless to say, there are a great many factors to consider such as land area available for energy versus food, crop yields, population growth, water availability, production costs, and the ability of the government for a country to properly manage forest growth and waste. Fortunately, the data for "Energy crops" were reassessed by Searle and Malins[12] to normalize them using a consistent criteria for energy crop yields, land availability, production costs, and good governance. The normalized results for "Energy crops" and other data for

Table 8.4 Currently Used and Potential Biomass for the United States, EJ/year

Feedstock, EJ/year	2040	2040
Currently Used Resources	Potential: Base Case	Potential: High-Yield Scenario
Surplus Forest Growth	1.0	1.0
Energy Crops	1.4	1.4
Agricultural Residues and Waste	0.5	0.5
Total Currently Used	2.9	2.9
Potential: Base-Case Scenario		
Surplus Forest Growth	0.7	0.5
Energy Crops	3.9	7.1
Agricultural Residues and Waste	2.2	2.3
Total (Case Case Scenario)	6.8	9.9
Total (Currently Used + Potential)	9.6	12.8

the ten studies are shown in Table 8.5 and data for North America, where available, are shown in Table 8.6. The original report data for "Energy crops" before normalization are shown in parentheses. Details for each study are not given here, but can be found in the individual reports. As expected, regions with either hot and humid climates, large land area, or both had the greatest potential for biomass energy.

Table 8.5 Global Biomass Potential for 2050, EJ/Year[12]

EJ/year	Surplus Forest Growth	Energy Crops	Agricultural Residues and Waste	Total
Fischer and Schrattenholzer[13]	100	45–49 (147–207)	143	288–292
Yamamoto et al.[14]		81 (110)	72	153
Wolf et al.[15]		75–111 (162–360)		75–111
Hoogwijk et al.[16]		97–104 (123–496)		97–104
Field et al.[17]		79 (27)		79
Smeets et al.[18]	59–103	65–87 (215–1,272)	76–96	200–286
Buchmann et al.[19]		76 (34–120)		76
D.P. Van Vuuren et al.[20]		89 (150)		89
Beringer et al.[21]		76–78 (26–116)		76–78
M. Parikka[22]	42	37	25	104
Average	74	76	82	232

Table 8.6 Normalized North America Biomass Potential for 2050, EJ/Year[12]

EJ/year	Surplus Forest Growth	Energy Crops	Agricultural Residues and Waste	Total
Yamamoto et al.[14]		19		19
Hoogwijk et al.[16] [(a)]		14.5–15.8		14.5–15.8
Smeets et al.[18]	5–11	6–12	14–19	25–42
Buchmann et al.[19]	2.2	2.3		
D.P. Van Vuuren et al.[20]	6.8	9.9		16.7
Beringer et al.[21]		8		
M. Parikka[22]	12.8	4.1	3.0	19.9
Average	7.5	9.6	9.8	26.9

a. *Note*: From Hoogwijk et al.; they also break out North America into 8.6–9.5 EJ/year for the United States, 3.4–6.3 EJ/year for Canada, and 0–2.5 EJ/year for Central America.

As illustrated by Table 8.5, the normalization procedure narrows the range for "Energy crops" from 26 to 1,272 EJ/year to 37–111 EJ/year, with an average of about 76 EJ/year. Considering all three categories, the total biomass energy potential projected for 2050 is 232 EJ/year with contributions of 74, 76, and 82 EJ/year from "Surplus forest growth," "Energy crops," and "Agricultural Residues and waste," respectively.

Table 8.6 shows that for North America, the total biomass energy potential is 26.9 EJ/year with contributions of 7.5, 9.6, and 9.8 EJ/year from "Surplus forest growth," "Energy crops," and "Agricultural Residues and waste," respectively. The report of Hoogwijk et al.[16] breaks out the range of 14.5–15.8 EJ/year for North America into contributions of 8.6–9.5 EJ/year from the United States, 3.4–6.3 EJ/year from Canada, and 0–2.5 EJ/year from Central America. The range for the United States is in reasonable agreement with the range of 9.6–12.8 EJ/year shown in Table 8.4 for the year 2040.

Using the same factors that allowed the conversion of million tons of dry biomass in Table 8.2 to EJ/year in Table IV for the United States, the global biomass potential shown in Table 8.5 for 2050 can be converted from EJ/year to a total of 2.9 trillion gallons of ethanol with about 0.93 trillion from "Surplus forest growth," 0.95 trillion from "Energy crops," and 1.0 trillion from "Agricultural residues and waste." Like the US projections, however, the use of global biomass resources will compete with the alternate use of biomass to generate electricity or provide heating. Converting this 2.9 trillion gallons ethanol to the equivalent energy for gasoline results in about 1.8 trillion gge. According to the EIA, in 2016 the world used 26,043 thousand barrels per day of gasoline.[23] This is equivalent to about 0.4 trillion gallons per year of gasoline so, even with competition from electricity and heating, the worldwide biomass potential is substantial enough to replace gasoline with ethanol.

In terms of production, Table 8.1 shown earlier, indicates that the United States and Brazil have dominated the production of ethanol from biomass over the last 12 years. Extracting only the data for 2017 and 2018, shown again in what follows, we see that the United States accounted for 58% and 56% of the ethanol production in 2017 and 2018 while Brazil accounted for 26% and 28% in 2017 and 2018. Together, these two countries have dominated global ethanal production with a total of 84% for both years. The United States makes ethanol primarily from corn grain, or corn kernels, while Brazil makes ethanol primarily from sugarcane.

Table 8.1a Annual Ethanol Production Growth over the Last 12 Years (millions of gallons)[6,7]

Region	United States	Brazil	European Union	China	Canada	Rest of World	World
2016	15,329	7,295	1,377	845	436	1,301	26,583
2017	15,800	7,060	1,415	875	450	1,450	27,050
2018	16,061	7,920	1,430	1,050	480	1,629	28,570

Table 8.7 Annual Ethanol Consumption for the United States and Brazil (millions of gallons)[24,25]

Region	United States	Brazil
2016	14,356	6,795
2017	14,485	7,141
2018	14,382	7,587

Table 8.7 shows ethanol consumption for the United States and Brazil from 2016 to 2018 using data from the EIA[24] and the USDA (US Department of Agriculture) Foreign Agricultural Service.[25] Comparing Table 8.1.a and Table 8.7 shows that, overall, the US exports ethanol while for Brazil, production and consumption are about the same.

Although the United States was is a net exporter of ethanol, it did import 36 million gallons in 2016, 99% of which was sugarcane ethanol from Brazil;[26] the remainder came from Canada. The United States exported more than a billion gallons in 2016 to 41 countries, and the four largest trading partners were Brazil (26%), Canada (25%), China (17%), and India (8%).

8.3. Overview of Technology

The fermentation of sugar to make alcohol has been around for many centuries. One review states that ethanol residue was found on 9,000 year old pottery in China.[27] In addition, evidence of alcohol has been discovered in the ancient Middle East (~5400 BC), ancient Egypt (~3000 BC), and even Mexico (~2000 BC). Typically, fermentation to beer and wine will not result in alcohol contents greater than about 15%, limited by the yeast which becomes inactive when alcohol levels go above about 10%.

Of course, we know that many alcohols, such as scotch, whiskey, and vodka have much higher alcohol contents, and this higher alcohol concentration is obtained through distillation. As ethanol boils at 78.37°C (173.1°F) and water boils at 100°C (212°F), distillation can be used to raise its alcohol content. Basically, distillation is a process of heating the alcohol to transform it from liquid to gas, collect the gas as it rises overhead from the boiling pot, and then condense it back to a liquid. According to one review, Greek alchemists working in Alexandria carried out distillations during the first century AD.[27] And, in 1796, Johann Tobias Lowitz obtained pure ethanol by filtering distilled ethanol through activated charcoal. Pure ethanol, or anhydrous ethanol (containing no water), is what is needed to make ethanol blends with gasoline.

There are two basic routes to make ethanol which can be used as fuel, namely biochemical and thermochemical routes to convert the biomass. The thermochemical route first involves a process known as gasification, such that the biomass is converted into CO (carbon monoxide) and H_2 (hydrogen). After cleaning the gases of impurities, this gas mixture, also known as synthesis gas or syngas, can be converted directly to ethanol through a catalytic process. The basic chemical equation to describe this is shown in this equation:

$$2CO + 4H_2 \rightarrow C_2H_5OH \text{ (ethanol)} + H_2O \quad (1)$$

In addition, once the biomass is converted to syngas, it can be converted directly to gasoline and diesel fuel using the Fischer-Tropsch process or converted to methanol, and subsequently converted to gasoline using the methanol-to-gasoline (MTG) process. The Fischer-Tropsch process is discussed in Chapter 12.

This chapter, however, focuses on the biochemical route to ethanol, such that sugars are made into ethanol by fermentation. For the technology overview, three different classes of biomass are considered. Food crops containing starch, such as wheat, corn, and potatoes, are one type of biomass. There are two major routes to transform these food crops into ethanol, namely dry and wet milling. Food crops containing readily fermentable sugars, such as sugarcane, sugar beets, and sorghum, are a second type of biomass. The last type of biomass to be considered is cellulosic biomass, which uses wood, grass, or inedible parts of the food crops. Cellulose can be broken down into glucose, which is then fermented into ethanol. Some examples of cellulosic biomass include corn stover, corn cobs, switchgrass, and sugarcane bagasse (the pulpy residue left after extraction of juice).

To get an idea of the differences in food crops containing starch and their cellulosic counterparts, consider Table 8.8, taken from one reference.[28] Here, corn grain refers to the corn kernels and corn stover refers to the stalks and leaves left from the corn harvest. As the table illustrates, the corn grain has a high level of starch, which can

Table 8.8 Compositions of Corn Grain, Corn Cob, and Corn Stover (dry basis)[28]

Type	Corn Grain, %	Cobb, %	Stover, %
Starch	71.7	n/m	n/m
Cellulose	2.4	42.0	36.0
Hemicellulose	5.5	33.0	26.0
Protein	10.3	n/m	5.0
Oil	43	n/m	n/m
Lignin	0.2	18.0	19.0
Ash	1.4	1.5	12.0
Other	4.2	5.5	2.0
Total	100.0	100.0	100.0
Best case Ethanol Yield (gallons/ton)	135	128	105

Note: "n/m" means not measured

easily be converted into sugar and then ethanol, but the cob and stover do not contain starch. Instead, they have high levels of cellulose and hemicellulose, which require additional processing to make the ethanol.

The use of corn grain to make ethanol is discussed first. According to the Renewable Fuels Association,[29] roughly 90% of the grain ethanol produced in the United States today comes from the dry milling process, with the remaining 10% coming from wet milling. The main difference between the two processes is in the initial treatment of the grain. In dry milling, the first step is to grind the corn grain kernel into "meal" (or coarse flour). In wet milling, the corn is soaked up to 48 hours to assist in separating the parts of the corn kernel.[30] The purpose of grinding is to break the tough corn kernel coating and increase the surface area of the starch.

After milling, the corn is mixed, or slurried, with heated water to form a "mash," a fermentable starchy mixture. This process is also known as liquefaction, and an enzyme, such as α-amylase, is added to this slurry to catalyze the hydrolysis (chemical reaction with water) of starch into glucose. After this, another enzyme called glucoamylase is added to the corn mash to complete the conversion of starch into glucose, a process step also called saccharification. Overall, the conversion of starch into glucose through liquefaction and saccharification is shown in the equation, such that $C_6H_{10}O_5$ is the starch and $C_6H_{12}O_6$ is glucose.

$$\left(C_6H_{10}O_5\right)_n + nH_2O \xrightarrow{\text{enzyme}} nC_6H_{12}O_6 \tag{2}$$

The final step in the chemistry to make ethanol is called fermentation, such that yeast is added to the corn mash to convert the glucose into ethanol and carbon dioxide (CO_2). The chemistry of this step is shown here, with C_2H_5OH the chemical formula for ethanol.

$$C_6H_{12}O_6 \xrightarrow{\text{yeast}} 2\ C_2H_5OH + 2\ CO_2 \tag{3}$$

While fermentation is the final step in the chemical transformation of starch to ethanol, some more processing is needed to make 100% pure ethanol. Recall from the earlier discussion about fermentation to make beer and wine that the alcohol content is limited by the yeast to about 15%. Also, the fermented slurry has some solid material, and this is removed using a centrifuge. The solids from the centrifuge are recovered and part of these, known as distiller's dry grain solids, or DDGS, are used to feed farm animals. The liquid part is distilled and dehydrated to remove water and produce pure ethanol. Also, a denaturant such as gasoline is normally added to make the ethanol undrinkable.

In contrast to the United States, Brazil makes most of its ethanol from sugarcane, a food crop containing readily fermentable sugars. A sugarcane plant is composed of the stalk, leaves, and the top. Average compositions for the sugarcane stalk and cellulosic bagasse are shown in Table 8.9.[31] Comparing Tables 8.8 and 8.9 show that sugarcane contains sugars, mostly sucrose, compared to starch for the corn grain while the cellulosic content of the sugarcane bagasse is similar to the corn stover.

Table 8.9 Average Composition of Sugarcane Stalk and Bagasse[31]

Type	Stalk, %	Bagasse (dry basis)%; water content typically 48%–52%
Water	74.5	
Sugars	14.0	
Sucrose	12.5	
Glucose	0.9	
Fructose	0.6	
Fibers	10.0	
Cellulose	5.5	43.0
Lignin	2.0	23.1
Hemicellulose	2.0	25.4
Gums	0.5	
Ash	0.5	2.9
Extractives		4.8

Because the sugarcane already contains sugars, the process to make ethanol is easier than from corn grain. Basically, the sugarcane is crushed to get out the soluble sugars, which are then fermented into ethanol. The pulpy residue left after juice extraction, bagasse, can be used as biomass feed for a cellulosic ethanol plant. Comparing sugarcane to ethanol from corn grain, the liquefaction and saccharification steps are not needed, since the juice already contains sugars (sucrose, glucose, and fructose). After fermentation, the process is similar to ethanol from corn, such that solids are removed and the liquid part is distilled and dehydrated to make pure ethanol.

One additional difference is that the sugar in sugarcane is mostly sucrose, which is a dimer (two molecules linked together) of glucose and fructose. Without going into the chemistry too much, glucose and fructose have the same molecular formula ($C_6H_{12}O_6$) but glucose has a six member ring and fructose has a five member ring structure. Since sucrose is a dimer of glucose and fructose, it must first be broken into glucose before fermentation. This is shown in this chemical equation:

$$C_{12}H_{22}O_{11} + H_2O + \rightarrow 2\ C_6H_{12}O_6 \qquad (4)$$

After this, fermentation of the glucose occurs as with corn grain, shown earlier in equation (3).

For the conversion of cellulosic biomass, one review article[28] notes six major steps including (1) feedstock preparation, (2) pretreatment to release cellulose from the lignin shield, (3) saccharification (breaking down of the cellulose and hemicellulose by hydrolysis to sugars), (4) fermentation of sugar to ethanol, (5) distillation to separate the ethanol from the dilute aqueous solution and (6) combusting the residues to provide energy for the process.

Feedstock preparation normally involves washing to remove undesirable material and grinding to prepare the feedstock for the pretreatment phase.

For pretreatment, it is useful to first review what cellulose is. Cellulose is an insoluble material that is part of the plant wall structure. Interesting enough, both cellulose and starch are types of carbohydrates and even have the same chemical formula ($(C_6H_{10}O_5)_n$). In fact, both cellulose and starch are polysaccharides, meaning they have repeating structures of glucose. They differ, however, in the nature in which these glucose structures repeat and, in such a way that you cannot eat cellulose. Your body does have enzymes to break down starch into glucose but not cellulose. However, there are enzymes to break down the cellulose into glucose, but pretreatment is needed to break the physical structure of the plant wall and make the cellulose accessible to these enzymes, as well as the hemicellulose and lignin in the plant wall (see Tables 8.8 and 8.9). Depending on the cellulosic biomass, different pretreatments are available, such as the use of steam and/or acids. Also, the choice of pretreatment is important in minimizing the operating costs of the process.

For the third step, the saccharification of the cellulose and hemicellulose into sugars, the basic chemical reaction is described by equation (2), except different enzymes are used to perform the hydrolysis. After this is done, fermentation of sugar to ethanol and distillation are done in much the same manner as with corn grain or sugarcane. The lignin, which provided rigidity to the plant walls, can be burned to generate steam, and consequently electricity, to reduce operating costs of the ethanol process.

Figure 8.1 summarizes and compares the steps in processing these three types of biomass, including food crops containing starch (corn grain), food crops containing readily fermentable sugars (sugarcane), and cellulosic biomass.

To evaluate ethanol yields from different types of biomass, there are two important aspects to consider, namely the crop yield in tonnes per hectare (or tons per acre) and the conversion efficiency of ethanol in liters per dry tonne of biomass (or gallons per ton). The crop yield is affected by location and climate while the conversion efficiency is related to the quality of the biomass to make fermentable sugar as well as the process used to make it.

For the crop yield, Table 8.10 shows data in two year increments from 1995 to 2017 for corn and sugarcane. Data for corn were obtained from the USDA[32] while data for sugarcane were obtained from Brazil's UNICA, or the União da Indústria de Cana de Açúcar (Brazilian Sugarcane Industry Association).[33] As can be seen, sugarcane has a greater yield per land area, currently about five to six times that of corn. However, while the yield for sugarcane has been relatively constant over this 22 year period, the yield for corn has been increasing. Relative to 1995, the corn yield per land area for 2017 is about 56% greater. According to one source,[34] corn production has been steadily increasing since the 1930s because scientists have been breeding

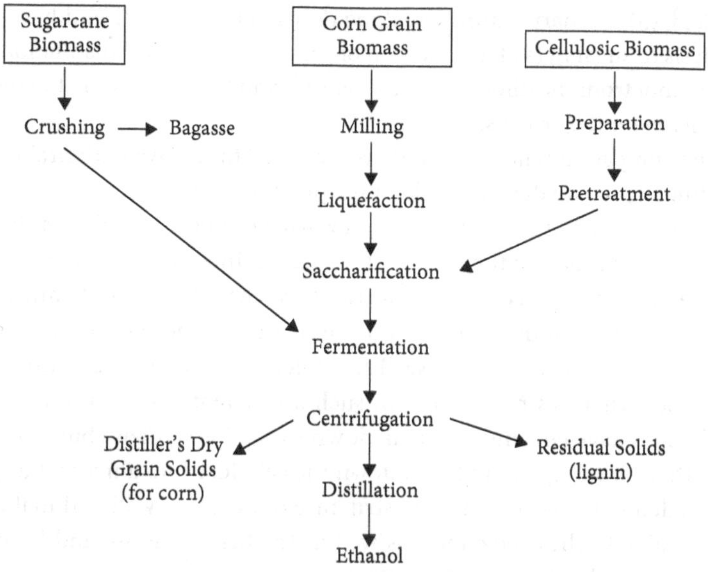

Figure 8.1 Biochemical Conversion Steps for Sugarcane, Corn Grain, and Cellulosic Biomass

Table 8.10 US Corn and Brazil Sugarcane Crop Yields

Year	US, Bushels Corn/ Acre[32]	US, Tonnes Corn/ Hectare[32]	Brazil, Tonnes Sugarcane/Hectare[33]
2017	176.6	11.1	62.7
2015	168.4	10.6	65.5
2013	158.1	9.9	63.7
2011	146.8	9.2	58.2
2009	164.4	10.3	68.1
2007	150.7	9.5	69.9
2005	147.9	9.3	66.2
2003	142.2	8.9	66.7
2001	138.2	8.7	58.3
1999	133.8	8.4	61.7
1997	126.7	8.0	62.0
1995	113.5	7.1	53.7

hybrid types of corn with bigger ears that could bunch together more closely. In addition, there have been improvements in fertilizer, chemical pesticides, and agricultural tools that have helped increase yields, and allowed corn to be grown in parts of the United States that were previously not possible.

It is also worth noting that not all corn or sugarcane is used to produce ethanol. In the United States, the 2018 harvest yielded 14.6 billion bushels, and about 38% (5.6 billion bushels) was used to produce ethanol.[32] In Brazil, the 2018 harvest yielded 620.8 million tonnes,[33] and about 61% was used to produce ethanol.[35] In 2017, 53.6% of Brazil's sugarcane was used to produce ethanol, and the higher percentage for 2018 was attributed to surplus sugar on the world market, making sugar a less attractive market than using it to produce ethanol.

Table 8.11 shows the second aspect, the conversion efficiency for ethanol per mass of biomass. Since the focus for this chapter is on corn and sugarcane to ethanol, 5 years of data are shown for these two types of biomass. Some cellulosic types of biomass are shown for comparison.

Comparing corn to sugarcane, Table 8.11 shows that the conversion efficiency for corn has steadily increased from 2014 to 2018 while that for sugarcane has remained

Table 8.11 Ethanol Yields for Different Biomass Feedstocks

Crop	Country	Crop Yield, tonnes/ha	Conversion Efficiency, liters/tonne	Yield, liters/ha	Yield, gallons/bushel	Year
Corn[9]	US	10.7	412.8	4,431	2.77	2014
Corn[9]	US	10.6	423.2	4,474	2.84	2015
Corn[9]	US	11.0	420.3	4,606	2.82	2016
Corn[9]	US	11.1	420.3	4,658	2.82	2017
Corn[9]	US	11.1	427.7	4,736	2.87	2018
Sugarcane[36]	Brazil	63.7	92.7	5,900		2013
Sugarcane[36]	Brazil	60.6	99.0	6,000		2014
Sugarcane[36]	Brazil	65.5	91.7	6,000		2015
Sugarcane[36]	Brazil	63.6	94.4	6,000		2016
Sugarcane[36]	Brazil	62.7	95.7	6,000		2017
Corn Stover[37]	US		330.1		2.22	2015
Corn Stover[38]	US	6.9	326.8	2,255	2.19	2014
Corn Stover[38]	US	6.58	229.6	1,511	1.54	2010
Sugarcane Bagasse[37]	Brazil		325.5			2015
Switchgrass[38]	US	8.17	240.9	1,968	1.59	2014
Switchgrass[38]	US	7.16	214.0	1,532	1.41	2012

Note: 56 lb/bushel was assumed for corn grain and corn stover, and 55 lb/bushel was assumed for switchgrass.

Table 8.12 Energy Balance for Corn, Corn Stover, and Sugarcane

MJ/l (mega Joules/liter)	Corn[39]	Corn Stover[39]	Sugarcane[40]
Total Energy Used	14.0	9.2	2.9
Farming	2.9	2.9	2.1
Transportation to ethanol plant	0.2	0.2	0.5
Ethanol Plant	10.6	5.9	0.3
Ethanol Distribution	0.3	0.3	
Byproduct Credit	4.6	4.1	3.4
Ethanol energy output	21.3	21.3	21.3
Energy Ratio with byproduct	1.9	2.7	8.5
Energy Balance (energy out – energy in)	11.9	16.2	21.8

relatively constant and is several factors less than corn. The conversion efficiency for sugarcane is lower largely because the stalk contains only around 14% sugars (Table 8.9) while the corn grain contains around 72% starch (Table 8.8). Nevertheless, sugarcane has a greater yield per land area than corn, currently about 27% higher. So, even though corn has a higher conversion efficiency, sugarcane has a greater crop yield for a given land area, resulting in an overall higher yield of ethanol per a unit of land area.

The cellulosic biomass types included in the table have lower yields than either corn or sugarcane.

Ethanol from sugarcane also benefits from a better energy balance than corn ethanol. Table 8.12 shows energy balances for corn, corn stover, and sugarcane. The ratio of energy produced to energy used is more than 4 times greater for sugarcane, 8.5 versus 1.9. Here, the energy ratios include the energy from byproducts. One reason for this higher ratio is that starch must first be converted into sugar before it is fermented into alcohol, which is not the case for sugarcane. This reduces the cost of the ethanol plant, as shown in the table. Another factor reducing the ethanol plant cost is that the sugarcane residues are burned to generate energy to operate the plant and generate electricity that can be sold to utility companies. These factors, coupled with the greater ethanol yield per land area, give sugarcane a better energy balance. Corn stover also has a better energy balance than corn, also primarily due to less energy use at the ethanol plant.

8.4. Capital Cost, Operating Cost, Well-to-Wheels Levelized Cost of Electricity, and Well-to-Wheels Levelized Cost of Fuel

The equations and methodology used to calculate levelized cost of fuel (LCOF), and specific input used for corn grain or corn stover to make ethanol, are detailed in

Appendix A. Calculations were based on an ethanol plant size of 100 million gallons per year. For corn grain, the dry milling process was used since 90% of the grain ethanol produced in the United States today comes from the dry milling process and only 10% from wet milling. The units for LCOF are in $/gallon of gasoline equivalent, or $/gge.

To determine the cost for a gge for ethanol, the cost of producing a gallon of ethanol has to be corrected to the cost of producing the same amount of energy in a gallon of gasoline. As the energy content of ethanol is lower than gasoline, 75,700 BTU versus 124,238 BTU for gasoline, a gge for ethanol is determined by adjusting the LCOF for ethanol using the energy ratio of gasoline to ethanol, or 1.64. In other words, the cost of making one gge of ethanol is 1.64 times that of making one gallon of ethanol.

Transportation fuel in the form of ethanol, derived from either corn grain or corn stover, is created over several steps including (1) planting, growing, and harvesting the corn or corn stover, (2) transporting the corn or stover to the ethanol plant, (3) the ethanol plant, (4) transporting the ethanol product to market, (5) taxes (both federal and state), and (6) marketing cost at the filling station. The cost for these different steps yields the "Well-to-Wheels" LCOF in $/gge. As described in Appendix A, (5) taxes are included in the LCOF equation.

For ethanol produced from corn grain, the cost of steps (1) and (2) are assumed to be the delivered cost of corn, shown in Table 8.13 for the last 10 years.[41] As the table illustrates, the price of corn has had wide variation, from a low of $3.30/bushel in 2017 to a high of $6.95/bushel in 2012. Also shown in the table are the average sales price of ethanol as well as the value of DDGS, a product used to feed farm animals that offsets part of the cost for operating the ethanol plant, step (3). For the base case study, the prices of corn and DDGS are set to the annual average values for 2018.

Table 8.13 Annual Average Prices for Ethanol, DDGS, and Corn[41]

Year Average	Ethanol per Gallon	DDGS per Ton	Corn per Bushel	Corn per Ton
2009	$1.63	$111.76	$3.56	$127.01
2010	$1.77	$116.96	$3.98	$142.15
2011	$2.56	$195.83	$6.61	$236.25
2012	$2.24	$238.20	$6.95	$248.39
2013	$2.34	$231.40	$6.16	$220.07
2014	$2.11	$160.78	$4.06	$144.95
2015	$1.43	$148.31	$3.62	$129.22
2016	$1.43	$120.97	$3.33	$118.93
2017	$1.43	$103.55	$3.30	$117.75
2018	$1.32	$142.30	$3.39	$120.90

The cost of operating the dry mill ethanol plant includes the capital cost of building the plant, the fixed and variable operating costs, the capacity factor, and the value of any coproducts. For the capacity factor, according to the Renewable Fuels Association (RFA)[42] the existing US production capacity for 2018 was 16,501 million gallons per year (MGY) while the operating production was 15,975 MGY. This yields a capacity factor of 96.8%. The capacity factor is defined as the percent of the rated capacity used each year.

For the capital cost to build the dry mill ethanol plant, there are several studies available. A 2002 USDA study showed that, for 21 plants surveyed, the cost for new plants ranged from \$1.05 to \$3.00 per gallon of installed capacity.[43] Another study cited a cost of \$1.30 to \$1.40 per gallon of capacity for 2002, but increasing to \$2.00 to \$2.30 per gallon in 2007.[44] The increasing capital cost was attributed to limited worker availability, as well as rising costs of construction material and production equipment. Similar numbers were reported in 2010 by the US DOE, with the capital cost of a dry mill ethanol plant as \$2.00 to \$2.25 per gallon of capacity.[45] A 2016 report from the University of Illinois gave a capital cost of \$2.11 per gallon of capacity,[46] as did a model from Iowa State University used to track ethanol profitability.[41] Both the University of Illinois and Iowa State studies assume a plant capacity of 100 million gallons of ethanol production per year and a yield of 2.80 gallons of ethanol per bushel of corn processed. Based on these two more recent studies, the capital cost is set to \$2.11 per gallon of installed capacity, so a 100 million gallon plant would have a capital cost of \$211 million. In terms of barrels per day of capacity, \$2.11 per gallon is equivalent to about \$32,000/BPD (barrels per day). This can be compared to the average capital cost for a refinery, shown in Chapter 1, of \$40,000/BPD.

Other details for the LCOF calculation are given in Appendix A. Since 10 years of data were available for the price of corn and the value of DDGS, all 10 years were used to calculate LCOF. Results are shown in Table 8.14. The LCOF for ethanol had a range of \$1.58 to \$2.51/gallon of ethanol produced, or \$2.60 to \$4.12 on a gge basis. For 2018, an LCOF of \$1.58/gallon was determined and, since Table 8.13 shows a sales price of \$1.32/gallon for ethanol, this means the production cost exceeded its sale value. This was also true for 2015 to 2017, although there were nice profits for 2013 and 2014. One explanation put forth for decreasing profits is that the value of ethanol has decreased because of increased production has overwhelmed the domestic and export use of ethanol.[47]

A sensitivity study was made for the dry mill ethanol plant by varying capital cost, the combined fixed and variable operating and maintenance (O&M) costs, and the cost of corn. Details are provided in Appendix A and results are shown in Table 8.15, where the base case was based on data for 2018. The greatest influence on the LCOF for ethanol was the cost of corn. Increasing the cost of corn from the base case value of \$3.39/bushel to the 2012 value of \$6.95/bushel increased the LCOF by about \$2.00/gallon.

We next turn to cellulosic ethanol by examining corn stover as a feedstock. The "Well-to-Wheels" steps are basically identical except that steps (1) and (2) involve the

Table 8.14 LCOF for Dry Mill Ethanol for a 100 Million Gallon per Year Plant and a Capital Cost of $211 Million

Year	Corn Cost, $/bushel	DDGS Value, $/ton	Levelized Capital Cost, $/gallon	LCOF, $/ gallon of ethanol	LCOF, gge
2018	$3.39	$142.30	$0.27	$1.58	$2.60
2017	$3.30	$103.55	$0.27	$1.68	$2.76
2016	$3.33	$120.97	$0.27	$1.63	$2.68
2015	$3.62	$148.31	$0.27	$1.64	$2.69
2014	$4.06	$160.78	$0.27	$1.75	$2.87
2013	$6.16	$231.40	$0.27	$2.23	$3.66
2012	$6.95	$238.20	$0.27	$2.48	$4.07
2011	$6.61	$195.83	$0.27	$2.51	$4.12
2010	$3.98	$116.96	$0.27	$1.87	$3.07
2009	$3.56	$111.76	$0.27	$1.75	$2.86

Table 8.15 Sensitivity Study for a Dry Mill Ethanol Plant

Cases	Capital Cost, MM$	Corn Cost, $/bushel	Combined O&M, $/gallon	LCOF, $/ gallon of ethanol	LCOF, gge
Base Case	$211	$3.39	$0.30	$1.58	$2.60
Capex = $1.05/ gallon capacity	$105	$3.39	$0.30	$1.45	$2.37
Capex = $1.05/ gallon capacity	$300	$3.39	$0.30	$1.69	$2.78
O&M = $0.15/ gallon	$211	$3.39	$0.15	$1.44	$2.36
O&M = $0.45/ gallon	$211	$3.39	$0.45	$1.74	$2.85
Corn Cost = $3.30/ bushel	$211	$3.30	$0.30	$1.55	$2.54
Corn Cost = $6.95/ bushel	$211	$6.95	$0.30	$2.81	$4.62

delivered cost of corn stover instead of corn grain and the ethanol plant for step (3) is different. Otherwise, the cost of transporting the ethanol product to market and the federal and state taxes are the same.

Results for five studies were used to establish the various costs for cellulosic ethanol from corn stover, two which are National Renewable Energy Laboratory (NREL) studies and three which take data from actual commercial plants. Results for the LCOF calculations, and important input data, are shown in Table 8.16.

As the table illustrates, the capital cost for a cellulosic ethanol plant is quite high, with the average more than four times that for corn grain. As a result, the average LCOF is $2.94/gallon, almost twice that of the 2018 value for corn grain of $1.58/gallon.

It is probably no surprise, then, ethanol produced from cellulosic biomass is struggling in the United States. The DuPont plant, located in Nevada, Iowa, was sold to the German company Verbio Vereinigte, which will use the plant to make renewable natural gas.[53] The Abengoa plant in Hugoton, Kansas, was sold in 2016 to Synata Bio Inc. because the parent company went bankrupt, and it is not clear if or when production will resume.[54] Finally, the Poet Project Liberty plant in Emmetsburg, Iowa, has been slow to meet its 20 million gallon per year target for ethanol, reportedly due to difficulty in the chemical pretreatment step used to perform enzymatic saccharification of the corn stover.[55]

A sensitivity study was made for cellulosic ethanol using corn stover as a feed by varying capital cost, O&M expenses, and the price of corn stover. Details are provided in Appendix A and results are shown in Table 8.17, where the base case was based on the data average in Table 8.16.

All three variables had a nearly equivalent effect on the LCOF. An interesting variable is the price for corn stover, the cellulosic biomass feedstock. Although on face value, this would seem to be a feedstock with very little cost since it is not rich in starch or used as food, there are actually costs associated with harvesting, collecting,

Table 8.16 LCOF for a Corn Stover Cellulosic Ethanol Plant

Data Source	Capital Cost	Corn Stover Cost	Liquid Product Yield	Capital Cost	Ethanol Yield	Combined O&M	LCOF, $/ gallon of ethanol	LCOF, gge
	$/ gallon of capacity	$/ton	MM gallons/ year	$/BPD	Gallons/ dry ton	$/gge	$/gallon	$/gge
NREL[48]	$4.87	$65.30	56	$74,596	72.5	$0.97	$2.68	$4.08
NREL[49]	$6.93	$58.50	59	$106,108	75.8	$0.64	$2.50	$3.80
DuPont[50]	$6.67	$60.00	30	$102,200	80.0	$0.91	$2.82	$4.29
Poet[51]	$13.75	$70.00	20	$210,788	71.2	$0.91	$3.56	$5.42
Abengoa[52]	$9.18	$60.00	25	$140,688	68.5	$1.09	$3.15	$4.78
Average	$8.28	$62.76	38	$126,876	73.6	$0.90	$2.94	$4.47

Table 8.17 Sensitivity Study for a Corn Stover Cellulosic Ethanol Plant

Cases	Capital Cost, MM$	Corn Stover Cost, $/ton	Combined O&M, $/gallon	LCOF, $/gallon of ethanol	LCOF, gge
Base Case	$316	$62.76	$0.90	$2.94	$4.47
Capex $5/gallon of capacity	$190	$62.76	$0.90	$2.53	$3.84
Capex $14/gallon of capacity	$532	$62.76	$0.90	$3.64	$5.54
Combined O&M is $0.45/gallon	$316	$62.76	$0.45	$2.52	$3.83
Combined O&M is $1.35/gallon	$316	$62.76	$1.35	$3.42	$5.20
Price of corn stover is $60/ton	$316	$60.00	$0.90	$2.90	$4.41
Price of corn stover is $100/ton	$316	$100.00	$0.90	$3.44	$5.23

and storing the corn stover. One source provides details for these steps. For example, harvesting corn stover must be accompanied by adding fertilizer to replace nutrients removed with the stover.[56] And, harvesting the stover requires some capital intensive operations, such as mowing, raking, and baling which require expensive equipment. Based on their analysis, the breakeven prices for stover range from $45.50 to $74.70 per ton. Another study provided a cost range of $40 to $80 per ton.[57] In addition, there are transportation costs associated with delivery of the corn stover, on the order of $20/ton, giving an overall delivered cost of $60 to $100 per ton. Looking back at Table 8.13, in 2018 corn grain was priced at $3.39/bushel, equivalent to $121/ton. Therefore, a delivered cost of corn stover as high as $100 per ton will cause a significant disadvantage in the viability of cellulosic ethanol over ethanol produced from corn grain.

8.5. Cost of Energy

The cost of energy for corn grain and corn stover is calculated in units of $/lb, $/ton and $/MM BTU. A study from the Penn State College of Agricultural Sciences[58] reports the energy content of dry corn grain as 8,000 to 8,500 BTU per pound. For corn stover, a different study[59] gives a range of 7,180–8,985 BTU per pound of dry matter. A Canadian report[60] gives values for both corn grain and corn stover as 7,313 and 7,614 BTU per pound, respectively. Taken together, there is not a significant difference in the energy content of corn grain and corn stover, so a nominal average of 8,000 BTU/lb is used to represent both.

In the economic analysis for dry mill corn grain, Table 8.14, a value of $3.39/bushel was used for corn grain. At 56 lb/bushel, this is equivalent to $0.06/lb or $121/ton. And, for the corn stover economic analysis shown in Table 8.16, the average cost for corn stover about $63/ton, or $0.03/lb. Using an energy content of 8,000 BTU/lb for both, the cost on an energy content basis is $7.50/MM BTU for corn grain and $3.75/MM BTU for corn stover.

For the nonrenewable fossil fuels discussed in earlier chapters, the cost of crude oil was given as $11.20/MM BTU (Chapter 1), natural gas as $3.17/MM BTU (Chapter 2), coal as $2.06/MM BTU (Chapter 3), and $0.30/MM BTU for uranium (Chapter 4). This means the cost of corn grain is more expensive than natural gas, coal, and uranium while corn stover is more expensive than coal and uranium but about the same as natural gas.

8.6. Capacity Factor

For the economic analysis discussed earlier, the capacity factor used for the dry mill ethanol plant was 96.8%. This value was calculated from US data for 2018 reported by the Renewable Fuels Association (RFA).[42] Existing US production capacity was 16,501 million gallons per year (MGY) while the operating production was 15,975 MGY.

For the cellulosic ethanol analysis, the capacity factor was assumed to be 1.0, or 100%, meaning that the operating production would equal the nameplate capacity. However, this value was assumed as there is not sufficient commercial operating data to calculate one.

Unlike other technologies, such that operating at nameplate capacity is limited by factors such as routine and unscheduled maintenance, variation in energy input (such as wind and solar), and electrical demand, ethanol plants operate near their design. According to the EIA,[61] nameplate capacity is not really a limit for many ethanol plants. And, by using more efficient operating techniques, many plants can exceed their nameplate capacity, if market conditions provide an incentive to do so. Indeed, for 2016 ethanol production was slightly more than 100% of the reported nameplate capacity.

8.7. Efficiency: Fraction of Energy Converted to Work

Efficiency, or thermal efficiency is the percentage of work done by the energy in the fuel sent to the engine. In Chapter 1, the efficiency was discussed for gasoline and diesel, refined from crude oil. Although theoretical efficiencies of 52% are possible, typical values for gasoline and diesel are much less than this. For Chapter 1, 35% and 45% were used for gasoline and diesel, respectively.

The question, then, is whether the addition of ethanol to gasoline will improve, or hurt, the thermal efficiency of the engine. Four papers were used to analyze typical results, especially for 20% ethanol blends (E20). In one study,[62] the addition of 20% ethanol improved thermal efficiency by 0 to 1%, depending on the spark ignition timing. The improvements were attributed to an increase in vapor pressure for the blend. Another study[63] showed little difference in thermal efficiencies for blends with 10, 20, and 30% ethanol; in fact, the 20% blend was as much as 1% lower in efficiency. This study was made on a conventional spark ignition engine. A third study[64] quantified the thermal efficiencies for pure gasoline, 5% ethanol and 10% ethanol, reporting efficiencies of 33.35%, 33.11%, and 32.27% at 2,000 rpm, respectively.

In contrast to these three studies, another study evaluated performance with a traditional spark-ignited engine and one with engine-controlled spark ignition.[65] While the traditional engine showed no significant impact of ethanol content on thermal efficiency, the controlled spark ignition showed as much as a 5% improvement in thermal efficiency.

Therefore, the answer as to the impact of ethanol on thermal efficiency appears to be whether or not the engine has been adapted to optimize the blended properties of the ethanol-gasoline blend. Based on these results, we would expect the thermal efficiency for gasoline-ethanol blends in a typical gasoline engine to be the same, or about 35%, as that for pure gasoline while an engine optimized for the 20% ethanol blend may possibly achieve efficiencies around 40%.

8.8. Energy Balance

Most biomass to ethanol reports talk about an energy ratio, also sometimes called the energy return on investment. It is defined as the ratio of energy in a gallon (or liter) of ethanol to the nonrenewable energy used to produce it. For example, Table 8.12 shown earlier gives an energy ratio of 1.9 for corn grain and 2.7 for corn stover. The same table also reports a net energy gain from the energy balance. This is somewhat deceiving, however, because it gives the impression that producing ethanol is an overall gain in energy. That is because it compares the energy in the ethanol relative to the nonrenewable energy used to produce it, but does not include the solar energy that also went into making the corn grain or stover.

In Chapter 6 for wind energy, the energy balance was the ratio of the electrical energy delivered to the customer relative to the kinetic energy contained in the wind flowing into the wind turbine. Likewise, for Chapter 7 on solar energy, the energy balance was the ratio of the electrical energy delivered to the customer relative to the energy contained in the solar radiation coming to the solar plant. Although the sources of energy for these two energy types are renewable, a poor energy balance still has an impact as it pertains to the capital cost and land area for the plant. In the case of an onshore wind turbine, the energy balance was calculated to be about 24% while that for a solar photovoltaic plant was calculated to be only 15%. For ethanol from corn grain and corn stover, the energy balance will also include the renewable solar energy that went in their production.

To assess the nonrenewable energy that goes into the energy balance, a 2015 USDA report[66] was used. Data taken from their Table 2 are shown in what follows, given in units of BTU per gallon of ethanol produced. The category labeled corn production includes all the nonrenewable energy that go into corn production, such as tilling, seeding, fertilizing, irrigating, and harvesting. To also determine the solar energy contained in the corn, the total starting energy in the corn grain or stover was calculated using the yields shown in the table (gallons/bushel), the assumption of 8,000 BTU/lb for energy content, and a bushel mass of 56 lb. This leads to a starting energy content of 162,319 BTU/gallon for the corn grain and 219,178 BTU/gallon for the corn stover. That means, relative to the energy content for the ethanol, it takes about 2.1 BTUs of energy to make one BTU of ethanol for corn grain and 2.9 BTUs to make one BTU of ethanol for corn stover.

The definition for the Energy Balance is defined as the percent energy derived from the feedstock. Thinking of this in terms of a "Well-to-Wheels" approach, the starting fuel is the biomass and the final use is ethanol blended with gasoline that powers an internal combustion engine. As discussed in the previous section, the thermal efficiency for a 20% ethanol-gasoline blend (E20) is anywhere from 35%–40%, depending on whether or not the engine has been optimized for the ethanol blend. For this energy balance, we assume 35% efficiency.

The energy balance, shown next, then is the ratio of the ethanol energy that goes towards powering the vehicle versus the original energy contained in the biomass (either corn grain or corn stover).

$$\text{Energy Balance} = \frac{[\text{Energy from ethanol that powers vehicle}]}{[\text{Energy contained in the biomass}]}$$

Using the data from Table 8.18, the energy balance was made for both corn grain and corn stover, shown here:

Ethanol from Corn Grain

162,319 (biomass) → 112,147 (energy used) → 95,556 (less byproduct credit) → 76,300 (other energy losses) → 26,705 (ethanol energy that powers vehicle)

Ethanol from Corn Stover

219,178 (biomass) → 186,074 (energy used) → 171,357 (less byproduct credit) → 76,300 (other energy losses) → 26,705 (ethanol energy that powers vehicle)

From this, we see that 16.5% of the original biomass energy from corn grain powers the vehicle versus 12.2% for the corn stover. It is worth noting that byproducts such as DDGS made from the corn grain and electricity made from the corn stover are, like

Table 8.18 Data from USDA Report[66] for Ethanol from Dry Mill Corn Grain and Corn Stover with 50% of Electricity Generated by the Cellulosic Plant

	Dry Mill Corn Grain	Corn Stover with 50% External Power Replacement
	BTU/gallon ethanol	BTU/gallon ethanol
Corn Production	9,007	9,007
Corn transport	701	701
Ethanol Conversion	38,141	21,073
Ethanol Distribution	993	993
Farm Machinery	1,330	1,330
Total energy used	50,172	33,104
Byproduct Credit (BPC)	16,591	14,717
Energy used less BPC	33,581	18,387
Ethanol Energy output	76,300	76,300
Energy Ratio, Ethanol Energy output/Energy used less BPC	2.3	4.1
Yield, gallons/bushel	2.76	2.04
Starting energy in corn grain or stover, BTU/lb	8,000	8,000
Bushel mass, lb/bushel	56	56
Starting energy in corn grain or stover, BTU/bushel	448,000	448,000
Starting energy in corn grain or stover, BTU/gallon	162,319	219,178

ethanol, positive energy credits. In spite of these low energy balances relative to fossil fuel plants, ethanol is still produced from renewable biomass, unlike fossil fuels that, once consumed, cannot be replaced.

8.9. Maturity: Experience

The concept of using ethanol as a fuel, either by itself or in a blend, is not new. Following a review on the history of biofuels,[67] the first spark-ignition piston engine using alcohol was invented by Samuel Morey in 1826, and the Otto-cycle engine invented by Nikolaus Otto in 1860 used ethanol. During the 1900s, Germany created a large-scale biofuels industry when Kaiser Wilhelm encouraged the use of ethanol fuel, made from potatoes. In 1906, 27 million gallons were produced, but the operation was small, spread out over 72,000 distilleries. And, the automotive pioneer Henry Ford stated in 1925 "The fuel of the future … is going to come from fruit like that sumac out by the road, or from apples, weeds, sawdust—almost anything. There is fuel in every bit of vegetable matter that can be fermented."

The first biofuel plant built in Brazil started in 1927, using sugarcane to make ethanol. And, by 1937, ethanol accounted for 7% of Brazil's fuel consumption. In the United States, ethanol plants opened for World War II but, following the war, cheap oil from the Middle East basically destroyed the ethanol market for fuel. During the 1973 oil embargo, when oil prices as much as quadrupled, the United States again looked back at corn-based ethanol. Another incentive for ethanol was environmental concerns with leaded gasoline and the removal of lead created a need for octane increases to gasoline. To promote development, President Jimmy Carter offered tax incentives for ethanol production, and by 1984 there were 163 ethanol plants in the United States producing around 600 million gallons. However, the worldwide drop in oil prices basically stopped the ethanol industry.

Since then, federal and state subsidies have helped keep the ethanol market afloat.[68] In the United States, corn is the primary biomass feedstock for ethanol because it is abundant and easily made into ethanol. A main regulatory driver in the United States is the Renewable Fuel Standard (RFS). The first RFS regulation, also called RFS1, was passed into law in 2005 and required that 4 billion gallons of biofuel be used in 2006, with an increase to 7.5 billion gallons by 2012.[69] This regulation was broadened in 2007 by President George W. Bush, and the so-called RFS2 required the use of 9 billion gallons in 2008 with a schedule to reach 36 billion gallons by 2022. In addition, the amount of ethanol from corn grain (or corn-starch ethanol) is limited to 15 billion gallons, with the remainder being made from cellulosic biomass.

For ethanol produced from cellulosic biomass, the first commercial plant in the world started construction in April 2011 in Crescentino, Italy, and officially opened in October 2013.[70] The plant produces around 20 million gallons of ethanol per year. In the United States, the first commercial cellulosic ethanol plant started operation in September 2014, in Emmetsburg, Iowa.[71] The plant, called Project Liberty, is designed to process 770 tons per day of corn stover and produce around 20 million gallons per year.

Table 8.1, shown again in what follows for just the United States, shows that ethanol production in 2018 was about 16 billion gallons. Unfortunately, according to data from the Environmental Protection Agency (EPA), cellulosic ethanol production in 2017 was 10 million gallons and dropped to 8.2 million gallons in 2018.[72] For 2014, 2015, and 2016 the numbers were 0.7, 2.2, and 3.8 million gallons, respectively. That is far short of the target of 7 billion gallons for 2018 and certainly not on schedule to reach the 21 billion gallons by 2022, with a total of 36 billion when corn grain ethanol is included. And, most of this cellulosic ethanol is coming from facilities that co-process corn grain and corn fiber. The Quad County Corn Processors plant, located

Table 8.1b Annual Ethanol Production Growth over the Last 12 Years (millions of gallons)[6,7]

Region	United States
2007	6,521
2008	9,309
2009	10,938
2010	13,298
2011	13,948
2012	13,300
2013	13,300
2014	14,313
2015	14,807
2016	15,329
2017	15,800
2018	16,061

in Galva, Iowa, uses their "Cellerate" process while Edeniq has six plants using their "Intellulose" process. As these two companies accounted for about 10 million gallons of production in 2017,[55] it is clear almost all cellulosic ethanol is coming from these two companies. That means that there is almost no cellulosic ethanol being produced by plants exclusively dedicated to corn stover, or other cellulosic material, and that the Poet Project Liberty plant in Emmetsburg, Iowa, is nowhere near their target of 20 million gallon per year. So, basically the United States has already met their 15 billion gallon limit for corn grain alcohol but is far short and well behind schedule to achieve the RFS2 target for 2022.

To make matters worse for cellulosic ethanol in the United States, the DuPont plant, located in Nevada, Iowa, was sold to the German company Verbio Vereinigte, which will use the plant to make renewable natural gas and the Abengoa plant in Hugoton, Kansas, was sold in 2016 to Synata Bio Inc. because the parent company went bankrupt. And, outside the United States, the world's first cellulosic ethanol plant in Crescentino, Italy, shut down in November, 2017.[73]

To summarize, ethanol has been used as a fuel for vehicles at least as early as 1826. Certainly, the use of corn grain to make ethanol has been around for centuries, but plants built specifically to produce commercial quantities of ethanol for vehicles have been around since at least 1927, when Brazil built a biofuel plant using sugarcane. In the United States, the first commercial ethanol plants started operation in the early 1940s to support the war effort. For ethanol produced from cellulosic biomass, the first commercial plant in the world started operation in October 2013 in Italy and in

September 2014 for the United States. Thus, the concept of using ethanol as a fuel has been around almost 200 years, plants making commercial quantities of ethanol from starch or sugar plants for about 90 years, and plants making commercial quantities of ethanol from cellulosic biomass for only 6 years.

8.10. Infrastructure

A first impression of ethanol is that it fits marvelously into the same infrastructure that already exists for gasoline and diesel, discussed in Chapter 1 for petroleum crude oil. Citing information in that chapter from the National Association of Convenience Stores (NACS),[74] the United States had more than 150,000 filling stations in 2019 and, according to an Information Handling Services (IHS) report,[75] there were about 278 million light vehicles with an average age of 11.8 years in 2019. Thus, there are filling stations and vehicles to use the produced ethanol, at least for common blends such as E10 and E20.

However, there are some new twists for ethanol, relative to gasoline and diesel, which will require some upgrades to the infrastructure. For example, as discussed earlier, ethanol is naturally attracted to water, and is miscible with water at any concentration. Thus, once a gasoline-ethanol blend is made, there is danger of phase separation, such that ethanol separates from the gasoline into the water phase. Pipelines that transport gasoline and diesel are not perfectly airtight, so moisture that enters the pipeline with the air can cause the ethanol to separate from the gasoline-ethanol blend. Also, since ethanol is a great solvent, when ethanol is transported through existing multiuse pipelines, it can dissolve polymers in the pipelines, thus contaminating the ethanol.[28]

Another issue is having a fleet of vehicles that can use the E85 blend, defined as a flexible fuel containing 51%–83% ethanol. A flexible-fuel vehicle (FFV) is needed to use this blend, such that the FFV has an internal combustion engine capable of running on more than one type of fuel. These FFVs have upgraded fuel systems with large fuel pumps and injectors that enable them to accommodate the greater fuel volume required for the same energy content as gasoline, as well as fuel tanks and lines are composed of ethanol-compatible materials.[28]

Therefore, full utilization of a large ethanol market will require infrastructure upgrades with respect to pipelines and vehicles.

Currently, 68% of US petroleum products are transported via pipeline, with barges (27%), trucks (3%), and rails (2%) making up the rest.[28] For comparison, a 12-inch pipeline can move around 4.2 million gallons of product per day, versus about 1.3 million gallons for the inland barge, 33,000 gallons for one railroad car, and 8,000 gallons for one truck. So pipeline transportation is very efficient. In addition, cost is an important factor. For a transportation distance of 1,000 miles, a pipeline will cost $0.015-0.025/gallon compared to $0.04–0.05/gallon for a barge, $0.075–0.125/gallon for a train, and $0.30–0.40/gallon for a truck. Since large ethanol plants are mostly located in the central US corn-producing states, transportation costs are a large factor to bring gasoline-ethanol blends to the east and west coasts.

A first pipeline dedicated to ethanol transportation began operation in December 2008, when a Kinder Morgan-built pipeline began transporting commercial batches of denatured ethanol in its 105 mile, 16-inch Central Florida Pipeline (CFPL) from Tampa to Orlando.[76] About $10 million was used to modify an existing pipeline including chemical cleaning, replacing parts of the pipeline that were incompatible with ethanol, and expanding storage capacity at their Orlando terminal to receive ethanol shipments. In addition, a proprietary additive was made to prevent ethanol from damaging the pipeline steel. Since then, Kinder Morgan also completed a dedicated ethanol pipeline between Carteret and Linden, New Jersey.[77] This 16-inch ethanol pipeline can handle as much as 36,000 barrels per day (about 1.5 million gallons). And, a $3.5 billion project to build a 1,700 mile dedicated ethanol pipeline was planned by Magellan Midstream Partners and Poet (the largest US ethanol producer) that would bring ethanol from the Midwest to the Northeast.[78] However, the two companies said the DOE loan guarantee program was critical for the project to move forward. In 2012, citing slim prospects for getting the DOE loan as well as losing their co-sponsor Magellan Midstream, Poet decided to stop the project.[79]

As mentioned, to use high levels of ethanol in a gasoline-ethanol blend, such as E85, FFVs are needed. These vehicles have internal combustion engines, but are designed to run on gasoline-ethanol blends up to E85, as well as gasoline. As of 2017, there were more than 21 million FFVs in the United States,[80] which have a fuel system and powertrain that is compatible with ethanol. That is, any part of the engine that will come in contact with ethanol, such as fuel tank, fuel lines, and fuel injectors, has been upgraded to allow for ethanol's corrosive properties. In contrast to the United States, Brazil has been more aggressive in embracing the ethanol economy, and had 20 million FFVs in 2013, about 87% of their market.[81] In addition, these FFVs can use E100, pure ethanol fuel! If you do own an FFV in the United States, as of 2018 there were around 4,500 E85 ethanol stations.[42]

In terms of ethanol plants, according to the Renewable Fuels Association[42] the United States had 210 ethanol plants operating in 2019, spread across 27 states. As the ethanol production was 16,061 million gallons in 2018, the average production for each plant is about 76 million gallons per year (MGY). In terms of production capacity, the top five states were Iowa (44 plants, 4,328 MGY), Nebraska (26 plants, 2,239 MGY), Illinois (14 plants, 1,787 MGY), Minnesota (22 plants, 1,297 MGY), and Indiana (14 plants, 1,198 MGY), accounting for about 68% of the production capacity. Not surprisingly, these are also the top five corn-producing states.

As mentioned in the previous section, there is very little infrastructure for the production of cellulosic ethanol. The 30 MGY DuPont plant in Nevada, IA and the 25 MGY Abengoa plant in Hugoton, Kansas, have been shut down. As this time, there is only one significantly sized plant dedicated exclusively to cellulosic feedstock, the 20 MGY Poet Project Liberty plant in Emmetsburg, Iowa, which processes corn stover. However, they have reportedly had difficulty meeting their production target because of problems in the chemical pretreatment step used to perform enzymatic saccharification of the corn stover. Some plants now produce cellulosic ethanol by co-processing corn grain and corn fiber, notably the Quad County Corn Processors plant, located in Galva, IA, and six Edeniq plants, four in Iowa, one in Nebraska, and one in California.

In summary, people in the United States who drive gasoline-powered automobiles can buy E10 or E20 blends with ethanol and use one of the more than 150,000 filling stations to refuel a fleet of 278 million light vehicles. The likely route for ethanol to reach the filling station is that the ethanol produced at the biorefinery will be shipped by either truck or rail to a terminal for blending and storage, and then be shipped by truck to the filling station. This is because, at this time, there are only two pipelines dedicated to ethanol transportation, with no current plans to add more. Starch-based ethanol production is still the primary method used at the 210 US ethanol plants that make this ethanol.

For those drivers in the United States that want to use E85, there are more than 21 million flexible fuel vehicles as of 2017 that can use this blend. For refueling, in 2018 there were around 4,500 E85 stations, compared to more than 150,000 gasoline filling stations.

8.11. Footprint and Energy Density

The footprint for ethanol, regardless of whether it is made from corn grain or corn stover, includes all the land area needed to (1) plant, grow, and harvest the corn grain or stover, (2) transport this biomass feedstock to the ethanol plant, (3) build and operate the ethanol plant, (4) transport the ethanol product to the fuel terminal for blending and (5) transport the gasoline-ethanol blend to the fueling station.

For Step (1), in 2018 the USDA reports that of the 81,740,000 acres of corn harvested, 38.8% or 31,715,120 acres were used for corn grain to ethanol.[32] This acreage produced 5.595 billion bushels of corn from which 16,061 million gallons of ethanol were produced, equivalent to 2.87 gallons of ethanol per bushel. The yield of ethanol per acre, then, is 506 gallons/(acre-year).

For cellulosic ethanol, an average of five data points for corn stover from a report by Zhang et al.[38] gives a yield of 1,320 liters per hectare, equivalent to 141 gallons/acre. Also, a report from Purdue University[57] estimates a yield of 1.5 dry tons of corn stover per acre. Earlier, a 2016 DOE report[8] reported a yield of 85 gallons of ethanol per dry ton of corn stover. Combining these two values give a yield of 128 gallons/acre. An average of these two cases results in a yield of 135 gallons/(acre-year).

Steps (2), (4), and (5) involve transportation. Transporting the biomass feedstock to the ethanol plant, Step (2), mostly occurs by truck. As mentioned earlier, the likely route for ethanol to reach the filling station is that the ethanol produced at the biorefinery will be shipped by either truck or rail to a terminal for blending and storage (Step (4)), and then be shipped by truck to the filling station (Step (5)). Trains move over tracks and trucks move over roads that already exist, and which can be used for

Table 8.19 Plant Footprints and Capacities for Ten Ethanol Plants

Plant	State	Feed	Capacity, MM Gallons/year	Plant Size, acres	Gallons/ (acre-year
Archer Daniels Midland, Decatur[82]	IL	Corn	365	1,125	324,444
Marquis Energy[83]	WI	Corn	75	240	312,500
Cargill Inc.—Blair[84]	NE	Corn	195	600	325,000
Valero Renewable Fuels — Charles City[85]	IA	Corn	120	225	533,333
Valero Renewable Fuels— Fort Dodge[86]	IA	Corn	120	270	444,444
Center Ethanol Company[87]	IL	Corn	54	57	947,368
Little Sioux Corn Processors LP[88]	IA	Corn	40	80	500,000
Absolute Energy[89]	IA	Corn	100	208	480,769
Average, Corn					483,482
Poet-DSM Advanced Biofuels LLC—Project Liberty[90]	IA	Corn Stover	20	27	740,741
Abengoa[91]	KS	Corn Stover	25	400	62,500
Average, Corn Stover					401,620

other purposes. Therefore, the land area for these steps will not be considered in the footprint.

For Step (3), the footprint of the ethanol plant, a small sample of plants were examined, and these are shown in Table 8.19 for both corn grain and corn stover.

For corn grain ethanol plants, the ethanol production per acre in 1 year varies from about 310,000 to 950,000 gallons, with an average of 483,000 gallons. Although only two data points were found for corn stover cellulosic ethanol plants, these data had an average of 402,000 gallons, although the two footprints were quite different. It is possible the Abengoa plant was built with expansion in mind, or that the 400-acre site provides a buffer zone for other reasons. And, in general, these footprints depend on several factors, such as whether they were built with the intention of expansion and how much onsite storage there is for the corn feed and product. Nevertheless, using a rough average of 500,000 gallons/(acre-year), we see that the average production is about 1,000 times larger than the land area needed to grow corn grain and about 3,000 times larger than the land area needed to harvest corn stover. Therefore, we can conclude that the size of the ethanol plant is insignificant compared to the land area needed to grow the biomass feedstock, and the footprint can be simply calculated from the biomass yield per acre.

The energy density can be derived from the values of 506 gallons/(acre-year) for corn grain and 135 gallons/(acre-year) for corn stover. Following the section on capacity factor, a capacity factor of 96.8% is used for corn grain ethanol and 100% for cellulosic ethanol. Recognizing that the energy content for ethanol is 75,700 BTU and there are 640 acres in a square mile, we can convert these two footprints to units of million BTU/(year-square mile). Doing this gives energy densities of 23.7 and 6.5 billion BTU/(year-square mile) for corn grain and corn stover, respectively. These energy densities can also be written as 23.7×10^9 and 6.5×10^9 BTU/(year-square mile).

Although not directly applicable to ethanol from biomass, these energy densities are converted to units of gigawatt-hours per square mile per year, or GWh/(square mile-year), in order to have a common basis for comparison to energy densities for other technologies. The conversion from BTU to Watt-hour, or Wh, uses this equation:

$$1 \text{ BTU} = 0.293071 \text{ Wh}$$

With this conversion and recognizing that a gigawatt (GW) is equivalent to one billion Watts (or 1×10^9), we can calculate the energy density for ethanol from corn grain as 7.0 GWh/(year-square mile) and the energy density for ethanol from corn stover as 1.9 GWh/(year-square mile).

Comparing these to other technologies, Chapter 5 showed a calculated energy density of 273 GWh/(year-square mile) for a run-of-river hydroelectric plant and 32 GWh/(year-square mile) for a storage dam hydroelectric plant. For an onshore wind farm, the calculated energy density was 51.8 GWh/(year-square mile) while an offshore wind farm was 67.3 GWh/(year-square mile). If the onshore wind farm considers only the directly impacted area, the energy density was 2,466.7 GWh/(square mile-year). For solar energy, Chapter 7 showed that the calculated energy density for a solar PV plant was 172.6 GWh/(year-square mile) and the energy density for a concentrating solar power (CSP) plant was 306.0 GWh/(year-square mile). Therefore, for renewable energy, the land area needed to grow crops to make energy in the form of ethanol gives a much lower return on energy than land area for other energy types.

8.12. Environmental Issues

Ethanol, like gasoline, is a flammable liquid, meaning the flash point is below 100°F (37.8°C). The flash point is the temperature at which the liquid has enough vapor over it to ignite in air and, for pure ethanol, this temperature is 61.9°F (16.6°C). Thus, ethanol is flammable at typical ambient temperatures.

Not surprisingly then, there is danger in transporting ethanol. For example, in September 2015, an ethanol train derailed and caught fire in South Dakota.[92] Of this 98-car train, three derailed cars leaked their ethanol cargo, causing a pasture fire.

However, according to the article, train derailments are not common and more than 99% of ethanol is delivered without any issues. Nevertheless, there is inherent danger in the transportation of a flammable liquid.

Of course, a safer route for ethanol transportation is to use pipelines, but the tendency of ethanol to separate from gasoline when water is present prevents the use of pipelines already used for petroleum products; thus, ethanol is primarily transported by rail, which is more expensive than pipelines. As mentioned in the section on Infrastructure, for a transportation distance of 1,000 miles, a pipeline will cost \$0.015–0.025/gallon while rail transfer will cost \$0.075–0.125/gallon. In addition, research in 2007 showed that pipeline steels are vulnerable to stress corrosion cracking, and that oxygen helps cause this to happen.[93] If this type of corrosion does happen, it can result in pipeline failure. Fortunately, research has shown that oxygen scavengers will mitigate stress corrosion cracking.

Another environmental issue of ethanol is the effect it has on land and waterways, especially when growing corn. Growing corn means the use of nitrogen, phosphorus, and potassium, and these nutrients can run off into the waterways, thus polluting the water.[94] The use of pesticides can also cause pollution issues.

Another factor is the enormous land use devoted to corn growing to support the production of ethanol. As mentioned in the previous section, ethanol from corn grain or stover has a larger footprint than other renewable energy types. And, if the use of land for the production of ethanol increases, this same land is not devoted to growing food crops. Or, alternatively, increasing the use of land dedicated to biomass growth could lead to an increase in deforestation, if more land area is needed to support both food crops and ethanol production.

To put some of this in perspective, according to the USDA,[32] the US corn yield in 2018 was 14.42 billion bushels, 5.59 billion of which were used for ethanol. This supported the production of 16,061 million gallons of ethanol made in 2018. This means that 38.8% of the corn production was used to make ethanol. According to the EIA,[95] the United States consumed 142.86 billion gallons of gasoline in 2018. So, the amount of ethanol represents 11% of the gasoline consumed. The USDA also reports that one third of the US corn crop is used for feeding cattle, hogs, and poultry,[96] so over 70% of the corn production is used to make ethanol and animals feed, thus not used to make food. Therefore, if ethanol demand increases, either through mandated regulations or market forces, even more land will have to be used to meet ethanol demand.

Not surprisingly, ethanol production can also affect food prices. As Table 8.13 shows, the price for a bushel of corn increased dramatically from \$3.56 in 2009 to \$6.61 in 2011; the price remained over \$6 per bushel until 2014, when the price dropped to \$4.06. One source blamed this increase on ethanol production, stating that "increased production of ethanol has a large impact on corn prices, not only because it's a major source of demand, but also because the demand is fixed.[97]" Continuing this argument, they claim that in a free market, if the price of corn goes up, demand will go down. However, having a federal mandate requires "the same amount of ethanol no matter how expensive corn is."

Another aspect about ethanol, which could hurt the environment, is the effect on the vehicle fuel system. Many types of engines are not designed to resist the corrosive effect of ethanol as a fuel.[98] Some of the problems, for example, that can occur from the use of 10% ethanol-gasoline blends (E10) include water in the fuel tank, decreased fuel efficiency due to the lower energy content of ethanol, decreased life of engine parts, damage to metal, rubber, and plastic parts of the fuel system, fuel loss through fuel lines, and clogging of the carburetor and fuel lines. And some of these issues will result in vapor loss into the atmosphere, thereby polluting the air.

The issue of decreased fuel efficiency makes it difficult to understand the benefits of using ethanol instead of gasoline as a fuel. Ethanol is expected to reduce both CO_2 and NO_x emissions; CO_2 is addressed specifically in the next section. This is because ethanol is a less carbon-intensive type of fuel and, also, some CO_2 is captured when growing the crops. However, ethanol has about 35% less energy than gasoline so, for example, an E10 blend would have 3% to 4% less energy than pure gasoline and this lower energy content will hurt fuel economy. A Consumer Reports study for a Chevrolet Tahoe running on E15 and E85 showed the fuel economy drop of 21 to 15 mpg on the highway and 9 to 7 mpg in the city.[94] Having a poorer fuel economy means vehicles will use more gallons of fuel, thus offsetting some of the air quality gains.

Studies on emissions from the use of ethanol-gasoline blends are mixed. One study that examined sixteen different vehicles with blends up to 20% ethanol did not show any statistically significant difference for emissions of volatile organic compounds (VOCs) or nitrogen oxides (NO_x).[99] Another study also showed mixed results for NO_x emissions using spark ignition engines.[100]

It is well known that NO_x is a major component in the formation of ozone. According to the California Air Resources Board (CARB),[94] the use of ethanol was found to increase emissions by as much as 72% in Southern California, 48% in Sacramento, and 55% in San Jose. Even more surprising is what happened in Sao Paulo, Brazil, in 2014, when price increases for ethanol, due to price increases in sugarcane, made people increase their use of ethanol-free gasoline. Even though gasoline created more NO_x, ozone went down due to a quenching effect, according to a Scientific American article.[101]

In conclusion, although ethanol is thought to be an improvement relative to gasoline because it helps to reduce our dependence on foreign oil, reduces the use of fossil fuels, and is renewable, there are significant issues concerning flammability, pipeline corrosion, increase in land usage and the effect on food prices, and mixed results with respect to the effect of ethanol on emissions.

8.13. CO_2 Production and the Cost of Capture and Sequestration

The mass of CO_2 produced during the process of making ethanol from corn grain or corn stover is not a straightforward calculation. Following a paper published by Wang et al.,[102] the "Well-to-Wheels" life cycle analysis for CO_2 from corn biomass to ethanol includes CO_2 made during farming, fermentation, transportation and delivery, and ethanol combustion, as well as credits for DDGS and electricity generation. And, unlike the burning of fossil fuels, ethanol has positive contributions from plant growth.

This has been referred to as the "biogenic carbon cycle," such that plants remove CO_2 from the atmosphere through photosynthesis and emit oxygen in return. Another, albeit more complicated aspect, of the CO_2 analysis is called "land use change," or LUC, such that switching crop growth to corn causes soil organic carbon (SOC) to either decrease or increase, depending on the crop. According to Wang et al.,[102] for land converted from cropland-pasture to corn the SOC will decrease, and CO_2 is released to the atmosphere.

Data taken from this reference, and shown in Table 8.20, were generated with the GREET (Greenhouse gases, Regulated Emissions, and Energy use in Transportation) model. As the table illustrates, CO_2 emissions from automotive combustion for corn grain and stover are quite low, offset by the removal of CO_2 via photosynthesis. In other words, the biogenic carbon cycle offsets greenhouse gas (GHG) emissions from automotive combustion. In contrast, gasoline which is produced from petroleum crude oil releases carbon that has been stored in this fossil fuel for millions of year, but there is no offset from photosynthesis.

As the table illustrates, CO_2 emissions generated from corn grain ethanol are only 66% of those from gasoline while those from corn stover ethanol are only 5%! For

Table 8.20 CO_2 Emissions for Gasoline, Corn Grain Ethanol, and Corn Stover Ethanol (Data from Wang et al.[83])

Activity	Gasoline, lb CO_2/MM BTU	Corn Grain Ethanol lb CO_2/MM BTU	Corn Stover Ethanol lb CO_2/MM BTU
Total	218.6	144.2	11.6
Crude Oil Recovery	14.0		
Refining	25.6		
Transportation & Delivery	7.0	9.3	7.0
Combustion in Vehicle	172.1	2.3	2.3
Fertilizer Production		23.3	14.0
Fertilizer N_2O Production		39.5	
Farming		9.3	7.0
Ethanol Production		72.1	23.3
Land Use Change		20.9	−2.3
Distiller's Dry Grain Solids (Credit)		−32.6	
Electricity (Credit)			−39.5

ethanol from corn grain, farming and ethanol production are the main contributors, with a nice CO_2 offset from the production of DDGS. Likewise, for ethanol from corn stover, farming and ethanol production are the main contributors, offset by production of electricity. The eventual combustion of the ethanol in a vehicle is quite small compared to gasoline combustion, as this is offset by positive contributions from plant growth.

It is worth mentioning that the "biogenic carbon cycle" with ethanol only works if there is good land stewardship, balancing biomass growth to capture carbon dioxide and offset carbon dioxide made from ethanol production and use. If, for example, there is increased corn demand to meet the needs for ethanol, food, and animal feed, and if subsequently there is an increased conversion of forests to land devoted to crop growth, GHG emissions could increase because of a disruption in the biogenic cycle.

Since this chapter focuses on the use of corn grain and stover to make ethanol in the form of transportation fuel, the application of an average heat rate to generate electricity is not appropriate. Therefore, there is no conversion of the CO_2 production values from lb CO_2/MM BTU to units of lb CO_2/kWh and g CO_2/kWh. The generation of electricity from biomass is discussed in the next chapter, Chapter 9. In summary, the total CO_2 generated from corn grain is estimated at 144.2 lb CO_2/MM BTU and that for corn stover is 11.6 lb CO_2/MM BTU.

The CO_2 made at the ethanol plants were 72.1 and 23.3 lb CO_2/MM BTU for corn grain and stover, so if capturing this CO_2 is economical and feasible, it would reduce corn grain CO_2 emissions by about 50% and virtually eliminate any CO_2 emissions for the corn stover. And the CO_2 produced from fermentation of corn is of very high purity, although water should be removed before it is compressed and stored. The Archer Daniels Midland Company actually started carbon capture and sequestration (CCS) at their 350 million gallons per year Decatur, Illinois, corn grain plant in 2017.[103] The plant, which produces one million tonnes of CO_2 per year, stores the CO_2 2 km (~1.25 miles) underground in the Mount Simon Sandstone, a saline aquifer. Touted as the world's first large-scale biofuel CCS project, the total cost to build the plant was $208 million.

One study estimates overall CCS costs of $42 to $60 per metric tonne, which would recover capital and operating costs, plus a reasonable profit.[104]

8.14. Chapter Summary

As Table 8.1 shows, there has been rapid growth in the US ethanol production, increasing from 6.5 billion gallons in 2007 to 16 billion gallons in 2018. This has happened in large part from Renewable Fuel Standards (RFS) passed by the US government. Ethanol from corn grain, which has the renewable identification number (RIN) label of D6, was mandated to achieve a production level of 15 billion gallons by 2015.[105] So, this mandate has been achieved. It is worth noting that the mandate was also accompanied by tax subsidies in the form of a "blender's credit." Referred to as the Volumetric Ethanol Excise Tax Credit (VEETC), it initially gave a tax credit of $0.51 for every gallon of ethanol blended, but was reduced to $0.45 in 2008. Since the federal tax for gasoline is $0.184 per gallon, this was a very nice credit indeed. Although the tax credit expired in 2012,[106] it certainly promoted ethanol as a fuel, and there are now 210 ethanol plants in the United States spread across 27 states. And, as the United States consumed 142.86 billion gallons of gasoline in 2018, the 16 billion gallons of ethanol helped reduce our dependence on oil and oil imports by replacing 11% of the gasoline with ethanol.

Unfortunately, the story is not as successful for ethanol produced from cellulosic biomass, with the RIN label D3. RFS2 originally mandated a production level of 7 billion gallons of cellulosic ethanol in 2018 and a final target of 16 billion gallons in 2022.[105] And, a tax credit up to $1.01 per gallon of cellulosic ethanol was made available to help promote this technology.[107] However, cellulosic ethanol production numbers from 2014 to 2018 were 0.7, 2.2, 3.8, 10.0, and 8.2 million gallons. And, in 2017 and 2018, most of this production came from plants that co-process corn grain and corn fiber, rather than operating exclusively with cellulosic biomass feedstock. At this time, the only major commercial-scale plant exclusively dedicated to cellulosic biomass feedstock is the Poet Project Liberty plant in Emmetsburg, Iowa, but they are currently not close to producing their nameplate capacity of 20 million gallons per year (20 MGY). Two other major US cellulosic plants, the 30 MGY DuPont plant in Nevada, Iowa, and the 25 MGY Abengoa plant in Hugoton, Kansas, have shut down. And, outside the United States, the world's first cellulosic ethanol plant in Crescentino, Italy, shut down in November 2017.

Certainly a problem with cellulosic ethanol is the plant capital cost and the LCOF. The cost of building a plant is $8.3 per gallon of installed capacity versus $2.11 per gallon for a corn grain ethanol plant. And, the LCOF for cellulosic ethanol was calculated to be $2.94 per gallon of ethanol versus $1.58 for corn grain ethanol. When these numbers are converted to a gasoline gallon equivalent (gge), they are $4.47 and $2.60, respectively. So, cellulosic ethanol is greatly disadvantaged relative to the LCOF for corn grain ethanol as well as that for conventional gasoline, which was calculated to be $2.25 per gge (see Chapter 1). In fact, ethanol from corn grain is also at a disadvantage compared to gasoline. Moreover, combining Tables 8.13 and 8.14 show that the sales price of ethanol over the last 10 years has not always been sufficient to recover the cost of producing the ethanol.

A life cycle analysis of CO_2 production gave values of 144.2 lb CO_2/MM BTU for ethanol from corn grain and 11.6 lb CO_2/MM BTU for ethanol from corn stover, substantially smaller than the value of 218.6 lb CO_2/MM BTU for gasoline. Ethanol produced from biomass benefits from the fact that combustion is nearly carbon neutral, as CO_2 generated by combustion in the vehicle is offset by crop growth using CO_2 for photosynthesis, the so-called "biogenic carbon cycle." This advantage only remains valid if good land stewardship is used to balance biomass growth that consumes CO_2 with CO_2 produced from ethanol combustion. If increased corn demand results in the conversion of forests to land for crop growth, GHG emissions could increase because of a disruption in the biogenic cycle.

For corn grain, the CO_2 produced at the ethanol plant represents about 50% of all the CO_2 produced in the life cycle analysis. One company, the Archer Daniels Midland Company, is now performing carbon capture and sequestration (CCS) at their 350 MGY Decatur, Illinois, corn grain plant. The CCS plant started operation in 2017, and captures one million tonnes of CO_2 per year which is stored in a saline aquifer 2 km ((~1.25 miles) underground.

The capacity factor for corn grain ethanol production is 96.8%, based on 2018 operating data. Thus, unlike other technologies such that operating at nameplate capacity is limited by factors such as routine and unscheduled maintenance, variation in energy input, and electrical demand, ethanol plants operate very near to their design capacity.

The footprints and energy densities for the production of ethanol is controlled by the land area needed to grow the plant feedstock, as the land area for the ethanol plant is insignificant by comparison. For ethanol from corn grain, an energy density of 7.0 GWh/(year-square mile) was calculated and for ethanol from corn stover, an energy density of 1.9 GWh/(year-square mile) was calculated. These were found to be much smaller than other technologies, such as hydroelectric, solar energy, wind farms, petroleum crude oil, and natural gas. The reason for this is the large land area needed to grow crops, giving it a much lower return on energy than land area for other energy types.

For ethanol from corn grain, the overall energy balance showed that 16.5% of the original biomass energy from the corn grain was used to power the vehicle; this compares to 12.2% for the corn stover. For this analysis, energy credits for the DDGS from corn grain and electricity generated with corn stover were not included. It is worth noting that many ethanol reports talk about a net energy gain in the energy balance. This is because they consider the energy made in the form of ethanol and byproducts to the nonrenewable energy needed to produce it, but do not include the solar energy that also went in making the corn grain or stover biomass.

In terms of experience and maturity, ethanol has been used as a fuel for vehicles at least since 1826. Plants built specifically for make commercial quantities of ethanol have been around since at least 1927, when Brazil built a biofuel plant using sugarcane. And, in the United States, the first commercial ethanol plants started operation in the early 1940s while cellulosic ethanol plants have been around since only 2013 (Italy) and 2014 (United States). Thus, the concept of using ethanol as a fuel has been around almost 200 years, plants making commercial quantities of ethanol from starch or sugar plants for about 90 years, and plants making commercial quantities of ethanol from cellulosic biomass for only 6 years.

Finally, the main environmental issues for ethanol are its flammability during transportation, pipeline corrosion, and the increase in land usage and possible effect on food prices. And, while ethanol does show a reduction in CO_2 emissions when compared to gasoline, various studies show mixed results when other emissions, such as volatile organic compounds (VOCs) and nitrogen oxides (NO_x) are evaluated.

References

1. "Biofuels: Ethanol and Biodiesel," Energy Information Administration, October 23, 2019, https://www.eia.gov/energyexplained/index.cfm?page=biofuel_home
2. "Energy Efficiency and Renewable Energy," US Department of Energy, 2017, https://www.fueleconomy.gov/feg/ethanol.shtml
3. "Water Tolerance of Petrol-Ethanol Blends, Zlata Mužíková et al., goriva i maziva 47 (1), 34–53, 2008.
4. "Ethanol-Water Phase Separation White Paper," Samir Jain, August 2015, http://national-petroleum.net/Ethanol-Water-Phase-Separation-facts.pdf
5. "Alternative Fuels Data Center," US Department of Energy, Energy Efficiency and Renewable Energy, 2020, https://www.afdc.energy.gov/fuels/ethanol_fuel_basics.html
6. "Renewable Fuels Association Industry Statistics," World Fuel Ethanol Production, 2019, http://www.ethanolrfa.org/resources/industry/statistics/#1454099103927-61e598f7-7643
7. "Alternative Fuels Data Center," US Department of Energy, Energy Efficiency & Renewable Energy, 2019, https://www.afdc.energy.gov/data/categories/fuels-infrastructure
8. "2016 Billion-Ton Report, Advancing Domestic Resources for a Thriving Bioeconomy," Volume 1, US Department of Energy, July 2016, https://energy.gov/eere/bioenergy/downloads/2016-billion-ton-report-advancing-domestic-resources-thriving-bioeconomy
9. "U.S. Corn for Fuel Ethanol, Feed and Other Uses," Alternative Fuels Data Center, US Department of Energy, Energy Efficiency & Renewable Energy, 2019, https://afdc.energy.gov/data/10339
10. "How Much Gasoline Does the United States Consume?," Energy Information Administration, March 3, 2020, https://www.eia.gov/tools/faqs/faq.php?id=23&t=10
11. "The Promise of Biomass: Clean Power and Fuel—If Handled Right," Union of Concerned Scientists, September, 2012, http://www.ucsusa.org/sites/default/files/legacy/assets/documents/clean_vehicles/Biomass-Resource-Assessment.pdf
12. "A Reassessment of Global Bioenergy Potential in 2050," Stephanie Searle and Chris Malins, GCB Bioenergy 7, 328–336, 2015, https://onlinelibrary.wiley.com/doi/full/10.1111/gcbb.12141
13. "Global Bioenergy Potentials through 2050," G. Fischer and L. Schrattenholzer, Biomass and Bioenergy 20, 151–159, 2001, http://pure.iiasa.ac.at/id/eprint/6527/1/RR-01-09.pdf
14. "Evaluation of Bioenergy Potential with a Multi-Regional Global-Land-Use-and-Energy Model," H. Yamamoto et al., Biomass and Bioenergy 21, 185–203, 2001, https://np-net.pbworks.com/f/Yamamoto+et+al+(2001)+Bio-energy+potential+-+global+land+use+-models,+Biomass+and+Bioenergy.pdf
15. "Exploratory Study on the Land Area Required for Global Food Supply and the Potential Global Production of Bioenergy," J. Wolf et al., Agricultural Systems, 76, 841-861, 2003, https://www.researchgate.net/publication/222567956_Exploratory_study_on_the_land_area_required_for_global_food_supply_and_the_potential_global_production_of_bioenergy
16. "Potential of Biomass Energy out to 2100, for Four IPCC SRES Land-Use Scenarios," M. Hoogwijk et al., Biomass and Bioenergy 29, 225–257, 2005, https://www.academia.edu/17352842/Potential_of_biomass_energy_out_to_2100_for_four_IPCC_SRES_land-use_scenarios
17. "Biomass Energy: The Scale of the Potential Resource," C.B. Field et al., Trends in Ecology and Evolution 23, 65–72, 2008, http://www.cas.miamioh.edu/~stevenmh/Field%20et%20al%202008.pdf

18. "A Bottom-Up Assessment and Review of Global Bio-Energy Potentials to 2050," E. Smeets et al., Progress in Energy and Combustion Science 33 (2007) 56–106, August 7, 2006, https://dspace.library.uu.nl/bitstream/handle/1874/21670/NWS-E-2006-202.pdf?sequence=1&isAllowed=y
19. "Future Bioenergy and Sustainable Land Use," H.J. Buchmann et al., WBGU, 1–393, editors R. Schubert et al., pp. 1–393, London: Earthscan, 2009, https://www.cbd.int/doc/biofuel/wbgu-bioenergy-SDM-en-20090603.pdf
20. "Future Bio-Energy Potential under Various Natural Constraints," D.P. Van Vuuren et al., Energy Policy 37, 4220–4230, 2009.
21. "Bioenergy Production Potential of Global Biomass Plantations under Environmental and Agricultural Constraints," T. Beringer et al., Global Change Biology Bioenergy 3, 299–312, 2011, https://onlinelibrary.wiley.com/doi/full/10.1111/j.1757-1707.2010.01088.x
22. "Global Biomass Fuel Resources," Matti Parikka, Biomass and Bioenergy 27, 613–620, 2004, http://citeseerx.ist.psu.edu/viewdoc/download?doi=10.1.1.584.5&rep=rep1&type=pdf
23. "Consumption of Motor Gasoline, World, Annual for 2016," Energy Information Administration, https://www.eia.gov/opendata/qb.php?category=2135028&sdid=INTL.62-2-WORL-TBPD.A
24. "Table 10.3 Fuel Ethanol Overview," Energy Information Administration, March 2020, https://www.eia.gov/totalenergy/data/monthly/pdf/sec10_7.pdf
25. "Brazil: Biofuels Annual 2018," USDA Foreign Agricultural Service, August 10, 2018, https://gain.fas.usda.gov/Recent%20GAIN%20Publications/Biofuels%20Annual_Sao%20Paulo%20ATO_Brazil_8-10-2018.pdf
26. "2016 Bioenergy Industry Status Report," Kristi Moriarty et al., March 2018, National Renewable Energy Laboratory, US Department of Energy, NREL/TP-5400-70397, https://www.nrel.gov/docs/fy18osti/70397.pdf
27. "Ethanol History—From Alcohol to Car Fuel," November 2010, http://www.ethanolhistory.com/
28. "Liquid Transportation Fuels from Coal and Biomass: Technological Status, Costs, and Environmental Impacts," America's Energy Future Panel on Alternative Liquid Transportation Fuels; National Academy of Sciences; National Academy of Engineering; National Research Council, ISBN 978-0-309-13712-6, 2009.
29. "How Ethanol Is Made?," Renewable Fuels Association, 2020, http://www.ethanolrfa.org/how-ethanol-is-made/
30. "Ethanol Production—Dry versus Wet Grind Processing," North Dakota State University, February 2008, https://www.ag.ndsu.edu/energy/biofuels/energy-briefs/ethanol-production-dry-versus-wet-grind-processing
31. "Sugarcane Biomass Composition for the Industrial Simulations in the Virtual Sugarcane Biorefinery (VSB)," T. Junquerira et al., Centro Nacional de Pesquisa em Energia e Materiais, MeT 22/2015, 2015, http://8k5sc3kntvi25pnsk2f69jf1.wpengine.netdna-cdn.com/wp-content/uploads/2016/08/MeT-222015.pdf
32. "QuickStatsforCorn,"UnitedStatesDepartmentofAgriculture,NationalAgriculturalStatistics Service, 2018, https://www.nass.usda.gov/Quick_Stats/Lite/index.php#5A3912FF-C666-340A-A898-784852C1B404
33. "Production Data," UNICA (União da Indústria de Cana de Açúcar or Brazilian Sugarcane Industry Association), 2019, http://www.unicadata.com.br/listagem.php?idMn=63
34. "A Brief History of U.S. Corn, in One Chart," Brad Plumer, Washington Post, August 16, 2012, https://www.washingtonpost.com/news/wonk/wp/2012/08/16/a-brief-history-of-u-s-corn-in-one-chart/
35. "Brazil Biofuels Annual 2018," USDA Foreign Agricultural Service Gain (Global Agricultural Information Network) Report, August 10, 2018, https://gain.fas.usda.gov/

Recent%20GAIN%20Publications/Biofuels%20Annual_Sao%20Paulo%20ATO_Brazil_8-10-2018.pdf
36. "Brazilian Sugarcane Sector: Recent Developments and the Path Ahead," Leticia Phillips, USDA's 94th Annual Agricultural Outlook Forum, February 23, 2018, https://www.usda.gov/oce/forum/2018/speeches/Leticia_Phillips.pdf
37. "BETO 2015 Peer Review," Kristi Moriarty, Biofuels Information Center, National Renewable Energy Laboratory, March 24, 2015, https://www.energy.gov/sites/prod/files/2015/04/f22/sustainability_and_strategic_analysis_moriarty_6301.pdf
38. "Diverse Lignocellulosic Feedstocks Can Achieve High Field-Scale Ethanol Yields While Providing Flexibility for the Biorefinery and Landscape-Level Environmental Benefits," Yaoping Zhang et al., Global Change Biology Bioenergy, June 4, 2018, https://onlinelibrary.wiley.com/doi/epdf/10.1111/gcbb.12533
39. "2015 Energy Balance for the Corn-Ethanol Industry," US Department of Agricultural (USDA), February 2016, https://www.usda.gov/oce/reports/energy/2015EnergyBalanceCornEthanol.pdf
40. Joaquim E. A. Seabra et al., "Greenhouse Gases Emissions Related to Sugarcane Ethanol," In Luis Augusto Barbosa Cortez (ed.), *Sugarcane bioethanol—R&D for Productivity and Sustainability*, São Paulo: Editora Edgard Blücher, 2014, p. 291–300, https://openaccess.blucher.com.br/article-details/19248
41. "Tracking Ethanol Profitability," Iowa State University Ag Decision Maker, D1-10, Ethanol Profitability, Don Hofstrand, Excel Model, 2019, https://www.extension.iastate.edu/agdm/energy/html/d1-10.html
42. "2019 Ethanol Industry Outlook," Renewable Fuels Association, 2019, https://ethanolrfa.org/wp-content/uploads/2019/02/RFA2019Outlook.pdf
43. "USDA's 2002 Ethanol Cost-of-Production Survey," Hosein Shapouri and Paul Gallagher, US Department of Agriculture, July 2005, https://lib.dr.iastate.edu/cgi/viewcontent.cgi?referer=https://www.google.com/&httpsredir=1&article=1021&context=econ_reportspapers
44. "Ethanol Plant Construct. Costs Are on the Rise," Flow Control, August 10, 2007, https://www.flowcontrolnetwork.com/ethanol-plant-construct-costs-are-on-the-rise/
45. "Current State of the U.S. Ethanol Industry," US Department of Energy Office of Biomass Programs, Fulfillment of Subcontract No. 02-5025, November 30, 2010, https://www.energy.gov/sites/prod/files/2014/04/f14/current_state_of_the_us_ethanol_industry.pdf
46. "The Profitability of Ethanol Production in 2016," Scott Irwin, Department of Agricultural and Consumer Economics, University of Illinois, February 1, 2017, https://farmdocdaily.illinois.edu/2017/02/the-profitability-of-ethanol-production-in-2016.html
47. "What Happened to the Profitability of Ethanol Production in 2017?," Scott Irwin, Department of Agricultural and Consumer Economics, University of Illinois, Farmdocdaily, March 14, 2018, https://farmdocdaily.illinois.edu/2018/03/profitability-of-ethanol-production-in-2017.html
48. "Biochemical Production of Ethanol from Corn Stover: 2008 State of Technology Model," D. Humbird and A. Aden, Technical Report NREL/TP-510-46214, August 2009, https://www.nrel.gov/docs/fy09osti/46214.pdf
49. "Process Design and Economics for Biochemical Conversion of Lignocellulosic Biomass to Ethanol," D. Humbird et al., Technical Report NREL/TP-5100-47764, May 2011, https://www.nrel.gov/docs/fy11osti/47764.pdf
50. "DuPont Advances Commercialization of Cellulosic Ethanol with Iowa Biorefinery Groundbreaking," Nevada, IA, November 2012, http://biosciences.dupont.com/media/news-archive/news/2012/dupont-advances-commercialization-of-cellulosic-ethanol-with-iowa-biorefinery-groundbreaking/archive/news/2012/

dupont-advances-commercialization-of-cellulosic-ethanol-with-iowa-biorefinery-groundbreaking/

51. "First Commercial-Scale Cellulosic Ethanol Plant in the U.S. Opens for Business," Emmetsburg, IA, Project Liberty, September 3, 2014, http://poet.com/pr/first-commercial-scale-cellulosic-plant
52. "Abengoa Celebrates Grand Opening of Cellulosic Ethanol Plant," Erin Voegele, Hugoton, KS, October 17, 2014, http://biomassmagazine.com/articles/11068/abengoa-celebrates-grand-opening-of-cellulosic-ethanol-plant
53. "DuPont Sells Iowa Ethanol Plant to German Company; It Will Soon Make Renewable Natural Gas," Donnelle Eller, Des Moines Register, November 9, 2018, https://www.desmoinesregister.com/story/money/agriculture/2018/11/08/dupont-cellulosic-ethanol-plant-nevada-sold-german-company-verbio-north-america-claus-sauter/1938321002/
54. "Is Cellulosic Ethanol Dead? Despite Setbacks, Signs of Progress," November 9, 2017, Environmental and Energy Study Institute (EESI), https://www.eesi.org/articles/view/is-cellulosic-ethanol-dead-despite-setbacks-signs-of-progress
55. "Zero to 10 Million in 5 years," Susanne Retka Schill, June 26, 2018, Ethanol Producer Magazine, http://www.ethanolproducer.com/articles/15344/zero-to-10-million-in-5-years
56. "To Harvest Stover or Not: Is It Worth it?," Madhu Khanna and Nick Paulson, Department of Agricultural and Consumer Economics University of Illinois, Farmdoc Daily, February 10, 2016, https://farmdocdaily.illinois.edu/wp-content/uploads/2016/04/fdd180216.pdf
57. "Corn Stover for Bioenergy Production: Cost Estimates and Farmer Supply Response," Jena Thompson and Wallace E. Tyner, Department of Agricultural Economics Purdue University, September 2011, https://www.extension.purdue.edu/extmedia/EC/RE-3-W.pdf
58. "Burning Shelled Corn—A Renewable Fuel Source," Dennis E. Buffington, Penn State College of Agricultural Sciences, July 2006, http://www.neo.ne.gov/neq_online/july2006/cornbtu.pdf
59. "Variation in Corn Stover Composition and Energy Content with Crop Maturity," L.O. Pordesimo et al., Biomass and Bioenergy 28 (4), 366–374, April 2005, http://www.sciencedirect.com/science/article/pii/S0961953404001795
60. "Ash Content and Calorific Energy of Corn Stover Components in Eastern Canada," Pierre-Luc Lizotte et al., Energies 8, 4827–4838. 2015, https://pdfs.semanticscholar.org/5ec4/254e7092ee2c1189f6c35a147ff461665f80.pdf
61. "U.S. Ethanol Plant Capacity Increases for Third Consecutive Year," Energy Information Administration, August 10, 2016, https://www.eia.gov/todayinenergy/detail.php?id=27452
62. "Combustion Performance of Bio-Ethanol at Various Blend Ratios in a Gasoline Direct Injection Engine," Dale Turner et al., Fuel 90, 1999–2006, 2011, https://www.birmingham.ac.uk/Documents/college-eps/mechanical/research/dmf-engine-performance/combustion-performance-of-bio-ethanol-at-various-blend-ratios-in-a-gasoline-direct-injection-engine-.pdf
63. "Experimental Investigation of Ethanol Blends with Gasoline on SI Engine," Gaurav tiwari, Int. Journal of Engineering Research and Applications 4 (10 Part 5), 108–114, October 2014, http://www.ijera.com/papers/Vol4_issue10/Part%20-%205/P41005108114.pdf
64. "Evaluating Environmental Effects of Bioethanol-Gasoline Blends in a SI Engine," Ahmet Necati Özsezen, Journal of FCE 4, December 2016, http://dergipark.gov.tr/download/article-file/303216
65. "Effect of Ethanol Content on Thermal Efficiency of a Spark-Ignition Light-Duty Engine," Luigi De Simio et al., International Scholarly Research Network, ISRN Renewable Energy 2012, Article ID 219703, https://www.hindawi.com/archive/2012/219703/ref/

66. "2015 Energy Balance for the Corn-Ethanol Industry," USDA study, February 2016, https://www.usda.gov/oce/reports/energy/2015EnergyBalanceCornEthanol.pdf
67. "History of Biofuels," Ron Kolb, Bioenergy Connection, 2014, http://bioenergyconnection.org/article/short-history-biofuels
68. "History of Ethanol Production and Policy," Cole Gustafson, North Dakota State University, 2008, https://www.ag.ndsu.edu/energy/biofuels/energy-briefs/history-of-ethanol-production-and-policy
69. "Renewable Fuel Standard," US Department of Energy, Alternative Fuels Data Center, 2007, https://www.afdc.energy.gov/laws/RFS
70. "World's First Commercial Cellulosic Ethanol Plant," April 30, 2011, Innovative Industry, http://www.innovativeindustry.net/worlds-first-commercial-cellulosic-ethanol-plant
71. "Commercial-Scale Cellulosic Ethanol Plant Opens," September 5, 2014, Energy Information Administration, https://www.eia.gov/todayinenergy/detail.php?id=17851
72. "RINs Generated Transactions, Generation Summary Report," US Environmental Protection Agency, 2018, https://www.epa.gov/fuels-registration-reporting-and-compliance-help/rins-generated-transactions
73. "World's 'First' Commercial Second-Generation Bioethanol Facility 'Shuts Down,'" Biofuels International, November 1, 2017, https://biofuels-news.com/news/worlds-first-commercial-second-generation-bioethanol-facility-shuts-down/
74. "U.S. Convenience Store Count," NACS (National Association of Convenience Stores), December 31, 2019, https://www.convenience.org/Research/FactSheets/ScopeofIndustry/IndustryStoreCount
75. "IHS Markit: Average Age of Cars and Light Trucks in US Rises again in 2019 to 11.8 Years," Green Car Congress, June 28, 2019, https://www.greencarcongress.com/2019/06/20190628-ihsmarkit.html
76. "Kinder Morgan Central Florida Pipeline Ethanol Project," Alternative Fuels Data Center, June 2009, https://www.afdc.energy.gov/pdfs/km_cfpl_ethanol_pipeline_fact_sheet.pdf
77. "Kinder Morgan Completes Dedicated Ethanol Pipeline Between Carteret and Linden, N.J.," April 3, 2012, http://ir.kindermorgan.com/press-release/all/kinder-morgan-completes-dedicated-ethanol-pipeline-between-carteret-and-linden-nj
78. "Magellan, Poet to Study U.S. Ethanol Pipeline," March 17, 2009, Reuters, https://www.reuters.com/article/us-magellan-poet-ethanol/magellan-poet-to-study-u-s-ethanol-pipeline-idUSTRE52G23Y20090317
79. "Poet Shelves $3.5B Ethanol Pipeline Due to Lack of Funds," Kelly Rizzetta, Law360, January 2012, https://www.law360.com/articles/302096/poet-shelves-3-5b-ethanol-pipeline-due-to-lack-of-funds
80. "Flexible Fuel Vehicles," Alternative Fuels Data Center, 2017, https://www.afdc.energy.gov/vehicles/flexible_fuel.html
81. "Brasil chega aos 20 milhões de motores flex, diz Anfavea" [Brazil reaches 20 million flex fuel cars], Fernando Calmon, UOL Carros, June 28, 2013, http://carros.uol.com.br/noticias/redacao/2013/06/28/brasil-chega-aos-20-milhoes-de-motores-flex-diz-anfavea.htm
82. "Archer Daniels Midland Decatur Plant," Illinois, Land Use Data Base, 2019, http://clui.org/ludb/site/archer-daniels-midland-decatur-plant
83. "Marquis Energy-Wisconsin," Biofuels Journal, 1st Quarter 2011, http://www.bluetoad.com/display_article.php?id=664743
84. "Big Cargill Corn Plant Feeds Green Economy," Christine Stebbins, Reuters News, September 29, 2010, https://www.reuters.com/article/us-usa-biorefinery-cargill/big-cargill-corn-plant-feeds-green-economy-idUSTRE68S4Y020100929?pageNumber=1
85. "Valero Charles City Ethanol Plant," 2020, https://www.valero.com/en-us/Pages/CharlesCity.aspx

86. "Valero Fort Dodge Ethanol Plant," 2020, https://www.valero.com/en-us/Pages/FortDodge.aspx
87. "Center Ethanol Company," 2018, http://www.centerethanol.com/
88. "Little Sioux Corn Processors, L.L.C.," US Securities and Exchange Commission, Washington, DC, Form 10-SB, General Form for Registration of Securities of Small Business Issuers, March 25, 2004, https://www.sec.gov/Archives/edgar/data/1229899/000104746904012183/a2131560z10sb12ga.htm
89. "Absolute Energy Breaks Ground on Ethanol Plant," Jan Horgen, Globe Gazette, August 29, 2006, https://globegazette.com/community/mcpress/news/local/absolute-energy-breaks-ground-on-ethanol-plant/article_b2af3442-1f6c-586c-afcc-d3e9440cb022.html
90. "Project Liberty Taking Shape," Larry Kershner, November 12, 2012, Governor's Biofuel Coalition, http://www.governorsbiofuelscoalition.org/project-liberty-taking-shape/
91. "Abengoa's Hugoton Cellulosic Ethanol Project Goes on the Block," Jim Lane, Biofuels Digest, July 18, 2016, http://www.biofuelsdigest.com/bdigest/2016/07/18/abengoas-hugoton-cellulosic-ethanol-project-goes-on-the-block/
92. "Ethanol Train Derails, Catches Fire in South Dakota," Evan Hendershot, Grand Forks Herald, September 19, 2015, http://www.grandforksherald.com/news/3843037-ethanol-train-derails-catches-fire-south-dakota
93. "Ethanol Pipelines," Oil & Gas Journal, February 18, 2008, http://www.ogj.com/articles/print/volume-106/issue-7/regular-features/journally-speaking/ethanol-pipelines.html
94. "The Environmental Costs of Ethanol," Max Gorders and H. Sterling Burnett, National Center for Policy Analysis, Brief Analyses | Energy and Natural Resources, No. 591, August 2, 2007, http://www.ncpathinktank.org/pub/ba591
95. "How Much Gasoline Does the United States Consume?," Energy Information Administration, March 3, 2020, https://www.eia.gov/tools/faqs/faq.php?id=23&t=10
96. "Corn Is America's Largest Crop in 2019," Tom Capehart and Susan Proper, Economic Research Service in Research and Science, July 29, 2019, US Department of Agriculture, https://www.usda.gov/media/blog/2019/07/29/corn-americas-largest-crop-2019
97. "Ethanol Blamed for Record Food Prices," Kevin Bullis, Sustainable Energy, March 23, 2011, https://www.technologyreview.com/s/423385/ethanol-blamed-for-record-food-prices/
98. "Gas-Caused (E10) Engine Damage and Performance Issues," Fuel Testers, 2009, http://www.fuel-testers.com/list_e10_engine_damage.html
99. "Effects of Mid-Level Ethanol Blends on Conventional Vehicle Emissions," Keith Knoll et al., National Renewable Energy Laboratory, Presented at the 2009 SAE Powertrain, Fuels, and Lubricants Meeting, 2–4 November 2009, San Antonio, Texas, https://www.nrel.gov/docs/fy10osti/46570.pdf
100. "Effect of Ethanol-Gasoline Blend on NO_x Emission in SI Engine," B.M. Masum et al., Renewable and Sustainable Energy Reviews 24, 209–222, 2013, http://www.academia.edu/4783015/Effect_of_ethanol_gasoline_blend_on_NOx_emission_in_SI_engine
101. "Ethanol Fuels Ozone Pollution: Shifts in the Use of Gasoline and Ethanol to Fuel Vehicles in Sao Paulo Created a Unique Atmospheric Chemistry Experiment," Mark Peplow, Scientific American, April 30, 2014, https://www.scientificamerican.com/article/ethanol-fuels-ozone-pollution/
102. "Well-to-Wheels Energy Use and Greenhouse Gas Emissions of Ethanol from Corn, Sugarcane and Cellulosic Biomass for US Use," Michael Wang et al., Environmental Research Letters 7, 045905, 2012, http://iopscience.iop.org/article/10.1088/1748-9326/7/4/045905/pdf
103. "ADM Starts Commercial-Scale CCS at Decatur Ethanol Plant," Alan Sherrard, Biofuels & Oils, April 20, 2017, https://bioenergyinternational.com/biofuels-oils/adm-starts-commercial-scale-ccs-decatur-ethanol-plant

104. "Capturing and Utilizing CO2 from Ethanol: Adding Economic Value and Jobs to Rural Economies and Communities While Reducing Emissions," white paper prepared by the State of Kansas CO2-EOR Deployment Work Group, December 2017, http://www.kgs.ku.edu/PRS/ICKan/2018/March/WhitePaper_EthanolCO2Capture_Dec2017_Final2.pdf
105. "An Introduction to the Renewable Fuel Standard & The RIN Credit Program," Ecoengineers, September 21–23, 2015, https://cleancities.energy.gov/files/u/news_events/document/document_url/84/2_-Session_0_-_RIN_101_-_FINAL.pdf
106. "After Three Decades, Tax Credit for Ethanol Expires," Robert Pear, January 1, 2012, New York Times, https://www.nytimes.com/2012/01/02/business/energy-environment/after-three-decades-federal-tax-credit-for-ethanol-expires.html
107. "Tax Policy: Driving Innovation and Making Ethanol Accessible," Renewable Fuels Association, 2020, https://ethanolrfa.org/tax/

9

Biomass

A Renewable Energy Type

9.1. Foreword

This chapter focuses on the use of the energy in biomass to make electricity. Chapter 8 examined the use of biomass energy to make ethanol, or bioethanol. For both of these chapters, biomass is defined as any organic material that can be used as a fuel.

In general, biomass can come from both plants and animals and is therefore a renewable source of energy. In the case of plants, energy is stored from the sun via photosynthesis, in which plants make food from carbon dioxide (CO_2) and water. Biomass can take on many forms, such as wood, agricultural crops, agricultural waste, and solid waste. Although not discussed here, vegetable oils, animal fats, recycled greases, and soybean oil can be used to make biodiesel.

To be sure, there is nothing new about harnessing the energy contained in biomass. As soon as man discovered how to make fire, biomass in the form of wood products was used for heating and cooking. And wood, like coal, has been used for several centuries to power steam engines. An advantage of biomass, relative to fossil fuels, is that it is a renewable energy source. The energy in the biomass comes from the sun and, when biomass is burned to release heat, the CO_2 produced will be consumed when new biomass is grown. This is referred to as the "biogenic carbon cycle," such that plants remove CO_2 in the atmosphere through photosynthesis, thereby recycling the CO_2 made from combustion.

It was in the 1970s that biomass was actively considered as an alternative energy source due, in part, to the increase in fossil fuels prices.[1] In the 1980s, generators specifically made to burn wood were built, thereby allowing the use of wood waste to produce electricity. Wood waste includes sawmill residues, forest thinning and residue, agricultural by-products, yard trimmings, and common household wood waste.[1]

Since 1970, there has been growth in generating electricity from biomass, with wood waste being the primary fuel source. Yearly total data, displayed in 5-year increments up to 2015 and yearly after that, are shown in Table 9.1 for biomass electricity generation. The data in this table were taken from the Energy Information Administration (EIA) monthly energy review for 2019.[2] As is evident, there was no significant generation before 1990 but, since then there has been steady growth to 2015, but little growth since then.

There are different methods to convert biomass into electricity,[3,4] and four are briefly described here. The first one is to burn the biomass to make heat, use this heat

The Changing Energy Mix. Paul F. Meier, Oxford University Press (2020). © Oxford University Press.
DOI: 10.1093/oso/9780190098391.001.0001.

Table 9.1 US Electricity Generation (GWh, gigawatt-hour)[2]

Date	Biomass Wood	Biomass Waste	Total Biomass
1950	390	NA	390
1955	276	NA	276
1960	140	NA	140
1965	269	NA	269
1970	136	220	356
1975	18	174	192
1980	275	158	433
1985	743	640	1,383
1990	32,522	13,260	45,782
1995	36,521	20,405	56,926
2000	37,595	23,131	60,726
2005	38,856	15,420	54,276
2010	37,172	18,917	56,089
2015	41,929	21,703	63,632
2016	40,947	21,813	62,760
2017	41,152	21,609	62,761
2018	41,411	21,354	62,765

Note: NA means not available and one GWh is equivalent to one billion watt-hours.

to convert water into steam, and then send it to a steam turbine to generate electricity. In this respect, the electricity generation is similar to that of a fossil fuel. Another method is to gasify the biomass to make synthesis gas ($CO + H_2$) and there are two basic approaches for this. One way to make the synthesis gas is to heat the biomass with less oxygen than needed for complete combustion. Another way is to use pyrolysis by heating at high temperature in the absence of oxygen. For both cases, once the synthesis gas is produced, it can be compressed and combusted, causing gas expansion that spins the gas turbine blades to generate electricity. After this, the heat from combustion can be used to make steam, which powers the steam turbine to generate electricity. This approach is similar to natural gas combined cycle (NGCC), such that both a gas and steam turbine are used to generate electricity, making more efficient use of the heat generated.

The fourth approach, not yet commercial, is to use fuel cells to convert biomass into electricity by using a catalyst activated by solar or thermal energy.[5] Unfortunately, the fuel cell is susceptible to impurities, so the biomass gas must be clean for this approach to work.

For the "Overview of Technology" and "Levelized Cost of Electricity" (LCOE) sections, two technologies are explored in more detail, including biomass bubbling fluidized bed (BFB) and biomass combined cycle (BCC). BFB is an example of the technology which uses heat to convert water into steam, and then generates electricity with a steam turbine. BCC is an example of the technology which makes synthesis gas, and then generates electricity with both gas and steam turbines.

9.2. Proven Reserves

The potential biomass in the United States and world was discussed in the previous chapter, Chapter 8, for converting biomass to ethanol. Now we consider using these same biomass reserves for making electricity. As discussed in the previous chapter, the amount of proven, or potential, reserves for biomass is difficult to assess.

There are a great number of factors to consider when evaluating the potential for biomass. Examples of these factors include which type of plants and waste to include as biomass, what land is available for energy versus food crops, what are the economic conditions for using the land, how will worldwide food consumption compete with biomass for energy, what new land can be introduced to make biomass, what increase in yields due to plant varieties and agricultural practices can be expected, how will loss of land due to soil degradation and increased land utilization for nonagricultural purposes impact biomass production, how will nature conservation affect land use, what land will be needed for flood protection, and what will be the use of biomass resources as raw materials in competing industries.[6]

To make an estimate of the potential biomass that could be used to generate electricity in the United States, data from a 2016 Department of Energy (DOE) Report[7] were used. This same report was used in Chapter 8 to estimate biomass that could be used to produce ethanol, and the report suggests that one billion or more tons of biomass could be available in the United States by 2030 and as much as one and a half billion by 2040, depending on crop yields, climate change impacts, harvest issues, and systems integration of production, harvest, and conversion. The report makes many projections based on the price of biomass per ton, and based on annual yield improvements for crops. Table 9.2 shows results with biomass priced at \$60/ton. The base case scenario assumes a 1% annual yield improvement in energy crops while the high-yield case assumes 3% growth. Energy crops are defined as plants cultivated solely for the purpose of producing non-food energy.

In this table, "Forestry Resources" are defined as whole-tree biomass and residues; "Agricultural Resources" including corn, hay, soybeans, wheat, cotton, sorghum, barley, oats, and rice; "Agricultural Residues" are defined as corn stover, cereal straws, sorghum stubble rice hulls, wheat dust, and sugarcane trash (bagasse); "Energy Crops" are defined as sorghum, energy cane (perennial tropical grass), eucalyptus, miscanthus (perennial bamboo-like grasses), pine trees, switchgrass (perennial

Table 9.2 Currently Used and Potential Biomass for the United States Available at $60 per Dry Ton (million dry tons) per Year[7]

Feedstock, Million Dry Tons	2017	2022	2030	2040
Currently Used Resources				
Forestry Resources	154	154	154	154
Agricultural Resources	144	144	144	144
Waste Resources	68	68	68	68
Total Currently Used	365	365	365	365
Potential: Base-Case Scenario				
Forestry Resources (all timberland)	103	109	97	97
Forestry Resources (no federal timberland)	84	88	77	80
Agricultural Residues	104	123	149	176
Energy Crops		78	239	411
Waste Resources	137	139	140	142
Total (Base Case Scenario)	343	449	625	826
Total (Currently Used + Potential)	709	814	991	1,192
Potential: High-Yield Scenario				
Forestry Resources (all timberland)	95	99	87	76
Forestry Resources (no federal timberland)	78	81	71	66
Agricultural Residues	105	135	174	200
Energy Crops		110	380	736
Waste Resources	137	139	140	142
Total (High-Yield Scenario)	337	483	782	1,154
Total (Currently Used + Potential)	702	848	1,147	1,520

native grass), and willow; and "Waste Resources" are defined as secondary and tertiary wastes from processing agricultural and forestry products, such agricultural secondary wastes, municipal solid waste (MSW), manure, and animal fats.

Next, an estimate is made for the potential electricity generation from the biomass shown in Table 9.2. To do this, some assumptions must be made with respect to the heat content for different types of biomass as well as the efficiency of converting thermal energy into electricity. This same 2016 DOE report gives an efficiency of 25% for the categories of "Waste Resources" and "Forestry Resources," and this efficiency is also assumed for the categories of "Agricultural Residues" and "Energy Crops."

Efficiency can also be defined in terms of a heat rate, the amount of energy needed to generate one kilowatt-hour (kWh) of electricity. As one kWh of electricity is equivalent to 3,412 BTU, a 25% efficiency means the heat rate is 13,648 BTU/kWh. In other words, it takes 13,648 BTU of energy to generate one kWh of electricity.

For the heat content, Table A-2 of this DOE report assigns a value of 4,900 BTU/lb for "Waste Resources," such as MSW and other waste biomass. "Forestry Resources" were assigned a heat content of 6,500 BTU/lb. The report did not provide a heat content for "Agricultural Resources" and "Energy Crops," so these are obtained from other sources. Assuming dry corn grain as a typical "Energy Crop," a study from the Penn State College of Agricultural Sciences[8] reports the energy content of dry corn grain as 8,000 to 8,500 BTU per pound. Assuming dry corn stover as a typical "Agricultural Resource," another study[9] gives a range of 7,180–8,985 BTU per pound of dry matter. A Canadian report[10] gives values for both corn grain and corn stover as 7,313 and 7,614 BTU per pound, respectively. Taken together, there is not a significant difference in the energy content of corn grain and corn stover, so a nominal average of 8,000 BTU/lb is used to represent both. Therefore, "Agricultural Resources" and "Energy Crops" are assigned a heat content of 8,000 BTU/lb.

Using these heat contents and a heat rate of 13,648 BTU/kWh, the data of Table 9.2 were converted to potential electricity generation, shown in Table 9.3.

Before looking at the potential for 2030 and 2040, we first consider what the DOE report gives as "Currently Used Resources." Table 9.2 breaks this down as 154 million dry tons of "Forestry Resources," 144 million dry tons of "Agricultural Resources," and 68 million dry tons of "Waste Resources." Table 9.3 shows what would happen if all biomass resources were used to generate only electricity, resulting in a value of 364.3 TWh. However, not all current biomass resources go to generating electricity, and the DOE report shows that of the available biomass energy for 2017, in terms of trillion BTUs, 433 went to residential, 156 to commercial, 2,547 to industry, 1,439 to transportation, and 510 to electricity, for a total of 5,085 trillion BTUs. For transportation, the biomass was used to produce ethanol, as discussed in Chapter 8. From Table 8.1 in Chapter 8, 15.8 billion gallons of ethanol were produced in 2017 and, with an energy content of 75,700 BTU/gallon, this amounts to 1,200 trillion BTUs, in reasonable agreement with the 1,439 trillion BTUs cited earlier for transportation. And, some of the waste biomass energy was used to make biodiesel, which is not included in the 15.8 billion gallons of ethanol.

Of the 5,085 trillion BTUs for 2017, about 10% was used to generate electricity. Taking 10% of the "Currently Used" from Table 9.3 gives 36.4 TWh, or 36,400 GWh, which is not in very good agreement with the value of 62,761 GWh shown in Table 9.1 for 2017. However, 11,543 GWh came from landfill gas in 2017, which is not in the form of solid biomass. Adjusting for this results in 51,218 GWh from solid biomass for 2017. Also, most of the 2017 electricity generation for biomass came from woody biomass, so it is possible the heat content of the biomass used to generate electricity was closer to a value of 8,000 BTU/lb. If 8,000 BTU/lb is assumed for the heat content of all biomass used to generate electricity, the calculated electricity generation increases to 42,910 GWh, which is in better agreement. So, the conversion "Currently Used" biomass from mass to electricity results in a value that is lower than that

Table 9.3 Currently Used and Potential Biomass Electricity in Units of Terawatt-Hours for the United States Available at $60 per Dry Ton (million dry tons) per Year

Feedstock, Million Dry Tons	Heat Content, BTU/lb	2030, Electricity, TWh	2040, Electricity, TWh
Currently Used Resources			
Forestry Resources	6,500	146.7	146.7
Agricultural Resources	8,000	168.8	168.8
Waste Resources	4,900	48.8	48.8
Total Currently Used		364.3	364.3
Potential: Base-Case Scenario			
Forestry Resources (all timberland)	6,500	92.4	92.4
Forestry Resources (no federal timberland)	6,500	73.3	76.2
Agricultural Residues	8,000	174.7	206.3
Energy Crops	8,000	280.2	481.8
Waste Resources	4,900	100.5	102.0
Total (Base Case Scenario)		647.8	882.5
Total (Currently Used + Potential)		1,012.1	1,246.8
Potential: High-Yield Scenario			
Forestry Resources (all timberland)	6,500	82.9	72.4
Forestry Resources (no federal timberland)	6,500	67.6	62.9
Agricultural Residues	8,000	204.0	234.5
Energy Crops	8,000	445.5	862.8
Waste Resources	4,900	100.5	102.0
Total (High-Yield Scenario)		832.9	1,271.7
Total (Currently Used + Potential)		1,197.2	1,636.0

Note: A TWh is equivalent to one trillion watt-hours.

reported for 2017, even after adjustments for landfill gas and energy content. Using a conservative approach is probably good, however, because the estimates for potential biomass growth likely are optimistic, especially those for the "High-Yield Scenario."

So, if all the biomass was used to generate electricity, and not used for ethanol or heating, the projected potentials for 2030 and 2040 are 1,197 and 1,636 TWh, respectively. And, if only 10% of the biomass is used to generate electricity, generation would be 120 and 164 TWh for 2030 and 2040, respectively. To put this in perspective, in 2017 the United States generated a total of 4,058 TWh from all energy types.[11] So, while biomass could not replace all electricity use, it could certainly have an important impact, even at the 10% level with other biomass going to transportation fuel and heating.

Turning now to the global potential for biomass, there are quite a few reports available that project the potential biomass energy for the year 2050. The ten different reports examined, which are the same that were examined in Chapter 8, showed a very wide range of potential biomass energy in the year 2050, 26 to 1,471 Exajoule/year (EJ/year). We deal later with the conversion of EJ to potential electricity generation but for now, one Exajoule is one quintillion Joules, or 1×10^{18} Joules. Like a British Thermal Unit (BTU), a Joule is a unit of energy and can be defined in the basic units of kg-m^2/s^2. To convert from BTU to Joules, the following equation can be used.

$$1\ \text{BTU} = 1.0548 \times 10^3\ \text{J}$$

Some of these reports only consider the category of "Energy crops" while others also consider "Surplus forest growth" and "Agricultural residues and waste." Needless to say, there are a great many factors to consider such as land area available for energy versus food, crop yields, population growth, water availability, production costs, and the ability of the government for a country to properly manage forest growth and waste. Fortunately, the data for "Energy crops" were reassessed by Searle and Malins[12] to normalize them using a consistent criteria for energy crop yields, land availability, production costs, and good governance. The normalized results for "Energy crops" and other data for the ten studies are shown in Table 9.4. The original report data for "Energy crops" before normalization are shown in parentheses. Details for each study are not given here, but can be found in the individual reports. As expected, regions with either hot and humid climates, large land area, or both had the greatest potential for biomass energy.

As illustrated by Table 9.4, the normalization procedure narrows the range for "Energy crops" from 26–1,272 EJ/year to 37–111 EJ/year, with an average of about 76 EJ/year. Considering all three categories, the total biomass energy potential projected for 2050 is 232 EJ/year with contributions of 74, 76, and 82 EJ/year from "Surplus forest growth," "Energy crops," and "Agricultural Residues and waste," respectively.

To put the potential biomass reserves in perspective, it is compared with the current energy consumption is in the world. Including all sectors, primarily heating, transportation, and electricity, an EIA report says the world used 575 quadrillion BTUs (575×10^{15} BTUs) in 2015.[23] Converting this to units of EJ so it can be compared to the biomass potential shown in Table 9.4 results in a worldwide consumption of 607 Exajoules for 2015, or 607 EJ/year. Consumption is expected to increase to 699 EJ/year by 2030 and 776 EJ/years by 2040. So, the projected biomass energy of 232 EJ/year shown in Table 9.4 represents 38% of the worldwide energy consumption in 2015, and 33% and 30% of that projected for 2030 and 2040, respectively. In other words, biomass energy could make a great contribution to worldwide energy needs.

Table 9.4 Global Biomass Potential for 2050, EJ/year[12]

EJ/year	Surplus Forest Growth	Energy Crops	Agricultural Residues and Waste	Total
Fischer and Schrattenholzer[13]	100	45–49 (147–207)	143	288–292
Yamamoto et al.[14]		81 (110)	72	153
Wolf et al.[15]		75–111 (162–360)		75–111
Hoogwijk et al.[16]		97–104 (123–496)		97–104
Field et al.[17]		79 (27)		79
Smeets et al.[18]	59–103	65–87 (215–1,272)	76–96	200–286
Buchmann et al.[19]		76 (34–120)		76
D.P. Van Vuuren et al.[20]		89 (150)		89
Beringer et al.[21]		76–78 (26–116)		76–78
M. Parikka[22]	42	37	25	104
Average	74	76	82	232

As was done for the United States, the global biomass potential of 232 EJ/year can also be converted to electricity generation, by converting Joules to BTU and using the same heat rate of 13,648 BTU/kWh. Doing this results in a value of 16,116 TWh/year. In this same EIA report,[23] worldwide electricity generation was given as 23,400 TWh for 2015, projected to increase to 29,400 TWh by 2030 and 34,000 TWh by 2040. So 16,116 TWh/year would be a significant contribution to current, and future, worldwide electricity generation. Realistically, heating and transportation compete with biomass use, and currently electricity generation represents 13% of biomass energy use. Using 13% of the potential biomass for electricity generation gives a generation of about 2,095 TWh/year.

Before leaving the discussion on the worldwide biomass potential, it is also interesting to consider how this potential is distributed around the world. Table 9.5 was made using data reported by the International Institute for Environment and Development.[6] Examining total potential by region, it is seen that North America, Latin America, Asia, and Africa all have similar potentials. That these numbers are greater than those for Europe, the Middle East, and the former Soviet Union countries is due to differences in total land mass and climate. Needless to say, the various factors to consider when evaluating biomass potential vary by country and region. For example, the competition of land use for the production of food crops versus biomass for energy may be more critical in Africa and Asia than North America and Europe. Likewise, the availability of new land to make biomass, maintain land for flood protection, and competition of biomass for making electricity versus heating or transportation fuels likely vary by region.

Table 9.5 Biomass Potential by Region[6]

Biomass Potential	North America	Latin America	Asia	Africa	Europe	Middle East	Former Soviet Union	World
Distribution	19.1%	20.7%	20.8%	20.6%	8.6%	0.7%	9.6%	100.0%

Table 9.6 Biomass Electricity Generation for Top Ten Countries in 2017, GWh[24]

Country	2017	% of Total
World	495,395	100.0%
Top Ten Total	358,232	72.3%
United States	68,857	13.9%
China	59,992	12.1%
Brazil	51,273	10.3%
Germany	50,929	10.3%
United Kingdom	31,870	6.4%
Thailand	25,519	5.2%
Japan	21,367	4.3%
Italy	19,378	3.9%
India	16,972	3.4%
Sweden	12,076	2.4%

In terms of current electricity generation and installed capacity, data were downloaded from the International Renewable Energy Agency (IRENA).[24] Generation data were available for 2017 while installed capacity data were available for both 2017 and 2018. For electricity generation, Table 9.6 shows the top ten countries for generation, along with the world total. The top ten countries control over 72% of the biomass electricity generation, with the United States, China, Brazil, and Germany as leaders. The table shows the US generation for 2017 as 68,857 GWh, which is 10% larger than the value of 62,761 shown earlier in Table 9.1, but it is not known why the data from IRENA are different than those reported by the EIA.

For installed capacity, Table 9.7 shows data for both 2017 and 2018, but the rankings are based on 2018 data. The top ten countries control 70% of the installed capacity with Brazil, China, and the United States as leaders.

To summarize, if all the biomass projected to be available in the United States by 2040 was used to generate electricity, it would amount to 1,636 TWh. However, currently only 10% of the biomass is used to generate electricity, with the other

Table 9.7 Biomass Installed Capacity for Top Ten Countries in 2017 and 2018, MW[24]

Country	2017	2018	% of Total
World	111,665	117,828	100.0%
Top Ten Total	78,104	82,677	70.2%
Brazil	14,559	14,782	12.5%
China	11,234	13,235	11.2%
United States	12,838	12,712	10.8%
India	9,531	10,271	8.7%
Germany	9,003	9,457	8.0%
United Kingdom	5,508	6,418	5.4%
Sweden	4,822	4,822	4.1%
Thailand	3,824	4,095	3.5%
Italy	3,451	3,511	3.0%
Canada	3,335	3,375	2.9%

biomass used for heating and transportation fuels. If this percentage is employed, generation would be 164 TWh. Similarly, the global biomass potential for 2050 would result in a generation of 16,116 TWh if all the biomass was used to generate electricity. However, currently only 13% of biomass is used globally for electricity generation and, if this percentage is employed, generation would be 2,095 TWh. For global electricity generation in 2017, the top ten countries controlled over 72% of the biomass electricity generation, with the United States, China, Brazil, and Germany as leaders. For installed capacity in 2018, the top ten countries controlled 70% of the installed capacity with Brazil, China, and the United States as leaders.

9.3. Overview of Technology

The most common approach to make electricity from biomass is direct combustion of the biomass to make high-pressure steam in a boiler. Next, the steam expands through a steam turbine to drive the electrical generator. In some biomass applications, excess and spent steam can further be used to heat buildings, a process called "combined heat and power" (CHP). Although this technology is not discussed here, using CHP increases the overall energy efficiency to about 75%–80%, compared to a process making only electricity with an efficiency of about 20%–25%.[25] It is also possible to use an approach called co-firing, in which biomass can be added to supplement a facility that uses another fuel, such as coal.

For direct combustion systems, there are two basic approaches. In one approach, the fuel sits on a grate, and air is passed through the grate to combust the fuel and generate char. Since the fuel is stationary, this type of furnace is often called a "fixed-bed" furnace. According to one review, this technology is used for biomass power plants up to 50 MW.[26]

Another approach for direct combustion is to use a fluidized bed, in which the air passes through nozzles at a velocity high enough to overcome gravity of the particles and fluidize the bed. For this approach, there two basic methods, namely a BFB and a circulating fluidized bed (CFB). For the BFB, fuel is added to the bottom of the bed, normally formed with sand and fluidized with air. For the BFB, the air velocity is just enough to overcome gravity and fluidize the particles, a velocity known as the minimum fluidization velocity. For this method, the typical combustion temperature is between 800°C and 950°C (~1,470–1,740°F) and the fluidization velocity around 1 m/s (meter/second).

Another method, used for larger applications, is the CFB. For this, fluidization velocities are much higher than BFB, as much as 4.5–6.7 m/s (15–22 ft/s).[26] These velocities are so high that some particles could be entrained and leave the combustion furnace; however, a device called a cyclone is used to capture these particles and return them to the furnace. Like BFB, sand is usually used to make the fluidized bed.

Other than direct combustion, gasification can also be used to make electricity. For this approach, the biomass is first made into synthesis gas ($CO + H_2$) and then this gas is used to drive the gas turbine to generate electricity. Next, the synthesis gas is combusted to make steam and generate more electricity in a steam turbine. This is essentially the same as the combined cycle approach discussed in Chapter 2 for natural gas, and the use of two turbines, gas and steam turbines, improves the efficiency (i.e., what percentage of the energy is used to make electricity). According to a US DOE review,[27] making electricity from direct combustion yields efficiencies of only 20%–25% while gasification can achieve efficiencies of 30%–40%. While gasification, or combined cycle technology, is proven in many natural gas plants, the efficiency and reliability of biomass to-gasification still needs to be established.[26]

It is worth noting here that with the gasification approach, once the synthesis gas is made, it is also possible to make liquid fuels via the Fischer-Tropsch synthesis. This is the subject of Chapter 12.

The thermal efficiencies for direct combustion of biomass to electricity, only 20%–25%, are low compared to other technologies that are used to generate electricity. In the Introduction, the thermal efficiency was defined as in Equation (1).

$$\eta = \frac{W}{Q_H} = 1 - \frac{T_C}{T_H} \tag{1}$$

Here:

W = net work taken from steam (for example, to make electricity)

Q_H = heat added to the system

T_C = temperature of the cold liquid (water)
T_H = temperature of the hot gas (steam)

So for direct combustion of biomass, only 20%–25% of the energy contained in the biomass ends up making electricity. However, for the BFB typical combustion temperatures are between 800°C and 950°C. Using 850°C for the hot temperature and assuming the cold temperature is room temperature (about 25°C), the thermal efficiency would be a very acceptable 74%. Thus, for biomass combustion to electricity, Equation (1) only tells part of the story.

An NREL (National Renewable Energy Laboratory) report[28] describes several factors not considered in the simplistic Equation (1). One factor is the loss of heat with the flue gas, such that the combustion gases (i.e., flue gases) leave the furnace at much higher temperatures than the temperature at which it was fed. A second important factor is the moisture in the fuel, which absorbs heat both through heating and through vaporization. Also while Equation (1) assumes 100% of the heat in the biomass fuel is available to make electricity, in reality some biomass does not undergo complete combustion and thus does not achieve the full release of its heat content. Finally, since the furnace has a temperature much higher than its surroundings, furnaces also lose heat.

A European Union report[29] summarizes data for 28 plants in eleven countries that use wood chips as the biomass fuel. The average data, organized by plant size, show that for plants sized from 5–15 MW, the average efficiency for making electricity was 24%. When the plant size ranged from 15–25 MW, the efficiency increased to 29% and, for plants larger than 25 MW, the average efficiency was 32%. So thermal efficiency improved as the power plant size increased. To some extent, since biomass is renewable, thermal efficiency is not as critical as for nonrenewable fossil fuels. However, a low thermal efficiency affects the plant size and amount of feedstock needed to achieve a certain electricity generation target, thus affecting capital and operating cost.

9.4. Capital Cost, Operating Cost, Well-to-Wheels Levelized Cost of Electricity, and Well-to-Wheels Levelized Cost of Fuel

The equations and methodology used to calculate LCOE, and specific input used for a biomass BFB and BCC are detailed in Appendix A. The units for LCOE are in \$/megawatt-hour, or \$/MWh.

Input for the BFB calculations came from 2019 EIA data,[30] 2017 data from Lazard,[31] and NREL 2017 data.[32] To show the change in capital and operating costs over the previous 3–4 years, the LCOE was also calculated using EIA data from 2015[33] and 2014 data from Lazard.[34] This had further importance for the comparison to the LCOE calculation for BCC, as the most recent data for this technology was from 2013.

This technology, also known as BIGCC (biomass integrated combined cycle plants), is mostly in the research and demonstration plant phase, with little large-scale commercial success.

For a biomass power plant, the "Well-to-Wheels" costs include (1) the cost of harvesting (and perhaps growing), preparing, and transporting the biomass to the power plant, (2) the cost of constructing the biomass plant, (3) power plant generating costs, (4) the cost of connecting the power plant to the electrical grid, (5) transmission and distribution costs, and (6) taxes (both federal and state).

Table 9.8 shows the key input used to make the LCOE calculations. The EIA study assumed a plant size of 50 MW for the BFB technology and 20 MW for the BCC technology, relatively small compared to coal-fired, natural gas, and nuclear power plants.

The most notable difference between the three references is the capacity factor, which is high for the EIA and Lazard studies but low for the NREL study. This is likely due to the EIA and Lazard studies assuming a performance similar to a coal-fired power plant, but the NREL study calculated a capacity factor from 2014 US operating data. For that year, 13.4 GW of capacity generated 64.3 TWh of electricity, giving a capacity factor of 55%. Of the 13.4 GW of biomass capacity, 61.9% used wood and wood-derived fuels, 15.7% used landfill gas, 16.4% used MSW, and 6.0% used other waste biomass.

Since wood and wood-derived fuels had the highest percent use, that is the basis for the fuel used for the LCOE calculations. The 2013 EIA report[33] used a wood fuel cost of $2.40/MM BTU. And, while the report does not specify the wood type or the disposition of the wood (chips, pellets, briquettes, etc.), it does give the water and heat

Table 9.8 Input Data for the LCOE Calculations

	Capacity Factor, %	Heat Rate, BTU/kWh	Capital Cost, $/kW	Fixed O&M, $/kW-yr	Variable O&M, $/MWh
Biomass Bubbling Fluidized Bed (BFB)					
EIA 2019 (50 MW plant)[30]	83	13,500	3,900	114.39	5.70
Lazard 2017[31]	80–85	14,500	1,700–4,000	50.00	10.00
NREL, 2017[32]	55	13,500	3,679	107.22	5.34
EIA 2015 (50 MW plant)[33]	83	13,500	4,114	105.63	5.26
Lazard 2014[34]	85–90	14,500	3,000–4,000	95.00	15.00
Biomass Combined Cycle (BCC)					
EIA[33]	0.83	12,350	8,180	356.07	17.49

content as 17.27% and 6,853 BTU/lb, respectively. On this basis, the dry heat content is 8,284 BTU/lb and, at $2.40/MM BTU, this translates into a cost of $40/dry ton of biomass.

This value seems too low, as a value of $7.50/MM BTU at $121/dry ton was used in Chapter 8 for corn grain and a value of $3.75/MM BTU at $63/dry ton was used for corn stover biomass. Indeed, just the transportation cost for corn stover is about $20/ton.[35] And a California report which evaluated the delivery of wood biomass to the power plant found that the cost was dependent on distance, with $10 per dry ton for a distance of 20 miles, $20 for a distance of 41 miles, and $30 for a distance of 61 miles.[36] Cost depended on whether the road used was a US highway, paved double lane, gravel single lane, or dirt single lane, but worked out to about $0.05/mile.

Also, the cost of biomass can vary a lot, as some power plants use feedstock blending to control the overall average feedstock price. For example, a model feed blend used in a DOE study[37] contained 45% pine wood ($99.49/dry ton), 32% logging residues ($67.51/dry ton), 3% switchgrass ($66.68/dry ton), and 20% waste ($58.12/dry ton), giving an average delivered cost of $80/dry ton. So, for the LCOE calculations, a delivered cost of $80/dry ton was used. Calculating a blended cost for the 2014 biomass feedstock data given earlier results in an average cost close to $80/dry ton and, moreover, this value is in between the cost of corn gran and corn stover. Later, the sensitivity study examines the impact of feedstock cost on the LCOE.

The costs of transmitting and distributing electricity were 0.9 cents/kWh (or $9/MWh) and 2.6 cents/kWh (or $26/MWh), taken from the EIA.[38] Here, transmission is defined as the high-voltage movement of electricity from the power station while distribution is the cost of electrical lines that take power from a substation to the customer. The cost of transmission and distribution is essential to get a true "Well-to-Wheels" analysis of the total cost.

Table 9.9 shows the LCOE for the cost of generating electricity and the additional cost of transmission and distribution. The average for the more recent data did not include the NREL study because of the low capacity factor. Although that capacity factor is based on actual operating data, it made for a better direct comparison with the BCC calculation to exclude it. Also, the current low capacity factor for commercial biomass plants may have more to do with current economics and demand for electricity, rather than down time due to normal maintenance issues.

This table illustrates several things. First, there has not been a great change in capital cost for the BFB technology, as average data from 2017–2019 are only about $500/kW less than those data from 2014–2015. Also, a comparison of the average BFB results, which have a high capacity factor of 83%, to the NREL study, which had a capacity factor of 55% based on annual operating data for US biomass power plants, shows the large impact of plant utilization on the LCOE. Lowering the capacity factor from 83% to 55% increased the LCOE by about $32/MWh. Finally, the BCC technology has a capital cost more than double that of the BFB technology, causing it to have a LCOE about $86/MWh higher, or 8.6¢/kWh.

Table 9.9 Economic Analysis for Biomass BFB and BCC

Technology	Capacity Factor	Capital Cost	Levelized Capital Cost; with MACRS Depreciation	Fixed O&M	Variable O&M (including fuel)	Transmission Investment	Total System Levelized Cost (LCOE) to Generate	LCOE Including Transmission and Distribution
		$/kW	$/MWh	$/MWh	$/MWh	$/MWh	$/MWh	$/MWh
BFB—EIA 2019[30]	0.83	$3,900	$37.3	$15.7	$73.2	$1.2	$135.5	$170.5
BFB—Lazard 2017[31]	0.85	$1,700	$19.3	$6.7	$82.5	$1.2	$109.7	$144.7
BFB—Lazard 2017[31]	0.80	$4,000	$48.3	$7.1	$82.5	$1.2	$139.2	$174.2
Average	0.83	$3,200	$35.0	$9.8	$79.4	$1.2	$128.1	$163.1
BFB—NREL 2017[32]	0.55	$3,679	$64.6	$22.3	$72.8	$1.2	$160.9	$195.9
BFB—EIA 2015[33]	0.83	$4,114	$47.9	$14.5	$72.8	$1.2	$136.4	$171.4
BFB—Lazard 2014[34]	0.90	$3,000	$32.2	$12.0	$87.5	$1.2	$133.0	$168.0
BFB—Lazard 2014[34]	0.85	$4,000	$45.5	$12.8	$87.5	$1.2	$146.9	$181.9
Average	0.86	$3,705	$41.9	$13.1	$82.6	$1.2	$138.8	$173.8
BCC—EIA 2015[33]	0.83	$8,180	$95.2	$49.0	$79.2	$1.2	$224.7	$259.7

A sensitivity study was made for both biomass technologies by varying the capital cost, the feedstock cost, and the capacity factor. The base case for the BFB technology used data from the 2019 EIA case while the base case for the BCC technology used the 2015 EIA case.

For the BFB technology, the capital cost examined a range of $500 to $8,000/kW while the BCC technology examined a capital cost range of $4,000 to $12,000/kW. For the capacity factor, both technologies examined a range of 55% to 90%. For the cost biomass delivered to the power plant, both technologies examined a range of $60 to $100/dry ton. Details on how these ranges were chosen are given in Appendix A. The results of the sensitivity study are shown in Table 9.10.

For the BFB, the range chosen for capital cost, which was based on a global data from a 2017 IRENA report,[39] had the greatest effect on the LCOE, although capacity

Table 9.10 Sensitivity Study for Biomass Bubbling Fluidized Bed (BFB) and Biomass Combined Cycle (BCC)

	Capacity Factor	Capital Cost	Fixed and Variable O&M (including fuel cost)	Total System Levelized Cost (LCOE) to Generate	Total LCOE Cost to Generate, Transmit, and Distribute
Units		$/kW	$/MWh	$/MWh	$/MWh
Base Case: BFB—EIA 2019	0.83	$3,900	$88.9	$135.5	$170.5
Capex = $500/kW	0.83	$500	$88.9	$96.0	$131.0
Capex = $8,000/kW	0.83	$8,000	$88.9	$183.3	$218.3
Capacity Factor = 0.55	0.55	$3,900	$96.9	$166.7	$201.7
Capacity Factor = 0.90	0.90	$3,900	$87.7	$130.8	$165.8
Feed Cost = $60/ton	0.83	$3,900	$72.0	$118.7	$153.7
Feed Cost = $100/ton	0.83	$3,900	$105.8	$152.4	$187.4
Base Case: BCC—EIA 2015	0.83	$8,180	$128.2	$224.7	$259.7
Capex = $4,000/kW	0.83	$4,000	$128.2	$176.0	$211.0
Capex = $12,000/kW	0.83	$12,000	$128.2	$269.1	$304.1
Capacity Factor = 0.55	0.55	$8,180	$153.1	$298.1	$333.1
Capacity Factor = 0.90	0.90	$8,180	$124.4	$213.4	$248.4
Feed Cost = $60/ton	0.83	$8,180	$112.8	$209.2	$244.2
Feed Cost = $100/ton	0.83	$8,180	$146.7	$240.1	$275.1

factor and biomass feed cost also had strong impacts on LCOE. Nevertheless, even with the broad ranges for these three variables, the largest calculated LCOE was still less than the base case for the BCC technology. Capital costs for this technology are too high and, because of the combined cycle system with two turbines and generators, it is difficult to significantly reduce the capital cost difference with the BFB technology.

9.5. Cost of Energy

The cost of energy for biomass to electricity depends on what type of biomass is used, the heat content of the biomass, and the delivery cost. In Section 2, which discussed proven reserves, five different biomass categories were identified including "Forestry Resources" (whole-tree biomass and residues), "Agricultural Resources" (corn, hay, soybeans, wheat, cotton, sorghum, barley, oats, and rice), "Agricultural Residues" (corn stover, cereal straws, sorghum stubble rice hulls, wheat dust, and sugarcane bagasse), "Energy Crops" (sorghum, energy cane, eucalyptus, miscanthus, pine trees, switchgrass, and willow), and "Waste Resources" (secondary and tertiary wastes from processing agricultural and forestry products, such agricultural secondary wastes, MSW, manure, and animal fats). Table 9.11 shows the heat content range and cost per dry ton for a sampling of biomass types. The heat content ranges came from a Congressional Research Service report,[40] and the cost per dry ton were taken from a variety of sources. From the cost and average heat content, the cost of energy is calculated in units of $/MM BTU ($ per million BTUs).

As the table illustrates, there is a wide range for this sampling of biomass, from $1.36 to $7.56/MM BTU. According to Table 9.1, about two-thirds of electricity from biomass comes from wood and wood waste. So the category "Forestry Resources," with a range of about $4 to $6/MM BTU, represents the typical cost for a biomass electric plant.

And, it may surprise some to see "Cattle Manure" in the table as a "Waste Resource." While this chapter has focused on solid biomass to electricity, it is also possible to make electricity from methane, as described in Chapter 2 for natural gas. In the case of manure, a process called anaerobic digestion, a natural process using microorganisms, breaks down the manure into various gases, referred to as biogas. Biogas is primarily composed of methane and CO_2, along with small amounts of water vapor, nitrogen, oxygen, hydrogen, and hydrogen sulfide. After the methane is isolated, it can be used as a fuel to generate electricity. And, as the table shows, the cost for the manure is quite low. And, interesting enough, the LCOE can be low as well. One study[48] reports an LCOE of $66/MWh, similar to a California study[49] giving a range of $67 to $77/MWh. A 2017 report from the United Kingdom.[50] reports an even lower range for LCOE, $23 to $53/MWh. However, a 2014 global status report from the Renewable Energy Policy Network for the 21st Century (REN21)[51] reports

Table 9.11 Cost of Energy for Various Biomass Types

Biomass Category / Biomass Type	Heat Content Range, BTU/lb (dry)	Heat Content Average, BTU/lb (dry)	$/ton (dry)	$/MM BTU
Forestry Resources				
Willow	7,983–8,497	8,240	$86 (added $20 for delivery)[41]	$5.22
Poplar	8,183–8,491	8,337	$82.43[42]	$4.94
Pine	8,000–9,120	8,560	$99.49[37]	$5.81
Pine Chips	8,000–9,120	8,560	$99.49[37]	$5.81
Mill Residue	7,000–10,000	8,500	$67.51[37]	$3.97
Agricultural Resources				
Corn Grain	8,000	8,000	$121[43]	$7.56
Agricultural Residues				
Corn Stover	7,587–7,967	7,777	$63[44]	$4.05
Wheat Straw	6,964–8,148	7,556	$66.66[45]	$4.41
Sugarcane Bagasse	7,450—8,349	7,900	$93.76[46]	$5.93
Energy Crops				
Miscanthus	7,781–8,417	8,099		
Switchgrass	7,754–8,233	7,994	$66.68[37]	$4.17
Sorghum	7,476–8,184	7,830		
Sugarcane	7,450–8,349	7,900	$28[47]	$1.77
Waste Resources				
Cattle Manure	8,500	8,500	$23.1 (added $20 for delivery)[48]	$1.36
Municipal Solid Waste	5,100	5,100	$58.12[37]	$5.70

the very wide range of $40 to $190/MWh. So, excepting the upper range from REN21 report, electricity from cow manure via biogas appears to have an LCOE much lower than the BFB and BCC technologies.

Comparing "Forestry Resources" to other energy sources, the range of $4 to $6/MM BTU is more expensive than natural gas at $3.17/MM BTU (Chapter 2) and coal at $2.06/MM BTU (Chapter 3), but less expensive than crude oil at $11.20/MM BTU (Chapter 1).

9.6. Capacity Factor

According to the EIA,[52] for the years 2013 to 2018, the capacity factors for landfill gas and MSW have ranged from 68.0% to 73.3% while those for wood and wood waste biomass have ranged from 55.3% to 58.9%. Since the United States has few, if any, BCC commercial plants in operation, these capacity factors are mostly based on BFB technology or anaerobic digestion of biogas to electricity.

On the other hand, a global renewables energy report for 2016[53] shows the capacity factor for the United States as an average of 93% with a range of 89 to 96%. In the same report, capacity factors for China ranged from 21% to 95% with an average of 62%, and the lowest weighted average was in Central America and Caribbean at 32%. As this report shows the correct value for the gigawatts (GW) of installed capacity for biomass power plants in the United States, the discrepancy must come from reconciling electricity generation.

A global renewables energy report for 2014[51] gave a capacity factor range of 50% to 90% for solid biomass to electricity, 40% to 80% for electricity from gasification, and 50% to 90% for biogas to electricity via anaerobic digestion. As the BCC (i.e., gasification) technology is mostly in the research and demonstration plant phase, with little commercial success, it is possible the 40% to 80% capacity factors from this report are based on coal plants using the same technological approach.

So, while economic studies typically assume high values for biomass power plant capacity factors, actual commercial operating data indicate much lower values. Indeed, the calculations in this chapter for the LCOE used capacity factors from 80% to 90%. Normally, for coal-fired and natural gas plants, a capacity factor is less than 100% because of routine or unexpected maintenance, as the supply of fuel is normally not a factor. Alternatively, capacity factors can be much lower than 100% if the electricity is not needed, the price of electricity is too low to make a profit, or there are problems getting the feedstock to the power plant. For example, combustion turbines using natural gas as a fuel have capacity factors from 10% to 30%, because they are used primarily to meet peak electrical loads during the summer months. And, solar and wind power plants have low capacity factors because the solar and wind energy input is not constant, and therefore the plant cannot achieve its rated capacity.

While the literature did not provide specific information, biomass power plants capacity factors probably suffer from all three factors, namely low demand, unfavorable economics, or issues getting the biomass feedstock to the plant.

9.7. Efficiency: Fraction of Energy Converted to Work

The thermal efficiency (the fraction of energy converted to electricity) can be calculated using the BTU content of one kilowatt-hour (kWh) of electricity, or 3,412 BTU and dividing this by the heat rate. The heat rate is the amount of energy used, in this case biomass, to make one kWh of electricity. This simple equation is shown here:

$$\begin{aligned}\text{Thermal Efficiency} &= \frac{[\text{Final energy in the form of electricity}]}{[\text{Fuel energy input}]}\\ &= \frac{[3{,}412\ \text{BTU/kWh}]}{[\text{Heat Rate, BTU/kWh}]}\end{aligned} \tag{2}$$

Table 9.8 showed the different heat rates used as input for the LCOE calculations. For the biomass BFB technology, the heat rates taken from the EIA[30] and the NREL[32] studies were the same, 13,500 BTU/kWh while that for the study by Lazard[31] was 14,500 BTU/kWh. Therefore, the thermal efficiency calculated for the EIA and NREL studies was 25% while that for the Lazard study was 24%. For the BCC technology, with a heat rate of 12,350 BTU/kWh, the thermal efficiency is a little better, 28%.

As mentioned earlier in this chapter, a European Union report[29] used data from 28 plants in eleven countries, all using wood chips as the fuel, to show that the thermal efficiency improved with increasing plant size. For plants sized from 5–15 MW, the average efficiency for making electricity was 24%, increasing to 29% for plants sized from 15–25 MW, and further increasing to 32% for plants larger than 25 MW.

Still, these thermal efficiencies are low compared to coal. For example, coal plants using subcritical technology have efficiencies around 34%, improving to about 39% for supercritical, and further improving to 43% for ultrasupercritical. One issue with biomass is moisture content, as this reduces the thermal efficiency. If the biomass contains water, some energy is needed to evaporate this water, thereby reducing the biomass energy available to produce electricity. The EIA study[33] cites a moisture content of 17.27% for the wood biomass used as fuel, so evaporation of this water reduces the overall thermal efficiency.

Also, biomass typically contains less energy on a mass basis than coal, ranging from 8,000 to 8,500 BTU/lb for wood and wood waste versus 12,712 BTU/lb for bituminous coal, so biomass also has less heat to create steam.

It is also surprising that the BCC technology had a thermal efficiency only 3% greater than the BFB technology. In the case of natural gas, a combined cycle plant has an efficiency around 45% versus 36% for a combustion turbine plant. Since a combined cycle plant uses heat from the gas turbine to produce steam and power the steam turbine, a thermal efficiency improvement of more than 3% would be expected.

Other studies do show this improvement. For example, a US DOE review[27] states that making electricity with a steam turbine, like the BFB technology, yields thermal efficiencies of only 20%–25% while the BCC technology can achieve efficiencies of 30%–40%. Also, a report from the Institute for Sustainable Economic Development[54] cites an efficiency of 29% for steam turbine systems versus 42% for the BCC technology. Adding carbon capture and sequestration (CCS) lowered the thermal efficiency for the steam turbine from 29% to 20% while adding CCS to the

BCC technology lowered the thermal efficiency from 42% to 30%.[54] These reductions happen because some of the biomass thermal energy is now being used to capture and sequester CO_2.

In summary, biomass power plants using only a steam turbine, like the BFB technology, have thermal efficiencies ranging from 24% to 32%, depending on the plant size. The thermal efficiency also depends on the biomass moisture content, which decrease with increasing moisture. Biomass power plants using combined cycle technology, like the BCC technology, have thermal efficiencies ranging from 30% to 40%. The addition of CCS reduces these thermal efficiencies by about 10%.

9.8. Energy Balance

For biomass, the energy balance is defined by this equation:

$$\text{Energy Balance} = \frac{[\text{Electrical energy from biomass delivered to the customer}]}{[\text{Energy contained in the biomass}]} \qquad (3)$$

For the conversion of biomass energy to electricity, there are several steps in the "Well-to-Wheels" analysis including the energy used to (1) harvest, perhaps grow, and prepare the feedstock, (2) transport the biomass to the electric plant, (3) construct the power plant, (4) generate the electricity, and then (5) distribute the electricity through the grid to your home.

To illustrate what is involved in the production and preparation of the feedstock, one 2013 paper,[55] which studied the use of poplar trees for the biomass fuel, gave the following steps: production of herbicides, use of tractors, soil cultivation, harvesting, turning the wood into chips, storage before transport, and stump removal.

An NREL report[56] performed a life cycle assessment for a BCC system. In the report, of the starting 43 units of energy, one unit of fossil fuel energy was consumed to produce 42 units of biomass energy sent to the plant. Of this, 0.76 units were for feedstock production and preparation, 0.16 units was to transport the biomass to the plant, and 0.08 units were for power plant construction. From these 42 units of biomass energy delivered to the power plant, 15.6 units were produced in the form of electricity. From these data, we can see that prior to the arrival of the biomass at the power plant, one unit out of 43 was consumed, or 2.3%. The thermal efficiency can be calculated by the electricity produced (15.6 units) and the energy in the biomass sent to the plant (42 units), or 37.1%. Another NREL report,[57] which considers both direct combustion using a BFB and BCC, gives energy losses for feedstock production and preparation, biomass transport to the plant, and power plant construction as 2.8% for BFB and 5.6% for BCC.

A study from the University of Wisconsin,[58] which examined the life cycle of wood pellet production from small diameter pine trees, showed an energy loss of 4.7%

before the wood pellets arrive at the plant. Finally, a report that studied BCC technology using wood chips produced from poplar trees as fuel performed an energy balance over a 16-year period.[55] In this study, the energy consumed in what they define as "cradle-to-plant" included land preparation, wood cutting, harvesting, chipping wood, stump removal, and regional transport. For the regional transport, a distance of 50 km (~31 miles) was assumed. The paper, which gives the results in units of GJ/ha (Giga Joules/hectare), shows a total of 1469.1 GJ/ha in the wood, 51.3 GJ/ha consumed in getting the wood to the power plant, and 546.5 GJ/ha of electric energy produced. Thus, of the total energy, 3.5% was consumed to get the wood to the plant and the thermal efficiency was calculated to be 37.2%.

Summarizing these several studies, the energy used to (1) harvest, grow, and prepare the feedstock (2) transport the biomass to the electric plant, and (3) construct the power plant was 2.8% for BFB technology and 2.3%, 5.6%, and 3.5% for BCC technology. The energy loss of 4.7% for the production of wood pellets from pine trees can be applied to either technology. Using an average of the three studies for BCC gives a value of 3.8%, so steps (1), (2), and (3) are assigned values of 2.8% for BFB and 3.8% for BCC.

In order to (4) make the electricity, thermal efficiencies were assumed using an average of the data presented in the previous section. For BFB, the thermal efficiency is set to 28% and for BCC the thermal efficiency is set to 35%.

It then remains to determine the energy needed to (5) distribute the electricity through the grid to your home. Transmission and distribution losses are affected by the voltage of the power line and the distance the electricity is transmitted. Historical data from the state of California shows annual average transmission losses around 5 to 7% over the time period of 2002 to 2008.[59] Similarly, EIA data for July 2016[60] show that of the electricity generated, 7% is lost in transmission and distribution. Based on these studies, a value of 7% is used for the amount of electricity lost in transmission and distribution after leaving the biomass power plant.

Putting these results together for the BFB technology, 2.8% of the biomass energy is consumed (1) producing and preparing the biomass, (2) transporting the biomass to the electric plant, and (3) constructing the power plant. For BCC, this value is 3.8%. To make the electricity, step (4), the thermal efficiencies are 28% for BFB and 35% for BCC. And, to (5) distribute the electricity through the grid to your home, 7% of the energy is consumed. Using 100 units of energy as a reference, we can arrive at these equations:

Biomass Direct Combustion Bubbling Fluidized Bed (BFB)

100 (forest) → 97.2 (preparation, transporting, and constructing power plant) → 27.2 (generate electricity) → 25.3 (transmission and distribution)

Biomass Gasification Combined Cycle (BCC)

100 (forest) → 96.2 (preparation, transporting, and constructing power plant) → 33.7 (generate electricity) → 31.3 (transmission and distribution)

9.9. Maturity: Experience

There is certainly nothing new about harnessing the energy contained in biomass. As soon as man discovered how to make fire, biomass wood fuel was used for heating and cooking. But when was biomass first used to generate electricity?

Before biomass plants were built exclusively to produce electricity, paper and wood mills generated electricity onsite to provide some, or all, of their electricity needs. According to biomass power plant data from the EIA,[61] four mills in the United States were doing this before 1940. These included the Port Townsend Paper Company in Washington state (3 MW), the WestRock Corporation in Panama City, Florida (4 MW), the Scotia Humboldt Redwood Company in California (7.5 MW), and the Rayonier Advanced Materials plant in Fernandina, Florida (7.5 MW). All of these plants use CHP, where heat from wood waste is used to generate steam which can be used to operate the mill and generate electricity.

The first fluidized bed boiler used for commercial power production was built on French Island in La Crosse, Wisconsin, starting operation in 1940.[62] However, the French Island plant was originally built as a coal-fired plant, converted to burn oil in 1972, and then converted to biomass wood waste in the early 1980s. Of the plant capacity of 140 MW, 18 MW are currently powered by biomass. Likewise, the Wilmarth 25 MW power plant in Blue Earth, Minnesota, started operation in 1948 using coal, but converted to wood biomass in 1987.[63]

Also, part of the problem establishing when the first biomass power plant was built, either in the United States or the world, is that coal-fired plants can be converted to biomass. Therefore, some of the first biomass power plants may not have been "greenfield," or "grass-roots" (i.e., new construction), plants, but coal plants converted to biomass. As one author explains,[64] there are basically two types of coal-fired power plants, namely fluidized bed and pulverized coal. A fluidized bed coal-fired plant can be converted to use wood chips by adding a wood handling and processing system. However, wood chips do not work with pulverized coal plants, as very fine particles are needed. But wood pellets can be milled to the same particle size specification as coal, after the addition of a storage and handling system.

Another way to evaluate the maturity and growth of biomass for generating electricity is to look at generation as a function of time. Data from Table 9.1 are shown graphically in Figure 9.1. It is pretty clear from this graphic that in the United States, significant electricity generation from biomass did not begin until after 1985.

The biomass power plant industry, as well as other renewable energy types, were helped by the Public Utilities Regulatory Policies Act (PURPA) of 1978.[65] In 1973, there was an oil crisis created when OPEC (Organization of Arab Petroleum Exporting Countries) imposed an oil embargo. As a response, the US government passed PURPA, meant to promote domestic and renewable energy. Biomass was found to be attractive since it is considered to be CO_2 neutral.

Although the data do not go as far back in time as Figure 9.1, Figure 9.2 shows global electricity generation from biomass since 1980. These data do not show the

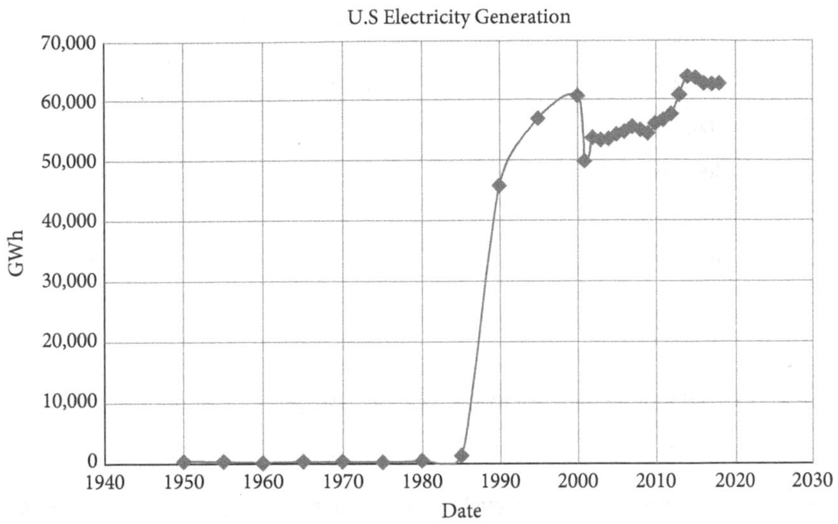

Figure 9.1 US Biomass Electricity Generation (GWh)[2]

large jump in 1985 seen in the United States, but rather steady growth over the last four decades. The rate of growth has increased since 2009.

Comparing US and global electricity generation from biomass, the United States generated 1,383 GWh in 1985, growing to 62,762 GWh in 2017 while global generation was 31,070 GWh in 1985, growing to 495,395 GWh in 2017. Therefore, US generation has increased by a factor of 45 and global generation by a factor of 16. Both represent impressive growth.

Of the two technologies examined in this chapter, one is direct combustion using a BFB. In this reactor, air passes through nozzles at a velocity high enough to overcome gravity of the particles and fluidize the bed. According to a review paper by Koornneef et al.,[67] this technology has been around since 1981, originally developed for coal combustion.

The other technology examined in this chapter is BCC. As a technology field by itself, gasification has been around a long time. Using a review published by NREL,[68] we find that the first patent obtained that deals with gasification was given to Robert Gardner in 1788. And, biomass gasification of wood was first conceived in 1798 by Philippe Lebon. Commercial use of gasification began in Germany prior to World War II, with the development of processes to make synthesis gas, or syngas (carbon monoxide and hydrogen), from coal. A process to convert coal to liquid fuels, primarily diesel, was invented by Hans Fischer and Franz Tropsch in 1925. This is the topic of Chapter 12. Following World War II, the South African Coal, Oil, and Gas Corporation Limited (SASOL) constructed a gasification plant in 1950 to make diesel, gasoline, liquefied petroleum gas, and chemicals from coal. And, in 1984, the first integrated gasification combined cycle (IGCC) demonstration plant was built near Barstow, California to produce electricity from coal gasification.

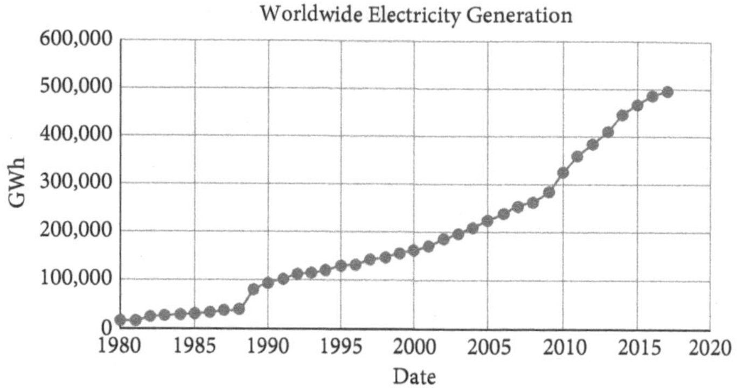

Figure 9.2 Global Biomass Electricity Generation (GWh)[24,65]

The first gasification combined cycle plant using biomass was the Värnamo IGCC Demonstration Plant in Sweden. This 6 MW electric plant, which also produced 9 MW of heat, finished construction in 1993 and operated from 1993–1999[69] using wood chips as the fuel. After 3,600 hours of operation, the demonstration program finished in 2000, and the plant was shut down. Another BCC demonstration plant was the 8 MW ARBRE (arable biomass renewable energy) plant in Selby, United Kingdom.[70] Construction began in 1998, but the plant went bankrupt in 2002; also, there were technical problems with the gasification technology.[71] Currently, there do not appear to be any commercial scale BCC plants.

In May 2019, the Drax power plant in North Yorkshire, United Kingdom, became the first plant in the world using biomass to do CCS.[72] The 3.96 GW plant has six units each generating 660 MW, and four operate exclusively with biomass while the other two use coal. The CCS pilot project is designed to capture one tonne of CO_2 per day using an organic solvent to separate the CO_2 from the flue gas.

The electric grid, which includes both transmission and distribution, is also an important component for any electric power plant. Transmission and distribution includes a step-up transformer, transmission lines, and a substation step-down transformer. The first electric power distribution system was built in 1882 in New York City, overseen by none other than Thomas Edison.[73] Between this time and the 1930's, the electrical grid system grew rapidly, roughly doubling every 6 years. Today, what we take for granted, involves more than 180,000 miles of high-voltage transmission lines connecting about 7,000 power plants.[73]

To summarize, biomass power plants started appearing in the early 1980s. Some of these first plants were converted from coal while others were new plants using the BFB technology. The BCC technology appeared with demonstration plants (less than 10 MW) in the 1990s, but the technology has not led to commercial-scale power plants. And, 2019 saw the first application of CCS technology to a biomass power plant. The electrical grid system was built in 1882 followed by rapid development through 1930s. Thus there are almost 40 years of experience with biomass power plants, and about 20 years of experience with biomass gasification BCC power plants, although this experience is only with demonstration plants. The application of CCS to a biomass plant has less than 1 year of experience. Finally, there are more than 135 years of experience with electric grids.

9.10. Infrastructure

Table 9.6, shown earlier, gives the 2017 global electricity generation from biomass as 495,395 GWh.[24] Of this, the United States generated 68,857 GWh, 13.9% of the global generation. And, according to Table 9.7, the global biomass plant installed capacity for 2017 was 111,665 MW, so the average capacity factor for 2017 was 51%. This capacity increased to 117,828 MW in 2018, but generation data were not available to calculate a capacity factor.

Table 9.7 also shows the US biomass plant installed capacity as 12,838 MW in 2017 and 12,712 MW in 2018, based on data from the IRENA.[24] Based on the IRENA reported generation, the average US capacity factor for 2017 was 61%. An Electric Power Annual 2017 report, published by the EIA,[11] presents different values for both capacity and generation. This report gives an installed capacity of 15,928 MW and generation of 62,761 GWh for 2017, resulting in a capacity factor of 45%. The total installed capacity in the United States for all power plants was 1,203,092 MW, so biomass made up 1.3% of the capacity while total generation was 4,058,258 GWh, with biomass making up 1.5% of the generation. As a percent of renewables, including hydroelectric, the biomass installed capacity was 6.3% and generation was 8.9%. It is not known why the IRENA and EIA numbers differ. One possible reason is that some coal-fired plants blend in some percentage of biomass as a co-feed, and this can cause confusion in counting both capacity and generation.

The EIA also provides a complete list of biomass power plants operating in the United States for 2015.[61] For this year, there were 758 biomass power plants including 354 that use landfill gas, 71 that use MSW, 105 that use "other waste biomass," and 228 that use wood and wood waste. These 758 plants are a sizable increase from 2003, when there were only 485 power plants.[74] Using a US installed capacity of 15,928 MW based on the EIA data in the previous paragraph, these 758 plants have an average capacity of about 21 MW, much smaller than typical natural gas, coal, and nuclear power plants.

Almost all US states have biomass power plants, 48 in all plus the District of Columbia, and the top six states are California with 114, Pennsylvania with 38, Florida with 37, New York with 37, and Michigan with 36. For installed capacity, the top five states in 2017 were Florida with 1.54 GW, California with 1.42 GW, Virginia with 1.02 GW, Georgia with 0.96 GW, and Maine with 0.77 GW.[75]

A Bioenergy International report[76] states that worldwide in 2018, 3,800 biomass power plants had an installed capacity of 60 GW, an average of about 16 MW per unit and similar to that for the United States. Since the global capacity in 2018 was 117,828

MW, or 117.8 GW, these 3,800 units are only about half of the total. Based on a proportional estimate, there are more than 7,400 biomass power plants in the world. This same report indicates that 300 new biomass power plants started operation in 2018. As the global capacity increased from 111,665 MW in 2017 to 117,828 MW in 2018, a difference of 6,163 MW, these 300 new plants had an average capacity of 20.5 MW, not much greater than the units already in operation.

Given the small size of most biomass power plants, it seems unlikely that new transmission lines will be built to accommodate them, so electricity from these plants will likely be transmitted around the United States into the more than 180,000 miles of high-voltage transmission lines already in our grid.[73]

9.11. Footprint and Energy Density

Footprint is a commonly used term to describe the land area needed to deliver energy for a certain energy type For biomass to electricity, the footprint includes the area to (1) grow and harvest the biomass, (2) transport the biomass to the power plant, (3) make electricity at the power plant, and (4) transmit and distribute the electricity via the grid to commercial customers.

For the first step, the land area needed to grow and harvest the biomass, the footprint can be difficult to assess. If the biomass fuel is, for example, MSW, a euphemism for garbage, it can be argued that no land area is needed since the garbage already exists and does not need to be "grown." In fact, burning it saves land area as it would otherwise be sent to a landfill. The same could be said for wood waste from a lumber factory, as well as corn stover and sugarcane bagasse, the waste from harvesting corn and sugarcane. On the other hand, if trees are grown specifically for the purpose of providing fuel to the biomass power plant, then it is entirely fair to include this land area. In this case we expect the land area needed to grow the biomass feedstock to dominate the land area needed for the power plant, as was the case in Chapter 8 for the land area needed to grow corn grain compared to land area for the ethanol plant. Because the biomass could come from either waste or wood grown specifically for the biomass plant, the footprint is determined for the land area needed to grow and harvest the biomass and the land area needed for the power plant.

One research paper[77] reports a yield for wood biomass from poplar and willow as 6.7 tons/[acres-year]. An MIT review[78] is less optimistic, citing a Swedish wood crop yield of 1.9 tons/[acres-year] but growing to 4.5 tons/[acres-year] in the future. A European Union report[79] reports a yield of 2.0 tons/[acres-year] for forest timber and residues, considered to be the maximum amount that could be harvested without degrading continued productivity. A report to examine the feasibility of a 1,000 MW biomass-fired power plant in the Netherlands[80] estimates wood pellet yields of 0.8 to 1.1 tons/[acres-year]. A Colgate University study[81] reports yields for willow stems at 3.7 to 5.1 tons/[acres-year]. Finally, a UK study, using a short rotation coppice for willow trees harvested on a 2- to 5-year cycle, reported a yield of 9.3 tons/

[acres-year].[82] Coppicing is a method where the tree is cut to ground level to stimulate their growth. Thus, estimates for these six reports range from 0.8 to 9.3 tons/[acres-year].

It is difficult to reconcile this wide range of yields per land area, but yield is certainly affected by how well the land is managed, regional differences in terrain and climate, moisture content, and what type of wood is planted. A simple average of the data in the previous paragraph gives a value of about 4.4 tons/[acres-year].

Knowing the yield per acre is only part of the footprint to grow and harvest the biomass because the biomass needs for the power plant are affected by several other issues. In order to know the mass flow of biomass to the plant to meet a certain target of electricity generation, the energy content of the biomass, the thermal efficiency in using this energy, and the capacity factor must be known. The energy content of the biomass is dependent on the type of biomass chosen and the moisture content. The thermal efficiency depends on the power plant process and the capacity factor depends on how many hours per year the plant is expected to operate. Unlike fossil fuel and nuclear power plants, where the supply of feed to the plant is expected to be fairly constant, this is not always the case for biomass power plants. As was discussed earlier in this chapter, from 2013 to 2018, wood and wood waste biomass plants in the United States had capacity factors ranging from 55.3% to 58.9%. But, it is unlikely you could justify building a biomass power plant with this capacity factor, as Table 9.10 showed that lowering the capacity factor from 83% to 55% increased the LCOE from $170/MWh to $202/MWh.

In order to determine the land area needed to provide biomass for a power plant, it is best to plan on what is needed to operate the plant at the lowest cost possible rather than the current commercial realities. For example, the feasibility study for the 1,000 MW biomass-fired power plant in the Netherlands[80] assumed an energy content of 7,300 BTU/lb for the wood pellet feed, a thermal efficiency of 40%, and a capacity factor of 80%. To meet this level of operation requires a feed rate of 3.7 million tonnes per year, or 4.1 million tons/year. So, with the 80% capacity factor built into the expected operation, this means that 4,100 tons are needed per year for each MW of capacity. Using an average biomass yield of 4.4 tons/[acres-year], it would take about 930,000 acres or 1,450 square miles, to provide this much biomass. To put this in perspective, the state of Rhode Island has 1,545 square miles, so a land area the size of Rhode Island is needed to provide the annual biomass supply for this 1,000 MW power plant! Dividing the power plant size by the acres needed gives a value of 0.0011 MW/acre. Stated differently, 930 acres or about 1 ½ square miles are needed to supply biomass for every one MW of electric plant production.

A thermal efficiency of 40% is optimistic, however, and in the energy balance section a thermal efficiency of 28% was used for the biomass BFB technology. Using this thermal efficiency and assuming a 100 MW plant using wood chips with a heat content of 6,500 BTU/lb and operating at a capacity factor of 80%, 8,540,000 MM BTU (million BTUs) are needed each year to meet the thermal energy needs. To meet this thermal energy input, the feed rate must be 657,000 tons/year, or 6,570 tons/MW.

With this lower thermal efficiency than the previous paragraph, the footprint is 0.0007 MW/acre.

For the next step of (2) transporting the biomass to the power plant, this transport occurs on existing railways, roads, and waterways. As this infrastructure would exist without the biomass power plant, it is assumed that the land area for transporting biomass to the power plant is not consequential in the calculations.

Next in the biomass power plant footprint is (3) making electricity at the power plant. Finding pertinent data for biomass power plants proved somewhat difficult. First, many biomass power plants are not grass roots (i.e., new) units but rather have been converted from coal. Therefore, these units do not necessarily demonstrate the optimal land area if the plant was built specifically for biomass. Second, biomass plants are built with onsite biomass storage, so the amount of storage chosen may vary from plant-to-plant. Finally, for the grass roots plants, some of these may have been built with plans for future expansion. Nevertheless, ten different plants are profiled in Table 9.12. Except for two feasibility studies, all of these are, or were, commercially operating plants. The ARBRE plant in the United Kingdom was the only footprint found for a combined cycle plant but, because the plant was only 8 MW, it was not included in the averages.

Intuitively, a biomass combined cycle (BCC) plant should have a larger footprint than direct combustion, since additional equipment is needed to gasify the biomass and a gas turbine is needed. Although the MW/acre for the ARBRE combined cycle plant was less than the average, it still fell within the range for the ten sets of direction combustion data; accordingly, there are not enough data to differentiate between the two technologies and the same average footprint, 1.42 MW/acre, is used for both.

Now that the electricity has been made, the next step is to (4) transmit and distribute it to customers via the grid. However, the land area needed for step-up and step-down transformers is very small and can be ignored. Likewise, the metal towers and electrical lines take up very little land area and can coexist with farming, pasture land, and roadways. Therefore, the important land footprints are the land area needed to (1) grow and harvest the biomass and the land area needed to (3) make electricity at the power plant.

Summarizing, for both the BFB and BCC technologies, a footprint of 0.0007 MW/acre is used for the biomass growth and harvest and a footprint of 1.42 MW/acre is used for the power plant.

We can convert the units of MW/acre to units of billion BTU/(year-square mile). A 1 MW plant operating at 100% capacity for 1 year will produce 8,760 MWh (1 MW * 24 hour/day * 365 days/year). Also, there are 640 acres per square mile and 3.412×10^6 BTU/MWh. For the capacity factors, Table 9.8 used in the economic calculations cited a capacity factor of 83% for both biomass BFB and BCC.[30,33]

From these values, the biomass production footprint of 0.0007 MW/acre is equivalent to an energy density of 11.1 billion BTU/(year-square mile), or 11.1×10^9. And the biomass power plant footprint of 1.42 MW/acre is equivalent to an energy density

Table 9.12 Biomass Power Plants

Name	MW	Acre	MW/ Acre	Fuel	Technology	Grass Roots	Start Date
Atikokan, Ontario Canada[83]	205	741	0.28	Wood pellets	Direct combustion	No	September 2014
Thunder Bay Ontario, Canada[84]	150	131	1.15	Wood pellets	Direct combustion	No	February 2015
Nacogdoches Sacul, Texas[85]	100	165	0.61	Wood residue	Direct Combustion	Yes	Mid 2012
Drax, North Yorkshire, United Kingdom[86]	740	350	2.11	Wood pellets	Direct combustion	No	2013
The Netherlands[80]	1,000	124	8.06	Wood pellets	Direct combustion and gasification	Yes	Feasibility study
Covanta Mendota, California[87]	25	80	0.31	Wood waste	Direct combustion	Yes	2007 (now closed)
Covanta Honolulu, Hawaii[88]	84	28	3.0	MSW	Direct combustion	No	2012
Gainesville, Florida[89]	102.5	131	0.78	Chipped wood	Direct combustion	Yes	December 2013
Palm Beach, Florida[90]	95	24	3.96	MSW	Direct combustion	Yes	July 2015
Kozani, Greece[91]	25	6.2	4.0	Wood and Waste	Direct Combustion	Yes	Feasibility study
ARBRE, Selby, United Kingdom[82]	8	11.1	0.72	Wood	Combined Cycle	Yes	1998 (now closed)
Average (not including ARBRE)	252.7	178.0	1.42				

a. *Note*: MSW is municipal solid waste

of 22,545 billion BTU/(year-square mile), or 22,545 × 10^9. As is evident, if the land area needed to grow and harvest the biomass fuel is included in the footprint and energy density calculations, it dominates the size of the power plant by a factor of about 2,000.

In order to compare energy densities for different technologies on a common basis of gigawatt-hours per square mile per year, or GWh/(square mile-year), energy was converted from BTU to Watt-hour, or Wh, using this equation:

$$1 \text{ BTU} = 0.293071 \text{ Wh}$$

Using this conversion and recognizing that a gigawatt (GW) is equivalent to one billion Watts (or 1×10^9), the calculated energy density for the land area needed to grow and harvest the biomass is 3.3 GWh/(year-square mile) and the energy density for the biomass power plant is 6,600 GWh/(year-square mile).

Assuming a 1,000 MW biomass power plant, the footprint is only 1.33 square miles but, if the land area needed to produce the biomass fuel is included, the footprint is 2,690 square miles. The state of Delaware has a land area of 2,489 square miles, so a land area greater than the state of Delaware is needed to provide the annual biomass supply for this 1,000 MW power plant!

According to the International Energy Agency,[92] the US electric power consumption per capita for 1 year was 12.8 MWh in 2016. Using a capacity factor of 83% for either BFB or BCC, a simple calculation shows that 1 MW will power about 570 homes for 1 year, so a 1,000 MW (1 GW) plant will power about 570,000 homes! If, however, the biomass power plant does not meet this high level of operation and the capacity factor is 50%, the plant will power about 340,000 homes.

9.12. Environmental Issues

At first glance, making electricity from biomass appears to have many environmental benefits. Since there is, in principle, an unlimited supply of biomass that can be grown and burned to make electricity, this type of energy is renewable. If waste biomass is used, such as MSW, wood waste, and crop waste, it will eliminate waste that would otherwise go to the landfill. And during photosynthesis, chlorophyll in plants and trees uses CO_2 and water to produce chemical energy. Therefore, relative to fossil fuels such as natural gas and coal, biomass has the potential to be carbon-neutral, consuming the CO_2 it produces when new biomass is grown.

However, reaping these benefits will require balancing biomass growth with consumption to produce electricity. This section considers five different possible issues with biomass including deforestation, CO_2 emissions, air emission from electricity generation, water use, and fertilizers and pesticides.

First we consider the issue of deforestation, broadly defined as the clearing of forests, and CO_2 emissions together. According to one report,[93] meeting energy demands by using biomass for energy rather than food will require a huge conversion of agricultural and forest lands to grow crops on a commercial scale. According to satellite data, 40% of the earth is already used for agriculture. Another report[94] states that without sustainable harvesting practices and responsible land use, the removal of forest to produce biomass energy can actually increase greenhouse gases. Both reports[93,94] state that 25% to 30% of greenhouse gases released each year are from deforestation.

Conceptually, the idea of using biomass to generate electricity is that the CO_2 produced will be used in new biomass growth. However, according to the Intergovernmental Panel on Climate Change,[95] this process may actually take decades. According to this report, one 50 MW biomass power plant burns about 650,000 tons of trees per year, which works out to 13,000 tons/(MW-year). This estimate is higher than that for a Netherlands 1,000 MB biomass power plant feasibility study,[80] which was said to require 4,100 tons/(MW-year) Also, an EIA report (EIA)[33] used values of 7,600 tons / (MW-year) for a BCC plant and 12,100 tons / (MW-year) for BFB. Based on these studies, to support only one MW of power for 1 year requires from about 4,000 to 13,000 tons of wood biomass per year, depending on the energy and moisture content of the wood. And, depending on the type of wood, it takes decades to grow back. To cite a few examples, the white pine tree grows at a rate of 8 to 12 inches per year[96] and, at an Australia plantation used to produce wood for lumber, the final harvest occurs at 30 to 35 years.[97] A fast growing tree, the hybrid poplar, takes about 5 years to grow to a harvestable height, growing at a rate of 10 feet per year.[98] And, the eastern cottonwood can grow 10 to 15 feet per year. Thus, if biomass power plants are to operate CO_2-neutral, it will require a balance between the harvesting rate and the growth and regrowth rate.[93] This may be a difficult balance to achieve.

Next, we consider air emissions. Whether a fossil fuel or biomass fuel is burned to make electricity, this combustion process creates exhaust gases, or flue gases, that can become air pollutants if not sequestered at the plant. Naturally, the type and amount of pollutants depend on the biomass fuel but it can normally be expected to include nitrogen oxides (NO_x), sulfur dioxide (SO_2), carbon monoxide (CO), and particulate matter (PM).[94] Consider, for example, the composition of wood used in the EIA report.[33] From their Table 2.3, the biomass contained 41.6% carbon, 4.8% hydrogen, 0.4% nitrogen, 0.01% sulfur, and 34% oxygen, the remainder being ash and water. The nitrogen in this fuel, once burned, will make NO_x and, if not sequestered at the power plant can lead to ground level ozone and the formation of acid rain in the atmosphere. Another possible issue is PM, defined as a mixture of very small particles that get into the air after combustion. According to one report,[95] biomass emits as much PM as coal and 1.5 times as much carbon monoxide and 1.5 times as much CO_2. While this may seem surprising, consider that in Chapter 3, the energy content for bituminous coal was referenced to be 12,712 BTU/lb. As the heating value of the wood used in the EIA report[33] was cited as 6,853 BTU/lb, the lower heat content for the biomass relative to coal means that more biomass must be burned to make an equivalent amount of electricity (assuming equivalent thermal efficiencies). The issue of potential air emission pollutants can be controlled using filters and electrostatic precipitators.

With respect to water use, making steam to run the steam turbine and generate electricity requires water. One report suggests that it takes between 20,000 and 50,000 gallons per metawatt-hour (MWh).[94] When this water is released back into the lake or river, it is normally at a higher temperature, which could affect the local ecosystem.

Also, if fertilizers and pesticides are used to grow the biomass, there is always the potential to have runoff of the chemicals into the lake or river. These fertilizers are known to increase algae growth.

Therefore, the renewable and CO_2-neutral benefits of using biomass to produce electricity require a careful balance between the consumption and growth of the biomass fuel. It also requires good stewardship on the part of the power plant to sequester potential air pollutants and prevent the return of water to lakes and rivers in a state that can affect the local ecosystem.

9.13. CO_2 Production and the Cost of Capture and Sequestration

An attractive feature of using biomass as a power plant fuel is that it is renewable and the growth of the biomass, through photosynthesis, negates most of the CO_2 produced from combustion. However, when considering a biomass power plant from a life cycle analysis, which includes feedstock production, transportation of the biomass to the power plant, and power plant construction, the net CO_2 production is still slightly positive, albeit it much less than a fossil fuel power plant.

In a detailed report by the NREL[56] which examined a biomass power plant using the technology of biomass gasification combined cycle (BCC), the net production of CO_2 was reported to be 46 g CO_2 per kWh of electricity, or 46 g CO_2/kWh. This total was broken down into 28 g CO_2/kWh for feedstock production, 6 for transportation, and 12 for power plant construction. The 890 g CO_2/kWh produced from plant operation (gasification), representing 95% of the total, is assumed to be captured by new tree growth through the "biogenic carbon cycle," such that plants remove CO_2 from the atmosphere through photosynthesis and emit oxygen in return.

An overview paper[99] examined nineteen studies on the life cycle analysis of biomass power plants and showed broad ranges of CO_2 production. For the direct combustion BFB, the analyses showed a range of 8.5 to 118 g CO_2/kWh. And, for biomass gasification combined cycle (BCC), the range was similar at 17 to 117 g CO_2/kWh. A World Nuclear Association report[100] gave a range for biomass CO_2 production as 10 to 101 g CO_2/kWh with a mean of 45. The type of technology was not specified, however. Taking the entirety of these data, we assume the CO_2 production is about the same for BFB and BCC and use a value of 46 g CO_2/kWh, or 0.101 lb CO_2/kWh, to represent both technologies.

Carbon dioxide production is commonly listed in the units of lb CO_2/MM BTU, lb CO_2/kWh, and g CO_2/kWh. To convert from units of lb CO_2/kWh to lb CO_2/MM BTU (lb CO_2/million BTU), we use the heat rates cited earlier as 13,500 BTU/kWh for BFB and 12,350 BTU/kWh for BCC. The following equation shows the calculation for BFB, resulting in a value of 7.5 lb CO_2/MM BTU. If, instead, we use the heat rate for BCC, the calculated value is 8.2 lb CO_2/MM BTU.

$$lb\ CO_2\ /\ MM\ BTU = \frac{0.101 * 1 \times 10^6}{13{,}500} = 7.5\ lb\ CO_2\ /\ MM\ BTU$$

Turning now to the discussion of CCS, it is clear that the CO_2 produced by feedstock production (tilling, seeding, fertilizing, irrigating, and harvesting), transportation of the biomass to the power plant, and the power plant construction cannot practically be captured and sequestered using CCS technology. Although photosynthetic growth of new biomass should eventually extract CO_2 that was made by the power plant, there is, nevertheless, interest in using CCS at a biomass power plant because it would actually lead, overall, to negative CO_2 emissions. That is, capturing the CO_2 produced at the power plant using CCS coupled with the removal of CO_2 from the atmosphere during the photosynthetic growth of the biomass is much greater than the CO_2 produced by feedstock production, transportation of the biomass to the power plant, and the power plant construction.

As an example, the earlier referenced NREL report[56] shows that for the 936 g CO_2/kWh total CO_2 production, 890 g CO_2/kWh or 95% will eventually be removed through the biogenic carbon cycle. However, if CCS is also added at the biomass power plant to remove most of this 890 g CO_2/kWh, the combination of CCS with biogenic removal would overall result in significant negative CO_2 emissions since the other CO_2 produced only amounted to 46 g CO_2/kWh.

Chapters 2 and 3 for natural gas and coal gave a cost of \$59 to \$77/ton of CO_2 captured by CCS. Unfortunately, estimates for CCS applied to a biomass power plant are much higher. Although the application of CCS to natural gas and coal power plants is certainly not a mature science, these concepts have been more fully studied and these power plants are generally much larger and therefore benefit from economy of scale. The CO_2, once captured, will likely be sent by pipeline to some location for underground storage. Since natural gas and coal power plants are much larger on average than biomass power plants, they should require less pipeline infrastructure. For example, in Chapter 2 the EIA listed 1,820 operating natural gas power plants in 2017 with a capacity of 522,378, giving an average of 287 MW per unit. But the average capacity for biomass power plants in the United States is about 21 MW, as discussed earlier in the Section 10 on infrastructure. So the small scale of biomass power plants means they will need more miles of pipeline to transport the CO_2 to the sequestration location.

In terms of cost, a European report[101] calculated a CCS cost of \$90 to \$180 per ton of CO_2. And, a United Nations Intergovernmental Panel on Climate Change (IPCC) report[102] suggests the broad cost range of \$55 to \$230/ton of CO_2 captured. A simple average of these data gives a cost of \$140/ton of CO_2 captured, roughly double that for natural gas and coal.

To convert this to a cost in terms of \$/MWh, consider a 1,000 MW power plant operating at a capacity factor of 83%, which will produce 19,920 MWh/day (megawatt-hours/day). Assuming the previous information from the NREL report, the plant production of CO_2 is 890 g CO_2/kWh or 1,962 lb CO_2/MWh. Based on these values, the plant will generate about 19,540 tons per day of CO_2 emissions. At \$140/ton for CCS, the daily cost would be \$2.74 million, or about \$137/MWh! Considering the economic analyses of Table 9.9, showing a LCOE of \$163.1 for BFB technology and \$259.7 for BCC technology, the cost of CCS is quite significant.

As discussed in Chapters 2 and 3 for natural gas and coal, amine scrubbing is the most common process used for capturing CO_2 from the flue gas of a power plant.

The CO_2 reacts with the amine in an absorber and is then later separated by stripping. Typical amines include monoethanolamine (MEA) and methyldiethanolamine (MDEA). Once concentrated from the flue gas, the CO_2 is liberated from the amine solvent and then compressed for transportation and storage. Since this is expected to be more expensive for a biomass power plant, there is an incentive to look at other approaches for CO_2 capture. One such approach[103] advances a concept in which the CO_2 is captured in biomass plants larger than 100 MW using calcium oxide (CaO) as the sorbent. The product, calcium carbonate ($CaCO_3$) would then later be burned to produce the CO_2 for permanent storage (CCS). The authors claim the process can achieve 84% removal at a cost of 43 €/ton CO_2. Assuming a conversion rate of 1.25 $/€ gives a cost of about $54/ton of CO_2 captured. This translates to a value of $53/MWh, which is a lot more palatable than $137/MWh.

9.14. Chapter Summary

Over the last four decades, growth in the use of biomass to generate electricity has been impressive. In the United States, electricity generation grew from 1,383 GWh in 1985 to 62,762 GWh in 2017 and global generation grew from 31,070 GWh in 1985 to 495,395 GWh in 2017. And while the US growth has been mostly level since 2000, global generation continues to grow.

There are some large biomass power plants including the 740 MW Ironbridge plant in the United Kingdom, the 205 MW Polaniec plant in Poland, the 140 MW New Hope plant in the United States, and the 250 MW Zhanjiang plant in the Chinese province of Guangdong.[86,104] And Finland has six biomass power plants including the 265 MW Alholmens Kraft, the 160 MW Kymijärvi II plant, the 140 MW Vaasa plant, the 140 MW Wisapower plant, the 125 MW Kaukaan Voima plant, and the 125 MW Seinäjoki plant.[86]

On the other hand, the BCC technology has had limited success, with only a few demonstration plants but little, or no, commercial success. Also, biomass power plants suffer from high capital cost, low thermal efficiencies, and low capacity factors. For example, the capital cost for a biomass BFB plant is about $3,200/kW and the BCC plant is about $8,200/kW versus less than $1,000/kW for a natural gas plant. With respect to biomass capacity factors, while economic studies use values greater than 80%, a typical commercial value in the United States for a plant powered by wood and wood waste is 55% and globally this value ranges from 50% to 90%. Although the reason for these lower capacity factors is not known, it is likely due to the availability and logistics of providing the biomass feed.

In terms of the potential for biomass in the United States, estimates for 2030 and 2040 suggest from 1 to 1 ½ billion tons per year, which would result in the generation of about 1,200 to 1,600 TWh (terawatt-hours) if all the biomass was used to generate electricity. As the United States generated 4,058 TWh in 2017, biomass could certainly have a big impact for electricity generation. Global potential for biomass by 2050 could be as much as 232 EJ/year (Exajoule, or 1×10^{18} Joules), which would generate about 16,000 TWh/year. Worldwide electricity generation was 23,400 TWh in 2015 and projected to increase to 34,000 TWh by 2040, so 16,000 TWh/year would be a significant contribution to current, and future, worldwide electricity generation. Realistically, heating, transportation, and cooking compete with biomass use, and currently electricity generation represents 13% of biomass energy use. Using 13% of the potential biomass for electricity generation gives a generation of about 2,095 TWh/year.

In terms of current electricity generation, the United States, China, Brazil, and Germany lead the world generating 13.9%, 12.1%, 10.3%, and 10.3%, respectively, of the global total. And, in terms of installed capacity, Brazil, China, and the United States lead the world with 12.5%, 11.2%, and 10.8%, respectively, of the global total.

The "Well-to-Wheels" economic analysis showed an average LCOE, including transmission and distribution, of $163.1/MWh for BFB and $259.7/MWh for BCC. This large difference is due primarily to the capital cost of construction, which is more than double for BCC. Adding CCS at the biomass power plant, which would lead to an overall negative CO_2 footprint, increases the cost to generate electricity by about 13.7¢ per kWh and possibly less if new technologies are commercially viable. In terms of mass, the cost of CCS is $140/ton.

For a 1,000 MW biomass power plant operating at a capacity factor of 83%, the CO_2 generation is about 19,540 tons CO_2/day. In other units, this is equivalent to 7.5 lb CO_2/MM BTU for BFB and 8.2 lb CO_2/MM BTU for BCC. It can also be given in units of 0.101 lb CO_2/kWh and 46 g CO_2/kWh.

Using wood as an example for biomass fuel, the cost on the basis of energy content ranges from $4 to $6/MM BTU. This is more expensive than natural gas at $3.17/MM BTU (Chapter 2) and coal at $2.06/MM BTU (Chapter 3), but less expensive than crude oil at $11.20/MM BTU (Chapter 1).

In terms of experience and maturity, biomass power plants have been around since the early 1980's, with some of the first plants converted from coal operation as well as new, grass-roots, plants using the direct combustion BFB technology. Biomass combined cycle (BCC) technology started appearing in demonstration plants (less than 10 MW) in the 1990s, but the technology has not led to commercial-scale power plants. And, the first electrical grid system was built in 1882, followed by rapid development through the 1930s.

The energy density for only the biomass power plant is 6,600 GWh/(year-square mile). However, if the land area needed to grow and harvest the biomass is included, the energy density is much smaller at 3.3 GWh/(year-square mile). Therefore the energy density depends a lot on whether the biomass used is considered waste, with no footprint, or biomass grown specifically for power generation. Using a 1,000 MW biomass power plant as a reference point, the biomass power plant footprint is a tiny 1.33 square miles but, if the land area needed to produce the biomass fuel is included, the footprint is 2,690 square miles! The state of Delaware has a land area of 2,489 square miles, so a land area greater than the state of Delaware is needed to provide the annual biomass supply for this 1,000 MW power plant!

Finally, the main environmental issues for biomass are possible deforestation, CO_2 emissions, air emission from electricity generation, water use, and possible runoff of fertilizers and pesticides used to grow the biomass. Deforestation is very problematic, because it is the photosynthetic CO_2 consumption of the forest biomass that is needed to make biomass a nearly CO_2-neutral technology. This means that the use of biomass power on a large scale will require good land stewardship and careful control of biomass growth and consumption.

References

1. "Unfolding the Improvements of Biomass History," Alternative Energy Sources Information, 2015, http://www.alternativeenergysourcesinfo.com/biomass-history.html
2. "Monthly Energy Review," Energy Information Administration, July 2019, file:///C:/Users/paulf/Documents/Book/References_Figures_Tables_Revised/Biomass/mer2019.pdf
3. "How Electricity Is Generated from Biomass," Mathias Aarre Maehlum, Energy Informative, May 3, 2013, http://energyinformative.org/how-electricity-is-generated-from-biomass/
4. "Biomass for Electricity Generation," Department of Energy Federal Energy Management Program (FEMP), September 15, 2016, https://www.wbdg.org/resources/biomass-electricity-generation
5. "Solar-Induced Hybrid Fuel Cell Produces Electricity Directly from Biomass," Georgia Tech News Center, February 18, 2014, http://www.news.gatech.edu/2014/02/15/solar-induced-hybrid-fuel-cell-produces-electricity-directly-biomass
6. "Mapping Out Global Biomass Projections, Technological Developments and Policy," prepared for International Institute for Environment and Development (IIED) by Mairi Black and Goetz M. Richter, October 19–21, 2010, http://pubs.iied.org/pdfs/G02986.pdf
7. "2016 Billion-Ton Report, Advancing Domestic Resources for a Thriving Bioeconomy," Volume 1, Department of Energy, July 2016, https://energy.gov/eere/bioenergy/downloads/2016-billion-ton-report-advancing-domestic-resources-thriving-bioeconomy
8. "Burning Shelled Corn—A Renewable Fuel Source," Dennis E. Buffington, Penn State College of Agricultural Sciences, July 2006, http://www.neo.ne.gov/neq_online/july2006/cornbtu.pdf
9. "Variation in Corn Stover Composition and Energy Content with Crop Maturity," L.O. Pordesimo et al., Biomass and Bioenergy 28 (4), 366–374, April 2005, http://www.sciencedirect.com/science/article/pii/S0961953404001795
10. "Ash Content and Calorific Energy of Corn Stover Components in Eastern Canada," Pierre-Luc Lizotte et al., Energies 8, 4827–4838, 2015, https://pdfs.semanticscholar.org/5ec4/254e7092ee2c1189f6c35a147ff461665f80.pdf
11. "Electric Power Annual 2017," Energy Information Administration, October 2018, Revised December 2018, https://www.eia.gov/electricity/annual/pdf/epa.pdf
12. "A Reassessment of Global Bioenergy Potential in 2050," Stephanie Searle and Chris Malins, GCB Bioenergy 7, 328–336, 2015, https://onlinelibrary.wiley.com/doi/full/10.1111/gcbb.12141
13. "Global Bioenergy Potentials through 2050,", G. Fischer and L. Schrattenholzer, Biomass and Bioenergy 20, 151–159, 2001, http://pure.iiasa.ac.at/id/eprint/6527/1/RR-01-09.pdf
14. "Evaluation of Bioenergy Potential with a Multi-Regional Global-Land-Use-and-Energy Model," H. Yamamoto et al., Biomass and Bioenergy, 21, 185–203, 2001, https://np-net.

pbworks.com/f/Yamamoto+et+al+(2001)+Bio-energy+potential+-+global+land+use+-models,+Biomass+and+Bioenergy.pdf
15. "Exploratory Study on the Land Area Required for Global Food Supply and the Potential Global Production of Bioenergy," J. Wolf et al., Agricultural Systems, 76, 841–861, 2003, https://www.researchgate.net/publication/222567956_Exploratory_study_on_the_land_area_required_for_global_food_supply_and_the_potential_global_production_of_bioenergy
16. "Potential of Biomass Energy out to 2100, for Four IPCC SRES Land-Use Scenarios," M. Hoogwijk et al., Biomass and Bioenergy, 29, 225–257, 2005, https://www.academia.edu/17352842/Potential_of_biomass_energy_out_to_2100_for_four_IPCC_SRES_land-use_scenarios
17. "Biomass Energy: The Scale of the Potential Resource," C.B. Field et al., Trends in Ecology & Evolution, 23, 65–72, 2008, http://www.cas.miamioh.edu/~stevenmh/Field%20et%20al%202008.pdf
18. "A Bottom-Up Assessment and Review of Global Bio-Energy Potentials to 2050," E. Smeets et al., Progress in Energy and Combustion Science 33 (2007), 56–106, August 7, 2006, https://dspace.library.uu.nl/bitstream/handle/1874/21670/NWS-E-2006-202.pdf?sequence=1&isAllowed=y
19. "Future Bioenergy and Sustainable Land Use," H.J. Buchmann et al., Wissenschaftlicher Beirat Globale Umweltveränderungen, 1–393, 2009, editors R. Schubert R, S. Schellnhuber, and J. Schmid, pp. 1–393, Earthscan, London, https://www.cbd.int/doc/biofuel/wbgu-bioenergy-SDM-en-20090603.pdf
20. "Future Bio-Energy Potential under Various Natural Constraints," D.P. Van Vuuren et al., Energy Policy, 37, 4220–4230, 2009.
21. "Bioenergy Production Potential of Global Biomass Plantations under Environmental and Agricultural Constraints," T. Beringer et al., Global Change Biology Bioenergy, 3, 299–312, 2011, https://onlinelibrary.wiley.com/doi/full/10.1111/j.1757-1707.2010.01088.x
22. "Global Biomass Fuel Resources," Matti Parikka, Biomass and Bioenergy 27, 613–620, 2004, file:///C:/Users/paulf/Documents/Book/References_Figures_Tables_Revised/Ethanol/New%20folder/Global%20biomass%20fuel%20resources%202004.pdf
23. "International Energy Outlook 2017," Energy Information Administration, September 14, 2017, https://www.eia.gov/outlooks/ieo/pdf/0484(2017).pdf
24. "International Renewable Energy Agency (IRENA)," Renewable Electricity Capacity and Generation Statistics, 2018, http://resourceirena.irena.org/gateway/dashboard/?topic=4&subTopic=54
25. "Biomass Energy: Efficiency, Scale, and Sustainability," Biomass Energy Resource Center, August 14, 2005, file:///C:/Users/paulf/Documents/Book/References_Figures_Tables_Revised/Biomass/FSE-Policy.pdf
26. "Converting Biomass to Energy—A Guide for Developers and Investors," International Finance Corporation (IFC), June 2017, https://www.ifc.org/wps/wcm/connect/7a1813bc-b6e8-4139-a7fc-cee8c5c61f64/BioMass_report_06+2017.pdf?MOD=AJPERES
27. "Biomass Resources and Technology Options," John Scahill, 2003 Tribal Energy Program Project Review Meeting, US Department of Energy, November 20, 2003, https://energy.gov/sites/prod/files/2016/01/f29/biomass_scahill_tep_nov03.pdf
28. "Technical Manual for the SAM Biomass Power Generation Model," Jennie Jorgenson et al., National Renewable Energy Laboratory, Technical Report, NREL/TP-6A20-52688, September 2011, https://www.nrel.gov/docs/fy11osti/52688.pdf
29. "Report on Conversion Efficiency of Biomass," BASIS—Biomass Availability and Sustainability Information System, Version #2, July 2015, Intelligent Energy Europe, http://www.basisbioenergy.eu/fileadmin/BASIS/D3.5_Report_on_conversion_efficiency_of_biomass.pdf

30. "Cost and Performance Characteristics of New Generating Technologies," Annual Energy Outlook 2019, Energy Information Administration, Table 2, January 2019, https://www.eia.gov/outlooks/aeo/
31. "Lazard's Levelized Cost of Energy Analysis—Version 11.0," November 2017, https://www.lazard.com/media/450337/lazard-levelized-cost-of-energy-version-110.pdf
32. "Electricity Generation Baseline Report," Jeffrey Logan et al., National Renewable Energy Laboratory, NREL Report 67645, January 2017, https://www.nrel.gov/docs/fy17osti/67645.pdf
33. "Updated Capital Cost Estimates for Utility Scale Electricity Generating Plants," Energy Information Administration, April 2013, and "Levelized Cost and Levelized Avoided Cost of New Generation Resources in the Annual Energy Outlook 2015," Table 1, June 2015.
34. "Lazard's Levelized Cost of Energy Analysis—Version 8," September 2014, https://www.lazard.com/media/1777/levelized_cost_of_energy_-_version_80.pdf
35. "Corn Stover for Bioenergy Production: Cost Estimates and Farmer Supply Response," Jena Thompson and Wallace E. Tyner, Department of Agricultural Economics Purdue University, September 2011, https://www.extension.purdue.edu/extmedia/EC/RE-3-W.pdf
36. "Biomass Power Plant Feedstock Procurement: Modeling Transportation Cost Zones and the Potential for Competition," Anil R. Kizha et al., California Agriculture 69(3,184–190, July 1, 2015, http://calag.ucanr.edu/Archive/?article=ca.v069n03p184
37. "Advancing Systems and Technologies to Produce Cleaner Fuels Technology Assessments," Quadrennial Technology Review 2015, Chapter 7, US Department of Energy, https://www.energy.gov/sites/prod/files/2016/01/f28/QTR2015-7B-Biomass-Feedstocks-and-Logistics.pdf
38. "Annual Energy Outlook 2015 with projections to 2040," DOE/EIA-0383, Table A8, April 2015, https://www.eia.gov/outlooks/aeo/pdf/0383(2015).pdf
39. "Renewable Power Generation Costs in 2017," International Renewable Energy Agency (IRENA), 2018, file:///C:/Users/paulf/Documents/Book/References_Figures_Tables_Revised/Biomass/IRENA_2017_Power_Costs_2018.pdf
40. "Biomass Feedstocks for Biopower: Background and Selected Issues," Kelsi Bracmort, October 6, 2010, Congressional Research Service, https://biomassboard.gov/pdfs/crs_biopower_feedstocks.pdf
41. "Economic Comparative Advantage of Willow Biomass in the Northeast USA," Matthew Langholtz et al., October 11, 2018, Wiley, https://onlinelibrary.wiley.com/doi/10.1002/bbb.1939
42. "Poplar (Populus spp.) Trees for Biofuel Production," April 3, 2019, https://farm-energy.extension.org
43. "Tracking Ethanol Profitability," Iowa State University Ag Decision Maker, D1-10, Ethanol Profitability, Don Hofstrand, Excel Model, 2019, https://www.extension.iastate.edu/agdm/energy/html/d1-10.html
44. "From Chapter 8: Ethanol in this Book, Table 8.16," Corn stover price is an average of five studies.
45. "Economics of Baling Wheat Straw," Jordan Shockley, March 27, 2018, https://www.kygrains.info
46. "Case Study: Biomass Export Potential of Colombia," Wolter Elbersen and Rocio Diaz-Chavez et al., Workshop "Towards a Sustainable European Bioenergy Trade Strategy for 2020 and Beyond," Brussels, June 14, 2016, https://www.biotrade2020plus.eu/images/workshop_brussels_2016/08_Elbersen_Colombia.pdf
47. "Assessment of Sugarcane-Based Ethanol Production," Rubens Eliseu Nicula de Castro et al., Intechopen, November 5, 2018, https://www.intechopen.com/books/fuel-ethanol-production-from-sugarcane/assessment-of-sugarcane-based-ethanol-production

48. "Economic Analysis of Cow Manure Biogas as Energy Source for Electricity Power Generation in Small Scale Ranch," Arini Wrestaa et al., Energy Procedia 68, 122–131, 2015, file:///C:/Users/paulf/Documents/Book/References_Figures_Tables_Revised/Biomass/Economic_Analysis_of_Cow_Manure.pdf
49. "Biomethane from Dairy Waste: A Sourcebook for the Production and Use of Renewable Natural Gas in California, Chapter 8. Financial Analysis of Biomethane Production," Ken Krich et al., 2005, http://www.suscon.org/pdfs/cowpower/biomethaneSourcebook/Chapter_8.pdf
50. "Potential for Energy Production from Farm Wastes: Using Anaerobic Digestion in the UK: An Economic Comparison of Different Size Plants," Gabriel D. Oreggioni et al., Energies 2017, 10, September 13, 2017, https://www.mdpi.com/1996-1073/10/9/1396
51. "Renewables 2014—Global Status Report," Renewable Energy Policy Network for the 21st Century (REN21), 2014, file:///C:/Users/paulf/Documents/Book/References_Figures_Tables_Revised/Biomass/GSR2014_Full-Report_English.pdf
52. "Table 6.7.B. Capacity Factors for Utility Scale Generators Not Primarily Using Fossil Fuels, January 2013–November 2017, Electric Power Monthly, Energy Information Administration, https://www.eia.gov/electricity/monthly/epm_table_grapher.php?t=epmt_6_07_b
53. "Renewables 2016: Global Status Report," Renewable Energy Policy Network for the 21st Century, 2016, http://www.ren21.net/wp-content/uploads/2016/06/GSR_2016_Full_Report.pdf
54. "Cost-Effective Policy Instruments for Greenhouse Gas Emission Reduction and Fossil Fuel Substitution through Bioenergy Production in Austria," J. Schmidt et al., December 2010, http://pure.iiasa.ac.at/id/eprint/9753/1/XO-11-056.pdf
55. "Biomass Yield and Energy Balance of a Short-Rotation Poplar Coppice with Multiple Clones on Degraded Land during 16 Years," S.Y. Dillen et al., Biomass and Bioenergy 56, 157–165, September 2013, https://ac.els-cdn.com/S0961953413002110/1-s2.0-S0961953413002110-main.pdf?_tid=237d26b6-0af2-11e8-b603-00000aab0f6b&acdnat=1517889753_560a2395ef7d59850c9cb8f3ccd86521
56. "Life Cycle Assessment of a Biomass Gasification Combined-Cycle System," Margaret K. Mann and Pamela L. Spath, NREL Report 23076, December 1997, https://www.nrel.gov/docs/legosti/fy98/23076.pdf
57. "Biomass Power and Conventional Fossil Systems with and without CO_2 Sequestration—Comparing the Energy Balance, Greenhouse Gas Emissions and Economics," Pamela L. Spath and Margaret K. Mann, Report NREL/TP-510-32575, January 2004, https://www.nrel.gov/docs/fy04osti/32575.pdf
58. "Energy Balance of Wood Pellets," Pellet Fuels Institute and the University of Wisconsin, March 2011, http://www.gfc.state.ga.us/utilization/forest-biomass/sustainability/EnergyBalanceofWoodPellets-Mar2011.pdf
59. "A Review of Transmission Losses in Planning Studies," Lana Wong, California Energy Commission, Figure ES-1, August 2011.
60. "Monthly Energy Review," Energy Information Administration, July 2016, https://www.eia.gov/totalenergy/data/monthly/pdf/mer.pdf
61. "Electricity: Form EIA-860 Detailed Data," Energy Information Administration, 2015 data, https://www.eia.gov/electricity/data/eia860/
62. "French Island Generating Station," Xcel Energy, 2020, https://www.xcelenergy.com/energy_portfolio/electricity/power_plants/french_island
63. "Wilmarth Generating Station," 2015, http://www.energyjustice.net/map/displayfacility-64859.htm
64. "A Strategy for Converting Coal Fueled Power Plants to Biomass That Does Not Raise the Cost of Electricity and Creates Jobs: Why Using Wood Pellets Can Be Better Than Wood

Chips for Some Power Plant Conversions," William Strauss, September 2014, http://www.futuremetrics.info/wp-content/uploads/2014/09/Pulverized_Coal_to_Pulverized_Wood_Pellets.pdf
65. "Public Utility Regulatory Policies Act (PURPA)," Publication L. 95–617, November 9, 1978, https://www.usbr.gov/power/legislation/purpa.pdf
66. "The World Bank: World Development Indicators," 2020, http://www.tsp-data-portal.org/Historical-Electricity-Generation-Statistics#tspQvChart
67. "Development of Fluidized Bed Combustion—An Overview of Trends, Performance, and Cost," J. Koornneef et al., Progress in Energy and Combustion Science 33, 19–55, 2007, http://www.academia.edu/16259663/Development_of_fluidized_bed_combustion_An_overview_of_trends_performance_and_cost
68. "History of Gasification," National Renewable Energy Laboratory (NREL), US Department of Energy, 2020, https://www.netl.doe.gov/research/coal/energy-systems/gasification/gasifipedia/history-gasification
69. "Biomass IGCC at Värnamo, Sweden—Past and Future," Krister Ståhl et al., GCEP Energy Workshop, April 27, 2004, https://www.researchgate.net/publication/242292402_Biomass_IGCC_at_Varnamo_Sweden_past_and_future
70. "BIGCC system for New Zealand: An Overview and Perspective," Shusheng Pang and Jingge Li, New Zealand Journal of Forestry 51(2), January 2006, file:///C:/Users/paulf/Documents/Book/References_Figures_Tables_Revised/Biomass/BIGCC_New_Zealand.pdf
71. "Project ARBRE: Lessons for Bio-Energy Developers and Policy-Makers," Athena Piterou et al., Energy Policy, February 25, 2008, https://www.academia.edu/6199177/Project_ARBRE_Lessons_for_bio-energy_developers_and_policy-makers
72. "Drax's Great Biomass Carbon Capture Experiment," Heidi Vella, Power Technology, May 20, 2019, https://www.power-technology.com/features/draxs-carbon-capture/
73. "United States Electricity Industry Primer," Office of Electricity Delivery and Energy Reliability U.S. Department of Energy DOE/OE-0017, July 2015, file:///C:/Users/paulf/Documents/Book/Chapters_for_OUP/New_Gen_Documents/united-states-electricity-industry-primer.pdf
74. "U.S. Biomass Power, Dampened by Market Forces, Fights to Stay Ablaze," Sonal Pate, October 1, 2018, https://www.powermag.com/u-s-biomass-power-dampened-by-market-forces-fights-to-stay-ablaze/
75. "2017 Renewable Energy Data Book Including Data and Trends for Energy Storage and Electric Vehicles," US Department of Energy, Office of Energy Efficiency & Renewable Energy, January 2019, file:///C:/Users/paulf/Documents/Book/References_Figures_Tables_Revised/Biomass/2017_Renewable_Energy_72170.pdf
76. "Biomass Power Continues Global Growth Trajectory amid Support Policy Changes," Bioenergy International, January 28, 2019, https://bioenergyinternational.com/heat-power/26464
77. "Evaluation of Plant Biomass Resources Available for Replacement of Fossil Oil," Robert J Henry, Plant Biotechnology Journal 8(3): 288–293, April 2010, https://www.ncbi.nlm.nih.gov/pmc/articles/PMC2859252/
78. "Biomass Energy and Competition for Land," MIT Joint Program on the Science and Policy of Global Change, John Reilly and Sergey Paltsev, Report No. 145, April 2007, http://web.mit.edu/globalchange/www/MITJPSPGC_Rpt145.pdf
79. "A Calculation of the EU Bioenergy Land Footprint: Discussion Paper on Land Use Related To EU Bioenergy Targets for 2020 and an Outlook for 2030," Liesbeth de Schutter and Stefan Giljum, March 2014, https://www.foeeurope.org/sites/default/files/agrofuels/2015/foee_bioenergy_land_footprint_may2014.pdf

80. “Opportunities for a 1000 MWe Biomass-Fired Power Plant in the Netherlands,” 50461976-KPS/PIR 04-1114, W. Fleuren et al., August 29, 2005, http://www.greenpeace.nl/Global/nederland/report/2007/5/rapport-biomassa-kema.pdf
81. “Willow Biomass: An Assessment of the Ecological and Economic Feasibility of Growing Willow Biomass for Colgate University,” Jeremy Bennick et al., Report ENST 480, Spring 2008, http://www.colgate.edu/portaldata/imagegallerywww/3869/imagegallery/2_480enst_willowbiomass.pdf
82. “Life-Cycle Energy Densities and Land-Take Requirements of Various Power Generators: A UK Perspective,” K.M. Vincent et al., Journal of the Energy Institute 90 (2), 201–213, April 2017, https://www.sciencedirect.com/science/article/pii/S1743967115300921
83. “North America's Largest Biomass Power Plant Now Operating,” Bioenergy, September 18, 2014, http://www.sunwindenergy.com/bioenergy/north-americas-largest-biomass-power-plant-now-operating (plant acre size from Wikipedia), https://en.wikipedia.org/wiki/Atikokan_Generating_Stationhttps://en.wikipedia.org/wiki/Atikokan_Generating_Station
84. “Thunder Bay Generating Station,” 2018, https://en.wikipedia.org/wiki/Thunder_Bay_Generating_Station
85. “Largest US Biomass Plant Online in Texas,” Leon Walker, July 20, 2012, https://www.environmentalleader.com/2012/07/largest-us-biomass-plant-online-in-texas/
86. “Power from Waste—The World's Biggest Biomass Power Plants,” Power Technology, April 1, 2014, https://www.power-technology.com/features/featurepower-from-waste-the-worlds-biggest-biomass-power-plants-4205990/; “Buyer found for former Ironbridge power station,” BBC News, June 19, 2018, https://www.bbc.com/news/uk-england-shropshire-44532490
87. “Covanta Energy,” Biomass Power Association, 2020, https://www.usabiomass.org/pnw-members/covanta-energy/
88. “Covanta Launches $302 Million Expansion Project,” Anna Austin, Biomass Magazine, October 2009, http://biomassmagazine.com/articles/3444/covanta-launches-$302-million-expansion-project
89. “Gainesville Renewable Energy Center,” 2020, https://www.power-technology.com/projects/gainesville-center/
90. “TOP PLANT: Palm Beach Renewable Energy Facility 2, West Palm Beach, Florida,” Aaron Larson, Power Magazine, December 1, 2016, http://www.powermag.com/palm-beach-renewable-energy-facility-2-west-palm-beach-florida-2/
91. “Presentation of 25MW Biomass Power Plant in Kozani Area,” International Conference Regional Policies in Bioenergy—Kozani, Greece, November 21, 2012, http://www.bioenarea.eu/sites/www.bioenarea.eu/files/9_PPC_Renewables.pdf
92. “Electric Power Consumption (kWh per capita),” IEA Statistics for 2016, http://energyatlas.iea.org/#!/tellmap/-1118783123/1
93. “Working Group for Sustainable Biomass Utilisation Vision in East Asia (2008), ‘Environmental Aspects of Biomass Utilisation,’” in M.Sagisaka (ed.), Sustainable Biomass Utilisation Vision in East Asia, ERIA Research Project Report 2007-6-3, Chiba: IDE-JETRO, pp. 70–103, http://www.eria.org/publications/research_project_reports/images/pdf/PDF%20No.6-3/No.6-3-5%20Chap%204%20Biomass.pdf
94. “Negative Effects of Biomass,” Max Roman Dilthey, Sciencing, April 25, 2017, https://sciencing.com/negative-effects-biomass-19624.html
95. “The Harmful Impacts of Biomass Energy Generation: Undermining the Fight Against Global Warming,” Massachusetts Environmental Energy Alliance, March 15, 2014, http://massenvironmentalenergy.org/docs/biomass%20factsheet%20from%20MEEA.pdf
96. “Pine Factsheet,” Clemson College of Agriculture, Forestry and Life Sciences, October 19, 1999, https://hgic.clemson.edu/factsheet/pine/

97. "The Pine Plantation Rotation," 2016, http://www.forestrycorporation.com.au/__data/assets/pdf_file/0009/238473/pine-plantation-rotation.pdf
98. "10 Fastest Growing Trees & Plants In The World," April 25, 2014, Conservation Institute, https://www.conservationinstitute.org/10-fastest-growing-trees-plants-in-the-world/
99. "Life Cycle Assessment (LCA) of Electricity Generation Technologies: Overview, Comparability and Limitations," Roberto Turconi et al., Renewable and Sustainable Energy Reviews 28, 555–565, 2013, http://www.uni-obuda.hu/users/grollerg/LCA/hazidolgozathoz/lca-electricity%20generation%20technologies.pdf
100. "Comparison of Lifecycle Greenhouse Gas Emissions of Various Electricity Generation Sources," World Nuclear Association Report, July 2011, http://www.world-nuclear.org/uploadedFiles/org/WNA/Publications/Working_Group_Reports/comparison_of_lifecycle.pdf
101. "Six Problems with BECCS," Fern, September 2018, file:///C:/Users/paulf/Documents/Book/References_Figures_Tables/Conclusions/References/Fern%20BECCS%20briefing_0.pdf
102. "Extracting Carbon from Nature Can Aid Climate but Will Be Costly: U.N.," Alister Doyle, March 26, 2014, https://www.reuters.com/article/us-climatechange-ccs/extracting-carbon-from-nature-can-aid-climate-but-will-be-costly-u-n-idUSBREA2P1LK20140326
103. "Process and Cost Analysis of a Biomass Power Plant with In Situ Calcium Looping CO2 Capture Process," Dursun Can Ozcan et al., Ind. Eng. Chem. Res. 53 (26), 10721–10733, 2014, https://pubs.acs.org/doi/abs/10.1021/ie500606v
104. "World's Largest Biomass Power Plant Produces Electricity by Consuming Agricultural Residue," Zhang Huan, People's Daily Online, August 2, 2017, http://en.people.cn/n3/2017/0802/c90000-9250265.html

10
Geothermal
A Renewable Energy Type

10.1. Foreword

Anyone who has had the pleasure of visiting Yellowstone National Park, established as our first US National Park in 1872, has seen a splendid and awesome display of geothermal energy. Home to geysers, hot springs in which pressurized water erupts into spraying water and steam, some as regular as Old Faithful erupting about every hour, Yellowstone also hosts other examples of geothermal activity such as hot springs (geothermally heated groundwater), bubbling mud pots, and fumaroles emitting steam and toxic gases.

The word geothermal is self-explanatory, and originates from the Greek words of "ge" (γη) for earth and "thermos" (θερμος) for hot. It might surprise you, and strike fear in your heart, that the temperature of the earth's core is estimated at 4,000 to 7,000 K (Kelvin), or about 7,000 to 12,000°F.[1] And how do we know this? Well, unless you are Jules Verne's fictional character Otto Lidenbrock, who traveled there in *Journey to the Center of the Earth*, the temperature must be determined by other means. In the 1930s, a Danish seismologist named Inge Lehmann used data for "S-waves" and "P-waves" to determine that the earth's core is composed of two layers, a solid inner core and a molten outer core, both made up mostly of iron.[2] "P-waves," or primary waves, are fast moving waves that travel through all materials, including liquids, while "S-waves" or secondary waves, travel slower and can only move through rocks. The pressure can be estimated by knowing the depth to the core, around 6,400 kilometers or 4,000 miles. According to a Scientific American article, the temperature estimate of 4,000 to 7,000 K can then be made by inferring what temperature is necessary at the pressure of the earth's core to render iron in a molten state.[1] Direct measurement is not possible, as the deepest hole ever drilled was the Kola Superdeep Borehole in Russia, to a depth of 12.3 kilometers, or 7.7 miles.[2] By the way, I think it would have been a bit hot for our fictional Otto Lidenbrock.

The three things that keep the earth heated are (1) heat when earth was formed, (2) frictional heating caused by denser material sinking toward the earth's center, and (3) heat from the decay of radioactive elements.[1] According to a Physics World article,[3] radioactive decay of elements such as uranium and thorium account for about 50% of the heat.

Mankind has utilized geothermal energy for centuries. There is archaeological evidence showing that the first human use of geothermal resources in North America

The Changing Energy Mix. Paul F. Meier, Oxford University Press (2020). © Oxford University Press.
DOI: 10.1093/oso/9780190098391.001.0001.

occurred more than 10,000 years ago with the settlement of Paleo-Indians at hot springs.[4] These springs undoubtedly provided warmth and bathing, and may likely have also been considered good for healing.

For nonrenewable fossil fuels, including coal, natural gas, and crude oil, we owe their formation to geothermal energy and the intense pressure of the earth's interior. However, rather than waiting many millions of years for geothermal energy to repeat this process on dead organisms and plants, geothermal energy can be used more directly.

There are three general categories in which geothermal energy can be applied.[5] The first is called "Direct use and district heating systems." This is the use of hot springs or reservoirs for cooking and heating. In the case of heating, hot water near the surface of the earth can be piped directly into buildings, something very common in Iceland. Direct use can also include using this heat for industrial applications, such as gold mining, food dehydration, and milk pasteurizing. The second category is the use of "Geothermal heat pumps," such that heat pumps are used to transfer heat from the ground to heat buildings in winter, or to remove heat from the buildings into the ground during summer. The third category is the use of geothermal energy to generate electricity. In this application, power plants are built near the location of geothermal reservoirs and use hot water and/or steam to power the electric plant steam turbine and generate electricity. It is this third category that is the subject of this chapter.

10.2. Proven Reserves

The temperature of the earth's core is estimated at 4,000 to 7,000 K (Kelvins), or about 7,000 to 12,000°F. However, the deepest it has been possible to drill is 12.3 kilometers, or 7.7 miles. Therefore, estimating geothermal reserves for the world, or any country for that matter, has to do with how deep you are willing to drill, and at what cost, to harness the geothermal energy. In addition, the rate the temperature increases with depth, also called the geothermal gradient, varies with the geology beneath the earth's surface. Anyone who has traveled to Yellowstone National Park or Iceland can surmise that these areas have a sharp geothermal gradient, such that high temperatures can be achieved with shallow drilling.

Another factor to consider is how the geothermal energy is captured. Normally, geothermal power plants have been built in those locations where naturally occurring heat, water or steam, and rock permeability are suitable to generate electricity. However, enhanced or engineered geothermal systems (EGS) can generate electricity from geothermal energy in dry and perhaps impermeable rocks. EGS is discussed in more detail in the "Overview of Technology" section. Therefore, unlike fossil fuels, the amount of proven, or potential, reserves for geothermal energy is more difficult to assess.

A 2016 report by the Geothermal Energy Association[6] shows a worldwide potential of 211 GW (210,876 MW). However, as the report indicates, the estimates were either reported by the respective country's government or taken from published works, but are not necessarily an official estimate based on reservoir studies. And, for some of the smaller countries, the potential may be larger than these countries would ever need to develop. Nevertheless, these results are shown in Table 10.1, in which the top eleven countries are shown (it would not seem right to list potential geothermal energy without including Iceland, so the list was extended to eleven rather than ten). These eleven countries control about 66% of the worldwide geothermal potential.

Another Geothermal Energy Association report[7] states that at a depth of 6.5 km (~4 miles), the temperature is above the boiling point of water nearly everywhere in the United States. Therefore, the potential to generate electricity can be greater than that shown in Table 10.1 and it is possible to generate geothermal electric power in every state in the United States. According to the US Geological Survey,[7] known geothermal systems could generate 16,457 MW, resources yet to be discovered could raise this total to 73,286 MW from resources yet to be discovered, and the use of EGS could raise this to 727,900 MW, or 728 GW!

A 2006 report by the Massachusetts Institute of Technology (MIT)[8] suggests that the use of EGS in the United States could allow geothermal electric power generation to increase to 100,000 MW (100 GW) by 2050. Thus, both the USGS and MIT studies suggest the United States has a much greater potential than the 9,057 MW shown in Table 10.1.

Table 10.2 shows the worldwide growth of electric power from geothermal energy for the last seven decades. Data from 1950 to 2015 were taken from a report by Berrtani[9] and data for 2018 came from an article by Think Geoenergy.[10]

Table 10.1 Country Geothermal Potentials[6]

Country	Potential (MW)
World	210,876
Indonesia	29,000
Japan	23,000
Chile	16,000
Pakistan	12,000
Mexico	10,500
India	10,000
Kenya	10,000
United States	9,057
Taiwan	7,150
China	6,744
Iceland	5,650

Table 10.2 Total Worldwide Installed Capacity from 1950 up to 2018[9,10]

Year	Installed Capacity, MW
1950	200
1955	270
1960	386
1965	520
1970	720
1975	1,180
1980	2,110
1985	4,764
1990	5,834
1995	6,832
2000	7,972
2005	8,933
2010	10,897
2015	12,635
2018	14,369

While the table does show rapid growth since 1950, the total installed capacity reported for 2018 was only 14.4 GW, much smaller than other alternate energy technologies, such as the 539 GW for wind (Chapter 6) and the 386 GW for solar photovoltaic (Chapter 7).

Table 10.3 shows data for the top ten countries in the world in terms of installed capacity for 2018.[10] The top ten countries control 93.8% of the total installed capacity and the top three countries were the United States, Indonesia, and the Philippines with 25.0%, 13.6%, and 13.0%, respectively.

Although Table 10.3 does not show data for all the countries in the world with installed geothermal electric plants, there were 27 countries in total.[11] Comparing the countries to Table 10.1, the geothermal potential, it is interesting that Pakistan, India, and Taiwan have potential but currently no geothermal plants. Also, Chile and China have only tapped a little of their potential, 48 MW for Chile and 26 MW for China.

Earlier, two other general categories were described for geothermal energy, namely "Direct use and district heating systems" and "Geothermal heat pumps." Although the focus of this chapter is the use of geothermal energy to generate electricity, it is interesting to look at energy use for these other two categories. One paper lumps both of these together as a category called "Direct utilization."[12] Table 10.4 shows data from 2015 for the world total and the top ten countries. The geothermal energy used is

Table 10.3 Total Worldwide Installed Capacity for the Top Ten Countries[10]

Country/area	2018 Capacity, MW
World	14,369
United States	3,591
Indonesia	1,948
Philippines	1,868
Turkey	1,200
New Zealand	1,005
Mexico	951
Italy	944
Iceland	755
Kenya	676
Japan	542

Table 10.4 Direct Utilization of Geothermal Energy Worldwide for 2015[12]

Country	MW (thermal)	TJ/yr
World	70,329	587,786
China	17,870	174,352
United States	17,416	75,862
Sweden	5,600	51,920
Turkey	2,886	45,126
Germany	2,849	19,531
France	2,347	15,867
Japan	2,186	26,130
Iceland	2,040	26,717
Switzerland	1,733	11,837
Finland	1,560	18,000

distributed as 55.3% for heat pumps, 20.3% for bathing and swimming, 15.0% for space heating (of which 89% is for district heating), 4.5% for greenhouses, 2.0% for aquaculture pond and raceway heating, 1.8% for industrial process heating, 0.4% for snow melting. 0.4% for agricultural drying, and 0.3% for other uses. In the table, data are also shown in TJ, or Terajoules, with a TJ is equal to one trillion joules (1×10^{12} Joules). The top ten countries control 80% of the direct utilization of geothermal energy on a MW basis, with China and the United States controlling 50%. In another paper by the same authors,[13] 2015 data are presented for electricity generation, geothermal heat pumps, and other direct utilization applications. For 2015, the breakdown was 264,776 TJ/year for electricity generation, 325,028 TJ/year for geothermal heat pumps, and 262,758 TJ/year for other direct uses for a total of 852,562 TJ/year. On a percent basis, 31% of the total went to electricity generation, 38% to geothermal heat pumps, and 31% to other direct uses so each of these three categories is a significant user of geothermal energy.

10.3. Overview of Technology

A good overview on the history of geothermal energy in the United States can be found at an Energy.gov website.[4] Key points in this history are cited here. The first known human use of geothermal energy in North America occurred more than 10,000 years ago when Paleo-Indians settled at hot springs to use them for warmth, bathing, and healing. Paleo-Indians is a term used to classify the first people to enter and inhabit North America during the end late Pleistocene period.

More recently, in 1807, settlers founded a city next to hot springs in Arkansas, giving it the same name. Later, in 1830, a man named Asa Thompson charged a dollar for baths that used the geothermally heated spring-fed water. According to the review article, this is the first known commercial use of geothermal energy. In 1852, the Geysers, north of San Francisco, were developed into a spa at the Geysers Resort Hotel. And, in 1892, Boise, Idaho, piped water from hot springs to heat buildings, considered the world's first heating system using geothermal energy. This system is still in use today!

Of course, these are examples of direct utilization of geothermal energy. The first use of geothermal energy to generate electricity occurred in Italy in July 1904.[14] Piero Ginori Conti used dry steam at the Larderello field to generate electricity, but only enough power to light five lightbulbs. Later, however, Conti would build a commercial power plant in 1911 at the Valle del Diavolo in Larderello using dry steam, and this electricity was used by the Italian railway system.

In the United States, John D. Grant drilled a well at the Geysers in 1921 to generate electricity. Although his first attempt was unsuccessful, in 1922 he found success to start the first commercial geothermal power plant in the United States, producing 250 kW used to power the Geysers Resort Hotel.

In 1948, a professor at Ohio State University, Carl Nielsen, developed the first ground-source heat pump, which he used at his own home. And, in 1960, a large-scale geothermal electric plant began operation at the Geysers, producing 11 MW of power.

Some legislative acts also occurred to promote the use of geothermal energy. In 1970, the Geothermal Resources Council was formed to encourage development of geothermal resources worldwide. And, in 1972, the Geothermal Energy Association was formed. In 1974, the US government founded the Geothermal Energy Research Development and Demonstration (RD&D) Act, promoting public and private companies to invest in geothermal energy. This agency was changed to ERDA in 1975, the Energy Research and Development Administration. With this act, the Geo-Heat Center was formed at the Oregon Institute of Technology in Klamath Falls. Interestingly, these acts occurred before the formation of the US Department of Energy (DOE), formed in 1977.

The first successful demonstration of the binary technology, to be discussed later, occurred in 1981 when Ormat Technologies demonstrated this commercial technology in Imperial Valley, California. A demonstration of the EGS technology occurred in 2012 at the Geysers, and Ormat Technologies started the first EGS commercial plant in 2013 in Desert Peak, Nevada.[15] EGS is also discussed later, but basically is a technology employed to generate electricity using geothermal energy in which the rocks are dry and, perhaps, impermeable. EGS seeks to extract this heat by fracturing the subsurface and then injecting water.

There are different types of geothermal energy that can be used to make electricity. To generate electricity from geothermal energy, the approach is similar to other technologies discussed in this book such that high-pressure steam expands through a steam turbine to drive an electricity generator. According to a World Energy Resources 2016 report,[16] the geothermal industry recognizes three grades of extractive geothermal energy: high temperature (> 180°C), intermediate temperature (101 to 180°C) and low temperature (30 to 100°C). The easiest places to extract high temperature geothermal energy are regions close to the earth's surface. It should not be surprising, then, that in Table 10.1, the top three countries in terms of geothermal potential are Indonesia, Japan, and Chile since these countries are part of the so-called Ring of Fire, a ring of volcanoes around the Pacific Ocean. This "Ring of Fire" is also the source of energy for the many geothermal power plants in the Western United States There are many hot spots around the Ring of Fire, where geothermal energy finds its way to the surface of the earth in the form of volcanoes, fumaroles, hot springs, and geysers. A lot of geothermal activity is also near the earth's surface in Iceland and Yellowstone National Park.

Following an IRENA (International Renewable Energy Agency) 2017 report,[17] there are four basic types of geothermal power plants. Direct dry steam plants are the first type, and the most desirable because of their simplicity. For this type of plant, the dry steam is taken directly from below the surface and used to drive the steam turbine. After the steam is used in the turbine, it is condensed, cooled, and returned to below the earth. Typically, this type of plant uses steam at a temperature of 150°C or higher, and the steam must be at least 99.995% dry to avoid scaling and erosion of the turbine and piping. According to the IRENA report,[17] direct dry steam plants range in size from 8 MW to 140 MW. The first geothermal power plant, built by Conti in Italy, used direct dry steam and, according to one source,[18] direct dry steam plants account for about 50% of the geothermal capacity in the United States, and all are located in California.

The next type of geothermal power plant is a flash plant, and this is the most common type of geothermal electric plant in operation today.[17] Fluid, in the form of hot water and steam, is pumped at high pressure into a tank with a lower pressure, allowing the fluid to vaporize, or flash. The steam from the flashing is used to drive the turbine after which it is condensed back to a liquid. This liquid is either reinjected into the ground or flashed again at a lower pressure. Since lowering the pressure also cools the fluid, this technology works best when temperatures are greater than 180°C. The size of the plant varies with the amount of flashing. For a single flash plant, 0.2 to 80 MW is common, increasing to 2–110 MW for double flashing and 60–150 MW for triple flashing.

A third type of geothermal power is binary plants, in which the geothermal reservoir does not have a high enough temperature to use either dry steam or flash steam. In this technology, hot water from the reservoir is used with a heat exchanger to heat a second, or binary, liquid that boils at a lower temperature than water. Compared to dry steam and flash plants, the hot water from the reservoir stays in a closed-loop system, going through a heat exchanger to extract the heat and then being returned to the ground. It is this binary liquid which then powers the turbine to make electricity. This technology uses temperatures between 100 and 170°C, thus fitting the intermediate temperature range defined earlier. Although it is possible to use lower temperature reservoirs, the thermal efficiency will decrease. Binary power plants typically range from 1 MW to 50 MW.[17]

The last type of geothermal power plant is an EGS. The three technologies discussed earlier all rely on permeable aquifers that have water flowing through them such that hot water or steam can be extracted. If, however, we want to realize the geothermal potential that exists at greater depths, deeper drilling alone will not be sufficient because at great depths the rock may not be permeable and there may be no water flow. To harness this geothermal energy, then, artificial fractures in the rock are made by injecting water and small amounts of chemicals at high pressure. Needless to say, as is discussed in the next section, the capital cost for this type of plant is greater than the other three technologies. Once the injected water is heated, the flash or binary type technology is used to generate electricity. Since there is no natural flow of water, water must be continually reinjected to maintain production. As you might imagine, this technology offers some challenges compared to the other three more straightforward technologies. These challenges include locating the appropriate subsurface environment, safely and cost-effectively drilling the resource, creating the large network of fractures to create enough surface area to be economically feasible, and sustaining the system over time.[19]

Table 10.5 shows the twelve largest geothermal plants in the world, with data taken from several references.[20-27] The table illustrates several things. First, as Table 10.3 showed that the global geothermal capacity for 2018 was 14,369 MW, these twelve plants represent 6,095 MW of that total, or 42%. Table 10.3 also shows that, for 2018, the US installed geothermal capacity was 3,591 MW, so by itself the Geysers plant in California represents 42% of the US total. The Philippines and Indonesia, two

Table 10.5 Twelve Largest Geothermal Plants in the World[20–27]

Name	Location	Startup Year (first plant)	Capacity, MW	Number of Generating Units	Type of Technology
Geysers	Near San Francisco, CA	1960	1,517	22	Dry Steam
Larderello	Italy	1913	770	34	Dry Steam
Cerro Prieto	Baja, Mexico	1973	720	5	Flash
Olkaria	Kenya	1985	648	15	Dry Steam
Makban	Philippines	1979	460	6	Binary
CalEnergy	Calipatria, CA	1982	340	10	Flash
Sarulla	Indonesia	2017	330	3	Binary / Flash
Hellisheidi	Iceland	2006	303 MW electric, 400 MW thermal	7	Flash CHP
Tiwi	Philippines	1979	290	3	Flash
Darajat	Indonesia	1994	260	3	Dry Steam
Malitbog	Philippines	1996	230	1	Flash
Wayang	Indonesia	1999	227	2	Flash

a. CHP is combined heat and power

countries along the "Ring of Fire," each have three of the world's largest plants. Not surprisingly, none of these large plants use the EGS technology, and all but two use dry steam or flash technology.

Also, it is a little hard to keep up with Kenya. At their Olkaria geothermal field, there are now five plants after 79 MW came online in August at 2019 Olkaria V.[27] A second unit is expected to come online soon, adding another 79 MW. The current Olkaria total of 648 MW has undergone frequent growth since the first 45 MW of capacity was brought online at Olkaria I in 1985, followed by Olkaria II in 2010; Olkaria III in 2000, 2009, 2014, and 2016; and Olkaria IV in 2014.

And, there are many new geothermal power plants that have either recently started operation, or are planned for the near future. According to *Renewable Energy World*, between March and September 2016, a total of 44 new geothermal power projects began development throughout 23 countries, which will eventually add 1,562.5 MW.[28] This rate of growth exceeded development over the previous 2 years. One recent large addition is the Sarulla Geothermal Power Project in Indonesia, included in Table 10.5, which started operation in 2017.[25] There is also an ambitious project with a proposed 1,000 km subsea cable, referred to as IceLink, which would allow the transfer of 1,200 MW from Iceland to Great Britain, with an expected startup data of 2025.[29]

In addition, the Indian government has set an ambitious 1,000 MW target in the near future and 10,000 MW to be developed by 2030, while China plans to triple its production over the next 4 years, adding 530 MW of new geothermal power.[28] In addition to the rapid growth in Kenya, elsewhere in Africa, Ethiopia plans to add 1,000 MW with two plants at Corbetti and Tulu Moye within the next 8 years.[30] For a final example of these new projects, Turkey has reached a country total of 1,100 MW geothermal power with the start of its 33 MW Melih plant at the end of 2017.[31]

Compared to conventional approaches to geothermal electric power generation, EGS projects are mostly in the demonstration phase. In the United States, a government website[32] gives progress on several projects. For example, a DOE-funded project at Desert Peak in Nevada is being used to extend the life of unproductive wells by drilling deep to capture energy from hot rocks thousands of feet below the surface. This increased their power output by 38% at an existing geothermal field, adding 1.7 MW of power. Likewise, at the Geysers, stimulating hot rocks at a depth of 3.7 km was done at an abandoned well in this geothermal complex, adding another 5 MW of power. And, at the Newberry Volcano near Bend, Oregon, a new, or "greenfield" project, was completed in January 2013, circulating cold water through cracks in hot rocks 1.8 km (~6,000 feet) below the surface. The size of this demonstration plant is not known at this time.

Elsewhere in the World, a power plant was commissioned in June 2008 at Soultz-sous-Forêts, France, concluding 30 years of work.[33] Drilling was done to a depth of 5 km and generating 1.5 MW of electricity. And, in Germany, there are two EGS projects. At Insheim, a 4 MW plant started operation in 2012, drilling to a depth of 3.5 km and at Landau, a 2.9 MW plant started operation in 2007, drilling to a depth of 3.0 km. And, in 2013, the 10-year 1 MW Habanero pilot project in Australia was completed, and included six deep geothermal wells to a depth of 4.6 km with a maximum measured temperature of 264°C.[34]

Although the focus of this chapter is the use of geothermal energy to generate electricity, it would be remiss to not mention geothermal heat pumps in this overview of technology, especially since it makes up 38% of the global use for geothermal energy. In a report by World Energy Resources,[16] direct utilization of geothermal energy is broken down into six categories including heat pumps, space heating, greenhouse heating, aquaculture pond heating, agricultural drying, and industrial uses. Of these six categories, geothermal heat pumps are the most common type of direct utilization, representing 71% of installed capacity and 55% of the total energy usage (see Figures 6 and 7 in this report).

Anyone that has been inside a cave in either winter or summer will notice that the temperature is relatively constant. That is because below the surface of the earth at depths of about 10 to 300 feet, the earth's temperature remains relatively constant.[35] Since the temperature in the ground is warmer than air in winter and cooler than air in summer, geothermal heat pumps can take advantage of this temperature difference by circulating fluids through pipes buried in the ground. The concept is relatively simple. In winter, cold air in the house or building is pumped below the earth, heated through heat exchange, and then returned to heat the house or building. In summer, the process is reversed, and hot air from the house or building is cooled in the ground and then returned to the house or building.

There are two basic heat pump systems, namely "open loop" and "closed loop." The open loop system is the oldest and cheapest type of geothermal heat pump system, in use since the 1970s, and uses available groundwater.[36] Groundwater is the heat exchange fluid and circulates through the geothermal heat pump system. After circulating through the system and performing the heating or cooling, it is returned to the ground.

The second system is a closed loop system, such that a fluid such as water or antifreeze is used to transfer the heat. The heat pump circulates the fluid through tubing buried in the ground, or submerged in water. The closed loop can be disposed either vertically or horizontally in the ground, or even placed in a pond or lake. For loops disposed horizontally, trenches are dug at least 4 feet deep while for vertically disposed loops, holes are drilled about 100 to 400 deep, 20 feet apart.[35] In the case of placing the closed loop in a pond or lake, the pipes are placed at least 8 feet below the water surface to prevent freezing in the winter.

10.4. Capital Cost, Operating Cost, Well-to-Wheels Levelized Cost of Electricity, and Well-to-Wheels Levelized Cost of Fuel

The equations and methodology used to calculate levelized cost of electricity (LCOE), and specific input used for geothermal energy to make electricity, are detailed in Appendix A. Input for the calculations came from 2019 Energy Information Administration (EIA) data,[37] NREL (National Renewable Energy Laboratory) 2018 data,[38] and 2018 data from Lazard.[39] The units for LCOE are in $/megawatt-hour, or $/MWh.

For a geothermal power plant, the "Well-to-Wheels" costs include (1) the cost of constructing the geothermal power plant, (2) power plant generating costs, (3) cost of connecting the power plant to the electrical grid, (4) transmission and distribution costs, and (5) taxes (both federal and state).

Capital costs, fixed and variable operating costs, and the capacity factor were set according to the authors. Table 10.6 shows the change in capital cost over either a 3- or 4-year period. From 2013 to 2016, the NREL data show a 29% reduction in capital cost for a geothermal flash plant and a 13% reduction for a binary plant.[38,40] For the 4-year period from 2014 to 2018, the data for Lazard also show a 13% reduction in capital cost for the geothermal binary plant.[39,41] The capital cost reduction for the flash plant is similar to the 27% decrease in capital cost for an onshore wind farm for the period of 2015 to 2019, as discussed in Chapter 6. However, it is not as dramatic as the 48% decrease in solar photovoltaic plants from 2014 to 2018, as discussed in Chapter 7. Nevertheless, these limited data suggest that capital cost has decreased significantly for a geothermal flash plant, but not as much for a binary plant.

Table 10.6 Capital Cost for Flash and Binary Geothermal Plants

	Capital Cost
Flash	$/kW
NREL, 2016[38]	$4,229
NREL, 2013[40]	$5,992
Binary	
Lazard, 2018[39]	$4,000–$6,400
Lazard, 2014[41]	$4,600–$7,250
NREL, 2016[38]	$5,455
NREL, 2013[40]	$6,291

The costs of transmitting and distributing electricity were 0.9 cents/kWh (or $9/MWh) and 2.6 cents/kWh (or $26/MWh), set according to the EIA.[42] Here, transmission is defined as the high-voltage movement of electricity from the power station while distribution is the cost of electrical lines that take power from a substation to the customer. The cost of transmission and distribution is essential to get a true "Well-to-Wheels" analysis of the total cost.

Results for the different LCOE cases are shown in Table 10.7.

As the table illustrates, the capital cost, fixed O&M, and the LCOE are greater for the geothermal binary plant compared to the flash plant. In the case of LCOE, including transmission and distribution, the LCOE for the binary plant is $139.0/MWh versus $98.8/MWh for a flash plant. This is expected, as the binary is a more complicated system, using a secondary liquid to take the heat from the geothermal hot water before flashing and generating electricity at the turbine.

And the table also shows that an EGS plant is dramatically more expensive that their non-EGS counterpart. Applying the EGS technology to a flash plant increases the LCOE from $98.8/MWh to $261.2/MWh while applying EGS to a binary plant increases the LCOE from $139.0/MWh to $611.5/MWh. Capital costs are also significantly more expensive, as with EGS systems it may be necessary to drill to depths of 5 km (3.1 miles) or more, and the hot rocks have to be fractured to create permeability and water flow. However, these capital costs are likely to change, as the EGS technology has not reached the same commercial level of maturity as the flash and binary technologies.

Sensitivity studies for flash, binary, and EGS applied to binary technology were made by varying the capital cost, the fixed operating costs, and the capacity factor. The base case for each calculation was based on the 2016 NREL data.[38]

For the range in capital cost, a US DOE report gave possible capital cost reductions if larger geothermal plants are built and if improvements are made in exploration, drilling, and reservoir creation.[43] In the case of geothermal flash plants, this could

Table 10.7 Economic Analysis for Geothermal Flash, Binary, and EGS Technologies. For EGS, Cases Are Shown for EGS Applied to Both the Flash and Binary Technologies.

Energy Type	Capacity Factor	Capital Cost	Levelized Capital Cost; with MACRS depreciation	Fixed O&M	Variable O&M (including fuel)	Transmission Investment	Total System Levelized Cost (LCOE) to Generate	LCOE Including Transmission and Distribution
		$/kW	$/MWh	$/MWh	$/MWh	$/MWh	$/MWh	$/MWh
Flash								
EIA[37]	0.90	$2,787	$36.8	$15.5	$0.0	$1.4	$53.7	$88.7
NREL[38]	0.90	$4,229	$55.9	$16.6	$0.0	$1.4	$73.9	$108.9
Average	0.90	$3,508	$46.3	$16.1	$0.0	$1.4	$63.8	$98.8
Binary								
NREL[38]	0.80	$5,455	$81.1	$24.5	$0.0	$1.4	$107.0	$142.0
Lazard[39]	0.90	$4,000	$52.8	$25.0	$0.0	$1.4	$79.2	$114.2
Lazard[39]	0.85	$6,400	$89.5	$35.0	$0.0	$1.4	$125.9	$160.9
Average	0.85	$5,285	$74.5	$28.2	$0.0	$1.4	$104.0	$139.0
EGS								
EGS Flash—NREL[38]	0.90	$14,512	$191.7	$33.1	$0.0	$1.4	$226.2	$261.2
EGS Binary—NREL[38]	0.80	$32,268	$479.5	$95.6	$0.0	$1.4	$576.5	$611.5

result in a reduction from \$4,229/kW to \$3,319/kW while in the case of geothermal binary plants, this could result in a reduction from \$5,445/kW to \$4,273/kW. For the EGS technology applied to a binary geothermal plant, the capital cost could potentially be reduced from \$32,268/kW to \$5,509/kW.

The range for fixed operating and maintenance (O&M) costs were based on data from Lazard.[39] For a geothermal flash plant, the fixed O&M costs ranged from the base case value of \$16.6/MWh to \$26.6/MWh while that for the binary plant ranged from the base case value of \$24.5/MWh to \$34.5/MWh. For the EGS plant, the range given in the NREL data[38] was used, namely from the base case value of \$95.6/MWh to \$23.1/MWh.

Details for the range for the capacity factors is given in Appendix A and is also described in Section 6, so it is not repeated here. For the sensitivity study, the capacity factors range from the high values shown in Table 10.7 to the lowest observed US commercial value for each technology over the time period of 2008 to 2012, namely 76.9% for a geothermal flash plant and 49.3% for a binary plant. Since the EGS case is used in combination with the binary geothermal plant, the same lower value of 49.3% for the binary system is also used for the EGS technology.

The results of the sensitivity study are shown in Table 10.8. Not surprisingly, changing the EGS capital cost from \$32,268/kW to \$5,509/kW had a very large impact on LCOE. For the flash and binary plant cases, the reduction in capital cost resulted in a similar change to that for increasing the fixed O&M costs. And, for the binary plant case, reducing the capacity factor from 80% to 49.3% had a large impact, resulting in an LCOE increase of about \$66/MWh. For the EGS sensitivity study, even with a capital cost decrease from \$32,268/kW to \$5,509/kW, the LCOE was still greater than any of the flash and binary plant cases.

10.5. Cost of Energy

Unlike natural gas, coal, and nuclear electricity generation, geothermal power plants do not use fuel, generating the electricity solely from geothermal energy. Energy needed to operate and maintain the electric plant, also defined as the "parasitic" load, is taken from the electricity generated by the plant. The parasitic load for a geothermal power plant is generally much higher than a fossil fuel plant because of the low plant thermal efficiency, sometime less than 10%. Because of the low thermal efficiency, the energy needed to operate cooling water pumps, feed pumps, and well pumps consumes part of the electric power generated, resulting in high parasitic loads for geothermal plants.

10.6. Capacity Factor

The capacity factors shown in Table 10.7 for the LCOE calculations, ranging from 80% to 90%, are ideal numbers used for design and economic calculations. Commercial

Table 10.8 Sensitivity Study for Geothermal Flash, Binary, and EGS Technologies

	Capital Cost	Fixed O&M	Capacity Factor	Total System Levelized Cost (LCOE)	Total LCOE Cost to Generate, Transmit, and Distribute
Units	$/kW	$/MWh		$/MWh	$/MWh
Flash sensitivity case					
Base Case, NREL 2018	$4,229	$16.6	0.90	$73.9	$108.9
Capex is $3,319	$3,319	$16.6	0.90	$61.9	$96.9
Fixed O&M is $26.6/MWh	$4,229	$26.6	0.90	$83.9	$118.9
Capacity Factor is 0.769	$4,229	$19.4	0.769	$86.2	$121.2
Binary sensitivity case					
Base Case, NREL 2018	$5,455	$24.5	0.80	$107.0	$142.0
Capex is $4,273/kW	$4,273	$24.5	0.80	$89.4	$124.4
Fixed O&M is $34.5/MWh	$5,455	$34.5	0.80	$117.0	$152.0
Capacity Factor is 0.493	$5,455	$39.8	0.493	$172.8	$207.8
EGS sensitivity case					
Base Case, NREL 2018	$32,268	$95.6	0.80	$576.5	$611.5
Capex is $5,509/kW	$5,509	$95.6	0.80	$178.9	$213.9
Fixed O&M is $23.1/MWh	$32,268	$23.1	0.80	$504.0	$539.0
Capacity Factor is 0.493	$32,268	$155.1	0.493	$934.6	$969.6

values are, however, generally lower. One of the problems with capacity factors for commercial geothermal plants is that the flow rate of the water or steam, as well as the temperature of the hot fluid, may decline over the life of the geothermal field. This affects the plants' ability to meet their design capacity. Examining data from the EIA for commercial geothermal plants in the United States, the capacity factors were 78.1, 74.0, 74.3, 74.7, and 71.1% for the years of 2008 to 2012.[44] Breaking this down by technology, dry steam varied from 72.4 to 76.6%, flash systems varied from 76.9 to 89.3%, and binary systems varied from 49.3 to 69.7%. Although this report is a bit dated, more recent years for commercial US geothermal plants still have similar capacity factors. Although a breakdown by technology was not available, for the years 2013 to 2018 capacity factors were 73.6%, 74.0%, 74.3%, 73.9%, 74.0%, and 77.3%, respectively.[45]

Global commercial data for 2016 showed a wide ranges for the capacity factor.[46] For Asia, the capacity factor had a range of 41% to 93% with a weighted average of 83%, and the Pacific, where there is a lot of geothermal activity for countries like Indonesia and the Philippines, the range was from 60% to 80% with a weighted average of 80%. The lowest weighted average was in Central America and Caribbean at 58%.

One report on capacity factors indicates that a variation in energy output from the geothermal field is a factor.[44] As the temperature and flow rate of the steam or hot water declines over the life of the geothermal field, there is a decline in net electricity generation. One example used in this report showed that a decline in the fluid temperature from 165°C to 150°C resulted in a 22 to 27% decrease in capacity factor. While the reduction in energy produced is due to a reduction in thermal efficiency (to be discussed in the next section), it also affects the ability of the geothermal power plant to meet their design rating and therefore reduces the capacity factor.

A decline in temperature should have a greater effect on a binary plant compared to a flash steam plant. Recalling equation (1) from the Introduction, the maximum efficiency is related to the hot (T_H) and cold (T_C) fluid temperatures.

$$\eta = \frac{W}{Q_H} = 1 - \frac{T_C}{T_H} \quad (1)$$

In earlier discussions in the Overview of Technology, flash plants typically have inlet fluid temperatures greater than 180°C while binary plants have inlet fluid temperatures from 100 to 170°C. Consider, for example, a binary plant with an inlet fluid temperature of 150° and an outlet fluid temperature of 90°C. According to Equation (1), the thermal efficiency would be 14% (remembering to change the temperatures to Kelvin for the calculation). For a flash plant with an inlet temperature of 180°C and the same 90°C outlet temperature, the thermal efficiency would be 20%. Comparing that to a supercritical pulverized coal power plant with an inlet steam temperature from 540 to 580°C, it is easy to see why thermal efficiencies are much higher for coal and natural gas plants compared to geothermal power plants.

10.7. Efficiency: Fraction of Energy Converted to Work

For coal, natural gas, and nuclear energies, the thermal efficiency of the turbine used to generate electricity was calculated using the BTU content of one kilowatt-hour (kWh) of electricity, or 3,412 BTU, and dividing this by the heat rate. The heat rate is the amount of energy used to make one kWh of electricity. However, for geothermal electric power plants, as well as other renewables such as solar, wind, and hydroelectric, there is no fuel being consumed, and therefore energy used to make the electricity is not frequently measured. This is because renewable energy that is wasted in the process of making electricity has no fuel cost. If some of the geothermal heat energy is not converted into electricity, there was no alternate use, or economic value, for this energy.

However, there are still inefficiencies in the process of making electricity from renewable fuels because no mechanical device is perfect.

To calculate thermal efficiencies for renewable energy, the EIA[47] proposes an "Incident Energy Approach." In this approach, the amount of electricity made can be compared to the incident energy that is input to the mechanical device. In the case of geothermal power, this is the energy contained in the hot fluid at the surface of the geothermal well. From Table F1, they report an efficiency of 16% for geothermal power.

So why is the efficiency for geothermal power so low? By comparison, a natural gas combustion turbine with an efficiency of 35.9% (Chapter 2), a supercritical coal plant with an efficiency of 38.5% (Chapter 3), and a nuclear plant with an efficiency of 33% (Chapter 4) are much higher. As shown in the previous section, the maximum thermal efficiency is related to the hot (T_H) and cold (T_C) fluid temperatures according to equation (1). If, for example, a coal plant heats steam to 580°C and then cools it in a condenser to 90°C, the maximum efficiency is 57%. Now consider typical numbers for a geothermal binary plant, taken from a 2007 *Geothermics* article,[48] in which the steam temperature is 150°C and the condenser temperature is 50°C, giving a maximum efficiency of 24%. Clearly, the thermal efficiency of a geothermal plant is hurt by the temperature of the hot fluid, which is a function of the temperature of the steam or hot water extracted from the geothermal reservoir. The three temperature ranges recognized by the geothermal industry, namely high temperature (> 180°C), intermediate temperature (101 to 180°C) and low temperature (30 to 100°C) are much less than the temperature of steam, attained by combustion, for a coal, natural gas, or nuclear plant. Of course, there is no cost for the geothermal energy, compared to a fossil fuel, but the low thermal efficiency has an effect on the amount of electricity generated as well as the LCOE.

The EIA incident energy approach[47] paper states that the geothermal thermal efficiency was based on an informal survey of commercial plants. Next, we would like to establish typical thermal efficiencies for all three technologies profiled in the economic calculations, namely flash, binary, and enhanced geothermal system (EGS) plants. The overall thermal efficiency is affected by many parameters including the type of technology used, the plant MW size, the parasitic load of the plant, the temperature and flow rate of the steam or hot water, and the ambient conditions (which helps set T_C, the cold fluid temperature). In one research report,[49] it was shown that the thermal efficiency increases with increasing plant MW output, increasing well depth, and increasing well temperature.

Given that binary plants work with fluid temperatures less than dry steam or flash plants, we would expect binary plants to have lower thermal efficiencies. One worldwide review,[50] which examined commercial data from 94 plants, shows exactly this such that binary plants had thermal efficiencies ranging from 1% to 9% while flash plants had thermal efficiencies ranging from 6% to 18%. The authors found that thermal efficiencies were a function of the fluid enthalpy content, which is a direct function of the fluid temperature and flow rate. Another research report shows commercial thermal efficiencies for binary plants ranging from 7% to 12%.[51]

A paper from *Energy Science & Technology*[49] shows commercial data for different geothermal technologies including binary, single and double flash, and EGS plants. The data taken from this paper are shown in Table 10.9. From the data, we can calculate average thermal efficiencies for each geothermal system with 13.5% for binary, 10.0% for single-flash, 11.0% for double-flash, and 11.0% for EGS systems. Thus, although the reservoir temperatures for binary systems are less than those for flash systems, the commercially measured thermal efficiencies can be the equal, or even larger, than those for flash systems because of other factors affecting the thermal efficiency. This paper also shows that the EGS technology is able to have thermal efficiencies equal to other geothermal systems.

In summary, considering all of the information discussed here, there does not seem to be enough information to assign different thermal efficiencies to different types of geothermal technologies so, for the purpose of calculating an energy balance in the next section, the EIA value of 16% is used to represent all of the technologies.

10.8. Energy Balance

For a geothermal power plant, the energy balance is defined by this equation:

$$\text{Energy Balance} = \frac{\left[\begin{array}{c}\text{Electric energy from the geothermal plant delivered}\\ \text{to the customer}\end{array}\right]}{\left[\begin{array}{c}\text{Thermal energy contained in steam or hot water}\\ \text{flowing to the power plant}\end{array}\right]}$$

For the conversion of geothermal energy to electricity, there are several steps in the "Well-to-Wheels" analysis including the energy used to (1) drill the well and extract the steam or hot water from the geothermal field, (2) construct the power plant, (3) generate the electricity, and then (4) distribute the electricity through the grid to your home.

For the first two steps, (1) drilling the well and (2) constructing the power plant, information is available from two Argonne National Laboratory (ANL) reports where they used a GREET (Greenhouse Gases, Regulated Emissions, and Energy Use in Transportation) model to perform a life cycle analysis for geothermal systems. For what they refer to as the "plant cycle," which involves material production and plant construction for any power plant, they show a ratio of the energy used for the plant cycle to the energy output of the power plant. In their first report,[52] they give values of 1.16% and 7.52% for the flash and EGS technologies, respectively. Their second report[53] revises these two values, and also provides a value for binary geothermal systems. In this report, they show 2.17% for binary, 1.15% for flash, and 7.62% for EGS. Note that these ratios are to the energy output and not the energy input. The more recent values are used for the energy balance.

Table 10.9 Geothermal Plant Thermal Efficiencies[49]

Plant name	Type	Capacity (MW)	Maximum Depth (m)	Reservoir Enthalpy (kJ/kg)	Maximum temperature (°C)	Thermal Efficiency
Amatitlan, Guatemala	Binary	25.2	—	1300	285	11.3%
Ngawha, New Zealand	Binary	75	2300	975	320	16.1%
Wairakei, New Zealand	Binary	171	—	2750	240	18.0%
Rotokawa, New Zealand	Binary	34	2500	1550	300	20.0%
Chena Hot Springs, USA	Binary	0.21	915	306	73.3	8.0%
Brady Hot Springs, USA	Binary	4.33	—	750	108	8.0%
Heber, USA	Binary	6.87	—	702	165	13.2%
Mori, Japan	Single-Flash	50	3000	1199	260	4.8%
Miravalles III, Costa Rica	Single-Flash	27.5	—	1038	159	4.1%
Svartsengi, Iceland	Single-Flash	74.4	—	1148	240	16.2%
Nesjavellir, Iceland	Single-Flash	120	2200	1503	300	17.5%
Uenotai, Japan	Single-Flash	28.8	800	2350	300	7.2%
Onikobe, Japan	Single-Flash	12.5	500	1020	240	3.9%
Takigami, Japan	Single-Flash	28		925	198	6.7%
Olkaria, Kenya	Single-Flash	45	2500	2120	152	15.0%
Los Azufres, Mexico	Single-Flash	195	3500	2030	180	12.3%
Kizildere, Turkey	Single-Flash	20.4	2261	875	190	12.0%
Ahuachapan, El Salvador	Double-Flash	95	1500	1115	280	9.5%
Hellisheidi, Iceland	Double-Flash	210	2195	1365	300	5.4%
Hatchobaru, Japan	Double-Flash	110	2300	1068	300	7.4%
Cerro Prieto, Mexico	Double-Flash	720	4400	1396	315	21.6%

Continued

Table 10.9 *Continued*

Plant name	Type	Capacity (MW)	Maximum Depth (m)	Reservoir Enthalpy (kJ/kg)	Maximum temperature (°C)	Thermal Efficiency
Bruchsal, Germany	EGS	0.55	2542	1800	123	12.7%
Neustadt-Glewe, Germany	EGS	0.21	2250	—	99	7.5%
Altheim, Austria	EGS	1	2300	—	106	7.7%
Berlín, El Salvador	EGS	54	1250	1300	300	14.8%
Larderello, Italy	EGS	594.5	3000	2770	270	13.9%

For the third step, to (3) generate the electricity, the previous section set the geothermal thermal efficiency for making electricity at 16% regardless of geothermal technology type.

It then remains to determine the energy needed to (4) distribute the electricity through the grid to your home. Transmission and distribution losses are affected by the voltage of the power line and the distance the electricity is transmitted. Historical data from the state of California shows annual average transmission losses around 5 to 7% over the time period of 2002 to 2008.[54] Similarly, EIA data from July 2016[55] show that 7% of the electricity generated is lost in transmission and distribution. Based on these studies, a value of 7% is used for the amount of electricity lost in transmission and distribution after leaving the geothermal power plant.

Putting these results together we see that, for all three cases, only a small amount of energy is consumed (1) drilling the well and (2) constructing the power plant. Recall that the ANL report gives ratios of the energy consumed in steps (1) and (2) to the output energy, not the input, so these were determined iteratively to get the ratio of the so-called plant cycle to the energy output. As shown in the following equations, the poor thermal efficiency of the geothermal power plant is the main reason for the small total use of energy extracted from the geothermal field. For all three technologies, only about 15% of the original energy extracted from the geothermal field is delivered to customer in the form of electricity.

Flash Geothermal Power Plant

100 (geothermal field) $\rightarrow$ 99.8 (drill the well and construct the power plant) $\rightarrow$ 16.0 (generate electricity) $\rightarrow$ 14.9 (transmission and distribution)

Binary Geothermal Power Plant

100 (geothermal field) → 99.7 (drill the well and construct the power plant) → 15.9 (generate electricity) → 14.8 (transmission and distribution)

EGS Geothermal Power Plant

100 (geothermal field) → 98.9 (drill the well and construct the power plant) → 15.8 (generate electricity) → 14.7 (transmission and distribution)

10.9. Maturity: Experience

Man has been using geothermal energy for at least 10,000 years ago, when Paleo-Indians settled at hot springs to use them for warmth, bathing, and healing. However, the use of geothermal energy to generate electricity really starts in the early part of the last century.

The first dry steam plant was in 1904, when Piero Ginori Conti used dry steam to generate electricity, but only enough to light five lightbulbs. However, in 1911, he would build a commercial power plant at the Valle del Diavolo which was used by the Italian railway system. According to a book by DiPippo,[56] the plant generated 250 kW and was the first commercial generation of electricity from geothermal energy. Not long after, dry steam technology appeared at the Geysers in California in 1922, where they generated electricity for their resort hotel. The Geysers pathway to become a 1,517 MW commercial plant began later in 1960.

The same book by DiPippo[56] describes a flash steam plant built in Kamchatka in 1967, with 5 MW of power. And a larger flash steam plant, Cerro Prieto I, was built in Baja, Mexico, from 1973 to 1981 with four single flash units and one double flash. Although it is generally believed that the first geothermal binary power plant was put into operation in 1967 near the city of Petropavlovsk in Kamchatka (rated at 670 kW), there is evidence a 200 kW plant existed as early as 1952 in the Democratic Republic of the Congo.[56]

As for EGS systems, the earliest time period depends somewhat on how you define EGS technology. Hot Dry Rock (HDR) technology has been around as early as 1977. HDR uses rock formations, primarily those of granite, that have high temperature, low permeability, and no stored water. Permeability is created by hydraulic fracturing, also known as hydrofracking, since water is used. The first HDR project was the Fenton Hill project at Valles Caldera, New Mexico, in 1977. In phase I, a 60 kW plant was installed. The EGS, more broadly defined than HDR technology, had its early beginnings when a plant was commissioned in June 2008 at Soultz-sous-Forêts, France, concluding 30 years of work.[57] In addition, a 4 MW EGS plant started operation at Insheim, Germany in 2012, drilling to a depth of 3.5 km, and a 3.5 MW EGS plant started operation at Landau, Germany in 2007, drilling to a depth of 3.0 km. In the United States, an EGS demonstration plant started up in 2012 at the Geysers while

Ormat Technologies started the first commercial plant in 2013 at their Desert Peak, Nevada, facility. Understanding the difference between HDR and EGS really has to do with permeability and fluid content. HDR, with low permeability and no stored water, is a subset of EGS, which can also include geothermal systems with higher permeability and some water content.

While the focus of this chapter has been on the use of geothermal energy to make electricity, the electric grid is also an important component of the electric power industry. The electrical grid, used to accept power produced from various energy sources involves both transmission and distribution. Transmission and distribution includes a step-up transformer, transmission lines, and a substation step-down transformer. The first electric power distribution system was built in 1882 in New York City, overseen by none other than Thomas Edison.[58] Between this time and the 1930s, the electrical grid system grew rapidly, roughly doubling every 6 years. Today, what we take for granted, involves more than 180,000 miles of high-voltage transmission lines connecting about 7,000 power plants.[58]

In summary, geothermal power plants using dry steam technology started appearing as early as 1904, followed by binary systems in 1967 (and possibly as early as 1952) and flash steam systems in 1967. The concept of EGS has been around as early as 1977. Thus, there are more than a hundred years of experience in using geothermal energy to make electricity, but more recent advances in the technology (binary and flash) are about 50 years. EGS has been studied for about 40 years but currently there are no large commercial plants, only demonstration projects smaller than 5 MW. The first electrical grid system was built in 1882 followed by rapid development through 1930s. Thus there are more than 135 years of experience with electric grids.

10.10. Infrastructure

According to the IRENA, in 2017 the global electricity generation for geothermal plants was 85,978 GWh.[11] The global installed capacity was 14,369 MW,[10] which gives an overall capacity factor of 68%. For the United States, an *Electric Power Annual* 2017 report[59] gives data for both capacity and generation. In 2017, US geothermal plants had an installed capacity of 3,732.4 MW and generation of 15,927 GWh, which gives an overall capacity factor of 49%.

The total installed capacity in the United States for all power plants was 1,203,092 MW, so geothermal made up only 0.3% of the capacity while total generation was 4,058,258 GWh, with geothermal making up 0.4% of the generation. As a percent of renewables, including hydroelectric, the geothermal installed capacity was 1.5% and generation was 2.3%.

According to EIA electricity data,[60] the installed capacity of 3,732.4 MW came from 65 geothermal power plants operating in the United States, giving an average size of 57 MW. However, this average is heavily affected by the Geysers plant, with an installed capacity of 1,517 MW, about 41% of the US total. Excluding the Geysers, the average plant size is 35 MW. Table 10.10 shows data for the seven states that are generating electricity from geothermal plants, with California as the clear leader in terms of number of plants, percent of capacity, and electricity generated. With the exception of New Mexico, capacity factors for each state are about 50% or greater.

Worldwide, some countries have a high percentage of electricity generated from geothermal energy. Looking at 2017 IRENA data for geothermal electricity generation by country[11] and acquiring data for total electricity generation from a variety of sources,[61–65] there are seven countries that generate 5% or more of their electricity from geothermal energy including Kenya (51%), Iceland (27%), El Salvador (26%), New Zealand (18%), Nicaragua (16%), the Philippines (11%), and Indonesia (5%). By comparison, geothermal electricity generation only made up 0.3% of global generation and 0.4% of US generation. Data from 2017 show that 27 countries have geothermal power.[11]

For the last discussion of this section, we turn to the subject of transmission lines, critical infrastructure for any power plant. For fossil fuel–based power plants, for which the fuel can typically be delivered by pipeline, train, barge, and truck, the fuel does not control the plant location and it can be placed near existing transmission lines. However, for renewable energy such as geothermal, wind, and solar, power plants are built where the conditions are favorable for generating the electricity. If these locations happen to be near existing transmission lines, they can exploit the more than 180,000 miles of high-voltage transmission lines that exist in the grid.[58] If these plant locations are not near, or the grid infrastructure is already at capacity, then new transmission lines are needed.

Naturally, this adds cost and can be especially significant since most geothermal plants are small. Consider, for example, that the United States currently has 3,732

Table 10.10 Geothermal Plant Data for US States[60]

State	MW	Number of Plants	% of Capacity	% of Plants	Generation, GWh	Capacity Factor
California	2,807	32	73.2	49.2	11,560	47.0
Hawaii	51	1	1.3	1.5	323	72.3
Idaho	18	1	0.5	1.5	84	53.3
New Mexico	19	1	0.5	1.5	13	7.7
Nevada	820	25	21.4	38.5	3,292	45.9
Oregon	37	2	1.0	3.1	174	54.1
Utah	84	3	2.2	4.6	481	65.5
Total	3,835	65	100.0	100.0	15,927	47.4

Note: The total installed capacity for these 65 plants, 3,835 MW, is slightly different from the 3,732.4 MW from another EIA source.

MW in 65 power plants, an average of less than 60 MW per plant. This is much smaller than typical natural gas, coal, or nuclear plants, so the cost of a new transmission line can become a significant expense. A 2012 NREL report[66] shows that the cost of carrying electricity decreases as the line voltage increases. For example, for a transmission of 100 miles, a 230 kV line costs $0.43 million per MW of power compared to $0.29 million for a 345 kV line and $0.19 million for a 500 kV line. Over a distance of 600 miles, the values are $3.37, $1.74, and $0.94 million for 230 kV, 345 kV, and 500 kV lines, respectively. However, the higher voltage lines are designed to carry more electricity and, if they are underutilized, the cost effectiveness is lost.

Nevertheless, new transmission lines are being built for geothermal plants. For example, for the Raft River project in Idaho, rated at 18 MW, a 3.2 mile, 34.5 kV transmission line was built in 2007 to send geothermal electricity to Bridge, Idaho.[56] And, in Nevada, the One Nevada Transmission Line, which was built to connect the northern and southern power grids, has been expanded by 65 miles to reach Boulder City.[67] The purpose of this new transmission line is to accelerate the integration and exportation of Nevada's renewable energy resources, including geothermal, solar, and wind.

Around the world, other countries are investing in new transmission lines for geothermal energy. Three examples are highlighted here. In Kenya, a new 482 kilometer transmission line is built to transmit electricity from Olkaria to Mombasa.[68] In the Philippines, a new 183 kV transmission line, extending 80 kilometers, will send electricity from Leyte to Sumar.[69] And, in Turkey where geothermal power is rapidly expanding, they are spending $3.5 billion to build 14 new transmission lines from 2016 to 2019.[70]

10.11. Footprint and Energy Density

The footprint for a geothermal power plant is one of the more difficult to quantify. Since there is no fuel to be produced, as for natural gas, coal, nuclear, or biomass, the land area footprint starts at the geothermal field, and includes the area for the power plant. Thus, the land area footprint includes (1) the geothermal field and infrastructure needed to deliver the hot steam or water to the power plant, (2) the geothermal power plant, and (3) the infrastructure to transmit and distribute the electricity via the grid to commercial customers. Steps (1) and (2), generally speaking, occur at the same physical site, since hot flowing steam and water lose temperature and pressure over long pipelines.

If it were just the power plant, the issue of land area footprint would be easy and the footprint would be small. For the traditional approaches to geothermal power, dry steam, flash, and binary plants, the power plant equipment includes a main building, separator, cooling towers, condenser, heat exchangers, turbines, and the electric generator. However, accessing the energy of the geothermal field adds a production well to bring the steam or hot water to the earth's surface, pipes to deliver the fluids to the geothermal plant, and an injection well to return fluids back into the earth.

The question of how much area is contributed by the geothermal field is a difficult one to answer. For example, one power plant might be built at a geothermal field, with the intention of expanding to more plants in the future to use geothermal energy in the same field that is yet untapped. Also, as the production well ages, it may be necessary to eventually drill more production wells if the temperature and flow of the steam or hot water are reduced. Proper management of the reservoir by returning fluid into the earth via the injection well has a lot to do with the life of a production well. Considering the issue of geothermal field sustainability, an article by the American Geoscience Institute provides some qualitative statements.[71] Summarizing the article, the temperature of a geothermal reservoir, the fluid pressure, or the fluid flow may decrease over time. Although fluids can, and should, be reinjected into the well, this can also cool down the temperature of the reservoir. As a result, it may become necessary to drill additional production wells to maintain normal electricity generation levels.

Before looking at data for specific geothermal power plants, a review is made of general "rules-of-thumb." According to one US government website,[72] a geothermal field and power plant use 1 to 8 acres per megawatt (MW), or 0.125 to 1 MW/acre. And from the Geothermal Energy Association,[73] in which they consider 30 MW and 50 MW plants, total land area ranges from 53 to 367 acres, giving a range of 0.08 to 0.94 MW/acre. For this analysis, they state that 2 to 7 acres are for exploration (mainly road access and construction) and 51 to 360 acres for drilling operations (drilling and well field development, road improvement or construction, the power plant, installing wellfield equipment including pipelines, and installing transmission lines). For the power plants only, the 30 MW plant disturbed 15 acres (2 MW/acre) and the 50 MW plant disturbed 25 acres (also 2 MW/acre).

In an MIT study[74] data are shown for flash and binary plants, excluding wells, and a flash plant that includes wells and pipes. The results, shown in Table 10.11, have been adapted from their Table 8.2. This shows that the power plant footprint alone is about 2.9 to 3.2 MW/acre but when the geothermal field and associated infrastructure is included, the footprint drops to 0.54 MW/acre. Taking these three studies together, the footprint for only the power plant is about 2 to 3 MW/acre while including the geothermal field reduces the overall footprint to about 0.1 to 1.0 MW/acre.

The earlier-referenced report also talks about how geothermal power plants can harmoniously coexist with agriculture. Citing two examples from the Imperial Valley of California, they show photographs of the Heber binary plant amid fields of alfalfa. Here the power plant footprint is 4 MW/acre while the plant plus geothermal field has a footprint of 1.3 MW/acre. Similarly, the Heber flash plant has a total footprint of 2 MW/acre, also sitting in an agricultural environment.

Naturally, the issue of harmonious coexistence depends on the type of geothermal plant. For some dry steam regions, such as Iceland and even Yellowstone National Park, coexistence certainly is affected by the amount of geothermal activity at the earth's surface.

Table 10.11 Geothermal Plant Footprints[74]

Size, MW	Technology	m^2/MW	m^2	Acres	MW/acre
110	Flash, excluding wells	1260	138,600	34.2	3.21
20	Binary, excluding wells	1415	28,300	7.0	2.86
56	Flash, including wells and pipes	7460	417,760	103.2	0.54

In order to get more plant specific data, a search was made for the land area footprints of the world's twelve largest plants shown in Table 10.5. Unfortunately, footprints could not be found for the Darajat and Sarulla plants, so the Miravalles plant in Costa Rica and the Los Azufres plant in Mexico were added, as these are also large plants. The data are shown in Table 10.12 with a range of 0.02 to 0.27 MW/acre and an average of 0.11 MW/acre. For this small sample, there were five dry steam, two binary, and five flash plants. For these twelve plants, the trend for these data are that dry steam plants have the largest land footprint followed by flash and then binary plants. However, since the data set is so small no differentiation for the three technologies is made with respect to footprint and energy density.

It is worth noting that for some of the plants, the geothermal field used for the land area can support future additional capacity. The Olkaria site, for example, have had significant expansion since the first 45 MW were brought online in 1985, leading to the current installed capacity of 648 MW.

Also, the energy densities of Table 10.12 represent the total land area for the geothermal power plant including the plant structure, roads, wells, pipelines, and transmission lines. So, as was discussed in Chapter 6 for wind, there is a direct impact area disturbed by infrastructure as well as the total land area. Except for the plant structure and roads, other parts of the land area may be able to coexist with other activities, such as farming or ranching, depending on the nature of the geothermal field. For example, Kenya's Olkaria geothermal plants are in Hell's Gate National Park, which is also home to an abundance of wildlife (buffalo, zebras, gazelles, baboons, etc.) and birds.

The average for the data in Table 10.12, 0.11 MW/acre, is used for the combined footprint of (1) the geothermal field and infrastructure needed to deliver the geothermal source to the power plant and (2) the geothermal power plant. For the (2) geothermal power plant alone, a value of 2.5 MW/acre is used based on the range of 2 to 3 MW/acre given earlier.

Next, units of MW/acre are converted to units of billion BTU/(year-square mile). A 1 MW plant operating at 100% capacity for 1 year will produce 8,760 MWh (1 MW * 24 hour/day * 365 days/year). Also, there are 640 acres per square mile and 3.412×10^6 BTU/MWh. Since no differentiation was made for the footprint and energy density with respect to technology type, only one set of energy densities is calculated to

Table 10.12 Geothermal Plant Footprints

Name	Location	Startup Year (first plant)	Capacity, MW	Type of Technology	Acres	MW/ acre
Geysers[75]	Near San Francisco, CA	1960	1,517	Dry Steam	29,000	0.05
Larderello[76]	Italy	1913	770	Dry Steam	45,700	0.02
Cerro Prieto[77]	Baja, Mexico	1973	720	Flash	7,400	0.10
Olkaria[78]	Kenya	1985	648	Dry Steam	16,640	0.04
Makban[79]	Philippines	1979	460	Binary	1,730	0.27
CalEnergy[80]	Calipatria, CA	1982	340	Dry Steam	2,560	0.13
Hellisheidi[81]	Iceland	2006	300	Flash CHP	2,026	0.15
Tiwi[82]	Philippines	1979	290	Flash	2,965	0.10
Malitbog[83]	Philippines	1996	230	Flash	3,707	0.08
Wayang[84]	Indonesia	1999	227	Flash	9,884	0.03
Miravalles I–V[85]	Costa Rica	1994	163	Flash & Binary	5,189	0.06
Los Azufres[86,87]	Mexico	1982	195	Binary	1,169	0.26
Average						0.11

represent all three technologies. For the capacity factor, 85% was chosen using Table 10.7 as a guide.

From the entire land area, a value of 0.11 MW/acre is equivalent to an energy density of 1.79 trillion BTU/(year-square mile), or 1.79×10^{12}. For the geothermal electric plant alone, with a footprint of 2.5 MW/acre, the calculated energy density is 40.65 trillion BTU/(year-square mile), or 40.65×10^{12}.

Now that the electricity has been made, the next step is to (3) transmit and distribute it to customers via the grid. However, the land area needed for step-up and step-down transformers is very small and can be ignored. Likewise, the metal towers and electrical lines take up very little land area and can coexist with farming, pasture land, and roadways. This means that the energy density for a geothermal power plant is based either on all the land area used for the geothermal field and power plant or only the land area for the power plant structure. If the geothermal field can coexist with agriculture or ranching, then it is reasonable to base the energy density only on the land area for the geothermal plant structure. Alternatively, if the geothermal field land area cannot be used for other purposes because of emissions danger, seismic activity, or the danger of hot surface water or steam, then the energy density must also include the geothermal field land footprint.

In order to compare energy densities for different technologies on a common basis of gigawatt-hours per square mile per year, or GWh/(square mile-year), energy was converted from BTU to Watt-hour, or Wh, using this equation:

$$1 \text{ BTU} = 0.293071 \text{ Wh}$$

Using this conversion and recognizing that a gigawatt (GW) is equivalent to one billion Watts (or 1×10^9), the calculated energy density for the combined geothermal field plus power plant is 524.0 GWh/(year-square mile) and the energy density for the power plant alone is 11,910.2 GWh/(year-square mile).

Assuming a 1,000 MW geothermal power plant, the footprint for the geothermal field plus power plant is 16.7 square miles and for the power plant alone is 0.74 square miles. Compared to coal and nuclear power, the footprint of the combined geothermal field and power plant is larger, but smaller than natural gas, wind farms, and solar plants.

According to the International Energy Agency,[88] the US electric power consumption per capita for 1 year was 12.8 MWh in 2016. Using the capacity factor of 0.85 for a geothermal power plant, a simple calculation shows that 1 MW will power about 580 homes for 1 year, so a 1 GW plant will power about 580,000 homes!

10.12. Environmental Issues

Before discussing some environmental issues for geothermal power plants, it is useful to remember that they have positive benefits relative to electricity generation from fossil fuels. Certainly, they largely avoid air pollution that can occur from sulfur and nitrogen oxides that come from fossil fuels, as well as reducing greenhouse gas emissions. Also, there is no land use to provide fuel, as is the case with petroleum, coal, natural gas, and uranium, for which drilling and exploration or mining are needed.

There are some environmental issues for geothermal, including air emissions; noise; water use; the impact on natural phenomena, wildlife, and vegetation; and even seismic activity. The discussion of seismic activity is saved for last.

With respect to air emissions, these are not a concern for binary plants and only a small concern for dry steam and flash plants. For binary plants, the hot water from the reservoir exchanges heat with another fluid and is returned into the ground, so it is a closed system with no air emissions. For dry steam and flash plants, some of the steam or hot water from the reservoir is emitted into the air as water vapor, with the remainder sent back to the reservoir. The gases that can be released include sulfur dioxide (SO_2), carbon dioxide (CO_2), nitrous oxides (NO_x), hydrogen sulfide (H_2S), and particulate matter (PM). Nitrous oxides include NO_2 and NO, and are generally represented as NO_x. Particulate matter is a complex mixture of very small particles (micron-sized) that are emitted by power plants, and these particles can cause health effects. However, geothermal emissions resulting from water vapors are quite small compared to fossil fuel-based power plants. Table 10.13 shows some emission data taken from one report.[89] As can be seen, emissions are essentially zero for the binary technology, and very small for dry steam and flash technologies relative to fossil fuel plants.

Hydrogen sulfide (H_2S) is also a possible emission, and present in many types of geothermal activity including volcanic gases, hot springs, fumaroles, and geothermal steam and water. Any H_2S made at the geothermal power plant is now routinely treated at the plant and converted to elemental sulfur.[90]

Mercury is another possible emission, although it is not present in every geothermal resource.[89] Where it is present, equipment is used to sequester it, typically reducing potential emissions by 90% or more. As is the case with other emissions, this is not an issue for binary geothermal plants since all of the hot water sent through the heat exchanger is returned to the reservoir.

With respect to noise pollution, most of the noise produced at a power plant occurs during the drilling of a well.[90] Noise from the power plant itself is not an issue, the small amount coming from cooling fans and the turbine.

As for water use, it is very small compared to a coal or nuclear plant. Typically, dry steam and flash geothermal plants use about 20 liters of fresh water per MWh compared to 1,370 liters per MWh for a coal plant and about 1,700 liters for a nuclear plant.[90] A binary air-cooled plant does not use any fresh water. The only change to the water is to cool it, and then it is returned to the reservoir.

Geothermal plants also have very little impact on natural phenomena, wildlife, and vegetation. Power plants are normally not located near geysers, fumaroles, and hot springs, so that extracting steam or hot water from the reservoir does not harm these thermal activities.[90] In fact, most plants are located where there are no geothermal surface discharges, so it should be possible for the plant to coexist with agriculture or ranching and not affect wildlife.

Table 10.13 Air Emissions Summary from US Power Plants

kg / MWh	NO_x	SO_2	CO_2	Particulate Matter
Coal	1.95	4.71	993.8	1.01
Oil	1.81	5.44	758.4	NA
Natural Gas	1.34	1.0	550.0	0.06
EPA Average of All US Power Plants	1.34	2.74	631.6	NA
Geothermal Flash	0	0.16	27.2	0
Geothermal Binary	0	0	0	Negligible
Geothermal Dry Steam	0.0005	0.0001	40.3	Negligible

We finish this section with the interesting topic of the connection of EGS geothermal power plants with seismic activity. In 2006, the city of Basel, Switzerland, experienced an earthquake of 3.4, which damaged some property in the city.[91] What was supposed to be an environmentally friendly way to generate electricity for 10,000 homes was eventually abandoned when it was discovered that the underground fracturing of the rock to increase permeability was the cause of the earthquake. In this case, wells were drilled to a depth of three miles, or 4.8 kilometers.

This was not a singular event, and earthquakes have happened with other EGS tests.[92] EGS relies on creating artificial fractures in the rock by injecting water and small amounts of chemicals at high pressure. Because of this, it can also induce seismicity and earthquakes. At the German Landau EGS facility, a 2.7 magnitude earthquake occurred on August 15, 2009. The epicenter was essentially at a Landau drilling site, but no damage or injuries were reported. Similarly, the EGS project in Cooper Basin, Australia, experienced tremors up to 3.7 in magnitude in December 2003, when two wells were drilled 4.4 to a depth of kilometers. And, at the French Soultz-sous-Forêts EGS plant, a 2.9 magnitude quake occurred in 2003 when two test wells had been drilled to 5 km.

And these earthquakes are not restricted to EGS plants. At Hellisheidi, Iceland, a combined heat and power flash plant, two 3.8 earthquakes occurred on October 15, 2011.[93] The quakes were attributed to induced seismicity due to fluid injection into the geothermal field. According to this article, reinjection of wastewater into the geothermal reservoir is necessary to (a) protect the surrounding environment, (b) prevent contamination of groundwater reserves, and (c) maintain pressure in the system. For these two earthquakes, structural vibrations were felt as far away as 12 km in the village of Hveragerdi.

Of course, not all earthquakes cause damage. Following some general scales,[94] an earthquake of 2.5 or less is generally not felt. Those from 2.5 to 5.4 can be felt, but cause only minor damage. From 5.5 to 6.0, there can be slight damage to buildings. Thus, as long as the magnitude is less than 2.5, there should not be any significant issues.

A US Geological Survey focused on seismic activity data at The Geysers power plants in California to provide some insight into the cause of these earthquakes.[95] To date, the largest earthquake recorded at The Geysers was magnitude 4.5. Although there was poor documentation of seismic activity when the plant first started in 1960, high quality seismic data have been available since 1975. From these data, it has been demonstrated that increased steam production and fluid injection correlates positively with earthquake activity. Seismologists have proposed three mechanisms to explain the cause of these earthquakes. First, when steam and water, as well as heat, are extracted from the reservoir, this can cause the surrounding rock to contract, causing a stress that could induce earthquakes. Second, when water is reinjected back into the reservoir at depths of one to three kilometers, this water is relatively cool compared to the rock temperature, and this thermal contrast may induce an earthquake. As mentioned, reinjecting water is necessary to extend the life of the field. Third, it is also possible that the hydraulic pressure of the reinjected water fractures rock when it enters faults.

10.13. CO_2 Production and the Cost of Capture and Sequestration

Since a geothermal power plant does not use any fossil fuel to generate heat, instead using natural geothermal energy, no CO_2 is produced from combustion when the electricity is made. However, some CO_2 is produced during operation, as well as during plant construction. Accordingly, various "life cycle" studies have been made to assess CO_2 produced at different stages of the plant's life.

Two reports from the ANL[52,53] give weighted average data, based on 85% of the geothermal plants in the world, for three types of geothermal plants, namely an EGS, a flash plant, and a binary plant. For the EGS plant, binary technology is used to generate the electricity. For this type of plant, CO_2 emissions were determined to be 23 g CO_2/kWh for construction but zero from operation. Likewise, the binary plant generated 5.7 g CO_2/kWh from construction but zero from operation. The flash plant generated 4.1 g CO_2/kWh from construction and 98.9 CO_2/kWh from operation, for a total of 103.0 CO_2/kWh. Thus, for the flash plant, 4% of the CO_2 emissions came from construction. As mentioned, the hot water extracted from the ground for binary technology stays in a closed loop and is returned to the ground with no venting to the atmosphere. On the other hand, in a flash plant some noncondensable gases go into the atmosphere after they pass through the steam turbine. That is why only the flash plant generates CO_2 emissions from operation.

In a paper by Hondo,[96] a double flash geothermal power plant was modeled, with calculated values of 5.3 CO_2/kWh from construction and 9.7 CO_2/kWh from operation, for a total of 15.0 CO_2/kWh. Breaking this down further, the authors show for the 5.3 CO_2/kWh from construction, 2.0 are from building the foundations, 3.2 from machinery, and 0.1 from exploration. For the 9.7 CO_2/kWh from operation, 2.9 are from drilling additional wells, 2.3 are from general maintenance, and 4.5 are from replacing equipment. This report shows a sharp difference from the Argonne studies, with 35% of the CO_2 emissions coming construction. It is also in better agreement with the value of 27.2 CO_2/kWh shown earlier in Table 10.13.

There are, in fact, many life cycle analysis studies for geothermal power plants. The NREL performed a systematic study of more than 180 reports, finally selecting 29 studies that fit their criteria.[97] A summary of their report is shown in Table 10.14. As before, the EGS plant used binary technology to generate electricity.

As the table shows, there is a significant range between the minimum and maximum for each technology. Differences can be attributed to assumptions about the location and depth of the wells drilled to harness the geothermal energy. In the case of the flash plant, which also shows the largest range, the report states that most of the

Table 10.14 Life Cycle CO_2 Emissions from EGS (binary), Flash, and Binary Geothermal Power Plants[97]

	EGS Binary	Flash	Binary
Minimum (g CO_2eq/kWh)	16.9	15.0	5.6
Median (g CO_2eq/kWh)	32.0	47.0	11.3
Maximum (g CO_2eq/kWh)	79.0	245.2	97.2
Sample Size	18	9	8

variability in CO_2 emissions comes from plant operation and not construction. The variation among the different studies depends on the efficiency of the plant to condense the water taken from the ground and its composition.

Since the NREL study includes the other studies highlighted previously, we use the median values of Table 10.13 to establish emissions for each of the technologies. Thus, 32.0 g CO_2/kWh is used for the EGS plant, 47.0 g CO_2/kWh for the flash plant, and 11.3 g CO_2/kWh for the binary plant. Also, even though the EGS and binary plants use similar, or the same, binary technology to generate electricity, it is logical that CO_2 emissions would be higher from construction for the EGS plant since drilling is done to greater depths.

Carbon dioxide production is commonly listed in the units of lb CO_2/MM BTU. However, geothermal power plants generate electricity without the combustion of a fuel. Because of this, there is no heat rate to convert from CO_2 produced on a mass basis and that produced on an energy basis. The conversion from gCO_2/kWh to lb CO_2/kWh is a simple conversion, resulting in values of 0.071 lb CO_2/kWh for the EGS plant, 0.104 lb CO_2/kWh for the flash plant, and 0.025 lb CO_2/kWh for the binary plant.

For a 1,000 MW plant with a capacity factor of 85%, the plant will generate 20,400 MWh/day. For CO_2 produced on a kWh basis, the life cycle production of CO_2 is equivalent to 1,439,200 lb CO_2/day (720 tons CO_2/day) for EGS, 2,113,800 lb CO_2/day (1,057 tons CO_2/day) for flash, and 508,200 lb CO_2/day (254 tons CO_2/day) for binary. Compared to fossil fuel technologies, this CO_2 production is small. For example, a pulverized coal combustion plant produces 22,214 tons CO_2/day (Chapter 3) and a natural gas combined cycle plant produces 7,500 tons CO_2/day (Chapter 2).

It is possible that CO_2 produced from a dry steam or flash plant could be removed by CO_2 capture and sequestration, but only that from operation and not construction. If 35% of CO_2 emissions are from construction, that means that of the 1,057 tons CO_2 made per day for a flash plant, only about 700 tons CO_2 could be captured by CCS, about 10% of that for natural gas and 3% of that for coal. For EGS using binary technology, or a binary power plant, CCS is not practical as this CO_2 is only made during construction.

10.14. Chapter Summary

While the use of geothermal energy to generate electricity has grown rapidly worldwide since 1950, going from an installed capacity of only 200 MW to more than 14,369 MW in 2018, it is still a relatively small player compared to other types of energy. In the United States, for example, geothermal power plants only accounted for 0.4% of the total electricity generated in 2017, and 2.3% of all types of renewable energy (including hydroelectric). Only seven states generated any geothermal electricity, but two states really dominated this generation, California and Nevada. Overall, 72.6% of all the US electricity generated from geothermal energy came from the state California and 20.7% from Nevada. These percentages were similar to those for installed capacity with California having 73.2% and Nevada 21.4%.

Globally, geothermal electricity only made up 0.3% of generation, but 27 countries had some amount of geothermal electricity in 2017. Seven countries generated 5% or more of their electricity from geothermal energy including Kenya (51%), Iceland (27%), El Salvador (26%), New Zealand (18%), Nicaragua (16%), the Philippines (11%), and Indonesia (5%). Ethiopia and Turkey are also adding large amounts of geothermal power.

The use of geothermal power is expected to have continued growth, as there is potential for 211 GW worldwide. The top three countries in terms of geothermal potential are Indonesia, Japan, and Chile, and these countries are part of the so-called Ring of Fire, a ring of volcanoes around the Pacific Ocean.

The "Well-to-Wheels" economic analysis showed an average LCOE, including transmission and distribution, of $98.8/MWh for a flash plant and $139.0/MWh for a binary plant. For an EGS plant using flash technology, the levelized cost was $261.2/MWh and for an EGS plant using binary technology, the levelized cost was $611.5/MWh. The higher LCOE for EGS is due to greater drilling depths, as much as 5 km (3.1 miles), which results in significantly higher capital costs for construction. Operating cost are also much higher.

The high LCOE for EGS is one of the reasons this technology has not been used to build geothermal power plants with significant size, as the EGS plants currently have installed capacities of less than 5 MW. EGS also suffers from technical challenges, such as induced seismic activity, leading to small earthquakes, and maintaining high temperature steam production from the injection well to the production well.

For a 1,000 MW power plant operating at a capacity factor of 85%, the life cycle production of CO_2 is about 1,000 tons CO_2/day for a flash plant, about 250 tons CO_2/day for a binary plant, and 720 tons CO_2/day for an EGS plant. In other units, EGS emissions are 32.0 g CO_2/kWh, flash emissions are 47.0 g CO_2/kWh, and binary emissions are 11.3 g CO_2/kWh. Emissions are lower for binary and EGS technologies because the hot water taken from the geothermal reservoir is in a closed loop, and therefore no CO_2 is vented to the atmosphere. CO_2 emissions for these two technologies come from plant construction.

For a 1,000 MW geothermal plant with a capacity factor of 85%, the footprint for the geothermal field plus power plant is 16.7 square miles. For the power plant alone, the footprint is 0.74 square miles.

The overall energy balance for a flash geothermal power plant is 14.9%, meaning that about 15% of the geothermal energy taken from the reservoir ends up as energy at the distribution point as electricity. For binary and EGS power plants, the values are similar at 14.8% and 14.7%, respectively. The reason these numbers are so low is due to the lower thermal efficiency of geothermal plants, only about 16%.

In terms of experience and maturity, geothermal power plants using dry steam technology have been around since 1904, followed by binary systems and flash systems in 1967. The concept of EGS has been around as early as 1977. And, the first electrical grid system was built in 1882, followed by rapid development through the 1930s.

The energy density for the combined geothermal field plus power plant is 524 GWh/(year-square mile). However, the energy density for only the power plant is much greater at 11,910 GWh/(year-square mile).

Finally, the main environmental issue for geothermal plants is induced seismicity, leading to earthquakes. Other issues, such as air emissions, noise, water use, and the impact on natural phenomena (geysers, fumaroles, and hot springs), are minor. In fact, emissions are essentially zero for the binary technology, and very small relative to fossil fuel plants for dry steam and flash technologies. With respect to earthquakes, there have been a number of earthquakes resulting from geothermal power plants, with earthquake magnitudes as high as 3.8. These are apparently caused by drilling very deep wells, as much as 5 kilometers (or 3.1 miles), as well as fluid injection into the geothermal field. However, earthquakes ranging from 2.5 to 5.4 can be felt, but cause only minor damage and those less than 2.5 are generally not considered significant.

References

1. "Why Is the Earth's Core So Hot? And How Do Scientists Measure Its Temperature?," Scientific American, October 6, 1997, https://www.scientificamerican.com/article/why-is-the-earths-core-so/
2. "How We Know What Lies at the Earth's Core," Chris Baraniuk, August 13, 2015, BBC, http://www.bbc.com/earth/story/20150814-what-is-at-the-centre-of-earth
3. "Radioactive Decay Accounts for Half of Earth's Heat," July 19, 2011, Physics World, http://physicsworld.com/cws/article/news/2011/jul/19/radioactive-decay-accounts-for-half-of-earths-heat
4. "A History of Geothermal Energy in America," Energy.gov, 2013, https://energy.gov/eere/geothermal/history-geothermal-energy-america
5. "Use of Geothermal Energy," Energy Information Administration, March 25, 2020, https://www.eia.gov/energyexplained/index.cfm?page=geothermal_use
6. "2016 Annual U.S. & Global Geothermal Power Production Report," March 2016, Benjamin Matek, Geothermal Energy Association, http://geo-energy.org/reports/2016/2016%20Annual%20US%20Global%20Geothermal%20Power%20Production.pdf

7. "What Is the Potential of Using Geothermal Resources in the U.S.?," Leslie Blodgett, Geothermal Energy Association, 2014, https://www.geothermal.org/Policy_Committee/Documents/Geothermal_101-Basics_of_Geothermal_Energy.pdf
8. "The Future of Geothermal Energy: The Future of Impact of Enhanced Geothermal Systems (EGS) on the United States in the 21st Century," Massachusetts Institute of Technology, 2006, ISBN: 0-615-13438-6, http://geothermal.inel.gov/publications/future_of_geothermal_energy.pdf
9. "Geothermal Power Generation in the World 2010–2014 Update Report," Ruggero Bertani, Proceedings World Geothermal Congress 2015, Melbourne, Australia, April 19–25, 2015, https://pangea.stanford.edu/ERE/db/WGC/papers/WGC/2015/01001.pdf
10. "Global Geothermal Capacity Reaches 14,369 MW—Top 10 Geothermal Countries," Alexander Richter, Think Geoenergy, September 28, 2018, http://www.thinkgeoenergy.com/global-geothermal-capacity-reaches-14369-mw-top-10-geothermal-countries-oct-2018/
11. "International Renewable Energy Agency (IRENA)," Renewable Electricity Capacity and Generation Statistics, 2019, http://resourceirena.irena.org/gateway/dashboard/?topic=4&subTopic=54
12. "Direct Utilization of Geothermal Energy 2015 Worldwide Review," John W. Lund and Tonya L. Boyd, Proceedings World Geothermal Congress 2015, Melbourne, Australia, April 19–25, 2015, https://www.unionegeotermica.it/pdfiles/usi-diretti-energia-geotermica-nel-mondo.pdf
13. "Worldwide Geothermal Energy Utilization 2015," John W. Lund et al., GRC Transactions 39, 79–92, 2015, http://pubs.geothermal-library.org/lib/grc/1032136.pdf
14. "A History of Geothermal Energy," James Bratley, Clean Energy Ideas, August 24, 2016, https://www.clean-energy-ideas.com/geothermal/geothermal-energy/history-of-geothermal-energy
15. "First Commercial Success for Enhanced Geothermal Systems (EGS) Spells Exponential Growth for Geothermal Energy," Energy.gov, April 15, 2013, https://energy.gov/eere/geothermal/articles/first-commercial-success-enhanced-geothermal-systems-egs-spells-exponential
16. "World Energy Resources Geothermal | 2016," World Energy Council, https://www.world-energy.org/wp-content/uploads/2017/03/WEResources_Geothermal_2016.pdf
17. "Geothermal Power Technology Brief," IRENA (International Renewable Energy Agency), September 2017, http://www.irena.org/-/media/Files/IRENA/Agency/Publication/2017/Aug/IRENA_Geothermal_Power_2017.pdf
18. "What Is Geothermal Energy?," Leslie Blodgett, Geothermal Energy Association, 2014, http://geo-energy.org/Basics.aspx
19. "Chapter 4: Advancing Clean Electric Power Technologies, Geothermal Power, Technology Assessments," Quadrennial Technology Review, 2015, https://energy.gov/sites/prod/files/2015/12/f27/QTR2015-4I-Geothermal-Power.pdf
20. "The Top 10 Biggest Geothermal Power Plants in the World," Praveen Duddu, November 11, 2013, Power Technology, https://www.power-technology.com/features/feature-top-10-biggest-geothermal-power-plants-in-the-world/
21. "Largest Geothermal Power Plants in the World," World Atlas, April 25, 2017, https://www.worldatlas.com/articles/largest-geothermal-power-plants-in-the-world.html
22. "Geothermal Development in the Philippines: The Country Update," Ariel D. Fronda et al., Proceedings World Geothermal Congress 2015, Melbourne, Australia, April 19–25, 2015, http://large.stanford.edu/courses/2016/ph240/makalinao1/docs/01053.pdf
23. "Darajat Unit II/III Interface Debottlenecking Project," Wibisono Yamin et al., Proceedings World Geothermal Congress 2015, Melbourne, Australia, April 19–25, 2015, https://pangea.stanford.edu/ERE/db/WGC/papers/WGC/2015/25035.pdf

24. "Wayang Windu Geothermal Power Plant," Hiroshi Murakami, Fuji Electric Company, 2001, https://www.fujielectric.com/company/tech_archives/pdf/47-04/FER-47-04-102-2001.pdf
25. "330 MW Sarulla Geothermal Plant in Indonesia Completed with Third Unit Online," Alexander Richter, Think Geoenergy, May 8, 2018, http://www.thinkgeoenergy.com/330-mw-sarulla-geothermal-plant-in-indonesia-completed-with-third-unit-online/
26. "Olkaria Geothermal Expansion Project, Rift Valley Province, Kenya," Sonal Patel, Power Magazine, December 1, 2015, https://www.powermag.com/olkaria-geothermal-expansion-project-rift-valley-province-kenya/?pagenum=1
27. "ConstructionofUnit1ofOlkariaVGeothermalPlantinKenyacompleted,"DominicMandela, Construction Review Online, August 1, 2019, https://constructionreviewonline.com/2019/08/construction-of-unit-1-of-olkaria-v-geothermal-plant-in-kenya-completed/
28. "2017 Outlook: Geothermal Is Trending Upwards," Allie Nelson, Renewable Energy World, February 3, 2017, http://www.renewableenergyworld.com/articles/print/volume-20/issue-1/features/geothermal/2017-outlook-geothermal-is-trending-upwards.html
29. "IceLink Interconnector in Operation by 2025?," Askja Energy, April 17, 2018, https://askjaenergy.com/2018/04/17/icelink-in-operation-by-2025/
30. "Ethiopia to Add 1,000 MW Capacity with Two New Geothermal Power Plants," Xinhua, December 21, 2017, http://www.xinhuanet.com/english/2017-12/21/c_136841056.htm
31. "Turkey Reaches Milestone 1,100 MW of Installed Geothermal Power Generation Capacity," Alexander Richter, Think GeoEnergy, January 5, 2018, http://www.thinkgeoenergy.com/turkey-reaches-milestone-1100-mw-of-installed-geothermal-power-generation-capacity/
32. "Enhanced Geothermal Systems Demonstration Projects," Energy.gov, 2014, https://www.energy.gov/eere/geothermal/enhanced-geothermal-systems-demonstration-projects
33. "A Global Review of Enhanced Geothermal System (EGS)," Shyi-Min Lu, Renewable and Sustainable Energy Reviews 81 (Part 2), 2902–2921, January 2018, https://www.researchgate.net/publication/318032275_A_global_review_of_enhanced_geothermal_system_EGS
34. "Habanero Pilot Project—Australia's First EGS Power Plant," Tony Mills and Ben Humphreys, 35th New Zealand Geothermal Workshop: 2013 Proceedings, Rotorua, New Zealand, November 17–20, 2013, file:///C:/Users/paulf/Documents/Book/References_Figures_Tables_Revised/Geothermal/Mills_Final.pdf
35. "Geothermal Heat Pumps," Energy.gov, 2020, https://www.energy.gov/energysaver/heat-and-cool/heat-pump-systems/geothermal-heat-pumps
36. "Ground-Source Heat Pumps: Overview of Market Status, Barriers to Adoption, and Options for Overcoming Barriers," William Goetzler et al., Final Report prepared for the US Department of Energy Efficiency and Renewable Energy, Geothermal Technologies Program, Navigant Consulting, Inc., February 3, 2009, https://www1.eere.energy.gov/geothermal/pdfs/gshp_overview.pdf
37. "Cost and Performance Characteristics of New Generating Technologies," Annual Energy Outlook 2019, Energy Information Administration, Table 2, January, 2019, https://www.eia.gov/outlooks/aeo/
38. "2018 Annual Technology Baseline ATB Cost and Performance Data for Electricity Generation Technologies—Interim Data without Geothermal Updates," National Renewable Energy Laboratory, July 2, 2018, https://data.nrel.gov/submissions/89
39. "Lazard's Levelized Cost of Energy Analysis—Version 12.0," November 2018, https://www.lazard.com/media/450784/lazards-levelized-cost-of-energy-version-120-vfinal.pdf
40. "2015 Annual Technology Baseline ATB Cost and Performance Data for Electricity Generation Technologies—Interim Data without Geothermal Updates," National Renewable Energy Laboratory, 2015, https://atb.nrel.gov/electricity/archives.html

41. "Lazard's Levelized Cost of Energy Analysis—Version 8," September 2014, https://www.lazard.com/media/1777/levelized_cost_of_energy_-_version_80.pdf
42. "Annual Energy Outlook 2015 with projections to 2040," DOE/EIA-0383, Table A8, April 2015, https://www.eia.gov/outlooks/aeo/pdf/0383(2015).pdf
43. "Geo Vision: Harnessing the Heat beneath Our Feet," US Department of Energy, May 2019, https://www.energy.gov/sites/prod/files/2019/05/f63/GeoVision-full-report.pdf
44. "Geothermal Plant Capacity Factors," Greg Mines et al., Proceedings, Fortieth Workshop on Geothermal Reservoir Engineering, Stanford University, Stanford, California, January 26–28, 2015, https://pangea.stanford.edu/ERE/db/GeoConf/papers/SGW/2015/Nathwani.pdf
45. "Table 6.7.B. Capacity Factors for Utility Scale Generators Not Primarily Using Fossil Fuels, Electric Power Monthly, Energy Information Administration, January 2013–June 2019, https://www.eia.gov/electricity/monthly/epm_table_grapher.php?t=epmt_6_07_b
46. "Renewables 2016: Global Status Report," Renewable Energy Policy Network for the 21st Century, http://www.ren21.net/wp-content/uploads/2016/06/GSR_2016_Full_Report.pdf
47. "Alternatives for Estimating Energy Consumption," Appendix F, Annual Energy Review 2010, https://www.eia.gov/totalenergy/data/annual/pdf/sec17.pdf
48. "Ideal Thermal Efficiency for Geothermal Binary Plants," Ronald DiPippo, Geothermics 36 (2007) 276–285, May 4, 2007, https://www.sciencedirect.com/science/article/pii/S0375650507000375
49. "An Analysis of Thermodynamic Properties in Both Traditional and Enhanced Geothermal Systems to Compare Thermal Efficiencies," Samuel Martin et al., University of Technology Sydney, PAM Review: Energy Science & Technology, 4, 2017, https://epress.lib.uts.edu.au/student-journals/index.php/PAMR/article/view/1446/1542
50. "Efficiency of Geothermal Power Plants: A Worldwide Review," Hyungsul Moon and Sadiq J. Zarrouk, New Zealand Geothermal Workshop 2012 Proceedings, Auckland, New Zealand, November 19–21, 2012, https://www.geothermal-energy.org/pdf/IGAstandard/NZGW/2012/46654final00097.pdf
51. "Production of Electricity from Geothermal Energy," Geothermal Systems and Technologies, Chapter 7.2, Power from GE, 2013, http://www.geothermalcommunities.eu/assets/elearning/7.2.Power_from_GE.pdf
52. "Life-Cycle Analysis Results of Geothermal Systems in Comparison to Other Power Systems," J.L. Sullivan et al., Argonne National Laboratory, August 2010, ANL/ESD/10-5, http://www.evs.anl.gov/downloads/ANL_ESD_10-5.pdf
53. "Life-Cycle Analysis Results for Geothermal Systems in Comparison to Other Power Systems: Part II," J.L. Sullivan et al., Argonne National Laboratory, ANL/ESD/11-12, November 2011, http://www.anl.gov/energy-systems/publication/life-cycle-analysis-results-geothermal-systems-comparison-other-power
54. "A Review of Transmission Losses in Planning Studies," Lana Wong, California Energy Commission, Figure ES-1, August 2011.
55. "Monthly Energy Review," Energy Information Administration, July 2016, https://www.eia.gov/totalenergy/data/monthly/pdf/mer.pdf
56. "Geothermal Power Plants: Principles, Applications, Case Studies and Environmental Impact," 3rd Edition, 2012, Ronald DiPippo, Elsevier, Oxford. https://www.u-cursos.cl/usuario/c658fb0e38744551c1c51c640649db2e/mi_blog/r/Geothermal_Power_Plants.pdf
57. "The Soultz-sous-Forêts' Enhanced Geothermal System: A Granitic Basement Used as a Heat Exchanger to Produce Electricity," Béatrice A. Ledésert and Ronan L. Hébert, 2012, Géosciences et Environnement Cergy, Université de Cergy-Pontoise France, ISBN: 978-953-51-0278-6, http://www.intechopen.com/books/heat-exchangers-basics-design-

applications/the-soultz-sous-for-tsenhanced-geothermal-system-a-granitic-basement-used-as-a-heat-exchanger-to-pr

58. "United States Electricity Industry Primer," Office of Electricity Delivery and Energy Reliability U.S. Department of Energy DOE/OE-0017, July 2015, file:///C:/Users/paulf/Documents/Book/Chapters_for_OUP/New_Gen_Documents/united-states-electricity-industry-primer.pdf
59. "Electric Power Annual 2017," Energy Information Administration, October 2018, Revised December, 2018, https://www.eia.gov/electricity/annual/pdf/epa.pdf
60. "Electricity: Form EIA-860 detailed data," Energy Information Administration, 2015 data, https://www.eia.gov/electricity/data/eia860/
61. "BP Statistical Review of World Energy 2019 | 68th edition," https://www.bp.com/content/dam/bp/business-sites/en/global/corporate/pdfs/energy-economics/statistical-review/bp-stats-review-2019-electricity.pdf
62. "Energy Statistics in Iceland 2017," Orkustofnun, 2018, https://orkustofnun.is/gogn/os-onnur-rit/Orkutolur-2017-enska-A4.pdf
63. "Nicaragua, ClimateScope 2017, http://2017.global-climatescope.org/en/country/nicaragua/#/enabling-framework
64. "Electricity Generation by Fuel: El Salvador 1990–2016," International Energy Agency, 2016, https://www.iea.org/statistics/?country=ELSALVADOR&year=2016&category=Electricity&indicator=ElecGenByFuel&mode=chart&dataTable=ELECTRICITYANDHEAT
65. "Electricity Generation by Fuel: Kenya 1990–2016," International Energy Agency, 2016, https://www.iea.org/statistics/?country=KENYA&year=2016&category=Electricity&indicator=ElecGenByFuel&mode=chart&dataTable=ELECTRICITYANDHEAT
66. "Geothermal Power and Interconnection: The Economics of Getting to Market," David Hurlbut, Technical Report NREL/TP-6A20-54192, April 2012, https://www.nrel.gov/docs/fy12osti/54192.pdf
67. "Transmission Line Expansion Called 'Game Changer' for Clean Energy in Nevada," Las Vegas Review-Journal, Sean Whaley, January 12, 2016, https://www.reviewjournal.com/business/energy/transmission-line-expansion-called-game-changer-for-clean-energy-in-nevada/
68. "New Transmission Line to Deliver Geothermal Power to Port Town of Mombasa, Kenya," Alexander Richter, Think Geoenergy, February 20, 2017, http://www.thinkgeoenergy.com/new-transmission-line-to-deliver-geothermal-power-to-port-town-of-mombasa-kenya/
69. "New Transmission Lines Strengthening Power Delivery from Leyte Plants," Alexander Richter, Think Geoenergy, February 26, 2016, http://www.thinkgeoenergy.com/new-transmission-lines-strengthening-power-delivery-from-leyte-plants/
70. "Turkey Ramping Up Investment of $3.5 Billion in Transmission Lines," Francisco Rojas, Think Geoenergy, April 28, 2016, http://www.thinkgeoenergy.com/turkey-ramping-up-investment-of-3-5-billion-in-transmission-lines/
71. "Why Is Geothermal Energy a Renewable Resource? Can It Be Depleted?," Drew L. Siler, American Geoscience Institute, 2020, https://www.americangeosciences.org/critical-issues/faq/why-geothermal-energy-renewable-resource-can-it-be-depleted
72. "Geothermal Power Plants — Minimizing Land Use and Impact," Energy.gov, 2020, https://energy.gov/eere/geothermal/geothermal-power-plants-minimizing-land-use-and-impact
73. "Geothermal Basics—Environmental Benefits," Geothermal Energy Association, September 2012, https://www.geothermal.org/PDFs/Geothermal_Basics_QandA.pdf
74. The Future of Geothermal Energy the Future of Impact of Enhanced Geothermal Systems (EGS) on the United States in the 21st Century, 2006, http://geothermal.inel.gov/publications/future_of_geothermal_energy.pdf
75. "Geysers by the Numbers: The Geysers Geothermal Field 2016 Statistics," 2016, https://geysers.com/The-Geysers/Geysers-By-The-Numbers

76. "Geothermal Power Plants of Italy: A Technical Survey of Existing Installations," R. DiPippo, October 1978, https://www.osti.gov/servlets/purl/6070343
77. "Geothermal Power Plants of Mexico and Central America: A Technical Survey of Existing and Planned Installations," Ronald DiPippo, Department of Energy, 1978, https://books.google.com/books?id=W3CbssgJjVYC&pg=PA3&lpg=PA3&dq=Cerro+Prieto+geothermal+plant+acres&source=bl&ots=GQm5Xl_tdk&sig=9LwFVjV3vLD64fprxVDU
78. "Geothermal Energy Grows in Kenya," Amy Yee, February 23, 2018, New York Times, https://www.nytimes.com/2018/02/23/business/geothermal-energy-grows-in-kenya.html
79. "Mak-Ban Geothermal Field, Philippines: 30 Years of Commercial Operation," V. Capuno et al., Proceedings World Geothermal Congress 2010, Bali, Indonesia, April 25–29, 2010, https://www.geothermal-energy.org/pdf/IGAstandard/WGC/2010/0649.pdf2010
80. "A Path to Increasing Geothermal Development in California" Randy Keller, October 20, 2016, California Geothermal Forum, BHE Renewables, https://geothermal.org/PDFs/California_Geothermal_Forum/Session_2_1_Keller.pdf
81. "The Hellisheidi Geothermal Project—Financial Aspects of Geothermal Development, Einar Gunnlaugsson, Presented at Short Course on Geothermal Development and Geothermal Wells, Santa Tecla, El Salvador, March 11–17, 2012, https://orkustofnun.is/gogn/unu-gtp-sc/UNU-GTP-SC-14-12.pdf
82. "Tiwi Geothermal Field, Philippines: 30 Years of Commercial Operation," A. Menzies et al., Proceedings World Geothermal Congress 2010, Bali, Indonesia, April 25–29, 2010, https://www.geothermal-energy.org/pdf/IGAstandard/WGC/2010/0648.pdf
83. "Repeat Microgravity and Leveling Surveys at Leyte Geothermal Production Field, North Central Leyte, Philippines," Nilo A. Apuada et al., Proceedings, Thirtieth Workshop on Geothermal Reservoir Engineering Stanford University, Stanford, California, January 31–February 2, 2005, SGP-TR-176, https://pangea.stanford.edu/ERE/pdf/IGAstandard/SGW/2005/apuada.pdf
84. "Fifteen Years (Mid-Life Time) of Wayang Windu Geothermal Power Station Unit-1: An Operational Review," M. Purnanto and A. Purwakusumah, Proceedings World Geothermal Congress 2015, Melbourne, April 19–25, 2015, https://pangea.stanford.edu/ERE/db/WGC/papers/WGC/2015/26040.pdf
85. "Utilization of Geothermal Energy in Protected Areas of Costa Rica," Paul Moya et al., Workshop for Decision Makers on Geothermal Projects in Central America, San Salvador, November 26–December 2, 2006, https://orkustofnun.is/gogn/flytja/JHS-Skjol/El%20Salvador%202006/10_MoyaUtilization.pdf
86. "An Update of the Los Azufres Geothermal Field, after 21 Years of Exploitation," M. A. Torres-Rodríguez et al., Proceedings World Geothermal Congress, Antalya, Turkey, April 24–29, 2005, https://www.geothermal-energy.org/pdf/IGAstandard/WGC/2005/0916.pdf
87. "Los Azufres Geothermal Power Plant Mexico," Global Energy Observatory, January 1, 2015, http://globalenergyobservatory.org/geoid/4077
88. "Electric Power Consumption (kWh per capita)," IEA Statistics for 2016, http://energyatlas.iea.org/#!/tellmap/-1118783123/1
89. "Environmental Impacts of Geothermal Energy," Geothermal Communities, 2013, http://www.geothermalcommunities.eu/assets/elearning/8.1.GE%20vs%20Environment.pdf
90. "Characteristics, Development and Utilization of Geothermal Resources," John W. Lund, 2007, Geo-Heat Center, Oregon Institute of Technology, https://pdfs.semanticscholar.org/084d/ecf2a958f35fa12525118125340l600a5708.pdf
91. "Swiss Geothermal Power Plan Abandoned after Quakes Hit Basel," Adam Gabbatt, The Guardian, December 15, 2009, https://www.theguardian.com/world/2009/dec/15/swiss-geothermal-power-earthquakes-basel

92. "Assessing the Earthquake Risk of Enhanced Geothermal Systems," Sonal Patel, Power Magazine, December 1, 2009, http://www.powermag.com/assessing-the-earthquake-risk-of-enhanced-geothermal-systems/
93. "On the Effects of Induced Earthquakes Due to Fluid Injection at Hellisheidi Geothermal Power Plant, Iceland," B. Halldorsson et al., Earthquake Engineering Research Centre, University of Iceland, Selfoss, Iceland, 15th World Conference on Earthquake Engineering, 2012, http://www.iitk.ac.in/nicee/wcee/article/WCEE2012_4069.pdf
94. "Earthquake Magnitude Scale," 2020, http://www.geo.mtu.edu/UPSeis/magnitude.html
95. "Why Are There So Many Earthquakes in the Geysers Area in Northern California?," US Geological Survey, 2020, https://www.usgs.gov/faqs/why-are-there-so-many-earthquakes-geysers-area-northern-california?qt-news_science_products=7#qt-news_science_products
96. "Life Cycle GHG Emission Analysis of Power Generation Systems: Japanese Case," H. Hondo, Energy 30 (2005) 2042–2056.
97. "Systematic Review of Life Cycle Greenhouse Gas Emissions from Geothermal Electricity," Annika Eberle et al., National Renewable Energy Laboratory, Technical Report NREL/TP-6A20-68474, September 2017, https://www.nrel.gov/docs/fy17osti/68474.pdf

11
Hydrogen
A Renewable or Nonrenewable Energy Type

11.1. Foreword

With the exception of Chapter 1 ("Petroleum Crude Oil"), Chapter 8 ("Ethanol"), and Chapter 12 ("Fischer-Tropsch Synthesis"), the other chapters in this book discuss the use of energy to generate electricity. Although some crude oil is used to generate electricity, its main application is for making transportation fuels. And, for ethanol and the Fischer-Tropsch synthesis, the energy is used exclusively for transportation fuels.

However, with the emergence of electric cars, the separation between energy for electricity and energy for transportation fuels is disappearing. While conventional vehicles are powered by gasoline or diesel in an internal combustion engine (ICE), electric vehicles (EVs) are powered by electric motor(s) using electricity stored in a battery in the form of chemical energy.

An alternative to an EV is the fuel cell vehicle (FCV), which is also powered by an electric motor but uses onboard storage of a fuel instead of a battery. Simply stated, EVs use stored chemical energy inside the battery that is converted to electricity and FCVs use fuel external to the battery to produce electricity. In this case, that fuel is hydrogen and the use of hydrogen in an FCV is the topic of this chapter.

Before we understand the reasons for using an electric powered vehicle which uses hydrogen instead of a battery, it is important review EVs and understand the motivation for a hydrogen-powered FCV. Currently, there are quite a few EVs on the market, and Table 11.1 compares data from a review of 11 different EVs.[1] If you take the EPA-certified range in miles and divide by the kWh of energy stored in the battery, you get the calculated values in the last column. As shown, these range from 2.4 to 4.1 miles/kWh. So is this good or bad? Well, the equivalent energy content for an average gallon of gasoline is 33.4 kWh, or stated differently, 33.4 kWh is one gallon of gasoline equivalent (gge). Assuming your gasoline-powered vehicle gets 25 miles/gallon, the calculated efficiency for an ICE is only 0.75 miles/kWh. Another way to look at this is to take a rough average for EVs of 3.5 miles/kWh and multiply by 33.4 kWh/gallon, which gives a value of 117 miles/gallon. So EVs are much more efficient with their energy, about four times more efficient than an ICE in this example.

Another important aspect for the data in Table 11.1 is the vehicle driving range. With the exception of the Chevrolet Bolt EV and Tesla Models S and X, the vehicle ranges vary from 68 to 151 miles. It is important to note that the Chevrolet Bolt EV

The Changing Energy Mix. Paul F. Meier, Oxford University Press (2020). © Oxford University Press.
DOI: 10.1093/oso/9780190098391.001.0001.

Table 11.1 Electric Vehicle Data[1]

Vehicle	Description	EPA-Certified Range (miles)	Miles/ kWh
BMW i3	5-door hatchback, 22 kWh or 33 kWh battery pack	81 or 114	3.7, 3.5
Chevrolet Bolt EV	Compact crossover with 60kWh lithium-ion battery. Can be recharged in approximately 9 hours using Level Two 240-volt/32 amp charging.	238	4.0
Fiat 500e	All electric version of Fiat 500 powered by a 24 kWh lithium-ion battery. On a 240V circuit will charge from empty to full in about 4 hours.	87	3.6
Ford Focus Electric	Powered by 33.5 kWh of Li-ion batteries.	115	3.4
Kia Soul EV	All electric version of Kia's popular Soul. Has 27kWh battery pack.	111	4.1
Mercedes B250e	All-electric hatchback based on Mercedes' B-Class body style powered by a 36 kWh battery pack. Will recharge in approximately 3 hours on Level 2 charging.	87	2.4
Nissan LEAF	Five-seater hatchback, powered by a 40 kWh battery pack. Can be recharged in 8 hours on a 6.6 kW Level 2 charger.	151	3.8
Smart ED	Powered by a 17.6 kWh lithium ion battery pack, but does not feature fast charging.	68	3.9
Tesla Model S	4-door sedan with four choices of battery capacity: 60 kWh (210 miles), 75 kWh (249 miles), 90 kWh pack (294 miles), and 100 kWh (315 miles). Recharging at home will take 4 hours from a Tesla charging station.	210–335	3.5, 3.3, 3.3, 3.2
Tesla Model X	All-electric crossover that seats 7 adults. Has three battery packs including 75kWh (237 miles), 90kWh (257 miles), and 100kWh (295 miles)-mile range.	237–295	3.2, 2.9, 3.0
Volkswagen e-Golf	All electric version of Volkswagen's popular Golf hatchback, powered by a 36 kWh battery pack.	125	3.5

and Tesla models get about the same miles/kWh, so their greater driving range is due to the higher rated battery packs. Now let's compare that to a gasoline-powered vehicle. For example, my Subaru Outback gets about 30 miles per gallon on the highway, and has a gasoline tank capacity of 18 gallons. That means I have a range of 540 miles. Clearly, then, one drawback of EVs is that they are, for the most part, vehicles that can only be used for short commutes. Even the Chevrolet Bolt EV and Tesla models only have about half the range of my Subaru.

So, if you do take your EV on a road trip, you are going to want to recharge the battery, before it has no more charge, after about 150–200 miles with the Bolt EV or Tesla models, just as most of us look to refill our automobiles when we get to about ¼ of a tank of gasoline. When you pull your gasoline-powered vehicle into the filling station, you probably expect the stop to last about 5 to 10 minutes. However, charging an electric battery is a much longer process. Without going into a wieldy discussion about the theory of battery charging, we can again use Table 11.1 to make some estimates. Charging times are given for the Chevy Bolt, Fiat 500e, Mercedes B250e, and Nissan LEAF. If we take those times and calculate how many miles of range are added for each hour of charging we get 26, 22, 29, and 19 for the Bolt, Fiat, Mercedes, and LEAF. In other words, for every hour of charging, about 25 miles of range is added. And this is for a Level Two 240-volt charging system. If you simply use your 110 V Level One home outlet in your garage, the charging times will more than double according to Consumer Reports.[2] As this article also mentions, there are other home charging systems, such as Tesla's High Power Wall Connector, which will charge the battery faster. If you are on the road, there are also direct current (DC) fast chargers, which can power up to 80% of the battery's capacity in about 20 to 30 minutes. At this time, however, there are only around 300 DC fast chargers available in the United States. And the Tesla supercharger network has over 500 stations around the United States which can add 100 miles of range in as little as 20 minutes. Compare this to the National Association of Convenience Stores (NACS)[3] 2019 report that there are more than 150,000 gasoline and diesel filling stations in the United States and you can see there is currently a lot less infrastructure for EVs.

If you want to enjoy the increase in fuel efficiency with a vehicle powered by an electric motor, refueling time for EVs is a big problem when you go for a long trip. So an FCV becomes a useful alternative, because using hydrogen to power the electric motor would resolve the issue of refueling time.

In an effort to reduce the amount of crude oil used in the United States, the George W. Bush administration started a Freedom Car initiative in January 2002, which was later expanded in September 2003.[4] Partners in the initiative included the US Department of Energy (DOE), BP America, Chevron Corporation, ConocoPhillips, Exxon Mobil Corporation, Shell Hydrogen LLC, and the United States Council for Automotive Research (USCAR). The latter is a legal partnership between DaimlerChrysler Corporation, Ford Motor Company, and General Motors Corporation. A main goal of the partnership was to examine and advance technologies to produce affordable hydrogen FCVs, as well as developing a national infrastructure to support the use of hydrogen. As a representative of ConocoPhillips, I served on this partnership from 2006 to 2008 as a member of the team to examine technology for onboard hydrogen storage. This and other aspects of a hydrogen FCV are discussed later in Section 3, "Overview of Technology."

An important issue for the success of FCVs is the source for the hydrogen fuel. Although hydrogen is quite abundant in nature, in hydrocarbons and water, it does not exist in large quantities in the pure gaseous form. Therefore, the manufacture and distribution of hydrogen are very important elements for using hydrogen as a fuel in FCVs. Other than its use in an FCV, hydrogen is manufactured today to serve various purposes, and can be made from reforming natural gas, coal gasification, and biomass gasification. It can also be made by electrolysis of water. These manufacturing routes are discussed later in the overview of technology. Some of the uses for hydrogen include food processing, treating metals, the production of methanol, the production of hydrochloric acid, and cryogenics. On rare occasions, it is also used as rocket fuel. The greatest uses for hydrogen, though, are the manufacture of ammonia to make fertilizer and as a feedstock in various refinery processes.

11.2. Proven Reserves

A description of the fuel cell technology are detailed in the next section, but basically hydrogen is mixed with oxygen to make water and generate electric current, which is in turn used to power the vehicle electric motor. The hydrogen can be produced either by water electrolysis or as a product of reforming or gasifying fossil fuels, such as natural gas, coal, or biomass. Equation (1) shows the equation for water electrolysis, and at standard conditions, 25°C (77°F) and atmospheric pressure (1 bar of pressure), 285.8 kJ/mole of energy is needed to break the water molecule into hydrogen and oxygen. The reverse of Equation (1) is the fuel cell equation, so the same amount of energy is released when hydrogen and oxygen react to make water, assuming the same temperature and pressure.

$$H_2O + 285.8\ kJ/mole \rightarrow H_2 + \tfrac{1}{2}\, O_2 \qquad (1)$$

Since water is a product of the fuel cell, that means if hydrogen was made from water electrolysis, the worldwide reserves to make hydrogen are basically unlimited. Naturally, if the energy required to perform the electrolysis comes from fossil fuels, there is a limit from an energy standpoint, but renewable energy can also be used to perform the electrolysis.

Alternatively, if the hydrogen was made from natural gas or coal, there is a limit on the amount of hydrogen that can be made, as these are not renewable resources. Hydrogen can also be made from biomass, which is renewable. In this section, we examine the potential reserves for hydrogen from a standpoint of natural gas, coal, and biomass.

According to the Office of Energy Efficiency and Renewable Energy,[5] about 95% of the hydrogen currently produced in the United States is made from natural gas reforming. As we learned in Chapter 2 (see Table 2.1), natural gas contains about 95%

methane (CH_4). In that chapter, a British Petroleum World Energy report[6] was used to quantify world natural gas reserves of 6,832 trillion cubic feet and US reserves of 309 trillion cubic feet in 2017 (see Table 2.2). Reforming natural gas is also called steam methane reforming (SMR), as steam and methane are the reactants. Equation (2) shows this in equation form.

$$CH_4 + H_2O\,(+\text{ heat}) \rightarrow CO + 3H_2 \tag{2}$$

Reaction (2) takes place using a catalyst, and typical operating conditions are temperatures of 700°C to 1,000°C and pressures of 3 to 25 bar (44–363 psi).[5] The reaction is endothermic, meaning that heat must be added to make the reaction happen. To further increase the yield of hydrogen, a second reaction called the "water-gas shift reaction," or WGS, is used. This is shown in Equation (3).

$$CO + H_2O \rightarrow CO_2 + H_2\,(+\text{heat}) \tag{3}$$

This reaction is slightly exothermic, meaning it gives off heat. Combining Equations (2) and (3) gives the overall balance, shown in Equation (4). Consequently, one mole of methane makes four moles of hydrogen (H_2).

$$CH_4 + 2\,H_2O \rightarrow 4H_2 + CO_2 \tag{4}$$

Unfortunately, chemical reactions and processes are not 100% efficient, so in commercial operation the yield is not four moles of hydrogen for one mole of methane. Although the details are given in the next section, one production case for methane to hydrogen is given here so that an estimate can be made for the amount of hydrogen that can be produced from natural gas reserves. One such pathway is to produce the hydrogen at the refueling site, also called "distributed natural gas." Energy balance data from an NREL report[7] show that 159,000 BTU of natural gas energy plus 11,000 BTU of electrical energy will yield 116,000 BTU of hydrogen with a loss of 54,000 BTU of energy. The energy content chosen for the hydrogen product is the same energy content of a gallon of gasoline, or a gallon of gasoline equivalent (gge). This energy balance shows that 73% of the energy in the natural gas ends up as hydrogen energy.

Knowing that a gge of hydrogen is 116,000 BTU, the percent of natural gas energy going to hydrogen is 73%, and a cubic foot of natural gas is 1,032 BTU, it is possible to calculate the amount of hydrogen production from the worldwide and US natural gas reserves. Also, in terms of mass, a gge of hydrogen is 0.997 kg, which is rounded to 1 kg for the calculations. Assuming that all natural gas reserves are devoted to hydrogen production, worldwide reserves could yield 4.43×10^{13} kg of hydrogen, or 44,300 million metric tonnes (MMT). For the US natural gas reserves, the yield is 2.01×10^{12} kg of hydrogen, or 2,010 MMT.

Next, an estimate for the production of hydrogen from coal reserves is made. As we learned in Chapter 3, a BP World Energy report[6] was used to quantify world coal gas reserves of 1,140,904 million tons and US reserves of 276,587 million tons in 2017 (see Table 3.2). To manufacture hydrogen from coal, a process called coal gasification is used. Generally speaking, gasification is a process that uses oxygen to convert fossil fuels into carbon monoxide (CO), hydrogen (H_2), and carbon dioxide (CO_2). Although coal has small amounts of hydrogen, for example bituminous coal has 4 to 5% hydrogen, Equation (5) is shown in simplified form in which all of the coal is represented as carbon.

$$3C + O_2 + H_2O \rightarrow H_2 + 3CO \tag{5}$$

As was the case with SMR, the WGS reaction is used to further increase the yield of hydrogen, shown again in Equation (6).

$$CO + H_2O \rightarrow CO_2 + H_2 \tag{6}$$

Combining Equations (5) and (6) gives the overall balance, shown in Equation (7). Overall, the combined equation shows that three moles of coal yields four moles of hydrogen (H_2).

$$3C + O_2 + 4H_2O \rightarrow 4H_2 + 3CO_2 \tag{7}$$

As was the case with SMR, coal gasification is not a 100% efficient process. The same NREL report referenced earlier also describes a process referred to as "central coal," such that the coal gasification is done offsite before the hydrogen is delivered to the refueling station. For this case, coal is delivered by rail to the coal gasification plant and the produced hydrogen is sent to the refueling station by pipeline. Carbon capture and sequestration (CCS) is also part of the coal gasification plant. The NREL report[7] energy balance shows that 213,000 BTU coal and 19,000 BTU of electricity energy yield 116,000 BTU of hydrogen, with an energy loss of 116,000 BTU. This energy balance shows that 54.5% of the energy in the coal ends up as hydrogen energy.

Recalling from Chapter 3 that 12,712 BTU/lb was used to represent the energy content of bituminous coal, that a gge of hydrogen is 116,000 BTU or 1 kg, and that the percent of coal energy going to hydrogen is 54.5%, it is possible to calculate the amount of hydrogen production from worldwide and US coal reserves. Assuming that all coal reserves are devoted to hydrogen production, worldwide coal reserves could yield 1.36×10^{14} kg of hydrogen, or 136,000 million metric tonnes (MMT). For the United States, the yield is 3.30×10^{13} kg of hydrogen, or 33,000 MMT.

In Chapter 9 on biomass, the worldwide potential for biomass was given as 232 Exajoules/year (1×10^{18} Joules). Also, a typical heat content for wood and wood waste is 6,500 BTU/lb. In the United States, about two-thirds of electricity from biomass comes from wood and wood waste, so this heat content is used for the biomass heat content, recognizing that biomass can also come from agricultural resources and residues (such as corn and corn stover), energy crops (such as miscanthus), and waste (such as municipal solid waste). Working through the math, this translates into 16,915 million tons per year of biomass. Note, this is in units of tons (2,000 lb) and not metric tonnes (1,000 kg). Also in Chapter 9, the potential biomass for the United States was given as 1,147 million dry tons per year (see Table 9.2).

To manufacture hydrogen from biomass, gasification is used, and the chemistry is similar to that shown earlier in Equations (5) and (6). The NREL report used before[7] describes a process of "central biomass," such that the gasification is done offsite and the hydrogen is delivered to the refueling station by truck transport. Truck transport is used because it is expected to be lower volume than coal, and located within a 50-mile radius of the refueling station. As was the case with methane reforming and coal gasification, it is not a 100% efficient process for biomass. The NREL report[7] energy balance shows that 271,000 BTU of biomass, 36,000 BTU of electricity energy, and 7,000 BTU of natural gas yield 116,000 BTU of hydrogen, with an energy loss of 198,000 BTU. This energy balance shows that 42.8% of the energy in the biomass ends up as hydrogen energy.

Using the worldwide potential of 16,915 million tons/year and US potential of 1,147 million tons/year, the 42.8% energy efficiency, and an energy content of 6,500 BTU/lb for wood and wood waste, the worldwide biomass reserves could yield 8.151×10^{11} kg of hydrogen per year, or 815 million metric tonnes (MMT/year). For the United States, the yield is 5.53×10^{10} kg of hydrogen per year, or 55 MMT/year.

To summarize worldwide reserves, natural gas could yield 44,300 million metric tonnes (MMT) of hydrogen, coal could yield 136,000 MMT, and biomass could yield 815 MMT/year. For the United States, natural gas could yield 2,010 MMT, coal could yield 33,000 MMT, and biomass could yield 55 MMT/year. Note that the biomass is a yearly yield while the yields for natural gas and coal are based on total existing reserves.

To put these numbers in perspective, according to the Energy Information Administration (EIA),[8] the United States used 142.86 billion gallons of gasoline in 2018, which is about 9.3 million barrels per day (one barrel is 42 gallons). As mentioned, one kilogram (1 kg) of hydrogen has the same energy content as a gallon of gasoline (1 gge). Thus, for example, if we use the value calculated for the potential production of hydrogen from US natural gas reserves, 2.01×10^{12} kg of hydrogen, recognize that one kg of hydrogen is equivalent to one gge of gasoline, and divide by the annual US consumption of gasoline, we can calculate a 14-year supply of gge for hydrogen made from US natural gas reserves. For US coal reserves, the calculated value is a much larger supply, a 231 year supply. And, for US biomass reserves, the yearly potential production would be equivalent to 39% of the annual consumption, a supply of 4 to 5 months. Of course, once the natural gas or coal is consumed, it is lost forever, but the biomass is renewable.

And, although one kg of hydrogen is considered one gge on a gasoline equivalent energy content, fuel economy is expected to be better for a hydrogen-powered FCV.

According to a NREL report,[7] the estimated fuel economy for an FCV is 45 miles per hydrogen gge. By comparison, according to the EPA the fuel economy of new US cars and trucks in 2016 was 24.7 miles per gallon.[9] Although this was a record high, it is still only 55% of the estimated fuel economy for the hydrogen-powered FCV.

Needless to say, natural gas, coal, and biomass resources compete with other uses, such as electricity generation, home heating, and ethanol manufacture. These evaluations do show, however, the potential to make hydrogen from existing reserves of biomass and nonrenewable natural gas and coal.

Next, we examine current global and US hydrogen production. To understand the hydrogen market, it helps to understand terminology to distinguish between different types of hydrogen, namely "captive" and "merchant." Captive hydrogen is defined as hydrogen produced and used onsite, which is typical for a refinery. On the other hand, merchant hydrogen is produced at some central site, and then sold to a customer, transported by pipeline or truck. Hydrogen production data for 2014 are shown in Table 11.2, using data taken from a DOE report[10] and a research report.[11] As the table illustrates, US production was about 9 to 10 million metric tonnes per year (MMT) while worldwide production was about 62 MMT. Although these data are a little out-of-date, US hydrogen production in 2018 was about 10 MMT, with similar percentages for refinery use and fertilizer production.[12] Unfortunately, the more recent data did not provide the details shown in Table 11.2.

Table 11.2 Hydrogen Production for 2014 (million metric tonnes)[10,11]

	United States[10]	United States[10,11]	World[11]
Captive Production	5.26	5.86	55.39
Refineries	3.6	3.4	10.08
Ammonia	1.27	2.03	31.81
Methanol	0.11	0.43	13.5
Other	0.28		
Merchant	3.77	3.83	6.16
Refineries	2.64	3.19	4.95
Ammonia	0.87		
Other	0.26	0.64	1.21
Total	9.03	9.69	61.55
% Refineries	69.1	68.0	24.4
% Ammonia	23.7	20.9	51.7
% Methanol	1.2	4.4	21.9
% Other	6.0	6.6	2.0
% Total	100.0	100.0	100.0

Looking at percentages for the United States, almost 70% of the hydrogen was produced and consumed by refineries while about 22% went to ammonia. Worldwide, more than 50% went to ammonia production with refineries consuming about 25%. Considering just the merchant hydrogen, the two previous sets of data for the United States show 70 and 83%, respectively, going to refineries. According to a report by Brown,[13] the amount of merchant hydrogen has steadily increased since 1990 due to increases in hydrogen demands from refineries. The increase in demand is due to the requirement of lower sulfur levels for gasoline and diesel, to meet the Clean Air Act Amendment, as well as the use of heavier crude oils, which require more hydrogen for upgrading into lighter, more valuable, products. Worldwide, about 80% of merchant hydrogen is used in refineries.

If the entire US production of hydrogen, about 10 MMT, was devoted to powering FCVs, this 10 MMT is equivalent to the energy content of 10 billion gallons of gasoline (gge), only 7% of the 142.86 billion gallons of gasoline the United States consumed in 2018. Thus, it is clear a step change in hydrogen production is needed if this fuel is to make an impact on the transportation fuel market.

11.3. Overview of Technology

This section covers several topics including routes for producing hydrogen, fuel cell technology and FCVs, and options for storing hydrogen on board the FCV.

We start by discussing different routes for producing hydrogen. A 2013 NREL report[14] evaluated ten different routes for producing and delivering hydrogen. These routes include (1) distributed natural gas reforming, (2) distributed ethanol reforming, (3) distributed electrolysis using grid electricity, (4) central biomass gasification delivered by pipeline, (5) central biomass delivered by truck as a gas, (6) central biomass delivered by truck as a liquid and dispensed as a gas, (7) central biomass delivered as a liquid and dispensed as a cryo-compressed liquid, (8) central natural gas delivered by pipeline, (9) central electrolysis using wind electricity and delivered by pipeline, and (10) central coal gasification delivered by pipeline. "Distributed" means the hydrogen is produced at the refueling site while "central" means the hydrogen is produced offsite and then delivered to the refueling site (by pipeline or truck). Thus, options (1) to (3) involve producing hydrogen at the refueling site while (4) to (10) involve producing hydrogen offsite. To give a sampling of the different approaches and to be consistent with the proven reserves cases in the previous section, options (1), (5), (9), and (10) are reviewed. A simplified schematic depicting some of these routes, taken from a guide published by the California Fuel Cell Partnership,[15] is shown in Figure 11.1. This schematic presents a simple picture of how different sources of energy, such as natural gas, biomass, and electricity, can produce hydrogen for the FCV.

Option (1), distributed natural gas reforming was discussed in the previous section and is reviewed again here. According to the Office of Energy Efficiency and Renewable Energy,[5] about 95% of the hydrogen currently produced in the United

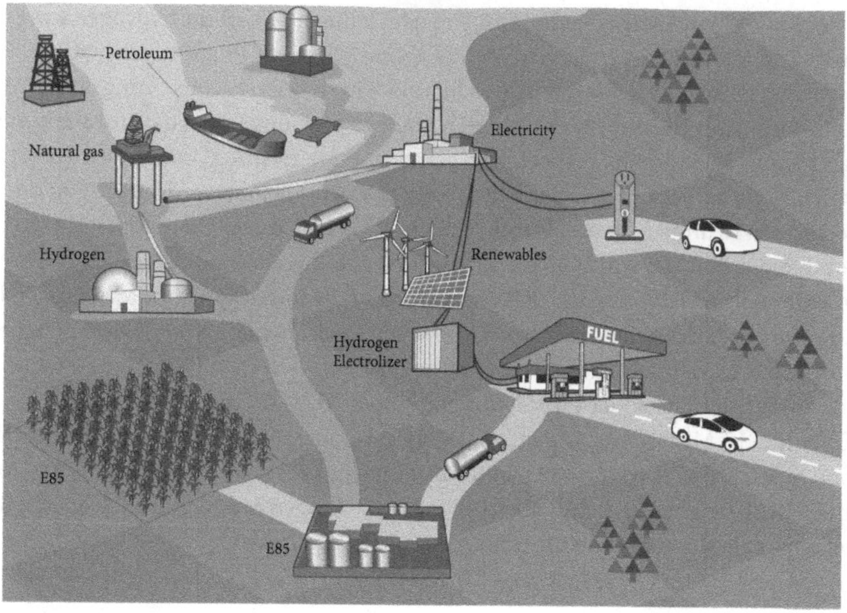

Figure 11.1 Schematic of Different Energy Sources Used to Produce Hydrogen (*Source*: California Fuel Cell Partnership[15])

States is made from natural gas reforming and, as you will see later from the economic evaluation, this is currently the most promising approach for producing hydrogen. Reforming natural gas is also called SMR, since steam and methane are the reactants. Equation (2), shown earlier, is shown again here:

$$CH_4 + H_2O\ (+\ \text{heat}) \rightarrow CO + 3H_2 \tag{2}$$

Reaction (2) takes place using a catalyst, and typical operating conditions are temperatures of 700°C to 1,000°C and pressures of 3 to 25 bar (44–363 psi).[5] The reaction is endothermic, meaning that heat must be added to make the reaction happen. To further increase the yield of hydrogen, the WGS is used. This is shown in Equation (3).

$$CO + H_2O \rightarrow CO_2 + H_2\ (+\ \text{heat}) \tag{3}$$

This reaction is slightly exothermic, meaning it gives off heat. Combining Equations (2) and (3) gives the overall balance, shown in Equation (4). Basically, it shows that one mole of methane makes four moles of hydrogen (H_2).

$$CH_4 + 2\ H_2O \rightarrow 4H_2 + CO_2 \tag{4}$$

In the NREL report,[14] 1,500 kg H_2/day is produced by SMR at the refueling site. The hydrogen is compressed to 875 bar (12,687 psi) and stored on site before being dispensed into the FCV at a pressure of 700 bar (~10,000 psi). In practice Equation (4) is not 100% efficient, and the energy balance shows that 73% of the energy in the natural gas ends up as hydrogen energy.

Option (10), central coal gasification delivered by pipeline, was also discussed in the previous section and is reviewed again here. For this option, coal gasification is used, such that oxygen is used to gasify the coal into hydrogen and carbon monoxide according to Equation (5). Although coal has small amounts of hydrogen, for example bituminous coal with 4 to 5% hydrogen, Equation (5) is shown in simplified form in which all of the coal is represented as carbon.

$$3C + O_2 + H_2O \rightarrow H_2 + 3CO \tag{5}$$

Following this, the yield of hydrogen is increased using the WGS reaction, shown again in Equation (6).

$$CO + H_2O \rightarrow CO_2 + H_2 \tag{6}$$

Combining Equations (5) and (6) gives the overall balance, shown in Equation (7). Overall, the combined equation shows that three moles of coal yields four moles of hydrogen (H_2).

$$3C + O_2 + 4H_2O \rightarrow 4H_2 + 3CO_2 \tag{7}$$

In the NREL report,[14] coal is delivered by rail to an offsite coal gasification plant where the hydrogen is made before being sent to the refueling station by pipeline. This central gasification plant is much larger than the distributed natural gas plant, producing 308,000 kg H_2/day. Also, in this scenario, CO_2 is captured and sequestered (CCS). The hydrogen is compressed to 69 bar (1,000 psi) and sent to the refueling station by pipeline. At the refueling station, hydrogen is further compressed to 875 bar (12,687 psi) and stored on site before being dispensed into the FCV at a pressure of 700 bar (~10,000 psi). Equation (7) is not 100% efficient, and the energy balance shows that 54.5% of the energy in the coal ends up as hydrogen energy.

In option (5), central biomass delivered by truck as a gas, 2,000 metric tons per day of biomass wood are used to produce 155,000 kg H_2/day. The biomass plant uses gasification and the WGS reaction, similar to that shown earlier in Equations (5) to (7). The biomass gasification plant is presumed to be within a 50-mile radius of the refueling station and the hydrogen is transported by truck in gaseous form to the refueling station. In the truck, the hydrogen is compressed to 260 bar (3,777 psi) and, at the refueling station, is further compressed to 875 bar (12,687 psi) and stored on site before being dispensed into the FCV at a pressure of 700 bar (~10,000 psi). The biomass gasification process is also not 100% efficient, and the energy balance shows that 42.8% of the energy in the biomass ends up as hydrogen energy.

The last option to be discussed is option (9), central electrolysis using wind electricity and delivered by pipeline to the refueling station. In this option, hydrogen is produced offsite at a facility using electrolysis with a design capacity of 52,300 kg H_2/day. Power for the electrolysis is provided from a wind farm. Naturally, for this case, the electricity could also come from other renewable energy sources, such as solar photovoltaic (PV), concentrating solar power (CSP), geothermal, or hydroelectric. For electrolysis, the chemistry is simple, with one molecule of water split into one mole of hydrogen and a half mole of oxygen.

$$H_2O + 285.8\ \text{kJ/mole} \rightarrow H_2 + \tfrac{1}{2}\,O_2 \qquad (8)$$

In this NREL scenario, the hydrogen is compressed to 69 bar (1,000 psi) and sent by pipeline to the refueling station. As was the case with the other technologies, the hydrogen is further compressed to 875 bar (12,687 psi) at the refueling station and stored on site before being dispensed into the FCV at a pressure of 700 bar (~10,000 psi).

According to the NREL report, the electrolysis system is based on a bipolar alkaline electrolyzer system. Normally, the alkaline material used for the electrolyte is potassium hydroxide (KOH). An electrolyte is a chemical that is dissolved in a polar solvent, such as water, and separates into cations and anions to make an electrically conducting solution. For KOH, the cation is K^+ and the anion is OH^-. Bipolar means that electrolytic cells are connected in series, to increase the total cell voltage. A good review on electrolytic systems is given by Smolinka et al.[16]

A final point about electrolysis is the difference compared to reforming and gasification. For Equations (4) and (7), the hydrogen comes from the natural gas, coal, or biomass, fuels already containing energy. In the case of natural gas and coal, this energy was formed over centuries of geothermal action while in the case of biomass, the energy comes from photosynthesis. These fuels contain energy which, minus losses, is transferred to the hydrogen. In Equation (8), however, electricity provides the energy to make the hydrogen fuel, as water is not a fuel. The reverse of Equation (8), shown as Equation (9), generates no more energy than the energy required to make hydrogen in Equation (8).

$$H_2 + \tfrac{1}{2}\,O_2 \rightarrow H_2O + 285.8\ \text{kJ/mole} \qquad (9)$$

And, as no system is perfect, a good deal of energy is lost in producing the hydrogen. According to the NREL report,[14] 176,000 BTU of electricity are required to make 116,000 BTU of hydrogen, for an energy efficiency of 65.9%. As this energy comes from a renewable energy source, the energy had no cost so the energy efficiency is less important than that for a fossil fuel source.

If the hydrogen is produced onsite, as is the case with option (1) for distributed natural gas reforming, the other cost associated with hydrogen are the refueling station costs which include compression, storage, and dispensing. NREL refers to these as CSD costs. If the hydrogen is produced offsite, as is the case with central biomass delivered by truck (options (5)), central electrolysis using wind electricity delivered by pipeline (option (9)), and central coal gasification delivered by pipeline (option (10)), there are also costs incurred for delivery as well as hydrogen lost along the route to the refueling station. Although the levelized cost of fuel (LCOF) is presented in the next section, Table 11.3 gives a preview of these results. As the table illustrates, distributed natural gas reforming is advantaged because hydrogen is produced at the refueling station, thereby avoiding the cost of delivery. The electrolysis route using wind electricity is the most expensive, more than triple that of distributed natural gas reforming. As shown in the next section, this route is highly dependent on the cost of electricity used to generate the hydrogen from electrolysis. All of the results in Table 11.3 are model projections, but some real data are also presented in the next section.

At this time, there are not many hydrogen refueling stations. According to the Alternative Fuels Data Center,[17] there are only 46 hydrogen stations in the United States with 41 in California, 1 in South Carolina, 1 in Massachusetts, 1 in Michigan, 1 in Hawaii, and 1 in Connecticut. Thus hydrogen has a long way to go to compete with the 150,000 gasoline and diesel filling stations reported by the National Association of Convenience Stores[3] in 2019. And according to FuelCellWorks,[18] as of January 2017 there were 274 hydrogen refueling stations worldwide, with 106 in Europe, 101 in Asia, 64 in North America, 2 in South America, and 1 in Australia. Using performance of existing US hydrogen stations, NREL reports an average refueling time of 3.8 minutes.[19] And, a National Petroleum Council report on hydrogen[20] gives a retail pump rate of 7 gallons per minute for gasoline and 7 kg per minute for hydrogen. Thus, refueling times for hydrogen are basically the same as those for gasoline.

We next turn our discussion to the FCV technology. There are several types of fuel cells including polymer electrolyte membrane (PEM), direct methanol, alkaline, phosphoric acid, molten carbonate, and solid oxide fuel cells (SOFCs).[21] According to the Alternate Fuels Data Center,[22] polymer electrolyte membrane (PEM) fuel cells

Table 11.3 Well-to-Wheels Hydrogen Costs

	Option 1: Distributed Natural Gas	Option 5: Central Biomass, Truck Delivery	Option 9: Central Wind, Pipeline Delivery	Option 10: Central Coal, Pipeline Delivery
H_2 Production Cost, \$/kg H_2	\$1.32	\$1.99	\$6.93	\$1.58
Delivery Cost, \$/kg H_2		\$1.88	\$2.20	\$2.20
Station Dispensing Costs (CSD), \$/kg H_2	\$2.05	\$1.30	\$1.61	\$1.61
Total, \$/kg H_2	\$3.37	\$5.17	\$10.74	\$5.39

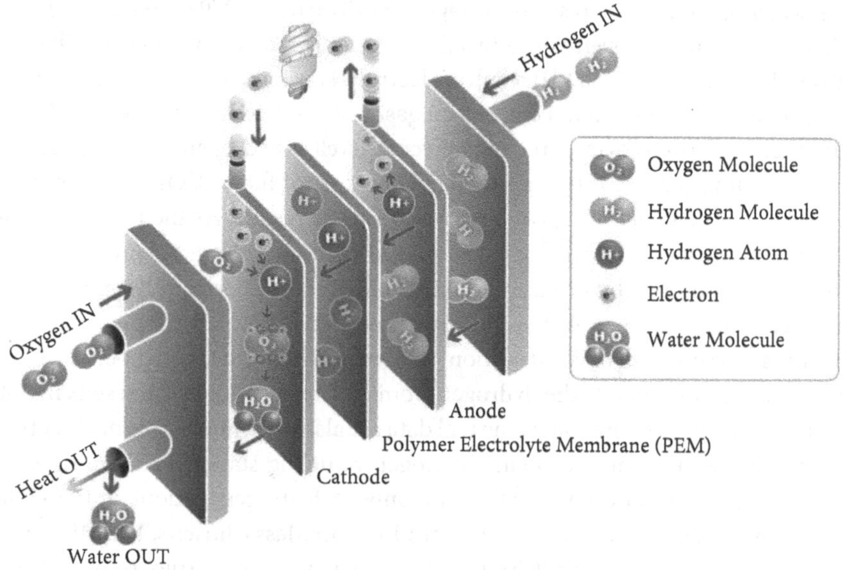

Figure 11.2 Schematic of a Polymer Electrolyte Membrane (PEM) Fuel Cell (Source: US Department of Energy[23])

are the most common type for FCVs, so our discussion focuses on that type of fuel cell. A schematic of a PEM fuel cell, taken from a DOE presentation, is shown in Figure 11.2.[23]

Like a battery, a fuel cell has three main parts including a cathode (positively charged), anode (negatively charged), and an electrolyte. Following Figure 2, hydrogen is introduced to the fuel cell at the anode and split into hydrogen ions (H^+) and electrons (e^-) using a noble-metal catalyst, typically platinum.[21] Oxygen is introduced at the cathode. The composition of the electrolyte membrane is beyond the scope of this chapter, but its purpose is to conduct the hydrogen ions, or protons, towards the anode to combine with oxygen to make water. The membrane is designed to allow protons to pass through but not electrons; these flow through the outer circuit to provide power to the electric motor of the FCV, but shown in the figure as a light bulb.

An important distinction to be made for a fuel cell is that the conversion of hydrogen and oxygen to water is a chemical reaction, or electrochemical reaction, and not a heat engine using heat from combustion. Recalling Equation (1) from the Introduction, a heat engine such as that used in an ICE or a fossil-fueled power plant is constrained by the so-called Carnot efficiency, named after the French scientist by that name. This equation is shown again here:

$$\eta = \frac{W}{Q_H} = 1 - \frac{T_C}{T_H}$$

Here:

W = net work taken from steam
Q_H = heat added to the system
T_C = temperature of the cold gas
T_H = temperature of the hot gas

If, for example, the ICE operates at a hot temperature of 500°C with a cold air temperature of 100°C, the maximum efficiency is 52% (don't forget to convert °C to absolute temperature in Kelvin, or K). In practice, the efficiency is less than 40% due to other factors. Without going into a lengthy discussion about the 2nd law of thermodynamics, it is not possible to take all of the heat added to the system (Q_H) and produce mechanical work (W), as some of that heat exits the system at the cold gas temperature (T_C). This results in an associated waste in energy and lowers the thermal efficiency.

In contrast, the FCV uses an electric motor, which is not constrained by the Carnot efficiency. The efficiency for a fuel cell can be described by Equation (10). In the case of hydrogen, the higher heating value (HHV) is 285.8 kJ/mole but the maximum electricity produced is only 237.2 kJ/mole, in which the difference of 48.6 kJ/mole is heat produced by the fuel cell.[24] This results in a maximum efficiency of 83%.

$$\text{Electrical Efficiency} = \frac{\text{Electricity Produced}}{\text{HHV of Fuel Used}} \qquad (10)$$

It is worth noting that the reaction of hydrogen and oxygen to make water can be interpreted in two ways, as the product of water can be considered as a liquid or vapor. If the produced water is not condensed to a liquid, there is a difference in energy known as the latent heat of condensation, which is 44 kJ/mole. Not considering the latent heat of condensation is known as using the lower heating value (LHV), and the LHV of the hydrogen fuel is 241.8 kJ/mole. As the electricity produced from the reaction of hydrogen and oxygen to make water vapor is 228.6 kJ/mole, this means the maximum efficiency based on an LHV basis is 94.5%.[24] Therefore, when fuel cell efficiencies are reported, it is important to know whether HHV or LHV was used.

The last topic for this section concerns the onboard storage of hydrogen on the FCV. This part of the technology is especially interesting to me as I served on the DOE USCAR onboard storage team from 2006 to 2008 while working for ConocoPhillips. There are several concepts for storing hydrogen on an FCV including high pressure hydrogen (350 and 700 bar), cryo-compressed hydrogen (hydrogen stored under pressure at very low temperatures), chemical storage, carbon-based storage, and metal hydrides. Each of these is discussed individually, but first it is important to understand hydrogen as a fuel as it compares to gasoline. As a liquid at room temperature gasoline, depending on its composition, has a density range of 0.71 to 0.77 kg/L (kilograms/liter). For the purpose of comparison, we use a density of 0.75 kg/L, or 2.84 kg/gallon. An average gallon of gasoline has an energy content of 33.4 kWh, so the energy content for gasoline can also be written as 11.76 kWh/kg.

In contrast, hydrogen is not a liquid at room temperature and, in liquid form, has a temperature of 20.3K (−252.9°C or −423.2°F). Absolute zero, or zero K, is the lowest temperature possible such that there is no motion or heat, so 20.3K is very cold indeed. At this temperature, the density of liquid hydrogen is 0.07 kg/L. Since this is not a very practical temperature for onboard FCV hydrogen storage, one concept is to use pressurized hydrogen stored at 700 bar. At 700 bar and room temperature, the density of pressurized hydrogen is only 0.04 kg/L. Recalling that 1 kg of hydrogen is equivalent to a gallon of gasoline (gge) and contains the same 33.4 kWh of energy, means that on a mass basis, hydrogen has an energy density 33.4 kW//kg compared to 11.76 kWh/kg for gasoline, or about 2.8 times higher. The hydrogen gas density can be converted from a liter basis to a gallon basis, resulting in a value of 0.15 kg/gallon. So, on a volume basis, hydrogen compressed to 700 bar is about 5 kWh/gallon compared to 33.4 kWh/gallon for gasoline, meaning the energy density for gasoline on a volume basis is about 6 ½ times larger. In summary, on a mass basis hydrogen stores about 2.8 times more energy but on a volume basis needs about 6 ½ times the volume to store the same amount of energy.

Using a 20-gallon tank as an example, gasoline with a density of 0.75 kg/L contains a mass of 56.8 kg, equivalent to 125 lb. A tank containing the same amount of energy in hydrogen at 700 bar form would occupy 132 gallons but weigh only 44 lb. Thus, as a fuel pressurized to 700 bar on an FCV, hydrogen is disadvantaged relative to volume and advantaged relative to mass.

Assuming your FCV should have a fuel storage system comparable in volume to that for a gasoline-powered vehicle, and further assuming that hydrogen has a fuel economy 2.5 to 3.0 times that of gasoline, you can use real data from the auto industry to set mass and volume targets for the hydrogen storage system. The DOE did just this,[25] setting original targets of 3.0 kWh/kg and 2.7 kWh/L (0.081 kg H_2/L) for the hydrogen storage system. Later, these were revised using FCV performance data to 2.5 kWh/kg and 2.3 kWh/L (0.070 kg H_2/L). The mass performance target of 2.5 kWh/kg translates to hydrogen being 7.5 wt% of the storage system.

With these targets, then, let's examine each of the possible onboard hydrogen storage concepts. For the present, the most mature technology for storing hydrogen on an FCV is compressed hydrogen, typically at 350 and 700 bar (5,000 and 10,000 psi) stored in tanks. If it makes you uncomfortable to think of having a high-pressure tank of hydrogen below you in an FCV, don't forget that we routinely drive around with 20 gallons of highly flammable gasoline every day. At 350 and 700 bar, the densities of hydrogen are 0.025 kg/L and 0.04 kg/L, respectively. Comparing this to the DOE volumetric target of 0.07 kg/L, it is clear that both systems fall short. Assuming a 20-gallon tank as comparable to most gasoline-power vehicles, you would have about 1.9 kg of hydrogen in a 350 bar tank and 3.0 kg of hydrogen in a 700 bar tank (there are 3.785 L per gallon). Using an estimated fuel economy of 45 miles per kg of hydrogen for an FCV means the driving range for a 350 bar tank is about 85 miles and for a 700 bar tank is 135 miles. By comparison, using the 24.7 miles per gallon EPA fuel economy for new US cars and trucks in 2016, you could travel 494 miles on that same 20-gallon tank using gasoline. Improving the fuel economy for the FCV or relaxing the constraint of having the same volume dedicated to fuel storage as a gasoline vehicle would improve the disparity, but it is clear that high-pressure hydrogen storage suffers from the low volume density of hydrogen. Also, without going into the details, there are parts of the storage system that cannot be used to store hydrogen, so the working densities are less than those earlier. According to a DOE report,[26] the hydrogen densities including the total system volume are 0.018 kg/L and 0.024 kg/L for the 350 and 700 bar systems, respectively. Comparing these to the densities exclusive of the system volume, 0.025 and 0.04 kg/L, we can calculate that about 60 to 70% of the system volume is used to store hydrogen, which we can refer to as a volumetric efficiency of 60 to 70%.

To calculate the hydrogen content on a mass basis, we need to know the storage system volume and mass, as the ability to meet the DOE target for mass must also include the system mass, and not only the mass of hydrogen. As there are various systems possible for hydrogen storage at 350 and 700 bar, the volume and mass of the storage system will vary. An Energy.gov publicaton[27] shows a range for 350 bar systems from 2.8 to 3.8 wt% with a nominal average of 3.2 wt% for hydrogen in the total system mass. Likewise, the publication shows a range for 700 bar systems from 2.5 to 4.4 wt% with a nominal average of 3.4 wt%. These also fall short of the DOE target of 7.5 wt% for hydrogen in the storage system.

Another concept is to use cryo-compressed hydrogen, for which the hydrogen is stored under pressure at very low temperatures. For example, a hydrogen storage system at 350 bar and a temperature of 150 K has a density of 0.04 kg/L, equivalent to the density of the 700 bar system with hydrogen at room temperature (room temperature is assumed to be 25°C, or 298K). Keeping the pressure at 350 bar but further decreasing the temperature to 50K raises the density to 0.075 kg/L, which seems to achieve the DOE target of 0.07 kg/L. However, storing hydrogen at low temperature comes with some storage issues. For example, a DOE report[28] discusses one cryo-compressed hydrogen storage system with a total system volume of 323 L, but only storing 151 L of hydrogen, giving it a volumetric efficiency of 47%. In other words, over 50% of the system is dedicated to storing the hydrogen, but does not contain any hydrogen, making the effective storage density only 0.035 kg/L.

As was the case for compressed hydrogen, we obtained the percent of hydrogen in the total system mass from an Energy.gov report[27] as 5.5 wt%, representing a range of 5 to 6 wt% for a variety of storage systems. Conditions for the cryo-compressed hydrogen were 500 bar at around 100K (−280°F). This too falls short of the DOE target of 7.5 wt% for hydrogen in the storage system, but is closer than compressed hydrogen.

In an effort to overcome the volume issues with hydrogen compared to gasoline, the DOE is pursuing other forms of onboard hydrogen storage including chemical storage, carbon-based storage, and metal hydrides. The metal hydride and carbon-based storage systems are reversible, meaning they can be recharged with hydrogen in the vehicle. The chemical storage is nonreversible, meaning it will be recharged outside the vehicle.

Chemical storage systems use materials that have covalent hydrogen bonds. Without getting too deep into the chemistry, a covalent bond is one in which the electrons are shared between the atoms. An example of such a system is shown as follows with ammonia borane, or NH_3BH_3, which is a solid at room temperature. The reaction to produce hydrogen proceeds according to Equation (11), such that the ammonia borane is heated to release the hydrogen. Based on the atomic weights of the compounds, the amount of hydrogen produced is 19.6 wt% of the ammonia borane.

$$NH_3BH_3 \rightarrow BN + 3H_2 \tag{11}$$

Even though heating is required to dehydrogenate the ammonia borane, the reaction is exothermic. That means that adding back hydrogen (rehydrogenation) requires energy and must be done outside the vehicle. Using a Savannah River National Laboratory (SRNL) annual review presentation,[29] in which data are given for the system mass and weight, the following values can be calculated and compared to the DOE targets:

Ammonia Borane: 4.6 wt% H_2, 0.041 kg H_2/L, 1.5 kWh/kg, and 1.4 kWh/L

DOE Targets: 7.5 wt% H_2, 0.070 kg H_2/L, 2.5 kWh/kg, and 2.3 kWh/L

Thus, although almost 20 wt% hydrogen is made from the ammonia borane, when the system mass is included the percent of hydrogen drops to 4.6 wt%.

Metal hydrides are another material category, storing hydrogen in solid form and having either ionic or covalent bonds. An ionic bond is one in which an atom gives its electron to another atom. An example of such a system is shown as follows with sodium tetrahydroaluminate, or $NaAlH_4$, a solid at room temperature. The reaction to produce hydrogen proceeds according to Equation (12), such that hydrogen is released when the compound is heated to 200°C. Based on the atomic weights of the compounds, the amount of hydrogen produced is 7.4 wt% of the starting hydride compound. Note that some hydrogen remains tied up with product NaH and is not released as hydrogen gas. Normally, this reaction is catalyzed with the addition of titanium, which helps with the hydrogen release and subsequent addition when refueling.

$$NaAlH_4 \rightarrow NaH + Al\,1.5\,H_2 \tag{12}$$

Again using the SRNL annual review presentation,[29] their data for the system mass and weight can be used to calculate the following values for comparison to the DOE targets:

$$NaAlH_4: 1.2 \text{ wt\% } H_2, 0.0115 \text{ kg } H_2/L, 0.4 \text{ kWh/kg, and } 0.4 \text{ kWh/L}$$

$$\text{DOE Targets: } 7.5 \text{ wt\% } H_2, 0.070 \text{ kg } H_2/L, 2.5 \text{ kWh/kg, and } 2.3 \text{ kWh/L}$$

Although almost 7.4 wt% hydrogen is made from this metal hydride, when the system mass is included the percent of hydrogen drops to 1.2 wt%.

The last materials to discuss are carbon-based storage systems. The chemistry of these systems is complex, and it would be easy to get lost in the details. The goal is to use materials, such as micro-porous activated carbons or metal organic framework (MOF) compounds, to store hydrogen using what is known as van der Waals forces. These van der Waals forces are weak intermolecular attractive forces resulting from electron-rich and electron-poor regions of different molecules. The term is named after Dutch physicist Johannes Diderik van der Waals who introduced this concept in 1873; however, the interaction is weaker than a chemical bond. Having weak attractive forces holding hydrogen to the carbon-based storage system would be an advantage for both releasing and adding back hydrogen during refueling.

An example of such a system is a metal organic framework-5 (MOF-5), shown in Figure 11.3. As the figure indicates, MOF-5 is formed from an organic species, 1,4-benzodicarboxylic acid, and inorganic species, Zn_4O. The figure shows the pore spaces formed by the compound, which are intended for hydrogen storage. Because the hydrogen is stored with van der Waals forces and not an actual chemical bond, it is not straightforward to write a chemical equation as was done for the chemical and metal hydride storage systems.

Once again using the SRNL annual review presentation,[29] their data for the system mass and weight can be used to calculate the following values for MOF-5 to compare to the DOE targets. For onboard storage, the system will operate at a pressure around 100 bar:

$$\text{MOF-5: } 3.1 \text{ wt\% } H_2, 0.021 \text{ kg } H_2/L, 1.1 \text{ kWh/kg, and } 0.7 \text{ kWh/L}$$

$$\text{DOE Targets: } 7.5 \text{ wt\% } H_2, 0.070 \text{ kg } H_2/L, 2.5 \text{ kWh/kg, and } 2.3 \text{ kWh/L}$$

A summary of these different hydrogen systems is shown in Table 11.4, along with the DOE targets. As the table illustrates, none of the systems currently meet the DOE targets, but the cryo-compressed and chemical storage systems have the best results on both a gravimetric (weight) basis and volumetric basis. Unfortunately, the chemical storage system has the disadvantage of not being reversible onboard the vehicle, which may negatively affect if from an economic standpoint. All of the other systems are reversible.

Without improvements in the gravimetric and volumetric capacities of these different systems, FCVs powered by hydrogen will be challenged with respect to driving range before refueling. Improvements in fuel economy as well as relaxing the

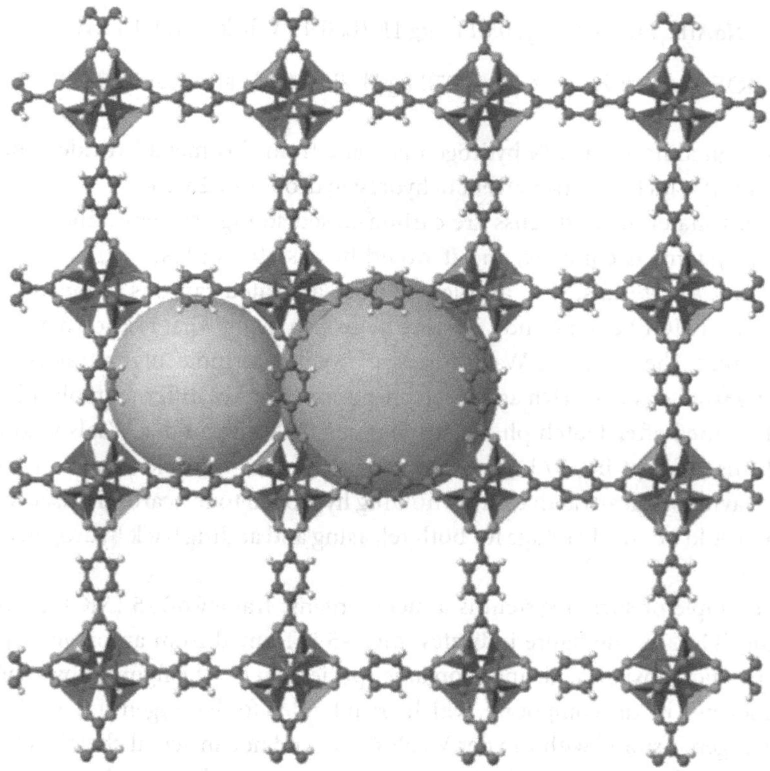

Figure 11.3 Metal Organic Framework-5 (MOF-5), formed with Zn_4O (zinc oxide) and 1,4-benzodicarboxylic acid ($C_{10}H_{10}O_4$). The large spheres embedded in the framework represent the pore size that can be used for hydrogen storage (*Source*: ChemTube3D[30])

Table 11.4 Hydrogen Storage Systems Summary[26,29]

System	Gravimetric Density, kg H_2/kg System	Gravimetric Energy Density, kWh/kg System	Volumetric Density, kg H_2/L System	Volumetric Energy Density, kWh/L System
DOE Targets	7.5%	2.5	0.070	2.3
350 bar at room temperature	3.2%	1.8	0.018	0.6
700 bar at room temperature	3.4%	1.4	0.024	0.8
Cryo-compressed at 500 bar and 100K	5.5%	2.3	0.035	1.4
Chemical Storage (NH_3BH_3)	4.6%	1.5	0.041	1.4
Metal Hydride ($NaAlH_4$)	1.2%	0.4	0.012	0.4
Carbon-based (MOF-5)	3.1%	1.1	0.021	0.7

constraint of having the same storage system volume as a gasoline-powered vehicle would help reduce this gap.

11.4. Capital Cost, Operating Cost, Well-to-Wheels Levelized Cost of Electricity, and Well-to-Wheels Levelized Cost of Fuel

The equations, methodology, and input used to calculate the LCOF for hydrogen are detailed in Appendix A. For hydrogen, the LCOF includes the cost of production, delivery, and onsite handling at the fueling station. Onsite handling includes hydrogen compression, storage, and dispensing.

Four cases were evaluated including (1) distributed natural gas reforming, (5) central biomass delivered by truck as a gas, (9) central electrolysis using wind electricity and delivered by pipeline, and (10) central coal gasification delivered by pipeline. The case numbers correspond to the earlier referenced NREL report[14] and LCOF was calculated for each of these four cases using economic data from this report. However, to be consistent with economic calculations made in other chapters of this book, the tax and discount rate were different than those used in the NREL report.

Another difference compared to the NREL report was the cost of energy used in the different cases. For Case (1), Case (5), and Case (10) the cost of each feed was taken from previous chapters using these same feeds. So, the feed cost for natural gas came from Chapter 2, the cost for biomass from Chapter 9, and the cost of coal from Chapter 3.

For Case (9), where electricity generated from a wind farm was used to perform electrolysis, the cost of electricity was based on the average US customers paid for electricity in 2018,[31] a value of \$0.1078/kWh. This value is also consistent with the levelized cost of electricity (LCOE) calculated for an onshore wind farm in Chapter 6, \$0.0998/kWh.

The units for LCOF are in \$/gallon of gasoline equivalent, or \$/gge. For hydrogen, 1 kg of hydrogen is equivalent in energy to one gallon of gasoline.

The cost of hydrogen was be calculated in terms of "Well-to-Wheels" cost. For distributed cases, for which the hydrogen is made at the refueling site, this included the (1) delivered cost of fuel to make the hydrogen, (2) the cost of production, (3) the cost of hydrogen compression, storage, and dispensing (CSD) at the refueling site, and (4) taxes (both federal and state). For central hydrogen, for which hydrogen is produced offsite, the cost included the (1) delivered cost of fuel to make the hydrogen, (2) the cost of production, (3) the cost of delivery (truck or pipeline), (4) the CSD costs, and (5) taxes.

The results for all four cases are shown in Table 11.5. There are several points that can be made from these results. Producing the hydrogen at the refueling site removes the need for delivery by either truck or pipeline, reducing the cost of hydrogen by about \$2/kg H_2. Although some of those costs are transferred to the refueling site in the form of compression, storage, and dispensing (CSD), overall the cost is lowered.

Table 11.5 Levelized Cost of Fuel (LCOF) for Hydrogen

	Case 1: Distributed Natural Gas	Case 5: Central Biomass, Truck Delivery	Case 9: Central Wind, Pipeline Delivery	Case 10: Central Coal, Pipeline Delivery
Feed Delivered Cost	$3.17/MM BTU	$72.6/dry tonne	$0.1078/kWh	$47.2/tonne
H_2 Product Yield, kg/day	1,500	155,200	52,300	307,700
Capital Cost (per kg H2), $/kg	$1,152	$1,296	$1,320	$2,497
Production, $/gge				
Fixed O&M	$0.19	$0.23	$0.23	$0.30
Non-Fuel Variable O&M	$0.12	$0.40	$0.06	$0.18
Feed Cost	$0.67	$0.98	$6.25	$0.37
Levelized Capital Cost	$0.34	$0.38	$0.39	$0.73
Total Cost of Production	$1.32	$1.99	$6.93	$1.58
Delivery, $/gge				
Levelized Cost of Delivery		$1.88	$2.20	$2.20
Compression, Storage, and Dispensing (CSD), $/gge				
Levelized Capital Cost	$1.11	$0.64	$0.77	$0.77
Other O&M	$0.54	$0.46	$0.41	$0.41
Energy & Fuel	$0.40	$0.20	$0.43	$0.43
Total Cost of CSD	$2.05	$1.30	$1.61	$1.61
LCOF	$3.37	$5.17	$10.74	$5.39

Note: O&M refers to Operations and Maintenance

Thus, if hydrogen can be safely produced and stored on site, this reduces the total cost for hydrogen.

The capital cost of production is the cost for building the plant and the H_2 product yield shows the plant capacity. For the onsite facility in Case (1), the rate is only 1,500 kg H_2/day. The capital cost for building the coal gasification plant, Case (10), is the highest by about double the other technologies, but the LCOF is similar to Case (5) and lower than Case (9) because of the lower feed cost. This higher capital cost is due, in part, to the inclusion of equipment for carbon capture and sequestration (CCS) at the coal gasification plant. The cost of pipeline delivery is the same for the wind and coal cases, but lower for truck delivery because of higher capital costs for building the pipeline.

Considering the overall total LCOF, Case (1) which has hydrogen produced at the refueling station by natural gas reforming is the lowest at \$3.37/kg H_2. On the other hand, Case (9) for which hydrogen is produced using electricity from wind power is the highest at \$10.74/kg H_2. As illustrated in the table, it is quite clear that the cost of hydrogen production, \$6.93/kg H_2, is the reason wind electricity-powered electrolysis has this high LCOF. This cost of electricity is one of the variables that is examined in the sensitivity study.

The data in Table 11.5 were calculated using NREL model data, along with current prices for the different feeds used to produce hydrogen. Before performing the sensitivity study, some actual data from commercially operating US hydrogen stations are examined. According to the US DOE Alternate Fuels Price Report for January 2018,[32] data received from eleven operating hydrogen stations reported an average price of \$15.01 per kg of hydrogen, or \$15.01/gge. In the same report, the average gasoline retail price for the same time period was \$2.50/gallon. In a report from the California Fuel Cell Partnership,[33] hydrogen fuel prices ranged from \$12.85 to more than \$16 per kg, with the most common price of \$13.99 per kg. Taking these two reports together, the average retail price for hydrogen is about \$14 per kg. As 41 of the 46 US hydrogen stations are located in California, it is also useful to compare the cost of hydrogen to the average retail price of gasoline in California which, according to the California Energy Commission[34] was \$3.20 in January 2018.

The earlier referenced NREL report[7] gave a FCV average of 45 miles per kg hydrogen versus 24.7 mpg for new US cars and trucks in 2016. Using a value of \$14/kg for hydrogen and \$3.20/gallon for gasoline, the average fuel economy is \$0.31/mile for hydrogen and \$0.13/mile for gasoline. Another way to look at this is that gasoline would have to cost \$7.70/gallon to be equivalent to a \$14/kg H_2 retail price.

In fairness, however, hydrogen refueling stations are in their infancy. According to one NREL report,[19] of 27 stations providing data for the second quarter of 2017, none of them had onsite natural gas reforming, the lowest cost as shown in Table 11.5. Also, maintenance costs were around \$6/kg hydrogen, due in part to low amounts dispensed and new station startup issues. Thus, as the fueling technology matures, we would certainly expect the average retail price to decrease from the current \$14/kg H_2. Also, there is reason to expect that FCVs will perform better than 45 miles per kg hydrogen, which will also lower the gap in fuel costs.

A sensitivity analysis was made for the high and low LCOF cases, namely Case (1), distributed natural gas reforming and Case (9), central electrolysis using wind electricity and delivered by pipeline. For both cases, the sensitivity analysis focuses on the cost of production. The rationale for choosing the variable ranges is described in Appendix A. For Case (1), variations in production capital cost, energy efficiency, and the price of natural gas fuel were examined. The price of natural gas is especially interesting, as data presented in Chapter 2 showed a range of \$3 to \$9/MM BTU over the last 14 years.

For Case (9), variations in production capital cost, energy efficiency, and the price of electricity were examined. The price of electricity is especially interesting, as a report from the DOE Office of Energy Efficiency[35] shows that the cost of onshore wind farm electricity is highly dependent on location, with the lowest cost projects in the so-called wind belt US interior, where wind speeds can top 8.5 meters per second, or 19 mph. The LCOE from plants in the US interior range from \$0.02 to \$0.06/kWh, compared to \$0.06 to \$0.12/kWh in other, less windy, parts of the United States.

Results for the sensitivity study are shown in Table 11.6. As the table illustrates, the capital cost and energy efficiency had a small impact on LCOF for Case (1) but the fuel cost, however, had a quite dramatic impact. Doubling the cost of natural gas increased the LCOF by \$0.6/gge and tripling the cost increased the LCOF by \$1.24/gge.

Table 11.6 Sensitivity Study for Distributed Natural Gas Reforming (Case (1)) and Central Electrolysis Using Wind Electricity and Delivered by Pipeline (Case (9))

	Energy Efficiency	Feed Cost	Capital Cost	Production Levelized Capital Cost	Total Cost of Production	LCOF
Units		\$/MM BTU	\$/kg H2	\$/gge	\$/gge	\$/gge
Case (1)						
Base Case	0.714	\$3.17	\$1,152	\$0.34	\$1.32	\$3.37
Capex is \$800/kg H_2	0.714	\$3.17	\$800	\$0.23	\$1.22	\$3.27
Capex is \$1,500/kg H_2	0.714	\$3.17	\$1,500	\$0.44	\$1.42	\$3.47
Energy efficiency is 0.6	0.600	\$3.17	\$1,152	\$0.34	\$1.45	\$3.50
Energy efficiency is 0.8	0.800	\$3.17	\$1,152	\$0.34	\$1.25	\$3.30
Natural gas is \$6/MM BTU	0.714	\$6.00	\$1,152	\$0.34	\$1.92	\$3.97
Natural gas is \$9/MM BTU	0.714	\$9.00	\$1,152	\$0.34	\$2.56	\$4.61
Case (9)		\$/kWh				
Base Case	0.719	\$0.1078	\$1,320	\$0.39	\$6.93	\$10.74
Capex is \$1,000/kg H_2	0.719	\$0.1078	\$1,000	\$0.29	\$6.83	\$10.65
Capex is \$2,000/kg H_2	0.719	\$0.1078	\$2,000	\$0.58	\$7.13	\$10.94
Energy efficiency is 0.6	0.600	\$0.1078	\$1,320	\$0.39	\$8.17	\$11.98
Energy efficiency is 0.8	0.800	\$0.1078	\$1,320	\$0.39	\$6.29	\$10.11
Electricity is \$0.04/kWh	0.719	\$0.0400	\$1,320	\$0.39	\$3.00	\$6.81
Electricity is \$0.08/kWh	0.719	\$0.0800	\$1,320	\$0.39	\$5.32	\$9.13

For Case (9), the cost of electricity had an even greater impact. Decreasing the cost of electricity from \$0.1078/kWh to \$0.08/kWh reduced the LCOF by \$1.61. Further decreasing the cost of electricity to \$0.04/kWh reduced the LCOF by \$3.93. Even with this sizable reduction, however, an LCOF of \$6.81/kg H_2 for inexpensive electricity at \$0.04/kWh is still about double that of the base case for distributed natural gas reforming and much greater than the average 2018 US price for gasoline at \$2.50/gallon.

11.5. Cost of Energy

The last section showed that using natural gas reforming to make hydrogen was the lowest cost option for providing hydrogen for FCVs. Also, using distributed production, for which the hydrogen is made at the refueling station, eliminates the need for a transport and delivery infrastructure, resulting in further cost reductions. Finally, according to the Office of Energy Efficiency and Renewable Energy,[5] about 95% of the hydrogen currently produced in the United States is made from natural gas reforming. Given this maturity and the low cost, it is likely that the early market for hydrogen production will rely heavily on natural gas reforming. For this reason, the cost of energy is evaluated with the assumption of natural gas as the fuel to produce hydrogen.

The Henry Hub spot price[36] for 2018 was \$3.17/MM BTU. To convert this to \$/lb, we use an energy content of 22,792 BTU/lb for natural gas, which gives a cost of \$0.07/lb for a mass basis.

11.6. Capacity Factor

The NREL report,[14] used as a source of data for the economic calculations, gave capacity factors for hydrogen production in each of the cases considered. These capacity factors are 0.89 (or 89%) for Case (1), distributed natural gas reforming, 0.90 for Case (5), central biomass delivered by truck as a gas, 0.90 for Case (9), central electrolysis using wind electricity and delivered by pipeline, and 0.90 for Case (10), central coal gasification delivered by pipeline.

For Cases (5) and (10), biomass and coal gasification were already discussed in Chapters 3 and 9. For coal gasification to electricity, also called integrated gasification combined cycle (IGCC), the capacity factor used in Chapter 3 was 0.85.[37] Likewise, the capacity factor used for biomass gasification combined cycle (BCC) was 0.83.[37] Thus, the NREL report, which used 0.90 for coal and biomass gasification is similar to these numbers.

For Case (1), natural gas reforming, a National Energy Technology Laboratory (NETL) report[38] on the Fischer-Tropsch synthesis, which uses natural gas reforming to make the synthesis gas ($CO + H_2$), gives a capacity factor of 0.90. A life cycle analysis report[39] looking at hydrogen production from natural gas steam reforming also gives a capacity factor of 0.90. In addition, the average 2018 capacity factor for US refineries was 0.931.[40] Since natural gas reforming is a type of refinery process, this value is also applicable. Thus, the 0.89 used in the NREL report is consistent with these capacity factors.

For Case (9), central electrolysis using wind electricity and delivered by pipeline, the NREL reported capacity factor of 0.90 deserves some explanation. In 2018, the average US capacity factor for wind-generated electricity was 0.374.[41] The capacity factor for wind is lower than fossil fuel-based technologies, since the wind has variable speeds and does not always produce electricity at the rated plant capacity. Thus, if hydrogen is directly produced from wind power, the NREL capacity factor of 0.90 is far too optimistic. However, in their report, they state that the capacity factor is not based on direct use of wind-generated electricity, but rather uses electricity purchased from the grid, with the cost reduced, or negated, using credits for electricity generated by wind power. If the electrolysis unit to make hydrogen was directly powered by the wind farm, the capacity factor would have to be much lower than the 0.90 used in their model calculations.

11.7. Efficiency: Fraction of Energy Converted to Work

Efficiency in this section is defined as the percent of energy contained in the hydrogen stored on the FCV that is used to power the vehicle. As mentioned earlier in the Overview of Technology, electrical efficiency in a hydrogen-powered FCV is not constrained by the Carnot efficiency, as is the case with internal combustion and heat engines, because the FCV uses an electric motor. According to Equation (10), the maximum efficiency for a hydrogen-powered FCV is 83%.

Unfortunately, few energy systems attain their maximum efficiency and, according to the DOE,[42] hydrogen FCVs which use electric motors typically use 40 to 60% of the fuel's energy. For the polymer electrolyte membranes (PEMs), which are commonly used in FCVs, the efficiency is typically 50 to 60%. The overall energy efficiency involves converting the energy contained in the hydrogen into electrical energy and, subsequently, converting that electrical energy into the mechanical work of powering the vehicle.

According to an NREL 2017 report[43] which evaluated on-road FCV data, energy efficiencies as high as 59% were observed. For the vehicles tested, efficiencies at ¼ power ranged from 51%–59%, with an average of 57%. At full power, efficiencies ranged from 30%–54%, with an average of 43%. Taking a simple average of ¼ and full power gives 50% for the energy efficiency.

By comparison, the thermal efficiencies for gasoline and diesel ICE were reported in Chapter 1 (Petroleum Crude Oil) as 35% and 45%, respectively. And, in Chapter 8 which discussed ethanol, the efficiency of a 10% blend of ethanol and gasoline was reported as 35%. Therefore the hydrogen-powered FCV is more efficient than the ICE.

In contrast, battery-powered electric vehicles (BEVs) have even higher energy efficiencies. One report[44] gives a range of BEV efficiency from a minimum of 71% to a maximum of 89%, giving a simple average of 80%.

In summary, for a hydrogen-powered FCV, about 50% of the energy contained in the hydrogen stored on the vehicle is used to power the vehicle. In contrast, BEVs have efficiencies of about 80% while ICEs are about 35% for gasoline and 45% for diesel.

11.8. Energy Balance

For a hydrogen-powered FCV, the energy balance is defined by the equation that follows. Hydrogen is an energy carrier, so it is the energy in the fuel source used to produce hydrogen that is the relevant reference point.

$$\text{Energy Balance} = \frac{[\text{Energy from hydrogen that powers the vehicle}]}{\left[\begin{array}{c}\text{Energy contained in the fuel source used to produce} \\ \text{the hydrogen}\end{array}\right]}$$

For the "Well-to-Wheels" analysis, there are several steps involved in the energy balance. These include (1) the energy to transport the fuel to the hydrogen production plant, (2) the energy used to produce the hydrogen, (3) the energy used to transport the hydrogen to the refueling station, if applicable, (4) the energy used to compress, store, and dispense (CSD) the hydrogen, and (5) the energy used to convert hydrogen into electricity and the mechanical energy to power the FCV.

The four cases examined for calculating the LCOF is also evaluated for the energy balance. These four cases include (1) distributed natural gas reforming, (5) central biomass delivered by truck as a gas, (9) central electrolysis using wind electricity and delivered by pipeline, and (10) central coal gasification delivered by pipeline. The same case numbers used in the NREL report are retained here.

For step (1) in the energy balance, transporting the fuel to the hydrogen production plant, the energy consumed is source specific. For Case (1), distributed natural gas reforming, we refer back to Chapter 2 for the information. According to a National Energy Technology Laboratory (NETL) report,[45] 91.9% of the original natural gas energy arrives at the hydrogen production plant.

For Case (5), central biomass delivered by truck to the refueling station, we refer to a NREL report[44] referenced in the chapter on biomass, Chapter 9, which shows that 2.3% of the biomass energy was consumed prior to the arrival of the biomass at the power plant, due to feedstock production and preparation as well as transporting the biomass to the plant. Thus, 97.7% of the original biomass energy arrives at the hydrogen production plant.

For Case (9), central electrolysis using wind electricity and delivered by pipeline to the refueling station, we refer to information discussed in the chapter on wind energy,

Chapter 6. For this case, the electricity generated from wind which is used with electrolysis to produce hydrogen has lost energy from that originally in the wind due to generation and transmission. That is, relative to the energy in the wind, energy is lost due to the wind turbine efficiency and energy lost by transmitting the electricity through the electric grid. An EIA report[47] gives an average wind turbine efficiency of 26%, and another EIA report[48] shows that of electricity generated in 2016, 7% was lost in transmission and distribution. From these data, we calculate that 24.2% of the original energy in the wind arrives at the hydrogen production plant.

Finally, for Case (10), central coal gasification delivered by pipeline, we refer to information discussed in the chapter on coal energy, Chapter 3. For this case, of the energy originally in the coal, some is lost due to extraction, processing, and transportation. For extraction and processing, a DOE report[49] reports that 1.5% of the coal energy is consumed due to extraction and processing. Another report[50] gives a loss of 0.4% of the coal energy due to transportation. Together, this means that 1.9% is lost altogether, so 98.1% of original energy in the coal arrives at the hydrogen production plant.

Summarizing step (1) in the energy balance, the energy consumed to transport the fuel the hydrogen production plant, for Cases (1), (5), (9), and (10), 91.9%, 97.7%, 24.2%, and 98.1% of the original energy arrives at the hydrogen production plant.

For steps (2) the energy used to produce the hydrogen, (3) the energy used to transport the hydrogen to the refueling station, if applicable, and (4) the energy used to compress, store, and dispense (CSD) the hydrogen, these are given in an NREL report as process energy efficiencies.[14] This part of Table 11.5 is shown again in Table 11.7. For example, of the 91.9% of the original energy in natural gas for Case 1 that arrives at the hydrogen production plant, 71.4% is made into hydrogen dispensed into the FCV.

While discussing efficiency, it is interesting to look at the energy efficiency for the electrolyzer system to make hydrogen from water, already contained in the 71.9% energy efficiency for Case (9). From the NREL report from which these data were extracted, they based the electrolyzer efficiency on a bi-polar alkaline system, with a higher heating value (HHV) efficiency of 67%.[14] In another NREL report,[51] current electrolyzer efficiencies were said to generally be in the range of 52% to 82% (HHV). Finally, in an NREL program review,[52] in which a polymer electrolyte membrane (PEM) electrolyzer was experimentally tested, the HHV efficiency was

Table 11.7 Process Energy Efficiencies for Four Hydrogen Production Cases[14]

	Case (1): Distributed Natural Gas	Case (5): Central Biomass, Truck Delivery	Case (9): Central Wind, Pipeline Delivery	Case (10): Central Coal, Pipeline Delivery
Process Energy Efficiency	0.714	0.460	0.719	0.568

measured as 67%. Thus, if electricity is used as the route to make hydrogen from water electrolysis, anywhere from 1/3 to ½ of the energy could be lost in this part of the process.

Finally, for step (5) in the energy balance, the energy used to convert hydrogen into electricity and the mechanical energy to power the FCV was given in the previous section as 50%.

Using 100 units of energy as a reference and putting these difference parts of the energy balance together, we can calculate an overall energy balance for each case, shown here:

Case (1): Distributed Natural Gas Reforming

100 (source) → 91.9 (preparation and transport to hydrogen production plant)
→ 65.6 (produce hydrogen at the refueling station and CSD)
→ 32.8 (hydrogen energy that powers FCV)

Case (5): Central Biomass Delivered by Truck as a Gas

100 (source) → 97.7 (preparation and transport to hydrogen production plant)
→ 44.9 (produce hydrogen, transport to refueling station and CSD)
→ 22.5 (hydrogen energy that powers FCV)

Case (9): Central Electrolysis using Wind Electricity Delivered by Pipeline as a Gas

100 (source) → 24.2 (preparation and transport to hydrogen production plant)
→ 17.4 (produce hydrogen, transport to refueling station and CSD)
→ 8.7 (hydrogen energy that powers FCV)

Case (10): Central Coal Gasification Delivered by Pipeline as a Gas

100 (source) → 98.1 (preparation and transport to hydrogen production plant)
→ 55.7 (produce hydrogen, transport to refueling station and CSD)
→ 27.9 (hydrogen energy that powers FCV)

While the focus of this chapter is on hydrogen-fueled FCVs, the energy efficiency for battery-powered vehicles (BEVs) is also calculated to compare to the previous cases. In the last section, BEVs were shown to have an average energy efficiency of about 80%. It then remains to calculate the efficiency for electricity generation, including the energy required to transmit and distribute the electricity through the grid. These calculations have already been made in earlier chapters for different energy sources. To compare directly to the previous cases, we need the energy balance to generate electricity from natural gas, biomass, wind, and coal. These were done previously in Chapters 2, 9, 6, and 3, respectively.

For the best natural gas case considered in Chapter 2, combined cycle, 44.7% of the original natural gas energy ended up as electricity. For biomass in Chapter 9, the better of the two cases analyzed, gasification combined cycle, 31.3% of the original biomass energy ended up as electricity. For wind in Chapter 6, 24.2% of the original wind energy ended up as electricity. And, for coal in Chapter 3, the best case was supercritical pulverized coal combustion, for which 35.1% of the original coal energy ended up as electricity.

Using a BEV efficiency of 80% with the previous numbers leads to overall "Well-to-Wheels" efficiencies of 35.8%, 25.0%, 19.4%, and 28.1% for natural gas, biomass, wind, and coal, respectively. Using the hydrogen-powered FCV cases calculated earlier, the BEV efficiency of 35.8% for natural gas compares to an FCV efficiency of 32.8%. For biomass, the BEV efficiency of 25.0% compares to an FCV efficiency of 22.5%. For wind, the BEV efficiency of 19.4% compares to an FCV efficiency of 8.7%. And, for coal, the BEV efficiency of 28.1% compares to an FCV efficiency of 27.9%. Thus, with the exception of wind energy, the energy balances for the BEV and the FCV are essentially the same.

11.9. Maturity: Experience

The concepts of both producing hydrogen from water via electrolysis and using hydrogen as a fuel have been around since the 1800s. Following a review on the history of hydrogen,[53] English scientists William Nicholson and Anthony Carlisle discovered they could produce hydrogen and oxygen from electric current in 1800, which was later called electrolysis. And, in 1838, the fuel cell effect was discovered by Swiss chemist Christian Friedrich, who combined hydrogen and oxygen to produce an electric current and water. This was demonstrated on a more practical scale in 1845 by English scientist William Grove, proving that electric current could be made from a reaction between hydrogen and oxygen, and earning him the title "father of the fuel cell."

As a fuel, Germany used hydrogen for ICEs as early as 1920. It is well known that NASA (National Aeronautics and Space Administration) used hydrogen in their space program for rocket propulsion and in fuel cells, to make electricity and water for the astronauts.

The first FCV in history was made by Harry Karl Ihrig, who in 1959 converted a farm tractor to run on more than one thousand alkaline fuel cells.[54] FCV activity grew in the 1970s, as environmental concerns about air pollution and high oil prices drove technology for clean air and energy efficiency.[55] During this decade, several German, Japanese, and US vehicle manufacturers began to experiment with FCVs. In 1994, Daimler Benz demonstrated a commercial FCV.[53] And, in 2000, Ballard Power Systems presented a commercial version of a polymer electrolyte membrane (PEM) fuel cell for automotive applications. The US initiative to develop a commercially viable hydrogen-powered FCV, the Freedom Car, was started in 2003 by President George W. Bush.

Summarizing, water hydrolysis to produce hydrogen has been around since 1800, fuel cells since 1845, and the first FCV since 1959. This means there are more than 200 years of experience in electrolysis, about 175 years of experience in fuel cells, and about 60 years of experience in FCVs.

11.10. Infrastructure

Unlike EVs, for which charging can be done at home as well as at recharging stations, hydrogen-powered FCVs require an infrastructure similar to what already exists for gasoline and diesel. In comparison to more than 150,000 gasoline and diesel filling stations in the United States,[3] there are only 46 hydrogen stations in the United States with 41 in California, 1 in South Carolina, 1 in Massachusetts, 1 in Michigan, 1 in Hawaii, and 1 in Connecticut.[17] Thus, the hydrogen-power FCV infrastructure is in its infancy. According to a National Renewable Energy Laboratory (NREL) presentation,[19] of the 27 stations reporting data, all reported using compressed hydrogen delivery, while one also used pipeline delivery, and three also used liquid delivery. Only one reported using onsite electrolysis and none reported using onsite natural gas reforming, currently the lowest cost option for hydrogen fuel cost. According to the California Fuel Cell Partnership,[56] 26 stations dispensed hydrogen as either 350 (5,000 psi) or 700 bar (10,000 psi) and one only at 700 bar. Thus, in the United States, hydrogen is mostly being made offsite and dispensed into the FCV as compressed hydrogen.

Worldwide, as of January 2017 there were 274 hydrogen refueling stations with 106 in Europe, 101 in Asia, 64 in North America, two in South America, and one in Australia.[18] A detailed worldwide study of how the hydrogen is produced was not made, but it is interesting that Denmark has nine hydrogen refueling stations, all of which use hydrogen produced by electrolysis using renewable energy.[57] Of these, six have the electrolysis onsite and three offsite and, for each station, the hydrogen is dispensed at 700 bar.

Another aspect of infrastructure are the FCVs to use the hydrogen fuel. According to a DOE overview,[58] there were more than 1,600 FCVs in the United States as of March 2017. And, according to one online magazine,[59] the number has continued to grow as the Toyota Mirai hydrogen fuel cell car topped 3,000 sold in California as of January 2018. This article says that the Mirai now accounts for 80% of the FCV sales in the United States, making it the leader by far. However, the market for hydrogen-powered FCVs is still small as, by comparison, Toyota sold over 100,000 Prius hybrids in 2017 alone and Nissan sold around 11,000 Leaf EVs.

In terms of performance, the DOE provides a comparison of the top three FCVs currently on the market,[60] shown in Table 11.8. Earlier in this chapter, a comparison

Table 11.8 Comparison of Fuel Cell Vehicles[60]

Vehicle Name	2018 Honda Clarity	2017 Hyundai Tucson	2018 Toyota Mirai
Combined City/Highway Fuel Economy (miles/kg H_2)	67	49	66
Range (miles)	366	265	312
Vehicle Class	Midsize	Small SUV	Subcompact

of hydrogen-powered FCVs to gasoline-powered vehicles was made using an NREL reported fuel economy of 45 miles per hydrogen gge.[7] The table shows that the Honda Clarity and Toyota Mirai do much better than that, and the Clarity is rated as a midsize vehicle. Dividing the range by the average fuel economy gives a value of about 5 kg of hydrogen, suggesting this is the typical fuel capacity for these vehicles. Since one kg of hydrogen is equivalent in energy to one gallon of gasoline, the equivalent gasoline-powered vehicle would carry five gallons, which would be considered a very small capacity. This shows how the higher fuel efficiency of hydrogen allows a reasonable driving range, even with a relatively low capacity for hydrogen.

11.11. Footprint and Energy Density

For the footprint and energy density, two cases were chosen, namely Case (1), distributed natural gas reforming and Case (9), central electrolysis using wind electricity and delivered by pipeline. According to the NREL report from which the data were taken,[14] hydrogen is produced at the hydrogen refueling site for Case (1) using a 1,500 kg H_2/day methane reformer with the WGS to further increase the yield of hydrogen. For Case (9), hydrogen is produced at a central production facility using a standalone electrolyzer with a design capacity of 52,300 kg H_2/day.

For both cases, there are four different steps that contribute to the footprint for hydrogen. For Case (1), these include the area needed to (1) extract natural gas from the ground, (2) process the natural gas in a gas processing plant, (3) transport the natural gas to the refueling station, and (4) produce, compress, store, and dispense hydrogen at the refueling station. For Case (9), in which we presume the wind farm that generates electricity is sited at the same location as the standalone electrolyzer, the steps include the area needed to (1) make electricity at the wind farm, (2) electrolyze water to hydrogen and oxygen, (3) transport the hydrogen by pipeline to the refueling station, and (4) compress, store, and dispense hydrogen at the refueling station.

For Case (1), steps (1) and (2) were already calculated in the chapter on natural gas, Chapter 2. The energy density to extract natural gas from the ground was calculated to be 159.3 billion BTU/(year-square mile) and the energy density to process the natural gas was calculated to be 0.76 quadrillion BTU/(year-square mile). Stated using scientific notation, these values are 159.3×10^9 and 0.76×10^{15}.

In order to compare energy densities for different technologies on a common basis of gigawatt-hours per square mile per year, or GWh/(square mile-year), energy was converted from BTU to Watt-hour, or Wh, using the following equation.

$$1 \text{ BTU} = 0.293071 \text{ Wh}$$

In these units, steps (1) and (2) are 46.7 GWh/(year-square mile) and 224,000 GWh/(year-square mile).

For step (3), transporting the natural gas to the refueling station, whether done by truck or pipeline, it is not expected to impact the overall footprint. If by truck, the highway infrastructure already exists and if by pipeline, this land area is small and pipelines can coexist with farming and pasture land.

For step (4), to produce, compress, store, and dispense hydrogen at the refueling station, we first refer to a Sandia National Laboratories report[61] that discusses one possible configuration for a hydrogen refueling station. In this configuration, for which as much as 1,300 kg of hydrogen can be stored onsite and 300 kg can be dispensed per day, the station occupies 18,000 ft^2 (1,672 m^2). Given that a kg of hydrogen contains 116,000 BTU of energy, using 300 kg/day of hydrogen dispensed leads to an energy density of 19.67 trillion BTU/(year-square mile), or 19.67×10^{12}. This can be converted to other units to give 5,766 GWh/(year-square mile).

The preceding configuration does not, however, include the onsite hydrogen reformer. An International Energy Agency (IEA) report[62] evaluated some small-scale reformers that produce hydrogen at either 50 Nm^3/hr (~660 kg/hr) or 100 Nm^3/hr (1,320 kg/hr), sizes that easily meet the 300 kg per day to be dispensed. Presumably, the reformer would not run continuously but rather would be used when the hydrogen storage tank needs refilling. For the reformers examined, the smaller units ranged from 9 to 10 m^2 and the larger ones from 10 to 18 m^2. As this area is negligible compared to the station area of 1,672 m^2 discussed previously, the addition of the reforming unit should not significantly affect the size of the hydrogen refueling station.

Summarizing Case (1), distributed natural gas reforming, the area needed to (1) extract natural gas from the ground is 46.7 GWh/(year-square mile), (2) process natural gas in a gas processing plant is 224,060 GWh/(year-square mile), (3) transport the natural gas to the refueling station is negligible, and (4) produce, compress, store, and dispense hydrogen at the refueling station is 5,766 GWh/(year-square mile). Based on these values, the area for extracting the natural gas from the ground dominates the overall energy density.

For Case (9), step (1), the land area needed to make electricity at the wind farm, was already calculated in the chapter on wind energy, Chapter 6. For an onshore wind farm, this was calculated to be 0.177 trillion BTU/(year-square mile), or 0.177×10^{12}. This can be converted 51.8 GWh/(year-square mile).

Step (2) is the footprint for the electrolyzer unit. In a brochure from Nel,[63] one company that sells electrolysis units, the Nel C-300 is listed having a total footprint of 87 m^2, or about 936 ft^2. The C-300 produces 300 Nm^3/hr of hydrogen, about 95,000 kg/day. Since Case (9) is designed with a standalone electrolyzer with a capacity of 52,300 kg H_2/day, this unit more than meets this goal. Given the land area for a wind farm, the footprint for this unit would be negligible, so we can conclude that the electrolyzer unit does not significantly affect the overall footprint.

For step (3), transporting the hydrogen by pipeline to the refueling station, the land area for this is not expected to impact the overall footprint. For pipelines, the land area is small and they can coexist with farming and pasture land.

Finally, for step (4), compressing, storing, and dispensing hydrogen at the refueling station, we expect the footprint to be the same as the previous case. This means the energy density is 19.67 trillion BTU/(year-square mile) or 5,766 GWh/(year-square mile).

Summarizing Case (9), central electrolysis using wind electricity and the hydrogen delivered by pipeline, the area needed to (1) make electricity at the wind farm is 51.8 GWh/(year-square mile). The areas to (2) electrolyze water to hydrogen and (3) transport hydrogen by pipeline to the refueling station are negligible. The area to (4) compress, store, and dispense hydrogen at the refueling station is 5,766 GWh/(year-square mile). Based on these values, the area to make electricity at the wind farm dominates the overall energy density.

It is clear then from these two cases, that the land area needed to harness the feedstocks to produce hydrogen, whether extracting natural gas from the ground or making electricity at the onshore wind farm, dominates the overall energy density.

11.12. Environmental Issues

Unlike gasoline and diesel-powered vehicles, hydrogen-powered FCVs have no emissions except water. However, considering the entire "Well-to-Wheels" process, there are harmful emissions depending on the fuel source used to produce the hydrogen. When fossil fuels such as natural gas and coal are used to produce the hydrogen, there can be emissions of sulfur dioxide (SO_2), nitrogen oxides (NO_x), and CO_2. Because FCVs have higher efficiencies than ICE vehicles, these emissions should be lower. Also, since these emissions are created at the hydrogen production plant, they could potentially be captured and sequestered, unlike emissions from a vehicle.

Considering the four cases to produce hydrogen discussed in this chapter, the amount and type of emissions vary according to the fuel source. While a brief review is given here, more detailed environment assessments can be found by referring to the chapter on each specific fuel source. The first case, Case (1), uses distributed natural gas reforming to produce the hydrogen. Referring to Chapter 2 on natural gas, the main environmental issues are hydrogen sulfide (H_2S) and nitrogen. The H_2S leaving the plant will be at a level that is not dangerous for people and the amount of

NO_x produced is about one-fifth of that made from coal or crude oil. If CO_2 capture and sequestration (CCS) become mandatory, it could be captured during hydrogen production. This is certainly more likely to be done if the methane reforming is done at a "central," offsite plant, rather than a small reforming unit onsite at the refueling station.

The second case examined was Case (5), central biomass delivered by truck as a gas. Referring to Chapter 9 on biomass, the main environmental issues are possible deforestation, CO_2 emissions, water use, and possible runoff of fertilizers and pesticides used to grow the biomass. Deforestation is especially problematic, because it is the photosynthetic CO_2 consumption of the forest biomass that is needed to make biomass a nearly CO_2-neutral technology. Failure to maintain careful control of biomass growth and consumption could result in CO_2 emissions becoming an issue.

The third case examined was Case (9), central electrolysis using wind electricity and delivered by pipeline. Referring to Chapter 6 on wind, the main environmental issues are land use, habitat disturbance, impact on birds and bats, soil erosion, noise, and visual impact. This approach to produce hydrogen is attractive since there are no harmful gas emissions since fossil fuels are not used.

The last case examined was Case (10), central coal gasification delivered by pipeline. Referring to Chapter 3 on coal, the main environmental issues are mine explosions, carbon monoxide (CO) poisoning, ash disposal, capture of sulfur and NO_x from the flue gas, wastewater disposal, and the destruction of land and wildlife habitats. If the sulfur and nitrogen oxides are captured and sequestered at the hydrogen production plant, these would not be an issue. Likewise, any CO_2 produced at the plant could be significantly reduced or eliminated with CCS.

Although the only emission from an FCV is water, there is some concern that this water could create treacherous road conditions in the form of fog and ice. To examine this further, we consider water production from both a hydrogen-powered FCV and a gasoline ICE. Earlier, we used NREL fuel economies of 45 miles per kg of hydrogen for an FCV and 24.7 miles per gallon for a gasoline-powered vehicle.[7] Assuming a simple composition of "-CH_2-" repeating units for gasoline, the equations for the ICE and FCV can be written thus:

Gasoline-Powered ICE

$$CH_2 + 3/2\ O_2 \rightarrow CO_2 + H_2O \tag{12}$$

Hydrogen-Powered FCV

$$H_2 + ½\ O_2 \rightarrow H_2O \tag{13}$$

To compare these equations on a mass basis, a density of 0.75 kg/L is assumed for gasoline, or 2.84 kg/gallon. Also, simple atomic weights are used for carbon (12), hydrogen (1), and oxygen (16). As 2.84 kg of CH_2 represents 0.203 moles, the rest of the equation for gasoline can be converted to mass. The equation for hydrogen was also converted to mass starting with one gge of hydrogen, or 1 kg H_2.

Gasoline-Powered Internal Combustion Engine

$$2.84 \text{ kg } CH_2 + 9.74 \text{ kg } O_2 \rightarrow 8.93 \text{ kg } CO_2 + 3.65 \text{ kg } H_2O \quad (14)$$

Hydrogen-Powered FCV

$$1 \text{ kg } H_2 + 8 \text{ kg } O_2 \rightarrow 9 \text{ kg } H_2O \quad (15)$$

Since the 2.84 kg of gasoline represents one gallon and a fuel efficiency of 24.7 miles per gallon was assumed, the water production is 0.12 kg/mile. For the FCV, the fuel economy is assumed to be 45 miles per kg of hydrogen, so the water production is 0.2 kg/mile. And, if we use the fuel economy of 66 mpg for the Toyota Mirai (Table 11.8), the water production is 0.14 kg/mile. Because of the higher fuel efficiency for the hydrogen-powered FCV, the water produced per mile is about the same for both types of vehicles.

Another possible concern, however, is the water exhaust temperature and the possibility of condensation into fog and ice, especially during winter. For a gasoline-powered ICE, typical operating temperatures are 500 to 700°C. For the hydrogen-powered FCV, however, typical operating temperatures of the fuel cell are 60 to 80°C.[64] Thus, although water production is about the same for both gasoline and hydrogen, the exhaust temperature for the water produced by the FCV will be cooler, which could be an issue during winter.

11.13. CO_2 Production and the Cost of Capture and Sequestration

Several studies report the mass of carbon dioxide equivalent (CO_2eq) per mile for FCVs using a variety of fuels to produce hydrogen. Since methane is used as a fuel source to produce hydrogen, the masses are given in effective global warming potentials (GWP) which include the effect of methane. According to the Environmental Protection Agency (EPA),[65] methane is estimated to have a GWP of 28 to 36 times that of CO_2. While methane in the atmosphere has a life of about a decade, compared to thousands of years for CO_2, it absorbs more solar energy than CO_2 and, will produce ozone which itself is a greenhouse gas. These factors result in methane have a higher GWP.

An Argonne National Laboratory (ANL) report[66] gives results for a life-cycle analysis (LCA) where hydrogen is produced from natural gas reforming. The LCA includes materials acquisition, transportation, and processing as well as plant construction and production distribution. In the report, they also break down the different components that contribute to the overall total. For natural gas reforming to hydrogen, the total is 286 g CO_2eq/mile driven. Of this, 20 g CO_2eq/mile come from the natural gas production, 220 from the methane reforming to hydrogen, and 46 from methane leakage in the overall process. By comparison, a gasoline ICE vehicle produces 480 g CO_2eq/mile. Also, a BEV using a natural gas combined cycle (NGCC) power plant to make the electricity produces 220 g CO_2eq/mile and a BEV using coal to make the electricity produces 520 g CO_2eq/mile.

Another LCA analysis[67] reports CO_2eq produced from natural gas reforming, coal gasification, and water electrolysis using wind-generated electricity. Here, the results are given in units of CO_2eq/kg H_2. The reported values are 11.9 kg CO_2eq/kg H_2 for natural gas reforming, 11.3 for coal gasification, and 0.97 for water electrolysis from wind electricity. Assuming a fuel economy of 45 miles per kg of hydrogen, these values can be converted to g CO_2eq/mile. Doing this gives calculated values of 260 g CO_2eq/mile, 250 g CO_2eq/mile, and 22 g CO_2eq/mile for natural gas, coal, and wind, respectively. Note that the value of 260 g CO_2eq/mile for natural gas reforming is similar to the value of 286 given in the previous paragraph.

Finally, the NREL report[14] used to as a data source to calculate the LCOF also reports the amount of CO_2eq produced per mile driven. In this chapter, we focused on four of the NREL reports ten routes to produce hydrogen, namely Case (1) distributed natural gas reforming, Case (5) central biomass delivered by truck as a gas, Case (9) central electrolysis using wind electricity and delivered by pipeline, and Case (10) central coal gasification delivered by pipeline. A range of values was given in this report based on assumed fuel economies of 48 and 68 mpgge (miles per gallon of gasoline equivalent). For Case (1), the values are 350 and 250 gCO_2eq/mile, for Case (5) 200 and 140 gCO_2eq/mile, for Case (9) 80 and 50 gCO_2eq/mile, and for Case (10) 200 and 130 gCO_2eq/mile. For Case (9), in which the hydrogen is made by electrolysis using wind-generated electricity, the small amounts of CO_2 generated are due mainly to the electric plant and pipeline construction. Case (5), which uses biomass to produce the hydrogen, will benefit from new biomass growth, reducing produced CO_2 via photosynthesis. Case (10), which uses coal gasification to produce hydrogen, includes carbon capture and sequestration (CCS) at the coal gasification plant. Otherwise, these values would be much larger.

As this chapter focuses on the use of hydrogen for transportation fuel, the application of an average heat rate to generate electricity is not appropriate. Therefore, there is no conversion of the CO_2 production values to units of lb CO_2/kWh and g CO_2/kWh. It is useful, however, to convert from units of g CO_2eq/mile to units of g CO_2eq/MM BTU and lb CO_2eq/MM BTU. From the previous discussion, we start with the following values for each case.

Case (1): Distributed natural gas reforming = 350 gCO_2eq/mile
Case (5): Central biomass delivered by truck as a gas = 200 gCO_2eq/mile
Case (9): Central electrolysis using wind electricity = 80 gCO_2eq/mile
Case (10): Central coal gasification = 200 gCO_2eq/mile

For comparison purposes, we also include:

Gasoline ICE = 480 g CO_2eq/mile
Battery electric vehicle using a natural gas combined cycle = 220 g CO_2eq/mile
Battery electric vehicle using coal gasification = 520 g CO_2eq/mile

To convert from units of g CO_2eq/mile to g CO_2eq/MM BTU, the fuel economy was assumed to be 45 miles per kg hydrogen for the FCV and 24.7 miles per gallon for the gasoline-powered ICE. For the BEV, Table 11.1 gave BEV fuel economies from 2.9 to 4.1 miles/kWh and a gallon of gasoline equivalent (gge) is 33.4 kWh. Using the lower value of 2.9 miles/kWh, this can be converted to a fuel economy of 96.9 miles/gge, which is rounded to 97 miles/gge. Also, the energy content for both a gallon of gasoline and one kg of hydrogen (one gge) is 116,000 BTU. The results are shown in Table 11.9, which also includes units of lb CO_2eq/MM BTU.

Table 11.9 illustrates several things. Not surprisingly, the use of wind-generated electricity for Case (9) gives low production of greenhouse gases (GHGs), as the only GHGs produced come from plant construction and storage and dispensing of the hydrogen. Case (10), central coal gasification, is better than Case (1), distributed natural gas reforming, because carbon capture and sequestration (CCS) is included for coal. If natural gas was used offsite to generate hydrogen with the inclusion of CCS, natural gas would produce less GHGs than coal. Interesting enough, in spite of poorer fuel economy, the GHGs produced per mile for the gasoline ICE are not worse than distributed natural gas, Case (1).

Also interesting is the high values for the two BEVs cases. However, this may not be surprising as the electricity for these two cases was generated by fossil fuels, namely natural gas and coal. If CCS is included at the natural gas or coal electric plant, these numbers would be significantly reduced. If you own a BEV, the GHGs produced depend on the mix of fuels that are used to generate electricity in your part of the

Table 11.9 Comparison of Fuel Cell Vehicles

	g CO_2eq/ mile	Fuel Economy, miles/gge	BTU/gge	g CO_2eq/ MM BTU	lb CO_2eq/ MM BTU
Case (1) FCV	350	45	116,000	135,776	299
Case (5) FCV	200	45	116,000	77,586	171
Case (9) FCV	80	45	116,000	31,034	68
Case (10) FCV	200	45	116,000	77,586	171
Gasoline ICE	480	24.7	116,000	102,207	225
BEV NGCC	220	97	116,000	183,966	406
BEV Coal Gasification	520	97	116,000	434,828	959

country. These indirect emissions (that is, not emitted from the vehicle) depend on your local mix for electricity generation, which could include varying percentages of renewables such as wind, hydroelectric, and solar as well as fossil fuels and nuclear power. These results make it clear that the advantage of reduced GHG emissions from an FCV or BEV depend on how the hydrogen or electricity is generated. And, if the hydrogen or electricity is generated from fossil fuels without the use of CCS, there may not be any real advantage in using hydrogen as a fuel to reduce GHGs.

11.14. Chapter Summary

The use of hydrogen in a FCV is in its infancy, with only 46 hydrogen stations currently in the United States, most of these in California. Of these stations, almost all are dispensing hydrogen at either 350 or 700 bar pressure, so novel approaches to onboard hydrogen storage are not yet practical from a commercial standpoint. Worldwide there were 274 refueling stations as of January 2017. Denmark is aggressively working to build a hydrogen infrastructure, and uses electrolysis powered by renewable energy to generate the hydrogen. In terms of FCVs on the road in the United States, the number of vehicles is currently greater than 3,000, much less than gasoline-powered and BEV. Relative to BEVs, a hydrogen-powered FCV eliminates the issue of long refueling time, because hydrogen refueling is equivalent to gasoline and diesel.

Calculated values for the LCOF range from about \$3.4 kg/H_2 to \$10.7/kg H_2. As one kg of hydrogen is equivalent in energy to one gallon of gasoline, this basically means that hydrogen costs from \$3.4 to \$10.7 per gallon of gasoline equivalent (gge). However, the current market price for hydrogen is about \$15/kg H_2, or \$15/gge. If the hydrogen infrastructure continues to grow, these prices should continue to decrease to more affordable levels, thus allowing it to compete with gasoline. Also, since the FCV gets about twice the miles per gge compared to a gasoline or diesel-powered vehicle, if an LCOF of \$3.4/kg H_2 can be commercially achieved, this is roughly equivalent to \$1.7 per gasoline gallon when the better fuel economy is taken into consideration.

As an energy type, hydrogen from fossil fuels such as natural gas or coal is a nonrenewable potential energy source. Although hydrogen on earth only exists in minute amounts in the atmosphere, it does exist in the form of fossil fuels, which are energy sources. If the hydrogen is made from electrolysis of water, it is a renewable potential energy carrier where renewable energy, such as wind, solar, or hydroelectric, have been converted to energy in the form of hydrogen.

A life cycle analysis of CO_2eq production for hydrogen-powered FCVs gave values ranging from 68 to 299 lb CO_2/MM BTU. The amount of greenhouse gas emissions depends strongly on the fuel that is used to produce the hydrogen. Interesting enough, in spite of poorer fuel economy, the gasoline ICE produces 225 lb CO_2/MM BTU, similar to the FCV cases where the hydrogen is made from natural gas, coal, or biomass. Not surprisingly, producing the hydrogen from wind-powered electrolysis had the lowest CO_2 emissions.

The capacity factors for hydrogen production were all presumed to be 90% in the model studies chosen for this chapter. When more hydrogen refueling stations and hydrogen generation plants are built, commercial operating data may change these values. Nevertheless, 90% is consistent with commercial electrical plants that generate electricity from natural gas (Chapter 2) and coal (Chapter 3).

The footprints and energy densities for hydrogen generation are controlled by the land area needed to harness the energy to produce hydrogen, whether extracting natural gas from the ground or making electricity at the onshore wind farm. For central electrolysis using wind electricity to make the hydrogen, the energy density was calculated to be 51.8 GWh/(year-square mile). For distributed natural gas reforming, the energy density was calculated to be 46.7 GWh/(year-square mile).

For hydrogen produced from natural gas reforming, the overall energy balance showed that 32.8% of the original natural gas energy was used to power the FCV. For hydrogen produced from biomass, electrolysis using wind electricity, and coal these energy balances were 22.5%, 8.7%, and 27.9%, respectively. The poor energy balance for producing hydrogen from wind electricity is due to the low turbine efficiency of 26%, meaning that generating electricity from wind captures only 26% of the original energy in the wind.

Finally, the main environmental issues are related to the fuel source used to produce the hydrogen. If a fossil fuel, such as natural gas or coal is used, and emissions generated are not captured at the hydrogen production plant, then producing hydrogen can generate SO_2, NO_x, and CO_2. If a renewable fuel, such as wind or solar is used, then there are no harmful gas emissions. Hydrogen FCVs will emit water, but the amount generated is similar to a gasoline-powered ICE. There is some concern, however, that water from an FCV could create treacherous road conditions in the form of fog and ice. Even though the amount of water generated from the tailpipe of an FCV is equivalent to that of a gasoline ICE, the lower exhaust temperature for the FCV could make condensation into fog and ice an issue.

References

1. "Find Plug-in Vehicles," Plug in America, 2018, https://pluginamerica.org/vehicles/?gclid=EAIaIQobChMImq_b5ND22QIVDLjACh0BKAbxEAAYAyAAEgLojPD_BwE&fwp_type=ev&fwp_class=cars-trucks
2. "Electric Cars 101: The Answers to All Your EV Questions" Plugging in to the Reality of EVs," Consumer Reports, March 29, 2017, https://www.consumerreports.org/hybrids-evs/electric-cars-101-the-answers-to-all-your-ev-questions/
3. "U.S. Convenience Store Count," NACS (National Association of Convenience Stores), December 31, 2019, https://www.convenience.org/Research/FactSheets/ScopeofIndustry/IndustryStoreCount

4. "Partnership Plan," March 2006, FreedomCar Fuel Partnership, https://www.hydrogen.energy.gov/pdfs/fc_fuel_partnership_plan.pdf
5. "Hydrogen Production: Natural Gas Reforming," Office of Energy Efficiency & Renewable Energy, Energy.gov, 2020, https://www.energy.gov/eere/fuelcells/hydrogen-production-natural-gas-reforming
6. "BP Statistical Review of World Energy," June 2018, https://www.bp.com/content/dam/bp/business-sites/en/global/corporate/pdfs/energy-economics/statistical-
7. "Hydrogen Pathways: Cost, Well-to-Wheels Energy Use, and Emissions for the Current Technology Status of Seven Hydrogen Production, Delivery, and Distribution Scenarios," Mark Ruth et al., National Renewable Energy, Technical Report NREL/TP-6A1-46612, September 2009, https://www.nrel.gov/docs/fy10osti/46612.pdf
8. "How Much Gasoline Does the United States Consume?," Energy Information Administration, March 3, 2020, https://www.eia.gov/tools/faqs/faq.php?id=23&t=10
9. "U.S. Vehicle Fuel Economy Rises to Record 24.7 mpg: EPA," David Shepardson and Nick Carey, Reuters News, January 11, 2018, https://www.reuters.com/article/us-autos-emissions/u-s-vehicle-fuel-economy-rises-to-record-24-7-mpg-epa-idUSKBN1F02BX
10. "Current U.S. Hydrogen Production," Fred Joseck et al., DOE Hydrogen and Fuel Cells Program Record, Record #: 14015, May 24, 2016, https://www.hydrogen.energy.gov/pdfs/14015_current_us_h2_production.pdf
11. "U.S. and World H2 Production 2014," Daryl R. Brown, March 2016, CryoGas International, https://www.researchgate.net/publication/301959555_US_and_World_H2_Production_2014
12. "DOE: 10 Million Metric Tons of Hydrogen Produced Annually in the US; 68% for Petroleum Processing," Green Car Congress, May 16, 2018, https://www.greencarcongress.com/2018/05/20180516-doeh2.html
13. "Hydrogen Production and Consumption in the U.S.; The Last 25 Years," Daryl R. Brown, CryoGas International 53(9),40–41, September 1, 2015, https://www.researchgate.net/publication/296332889_Hydrogen_Production_and_Consumption_in_the_US_The_Last_25_Years
14. "Hydrogen Pathways Updated Cost, Well-to-Wheels Energy Use, and Emissions for the Current Technology Status of Ten Hydrogen Production, Delivery, and Distribution Scenarios," T. Ramsden et al., National Renewable Energy Laboratory, Technical Report NREL/TP-6A10-60528, March 2013, https://www.nrel.gov/docs/fy14osti/60528.pdf
15. "Air, Climate, Energy, Water, and Security, California Fuel Cell Partnership, 2016, https://cafcp.org/sites/default/files/W2W-2016.pdf
16. "Overview on Water Electrolysis for Hydrogen Production and Storage: Results of the NOW Study," Tom Smolinka et al., Symposium—Water electrolysis and hydrogen as part of the future Renewable Energy System, May 10, 2012, http://www.hydrogennet.dk/fileadmin/user_upload/PDF-filer/Aktiviteter/Kommende_aktiviteter/Elektrolysesymposium/Tom_Smolinka.pdf
17. "Alternative Fueling Station Count by State," Alternative Fuels Data Center, US Department of Energy, Energy Efficiency & Renewable Energy, 2019, https://afdc.energy.gov/stations/states
18. "92 new hydrogen refueling stations worldwide in 2016," FuelCellsWorks, February 23, 2017, https://fuelcellsworks.com/news/92-new-hydrogen-refuelling-stations-worldwide-in-2016
19. "Performance of Existing Hydrogen Stations," Keith Wipke et al., National Renewable Energy Laboratory, 2017 Fuel Cell Seminar and Energy Exposition, November 8, 2017, NREL/PR-5400-70527, https://www.nrel.gov/docs/fy18osti/70527.pdf
20. "Advancing Technology for America's Transportation Future," National Petroleum Council, Chapter 15, Hydrogen, 2012.

21. "Types of Fuel Cells," Energy.gov, Energy Efficiency & Renewable Energy, 2020, https://www.energy.gov/eere/fuelcells/types-fuel-cells
22. "Fuel Cell Electric Vehicles," US Department of Energy, Energy Efficiency & Renewable Energy, Alternate Fuels Data Center, 2020, https://www.afdc.energy.gov/vehicles/fuel_cell.html
23. "Hydrogen and Fuel Cells Overview," Sunita Satyapal, US Department of Energy, DLA Worldwide Energy Conference, April 12, 2017, https://www.energy.gov/sites/prod/files/2017/06/f34/fcto-h2-fc-overview-dla-worldwide-energy-conf-2017-satyapal.pdf
24. "Hydrogen Production: Fundamentals and Case Study," K.W. Harrison et al., National Renewable Energy Laboratory, 18th World Hydrogen Energy Conference, May 16–21, 2010, https://www.nrel.gov/docs/fy10osti/47302.pdf
25. "Executive Summaries for the Hydrogen Storage Materials Centers of Excellence Chemical Hydrogen Storage CoE, Hydrogen Sorption CoE, and Metal Hydride CoE, Period of Performance: 2005–2010, Fuel Cell Technologies Program, Office of Energy Efficiency and Renewable Energy, US Department of Energy, April 2012, https://www.energy.gov/sites/prod/files/2014/03/f11/executive_summaries_h2_storage_coes.pdf
26. "Hydrogen Storage Tech Team Roadmap," July 2017, US Drive, https://www.energy.gov/sites/prod/files/2017/08/f36/hstt_roadmap_July2017.pdf
27. "Status of Hydrogen Storage Technologies," Office of Energy Efficiency & Renewable Energy, 2019, https://www.energy.gov/eere/fuelcells/status-hydrogen-storage-technologies
28. "Technical Assessment: Cryo-Compressed Hydrogen Storage for Vehicular Applications," October 30, 2006, US Department of Energy Hydrogen Program, https://www.hydrogen.energy.gov/pdfs/cryocomp_report.pdf
29. "2016 Annual Merit Review Presentation," D.L. Anton and T. Motyka, Savannah River National Laboratory (SRNL), June 9, 2016, Hydrogen Storage Engineering Center of Excellence, https://www.hydrogen.energy.gov/pdfs/review16/st004_anton_2016_o.pdf
30. "MOF-5 (or IRMOF-1) Metal Organic Framework," ChemTube3D, University of Liverpool, 2020, http://www.chemtube3d.com/solidstate/MOF-MOF5.html
31. "Electric Power Monthly, Energy Information Administration, Data for June 2018, Table 5.6.A. Average Price of Electricity to Ultimate Customers by End-Use Sector, https://www.eia.gov/electricity/monthly/epm_table_grapher.php?t=epmt_5_6_a
32. "Clean Cities Alternative Fuel Price Report," US Department of Energy, January 2018, https://www.afdc.energy.gov/uploads/publication/alternative_fuel_price_report_jan_2018.pdf
33. "Cost to Refill," California Fuel Cell Partnership, 2016, https://cafcp.org/content/cost-refill
34. "California Average Weekly Retail Gasoline Prices," California Energy Commission, January 2018, http://www.energy.ca.gov/almanac/transportation_data/gasoline/retail_gasoline_prices2.html#2017
35. "Average US Wind Price Falls to $20 per Megawatt-Hour," Emma Foehringer Merchant, Green Tech Media, August 24, 2018, https://www.greentechmedia.com/articles/read/us-wind-prices-20-megawatt-hour#gs.a7q2sr
36. "Natural Gas: Henry Hub Natural Gas Spot Price," Energy Information Administration, 2018, https://www.eia.gov/dnav/ng/hist/rngwhhdm.htm
37. "Updated Capital Cost Estimates for Utility Scale Electricity Generating Plants," Energy Information Administration, April 2013; and "Levelized Cost and Levelized Avoided Cost of New Generation Resources in the Annual Energy Outlook 2015," Table 1, June 2015.
38. "Analysis of Natural Gas-to Liquid Transportation Fuels via Fischer-Tropsch," September 13, 2013, Erik Shuster, DOE/NETL-2013/1597, National Energy Technology Laboratory (NETL), https://www.netl.doe.gov/energy-analyses/temp/FY14_AnalysisofNaturalGas-to-LiquidTransportationFuelsviaFischer-Tropsch_090113.pdf

39. "Module 14. 'Life Cycle Assessment (LCA)': 4 Steps of LCA, Approaches, Software, Databases, Subjectivity, Sensitivity Analysis, Application to a Classic Example," Program for North American Mobility in Higher Education, École Polytechnique de Montréal, 2003, www.polymtl.ca/namp/docweb/Modules_Web/M14_Part1_Tier3.ppt
40. "Petroleum & Other Liquids: Refinery Utilization and Capacity," Energy Information Administration, 2018 annual average, https://www.eia.gov/dnav/pet/pet_pnp_unc_dcu_nus_a.htm
41. "Table 6.7.B. Capacity Factors for Utility Scale Generators Not Primarily Using Fossil Fuels," Electric Power Monthly, January 2013–February 2019, https://www.eia.gov/electricity/monthly/epm_table_grapher.php?t=epmt_6_07_b
42. "Hydrogen Fuel Cells," Department of Energy Hydrogen Program, October 2006, https://www.hydrogen.energy.gov/pdfs/doe_fuelcell_factsheet.pdf
43. "Fuel Cell Electric Vehicle Evaluation," Jennifer Kurtz et al., DOE 2017 Annual Merit Review, National Renewable Energy Laboratory, Project ID TV001, https://www.hydrogen.energy.gov/pdfs/review17/tv001_kurtz_2017_o.pdf
44. "Comparison between Battery Electric Vehicles and Internal Combustion Engine Vehicles Fueled by Electrofuels from an Energy Efficiency and Cost Perspective," Tobias Gustafsson and Anders Johansson, Department of Energy and Environment, Chalmers University of Technology, Master's Thesis FRT 2015:02, Gothenburg, Sweden 2015, https://pdfs.semanticscholar.org/cfb3/aa91cb4b0358c73bb2690629663c0947411a.pdf
45. "Life Cycle Analysis of Natural Gas Extraction and Power Generation," T.J. Skone et al., National Energy Technology Laboratory (NETL), DOE Contract Number DE-FE0004001, Figure 4.3, May 29, 2014.
46. "Life Cycle Assessment of a Biomass Gasification Combined-Cycle System," Margaret K. Mann and Pamela L. Spath, December 1997, NREL Report 23076, https://www.nrel.gov/docs/legosti/fy98/23076.pdf
47. "Alternatives for Estimating Energy Consumption," Appendix F, Annual Energy Review 2010, https://www.eia.gov/totalenergy/data/annual/pdf/sec17.pdf
48. "Monthly Energy Review," Energy Information Administration, July 2016, https://www.eia.gov/totalenergy/data/monthly/pdf/mer.pdf
49. "Mining Industry Energy Bandwidth Study," June 2007, US DOE, Industrial Technologies Program.
50. "Railroad Transportation Energy Efficiency," Chris Barkan, University of Illinois at Urbana-Champaign, 2007, http://www.istc.illinois.edu/about/seminarpresentations/20091118.pdf
51. "Wind-to-Hydrogen Project: Electrolyzer Capital Cost Study," Genevieve Saur, National Renewable Energy Laboratory, Technical Report NREL/TP-550-44103, December 2008, https://www.nrel.gov/docs/fy09osti/44103.pdf
52. "Renewable Electrolysis Integrated System Development and Testing, Kevin W. Harrison, National Renewable Energy Laboratory, 2008 DOE Hydrogen Program Annual Merit Review, June 11, 2008, https://hydrogendoedev.nrel.gov/pdfs/review08/pdp_4_harrison.pdf
53. "The History of Hydrogen," James Jonas, Altenergy Magazine, April 1, 2009, https://www.altenergymag.com/article/2009/04/the-history-of-hydrogen/555/
54. "History of Hydrogen Powered Cars," January 30, 2016, http://fuel-efficient-vehicles.org/energy-news/?page_id=819
55. "Fuel Cell History," Fuel Cell Today, 2020, http://www.fuelcelltoday.com/history
56. "Station Status," California Fuel Cell Partnership, 2019, https://m.cafcp.org/
57. "With 9 Hydrogen Fuel Stations, Denmark Is 1st Country with Basic National Network," Mark Kane, March 14, 2016, https://insideevs.com/with-9-hydrogen-fuel-stations-denmark-is-1st-country-with-basic-national-network/

58. "Hydrogen and Fuel Cells Progress Overview," Sunita Satyapal, Houston, TX, May 23, 2017, https://www.energy.gov/sites/prod/files/2017/05/f34/fcto_may_2017_h2_scale_wkshp_satyapal.pdf
59. "Toyota Mirai Hydrogen Fuel-Cell Car Tops 3,000 California Sales," Stephen Edelstein, January 24, 2018, http://www.thedrive.com/tech/17924/toyota-mirai-hydrogen-fuel-cell-car-tops-3000-california-sales
60. "Compare Fuel Cell Vehicles," US Department of Energy, Energy Efficiency and Renewable Energy, March 2020, https://www.fueleconomy.gov/feg/fcv_sbs.shtml
61. "Safety, Codes and Standards for Hydrogen Installations: Hydrogen Fueling System: Footprint Metric Development," A.P. Harris et al., Sandia National Laboratories, Report 2014-3416, April 2014, http://energy.sandia.gov/wp-content/gallery/uploads/SAND_2014-3416-SCS-Metrics-Development_distribution.pdf
62. "IEA-HIA Task 23 Small-Scale Reformers for On-Site Hydrogen Supply," 2006–2011, Final report, Ingrid Schjølberg et al., International Energy Agency, http://ieahydrogen.org/pdfs/Task23_Final-Report_ISBN.aspx
63. "Nel Hydrogen Electrolyser: The World's Most Efficient and Reliable Electrolyser," January 2017, http://nelhydrogen.com/assets/uploads/2017/01/Nel_Electrolyser_brochure.pdf
64. "Fuel Cell Electric Vehicles and Hydrogen Infrastructure: Status 2012," Ulrich Eberle and Rittmar von Helmolt, Energy & Environmental Science 5(10), 8790–8798, July 2012, https://www.researchgate.net/publication/233987484_Fuel_cell_electric_vehicles_and_hydrogen_infrastructure_Status_2012
65. "Greenhouse Gas Emissions, Understanding Global Warming Potentials," US Environmental Protection Agency, January 2017, https://www.epa.gov/ghgemissions/understanding-global-warming-potentials
66. "XI.2 Life-Cycle Analysis of Vehicle and Fuel Systems with the GREET Model," Michael Wang et al., Argonne National Laboratory, 2012, https://www.hydrogen.energy.gov/pdfs/progress12/xi_2_wang_2012.pdf
67. "Life Cycle Assessment of Hydrogen Production Methods—A Review," R. Bhandari et al., June 2012, STE Research Report, http://www.fz-juelich.de/SharedDocs/Downloads/IEK/IEK-STE/DE/Publikationen/research_reports/2012/report_06_2012.pdf?__blob=publicationFile

12
Fischer-Tropsch Synthesis
A Renewable or Nonrenewable Energy Type

12.1. Foreword

Simply stated, the Fischer-Tropsch synthesis is a catalytic polymerization reaction that employs a metal catalyst to hydrogenate carbon monoxide (CO) into hydrocarbon chains. Polymerization is generally defined as a process to combine monomers, molecules that combine with others of the same type, to form a long chain molecule. Equation (1) shows this reaction for CO with "n" representing an integer.

$$(2n+1)H_2 + nCO \rightarrow C_nH_{(2n+2)} + nH_2O \tag{1}$$

If "n" equals one, the Equation (1) simply becomes the methanation reaction, shown in Equation (2). That is, CO is hydrogenated with hydrogen (H_2) to form methane.

$$3H_2 + CO \rightarrow CH_4 + H_2O \tag{2}$$

The methanation reaction has been around since 1902, when it was discovered by Paul Sabatier and Jean-Baptiste Senderens using nickel (Ni) and cobalt (Co) catalysts. In the case of the Fischer-Tropsch synthesis, the resulting product is a mixture of different hydrocarbon chain lengths in which "n" ranges from one to very large integers. Because the Fischer-Tropsch synthesis is a polymerization reaction, excepting methane, the selective formation of specific hydrocarbon molecules is not possible in a one step process.

The chemistry of the Fischer-Tropsch synthesis is discussed in more detail later. Briefly, CO is a relatively stable molecule that cannot be easily hydrogenated with H_2. However, when the CO is "activated" on a catalyst, it is much more reactive. Here, activation means that the catalyst metal has adsorbed the CO and made it more reactive by lowering the energy needed to initiate the reaction with H_2 (the so-called activation energy). Thus, Equation (1) is simplified and overall contains two main reactions, namely CO activation and dissociation (into C and O) on the metal catalyst and the subsequent formation of carbon-carbon (C-C) and carbon-hydrogen (C-H) bonds.

The Changing Energy Mix. Paul F. Meier, Oxford University Press (2020). © Oxford University Press.
DOI: 10.1093/oso/9780190098391.001.0001.

If you have not heard of the Fischer-Tropsch synthesis, you are probably not alone. However, it may surprise you to know that this reaction has been around since 1925. It was discovered by Germans Franz Fischer and Hans Tropsch at the Kaiser-Wilhelm-Institut für Kohlenforschung (that is, the institute for coal research). The feedstock to produce the CO and H_2 for Equation (1) is typically coal or natural gas, but biomass can also be used. In the case of Nazi Germany, which did not have vast resources of petroleum, this process was used in World War II to help meet fuel demands using their abundant supply of coal. At one point, they had 9 plants operating and producing about 14,000 barrels per day (BPD).[1]

After World War II, the United States collected information from the Germans and worked to commercialize this process. However, vast fields of cheap oil in the Middle East made refining crude oil to make transportation fuels much cheaper than the Fischer-Tropsch process, so support for commercialization was stopped. Ironically, another ill-fated regime, the apartheid government of South Africa, took up the mantle for the Fischer-Tropsch technology. Because of apartheid, they became isolated and many countries refused to sell them crude oil. With their vast coal reserves and little access to crude oil, they formed the South Africa Synthetic Oil Liquid (SASOL) company in 1950 to produce transportation fuels via the Fischer-Tropsch synthesis.

Today, there is renewed interest in the Fischer-Tropsch synthesis, more so using natural gas and biomass as fuels instead of coal. In the world, natural gas exists in some remote and stranded locations, so if the Fischer-Tropsch process was used to produce liquid hydrocarbons, they could be more easily transported to market than by building long and expensive pipelines. And, as discussed in Chapters 8 and 9, biomass is an abundant and renewable resource, therefore making it a good feedstock to produce liquid transportation fuels.

12.2. Proven Reserves

Unlike other forms of energy, such as fossil fuels (petroleum, coal, natural gas), nuclear, or renewables (wind, solar, hydroelectric, biomass), there are no readily available reserves of the H_2 and CO gases needed to make the Fischer-Tropsch (Fischer-Tropsch) liquids. We can, however, estimate the potential for making H_2 and CO, and subsequently the Fischer-Tropsch liquid products, based on the fuel used as input to the process. In this chapter, we focus on three approaches to making liquid transportation fuels namely, gas-to-liquids (GTL), coal-to-liquids (CTL), and biomass-to-liquids (BTL). In the case of GTL, the gas is natural gas.

The next section, "Overview of Technology," discusses the details and steps for each of these processes. Here it is sufficient to state that these steps include the Fischer-Tropsch synthesis, a polymerization reaction, followed by subsequent processing to make the desired liquid products with an acceptable level of quality. In general, since the Fischer-Tropsch makes excellent quality diesel fuel with a high cetane number, commercial processes focus on maximizing diesel transportation fuel with a lesser amount of gasoline. To help make the estimate of potential transportation fuels from natural gas (GTL), coal (CTL), and biomass (BTL), the yields of gasoline and diesel from each fuel are needed.

Although a variety of products are made in the process, commercial units are generally designated in terms of the liquid product total for gasoline and diesel. Therefore, the potential transportation fuels that can be made is based on the total of these two products, recognizing that other useful products are also generated.

Referring to the literature, one National Energy Technology Laboratory (NETL) study[2] reports a liquid yield of 69% diesel for a GTL plant having a total liquid yield of 50,000 BPD; gasoline makes up the other 31%. Similarly, another report[3] gives a diesel yield of 71% based on commercial data from the Oryx Fischer-Tropsch plant in Qatar, which has a total liquid yield of 34,000 BPD. Using the product yield given in each report and the natural gas feed rate needed to meet this yield, we can calculate the total liquid product yield per million standard cubic feet per day (SCFD) of natural gas. For the first paper, the value is 5,115 gallons/MM SCFD and for the second paper the value is 4,327 gallons/MM SCFD (Note that one barrel is equivalent to 42 gallons). For the purpose of making estimates, we use a value of 5,000 gallons/MM SCFD of natural gas.

In Chapter 2, the chapter discussing natural gas, a British Petroleum (BP) review of world energy[4] reported world reserves of 6,832 trillion cubic feet and US reserves of 309 trillion cubic feet in 2017 (see Table 2.2). Using the previous estimate of 5,000 gallons/MM SCFD, if all these worldwide reserves were used for making Fischer-Tropsch transportation fuels, the total potential would be 34.2 trillion gallons, while if all the US reserves were used, the total potential would be 1.5 trillion gallons.

Next, we look at the potential Fischer-Tropsch liquid transportation fuels from coal. An NETL report on CTL[5] gives a diesel yield of 69% for a plant producing 50,000 BPD of liquid transportation fuels. This 50,000 BPD plant had a coal feed rate of 21,719 tons/day, meaning that one ton of coal produced 96.7 gallons of liquid products. This number is rounded to 97 gallons/ton.

The same BP review of world energy[4] was also used for coal reserves in Chapter 3 (see Table 3.2). The review gives world reserves for 2017 as 1,140,904 million tons and US reserves as 276,587 million tons. Using a liquid product yield of 97 gallons per ton, if all these worldwide reserves were used for making Fischer-Tropsch transportation fuels, the total potential would be 110.7 trillion gallons, while if all the US reserves were used, the total potential would be 26.8 trillion gallons.

Lastly, we look at the potential Fischer-Tropsch liquid transportation fuels from biomass. An NETL report on BTL[5] gives a diesel yield of 69% for a plant producing 5,000 BPD of liquid transportation fuels from switchgrass. This 5,000 BPD plant has a biomass feed rate of 4,084 tons/day, meaning that one ton of biomass produced 51.4 gallons per ton. A National Renewable Energy Laboratory (NREL) report[6] reported a similar 68% diesel yield, producing 2,479 BPD of liquid transportation fuels with a biomass feed rate of 2,205 tons/day. Here, the biomass is corn stover, and the yield is 47.2 gallons/ton. The average yield from these two reports, 49.3 gallons/ton, is used to estimate the potential yield from biomass. Note that the yield of 49.3 gallons/ton is only about half of that for coal, 97 gallons/ton. This is because the carbon content of biomass is quite a bit lower than coal. For example, in Chapter 3 the carbon content for bituminous coal was given as 80%–90% (see Table 3.1). On the other hand, the composition of wood biomass used in one Energy Information Administration (EIA) report[7] is listed as having a carbon content of only 41.6%.

In Chapter 9 on biomass, the worldwide potential was given as 232 Exajoules/year (1×10^{18} Joules) and the US potential was given as 1,147 million dry tons/year (see Table 9.2). In Chapter 11 on H_2, a typical heat content for wood and wood waste, 6,500 BTU/lb, was used to convert the worldwide potential from Exajoules to million tons per year, resulting in a value of 16,915 million tons per year. Using a liquid product yield of 49.3 gallons per ton, if all these worldwide reserves were used for making Fischer-Tropsch transportation fuels, the total potential would be 0.83 trillion gallons per year, while if all the US reserves were used, the total potential would be 0.057 trillion gallons per year.

In summary, the worldwide potential for making Fischer-Tropsch transportation fuels from natural gas is 34.2 trillion gallons, from coal is 110.7 trillion gallons, and from biomass is 0.83 trillion gallons per year. For the United States, the potential from natural gas is 1.5 trillion gallons, from coal is 26.8 trillion gallons, and from biomass is 0.057 trillion gallons per year. It is worth noting that the biomass yield is given on an annual basis and is made from renewable fuel while the yields from natural gas and coal are based on total reserves, so there is a limit on the amount of transportation fuels that can be made as these resources are not renewable.

To help put these numbers in perspective, according to the EIA,[8] the annual consumption of gasoline for the United States in 2018 was 142.86 billion gallons, which is about 9.3 million BPD (a barrel is 42 gallons). Another EIA reference[9] gives the 2018 diesel consumption as 63.6 billion gallons, about 4.15 million BPD. Combined, this equals an annual consumption of 206.5 billion gallons, about 13.5 million BPD.

The current US consumption level for gasoline and diesel represents 14% and 0.8%, respectively, of the total potential from natural gas and coal. For biomass, the potential production equals about 28% of the annual US transportation fuel consumption. Of course, natural gas, coal, and biomass resources compete with other uses, such as electricity generation, home heating, and ethanol manufacture. Another way to look at this is that if we were to use all the US reserves for natural gas and coal to only produce Fischer-Tropsch transportation fuels, the natural gas would supply about 7 years and coal about 130 years, based on current transportation fuel consumption levels.

Next, we turn to current levels of commercial Fischer-Tropsch plant production. Unfortunately, while the technology has been around many years, there are currently only a small number of commercial plants operating in the world, due primarily to high capital costs and the still relatively low cost of crude oil. Although there are a number of research and demonstration plants worldwide, Table 12.1 shows only commercial GTL and CTL plants currently in operation, or planned to start operation in the case of the Chinese Shenhua plant. The table shows the locations, startup dates,

Table 12.1 Worldwide Fischer-Tropsch Plants

Name and Location	Startup Date	Company	Liquid Products, BPD	Cost, $billion	$/BPD
GTL Plants					
Pearl, Qatar[3,10]	2010	Shell	140,000	19	136,000
Oryx, Ras Laffan Qatar[3]	2007	Sasol/Qatar Petroleum	34,000	1.2	35,000
Bintulu, Malaysia[3,10]	1993	Shell MDS	14,700	1	68,000
Mossel Bay, South Africa[3]	1992	PetroSA	45,000	4	89,000
Escravos, Nigeria[12,13]	2014	Chevron and Nigerian National Petroleum Company	34,000	10	294,000
GTL Total			267,700		
CTL Plants					
Secunda II, South Africa[14]	1980	Sasol	40,000	3.2	$40,000
Secunda III, South Africa[14]	1984	Sasol	40,000	2.5	31,000
Shenhua, Ningxia, China[11]	2018	Synfuels China	100,000	8.73	87,000
CTL Total			180,000		

company, yields of liquid transportation fuels, the cost, and the cost on a BPD basis. The capital costs were taken directly from the referenced work and were not adjusted to present day dollars, and the BPD values were rounded to nearest thousand. There are some small-scale BTL plants, such as the one operating in Oklahoma City with a daily output of 200 BPD,[15] but at the time this chapter was written, there were no BTL plants with capacities greater than 5,000 BPD.

As the table illustrates, the total for commercial GTL technology is less than 270,000 BPD, about the typical size for one US refinery. Likewise, the CTL total is 180,000 BPD. Thus, production levels are quite small compared to traditional refining methods to produce gasoline and diesel. Clearly, Qatar and South Africa dominate the commercial business, but when the Chinese plant starts operation, China will become a significant participant.

Finally, Table 12.1 shows that the capital costs for all these commercial plants are greater one billion dollars. And, on a cost per barrel basis, the average for a GTL plant is about $124,000/BPD and for a CTL plant is about $53,000/BPD. These values, and especially the one for GTL, are quite large compared to that for conventional refining, which is about $40,000/BPD (see Table A.1.1 in Appendix A). Economic calculations for the levelized cost of fuel (LCOF) are given later in this chapter, but these high capital costs are a major impediment to the construction of new commercial Fischer-Tropsch plants.

12.3. Overview of Technology

The Fischer-Tropsch (Fischer-Tropsch) synthesis is a catalytic polymerization reaction that uses metal catalysts to polymerize CO into long hydrocarbon chains. Polymerization is a chemical field that has been around a long time and the foundation for many processes that produce plastics. Polymerization is a type of chemical reaction such that monomers (molecules that can combine with others of the same type) react to form long chain molecules. The scope, complexity, and variety of mechanisms offered up for the Fischer-Tropsch synthesis are beyond the scope of this book, but a simple explanation of the polymerization reaction is as follows. First, CO adsorbs on the catalyst surface and subsequently, it is believed that the CO disassociates and is hydrogenated to form a "$-CH_2-$" monomer. Next, chain growth continues when additional CO molecules are inserted into the chain and hydrogenated, to form longer chains. This process continues and will eventually terminate, followed by product desorption from the catalyst surface. The longest molecule formed, as well as the average chain length, depends on the catalyst and operating conditions.

Another interesting aspect of the chemistry is that the Fischer-Tropsch synthesis makes straight chain molecules with no branching to form isomers, assuming there are no other catalytic influences added. Using Figure 12.1 as an example, a molecule with eight carbon atoms made by the Fischer-Tropsch synthesis will be the straight chain normal-octane (n-C_8H_{18}) such that the carbons line up with no branching from the main chain, in contrast to 3-methylheptane (also C_8H_{18}). In Figure 12.1, it is understood that the vertices are carbon atoms and that each carbon atom is fully saturated with hydrogen atoms.

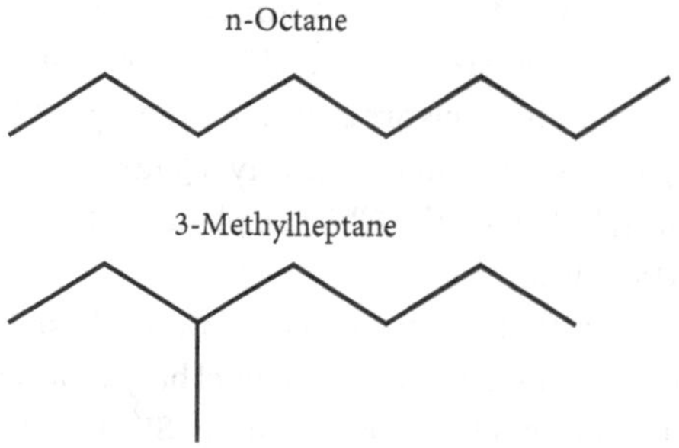

Figure 12.1 Structure of n-Octane and Its Isomer 3-Methylheptane

As shown earlier, Equation (1) gives a general overall chemical equation for this process, for which "n" is an integer.

$$(2n+1)H_2 + nCO \rightarrow C_nH_{(2n+2)} + nH_2O \qquad (1)$$

In 1956, Robert Anderson recognized that the products from the Fischer-Tropsch synthesis followed a Schulz-Flory kinetic distribution,[16] so the equation to describe the kinetic distribution from the reaction is generally referred to as the Anderson-Schulz-Flory distribution, shown in Equation (3).

$$W_n = n\,\alpha^{n-1}(1-\alpha)^2 \qquad (3)$$

Here, "W" is the weight fraction, "n" is the carbon number, and "α" is the probability of chain growth. Two examples of this distribution are shown in Figure 12.2. Here, "α" was chosen to maximize gasoline or diesel.

Defining gasoline as having carbon numbers ranging from C_5 to C_{11}, a maximum gasoline yield of 47.6 wt% was calculated at an "α" value of 0.76. At this value, the diesel yield was 17.1 wt%, with diesel defined as C_{12} to C_{25}. For maximum diesel yield, a value of 41.0 wt% was obtained at an "α" of 0.89; the gasoline yield was 29.0 wt% for this value.

Recalling earlier discussions, however, typical commercial plants have a liquid yield of around 70% for diesel and 30% for gasoline. The reason for this is that commercial plants employ additional chemistry downstream of the Fischer-Tropsch process.

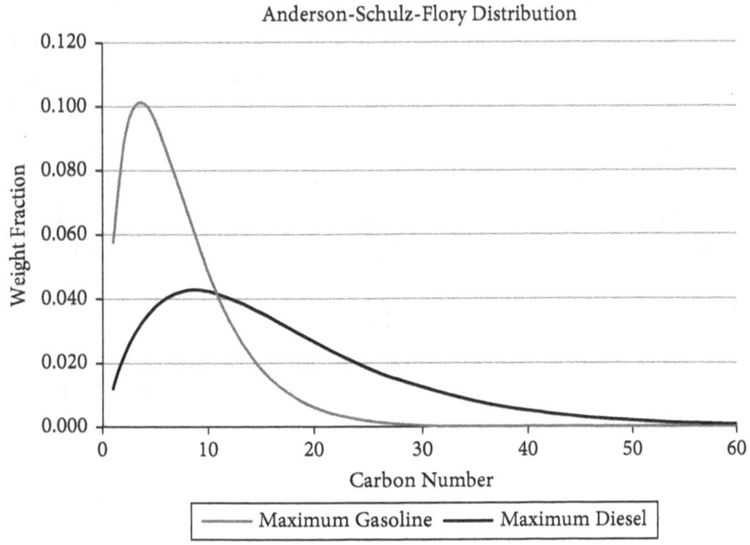

Figure 12.2 Anderson-Schulz-Flory Distribution for Maximum Gasoline and Maximum Diesel

Without getting too deep into the chemistry, diesel quality is rated by cetane number, and a good quality cetane number is obtained with linear hydrocarbons, which the Fischer-Tropsch process makes. On the other hand, gasoline is rated by octane number, and a good octane number is obtained with branched hydrocarbons, as well as olefins, aromatics, and naphthenes. Aromatics and naphthenes are ringed carbon molecules. Thus, the Fischer-Tropsch process does not produce high quality octane gasoline.

Therefore, one downstream process used is the isomerization of the naphtha stream (unfinished gasoline is called naphtha) from straight chain hydrocarbons (paraffins) into branched hydrocarbons (isoparaffins). Another typical downstream process is reforming, used to change the naphtha into some aromatic compounds. Both processes increase the octane value of the gasoline.

For the diesel, the cetane number is already excellent. However, to increase the yield of diesel, carbon numbers larger than diesel are hydrocracked to increase the diesel yield, a process which also increases the gasoline yield. Hydrocracking is a catalytic process such that H_2 is used with a catalyst to break carbon-carbon bonds to make smaller molecules.

Although beyond the scope of this book, it is worth noting that there has been extensive research to find Fischer-Tropsch catalysts that will defeat the Anderson-Schulz-Flory distribution and make more selective product distributions. In fact, that is how my research career began in 1978, trying to make C_2 to C_4 olefins in high selectivity. While it is possible to make methane with high selectivity (not a useful process if you are starting with natural gas), high selectivity of other hydrocarbon molecules has remained elusive. Therefore, it is more common to combine the Fischer-Tropsch synthesis with secondary processes to arrive at the desired product distribution. That means in order to achieve the 70% diesel yield many commercial plants strive for, the normal approach is to have a chain growth probability ("α") 0.9 or greater, recognizing that the Fischer-Tropsch synthesis combined with hydrocracking will result in a diesel yield greater than possible with the Fischer-Tropsch synthesis alone.

Turning to choice of feedstock, this chapter examines the production of Fischer-Tropsch products from natural gas (GTL), coal (CTL), and biomass (BTL). Referring back to Equation (1), the reactants for the Fischer-Tropsch are H_2 and CO with a molar ratio of H_2 to CO greater than two (H_2/CO). For natural gas, which is typically about 95% methane, the molar ratio is already two. However, for coal and biomass, the ratio is much lower. In Chapter 3 on coal, bituminous coal was shown to have 80 to 90 wt% carbon and 4 to 5 wt% H_2. Assuming 85 wt% carbon and 5 wt% H_2 and simple atomic weights of 12 and 2 for carbon and H_2, the H_2/CO ratio is 0.35. Likewise, in Chapter 9 for biomass, a typical composition of wood was shown to have 41.6 wt% carbon and 4.8 wt% H_2, leading to a H_2/CO molar ratio of about 0.69.

In general, coal and biomass have low H_2-to-carbon ratios, so to make the H_2 and CO reactants for the Fischer-Tropsch synthesis with the proper molar ratio, it is necessary to either add H_2, reject carbon, or both. To make the starting mixture of CO and H_2, also called synthesis gas or syngas, different processes are employed for natural gas versus coal and biomass.

There are at least three ways to make syngas from natural gas, using water and/or oxygen. The first is steam reforming, also called steam methane reforming (SMR), such that methane reacts with water according to Equation (4). This reaction is endothermic, meaning that heat must be added to make the reaction happen.

$$CH_4 + H_2O\ (+\ \text{heat}) \rightarrow CO + 3H_2 \tag{4}$$

The second is called partial oxidation, for which oxygen is burned with methane according to Equation (5). It is referred to as "partial," because the oxidation is not carried out to the full extent of making carbon dioxide (CO_2). This reaction is exothermic, meaning that heat is released from the reaction. This approach has the added expense of an air separation unit (ASU), to cryogenically separate oxygen from air. Although it is conceptually possible to use air instead of pure oxygen, this adds the additional complication that about 80% of the feed gas, in the form of nitrogen, does not participate in the reaction, thus roughly quadrupling the reactor size.

$$CH_4 + \tfrac{1}{2}\,O_2 \rightarrow CO + 2H_2 (+\ \text{heat}) \tag{5}$$

A balanced approach to these two methods is called autothermal reforming (ATR), which involves combining steam reforming with partial oxidation. An example of this is shown in Equation (6), in which one part of Equation (4) has been combined with two parts of Equation (5). ATR is exothermic, but less so than partial oxidation, and thus temperature control of the reactor is easier.

$$3CH_4 + H_2O + O_2 \rightarrow 3CO + 7H_2 \tag{6}$$

Because a molar ratio of about two for H_2-to-CO is desired, partial oxidation or ATR is normally used for GTL.

For CTL and BTL, a process called gasification is normally used. Generally speaking, gasification is a process that uses oxygen to convert solid fuels, such as coal and biomass, into CO, H_2, and CO_2. Although coal and biomass have small amounts of H_2, Equation (7) shows a simplified equation where only carbon is considered as the reactant to make the syngas. Like partial oxidation, gasification requires pure oxygen.

$$3C + O_2 + H_2O \rightarrow H_2 + 3CO \tag{7}$$

To increase the H_2-to-CO molar ratio to the desired value of about two, a second reaction, called the "water-gas shift reaction" or WGS, is employed. This is shown in Equation (8).

$$CO + H_2O \rightarrow CO_2 + H_2 \quad (8)$$

Based on these approaches to syngas production for GTL, CTL, and BTL, simplified flowsheets can be made for the overall processes, shown in Figures 12.3 and 12.4. These process flowsheets show the general product flow from producing syngas, the Fischer-Tropsch synthesis, product separation, and product upgrading. These simplified flowsheets do not show, however, many of the detailed steps in the process. For example, in the case of coal, it is necessary to remove and treat some of the bad things in coal, such as mercury, sulfur, and nitrogen. The flowsheets for natural gas, Figure 12.3, is shown using partial oxidation to make the syngas, recognizing that this step could be replaced with steam reforming or autothermal reforming.

The cryogenic air separation unit (ASU) to produce pure oxygen from air is an expensive component, making up about 30% of the cost of making the syngas and about 7 to 8% of the entire capital costs.[2,12] For this reason, there has been much research into reducing the cost for this part of the process. One approach is to use an oxygen transport membrane (OTM), such that oxygen is selectively transported across a ceramic membrane, leaving the nitrogen behind. Once transported, partial oxidation or autothermal reforming to make syngas could proceed as described earlier.

As a representative of Phillips 66, I participated in an OTM research alliance with four other companies, BP-Amoco, Praxair, Sasol, and Statoil from 1998 to 2001. Unfortunately, there are technical difficulties to transform the ceramic membrane

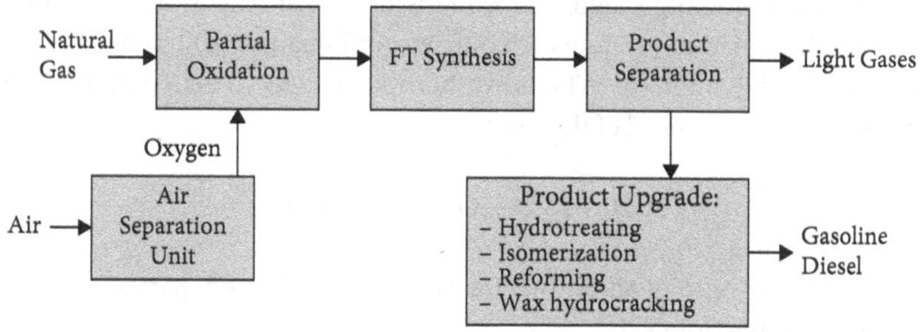

Figure 12.3 Simple Process Flowsheet for GTL Plant Using Partial Oxidation

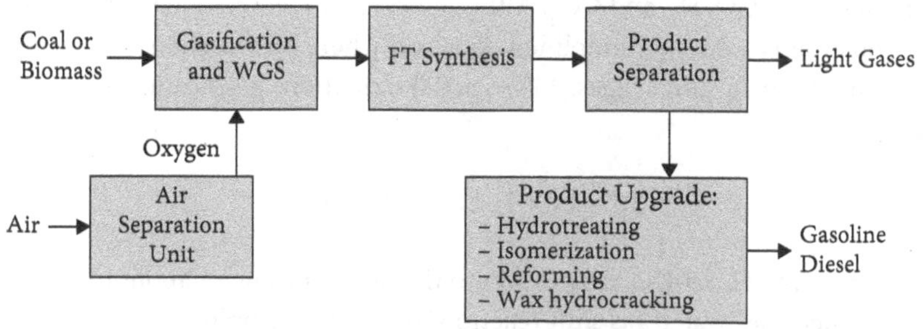

Figure 12.4 Simple Process Flowsheet for CTL or BTL Plant Using Gasification

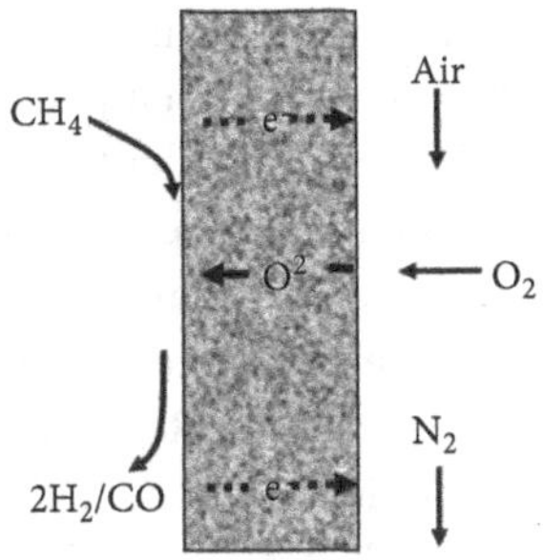

Figure 12.5 Schematic of Oxygen Transport Membrane (OTM)

into tubes that can be placed in a reactor, difficulties getting a high oxygen flux across the membrane, and keeping the cost low enough to compete with an ASU. As a result, this approach was never successfully commercialized.

The last topic for this section is the types of catalysts used in the Fischer-Tropsch synthesis. Indeed, entire books have been written on catalysts for the Fischer-Tropsch synthesis. In the field of catalysis, for which catalysts are materials to facilitate chemical reactions at lower temperatures and with unique selectivity to products, the Group VIIIB elements are important. Looking at a periodic table, the elements iron (Fe), cobalt (Co), and Nickel (Ni), as well as the atoms below them in their respective columns, including ruthenium (Ru), rhodium (Rh), palladium (Pd), osmium (Os) and iridium (Ir), and platinum (Pt), make up the Group VIIIB elements. Of these, iron, cobalt, ruthenium, and nickel have been used in the Fischer-Tropsch synthesis. However, ruthenium is scarce and expensive and nickel is known as a methanation catalyst, meaning it typically makes methane and not the long-chain hydrocarbon molecules desired for gasoline and diesel. As a result, only iron- and cobalt-based Fischer-Tropsch catalysts have found commercial application.[17] Iron generally makes more olefins and oxygenates than cobalt, which yields better quality diesel (paraffins) and can operate at lower pressure than iron catalysts.

Metal promoters, things added to enhance the performance of the basic metal catalyst, are also important in the Fischer-Tropsch synthesis. Of note are the alkali metals, lithium (Li), sodium (Na), potassium (K), and rubidium (Rb), which have the special property of increasing the selectivity to longer chain molecules.

12.4. Capital Cost, Operating Cost, Well-to-Wheels Levelized Cost of Electricity, and Well-to-Wheels Levelized Cost of Fuel

The equations and methodology used to calculate LCOF, and specific input used for the Fischer-Tropsch synthesis to make gasoline and diesel, are detailed in Appendix A. Three different Fischer-Tropsch approaches were evaluated, namely GTL, CTL, and BTL. In the case of GTL, the gas is natural gas.

The units for LCOF are in \$/gallon of gasoline equivalent, or \$/gge. It should be noted that for most commercial Fischer-Tropsch plants, the main product is diesel, with gasoline secondary. We could therefore calculate the LCOF in terms of \$/gallon of diesel. Since the energy contents of gasoline and diesel are 124,238 and 138,690 BTU/gallon (see Appendix B, Equations (28) and (29)), the LCOF for diesel would be about 90% of that for gasoline on an equivalent energy basis. However, to be consistent with other chapters, the LCOF is calculated based on gasoline.

As was the case with other technologies, the LCOF for the Fischer-Tropsch cases were analyzed in terms of "Well-to-Wheels" cost. For GTL, the "Well-to-Wheels" costs included (1) the cost of natural gas production and transportation to the Fischer-Tropsch plant, (2) the Fischer-Tropsch plant including post-Fischer-Tropsch processing units, (3) transportation of products to market, (4) taxes (both federal and state), and (5) marketing cost at the filling station. For CTL and BTL, the "Well-to-Wheels" costs included (1) the cost of coal or biomass production, preparation, and transportation to the Fischer-Tropsch plant, (2) the Fischer-Tropsch plant including post-Fischer-Tropsch processing units, (3) transportation of products to market, (4) taxes (both federal and state), and (5) marketing cost at the filling station.

The cost for step (1) is included in the delivered cost of the fuel, whether natural gas, coal, or biomass. For GTL, the cost of natural gas was \$3.17/MM BTU based on the Henry Hub spot price for 2018.[18] This same cost was also used in Chapter 2, the chapter for natural gas to electricity. For CTL, the price of coal was based on Illinois No. 6 bituminous coal. This value was \$2.06/MM BTU, taken from a 12-month average of 2018 EIA data[19] and the same value used in Chapter 3 for coal to electricity. For BTL, the price of biomass was based on a Department of Energy (DOE) study[20] and the same value used in Chapter 9 for biomass to electricity. That value is \$80/dry ton, or \$72.6/dry tonne.

For steps (3) and (5), transportation of gasoline and diesel to the market plus the marketing cost at the filling station, an April 2019 update by the EIA[21] reports values of \$0.32/gallon for gasoline and \$0.60/gallon for diesel. Since LCOF is calculated for gasoline, \$0.32/gallon is used. The taxes for step (4) are set to 38% for both federal and state.

Data for the other steps were taken from various published articles. For GTL, the data and calculated LCOF are shown in Table 12.2. With the exception of the NETL report, the other data are based on existing commercial units.

For CTL, the data and calculated LCOF are shown in Table 12.3. None of these data came from existing commercial units and all of the cases were designed with liquid product yields of 50,000 BPD. Two of the studies add cost for carbon capture and sequestration (CCS). Adding CCS increases the LCOF by \$0.06 to \$0.09, or about \$2.5 to \$4 per barrel (42 gallons in a barrel).

For BTL, the data and calculated LCOF are shown in Table 12.4. None of these data came from existing commercial units. Two of the studies added the cost of CCS, increasing the LCOF by \$0.10 to \$0.17, or about \$4 to \$7 per barrel. Although the feed costs were similar for the cases taken from these three studies, the biomass types were

Table 12.2 Economic Analysis for GTL Plants

Reference	Capacity Factor	Feed Cost	Liquid Product Yield	Levelized Capital Cost	Fixed O&M	Non-Fuel Variable O&M	Feed Cost	T&M (a)	LCOF
Units		$/MM BTU	BPD	$/gge	$/gge	$/gge	$/gge	$/gge	$/gge
NETL Report[2]	0.90	$3.17	50,000	$0.77	$0.37	$0.25	$0.64	$0.38	$2.42
Pearl Qatar[22]	0.90	$3.17	140,000	$1.22	$0.14	$0.07	$0.89	$0.38	$2.70
Bintulu, Malaysia[10]	0.90	$3.17	14,700	$0.55	$0.14	$0.07	$0.58	$0.38	$1.72
Mossel Bay, South Africa[3]	0.90	$3.17	45,000	$0.80	$0.14	$0.07	$0.62	$0.38	$2.01
Oryx, Qatar[3]	0.90	$3.17	34,000	$0.32	$0.14	$0.07	$0.75	$0.38	$1.66
Escravos, Nigeria[12]	0.90	$3.17	34,000	$1.62	$0.14	$0.07	$0.75	$0.38	$2.96
Average	0.90	$3.17	52,951	$0.88	$0.18	$0.10	$0.71	$0.38	$2.25

a. Transportation and marketing.

Table 12.3 Economic Analysis for CTL Plants

Reference	Capacity Factor	Feed Cost	Levelized Capital Cost	Fixed O&M	Non-Fuel Variable O&M	Feed Cost	T&M (a)	Power Credit	LCOF
Units		$/ton	$/gge	$/gge	$/gge	$/gge	$/gge	$/gge	$/gge
NETL Report (no CCS)[5]	0.90	$48.06	$0.68	$0.37	$0.08	$0.50	$0.38	-$0.04	$1.96
NETL Report (w/CCS)[5]	0.90	$48.06	$0.70	$0.38	$0.08	$0.49	$0.38	$0.00	$2.02
Princeton (no CCS)[23]	0.90	$54.03	$0.75	$0.28	$0.07	$0.61	$0.38	-$0.29	$1.81
Princeton (w/CCS)[23]	0.90	$54.03	$0.76	$0.29	$0.07	$0.61	$0.38	-$0.22	$1.90
NETL Report (no CCS)[24]	0.85	$54.08	$0.74	$0.31	$0.02	$0.63	$0.38	-$0.07	$2.01
Average (no CCS)	0.88	$52.06	$0.72	$0.32	$0.06	$0.58	$0.38	-$0.14	$1.92

a. Transportation and marketing.

different and included switchgrass, corn stover, and prairie grass. Compared to Tables 12.2 and 12.3 for GTL and CTL, the BTL cases had small liquid product yields, from about 2,500 to 5,000 BPD (BPD).

Considering Tables 12.2 to 12.4, the average LCOFs for GTL, CTL, and BTL are $2.25, $1.92, and $3.89, respectively. By comparison, the LCOF for petroleum crude oil is $2.25 (see Chapter 1, Table 1.6). Thus, CTL and GTL compare very well with petroleum crude oil.

The calculated values for LCOF are largely impacted by the feed cost which is, on average, 32% for GTL, 30% for CTL, and 39% for BTL. Thus, the fuel cost for all three technologies makes up about 1/3 of the LCOF.

Before discussing the sensitivity studies, it is instructive to look collectively at the capital costs for all three types of plants, shown in Table 12.5. On average, GTL and CTL plants cost about $100,000 per BPD installed capacity and BTL plants cost about $180,000. These values are significantly different compared to average installed refinery costs of $40,000/BPD (see Table A.1.1 in Appendix A). This is a large drawback for building new Fischer-Tropsch plants, because a typical GTL or CTL plant that produces 50,000 BPD of transportation fuels will cost five billion dollars at $100,000/BPD and a small 5,000 BPD BTL plant will cost $0.9 billion dollars at $180,000/BPD! When I last worked on Fischer-Tropsch R&D in the late 1990s and early 2000s, the

Table 12.4 Economic Analysis for BTL Plants

Reference	Capacity Factor	Feed Cost	Liquid Product Yield	Levelized Capital Cost	Fixed O&M	Non-Fuel Variable O&M	Feed Cost	T&M (a)	Power Credit	LCOF
Units		$/ton	BPD	$/gge	$/gge	$/gge	$/gge	$/gge	$/gge	$/gge
NETL Report (no CCS)[5] (b)	0.85	$85.60	5,000	$1.78	$0.89	$0.11	$1.66	$0.38	$0.00	$4.82
NETL Report (w/CCS)[5] (b)	0.85	$85.60	5,000	$1.88	$0.94	$0.11	$1.69	$0.38	$0.00	$4.99
NREL Report[6] (c)	0.85	$81.61	3,200	$1.55	$0.34	$0.29	$1.23	$0.38	–$0.13	$3.66
NREL Report[6] (d)	0.85	$81.61	2,479	$1.64	$0.38	$0.40	$1.59	$0.38	–$0.20	$4.19
Princeton (no CCS)[23] (e)	0.90	$85.04	4,410	$0.88	$0.33	$0.00	$1.54	$0.38	–$0.26	$2.87
Princeton (w/CCS)[23] (e)	0.90	$85.04	4,415	$0.90	$0.34	$0.00	$1.54	$0.38	–$0.19	$2.97
Average (no CCS)	0.86	$83.47	3,772	$1.46	$0.49	$0.20	$1.51	$0.38	–$0.15	$3.89

a. Transportation and marketing.

b. The biomass fuel is switchgrass.

c. Gasification is done at high temperature and the biomass fuel is corn stover.

d. Gasification is done at low temperature and the biomass fuel is corn stover.

e. The biomass fuel is prairie grass.

Table 12.5 Total Capital Costs and Cost on a $/BPD Basis for GTL, CTL, and BTL

	Total Project Investment	Liquid Product Yield	Capital Cost (per BBL product)
Units	**$ MM**	**BPD**	**$/BPD**
GTL Plants			
NETL Report[2]	$4,310	50,003	$86,189
Pearl, Qatar[22]	$19,000	140,000	$135,714
Bintulu, Malaysia[10]	$900	14,700	$61,225
Mossel Bay, South Africa[3]	$4,000	45,000	$88,889
Oryx, Qatar[3]	$1,200	34,000	$35,294
Escravos, Nigeria[12]	$6,120	34,000	$180,000
GTL Average			**$97,885**
CTL Plants			
NETL Report (no CCS)[5]	$5,125	50,000	$102,500
NETL Report (w/CCS)[5]	$5,310	50,000	$106,200
Princeton (no CCS)[23]	$4,878	50,000	$97,568
Princeton (w/CCS)[23]	$4,945	50,000	$98,908
NETL Report (no CCS)[24]	$4,528	49,992	$90,574
CTL Average			**$99,150**
BTL Plants			
NETL Report (no CCS)[5]	$1,086	5,000	$217,200
NETL Report (w/CCS)[5]	$1,148	5,000	$229,600
NREL Report[6]	$606	3,200	$189,350
NREL Report[6]	$498	2,479	$201,000
Princeton (no CCS)[23]	$503	4,410	$114,131
Princeton (w/CCS)[23]	$515	4,415	$116,726
BTL Average			**$178,001**

capital cost goal was $25,000/BPD, so capital cost reduction has a long way to go to meet this target. There is, however, large variability in the commercial data for GTL plants, ranging from ~$35,000/BPD for the Oryx plant in Qatar to $180,000/BPD for the Escravos plant in Nigeria, suggesting that capital costs are affected by location and that the $25,000/BPD could be achieved under the right circumstances.

For the sensitivity study, capital cost, feed cost, and the combined fixed and non-fuel variable operating (O&M) costs were examined. Using the previous table as a guide, the capital cost for commercial GTL plants ranged from $35,000/BPD to

$180,000/BPD with an average of about $100,000/BPD. Based on this, the base case used $100,000/BPD, with lower and upper limits of $35,000 and $180,000/BPD. For CTL, the data in this same table, which reports model studies and not commercial data, do not show much variation in capital cost. Therefore, the commercial data for GTL plants was used as a guide, so the same lower and upper limits of $35,000/BPD and $180,000/BPD used for GTL are also used for CTL. And for BTL, Table 12.5 was used to establish a base case of $180,000/BPD, with low and high ranges of $120,000/BPD and $240,000/BPD. The results for the three sensitivity studies are shown in Tables 12.6, 12.7 and 12.8.

For the GTL sensitivity study, both capital cost and the cost of natural gas had a great impact on LCOF. Over the last 14 years, the Henry Hub average annual prices for natural gas[18] have varied from a low of $2.52/MM BTU to a high $8.86/MM BTU. And, as Table 12.6 illustrates, changing the natural gas price from this low to high results in an LCOF difference of $1.32/gge, a quite significant change. Therefore, even if investors can tolerate the initial high capital costs, the price of fuel has a strong effect on the success of the plant. In fact, GTL is competitive with gasoline at the low-end cost for natural gas, a calculated LCOF of $1.99/gge compared to $2.25/gge for crude oil.

The sensitivity study for CTL indicates that it is not as dependent on feed cost as GTL, and the capital costs had the greatest impact on LCOF. As in the case of GTL, if investors can tolerate the high initial capital costs, CTL is quite competitive with a petroleum refinery at a capital cost of $100,000/BPD.

For BTL, the LCOF starts disadvantaged relative to GTL and CTL because of the higher capital cost. Both capital cost and the biomass feed cost have a strong impact on LCOF. Even at the low end of capital cost or the low end of feed cost, the LCOF for BTL is still significantly higher than the cost of gasoline from a refinery, $2.25/gge.

Table 12.6 Sensitivity Study for GTL Plant

Reference	Capacity Factor	Capital Cost	Feed Cost	Fixed and Non-Fuel O&M	LCOF
Units		$/BPD	$/MM BTU	$/gge	$/gge
Base Case	0.90	$100,000	$3.17	$0.21	$2.13
Low Capex	0.90	$35,000	$3.17	$0.21	$1.55
High Capex	0.90	$180,000	$3.17	$0.21	$2.85
Opex is $12/BBL	0.90	$100,000	$3.17	$0.29	$2.20
Opex is $18/BBL	0.90	$100,000	$3.17	$0.43	$2.35
Natural Gas is $2.5/MM BTU	0.90	$100,000	$2.50	$0.21	$1.99
Natural Gas is $6/MM BTU	0.90	$100,000	$6.00	$0.21	$2.70
Natural Gas is $9/MM BTU	0.90	$100,000	$9.00	$0.21	$3.31

Table 12.7 Sensitivity Study for CTL Plant

	Capacity Factor	Capital Cost	Feed Cost	Fixed and Non-Fuel O&M	LCOF
Units		$/BPD	$/ton	$/gge	$/gge
Base Case	0.90	$100,000	$48.06	$0.43	$1.97
Capex is $35,000/BBL	0.90	$35,000	$48.06	$0.16	$1.26
Capex is $180,000/BBL	0.90	$180,000	$48.06	$0.78	$2.85
Opex is 3% fixed and 1% variable	0.90	$100,000	$48.06	$0.29	$1.83
Opex is 7% fixed and 1% variable	0.90	$100,000	$48.06	$0.58	$2.12
Coal is $20/ton	0.90	$100,000	$20.00	$0.43	$1.68
Coal is $80/ton	0.90	$100,000	$80.00	$0.43	$2.30

Table 12.8 Sensitivity Study for BTL Plant

	Capacity Factor	Capital Cost	Feed Cost	Fixed Operating Cost	Total LCOF
Units		$/BPD	$/ton	$/gge	$/gge
Base Case	0.90	$180,000	$80	$0.65	$4.11
Capex is $120,000/BBL	0.90	$120,000	$80	$0.43	$3.38
Capex is $240,000/BBL	0.90	$240,000	$80	$0.87	$4.83
Fixed Opex is 3% of Capex	0.90	$180,000	$80	$0.39	$3.85
Fixed Opex is 7% of Capex	0.90	$180,000	$80	$0.91	$4.37
Biomass is $40/ton	0.90	$180,000	$40	$0.65	$3.33
Biomass is $120/ton	0.90	$180,000	$120	$0.65	$4.89

Although not shown in the table, at the low end of both capital cost and biomass feed cost, the LCOF is $2.61/gge. Therefore, BTL needs low capital cost and low feed cost to effectively compete with gasoline from a refinery.

12.5. Cost of Energy

The cost of energy is an important component of the economic analysis. Comparing natural gas and coal, Table 12.9 shows that over the last 14 years coal has been

Table 12.9 Historical Prices for Natural Gas[18] and Coal[19] (MM BTU = million BTUs)

Year	Natural Gas, $/MM BTU	Coal, $/MM BTU
2005	8.69	1.54
2006	6.73	1.69
2007	6.97	1.77
2008	8.86	2.07
2009	3.94	2.21
2010	4.37	2.27
2011	4.00	2.39
2012	2.75	2.38
2013	3.73	2.34
2014	4.37	2.37
2015	2.62	2.22
2016	2.52	2.11
2017	2.99	2.06
2018	3.17	2.06

relatively stable on an energy basis, while the price of natural gas has fluctuated, from a low of $2.52/MM BTU in 2016 to a high of $8.86/MM BTU in 2008. The prices for natural gas are Henry Hub historical prices[18] while those for coal are historical prices for Illinois No. 6 bituminous coal.[19] Currently, natural gas is about 50% more expensive on an energy basis.

These can be converted to a mass basis using an energy content of 22,792 BTU/lb for natural gas and 12,712 BTU/lb for bituminous coal. For natural gas, this results in a cost of $0.072/lb and for coal a cost of $0.026/lb, or $52/ton.

It is difficult to establish historical, and even current prices for biomass since there are so many different types. In Chapter 9 on the use of biomass to generate electricity, it was shown that in the United States about two-thirds of electricity comes from wood and wood waste biomass, so this type of biomass is currently dominant in electricity generation. A typical heat content for wood and wood waste is 6,500 BTU/lb. For the economic studies discussed in the previous section, the price of this type of biomass was $80/dry ton, based on a DOE study[20] and the same value used in Chapter 9 for biomass to electricity. Converting from $80/dry ton to an energy basis results in a cost of $6.15/MM BTU.

Therefore, at current prices, coal is less expensive than both natural gas and biomass and advantaged from a feed cost standpoint for the Fischer-Tropsch synthesis. As the previous section demonstrated, the price of natural gas needs to decrease to $2.5/MM BTU to have an LCOF equivalent to coal at its current price.

12.6. Capacity Factor

The capacity factor for a GTL plant was shown in Table 12.2 to be 0.90, or 90%. Likewise, for CTL and BTL, the average capacity factors were shown in Tables 12.3 and 12.4 to be 0.88 and 0.86, respectively. Based on the authors, the addition of CCS was not expected to affect the on-stream performance.

Unfortunately, all these data are model projections and not actual commercial performance. According to one author,[5] a high capacity factor was expected for both CTL and BTL because their construction includes a spare gasifier, thus limiting down time due to maintenance on this important piece of equipment. The reason BTL plants were shown with a lower capacity factor than CTL, 0.85 versus 0.90 for this author, was because the supply of biomass feedstock is expected to be less reliable than the supply of coal.

For one commercial example, the Mossel Bay plant in South Africa stated that, since start-up, the complex had been running with an average capacity factor above 0.90.[25]

Also, the operating capacity factor should not be much different than that for a crude oil refinery, as equipment and maintenance issues are similar. According to the EIA, the average capacity factor for US refineries in 2018 was 0.93,[26] and has improved almost every year since 2010 when it was 0.86.

Thus, despite limited commercial data, capacity factors of 0.90, 0.88, and 0.86 for GTL, CTL, and BTL, respectively, seem to be reasonable based on similar experiences with refineries.

12.7. Efficiency: Fraction of Energy Converted to Work

The thermal efficiency for a Fischer-Tropsch plant is defined as the percent of energy that ends up as diesel and gasoline transportation fuels, plus any surplus electricity generated, divided by the energy in the feed input to the plant, be it natural gas, coal, or biomass.

One thing that affects the overall thermal efficiency is the methodology used to produce the synthesis gas, which can include steam reforming, partial oxidation, autothermal reforming, and gasification, processes described earlier in the chapter. For a GTL plant, the preheat temperatures for the oxygen and natural gas can also affect efficiency. Since plants need energy to operate, the thermal efficiency is also directly affected by how much electricity is generated from steam produced in the process, electricity that reduces the need for the use of external electrical energy. Using a heat exchanger to recover product heat from the reformer or partial oxidation processes will also improve thermal efficiency. And, for CTL and BTL plants, the inclusion of CCS will lower the thermal efficiency, since energy is needed to capture and compress the CO_2.

Looking at literature data for GTL, one report[2] shows that 62.4% of the original energy goes to liquid transportation fuels, and 68.0% to products if electricity generation is included. Another GTL paper[1] gives an overall thermal efficiency of 60%. An assessment by the EIA[27] reports a thermal efficiency of 54%. Thus, for these studies, GTL thermal efficiencies range from 54% to 68%.

For CTL, one report shows a thermal efficiency of 52.4% without CCS included,[5] but adding CCS does not change the efficiency much, reported as 53.0%. Another CTL report[23] reports a thermal efficiency of 43.3% to liquid transportation fuels, increasing to 49.1% if the exportable electricity is included. Adding CCS drops this overall efficiency to 47.6% because CO_2 compression requirements for pipeline delivery increase electricity use. For these two studies, the thermal efficiencies are 52.4% and 49.1% without CCS and 53.0% and 47.6% with CCS.

For BTL, one report[5] gives a thermal efficiency of 44.3% without CCS and 43.7% with CCS. Another report[6] shows thermal efficiencies of 50.1% and 42.7%, neither including CCS. Finally, it was reported[23] that a BTL plant without CCS converted 46.2% of the input energy into liquid transportation fuels and 5.7% into exportable electricity, for an overall efficiency of 51.9%. Adding CCS reduced the overall efficiency to 50.3%. Thus, the thermal efficiency without CCS ranged from 42.7% to 51.9% and with CCS ranged from 43.7% to 50.3%.

Taking simple averages, a value of 61% is used for the thermal efficiency of GTL, 50.8% and 50.3% are used for CTL without and with CCS, and 47.3% and 47.0% are used for BTL without and with CCS.

Based on these data, a GTL plant has a thermal efficiency about 10% better than CTL and about 14% better than BTL. Adding CCS for either CTL or BTL drops the thermal efficiency by 0.3% to 0.5%.

There is also a thermal efficiency associated with operation of an internal combustion engine (ICE) vehicle and, in Chapter 1 (Petroleum Crude Oil) we learned these are typically 35% and 45% for gasoline and diesel ICEs, respectively. Although the composition of Fischer-Tropsch-derived gasoline and diesel are somewhat different than those generated from refining, the differences are small and not expected to affect these thermal efficiencies.

12.8. Energy Balance

For a Fischer-Tropsch plant making liquid transportation fuels, and perhaps exportable electricity, the energy balance is defined by this equation:

$$\text{Energy Balance} = \frac{[\text{Energy in the diesel or gasoline that powers the vehicle}]}{[\text{Energy contained in the fuel source}]}$$

For the "Well-to-Wheels" analysis, there are several steps involved in the energy balance. These include (1) producing and preparing the feedstock, (2) transporting the fuel to the Fischer-Tropsch plant, (3) operating the Fischer-Tropsch plant and producing liquid transportation fuels, (4) transporting the liquid transportation fuels to market, and (5) the energy used to power the vehicle.

If the Fischer-Tropsch plant uses natural gas, a GTL plant, the first step includes extraction and processing, such that raw natural gas is treated to remove water, heavy hydrocarbons, and hydrogen sulfide (H_2S). The second step is normally pipeline delivery. In Chapter 2 on natural gas, an NETL report[28] was used to evaluate the natural gas energy lost in steps (1) and (2). The report shows that 0.7% of the natural gas energy is lost during extraction, 6.0% from processing, and 1.4% from transporting to the electrical plant, for a total of 8.1%. Assuming the energy lost transporting the natural gas to a Fischer-Tropsch is like that for an electrical plant, this means that 91.9% of the original natural gas energy arrives at the Fischer-Tropsch plant. That is, 8.1% of the energy is lost in steps (1) and (2).

If the Fischer-Tropsch plant uses coal, a CTL plant, the first step includes extracting and processing the coal. Extraction can include blasting, drilling, and digging. Processing includes crushing and grinding the coal, and separating out worthless material (beneficiation). In Chapter 3 on coal, a DOE study[29] was used to calculate that 1.5% of the coal energy was consumed to extract and process the coal. To transport the coal, according to 2014 report using EIA[30] data, 69% of coal shipments were delivered by rail, 13% by water, 11% by truck, and 7% by conveyor belts and tramways. Assuming rail transportation, it was calculated in Chapter 3 that 0.4% of the coal energy was consumed to transport the coal. Thus, for coal, 1.9% of the energy was lost in steps (1) and (2), meaning that 98.1% of the original energy in the coal arrives at the Fischer-Tropsch plant.

The last case is for a Fischer-Tropsch plant that uses biomass, a BTL plant. Using wood chips for an example of the first step, production and preparation of the feedstock involves soil cultivation, harvesting, turning the wood into chips, storage before transport, and stump removal. A life cycle assessment made by the NREL[31] showed that 1.8% of the biomass energy was consumed for production and preparation and 0.6% for transporting the biomass to the electrical plant. Assuming the energy lost transporting the biomass, by truck, to a Fischer-Tropsch is similar to that for an electrical plant, this means 2.4% of the original biomass energy was lost in steps (1) and (2). Therefore, 97.6% of the original biomass energy arrives at the Fischer-Tropsch plant.

In summary, for steps (1) and (2), feedstock production and transportation, the amount of energy arriving at the Fischer-Tropsch plant is 91.9% for a GTL plant, 98.1% for a CTL plant, and 97.6% for a BTL plant.

The amount of energy consumed in step (3), Fischer-Tropsch plant operation, was established with the thermal efficiencies set in the last section. The thermal efficiencies are 61% for GTL, 50.8% for CTL without CCS and 50.3% for CTL with CCS, and 47.3% for BTL without CCS and 47.0% for CTL with CCS.

For the energy consumed in step (4), transporting the liquid transportation fuels to market, it depends on the location of the Fischer-Tropsch plant. Diesel and gasoline transportation fuels can be transported to market by a variety of methods. These include rail, water, truck, and pipelines. Earlier in this section, it was shown that 1.4% of natural gas energy was used transporting by pipeline, 0.4% of coal energy was used transporting by rail, and 0.6% of biomass energy was used transporting by truck. Based on these values, it seems reasonable to use a figure of 1% for the energy consumed transporting the diesel and gasoline to market.

Finally, it remains to determine the energy consumed in step (5), the energy used to power the vehicle. In the last section on thermal efficiencies, these were given as 35% for gasoline and 45% for diesel. Since the Fischer-Tropsch reaction makes linear hydrocarbons, resulting in high quality diesel properties, commercial plants are normally focused on making diesel fuel. Therefore, the energy balance is made using diesel to power the vehicle.

Using 100 units of energy as a reference and putting the different parts of the energy balance together, we can calculate an overall energy balance for each case, shown here:

Gas-to-Liquids

100 (source) → 93.3 (produce and prepare feedstock) → 91.9 (transport to Fischer Tropsch plant) → 56.1 (operate Fischer-Tropsch plant) → 55.5 (transport liquid transportation fuels to market) → 25.0 (diesel energy that powers vehicle)

Coal-to-Liquids (no CCS)

100 (source) → 98.5 (produce and prepare feedstock) → 98.1 (transport to Fischer Tropsch plant) → 49.8 (operate Fischer-Tropsch plant) → 49.3 (transport liquid transportation fuels to market) → 22.2 (diesel energy that powers vehicle)

Coal-to-Liquids (with CCS)

100 (source) → 98.5 (produce and prepare feedstock) → 98.1 (transport to Fischer Tropsch plant) → 49.3 (operate Fischer-Tropsch plant) → 48.9 (transport liquid transportation fuels to market) → 22.0 (diesel energy that powers vehicle)

Biomass-to-Liquids (no CCS)

100 (source) → 98.2 (produce and prepare feedstock) → 97.6 (transport to Fischer Tropsch plant) → 46.2 (operate Fischer-Tropsch plant) → 45.7 (transport liquid transportation fuels to market) → 20.6 (diesel energy that powers vehicle)

Biomass-to-Liquids (with CCS)

100 (source) → 98.2 (produce and prepare feedstock) → 97.6 (transport to Fischer Tropsch plant) → 45.9 (operate Fischer-Tropsch plant) → 45.4 (transport liquid transportation fuels to market) → 20.4 (diesel energy that powers vehicle)

From these calculations we see that only 25.0% of the original energy in the natural gas becomes diesel fuel energy that powers the vehicle. For coal and biomass, these numbers are less, 22.2% and 20.6%, respectively. With the addition of CCS, coal decreases 0.2% to 22.0% and biomass decreases 0.2% to 20.4%.

12.9. Maturity: Experience

The chemistry for converting CO and H_2 into liquid transportation fuels has been around since 1925, when it was discovered by Germans Franz Fischer and Hans Tropsch at the Kaiser-Wilhelm-Institut für Kohlenforschung (that is, the institute for coal research). In fact, the chemistry employing the reactants of CO and H_2 has been around even longer, as Paul Sabatier and Jean-Baptiste Senderens discovered the methanation reaction in 1902, using nickel and cobalt catalysts, catalysts also used by the Fischer-Tropsch synthesis.

In spite of this, commercial operation of this process is not yet widespread. It is true that shortly after the 1925 laboratory discovery Nazi Germany used Fischer-Tropsch technology on a commercial scale, with nine plants producing about 14,000 BPD of liquid transportation fuels from coal, or CTL.[1] Also, the South Africa company Sasol built the Sasol-1 plant in 1955, producing about 8,000 BPD of liquid fuels from brown coal, more commonly known as lignite.[1] However, these plants were not built because of their economic competitive nature with crude oil-derived transportation fuels, but rather because of political reasons. In the case of Nazi Germany, which did not have vast resources of petroleum, they used this process in World War II to help meet fuel demands using their abundant supply of coal. In the case of South Africa, their apartheid policy isolated them as many countries refused to sell them crude oil, so they also employed their vast coal reserves to operate a commercial CTL plant. Sasol-1 switched from coal to natural gas in 2004, thus becoming the first GTL plant. As shown earlier in Table 12.1, other CTL plants started operation in South Africa when Sasol II and Sasol III came on line in 1980 and 1984, respectively. These both have capacities of 40,000 BPD liquid fuels. In addition, the Shenhua plant in Ningxia, China, started operation in 2018, with a liquid fuel production rate of 100,000 BPD.

For GTL plants, Table 12.1 also shows that the first commercial scale plant started operation in 1992 with the PetroSA plant in Mossel Bay, South Africa, producing a liquid product yield of 45,000 BPD. The following year, in 1993, the Shell Bintulu, Malaysia, plant started operation with a liquid product yield of 14,700 BPD. Following these two plants, only three significantly-sized plants have started operation, with two Qatar plants in Oryx, Rass Laffan (34,000 BPD) in 2007 and the Pearl (140,000 BPD) in 2010. The third plant was built by Chevron and the Nigerian National Petroleum Company in Escravos, Nigeria, starting operation in 2014 with a liquid product yield of 34,000 BPD.

In the United States, interest in CTL technology began in earnest after World War II, when they learned about this German technology.[32] After performing research and setting up a governmental structure for a synthetic fuels program, the Bureau of Mines contracted with the Koppers Corporation to build a CTL plant in Louisiana, Missouri, starting operation in 1951 with a daily liquid product yield of 80 BPD. However, this unit was plagued with catalyst issues, and the unit was shut down in 1953. Subsequently, as crude oil prices and availability stabilized, the United States got out of the CTL business.

With the availability of cheap natural gas, interest in the United States was revived in the first decade of this century, and several small-scale GTL plants started operation. In 2002, British Petroleum opened a GTL plant in Nikiski, Alaska, with a liquid fuel yield of 300 BPD, but this plant closed in 2009.[33] And, in 2003, two other plants started operation including the 400 BPD Conoco plant in Ponca City, Oklahoma, and the 70 BPD Syntroleum plant in Tulsa, Oklahoma.[33] The Conoco plant stopped operation in 2009 and the Syntroleum plant stopped operation in 2006. A very large commercial scale GTL plant with a liquid yield of 140,000 BPD was planned by Shell to start operation near Baton Rouge in Louisiana, but this project was canceled in December 2013 because of high capital costs.[35] The plant costs were projected to be twenty billion dollars,[35] about $140,000 per barrel of liquid product yield.

For BTL, the introduction of commercial plants has been more recent, and on a smaller scale than CTL and GTL. Two plants started up in 2010, including the 5,000 BPD Choren plant in Freiberg, Germany, and the Velocys one BPD pilot plant in Gussing, Austria.[36] Both plants use dry biomass wood, but the Choren plant has since ceased operation. In 2015, another two plants started operation including the Solena Fuels plant in Essex, United Kingdom, producing 1,157 BPD of jet fuel from municipal and commercial waste, and the Syndiese plant in Nevada, producing 530 BPD of liquid fuels from forest and agricultural waste.[36] Finally, in 2016, Sierra Biofuels started a plant in Nevada, in the United States, producing 657 BPD of liquid fuels from municipal solid waste and in 2017, Red Rock Biofuels started a plant in Oregon, producing 1,100 BPD of liquid fuel from forest and saw mill wood waste.[36]

In conclusion, the basic chemistry for the Fischer-Tropsch synthesis has been around for almost 100 years, as the initial discovery was made in 1925. The first CTL plants started operation in Nazi Germany during World War II, so commercial CTL plants have been around for about 80 years. In the United States, the first CTL plant started operation in 1951, almost 70 years ago. The first GTL plant started operation in 1992 in South Africa, so GTL commercial plants have been around for about 28 years. On a small scale, GTL plants have operated in the United States since 2002, only about 18 years. The first BTL commercial plant began operation in 2010, about10 years ago. Thus, CTL plants have been around for about 80 years, GTL for about 28 years, and BTL for about 10 years.

12.10. Infrastructure

The existing infrastructure for commercial Fischer-Tropsch plants is still a short list. As shown in Table 12.10, there are only five GTL plants worldwide, with a total nameplate capacity of 267,700 BPD. For CTL, there are three plants worldwide with a total nameplate capacity of 260,000 BPD. And, for BTL, there are five plants with a total nameplate capacity of 3,445 BPD. For BTL, the number of plants and nameplate capacity is likely to change, as other plants are planned for the future. Nevertheless, the number of worldwide commercial plants for all three technological approaches is still small. And, for the United States, there are currently no GTL or CTL plants, and only three BTL plants with a total nameplate capacity of 2,287 BPD.

If and when the Fischer-Tropsch technology finds a foothold in the United States, the infrastructure to support the production of liquid transportation fuels from this process already exists. For any Fischer-Tropsch commercial plant, there is a need to transport the feedstock to the plant and to transport the liquid products from the plant.

If the Fischer-Tropsch plant uses coal, a CTL plant, there is already an immense infrastructure to transport coal. For rail, by which 69% of coal is delivered, a 2012 report indicates that large railroads operated on 95,264 miles of track.[37] For trucks, by which 11% of coal is delivered, the interstate highway system currently has 46,876 miles.[38]

If the Fischer-Tropsch plant uses natural gas, a GTL plant, there is also an immense system of natural gas pipelines, around 1.6 million miles.[39] If the Fischer-Tropsch plant uses biomass, a BTL plant, the biomass will be delivered by truck, rail, or even barge. Initially, BTL plants will likely be built close to their input fuel, but access to this fuel could change over time. Regardless, a BTL plant can have their biomass delivered using the same infrastructure used to deliver coal.

Once the Fischer-Tropsch plant produces the liquid transportation fuels, these can be transported using the same infrastructure that exists for refineries. For petroleum products, there are about 95,000 miles of existing pipelines, varying in size from 8 to 42 inches in diameter.[40] And there are many refueling stations to receive these gasoline and diesel transportation fuels. According to the National

Table 12.10 List of Commercial Fischer-Tropsch Plants Worldwide

Location	Company	Barrels per Day (BPD)
Gas-to-Liquids (GTL)		
Pearl, Qatar	Shell	140,000
Oryx, Ras Laffan Qatar	Sasol / Qatar Petroleum	34,000
Bintulu, Malaysia	Shell	14,700
Mossel Bay, South Africa	PetroSA	45,000
Escravos, Nigeria	Chevron and Nigerian National Petroleum Company	34,000
GTL Total		**267,700**
Coal-to-Liquids (CTL)		
Secunda II, South Africa	Sasol	80,000
Secunda III, South Africa	Sasol	80,000
Shenhua, Ningxia, China	Synfuels China	100,000
CTL Total		**260,000**
Biofuels-to-Liquids (BTL)		
Freiberg, Germany	Choren (dry biomass wood)	5,000 (no longer operating
Gussing, Austria	Velocys (dry biomass wood)	1
Essex, UK	Solena Fuels (municipal solid waste)	1,157
Nevada	Syndiese (forest and agricultural waste)	530
Reno, Nevada	Sierra Biofuels (municipal solid waste)	657
Lakeview, Oregon	Red Rock Biofuels (forest and saw mill wood waste)	1,100
BTL Total		**3,445**

Association of Convenience Stores (NACS),[41] there were more than 150,000 filling stations in 2019.

In summary, whether the Fischer-Tropsch technology uses CTL, GTL, or BTL, it will fit nicely into the existing infrastructure both in terms of feedstock delivery and liquid product transportation.

12.11. Footprint and Energy Density

The "Well-to-Wheels" footprint and energy density for any type of Fischer-Tropsch plant includes the area needed to (1) produce the feedstock, (2) transport the feedstock to the plant, (3) operate the Fischer-Tropsch plant, and (4) transport the produced liquid transportation fuels to the fueling station.

For a Fischer-Tropsch plant using coal, a CTL plant, we refer back to the Chapter 3 on coal for the land area. Combining underground and surface mining data for 2017 and using an energy content of 12,712 BTU/lb for bituminous coal leads to a calculated energy density of 0.11 quadrillion BTU/(year-square mile) or 0.11×10^{15} BTU/(year-square mile). Also, the transportation of coal to market occurs on existing railways, roads, and waterways, which would exist in the absence of coal transport. Therefore, for the steps of (1) producing the coal feedstock (mining and upgrading before transport) and (2) transporting the coal to the Fischer-Tropsch CTL plant, the energy density is 0.11×10^{15} BTU/(year-square mile).

For a GTL Fischer-Tropsch plant, we refer back to Chapter 2 on natural gas for the land area. Here is was shown that the land area needed to extract the natural gas dominates the land area needed for gas processing as well as the land area needed to transport the natural gas to the plant via pipeline. The energy density for gas extraction was calculated to be 159.3 billion BTU/(year-square mile). Therefore, for the steps of (1) producing the natural gas feedstock (extraction and gas processing) and (2) transporting the natural gas to the GTL plant, the energy density is 159.3 billion BTU/(year-square mile) or 159.3×10^{9} BTU/(year-square mile).

For a BTL Fischer-Tropsch plant, we refer back to the chapter on biomass, Chapter 9 for the land area. Certainly, there are many types of biomass including wood, wood waste, crops, manure, and municipal solid waste. In Chapter 9, the calculations were made using wood as the fuel source, so those numbers are used for the BTL calculations here. The energy density for biomass production from wood was calculated to be 11.1 billion BTU/(year-square mile), or 11.1×10^{9} BTU/(year-square mile). Similar to the case for coal and CTL, the footprint for transporting the biomass to the BTL plant will occur on existing railways, roads, and waterways. Since this infrastructure already exists, and would exist without a BTL Fischer-Tropsch plant, the land area for transporting biomass to the BTL plant is not consequential in the calculations. Therefore, for the steps of (1) producing the biomass feedstock and (2) transporting the biomass to the Fischer-Tropsch BTL plant, the energy density is 11.1×10^{9} BTU/(year-square mile).

Summarizing steps (1) and (2) for all three technologies, the energy density for CTL is 0.11×10^{15} BTU/(year-square mile), the energy density for GTL is 159.3×10^{9} BTU/(year-square mile), and the energy density for BTL is 11.1×10^{9} BTU/(year-square mile). This means that for the same amount of land area, coal mining produces 700 times the amount of energy compared to the extraction of natural gas and about 10,000 times the amount of energy compared to wood biomass.

The next step in the footprint and energy density calculation is step (3), the land area needed to operate the Fischer-Tropsch plant. As illustrated in Table 12.1, there are not many commercial Fischer-Tropsch plants operating in the world, so only a small sample size is available to determine the footprint and energy density for a typical Fischer-Tropsch plant.

Reported land areas for a given commercial Fischer-Tropsch plant were combined with the rated capacity for liquid transportation fuel production to determine the BPD of transportation fuel production per square mile. These data are shown in Table 12.11 for CTL, GTL, and BTL plants. The source for the data is shown in the first column of the table.

One difference worth noting, relative to Table 12.1, is the nameplate capacity for the two South Africa SASOL CTL plants, Secunda II and III. When these units were built in 1980 and 1984, their stated capacities were 40,000 BPD of liquid products for each plant. Since then, the capacities have increased to 80,000 BPD for both units, for a total of 160,000 BPD of liquid transportation fuels. Also, because of the small amount of available data for CTL plants, a paper giving a generic footprint for a 50,000 BPD plant was used.

With these data, Table 12.11 shows that GTL plants produce the highest amount of liquid transportation fuels per square mile, in spite of the low production rate for the Mossel Bay GTL plant in South Africa. As for any commercial plant, the land area used may include future plans for expansion. Nevertheless, the average BPD of liquid transportation fuels for the four GTL plants examined was about 107,000 BPD per square mile.

For CTL plants, the average production rate was much lower, about 47,000 BPD of liquid fuel product per square mile. And, for BTL plants, the average was lower still, about 36,000 BPD of liquid fuel product per square mile.

Examining differences for the three technologies, the lower production rates for BTL may suffer in part due to economy of scale, as these plants are more than a magnitude smaller than GTL or CTL. Another difference is that coal and biomass likely need more posttreatment processing, as these fuels are not as clean as natural gas. That is, coal and biomass have issues with nitrogen, sulfur, ash, and metals in the case of coal, which are not issues with natural gas. Finally, a GTL plant uses either partial oxidation or autothermal reforming (ATR) to produce the synthesis gas while CTL and BTL use gasification to make the synthesis gas.

Therefore, although only a small sample of commercial data were available, Table 12.11 indicates that a GTL plant will produce about 107,000 BPD of liquid transportation fuels per square mile while CTL and BTL plants will produce about 47,000 and 36,000 BPD per square mile.

In order to compare the energy density to operate the Fischer-Tropsch plant with the energy densities to produce the feedstock and transport it to the plant, the averages of Table 12.11 must be converted into units of BTU/(year-square mile). This poses two problems. The first problem is establishing the distribution of liquid transportation fuels into gasoline and diesel. Earlier, in the section on Proven Reserves, an NETL report[2] gave a liquid yield of 69% diesel and 31% gasoline for a GTL plant while another report[3] gave a liquid yield of 71% diesel and 29% gasoline for the Oryx GTL

Table 12.11 Land Area for Some Fischer-Tropsch Plants

Company	Plant Name and Location	Nameplate Capacity, BPD	Hectares	Acreage	Square Miles	BPD/ square mile
GTL Plants						
Shell[42]	Pearl, Qatar	140,000	230	568.3	0.89	157,652
Sasol and Qatar Petroleum[43]	Oryx, Ras Laffan, Qatar	34,000	72	177.9	0.28	122,305
PetroSA[44]	Mossel Bay, South Africa	45,000	410	1013.1	1.58	28,427
Chevron and Nigerian National Petroleum Company[45]	Escravos, Nigeria	34,000	74	182.9	0.29	119,000
GTL Average						106,846
CTL Plants						
SASOL[46]	Secunda II and III, South Africa	160,000	1295	3200.0	5.00	32,000
Synfuels China[47]	Shenhua, Ningxia, China	100,000	560	1383.8	2.16	46,250
Generic[48]		50,000	202.3	500.0	0.78	64,000
CTL Average						47,416
BTL Plants						
Choren[49]	Choren, Freiberg, Germany	5,000	32	79.1	0.12	40,469
Solena[50]	Solena Fuels Plant, Essex, UK	1,157	8.1	20.0	0.03	36,995
Sierra Biofuels[51]	Sierra Fuels BTL Plant, Nevada	657	5.83	14.4	0.02	29,187
BTL Average						35,550

plant. Given these two reports, a nominal average of 70% diesel and 30% gasoline is used for the distribution of liquid transportation products, regardless of whether the technology is GTL, CTL, or BTL. With this distribution, the BTU contents for gasoline and diesel can be combined to get an aggregate energy content for the liquid product blend. In Appendix B, gasoline was assigned an energy content of 124,238 BTU per gallon and diesel an energy content of 138,690 BTU per gallon.

The second problem is that a Fischer-Tropsch plant makes more than just gasoline and diesel, producing light gases (C_4 and lighter) as well as liquids and waxes with molecular weights greater than diesel. However, gasoline and diesel constitute the majority of the energy content for the products, so other products are ignored in this calculation.

As there are 42 gallons in a barrel, the energy content for one barrel of gasoline and diesel can be calculated as 5,217,996 and 5,824,980 BTU/barrel, respectively. Blending these with a distribution of 70% diesel and 30% gasoline gives a blended Fischer-Tropsch transportation fuel barrel energy content of 5,642,885 BTU/barrel. In order to convert BPD to a yearly basis, the same capacity factors used in the calculation of the LCOF are used here, namely 0.90, 0.88, and 0.86 for GTL, CTL, and BTL, respectively.

Using a capacity factor of 0.90 and an average production rate of 107,000 BPD of liquid transportation fuels per square mile for a GTL plant, the calculated energy content is 198.3 × 10^{12} BTU/(year-square mile). Likewise, a value of 85.2 × 10^{12} BTU/(year-square mile) was calculated for CTL and a value of 63.8 × 10^{12} BTU/(year-square mile) was calculated for BTL. Reviewing again the energy densities to produce the feedstock and deliver it to the Fischer-Tropsch plant, the values for GTL, CTL, and BTL were 159.3 × 10^9 BTU/(year-square mile), 110 × 10^{12} BTU/(year-square mile), and 11.1 × 10^9 BTU/(year-square mile), respectively. Comparing the energy density to produce the energy with the energy density for the Fischer-Tropsch plant, the energy densities are similar for a CTL plant while for GTL and BTL the energy density to produce either the natural gas or biomass fuel dominates the land area required for the Fischer-Tropsch plant.

The last step in the "Well-to-Wheels" footprint is to (4) transport the produced liquid transportation fuels to the fueling station. However, this land area is negligible compared to the land area needed to produce the fuel, whether coal, natural gas, or wood biomass. Gasoline and diesel are generally transported by pipeline, train, or truck to the fueling station. Pipelines do not contribute very much to land area and generally can coexist with farming and pasture land, trains move over tracks which can be used for other purposes, and trucks move over roads that already exist.

Although not directly applicable to liquid transportation fuels from the Fischer-Tropsch synthesis, the energy densities for these three Fischer-Tropsch plant types are calculated on a basis of gigawatt-hour per square mile per year, or GWh/(square mile-year), for comparison to other technologies. The conversion from BTU to Watt-hour, or Wh is shown here:

$$1 \text{ BTU} = 0.293071 \text{ Wh}$$

Using this conversion and recognizing that a gigawatt (GW) is equivalent to one billion Watts (or 1 × 10^9), the energy density to produce the feedstock for GTL, CTL, and BTL were calculated to be 46.7, 32,511, and 3.3 GWh/(year-square mile). For the Fischer-Tropsch plants, the energy density for GTL, CTL, and BTL were calculated to be 58,115, 24,960, and 18,684 GWh/(year-square mile). For GTL and BTL, it is clear that the land area to produce the feedstock dominates the land area needed for the Fischer-Tropsch plant. However, for CTL, the land area to produce the coal is similar to that for the CTL Fischer-Tropsch plant, so both contribute to the overall footprint.

12.12. Environmental Issues

The "Well-to-Wheels" steps examined in the last section for the footprint and energy density are also steps that can produce environmental issues. These are the environmental issues associated with (1) producing the feedstock, (2) transporting the feedstock to the plant, (3) operating the Fischer-Tropsch plant, and (4) transporting the produced liquid transportation fuels to the fueling station. In addition, there can be environmental issues associated with automotive combustion of the Fischer-Tropsch gasoline and diesel fuels.

For automotive combustion of Fischer-Tropsch fuels, whether they are produced from natural gas, coal, or biomass, they are essentially free of sulfur and nitrogen. That is because any sulfur or nitrogen in the feedstock is removed during the formation of the synthesis gas (i.e., $CO + H_2$). The Fischer-Tropsch synthesis produces almost entirely linear hydrocarbons, although subsequent product upgrades using hydrocracking, isomerization, and reforming also produce isoparaffins and cycloparaffins. Therefore, unlike gasoline and diesel produced by a refinery, the Fischer-Tropsch gasoline and diesel are cleaner burning fuels with essentially no sulfur or nitrogen compounds.

Environmental issues associated with (1) producing the feedstock and (3) operating the Fischer-Tropsch plant are similar to those discussed in previous chapters for which natural gas, coal, and biomass were used to produce electricity. In the case of the Fischer-Tropsch GTL process, the environmental issues for natural gas were discussed earlier in Chapter 2. Briefly, the main environmental issues for natural gas are H_2S and nitrogen, but only very small amounts of H_2S leave the Fischer-Tropsch plant and at levels that are not dangerous for people. Although the small amount of nitrogen in the natural gas produces nitrogen oxide (NO_x), the amount of NO_x produced is about one-fifth of that made from coal or crude oil.

In the case of the Fischer-Tropsch CTL process, the environmental issues for coal were discussed earlier in Chapter 3. The main environmental issues for producing the coal are mine accidents (such as explosions and the release of H_2S and destruction of habitats (especially for surface coal mining). The main environmental issues encountered at the Fischer-Tropsch plant are the formation of sulfur and NO_x, metal compounds, and ash. The sulfur and NO_x should be captured and sequestered at the Fischer-Tropsch plant as well as heavy metals, especially arsenic, lead, cadmium, selenium, and mercury. The ash should be either landfilled or recycled. For example, an NETL report[24] for CTL gives an equipment list that includes mercury capture, ash removal, flue gas desulfurization (FGD) systems, and selective catalytic reduction and selective noncatalytic reduction (SCR/SNCR) systems to remove nitrogen.

In the case of the Fischer-Tropsch BTL process, some environmental issues are similar to coal. The main environmental issues for producing the biomass from wood are deforestation, if growth and harvesting are not properly managed, and fertilizers and pesticides used to promote growth that can runoff into lakes or rivers. The main environmental issues encountered at the Fischer-Tropsch plant are the formation of sulfur and NO_x, H_2S, and ash which, like coal, should be captured and sequestered at the Fischer-Tropsch plant. For example, one paper[23] gives an equipment list that includes ash handling costs, gas scrubbing, and biomass treatment prior to gasification.

Transporting the feedstock to the Fischer-Tropsch plant, step (2) and transporting the produced liquid transportation fuels to the fueling station, step (4) should not pose any problems different than those involved using natural gas, coal, or biomass to produce electricity. As long as good stewardship is used during the transportation process, whether by pipeline, rail, truck or water barges, there should not be any significant environmental issues. Since transportation does not involve chemical or thermal processes, any issues should be solely related to spills.

12.13. CO_2 Production and the Cost of Capture and Sequestration

The production of CO_2 for a Fischer-Tropsch commercial process can be complicated, and depends on several factors. Referring back to Equation (1), shown again following this paragraph, we see that the Fischer-Tropsch synthesis should not directly produce any CO_2, but only the side product of water. However, this is an ideal equation and, in fact, CO_2 is produced during the synthesis. The reason for this is beyond the scope of this chapter, but the amount produced depends on catalysts, catalyst promoters, and operating conditions. For example, for cobalt catalysts, CO_2 production can be very low, while for iron catalysts, CO_2 is a main byproduct.

$$(2n+1)H_2 + nCO \rightarrow C_nH_{(2n+2)} + nH_2O \quad (1)$$

This is important, not only in terms of CO_2 production and its effect on greenhouse gas emissions, but also because the production of CO_2 is a waste of energy that is not converted into the desired transportation fuels of diesel and gasoline.

In addition, the formation of the synthesis gas (CO and H_2) can be a major contributor of CO_2. If the feedstock is natural gas (GTL), partial oxidation or autothermal reforming is used to make the synthesis gas. These were shown earlier in Equations (5) and (6), shown again at the end of this paragraph. Partial oxidation and autothermal reforming produce two and 7/3 moles of H_2 per mole of CO, respectively. As a ratio of two or greater is desired, the use of these methods gives the desired composition of synthesis gas, and does not require any adjustments that would result in CO_2 formation.

$$CH_4 + \tfrac{1}{2} O_2 \rightarrow CO + 2H_2 (+ \text{heat}) \quad (5)$$

$$3CH_4 + H_2O + O_2 \rightarrow 3CO + 7H_2 \quad (6)$$

For coal (CTL) and biomass (BTL), these feedstocks have lower H_2 content compared to natural gas. Therefore, for these fuels synthesis gas is normally made by gasification, shown earlier in Equation (7).

$$3C + O_2 + H_2O \rightarrow H_2 + 3CO \quad (7)$$

As the equation illustrates, the H_2 to CO ratio is too low for the Fischer-Tropsch synthesis, and therefore an additional step is needed to raise the H_2 content. This additional step is called the "water-gas shift reaction," or WGS, and is shown again in Equation (8).

$$CO + H_2O \rightarrow CO_2 + H_2 \quad (8)$$

As this equation illustrates, some of the CO is consumed to make H_2, but at the expense of also producing CO_2. The trend in CO_2 formation is reflected by the trend in the energy balances discussed earlier in Section 8. For GTL, 25% of the original natural gas energy ended up as energy that powered the vehicle. However, for CTL and BTL, 22.2% and 20.6% ended up as power. That is, the formation of CO_2 uses energy from the starting fuel, thus reducing the energy that is available to eventually power the vehicle. In Chapter 1 on petroleum crude oil, 30.3% of the original crude oil energy is used to power the vehicle, more than any of these Fischer-Tropsch approaches.

Because of these factors, we can anticipate in advance that a GTL-produced fuel will produce more CO_2 than a fuel from refining crude oil, and that CTL will produce more CO_2 than GTL or refining. For BTL, CO_2 production is more complicated because, in spite of the lower energy balance, CO_2 production is offset because of photosynthesis and the carbon cycle for biomass growth.

Looking at this from a "Well-to-Wheels" analysis, CO_2 can be produced during (1) feedstock production, (2) transportation of the feedstock to the Fischer-Tropsch plant, (3) operation of Fischer-Tropsch plant, (4) transportation of the produced liquid transportation fuels to the fueling station, and (5) combustion in the vehicle. All of these steps combined can also be referred to as a "Life Cycle Analysis," such that the environmental impact is examined through all stages of a process, in this case from obtaining the starting fuel until its use in the vehicle.

For the amount of CO_2 made from transportation fuels produced from petroleum crude oil, we refer back to Table 1.8 in Chapter 1. This table has been reproduced here as Table 12.12 and shows the CO_2 production levels and the percent of the total. Clearly, the combustion of the gasoline or diesel in the vehicle produces the most CO_2 in the process, about 80% of the total. Relative to the energy content of the gasoline and diesel, the amounts of CO_2 produced are 212.1 and 209.4 lb CO_2/MM BTU, respectively.

As mentioned, the amount of CO_2 produced from a Fischer-Tropsch process depends on the catalyst type, the operating conditions, and how the synthesis gas is made. It also depends on the thermal efficiency of the process, as well as whether or not CCS is included in the commercial plant design. In Section 7, the thermal

Table 12.12 Greenhouse Gas Emissions for Petroleum Transportation Fuels Sold or Distributed in the United States in Year 2005 (lb CO_2/MM BTU)

Life Cycle Stage	Conventional Gasoline		Conventional Diesel	
Step 1 (Feedstock Production)	16.1	8%	14.6	7%
Step 2 (Transportation to the plant)	3.1	1%	2.9	1%
Step 3 (Operation of the refinery)	21.6	10%	20.9	10%
Step 4 (Transportation to the fueling station)	2.4	1%	2.0	1%
Step 5 (Combustion in the vehicle)	168.9	80%	169.1	81%
Total Well-to-Wheels	212.1	100%	209.4	100%

efficiencies established for the different Fischer-Tropsch approaches were 61% for GTL, 50.8% for CTL without CCS and 50.3% for CTL with CCS, and 47.3% for BTL without CCS and 47.0% for CTL with CCS. This means that the amount of CO_2 produced for a Fischer-Tropsch process will depend on the feedstock selected for the commercial plant.

For example, an Argonne National Laboratory (ANL) report[52] calculates the amount of CO_2 produced from a Life Cycle Analysis with diesel fuel produced from GTL, CTL, and BTL. The Life Cycle Analysis described in the report used the same five steps described earlier. For the GTL plant, a plant thermal efficiency of 63% was used and for the CTL plant, the assumed thermal efficiency was 52%. A case including CCS was also considered, with a CO_2 capture rate of 85%. For the BTL plant, a thermal efficiency of 47% was used. Thus, although not identical, the thermal efficiencies used were similar to those assumed in this chapter. For the GTL case, the amount of CO_2 produced was 242.5 lb CO_2/MM BTU and for CTL, the amount of CO_2 produced was 485.0 lb CO_2/MM BTU; adding CCS reduced the CTL value to 264.6 lb CO_2/MM BTU. For BTL, two cases were examined, one using trees as the fuel source and one using forest residues. For these two cases, the CO_2 production was −11.0 and 44.1 lb CO_2/MM BTU. These low, and even negative values, are the result of photosynthesis and the carbon cycle for biomass. The CO_2 produced from making and using Fischer-Tropsch diesel and gasoline is largely negated during the photosynthesis process, because CO_2 is used with sunlight to make carbohydrates during biomass growth.

A second report was also used to quantify the production of CO_2 from natural gas (GTL) and coal (CTL).[53] Like the previous report, this one also examined the same five Life Cycle Analysis steps. For the CTL case, gasification was used to produce the synthesis gas while for the GTL case, partial oxidation was used. For GTL, three cases were considered, including a plant in the United States making gasoline and diesel using domestic natural gas, a case for which the diesel and gasoline were foreign-produced and then shipped to the United States via an ocean tanker, and a case for which the natural gas was foreign-produced and shipped to the United States in an LNG tanker. Relative to the base case of domestically produced fuels using domestic natural gas, the second case had an increase in CO_2 due to the longer distance for transportation while the third case had an increase in CO_2 due to liquefying, transporting, subsequent gasifying, and then pipeline delivery of the natural gas. The addition of CCS was not considered for any of these cases. For the CTL plant with no CCS, the CO_2 production was 377.6 lb CO_2/MM BTU and for the GTL case with domestically produced fuels using domestic natural gas, the CO_2 production was 229.3 lb CO_2/MM BTU. Importing foreign-produced GTL liquids increased this to 236.0 lb CO_2/MM BTU while bringing foreign natural gas via an LNG tanker to the United States increased this to 283.2 lb CO_2/MM BTU. This interesting comparison shows the impact of transportation on CO_2 production, either shipping the Fischer-Tropsch transportation fuels or shipping the natural gas feedstock from a foreign country.

A third example examines calculations from an NETL report[5] which considered both CTL and BTL for diesel. For both CTL and BTL, three cases were examined including a case with no CCS, a case referred to as "simple CCS," and a case referred to as "aggressive CCS." In the "simple CCS" case, CO_2 made from the water-gas shift reaction is captured and sequestered at the plant. For the "aggressive CCS" case, light hydrocarbons (C_4 and lighter) produced from the Fischer-Tropsch synthesis are converted into H_2 and CO_2 using autothermal reforming (ATR), and this CO_2 is subsequently captured. For the three CTL cases, the calculated CO_2 values were 518.1, 198.9, and 184.5 lb CO_2/MM BTU for no CCS, "simple CCS," and "aggressive CCS." For BTL, the CO_2 produced for the same three cases were −19.4, −463.0, and −540.1 CO_2/MM BTU. As expected, adding CCS dramatically reduces the CO_2 produced, and can result in negative CO_2 production when biomass is used as the feedstock. Of course, while the use of biomass with CCS results in significant levels of CO_2 removal, these cases are not necessarily economically viable.

Summarizing the GTL results, the two cases for which both the commercial plant and natural gas were in the United States showed calculated values of 242.5 lb CO_2/MM BTU and 229.3 lb CO_2/MM BTU. A simple average of these two studies gives a GTL CO_2 production level of 235.9 lb CO_2/MM BTU.

Summarizing the CTL results, there were three cases with no CCS with calculated values of 485.0, 337.6, and 518.1 lb CO_2/MM BTU. A simple average of these three studies gives a CTL CO_2 production level of 446.9 lb CO_2/MM BTU. Adding CCS reduced this to 264.6 lb CO_2/MM BTU for one case and 198.9 lb CO_2/MM BTU for another. A simple average of these two studies gives 231.8 lb CO_2/MM BTU.

Summarizing the BTL results with no CCS, assuming the use of forest residues, two studies provided calculated values of 44.1 and 19.4 lb CO_2/MM BTU. A simple average of these two values is 31.8 lb CO_2/MM BTU.

Using these simple averages, the CO_2 production for the three commercial approaches to Fischer-Tropsch technology were 235.9 lb CO_2/MM BTU for GTL, 446.9 lb CO_2/MM BTU for CTL, and 31.8 lb CO_2/MM BTU for BTL. Adding CCS for the CTL case lowers the CO_2 production to 231.8 lb CO_2/MM BTU. These can be compared to the value of 209.4 lb CO_2/MM BTU for refinery-produced diesel fuel. Compared to the refinery-produced diesel fuel, the GTL value is 12.7% higher and the CTL value is 113.4% higher. Adding CCS to the CTL case reduces this to 10.7%, essentially the same as for GTL. Thus, the poor energy efficiency of the Fischer-Tropsch commercial plant, which includes the process to make the synthesis gas and the Fischer-Tropsch reaction, results in higher greenhouse gas emissions compared to a refinery. For the BTL case, however, the CO_2 production is only 15.2% of the CO_2 produced from refinery-produced diesel, because the subsequent growth of the biomass consumes most of the CO_2 produced by combustion of the diesel in the vehicle.

12.14. Chapter Summary

The Fischer-Tropsch synthesis is a commercial approach to make transportation fuels from natural gas, coal, or biomass. From about 1990 to 2010, there was keen interest in GTL technology, because of the worldwide abundance of natural gas and the fact that some natural gas in the world is stranded because pipeline or Liquefied Natural Gas (LNG) transport are difficult to implement due to both geography and geopolitical reasons. Transforming the natural gas into transportation fuels such as diesel and gasoline would add a mechanism for monetizing the natural gas and facilitate these transportation issues. If nothing else, it does add a third option for natural gas transport to go along with pipelines and LNG.

However, GTL, CTL, and BTL technologies are plagued by high capital costs. Referring back to the averages shown in Table 12.5, commercial plants have capital cost of about $100,000/BPD of capacity for both GTL and CTL, and about $180,000/BPD for BTL. So, for example, if you plan to build a Fischer-Tropsch GTL plant with a goal of producing 50,000 BPD of diesel and gasoline, the capital cost for the plant will be around five billion dollars! And, for a BTL plant, this figure would closer to nine billion dollars! The last petroleum crude oil refinery built in the United States with significant capacity was the Marathon Garyville, Louisiana, refinery built in 1977.[54] And, the average cost for refineries using petroleum crude oil are about $40,000/BPD of capacity. Given the resistance to building new refineries, with a much lower capital cost, it is not surprising that the enthusiasm for building commercial GTL plants has waned. As an example, the construction of a 140,000 BPD commercial GTL near Baton Rouge in Louisiana has been canceled because of high capital costs, which were projected to be 20 billion dollars![35]

The LCOF on a gallon basis was $2.25 for GTL, $1.92 for CTL, and $3.89 for BTL. Compared to the LCOF of $2.25/gallon for petroleum crude oil, it shows that both GTL and CTL can compete with refinery-produced fuels if these high capital costs can be overcome, or at least accepted. And, if the natural gas costs are low because of its stranded nature and no other available options for use, the comparison improves. Lowering the cost of natural gas to $2.5/MM BTU reduces the LCOF to $1.99/gge.

A life cycle analysis of CO_2 production gave values of 235.9, 446.9, and 31.8 lb CO_2 /MM BTU for GTL, CTL, and BTL, respectively. This can be compared to a value of 209.4 lb CO_2/MM BTU for refinery-produced diesel. The high value for CTL is caused by coal gasification to make the synthesis gas ($CO + H_2$) for the synthesis gas, because the water-gas shift (WGS) reaction is needed to increase the amount of H_2, thus producing additional CO_2. This is not generally an issue for GTL, which uses partial oxidation or autothermal reforming to make the synthesis gas, and has sufficient H_2 for the Fischer-Tropsch reaction without utilizing WGS. The low value for BTL assumes the CO_2 produced by fuel combustion in the vehicle will be offset by photosynthesis when new biomass is grown.

The capacity factors for the GTL, CTL, and BTL technologies were similar, 90%, 88%, and 86%, respectively. These values are similar to those for petroleum crude oil refineries.

The footprints and energy densities for GTL and BTL are controlled by the land area needed to produce the feedstock for the Fischer-Tropsch plant. For CTL technology, the land area needed for producing the coal is similar to that for the Fischer-Tropsch commercial plant. For the three technologies, the energy densities were 46.7; 32,511; and 3.5 GWh/(year-square mile) for GTL, CTL, and BTL, respectively.

The overall energy balance for a GTL plant was 25%, meaning that only 25% of the original energy in the natural gas becomes diesel fuel that powers the vehicle. For coal and biomass, these numbers are even lower, 22.2% and 20.6%, respectively. These are not as good as a refinery, for which 30.3% of the original crude oil energy powers the vehicle. The production of CO_2 from making the synthesis gas and from the Fischer-Tropsch reaction is wasted energy that reduces the amount of energy that ends up in the transportation fuels.

Finally, the main environmental issues are similar to those for which the fuel type is used to generate electricity. That is, environmental issues related to a GTL plant are similar to those for which natural gas is used to generate electricity. Likewise, Fischer-Tropsch plants that use coal or biomass will have similar environmental issues as their corresponding electric plant.

References

1. "Gas to Liquids: Historical Development and Future Prospects," Olga Glebova, Oxford Institute for Energy Studies, NG 80, November 2013, https://www.oxfordenergy.org/wpcms/wp-content/uploads/2013/12/NG-80.pdf
2. "Analysis of Natural Gas-to-Liquid Transportation Fuels via Fischer-Tropsch," Erik Shuster, National Energy Technology Laboratory, September 13, 2013, DOE/NETL-2013/1597, http://www.academia.edu/30467535/Analysis_of_Natural_Gas-to_Liquid_Transportation_Fuels_via_Fischer-Tropsch_OFFICE_OF_FOSSIL_ENERGY
3. "Gas to Liquids Plants: Turning Louisiana Natural Gas into Marketable Liquid Fuels," E. O'Brien, Louisiana Department of Natural Resources/Technology Assessment Division, November 2013, http://www.dnr.louisiana.gov/assets/TAD/newsletters/2013/2013-11_topic_1.pdf

4. "BP Statistical Review of World Energy," June 2018, https://www.bp.com/content/dam/bp/business-sites/en/global/corporate/pdfs/energy-economics/statistical-
5. "Affordable, Low-Carbon Diesel Fuel from Domestic Coal and Biomass," Thomas J. Tarka, DOE/NETL-2009/1349, January 14, 2009, https://www.netl.doe.gov/energy-analyses/pubs/CBTL%20Final%20Report.pdf
6. "Techno-Economic Analysis of Biofuels Production Based on Gasification," R.M. Swanson et al., NREL Technical Report NREL/TP-6A20-46587, November 1, 2010, https://www.nrel.gov/docs/fy11osti/46587.pdf
7. "Updated Capital Cost Estimates for Utility Scale Electricity Generating Plants," Energy Information Administration, April 2013; and "Levelized Cost and Levelized Avoided Cost of New Generation Resources in the Annual Energy Outlook 2015," Table 1, June 2015.
8. "How Much Gasoline Does the United States Consume?," Energy Information Administration, March 3, 2020, https://www.eia.gov/tools/faqs/faq.php?id=23&t=10
9. "Short-Term Energy Outlook," Energy Information Administration, October 8, 2019, https://www.eia.gov/outlooks/steo/report/us_oil.php
10. "The Shell GTL Process: Towards a World Scale Project in Qatar: The Pearl Project," Arend Hoek; Shell Global Solutions International, Amsterdam; DGMK-Conference "Synthesis Gas Chemistry," October 4–6, 2006, Dresden, Bintulu, Malaysia GTL (1993), https://www.yumpu.com/en/document/view/5278891/the-shell-gtl-process-dgmk
11. "Synfuels China Demonstrates First Fischer-Tropsch Products from the New Shenhua Ningxia Coal-to-Liquids Plant," Syngas Chemistry, December 12, 2016, http://www.syngaschem.com/news/synfuels-china-ningxia-plant
12. "Gas to Liquids Conversion," A. de Klerk, University of Alberta, Natural Gas Conversion Technologies Workshop of ARPA-E, US DOE, Houston, January 13, 2012, https://www.arpa-e.energy.gov/sites/default/files/documents/files/De_Klerk_NatGas_Pres.pdf
13. "Chevron's $10bn Escravos GTL Plant Produces First Liquids," African Independent Television, 2015, http://aitonline.tv/post-chevron_s__10bn_escravos_gtl_plant_produces_first_liquids
14. "SASOL's Secunda CTL Plant: Costly to Build, but Now It's a Cash Cow," Gas-to-Liquids News, January 1, 2005, https://www.highbeam.com/doc/1P3-793476071.html
15. "Pivoting to renewable fuels Annual report and accounts 2017," Velocys, 2017, https://www.velocys.com/wp-content/uploads/2019/07/Velocys-AR-2017_online.pdf
16. "Catalysts for the Fischer-Tropsch Synthesis," R.B. Anderson, Chapter 2 in Catalysis, Vol. 4, pp. 29–255, New York: Van Nostrand, Reinhold, 1956.
17. "Fischer–Tropsch Synthesis: Catalysts and Chemistry Chapter," Jan van de Loosdrecht et al., in Comprehensive Inorganic Chemistry, Vol. 2, Chapter 7.20, Eds. Jan Reedijk and Kenneth Poeppelmeier, Oxford, Elsevier, August 2013, https://www.researchgate.net/publication/258685985_Fischer-Tropsch_Synthesis_Catalysts_and_Chemistry
18. "Natural Gas: Henry Hub Natural Gas Spot Price," Energy Information Administration, 2018, https://www.eia.gov/dnav/ng/hist/rngwhhdm.htm
19. "Table 9.9 Cost of Fossil-Fuel Receipts at Electric Generating Plants," Energy Information Administration/Monthly Energy Review, p. 163, March 2019, http://www.eia.gov/totalenergy/data/monthly/pdf/sec9_13.pdf
20. "Advancing Systems and Technologies to Produce Cleaner Fuels Technology Assessments," Quadrennial Technology Review 2015 Chapter 7, US Department of Energy, https://www.energy.gov/sites/prod/files/2016/01/f28/QTR2015-7B-Biomass-Feedstocks-and-Logistics.pdf
21. "Gasoline and Diesel Fuel Update," April 1, 2019, Energy Information Administration, http://www.eia.gov/petroleum/gasdiesel/

22. "Pearl GTL Still Generating Work," MEED (Middle East Business Intelligence), November 24, 2013, http://www.meed.com/sectors/oil-and-gas/gas/pearl-gtl-still-generating-work/3186978.fullarticle
23. "Fischer-Tropsch Fuels from Coal and Biomass," Thomas G. Kreutz et al. Princeton Environmental Institute, 25th Annual International Pittsburgh Coal Conference, August 21, 2008, https://acee.princeton.edu/wp-content/uploads/2016/10/Kreutz-et-al-PCC-2008-10-7-08.pdf
24. "Baseline Technical and Economic Assessment of a Commercial Scale Fischer-Tropsch Facility," DOE/NETL-2007/1260, Lawrence Van Bibber et al., April 9, 2007, https://www.netl.doe.gov/energy-analyses/pubs/Small-Scale%20F-T%20Liquids%20Facilities_Final.pdf
25. "Value Added Synthetic Fluids Key to Mossgas' Success," December 6, 1999, Oil & Gas Journal, https://www.ogj.com/articles/print/volume-97/issue-49/special-report/value-added-synthetic-fluids-key-to-mossgas-success.html
26. "Petroleum & Other Liquids: Refinery Utilization and Capacity," Energy Information Administration, 2018 annual average, https://www.eia.gov/dnav/pet/pet_pnp_unc_dcu_nus_a.htm
27. "Gas-To-Liquid (GTL) Technology Assessment in Support of AEO2013," Energy Information Administration, Biofuels and Emerging Technologies Team, January 7, 2013, https://www.eia.gov/outlooks/documentation/workshops/pdf/AEO2013_GTL_Assessment.pdf
28. Life Cycle Analysis of Natural Gas Extraction and Power Generation," T.J. Skone et al., National Energy Technology Laboratory (NETL), DOE Contract Number DE-FE0004001, Figure 4.3, May 29, 2014
29. "Mining Industry Energy Bandwidth Study," US DOE, Industrial Technologies Program, June 2007.
30. Railroads and Coal—Association of American Railroads," July 2016, https://www.aar.org/BackgroundPapers/Railroads%20and%20Coal.pdf
31. "Life Cycle Assessment of a Biomass Gasification Combined-Cycle System," Margaret K. Mann and Pamela L. Spath, NREL Report 23076, December 1997, https://www.nrel.gov/docs/legosti/fy98/23076.pdf
32. "Early Days of Coal Research," Office of Fossil Energy, 2020, https://www.energy.gov/fe/early-days-coal-research
33. "Opportunities for the Early Production of Fischer-Tropsch (F-T) Fuels in the U.S.—An Overview," John Shen et al., National Energy Technology Laboratory (NETL), 2002 DEER Workshop, San Diego, California, August 25–29, 2002, https://www.energy.gov/sites/prod/files/2014/03/f9/2002_deer_shen.pdf
34. "Commercial GTL Project Opens in Southeast Texas," Stephen Whitfield, Journal of Petroleum Technology, February 1, 2017, https://www.spe.org/en/jpt/jpt-article-detail/?art=2613
35. "Shell Halts $20 Billion Louisiana Gas-to-Liquids Project," Bradley Olson, Bloomberg News, December 5, 2013, https://www.bloomberg.com/news/articles/2013-12-05/shell-halts-20-billion-louisiana-gas-to-liquids-project
36. "Biomass to Liquid Transportation Fuel via FischerTropsch Synthesis—Technology Review and Current Scenario," Snehesh Shivananda Ail and S. Dasappa, Renewable and Sustainable Energy Reviews 58(2016), 267–286, December 17, 2015, http://cgpl.iisc.ernet.in/dasappa/img/pdf/journals/Biomass%20to%20liquid%20transportation%20fuel%20via%20Fischer%20Tropsch%20synthesis%20-%20Technology%20review%20and%20current%20scenario.pdf
37. "Class I Railroad Statistics," Association of American Railroads, May 3, 2016, https://www.aar.org/Documents/Railroad-Statistics.pdf

38. "Highway History," Federal Highway Administration, US Department of Transportation, December 18, 2018,https://www.fhwa.dot.gov/interstate/faq.cfm#question3
39. "Annual Distribution and Transmission Miles of Pipeline," November 28, 2018, American Gas Association, https://www.aga.org/research/data/distribution-and-transmission-miles-of-pipeline/
40. "How Pipelines Make the Oil Market Work—Their Networks, Operation and Regulation," A Memorandum Prepared for the Association of Oil Pipe Lines And the American Petroleum Institute's Pipeline Committee, Allegro Energy Group, December 2001.
41. "U.S. Convenience Store Count," NACS (National Association of Convenience Stores), December 31, 2019, https://www.convenience.org/Research/FactSheets/ScopeofIndustry/IndustryStoreCount
42. "From Bintulu Shell MDS to Pearl GTL in Qatar: Applying the Lessons of Eleven Years of Commercial GTL Experience to Develop a World Scale Plant," Lars Carlsson and Niels Fabricius, Gastech 2005, Bilbao, Spain, March 14–17, 2005, http://www.ivt.ntnu.no/ept/fag/tep4215/innhold/LNG%20Conferences/2005/SDS_TIF/050139.pdf
43. "Oryx GTL: How Technip Delivered Middle East's First GTL Plant," September 2011, http://www.technip.com/sites/default/files/technip/fields/oryx_gtl.pdf
44. "Value Added Synthetic Fluids Key to Mossgas' Success," Oil & Gas Journal, December 6, 1999, https://www.ogj.com/home/article/17229861/value-added-synthetic-fluids-key-to-mossgas-success
45. "Energy to Grow," Chevron Line Rider Magazine, November 2006, http://www.chevronretirees.org/sf-docs/default-source/line-rider/Line_Rider_Issue_10.pdf?sfvrsn=0
46. "Barriers and Opportunities Relating to the Production of Coal Liquids and Its Environmental Issues," Ari Geertsema, Summit on Energy Security, Purdue University, April 7, 2019, https://www.purdue.edu/energysummit/presentations/geertsema_purdue.ppt
47. "Visiting the World's Biggest Single Coal-to-Liquid Project in Yinchuan, China," Xing Zhang, IEA Clean Coal Centre, June 27, 2017, https://www.iea-coal.org/visiting-the-worlds-biggest-single-coal-to-liquid-project-in-yinchuan-china/
48. "Coal Gasification & Co-Production of Chemicals & Fuels Workshop on Gasification Technologies," Daniel Cicero, Technology Energy Technology Laboratory (NETL), Indianapolis, June 11–12, 2007, https://www.ems.psu.edu/~radovic/Cicero_NETL_2007.pdf
49. "The CHOREN BTL Process Status—Progress—Future Prospects," Matthias Rudloff, Brussels, September 15, 2008, http://www.google.com/url?sa=t&rct=j&q=&esrc=s&source=web&cd=1&ved=0ahUKEwiX0qeLxr3bAhVI6IMKHX7HDYEQFggnMAA&url=http%3A%2F%2Fwww.renew-fuel.com%2Fdownload.php%253Fdl%253Dmatthias_rudloff-choren-080915.pdf%2526kat%253D19&usg=AOvVaw2vwLt7gJiumWjgY7BT1Ei5
50. "Exploitation Plan and Business Models: Final Version," Report Number: WP7-31-03-16, April 15, 2016, http://www.fuelfromwaste.eu/download/FFW_D7%202_V26.pdf
51. "Final Environmental Assessment Fulcrum Sierra Biofuels LLC, Waste to Fuel Facilities in McCarran, Storey County, Nevada," Jointly Executed for Department of Defense Title III Program, Wright-Patterson AFB and USDA-Rural Development, Washington, DC, August 2014, http://docplayer.net/13007490-Final-environmental-assessment-fulcrum-sierra-biofuels-llc-waste-to-fuel-facilities-in-mccarran-storey-county-nevada.html
52. "Life-Cycle Energy and Greenhouse Gas: Results of Fischer-Tropsch Diesel Produced from Natural Gas, Coal, and Biomass," Michael Wang, May Wu, and Hong Huo, Center for Transportation Research, Argonne National Laboratory, 2007 SAE Government/Industry Meeting, Washington, DC, May 14–16, 2007, https://content.sierraclub.org/coal/sites/content.sierraclub.org.coal/files/elp/docs/us-general_argonne_ctl_ghg_study_2007-05-14.pdf

53. "Comparative Analysis of the Production Costs and Life-Cycle GHG Emissions of Fischer-Tropsch Liquid Fuels from Coal and Natural Gas," Paulina Jaramillo et al., Civil and Environmental Engineering Department, Tepper School of Business, and Department of Engineering and Public Policy, Carnegie Mellon University, Environmental Science & Technology 42(20), 7559–7565, August 5, 2008, https://pubs.acs.org/doi/pdf/10.1021/es8002074
54. "Frequently Asked Questions: When Was the Last Refinery Built in the United States," Energy Information Administration, January 1, 2019, http://www.eia.gov/tools/faqs/faq.cfm?id=29&t=6
55. "Development of Baseline Data and Analysis of Life Cycle Greenhouse Gas Emissions of Petroleum-Based Fuels," DOE/NETL-2009/1346, Timothy J. Skone and Kristin Gerdes, National Energy Technology Laboratory, November 26, 2008, https://www.netl.doe.gov/File%20Library/Research/Energy%20Analysis/Life%20Cycle%20Analysis/NETL-LCA-Petroleum-based-Fuels-Nov-2008.pdf

13
Generation of Electricity by US States

13.1. Foreword

This chapter explores how the fifty US states and the District of Columbia generate electricity. Electricity generation in gigawatt-hours (GWh) is shown in Table 13.1 for each state as well as for the entire United States. These data were taken from a 2017 US Energy Information Administration (EIA) annual electric power report.[1] Table 13.2 shows the same data using percentages instead of GWh. In these tables, "Other Gases" include propane gas and other waste gases from fossil fuels that are not methane (natural gas) and "Other Energy Sources" include batteries, hydrogen, purchased steam, sulfur, tire-derived fuels and other miscellaneous energy sources.

These two tables show two different categories for solar photovoltaic (PV), namely "Utility Scale" and "Estimated Distributed Solar." Utility scale refers to electricity that was generated by a large plant and then distributed to customers through the electric grid. However, solar PV electricity is a unique energy type that can also be generated with small-scale PV systems typically found on commercial and residential rooftops. This is referred to as "distributed" solar PV. As can be imagined, these numbers are more difficult to track and measure than those for a large-scale commercial plant delivering electricity to the electric grid. To estimate electricity generation for distributed solar PV, in 2015 the EIA started using data from electric utilities and third-party owners (TPOs).[2] TPOs are solar companies that install and maintain the rooftop PV system, often with no upfront costs to the homeowner, and the homeowner pays the TPO for the electricity generated. Even though the distributed solar PV data are not as well-measured as data for commercial utilities, they are nevertheless included in this chapter to evaluate state trends in electricity generation.

Examination of these tables shows what a very different country we are state-by-state. While some states remain heavily dependent on coal for electricity generation, other states have transitioned to natural gas. Still others have added significant amounts of renewables, defined as the sum of conventional hydroelectric, pumped storage, wind, solar PV (both utility-scale and distributed), solar thermal, geothermal, and biomass.

To examine these trends, we break down the data in Tables 13.1 and 13.2 to look at states making heavy use of coal, natural gas, and nuclear, and those states with significant amount of total renewable electricity. In addition, renewable electricity is examined in more detail by individually considering hydroelectric, wind, solar, geothermal, and biomass.

The Changing Energy Mix. Paul F. Meier, Oxford University Press (2020). © Oxford University Press.
DOI: 10.1093/oso/9780190098391.001.0001.

Table 13.1 2017 US Electricity Generation, Gigawatt-hours (GWh)

2017, GWh	Coal	Natural Gas	Other Gases	Nuclear	Petroleum Total	Conventional Hydroelectric	Pumped Storage	Wind
United States	1,205,835	1,296,415	12,469	804,950	21,390	300,333	−6,495	254,303
Connecticut	198	15,961	0	16,500	178	332	2	13
Maine	68	2,237	0	0	132	3,389	0	2,333
Massachusetts	1,136	22,153	0	5,047	266	1,037	−443	233
New Hampshire	287	3,580	0	9,991	105	1,413	0	412
Rhode Island	0	7,188	0	0	58	2	0	149
Vermont	0	2	0	0	7	1,280	0	305
New Jersey	1,216	37,708	219	34,033	156	14	−162	22
New York	770	47,272	0	42,167	598	30,145	−416	4,136
Pennsylvania	47,634	72,503	446	83,200	355	3,123	−605	3,591
Illinois	57,980	15,016	180	97,191	51	125	0	12,268
Indiana	72,385	17,976	1,991	0	125	306	0	5,089
Michigan	42,021	26,131	1,652	32,381	1,117	1,679	−675	5,191
Ohio	68,344	28,799	786	17,688	1,241	277	0	1,589
Wisconsin	35,852	13,655	0	9,649	144	2,657	0	1,641
Iowa	25,358	4,567	0	5,214	147	1,034	0	21,373
Kansas	19,390	2,147	0	10,648	55	29	0	18,598
Minnesota	22,782	6,708	0	13,904	35	1,258	0	11,137

Missouri	67,519	5,206	0	8,304	61	1,182	100	2,032
Nebraska	21,174	629	0	6,913	6	1,489	0	5,084
North Dakota	26,756	676	42	0	36	2,582	0	11,359
South Dakota	2,062	654	0	0	3	5,256	0	2,958
Delaware	359	6,723	271	0	25	0	0	5
District of Columbia	0	20	0	0	0	0	0	0
Florida	37,513	161,104	5	29,146	1,463	218	0	0
Georgia	32,487	52,723	0	33,709	282	2,410	−1,248	0
Maryland	8,514	6,729	0	15,107	102	1,965	0	561
North Carolina	34,460	38,590	0	42,374	251	3,818	0	471
South Carolina	18,152	17,156	0	54,345	98	1,835	−1,025	0
Virginia	10,726	44,507	0	30,554	456	1,116	−1,278	0
West Virginia	68,349	1,527	34	0	120	1,658	0	1,682
Alabama	31,440	52,984	9	42,652	36	9,237	0	0
Kentucky	57,237	10,380	0	0	498	4,506	0	0
Mississippi	4,628	46,158	0	7,365	11	0	0	0
Tennessee	27,715	10,269	12	31,818	124	8,691	−686	43
Arkansas	26,285	17,314	0	12,691	49	2,943	20	0
Louisiana	12,316	58,973	2,143	15,410	4,570	906	0	0

Continued

Table 13.1 *Continued*

2017, GWh	Coal	Natural Gas	Other Gases	Nuclear	Petroleum Total	Conventional Hydroelectric	Pumped Storage	Wind
Oklahoma	17,368	30,451	0	0	16	2,036	–118	23,599
Texas	134,648	204,512	2,472	38,581	164	1,062	0	67,061
Arizona	31,396	29,591	0	32,340	57	6,832	–46	570
Colorado	29,242	12,532	0	0	7	1,897	–327	9,315
Idaho	24	3,079	0	0	0	10,670	0	2,545
Montana	13,864	417	14	0	458	10,946	0	2,155
Nevada	1,866	26,626	0	0	9	1,813	0	361
New Mexico	18,414	9,132	0	0	41	193	0	4,595
Utah	26,390	5,871	16	0	38	1,294	0	858
Wyoming	40,069	779	360	0	44	1,124	0	4,321
California	291	88,350	1,408	17,901	46	42,363	407	12,823
Oregon	1,728	15,066	0	0	10	38,294	0	6,227
Washington	5,490	10,849	356	8,128	23	82,183	3	6,925
Alaska	556	3,235	0	0	881	1,644	0	142
Hawaii	1,376	0	52	0	6,634	66	0	532

2017, GWh	Utility-Scale Solar Total	Solar PV—Utility Scale	Estimated Distributed Solar PV	Solar Thermal	Geothermal	Biomass Total	Other Energy Sources	Total
United States	53,284	50,017	23,990	3,269	15,927	62,762	13,094	4,058,259
Connecticut	40	40	407	0	0	792	547	34,970
Maine	5	5	41	0	0	2,704	396	11,305
Massachusetts	788	788	1,516	0	0	1,160	827	33,720
New Hampshire	0	0	87	0	0	1,610	49	17,534
Rhode Island	14	14	55	0	0	203	0	7,669
Vermont	99	99	111	0	0	448	0	2,252
New Jersey	926	926	1,660	0	0	929	583	77,304
New York	183	183	1,182	0	0	2,286	924	129,247
Pennsylvania	70	70	357	0	0	2,438	884	213,996
Illinois	52	52	58	0	0	474	254	183,649
Indiana	278	278	35	0	0	473	306	98,964
Michigan	63	63	65	0	0	2,494	260	112,379
Ohio	105	105	129	0	0	727	–4	119,681
Wisconsin	21	21	56	0	0	1,460	27	65,162
Iowa	5	5	89	0	0	210	2	57,999
Kansas	5	5	17	0	0	57	5	50,951
Minnesota	596	596	59	0	0	1,933	396	58,808
Missouri	56	56	204	0	0	144	4	84,812

Continued

Table 13.1 *Continued*

2017, GWh	Utility-Scale Solar Total	Solar PV—Utility Scale	Estimated Distributed Solar PV	Solar Thermal	Geothermal	Biomass Total	Other Energy Sources	Total
Nebraska	15	15	5	0	0	97	0	35,412
North Dakota	0	0	0	0	0	2	51	41,504
South Dakota	2	2	1	0	0	0	0	10,936
Delaware	50	50	92	0	0	63	0	7,588
District of Columbia	0	0	52	0	0	47	0	119
Florida	877	855	293	22	0	5,009	3,078	238,706
Georgia	1,986	1,986	226	0	0	5,018	87	127,680
Maryland	267	267	735	0	0	536	322	34,838
North Carolina	5,114	5,114	186	0	0	2,812	577	128,653
South Carolina	80	80	138	0	0	2,400	41	93,220
Virginia	313	313	66	0	0	3,809	213	90,482
West Virginia	0	0	8	0	0	0	–14	73,364
Alabama	181	181	8	0	0	3,426	0	139,973
Kentucky	20	20	25	0	0	495	45	73,206
Mississippi	86	86	11	0	0	1,477	2	59,738
Tennessee	92	92	87	0	0	948	21	79,134
Arkansas	31	31	10	0	0	1,437	4	60,784
Louisiana	2	2	200	0	0	2,767	632	97,919
Oklahoma	33	33	6	0	0	299	48	73,738

Texas	2,189	2,189	476	0	0	1,577	528	453,270
Arizona	4,942	4,218	1,893	724	0	171	–1	107,745
Colorado	954	954	531	0	0	166	57	54,374
Idaho	459	459	16	0	84	465	68	17,410
Montana	14	14	15	0	0	21	332	28,236
Nevada	4,146	3,986	412	160	3,292	58	32	38,615
New Mexico	1,193	1,193	204	0	13	18	–1	33,802
Utah	2,211	2,211	289	0	481	78	175	37,701
Wyoming	0	0	5	0	0	0	46	46,748
California	24,353	21,989	10,605	2,364	11,560	5,808	836	216,751
Oregon	194	194	179	0	174	981	39	62,892
Washington	0	0	116	0	0	1,899	56	116,028
Alaska	0	0	2	0	0	44	–3	6,501
Hawaii	174	174	971	0	323	293	363	10,784

Table 13.2 2017 US Electricity Generation, %

2017, %	Coal	Natural Gas	Other Gases	Nuclear	Petroleum Total	Conventional Hydroelectric	Pumped Storage	Wind
United States	29.7	31.9	0.3	19.8	0.5	7.4	−0.2	6.3
Connecticut	0.6	45.6	0.0	47.2	0.5	0.9	0.0	0.0
Maine	0.6	19.8	0.0	0.0	1.2	30.0	0.0	20.6
Massachusetts	3.4	65.7	0.0	15.0	0.8	3.1	−1.3	0.7
New Hampshire	1.6	20.4	0.0	57.0	0.6	8.1	0.0	2.3
Rhode Island	0.0	93.7	0.0	0.0	0.8	0.0	0.0	1.9
Vermont	0.0	0.1	0.0	0.0	0.3	56.8	0.0	13.5
New Jersey	1.6	48.8	0.3	44.0	0.2	0.0	−0.2	0.0
New York	0.6	36.6	0.0	32.6	0.5	23.3	−0.3	3.2
Pennsylvania	22.3	33.9	0.2	38.9	0.2	1.5	−0.3	1.7
Illinois	31.6	8.2	0.1	52.9	0.0	0.1	0.0	6.7
Indiana	73.1	18.2	2.0	0.0	0.1	0.3	0.0	5.1
Michigan	37.4	23.3	1.5	28.8	1.0	1.5	−0.6	4.6
Ohio	57.1	24.1	0.7	14.8	1.0	0.2	0.0	1.3
Wisconsin	55.0	21.0	0.0	14.8	0.2	4.1	0.0	2.5
Iowa	43.7	7.9	0.0	9.0	0.3	1.8	0.0	36.9
Kansas	38.1	4.2	0.0	20.9	0.1	0.1	0.0	36.5

2017, %	Coal	Natural Gas	Other Gases	Nuclear	Petroleum Total	Conventional Hydroelectric	Pumped Storage	Wind
Minnesota	38.7	11.4	0.0	23.6	0.1	2.1	0.0	18.9
Missouri	79.6	6.1	0.0	9.8	0.1	1.4	0.1	2.4
Nebraska	59.8	1.8	0.0	19.5	0.0	4.2	0.0	14.4
North Dakota	64.5	1.6	0.1	0.0	0.1	6.2	0.0	27.4
South Dakota	18.9	6.0	0.0	0.0	0.0	48.1	0.0	27.0
Delaware	4.7	88.6	3.6	0.0	0.3	0.0	0.0	0.1
District of Columbia	0.0	16.8	0.0	0.0	0.0	0.0	0.0	0.0
Florida	15.7	67.5	0.0	12.2	0.6	0.1	0.0	0.0
Georgia	25.4	41.3	0.0	26.4	0.2	1.9	−1.0	0.0
Maryland	24.4	19.3	0.0	43.4	0.3	5.6	0.0	1.6
North Carolina	26.8	30.0	0.0	32.9	0.2	3.0	0.0	0.4
South Carolina	19.5	18.4	0.0	58.3	0.1	2.0	−1.1	0.0
Virginia	11.9	49.2	0.0	33.8	0.5	1.2	−1.4	0.0
West Virginia	93.2	2.1	0.0	0.0	0.2	2.3	0.0	2.3
Alabama	22.5	37.9	0.0	30.5	0.0	6.6	0.0	0.0
Kentucky	78.2	14.2	0.0	0.0	0.7	6.2	0.0	0.0
Mississippi	7.7	77.3	0.0	12.3	0.0	0.0	0.0	0.0
Tennessee	35.0	13.0	0.0	40.2	0.2	11.0	−0.9	0.1

Continued

Table 13.2 *Continued*

2017, %	Coal	Natural Gas	Other Gases	Nuclear	Petroleum Total	Conventional Hydroelectric	Pumped Storage	Wind
Arkansas	43.2	28.5	0.0	20.9	0.1	4.8	0.0	0.0
Louisiana	12.6	60.2	2.2	15.7	4.7	0.9	0.0	0.0
Oklahoma	23.6	41.3	0.0	0.0	0.0	2.8	−0.2	32.0
Texas	29.7	45.1	0.5	8.5	0.0	0.2	0.0	14.8
Arizona	29.1	27.5	0.0	30.0	0.1	6.3	0.0	0.5
Colorado	53.8	23.0	0.0	0.0	0.0	3.5	−0.6	17.1
Idaho	0.1	17.7	0.0	0.0	0.0	61.3	0.0	14.6
Montana	49.1	1.5	0.0	0.0	1.6	38.8	0.0	7.6
Nevada	4.8	69.0	0.0	0.0	0.0	4.7	0.0	0.9
New Mexico	54.5	27.0	0.0	0.0	0.1	0.6	0.0	13.6
Utah	70.0	15.6	0.0	0.0	0.1	3.4	0.0	2.3
Wyoming	85.7	1.7	0.8	0.0	0.1	2.4	0.0	9.2
California	0.1	40.8	0.6	8.3	0.0	19.5	0.2	5.9
Oregon	2.7	24.0	0.0	0.0	0.0	60.9	0.0	9.9
Washington	4.7	9.4	0.3	7.0	0.0	70.8	0.0	6.0
Alaska	8.6	49.8	0.0	0.0	13.6	25.3	0.0	2.2
Hawaii	12.8	0.0	0.5	0.0	61.5	0.6	0.0	4.9

2017, %	Utility-Scale Solar Total	Solar PV—Utility Scale	Estimated Distributed Solar PV	Solar Thermal	Geothermal	Biomass Total	Other Energy Sources	Total
United States	1.3	1.2	0.6	0.1	0.4	1.5	0.3	100.0
Connecticut	0.1	0.1	1.2	0.0	0.0	2.3	1.6	100.0
Maine	0.0	0.0	0.4	0.0	0.0	23.9	3.5	100.0
Massachusetts	2.3	2.3	4.5	0.0	0.0	3.4	2.5	100.0
New Hampshire	0.0	0.0	0.5	0.0	0.0	9.2	0.3	100.0
Rhode Island	0.2	0.2	0.7	0.0	0.0	2.6	0.0	100.0
Vermont	4.4	4.4	4.9	0.0	0.0	19.9	0.0	100.0
New Jersey	1.2	1.2	2.1	0.0	0.0	1.2	0.8	100.0
New York	0.1	0.1	0.9	0.0	0.0	1.8	0.7	100.0
Pennsylvania	0.0	0.0	0.2	0.0	0.0	1.1	0.4	100.0
Illinois	0.0	0.0	0.0	0.0	0.0	0.3	0.1	100.0
Indiana	0.3	0.3	0.0	0.0	0.0	0.5	0.3	100.0
Michigan	0.1	0.1	0.1	0.0	0.0	2.2	0.2	100.0
Ohio	0.1	0.1	0.1	0.0	0.0	0.6	0.0	100.0
Wisconsin	0.0	0.0	0.1	0.0	0.0	2.2	0.0	100.0
Iowa	0.0	0.0	0.2	0.0	0.0	0.4	0.0	100.0
Kansas	0.0	0.0	0.0	0.0	0.0	0.1	0.0	100.0
Minnesota	1.0	1.0	0.1	0.0	0.0	3.3	0.7	100.0
Missouri	0.1	0.1	0.2	0.0	0.0	0.2	0.0	100.0

Continued

Table 13.2 *Continued*

2017, %	Utility-Scale Solar Total	Solar PV—Utility Scale	Estimated Distributed Solar PV	Solar Thermal	Geothermal	Biomass Total	Other Energy Sources	Total
Nebraska	0.0	0.0	0.0	0.0	0.0	0.3	0.0	100.0
North Dakota	0.0	0.0	0.0	0.0	0.0	0.0	0.1	100.0
South Dakota	0.0	0.0	0.0	0.0	0.0	0.0	0.0	100.0
Delaware	0.7	0.7	1.2	0.0	0.0	0.8	0.0	100.0
District of Columbia	0.0	0.0	43.7	0.0	0.0	39.5	0.0	100.0
Florida	0.4	0.4	0.1	0.01	0.0	2.1	1.3	100.0
Georgia	1.6	1.6	0.2	0.0	0.0	3.9	0.1	100.0
Maryland	0.8	0.8	2.1	0.0	0.0	1.5	0.9	100.0
North Carolina	4.0	4.0	0.1	0.0	0.0	2.2	0.4	100.0
South Carolina	0.1	0.1	0.1	0.0	0.0	2.6	0.0	100.0
Virginia	0.3	0.3	0.1	0.0	0.0	4.2	0.2	100.0
West Virginia	0.0	0.0	0.0	0.0	0.0	0.0	0.0	100.0
Alabama	0.1	0.1	0.0	0.0	0.0	2.4	0.0	100.0
Kentucky	0.0	0.0	0.0	0.0	0.0	0.7	0.1	100.0
Mississippi	0.1	0.1	0.0	0.0	0.0	2.5	0.0	100.0
Tennessee	0.1	0.1	0.1	0.0	0.0	1.2	0.0	100.0
Arkansas	0.1	0.1	0.0	0.0	0.0	2.4	0.0	100.0
Louisiana	0.0	0.0	0.2	0.0	0.0	2.8	0.6	100.0
Oklahoma	0.0	0.0	0.0	0.0	0.0	0.4	0.1	100.0

Texas	0.5	0.5	0.1	0.0	0.0	0.3	0.1	100.0
Arizona	4.6	3.9	1.8	0.7	0.0	0.2	0.0	100.0
Colorado	1.8	1.8	1.0	0.0	0.0	0.3	0.1	100.0
Idaho	2.6	2.6	0.1	0.0	0.5	2.7	0.4	100.0
Montana	0.0	0.0	0.1	0.0	0.0	0.1	1.2	100.0
Nevada	10.7	10.3	1.1	0.4	8.5	0.2	0.1	100.0
New Mexico	3.5	3.5	0.6	0.0	0.0	0.1	0.0	100.0
Utah	5.9	5.9	0.8	0.0	1.3	0.2	0.5	100.0
Wyoming	0.0	0.0	0.0	0.0	0.0	0.0	0.1	100.0
California	11.2	10.1	4.9	1.1	5.3	2.7	0.4	100.0
Oregon	0.3	0.3	0.3	0.0	0.3	1.6	0.1	100.0
Washington	0.0	0.0	0.1	0.0	0.0	1.6	0.0	100.0
Alaska	0.0	0.0	0.0	0.0	0.0	0.7	0.0	100.0
Hawaii	1.6	1.6	9.0	0.0	3.0	2.7	3.4	100.0

Before proceeding, it is also useful to note that some states are net importers of electricity while others are net exporters. This can cause confusion in comparing states. For example, if a state generated 30% of its electricity from renewables but imported electricity generated from a coal plant in another state, the percentage of electricity from renewables based on total consumption is really less than that reported from generation alone.

Using sales data reported by the EIA,[3] Table 13.3 was compiled to show the percent of electricity that was generated relative to consumption. If the value is close to 100%, that indicates the state was self-sufficient. If, on the other hand, the number is significantly greater or less than 100%, that indicates the state was a net exporter or importer of electricity.

Examining Table 13.3, it is clear that Vermont and the District of Columbia imported a significant portion of their electricity. According to the EIA,[4] Vermont imports a lot of energy from Canada and neighboring states. Since they are next door to New Hampshire, which generated 63% more than it consumed, it is likely that Vermont also imported electricity from New Hampshire. Another reason Vermont imported electricity is that their Yankee nuclear power station was permanently shut down in 2014.

And, while our nation's capital does generate a high percentage of electricity from solar PV and biomass, the amount generated is very small and almost all of the electricity comes from outside the city.[5] There are no utility-size power plants in the District of Columbia, and most of the electricity generated is distributed from commercial and residential rooftop solar PV. A small amount of electricity is generated at the Blue Plains Advanced Wastewater Treatment Plant using biomass solids left from the wastewater treatment.

The table also illustrates that New Hampshire, Pennsylvania, North Dakota, West Virginia, Alabama, Montana, New Mexico, and Wyoming export about 50% or more of the electricity they generated.

One state that deserves special attention is Hawaii, which generated 62% of its electricity from petroleum and 13% from coal. Each of Hawaii's six main islands has its own separate electricity grid, and these grids are not connected by undersea electric transmission cables.[6] In 2017, renewable sources provided about 22% of the electricity. Because of its isolation, Hawaii depends heavily on imported petroleum, resulting in the very high electricity price of 32.76¢/kWh. This is rapidly changing, however, and a 2018 Scientific American article indicates the state legislature wants to reach 100% renewable energy by 2045.[7]

13.2. Coal

As Table 13.2 illustrates, in 2017 coal and natural gas were almost equal in terms of electricity generation, with natural gas about 2% larger. In 2015 however, electricity generation was about 0.5% larger for coal. Electricity generated from natural gas overtook coal in 2015, and is now the leading technology in the United States for all electricity generation. Nevertheless, although electricity from coal has been decreasing,

Table 13.3 % Generated of Total Consumed

2017, %	% Generated of Total		% Generated of Total
Connecticut	124.3	North Carolina	97.9
Maine	100.8	South Carolina	119.4
Massachusetts	64.2	Virginia	81.1
New Hampshire	162.5	West Virginia	231.4
Rhode Island	103.8	Alabama	162.3
Vermont	41.5	Kentucky	100.8
New Jersey	105.3	Mississippi	124.9
New York	89.1	Tennessee	81.4
Pennsylvania	149.7	Arkansas	131.9
Illinois	133.9	Louisiana	107.4
Indiana	100.0	Oklahoma	121.9
Michigan	110.3	Texas	112.8
Ohio	81.6	Arizona	138.8
Wisconsin	94.3	Colorado	99.2
Iowa	118.6	Idaho	73.2
Kansas	126.5	Montana	192.0
Minnesota	87.6	Nevada	105.3
Missouri	110.9	New Mexico	146.9
Nebraska	116.6	Utah	123.3
North Dakota	206.1	Wyoming	278.6
South Dakota	88.8	California	84.3
Delaware	68.2	Oregon	125.7
District of Columbia	1.1	Washington	126.2
Florida	102.4	Alaska	105.1
Georgia	95.7	Hawaii	115.7
Maryland	58.7		

some states remain heavily dependent on coal. Table 13.4 lists those states that produce more than the US average of 29.7%.

Of these twenty-one states in the table, thirteen produce more than 50% of their electricity from coal: West Virginia, Indiana, Kentucky, Ohio, Wisconsin, Colorado, Montana, New Mexico, Missouri, Utah, Nebraska, Wyoming, and North Dakota.

Table 13.4 2017 Electricity Generation from Coal, %

	%		%
United States	29.7		
Illinois	31.6	West Virginia	93.2
Indiana	73.1	Kentucky	78.2
Michigan	37.4	Tennessee	35.0
Ohio	57.1	Arkansas	43.2
Wisconsin	55.0	Texas	29.7
Iowa	43.7	Colorado	53.8
Kansas	38.1	Montana	49.1
Minnesota	38.7	New Mexico	54.5
Missouri	79.6	Utah	70.0
Nebraska	59.8	Wyoming	85.7
North Dakota	64.5		

And, if 75% is used as the cutoff, four states are left: Missouri, West Virginia, Kentucky, and Wyoming. Wyoming, West Virginia, and Kentucky are well-known coal mining states, so it is not surprising they are on this list. Looking only at plant count and not specific capacity, one source[8] shows that these four states have 177 coal plants operating, about 50% of the US total of 359 in 2017. In 2018, 14 GW of coal-fired plants were either retired or converted to natural gas,[9] so these numbers will undoubtedly change in the coming years. At a capacity factor of 90%, 14 GW will produce almost 95,000 GWh of electricity, or about 8% of the 2017 total for coal.

On the opposite side, fourteen states and the District of Columbia have less than 5% of their electricity from coal: Connecticut, Maine, Massachusetts, New Hampshire, Rhode Island, Vermont, New Jersey, New York, Delaware, Idaho, Nevada, California, Oregon, and Washington. And, of these fourteen states, seven get less than 1% of their electricity from coal: Connecticut, Maine, Rhode Island, Vermont, New York, Idaho, and California. In fact, Rhode Island and Vermont show zero electricity from coal. As mentioned earlier, however, Vermont does import a significant of its electricity from Canada and neighboring states.

13.3. Natural Gas

The 2017 US average for electricity from natural gas was 31.9% and eighteen states produced more than this average. These are shown in Table 13.5. Of these eighteen states, seven exceeded 60%: Massachusetts, Rhode Island, Delaware, Florida,

Table 13.5 2017 Electricity Generation from Natural Gas, %

	%		%
United States	31.9		
Connecticut	45.6	Virginia	49.2
Massachusetts	65.7	Alabama	37.9
Rhode Island	93.7	Mississippi	77.3
New Jersey	48.8	Louisiana	60.2
New York	36.6	Oklahoma	41.3
Pennsylvania	33.9	Texas	45.1
Delaware	88.6	Nevada	69.0
Florida	67.5	California	40.8
Georgia	41.3	Alaska	49.8

Mississippi, Louisiana, and Nevada. In fact, Rhode Island and Delaware produced 93.7% and 88.6%, respectively, of their electricity from natural gas.

13.4. Nuclear

The 2017 US average for electricity from nuclear energy was 19.8% and nineteen states produced at or more than this 19.8%. These are shown in Table 13.6. Of these nineteen states, seven exceeded 40%: Connecticut, New Hampshire, New Jersey, Illinois, Maryland, South Carolina, and Tennessee. In fact, New Hampshire, Illinois, and South Carolina produced more than 50% of their electricity from nuclear energy. Illinois has nuclear power plants at six locations with a total capacity of 12.4 GW while South Carolina has nuclear power plants at four locations with a total capacity of 6.9 GW. The largest nuclear power plant in the United States is the Palo Verde plant west of Phoenix, having three units with a total capacity of about 4 GW. On the other end of the scale, twenty states did not produce any electricity from nuclear energy.

According to the EIA,[10] of those nuclear power plant currently in operation, the earliest started operation in 1969 and the most recent in 1996. Therefore, unlike renewable energy and natural gas, new nuclear power plants are not being built so the numbers in Table 13.6 are not likely to change very much in the coming years.

13.5. Renewable Energy

In 2017, the US average for renewable energy was 17.3%, the majority of which was conventional hydroelectric at 7.4% and wind at 6.3%. Total renewable energy is

Table 13.6 2017 Electricity Generation from Nuclear, %

	%		%
United States	19.8		
Connecticut	47.2	Georgia	26.4
New Hampshire	57.0	Maryland	43.4
New Jersey	44.0	North Carolina	32.9
New York	32.6	South Carolina	58.3
Pennsylvania	38.9	Virginia	33.8
Illinois	52.9	Alabama	30.5
Michigan	28.8	Tennessee	40.2
Kansas	20.9	Arkansas	20.9
Minnesota	23.6	Arizona	30.0
Nebraska	19.5		

defined here as the sum of conventional hydroelectric, pumped storage, wind, solar PV (both utility-scale and distributed), solar thermal, geothermal, and biomass. Using 20% as a demarcation point, nineteen states and the District of Columbia generated more than 20% of their electricity from renewable energy types. Data are shown in Table 13.7.

In terms of diversity, California was the only state to generate electricity from each of these types of renewable energy. If a demarcation point of 30% is chosen instead, the list decreases to twelve states and the District of Columbia. Vermont generated virtually 100% of their electricity from renewables, mainly hydroelectric, wind, and biomass. Also, Maine, South Dakota, Idaho, Oregon, and Washington generated more than 70% of their electricity from renewables. Now, we take a closer look at each of these renewable energy types.

13.5.1. Conventional Hydroelectric

Vermont, South Dakota, Idaho, Oregon, and Washington all made more than 65% of their electricity from renewables and, for these five states, 50% or more came from hydroelectric plants. For Idaho, Oregon, and Washington, they are blessed with hydroelectric dams along the Columbia and Snake Rivers. Likewise, South Dakota generated a lot of hydroelectric power with its four dams on the Missouri River. Although Vermont imported a significant amount of its electricity, it also has 85 operating hydroelectric facilities with a generating capacity of 190 MW.[11]

Table 13.7 2017 Electricity Generation from Renewables, %

2017, %	Conventional Hydroelectric	Pumped Storage	Wind	Solar PV—Utility Scale	Estimated Distributed Solar PV	Solar Thermal	Geothermal	Biomass Total	Renewables
United States	7.4	–0.2	6.3	1.2	0.6	0.1	0.4	1.5	17.3
Maine	30.0	0.0	20.6	0.0	0.4	0.0	0.0	23.9	74.9
New Hampshire	8.1	0.0	2.3	0.0	0.5	0.0	0.0	9.2	20.1
Vermont	56.8	0.0	13.5	4.4	4.9	0.0	0.0	19.9	99.6
New York	23.3	–0.3	3.2	0.1	0.9	0.0	0.0	1.8	29.0
Iowa	1.8	0.0	36.9	0.0	0.2	0.0	0.0	0.4	39.2
Kansas	0.1	0.0	36.5	0.0	0.0	0.0	0.0	0.1	36.7
Minnesota	2.1	0.0	18.9	1.0	0.1	0.0	0.0	3.3	25.5
North Dakota	6.2	0.0	27.4	0.0	0.0	0.0	0.0	0.0	33.6
South Dakota	48.1	0.0	27.0	0.0	0.0	0.0	0.0	0.0	75.1
District of Columbia	0.0	0.0	0.0	0.0	43.7	0.0	0.0	39.5	83.2
Oklahoma	2.8	–0.2	32.0	0.0	0.0	0.0	0.0	0.4	35.1
Colorado	3.5	–0.6	17.1	1.8	1.0	0.0	0.0	0.3	23.1
Idaho	61.3	0.0	14.6	2.6	0.1	0.0	0.5	2.7	81.8
Montana	38.8	0.0	7.6	0.0	0.1	0.0	0.0	0.1	46.6

Continued

Table 13.7 *Continued*

2017, %	Conventional Hydroelectric	Pumped Storage	Wind	Solar PV—Utility Scale	Estimated Distributed Solar PV	Solar Thermal	Geothermal	Biomass Total	Renewables
Nevada	4.7	0.0	0.9	10.3	1.1	0.4	8.5	0.2	26.1
California	19.5	0.2	5.9	10.1	4.9	1.1	5.3	2.7	49.8
Oregon	60.9	0.0	9.9	0.3	0.3	0.0	0.3	1.6	73.2
Washington	70.8	0.0	6.0	0.0	0.1	0.0	0.0	1.6	78.5
Alaska	25.3	0.0	2.2	0.0	0.0	0.0	0.0	0.7	28.2
Hawaii	0.6	0.0	4.9	1.6	9.0	0.0	3.0	2.7	21.9

13.5.2. Wind

In 2017, fourteen states generated more than 10% of their electricity from wind: Maine, Vermont, Iowa, Kansas, Minnesota, Nebraska, North Dakota, South Dakota, Oklahoma, Texas, Colorado, Idaho, New Mexico, and Oregon, all landlocked states with the exception of Oregon, Maine, and Texas. Nebraska, Texas, and New Mexico make the "wind" list, even though they did not generate more than 20% of their total electricity from renewables.

Iowa, Kansas, and Oklahoma produced more than 30% of their electricity from wind with 36.9, 36.5, and 32.0%, respectively. To show how rapidly the use of wind is increasing, in 2015, Iowa, Kansas, and Oklahoma generated 31.5, 24.2, and 18.4% from wind, significant increases over just 2 years. Over these same 2 years, the US average increased from 4.7 to 6.3%. Although blessed with abundant wind, it is somewhat surprising to see Texas and Oklahoma on the list, states traditionally rich in natural gas and petroleum. Anyone who has driven highways through western Texas and Oklahoma, as well as Kansas, has seen a dramatic transformation of the landscape with a seemingly endless skyline of wind turbines. In terms of net generation, Texas led the country in 2017 with 67,061 GWh, 26% of the US total from wind.

As mentioned in Chapter 6 for wind, capacity has grown rapidly since the turn of the century, with installed wind capacity growing from 2.5 GW in 2000 to 84 GW in 2017! Further growth for wind depends, in part, on the construction of new transmission lines to transmit electricity, sometimes long distances, from sparsely populated states with great wind energy potential to large urban areas.

13.5.3. Solar

For utility-scale solar plants, the sum of solar PV and solar thermal (or concentrating solar plants), generated 1.3% of the electricity. As only a handful of states have solar thermal plants, most of this generation came from solar PV.

The twelve states exceeding this 1.3% were Massachusetts, Vermont, Georgia, North Carolina, Arizona, Colorado, Idaho, Nevada, New Mexico, Utah, California, and Hawaii. California and Nevada were the leaders with California generating 11.2% of its utility-scale electricity from solar PV plus solar thermal. And, except for a small amount of solar thermal generation in Florida and Nevada, California and Arizona were the only states with significant solar thermal electricity production. 72% of all solar thermal electricity generation came from California.

Considering distributed solar PV, electricity typically generated from commercial and residential rooftops, some additional states enter the picture. Referring to Table 13.8, the total distributed solar PV for the United States was only 0.6% in 2017. Sixteen states and the District of Columbia exceeded this total: Connecticut, Massachusetts, Rhode Island, Vermont, New Jersey, New York, Delaware, Maryland, Arizona, Colorado, Nevada, New Mexico, Utah, California, Oregon, and Hawaii.

Table 13.8 2017 Electricity Generation from Distributed Solar PV, %

2017, %	Estimated Distributed Solar PV	2017, %	Estimated Distributed Solar PV
United States	0.6		
Connecticut	1.2	Arizona	1.8
Massachusetts	4.5	Colorado	1.0
Rhode Island	0.7	Nevada	1.1
Vermont	4.9	New Mexico	0.6
New Jersey	2.1	Utah	0.8
New York	0.9	California	4.9
Delaware	1.2	Oregon	0.3
District of Columbia	43.7	Hawaii	9.0
Maryland	2.1		

And six states were at 2% or above for distributed solar PV: Massachusetts, Vermont, New Jersey, Maryland, California, and Hawaii. Vermont, Hawaii, and California top the list at 4.9%, 9.0%, and 4.9%, respectively. As the cost and installation of solar PV panels continues to decrease, there is little doubt that we will see further growth in this industry, especially in the sun-drenched states of the southwest.

13.5.4. Geothermal

Electricity generated from geothermal energy is a special case, because not every state has access to geothermal activity. For 2017, the United States generated 0.4% of its electricity from geothermal, and only seven states generated any electricity from geothermal. Referring to Table 13.1, these were Idaho, Nevada, New Mexico, Utah, California, Oregon, and Hawaii. On both a percentage basis and total electricity generation basis, California and Nevada were by far the leaders. Nevada generated 8.5% of its electricity from geothermal, 3,292 GWh and California generated 5.3% of its electricity from geothermal, 11,560 GWh. Together these two states combined for 93% of all US geothermal electricity, with California generating almost 73%.

13.5.5. Biomass

In 2017, 1.5% of all US electricity was generated from biomass. Of this, 66% came from biomass wood and the remaining 34% from biomass waste. Referring to Table

13.2, 25 states equaled or exceeded this value. Only three states exceeded 5%: Maine, New Hampshire, and Vermont. Maine and Vermont had very high generation from biomass, 23.9% and 19.9%, respectively. As these are tree-rich states, this is not surprising. If growth is to come mainly from biomass wood, and if biomass-generated electricity can compete with other renewable energy types, northern tree-rich states like Minnesota, Idaho, Oregon, and Washington may see future growth in biomass-generated electricity.

References

1. "Electric Power Annual 2017," Energy Information Administration, October 2018, Revised December 2018, https://www.eia.gov/electricity/annual/pdf/epa.pdf
2. "EIA Electricity Data Now Include Estimated Small-Scale Solar PV Capacity and Generation," December 2, 2015, https://www.eia.gov/todayinenergy/detail.php?id=23972
3. "State Historical Tables for 2017," Energy Information Administration, February 22, 2019, Excel data.
4. "Vermont: Profile Analysis," July 19, 2018, https://www.eia.gov/state/analysis.php?sid=VT
5. "District of Columbia: Profile Analysis," August 16, 2018, https://www.eia.gov/state/analysis.php?sid=DC
6. "Hawaii: Profile Analysis," November 15, 2018, https://www.eia.gov/state/analysis.php?sid=HI
7. "As Hawaii Aims for 100% Renewable Energy, Other States Watching Closely," John Fialka, E&E News, Scientific American, April 27, 2018, https://www.scientificamerican.com/article/as-hawaii-aims-for-100-renewable-energy-other-states-watching-closely/
8. "Existing U.S. Coal Plants," Global Energy Monitoring, 2017, https://www.gem.wiki/Existing_U.S._Coal_Plants
9. "US Coal Retirements in 2019 to Hit at Least 6 Gigawatts," Joshua S. Hill, Clean Technica, January 23, 2019, https://cleantechnica.com/2019/01/23/us-coal-retirements-in-2019-to-hit-at-least-6-gigawatts/
10. "Net Generation by Energy Source: Total (All Sectors), 2008–2018," Table 3.1.A, Energy Information Administration, https://www.eia.gov/electricity/monthly/epm_table_grapher.cfm?t=epmt_1_1
11. "Building Our Renewable Energy Future: Clean, Sustainable, Secure," Renewable Energy Vermont, 2020, https://www.revermont.org/go-renewable/electricity/hydro/

14
Conclusions

Twelve different energy technologies, eight for generating electricity and four for making transportation fuels, have been examined in this book. While the focus was on technologies for producing electricity and transportation fuels, it is worth noting that energy has other uses, such as heating, cooking, and various industrial uses. We also consume energy in the food we eat. There are some other renewable energy technologies that were not covered in this book. For example, it is possible to harness the mechanical energy of ocean tides and waves. In a manner similar to a wind turbine, a rotating underwater turbine can be used to generate electricity and, as well, ocean waves moving through peaks and troughs can be used to generate electricity. Although the uses of biomass to generate electricity and ethanol were discussed, the production of biodiesel from vegetable oil, animal fats, and waste cooking oil was not.

Twelve different aspects of each technology were examined to systematically compare the different energy types, and some of these are used in this chapter to provide some perspective as to the future of energy.

The first section, "The Changing Energy Mix," explores current trends in energy and some of the driving forces affecting these trends. The second section, "The Electric Vehicle," discusses a potentially critical component of our energy future that provides a bridge between the electricity sector and the transportation sector. The next section is titled "Reducing CO_2 Emissions." Whatever your position is on global warming, or climate change, there is no doubt that this is an important topic in the energy arena, so this section examines how current carbon dioxide (CO_2) emissions could be reduced by the use of renewable fuels. Following this, the section "Proven Reserves and Consumption" summarizes both nonrenewable and renewable energy types. Finally, "Footprint and Energy Density" summarizes this topic for each technology.

14.1. The Changing Energy Mix

The energy mix used for both electricity generation and transportation fuels is changing, and renewable energy types are growing in importance. Consider Figure 14.1, showing data for the US from a British Petroleum energy report for hydroelectric, wind, and solar from 1965 to 2018.[1] If you were born before 1989, you lived in a United States where there was no electricity generated from either wind or solar power and very little from geothermal and biomass. However, while electricity from hydroelectric has been relatively stable over this time period, the growth in both wind and solar electricity generation has been dramatic since 1989. In fact, in 2018 wind generated

The Changing Energy Mix. Paul F. Meier, Oxford University Press (2020). © Oxford University Press.
DOI: 10.1093/oso/9780190098391.001.0001.

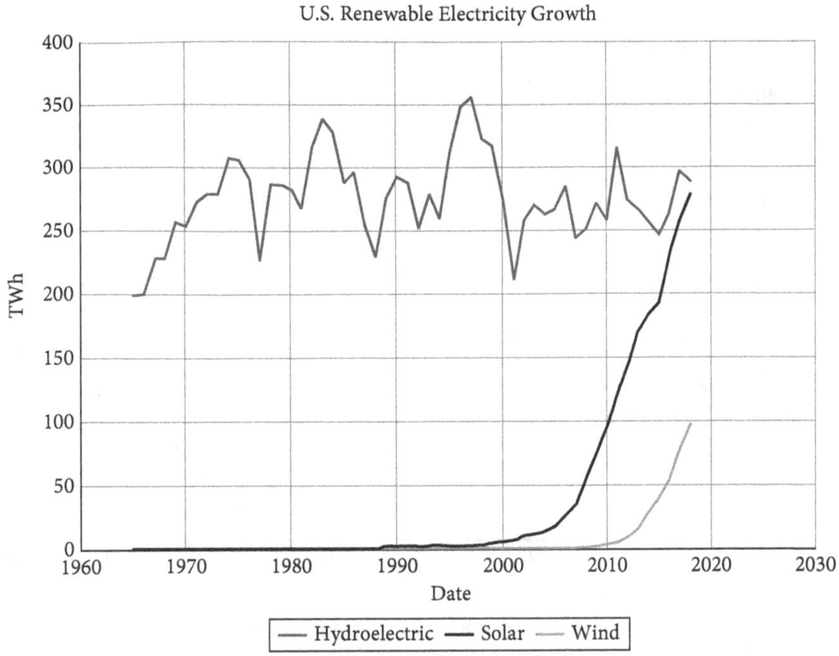

Figure 14.1 US Renewable Energy Growth

electricity was almost the same as hydroelectric, 277.7 TWh (terawatt-hours) from wind versus 288.7 TWh generated from hydroelectric. Energy from wind and solar made up 6.2% and 2.2% of the total US electricity generation in 2018, a remarkable increase from 20 years earlier in 1998, when wind and solar made up 0.08% and 0.01%, respectively. In fact, in 2017 there were fourteen states that generated more than 10% of their electricity from wind and three, Iowa, Kansas, and Oklahoma, generated more than 30% of their electricity from wind.

Likewise, the world did not see any electricity generation from wind until 1985 and none from solar until 1989. As Figure 14.2 shows, wind and solar, as well as hydroelectric, are experiencing rapid growth in the world. Solar and wind made up 2.2% and 4.8% of the electricity generation in 2018, a significant growth from 20 years earlier in 1998 when solar and wind were only 0.01% and 0.11%, respectively. And although hydroelectricity generation increased on a TWh basis, the percentage decreased from 17.9% in 1998 to 15.8% in 2018. This is because worldwide energy usage has been rapidly increasing, from 14,494 TWh in 1998 to 26,615 TWh in 2018, an increase of 84%.

The energy mix for fossil fuels used to generate electricity has also been changing in the United States, notably coal and natural gas. As Figure 14.3 illustrates, electricity from coal reached a peak in 2007 and has been steadily decreasing ever since, while electricity from natural gas is on the rise. Natural gas overtook coal in 2015, and is now the primary energy for producing electricity, 35.4% versus 27.9% for coal. This is

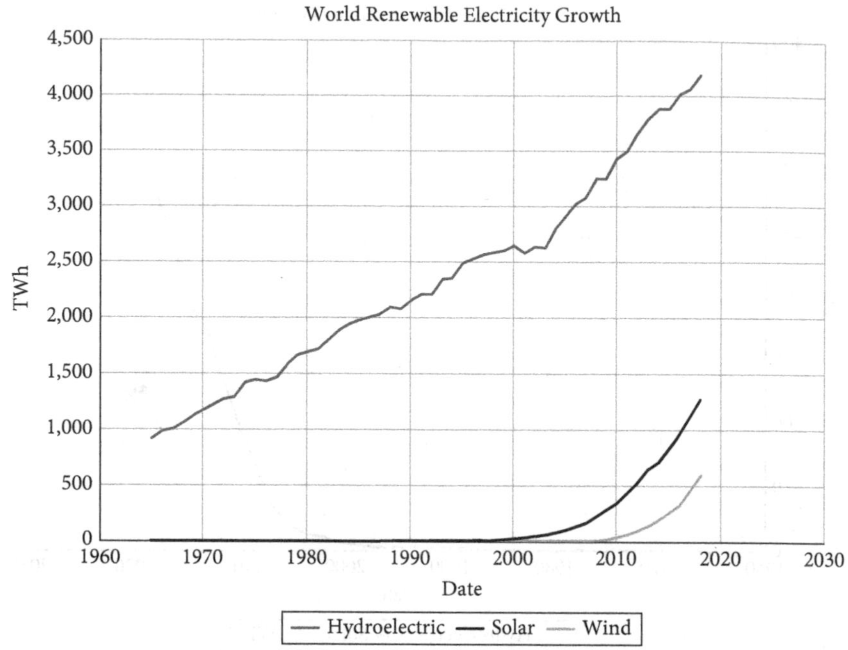

Figure 14.2 World Renewable Energy Growth

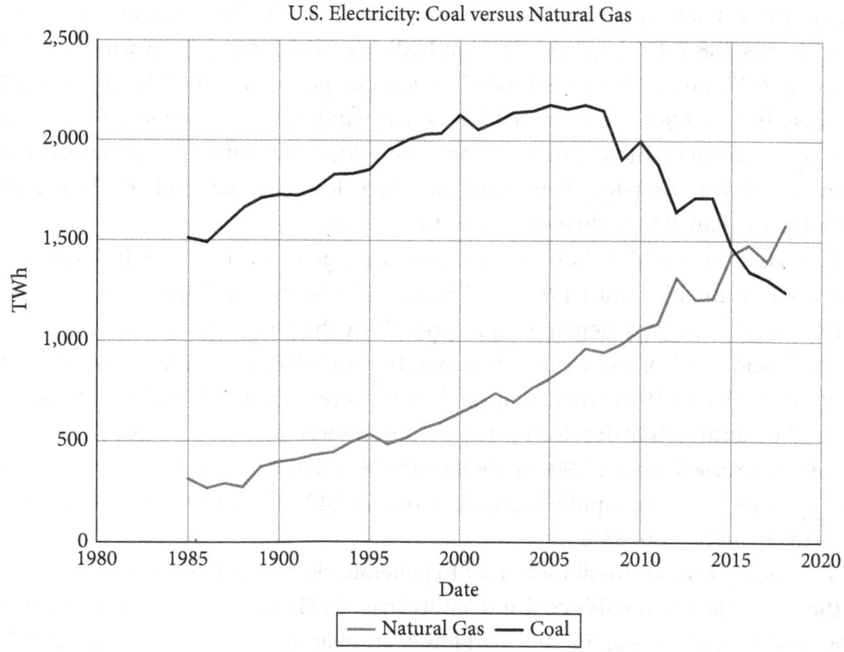

Figure 14.3 US Electricity: Coal versus Natural Gas

a big change from 20 years earlier in 1998, when coal was responsible for 52.6% of the US electricity generation versus 14.8% for natural gas.

To demonstrate the change in the energy mix over the last 20 years, Table 14.1 shows the main energy types used in the United States for electricity generation. In 1998, the traditional energy types of natural gas, coal, nuclear, and oil accounted for 89.4% of all electricity, decreasing to 83.0% in 2018. However, while nuclear has remained relatively constant on a percentage basis, electricity from coal and oil have decreased significantly, with natural gas now taking the leadership role. Overall electricity generation has increased 16% over these 20 years.

For renewables, hydroelectric has not changed much over the last 20 years, nor has biomass or geothermal. But clearly, wind and solar have made great strides in these two decades. And while their level of generation is still far behind natural gas, it is quite conceivable that wind will soon overtake hydroelectric. And if wind and solar continue the exponential growth shown in Figure 14.1, in 10 years their combined generation would surpass the current level of nuclear generation.

What are some of the driving forces behind this transformation? Some countries, including the United States, have regulations and subsidies to help reduce our dependence on fossil fuels, to create jobs, and to reduce greenhouse gases (GHGs). Reducing the use of fossil fuels has the benefits of greater energy independence and improving the environment, such as the reduction of sulfur oxides, nitrogen oxides, and CO_2. Greater energy independence can also be important in terms stabilizing electricity and transportation fuel costs, as there has been great fuel price variability in the past, especially for crude oil and natural gas. The reduction of sulfur oxides and nitrogen oxides is particularly important, as these are linked to ground level ozone and the formation of acid rain in the upper atmosphere. The reduction of GHGs in

Table 14.1 US Electricity Generation (percent of total in parentheses)

Terawatt-hours (TWh)	1998	2018
US Total[1]	3,855.8	4,460.8
Natural Gas[1]	571.2 (14.8%)	1,578.5 (35.4%)
Coal[1]	2,029.0 (52.6%)	1,245.8 (27.9%)
Nuclear[1]	709.2 (18.4%)	849.6 (19.0%)
Oil[1]	138.5 (3.6%)	26.4 (0.6%)
Wind[1]	3.1 (0.1%)	277.7 (6.2%)
Solar[1]	0.5 (0.0%)	97.1 (2.2%)
Hydroelectric[1]	322.1 (8.4%)	288.7 (6.5%)
Biomass[2]	58.8 (1.5%)	58.4 (1.3%)
Geothermal[2]	14.8 (0.4%)	16.0 (0.4%)

the United States is a topic of great importance or none at all depending on who you ask but, in Europe, there is great public pressure to reduce GHGs.

In the United States, federal subsidies have helped the development of renewable fuels, and are needed to help jump start new energy types that are disadvantaged with respect to infrastructure, technology not fully developed, and capital costs. Table 14.2 shows federal subsidies for three different years using data from an Energy Information Administration (EIA) report.[3] As the table illustrates, renewables received greater than 15 billion dollars of subsidies in 2010 and 2013, but that number decreased to less than $7 billion in 2016. Solar, wind, and biofuels received most of the 2016 subsidies, with 33.4% for solar, 18.9% for wind, and 42.1% for biofuels. Here biofuels refers primarily to ethanol and biodiesel.

It is also interesting that fossil fuels and nuclear continue to receive federal subsidies, more than one billion dollars for coal in each of these 3 years and about one and a half billion for nuclear in 2010 and 2013. Interesting, natural gas and petroleum liquids received close to three billion in subsidies for 2010 and 2013, but the value is negative for 2016. This negative value results from a change in the tax expenditures for oil and natural gas, which resulted in an inflow to the US Treasury instead of a subsidy.

Although this table does not provide a complete breakdown for each year and each technology, three different federal subsidy categories are defined including direct expenditures, tax expenditures, and research and development. Examples of tax expenditures include biodiesel producer tax credits, credits for residential energy efficiency, energy production credits, and energy investment credits. For 2016, the subsidies for all renewable energy included $909 million for direct expenditures, $5,316 million for tax expenditures, and $456 million for research and development.

Table 14.2 Federal Subsidies for Different Energy Technologies[3]

Millions of $	2010	2013	2016
Coal	1,062	1,104	1,262
Natural Gas and Petroleum Liquids	2,976	2,796	–773
Nuclear	1,537	1,390	365
Biomass	1,037	572	79
Geothermal	83	358	86
Hydroelectric	95	233	38
Solar	1,116	5,756	2,231
Wind	5,705	6,187	1,266
Other	410	280	169
Biofuels	7,340	1,878	2,813
Total Renewables	15,786	15,264	6,682

While there is some public pressure in the United States to reduce GHGs, the driving force is not as great as in Europe, so eventually energy technologies need to become competitive with more traditional forms of energy. And wind and solar have become more competitive. Using data presented in Chapter 6 and 7 for wind and solar, EIA data show significant reductions in capital cost for both technologies. In 2015, the capital cost for an onshore wind farm was $2,213/kW, but decreased to $1,624/kW in 2019, a reduction of 27%. For solar photovoltaics (PV), the capital cost was $4,183/kW in 2015, reducing to $1,783/kW in 2019, a reduction of 57%. One term used when discussing capital costs for new commercial technology is the "n^{th} plant," meaning that capital costs for new technologies are expected to decrease as the technology matures, eventually leveling out at the nth plant. Capital cost is decreasing for both of these technologies and it is not clear at this time if these technologies have reached the "n^{th}" plant status, so there could be future incremental decreases in the capital costs for both solar and wind.

While these capital costs are still greater than that for a natural gas combined cycle (NGCC) plant at $999/kW (see Chapter 2), renewable operating costs are less. For onshore wind, the operating cost is $13.5/MWh and for solar PV it is $10.3/MWh, versus $26.0/MWh for NGCC, of which $24.5/MWh is the cost of the natural gas fuel. The capital cost reduction for wind resulted in a decrease for levelized cost of electricity (LCOE) from ₵12.5/kWh in 2015 to ₵10.0/kWh in 2019 and the capital cost reduction for solar PV resulted in an LCOE decrease from ₵25.1/kWh to ₵13.4/kWh in 2019. Thus has helped narrow the gap with NGCC, having an LCOE of ₵7.6/kWh.

Figures 14.4a and 14.4b show the LCOE for different technologies, where the upper and lower ranges were determined in the sensitivity studies presented in each chapter and the circles represent the average. Figure 14.4b is shown with a narrower range on the y-axis in order to amplify the details for those technologies competing most closely with NGCC. The figures also show that geothermal flash technology is competitive with NGCC, even though the opportunity for this technology is mostly in the US western states.

Two other factors worth pointing out are that there are regional differences in technology costs and that the cost of CO_2 capture and sequestration (CCS) would affect these comparisons if CCS becomes mandatory in the future. As an example of regional differences in the LCOE, one report from the DOE Office of Energy Efficiency[4] shows that the cost of onshore wind farm electricity is dependent on location, with the lowest cost projects in the so-called wind belt US interior, where wind speeds can top 8.5 meters per second, or 19 mph. The LCOE from plants in the US interior range from ₵2 to ₵6/kWh, compared to ₵6 to ₵12/kWh in other, less windy, parts of the United States. Thus a wind farm in the US interior has an LCOE that is as good, or perhaps better, than the LCOE for NGCC.

With respect to CCS, NGCC and pulverized coal plants produce 0.73 and 2.08 lb CO_2/kWh compared to 0.060 and 0.088 lb CO_2/kWh for wind and solar PV plants. The small CO_2 emission levels for wind and solar are not made during electricity generation, but rather are part of the plant life cycle, and include things such as

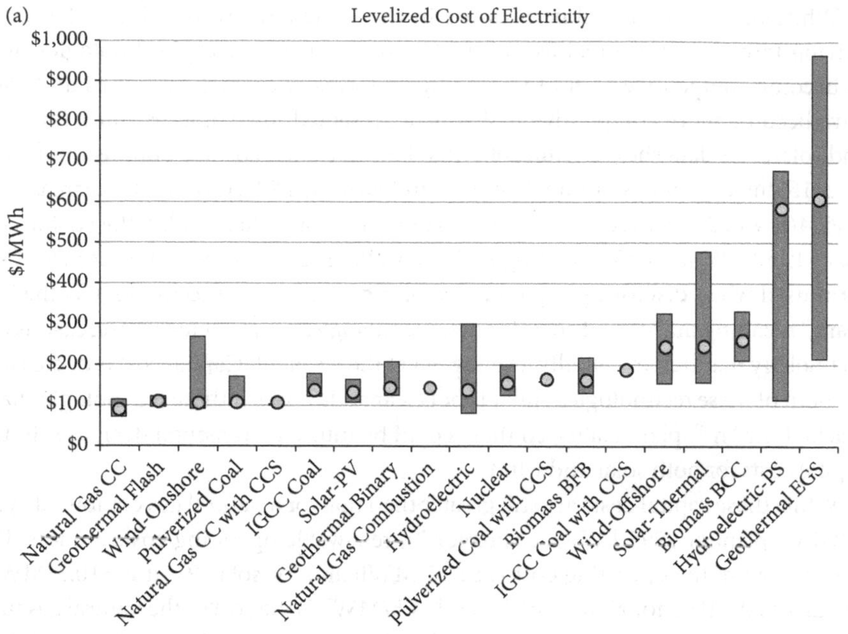

Figure 14.4a US Levelized Cost of Electricity (LCOE)

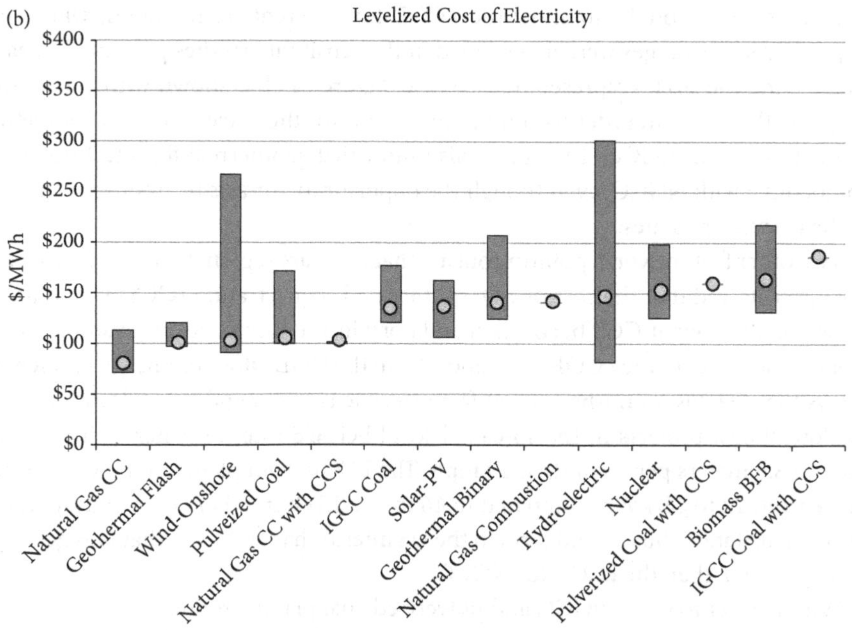

Figure 14.4b US Levelized Cost of Electricity (LCOE)

manufacturing materials transportation, installation, and maintenance. If CCS does become mandatory for NGCC and coal plants in the United States, the capture and sequester of this CO_2 adds about ¢2.8/kWh for an NGCC plant and about ¢5.8/kWh for a pulverized coal plant, thus making wind and solar even more attractive.

The energy mix is also changing for transportation fuels, but primarily with the addition of ethanol to gasoline and biodiesel to diesel. Biodiesel was not specifically discussed in a separate chapter in this book, but is a renewable fuel made from soybean oil, animal fats, and recycled grease. Table 14.3 shows a very busy table for the pool of US transportation fuels from 2008 to 2018. The table includes gasoline, jet fuel, diesel, ethanol, compressed natural gas (CNG), electricity used for electric vehicles (EVs), biodiesel, and hydrogen, with data taken from a variety of resources. CNG is primarily used in the operation of pipelines and compressors, but a small amount is used as vehicle fuel. This small amount is listed in a separate column, and is clearly a very small amount of the CNG total. The data of Table 14.3 are also shown as percentages of the total in Table 14.4.

These tables show that for gasoline-powered vehicles, CNG, electricity, and hydrogen have not yet made much of an impact. Looking at 2017, CNG for motor fuel, electricity and hydrogen were only 0.07%, 0.023%, and 0.0002% of the overall total. And, if we pool these fuels with gasoline and ethanol and focus only on the gasoline pool, CNG, electricity, and hydrogen make up 0.11%, 0.04%, and a very small 0.0003%, with gasoline and ethanol representing 89.9% and 9.9%, respectively. For diesel-powered vehicles, biodiesel makes up 0.74% of the entire transportation pool, and 3.2% of the diesel pool. So, at this time, the major changes to the gasoline and diesel market have been the introduction of ethanol and biodiesel, but CNG, electricity, and hydrogen have not yet made much of an impact.

This trend is also reflected by the vehicle market. In the United States in 2018, there were about 6,500 fuel cell vehicles (FCVs),[11] one million EVs,[12] and 175,000 natural gas vehicles (NGVs)[13] compared to more than 278 million gasoline vehicles (see Chapter 1).

The growth in ethanol as a transportation fuel is related to an event that happened more than 40 years ago, the oil embargo of 1973 when OPEC (Organization of Petroleum Exporting Countries) declared an oil embargo on countries supporting Israel during the Yom Kippur War, and crude oil prices as much as quadrupled. Following the rapid increase in crude oil prices, the United States started looking at corn-based ethanol as a fuel. Thus a driving force that has caused the growth in ethanol use, and to a lesser extent biodiesel, is energy independence and price stability rather than GHGs and public pressure.

Another driving force was the environmental concern with leaded gasoline, which boosted gasoline octane, so its removal created a need for additives to increase gasoline octane. The fuel additive tetraethyl lead (TEL) had been used since the 1920s but was phased out in the 1970s as part of the US Clean Air Act. Lead in air and soil was responsible for lead poisoning as well as poisoning the catalytic converter in your vehicle used to control emissions. To promote ethanol development, tax incentives were

Table 14.3 US Transportation Fuels by Year (million gallons)

Date	Gasoline[5]	Jet Fuel[5]	Diesel[5]	Ethanol[6]	CNG[7]	CNG as Motor Fuel[7]	Electricity, gge[8] (a)	Biodiesel[9]	Hydrogen[10]	Total
2008	138,182	23,886	60,868	9,309	4,990	113	0	304	0	237,652
2009	137,917	21,579	55,933	10,938	5,154	142	0	322	0	231,985
2010	137,857	22,172	58,564	13,298	5,184	126	0	260	0	237,461
2011	134,180	22,076	59,957	13,948	5,289	128	0.6	886	0	236,465
2012	133,463	21,701	57,594	13,300	5,622	124	3.3	899	0	232,706
2013	135,563	22,175	58,755	13,300	6,397	132	9.6	1,429	0	237,761
2014	136,756	22,715	62,029	14,313	5,477	138	18.9	1,417	0	242,864
2015	140,704	23,911	61,345	14,807	5,370	156	29	1,494	0.002	247,816
2016	143,222	24,984	59,734	15,329	5,459	167	41.3	2,085	0.10	251,021
2017	142,976	25,962	60,360	15,800	5,758	172	58.1	1,985	0.45	253,072
2018	143,013	26,355	63,639	16,061	6,294	189		1,904	0.91	257,456

a. The reference for electricity used electric vehicles was given in TWh, which were converted to gge using a value of 33.4 kWh/gge.

Table 14.4 US Transportation Fuels by Year (percent)

Date	Gasoline[5]	Jet Fuel[5]	Diesel[5]	Ethanol[6]	CNG[7]	CNG as Motor Fuel[7]	Electricity, gge[8]	Biodiesel[9]	Hydrogen[10]
2008	58.1	10.1	25.6	3.9	2.1	0.05		0.13	
2009	59.5	9.3	24.1	4.7	2.2	0.06		0.14	
2010	58.1	9.3	24.7	5.6	2.2	0.05		0.11	
2011	56.7	9.3	25.4	5.9	2.2	0.05	0.0003	0.37	
2012	57.4	9.3	24.7	5.7	2.4	0.05	0.0014	0.39	
2013	57.0	9.3	24.7	5.6	2.7	0.06	0.004	0.60	
2014	56.3	9.4	25.5	5.9	2.3	0.06	0.008	0.58	
2015	56.8	9.6	24.8	6.0	2.2	0.06	0.012	0.60	0.000001
2016	57.1	10.0	23.8	6.1	2.2	0.07	0.016	0.83	0.00004
2017	56.5	10.3	23.9	6.2	2.3	0.07	0.023	0.78	0.0002
2018	55.5	10.2	24.7	6.2	2.4	0.07		0.74	0.0004

offered during the 1980s. Crude oil prices dropped again in the 1980s and 1990s, so the ethanol industry came to a halt. However, crude oil prices were on the rise again in the first decade of this century, and interest in ethanol started again.

To drive the development of a US ethanol market, the first Renewable Fuel Standard, also called RFS1, was passed into law in 2005 and required that 4 billion gallons of biofuel be used in 2006, with an increase to 7.5 billion gallons by 2012. This regulation was broadened in 2007 and the so-called RFS2 required the use of 9 billion gallons in 2008 with a schedule to reach 36 billion gallons by 2022. In addition, the amount of ethanol from corn grain (or corn starch ethanol) is limited to 15 billion gallons, with the remainder being made from cellulosic biomass. Biodiesel also had a mandate of 0.5 billion gallons for 2009, growing to 1.28 billion gallons in 2013.[14] As shown in Table 14.3, corn grain ethanol met the 15 billion gallon target in 2016 and biodiesel met the 1.28 billion gallon target in 2013. Corn grain ethanol continues to meet this target, but there has not been much change over the last 3 years.

The enactment of RFS1 and RFS2 has been facilitated by federal subsidies, as shown in Table 14.5, although these have substantially decreased since 2010. For example, the Volumetric Ethanol Excise Tax Credit (VEETC) is a blending credit that was originally $0.51 per gallon of ethanol blended, but was reduced to $0.45 in 2008, and expired at the end of 2011. Similarly, there was a biodiesel credit of $1.00 per gallon blended which was established in 2004 but expired in 2017.

In spite of these government subsidies and mandates, however, the cellulosic ethanol market has not taken off. According to RFS2, the total for corn based and cellulosic ethanol was to be 36 billion gallons by 2022, of which 21 billion were to be

Table 14.5 Biofuels Federal Subsidies[3]

Millions of $	2010	2013	2016
Direct Expenditures	363	75	33
Tax Expenditures	6,890	1,740	2,690
Research and Development	86	62	90
Total	7,339	1,877	2,813

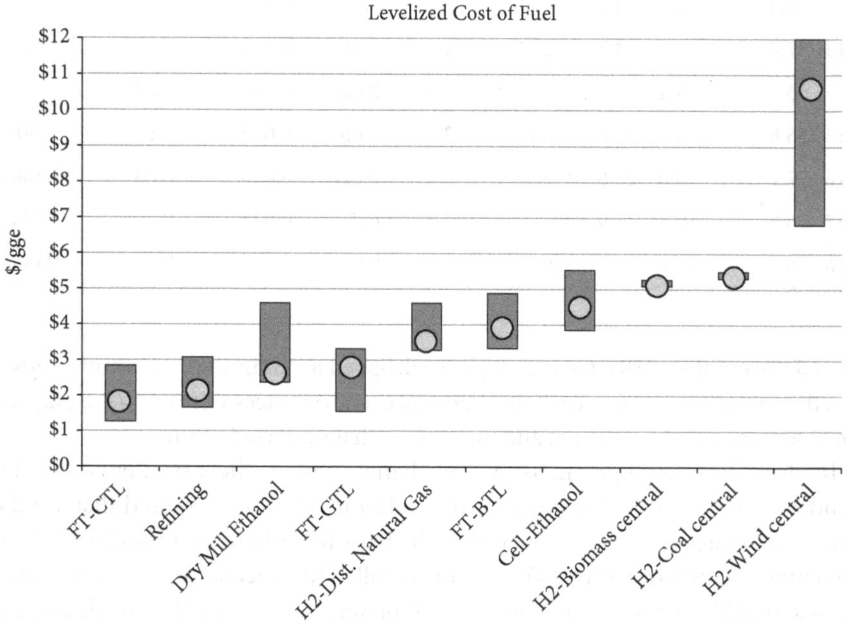

Figure 14.5 US Levelized Cost of Fuel (LCOF)

cellulosic ethanol. Only 10 million gallons were produced in 2017 and 8.2 million in 2018 (see Chapter 8). Faced with this reality, the EPA has revised the cellulosic mandate to 381 million gallons for 2019,[15] although this reduced value will very likely not be met. Most of what is currently being counted as cellulosic ethanol is coming from facilities that co-process corn grain and corn fiber. There is currently only one plant dedicated solely to cellulosic ethanol, the Poet Project Liberty plant in Emmetsburg, Iowa, and it is not yet meeting the 20 million gallon per year plant design target.

Cellulosic ethanol suffers from high capital cost and high operating costs, resulting in a high levelized cost of fuel (LCOF). Figure 14.5 shows results for different transportation fuels and cellulosic ethanol has an LCOF of $4.47/gge compared to $2.60/gge for corn grain ethanol and $2.25/gge for refinery gasoline. That means there is a price difference of greater than $2 per gallon compared to regular gasoline.

As mentioned, hydrogen as a transportation fuel has yet to make much of an impact. The best LCOF for hydrogen is \$3.37/gge, but this model-based calculation has yet to be realized in the real world, as the average price for 2018 was about \$15/gge.

Another potential competitor with refinery-produced gasoline and diesel is the Fischer-Tropsch (FT) technology using coal, or coal-to-liquids (CTL) and natural gas, or GTL. On an LCOF basis, it competes nicely with refinery-produced fuels, but there is a major impediment in capital costs, which are around \$100,000 per barrel per day (\$100,000/BPD) versus \$40,000/BPD for a refinery. A typical FT plant is 50,000 BPD, so this would require an initial investment of five billion dollars. Thus the high capital costs have held this technology back in spite of its favorable LCOF.

These data suggest that, for the near future, the US market will continue to use refinery-produced gasoline and diesel, blended with ethanol and biodiesel. The markets for CNG, hydrogen, and electricity are still quite small while the Fischer-Tropsch technology has been held back by high capital costs.

14.2. The Electric Vehicle

If the United States and world are ever going to have a transportation market that is fossil-fuel free, the choices are really the FCV using hydrogen or the EV. While hydrogen has the advantage of a rapid charging time equivalent to gasoline and a decent driving range, it will require a lot of new infrastructure, a reduction in hydrogen fuel cost, and some better ways to store hydrogen onboard the vehicle than high-pressure hydrogen.

The EV is interesting because it provides a bridge between transportation and electricity. As the capture and sequestering of CO_2 at the exhaust of a vehicle is not likely to become practical, the EV has the advantage that CO_2 made from electricity can be captured at the power plant in the case of a fossil fuel, or can be avoided by the use of renewable energy. It also has the advantage of fitting into the existing electrical industry infrastructure, but suffers from long recharging times and relatively short driving range. The driving range is typically less than 300 miles and more typically about 100 miles and charging times, even with a 240-volt charging system, are 4 to 8 hours. And both the FCV and EV are more expensive than gasoline-powered vehicles.

Where an EV has an advantage is in the cost of fuel per mile driven. According to the EIA, the average US cost of electricity was \$0.1078/kWh in 2018[16]. As there are 33.4 kWh/gge, this means the price of electricity as a fuel is \$3.60/gge. This can be compared with the LCOF of \$2.25/gallon for gasoline. However, the current corporate average fuel economy (CAFE) for a gasoline-powered vehicle is 25 mpg versus about 120 mpg for an EV (see Chapter 11). This means the driving cost per mile is \$0.03 for an EV versus \$0.09 for gasoline-powered. And if we use a typical retail price of \$3/gallon for gasoline, the cost per mile driven is \$0.12.

On a fuel cost basis, the EV makes sense but what about the capital cost of purchasing the EV? Table 14.6 show comparative data for EVs, plug-in hybrids, and a gasoline vehicle, all sedans with five seats.

If you choose to purchase the Chevy Bolt or Hyundai Ioniq EVs, you pay about $4,000 more than a plug-in hybrid and $6,000 to $14,000 more compared to a gasoline vehicle. Even if you ignore the cost of electricity used by the EV, taking an average difference of $10,000 for the EV versus the gasoline-powered vehicle, a price of $3/gallon of gasoline, and an average fuel economy of 25 mpg, means you would have to drive your EV more than 80,000 miles before breaking even. And if you assume a $5,000 difference between the plug-in hybrid and gasoline vehicle, you will drive your plug-in hybrid more than 40,000 miles before breaking even. Thus, if the decision is on cost alone, you will probably still purchase a gasoline vehicle. In addition, if you want to choose between an EV or plug-in hybrid, decisions based on economics and convenience will likely lead you to select a plug-in hybrid, which has a much greater driving range and is not constrained by the typical 4 to 8 hour charging time for an EV, assuming you have a Level 2, 240 volt charging station available. Otherwise, charging times at 120 volts are even longer.

Because of the cost for buying the EV, the limited driving range, and the battery recharging time, the EV market may not be ready to take off. Certainly, decreasing the cost of the EV and increasing the driving range through improved battery storage will change this.

One last consideration for this section is whether the electric market could support the demand for a US fleet of EVs. Assuming the entire US fleet of 278 million gasoline vehicles was converted to a fleet of EVs, how much electricity would be required? According to the US Department of Transportation Federal Highway

Table 14.6 Comparison of Electric, Plug-in Hybrid, and Gasoline Vehicles Including Manufacturer's Suggested Retail Price (MSRP) (but not including tax credits)[17]

Name	MSRP	Vehicle Type	EPA Range	Battery Size	Miles per Gallon (a)
Chevy Bolt	$37,500	Electric Vehicle	238 miles	60 kWh	132
Chevy Volt	$33,200	Plug-in Hybrid	53 miles electric but 420 miles with gasoline	18 kWh	43 city and 42 highway
Hyundai Ioniq	$29,500	Electric Vehicle	124 miles	28 kWh	148
Hyundai Ioniq	$25,000	Plug-in Hybrid	29 miles electric but 630 miles with gasoline	9 kWh	57 city and 59 highway
Toyota Camry	$23,645	Gasoline	650 miles		29 city and 41 highway

a. For the EVs, miles per gallon are shown on the basis of a gge, where 33.4 kWh is 1 gge.

Administration,[18] the average American now drives 13,476 miles per year. Using the assumption of 120 miles per gge and the conversion factor of 33.4 kWh per gge, the EV electricity economy is 3.6 miles/kWh. Assuming all of the 278 million vehicles drive the same 13,476 miles, the electricity demand would be 1,040,647 GWh. As the EIA reports a total electricity generation of 4,058,258 GWh for 2017,[19] the EV demand represents 25% of the amount of all electricity generated. Naturally, the transition of the entire 278 million vehicle fleet would be gradual, but eventually the electric power industry would have to increase generation to meet the demand for EV electricity.

14.3. Reducing CO_2 Emissions

This book is not intended to be political, but just raising a discussion about global warming, or the more euphemistic term climate change, can stir political feelings. While it appears that the European position on global warming is unequivocal, the position of the United States is much less clear. Different US administrations have taken different approaches to global warming, and you can get vastly different opinions from people based on their generation, income level, educational background, as well as other factors. Therefore it is not exactly clear what future position the United States will have with respect to global warming, or what plans there will be for reducing the production of CO_2. If you have an interest in the science between historical data for CO_2 and temperature as a function of years, two papers are referenced here.[20,21] In addition, if you have considered that the earth's warming is due to the sun burning hotter rather than GHG emissions, NASA has discounted this as the cause for the global warming trend that has occurred in the last 50 years.[22]

With these issues in mind, and the fact that the world is moving toward more strict control of CO_2 with or without of us, it is worth reviewing the available options for reducing CO_2 emissions within the context of the energy types discussed in this book. CO_2 emissions come from electricity generation, transportation fuel, and heating, and the two basic options to reduce emissions are to perform CCS at fossil fuel plants or replace fossil fuel energy with renewable energy.

Table 14.7 summarizes CO_2 emissions for each technology, taken from the appropriate chapter, as well as the cost of CCS and the percentage of CO_2 emissions that can reasonably be captured. As the table shows, values are listed in terms of CO_2 equivalent, or CO_2eq, because other GHGs, usually methane, can also be formed for some of the technologies. For example, methane can be formed with hydroelectric plants due to biomass decomposition in the reservoir. Methane is important to include, as it has a global warming potential (GWP) of 28 to 36 times that of CO_2.

Next, the data of Table 14.7 were combined with US data for electricity, heating, and transportation in order to calculate the tons of CO_2eq made in each sector for the year 2017. For electricity, 2017 data were taken from the EIA *Electric Power Annual*[23] and then CO_2 emissions were calculated using the appropriate value in Table 14.7 for

Table 14.7 CO_2 Production and the Cost of CCS

Energy Type	lb CO_2eq/ MM BTU	lb CO_2eq/ kWh	CO_2 from 1,000 MW Plant, tons	CO_2 Available for Capture	Cost of CCS
Fuel Technology					
Petroleum Crude Oil Refining	242.6	2.66		10%	$7/BBL ($90–120/ ton)
Fischer-Tropsch GTL	235.9				
Fischer-Tropsch CTL	446.9				$2.5–4/BBL
Fischer-Tropsch BTL	31.8				$4–7/BBL
Ethanol Corn Grain	144.2			50%	$46–66/ton
Ethanol Cellulosic	11.6			50%	
Hydrogen, Natural Gas Reforming	299			92%	
Hydrogen, Central Biomass	171			95%	
Hydrogen, Central Electrolysis	68			0%	
Hydrogen, Central Coal	171			92%	
Power Plant Technology					
Natural Gas Combined Cycle (CC)	112.4	0.73	7,500	92%	$30/MWh ($82/ton)
Pulverized Coal Combustion	236.4	2.08	22,214	92%	$58/MWh ($56/ton)
Nuclear	6.3	0.066	713	0%	
Hydroelectric, all types		0.165	743	0%	
Wind, all types		0.06	343	0%	
Solar Photovoltaic		0.088	233	0%	
Concentrating Solar Power		0.044	222	0%	
Biomass, all types	7.5	0.101	19,540	95%	$137/MWh ($140/ton)
Geothermal Flash		0.104	1,057	65%	
Geothermal Binary		0.025	254	0%	
Enhanced Geothermal Systems (EGS)		0.071	720	0%	

a. *Note*: CCS is carbon capture and sequestration.

b. *Note*: GTL, CTL, and BTL are gas-to-liquids, coal-to-liquids, and biomass-to-liquids.

lb CO_2eq/kWh. For heating, data were taken from the EIA *Annual Energy Review*[24] and CO_2 emissions were calculated using the appropriate value in Table 14.7 for lb CO_2eq/MM BTU. Likewise, for transportation, the CO_2 emissions were calculated using the appropriate value in Table 14.7 for lb CO_2eq/MM BTU along with the data shown earlier in Table 14.3. The results of these calculations are shown in Tables 14.8, 14.9 and 14.10 for electricity, heating, and transportation. Where values were not available for a particular technology, or where the fuel has different technology approaches, the value for lb CO_2eq/MM BTU and lb CO_2eq/kWh was assumed, as noted in the table. For example, the calculations for coal and natural gas were based on CO_2 production values for pulverized coal and NGCC, even though the emissions may be somewhat different if the coal plant used integrated gasification combined cycle (IGCC) or if the natural gas plant used combustion. In the case of using petroleum for electricity generation, the CO_2 emission value was not available, so the value for coal was used instead.

Summarizing the three tables, the estimate for the total CO_2eq produced in 2017 was about 6,400 million tons, with 28.2% from electricity, 20.9% from heating, and 50.9% from transportation. Note that CO_2eq was also produced from renewable energy types, such as wind and solar, because of CO_2 emissions produced during the life cycle of a plant from things such as manufacturing materials transportation, installation, and maintenance. Clearly, though, these renewables contributed very little to the CO_2 emissions. For example, fossil fuels produced 96.3% of all the CO_2 emissions for electricity.

Focusing first on electricity, the key to reducing CO_2 is to replace coal and natural gas with renewable energy. Although not shown here in tabular form, the last 10 years have seen little change in hydroelectric, geothermal, and biomass electricity, as almost all the increase in renewable electricity has come from solar and wind. If the data for wind shown in Figure 14.1 are fit to a polynomial trend line and extrapolated to the years 2030 and 2040, the amount of wind-generated electricity will increase from 256.9 TWh in 2017 to 932.0 TWh in 2030 and 1,883.3 TWh in 2040. Likewise, the value of solar-generated electricity for 2017 is 78.1 TWh in 2017, increasing to 458.0 TWh in 2030 and to 1,111.8 TWh in 2040.

In other words, if the current trends for solar and wind continue, the total for these two renewables will increase from 335.0 TWh in 2017 to 1,390.1 TWh in 2030 and 2,995.0 TWh in 2040, increases of 1,055 TWh and 2,660 TWh. Assuming the total electricity generated in 2017, 4,058 TWh, remains constant, solar and wind would represent 26% of the electricity mix by 2030 and 66% of the mix by 2040. By 2030, this would reduce the electricity CO_2 emissions from 1,820.2 million tons CO_2eq/year to 888.7 and by 2040, reduce it to less than 130 tons CO_2eq/year, as wind and solar would completely eliminate all fossil-fuel generated electricity.

Overall, the total CO_2 produced by the United States would decrease from 6,400 million tons to about 4,675 million tons, a reduction of 27%. Naturally, there are a lot of assumptions here, not the least of which is that electricity consumption would remain constant. It would also require new grid infrastructure to deliver this

Table 14.8 US CO_2 Production from Electricity in 2017

2017	Net Generation, GWh	CO_2 Available for Capture	Million Tons CO_2eq/Year
Coal (based on pulverized coal)	1,205,835	92%	1,254.1
Natural Gas (based on NGCC)	1,296,415	92%	473.2
Other Gases (based on NGCC)	12,469	92%	4.6
Nuclear	804,950	0%	26.6
Petroleum (based on coal)	21,390	92%	22.2
Conventional Hydroelectric	300,333	0%	24.8
Hydro-Pumped Storage (based on Hydroelectric)	−6,495	0%	−0.5
Wind	254,303	0%	7.6
Solar PV—Utility Scale	50,017	0%	2.2
Solar PV—Small	23,990	0%	1.1
Solar PV Total	74,007	0%	3.3
Solar Thermal	3,269	0%	0.1
Solar Total	77,276		3.3
Wood and Wood-derived Fuels	41,152	95%	2.1
Geothermal (based on flash)	15,927	65%	0.8
Other Biomass	21,609	95%	1.1
Other Energy Sources (based on wind)	13,094	0%	0.4
Total	4,058,258		1,820.2

wind and solar-generated electricity to market. Nevertheless, it demonstrates that if the current growth trend continues for wind and solar, US electricity could be CO_2-free by 2040. In addition, Chapter 6 on wind suggested a combined onshore and offshore potential of about 38,000 TWh and Chapter 7 on solar suggested a US potential of 2,019 TW which, at a capacity factor of 26%, would generate 4.6 million TWh. Therefore, the potential does exist for wind and solar to meet the generation levels projected for 2030 and 2040.

As for cost, the LCOE calculated for NGCC was $73.9/MWh, or ¢7.4/kWh. For wind the LCOE was ¢10.0/kWh and for solar PV was ¢13.4/kWh, ¢2.6/kWh and ¢6.0/kWh more expensive than NGCC, respectively. However, from 2015 to 2019, the LCOE for wind decreased from $125.4/MWh to $99.8/MWh, a decrease of ¢2.6/kWh and the LCOE for solar PV decreased from $250.7/MWh to $134.1/MWh, a decrease

Table 14.9 US CO_2 Production from Heating in 2017

Natural Gas	Billion cubic feet	MM BTU	Million Tons CO_2 eq/year
Residential	4,412.3	4,553,475,024	255.9
Commercial	3,164.1	3,265,328,496	183.5
Industrial	9,513.7	9,818,105,376	551.8
Coal	**Thousand tons**	**MM BTU**	
Residential	350.0	8,898,400	1.1
Commercial	1061.1	26,978,500	3.2
Industrial	50801.5	1,291,577,031	152.7
Petroleum	**Thousand BPD**	**MM BTU**	
Residential	516.5	1,093,451,670	132.6
Commercial	266.7	564,591,198	68.5
Total			1,349.2

Table 14.10 US CO_2 Production from Transportation Fuel in 2017

	Million Gallons	MM BTU	Million Tons CO_2eq/ year
Gasoline	142,976	17,763,052,288	1,883.8
Jet Fuel	25,962	3,600,669,780	368.3
Diesel	60,360	8,371,328,400	876.5
Ethanol	15,800	1,196,060,000	1.4
CNG	5,758	662,170,000	37.2
CNG as Motor Fuel	172	19,780,000	1.1
Biodiesel	1,985	275,299,650	28.8
Total	253,013	31,888,360,118	3,197.1

of ¢11.7/kWh. Therefore, it entirely possible that by 2030 and 2040, the LCOE for wind and solar could be equal, or nearly so, to NGCC.

An alternative approach is to use CCS to remove CO_2 from natural gas and coal plants instead of renewable energy. Table 14.7 shows the cost of CCS for NGCC and pulverized coal as $30/MWh and $58/MWh, respectively. This means that if CCS became required for natural gas and coal generated electricity, the LCOE would increase ¢3/kWh for NGCC and ¢5.8/kWh for pulverized coal. For the current LCOE values calculated for NGCC and wind, this would make the two equivalent in price. This suggests that for the two approaches, namely adding CCS to natural gas and coal plants versus replacement with wind and solar, replacement of fossil fuel plants with renewable energy is the better approach.

The biggest contributor to US CO_2 emissions is transportation, estimated to be around 50%. Only about 10% of CO_2 emissions from transportation occur at the refinery, the other 90% coming from the exhaust of our vehicles. So, unlike coal and natural gas power plants, where it is possible to do CCS and remove almost all CO_2 at the plant, CCS at a refinery would be, at most, 10% effective.

As it is unlikely that every vehicle in the United States could be equipped with some sort of CCS device, this means the only effective way to remove CO_2 from the transportation sector is to capture and sequester it before it enters the vehicle. At this time, the three main options are the hydrogen-fueled FCV, the EV, or a vehicle running on a high percentage of ethanol. Clearly, a CNG-powered vehicle still emits CO_2, albeit less than gasoline. In the case of ethanol, there is a fuel blend called E85, which contains 51% to 83% ethanol and is used in flexible fuel vehicles (FFVs). Since the life cycle production of corn grain ethanol produces 144.2 lb CO_2eq/MM BTU versus 212.1 lb CO_2eq/MM BTU for gasoline, every gallon of ethanol used reduces CO_2 by about 1/3. Assuming an average ethanol content of 75% for E85, and assuming all 142,976 million gasoline gallons used in 2017 were replaced with E85, the CO_2 from gasoline would be reduced from 1,884 million tons CO_2eq/year to 471 million tons CO_2eq/year, while that for ethanol would increase from 86.2 to 671.5 million tons, a net reduction of 827.6 million tons CO_2eq/year.

Overall, this would reduce transportation CO_2 emissions from 3,282 million tons CO_2eq/year to 2,454 million tons CO_2eq/year, a reduction of 25%. Needless to say this approach has some issues, the main one being the amount of corn grain needed to make this happen. In 2018, the US corn production was 14.60 billion bushels, 5.60 billion of which were used for ethanol or 38.4% (see Table 8.3 in Chapter 8). Replacing 75% of the 142,976 million gallons of gasoline used in 2017 with ethanol would require 107,232 million gallons of ethanol, and it would take 38 billion bushels of corn, more than 2 ½ times the total US corn production in 2017. Thus, for this option to work, it will require a breakthrough in corn production, a reduction in the cost for cellulosic ethanol, or a breakthrough in the use of some other form of biomass to make ethanol. As the LCOF for cellulosic ethanol was calculated to be \$4.47/gge versus \$2.60 for corn grain ethanol, there is still quite a difference in price. Even so, a United States using only E85 still produces a substantial amount of CO_2.

The other two main options are the hydrogen-fueled FCV or the EV. As Table 14.3 demonstrated, the hydrogen-fueled FCV is still struggling to gain traction, producing less than one million gge of transportation fuel versus a total for 2017 of 253,000 million gge. While this technology has the advantage of better fuel economy than gasoline, and an equivalent refueling time, fuel costs are still high and there are not many refueling stations. At this time, there are only 46 hydrogen stations in the United States, with 41 in California, and the average price for hydrogen fuel in

2018 was \$15/gge. Most of these stations use compressed natural gas delivered from a central location. Of the four options considered for hydrogen, distributed natural gas, central biomass, central coal, and central wind, only central wind is truly CO_2-free, as the others approaches produce CO_2. Unfortunately, for the calculated LCOF of each of these options, distributed natural gas was lowest at \$3.37/gge and central wind was the highest at \$10.74/gge. To truly become an attractive alternative to gasoline will require a lot of new infrastructure and a breakthrough in the cost of central wind hydrogen. It may also require some novel onboard storage methods as currently hydrogen is stored on the vehicle at 5,000 or 10,000 psi.

Therefore, in my opinion, the EV is the best future for reducing CO_2 from the transportation sector, especially using electricity made from renewable energy. Recharging time and driving range are the main impediments at this time. Currently there are over 500 Tesla supercharger network stations around the United States which can add 100 miles of range in as little as 20 minutes. Although this is considerably less than the 150,000 gasoline and diesel filling stations, it is still a lot more market penetration than the 46 hydrogen stations. With the hope that these two impediments can be overcome, estimates will be made for the reduction in CO_2 emissions and the cost to the driver.

In Chapter 11, the current corporate average fuel economy (CAFE) for a gasoline-powered vehicle was given as 25 mpg versus about 120 mpgee (miles per gasoline gallon equivalent) for an EV. That means the 142,976 million gallons of gasoline shown in Table 14.10 do not need to be replaced with an equivalent amount of electric fuel because of the better economy. A simple ratio shows that if all the gasoline were replaced with electricity, it would require about 30,000 million gge of electricity. Since there are 33.4 kWh/gge, this 30,000 million gge of electricity is equivalent to about 1,000 TWh. In 2018, the average US cost of electricity was \$0.1078/kWh, so 1,000 TWh would cost about \$108 billion. Using the LCOF for gasoline of \$2.25/gallon, the 142,976 million gallons would cost \$322 billion. It is worth noting that 1,000 TWh represents about 25% of the total electricity generated in 2017, 4,058 TWh. This would certainly impact the US electric market but, as discussed earlier, there is still great untapped potential for solar and wind, as well as other renewables such as hydroelectric, biomass, and geothermal.

Referring back to Table 14.6, the price difference between an EV and a Toyota Camry is about \$10,000, so the difference of purchasing an EV versus a Toyota Camry for the entire US fleet would cost \$2,780 billion, a very significant figure. Even if you ignore the cost of electricity used by the EV, the price difference of \$10,000 combined with an assumed price of \$3/gallon and an average fuel economy of 25 mpg means you would need to drive more than 80,000 miles before breaking even. Clearly, the EV must come down in price before it receives broad market penetration.

In terms of CO_2, if the EV is powered by electricity generated from wind and solar PV, the entire 1,883.8 million tons CO_2eq/year generated by gasoline-powered vehicles would be eliminated. Also, a transition from fossil fuel generated electricity to electricity from wind and solar could reduce CO_2 emissions from 1,820 million tons in 2017 to 889 tons in 2030 and less than 130 tons in 2040. Combining these two means that the total CO_2 produced in 2017 could be reduced from 6,400 million tons to 2,877 million tons by 2040.

Therefore, even with these dramatic changes to the electric power and transportation industry, there is still a lot of CO_2 being produced. In the transportation sector, other major contributors include jet fuel and diesel fuel. The combination of these two is 1,245 million tons CO_2eq per year. Presuming the EV can transition to larger vehicles, part of this CO_2 could be eliminated, but the transition to commercial electric planes is likely a long way in the future, if ever. Another major contributor is residential, industrial, and commercial heating, representing another 1,350 million tons of CO_2 per year. Heating accounts for a little more than 20% of the CO_2 emissions produced in the United States for 2017. Some options for CO_2 reduction in this sector include the geothermal heat pump, solar heating, and biomass furnaces. Geothermal heat pumps reduce heating costs by 30%–70% and cooling costs from 20%–50%, depending on location. Unfortunately, installation costs can exceed $20,000 so it is a large initial expense for the home owner. Solar heating can include solar hot water, passive solar heating and daylighting, and solar process heating and cooling. Solar hot water is the use of solar energy to heat water, essentially replacing your hot water heater. Passive solar heating and daylighting means the building is constructed with large windows on the wall receiving the most sunlight. In addition, materials that absorb and store solar heat can be selected for floors and walls. Daylighting simply means that sunlight is used to brighten the interior of the building, which can include windows placed near, or on, the roof. The effect of implementing these options in the heating sector is not explored here but, like every new technology, they can make incremental reductions in CO_2 emissions.

14.4. Proven Reserves and Consumption

This section summarizes proven reserves and current consumption for each technology. In the case of fossil fuels and nuclear energy, current consumption also determines the ultimate lifetime, pending the discovery of more resources. In the case of renewable energies, estimates of "proven reserves" were made based on various criteria. For example, potential reserves for solar energy were determined by using average solar insolation levels and land area for different countries. And, for wind energy, reserves were determined using various technical factors, such as turbine ratings, turbine heights and the wind speed at that height, and what land and ocean areas are available and appropriate for use. More details can be found in the individual chapters for each technology.

Although current consumption is used here to calculate the ultimate lifetime of nonrenewable energy types, growth in world energy use is expected that would further decrease these calculated lifetimes. Given growing energy use in countries like China, India, and Vietnam, as well as the increasing world population, it seems likely that consumption will increase even though more efficient use of energy will counteract this somewhat.

Here then are some projections for each of the twelve different technologies.

Petroleum Crude Oil

- Worldwide reserves are around 1.66–1.7 trillion barrels
- US reserves are 50 billion barrels, 2.9% of world total reserves
- Worldwide consumption for 2017 was 98.2 million barrels per day (MBPD)
- Assuming current consumption levels and no significant increase in reserves, there are about 47 years remaining for crude oil
- Improved oil development, such as deepwater drilling and hydraulic fracking would extend the life for crude oil
- Eight countries each have 100 billion or more proven barrels: Canada, Venezuela, Russia, Iran, Iraq, Kuwait, Saudi Arabia, and the United Arab Emirates, and these eight countries control 79.6% of the worldwide proven reserves based on a total of 1.7 trillion barrels

Natural Gas

- Worldwide reserves are 6,832 trillion cubic feet
- US reserves are 308.5 trillion cubic feet, 4.5% of world total reserves
- Worldwide consumption for 2017 was 129.6 trillion cubic feet
- Assuming current consumption levels and no significant increase in reserves, there are 53 years remaining
- Hydraulic fracking could extend the life for natural gas, while increased consumption using natural-gas power vehicles and the Fischer-Tropsch synthesis using natural gas would further shorten it
- Ten countries control 78.8% of reserves: Russia, Iran, Qatar, the United States, Saudi Arabia, Turkmenistan, United Arab Emirates, Venezuela, Nigeria, and China

Coal

- Worldwide reserves are 1,140,904 million tons
- US reserves are 276,587 million tons, 24.2% of world total reserves
- Worldwide consumption for 2017 was 8,434 million tons
- Assuming current consumption levels and no significant increase in reserves, there are 135 years remaining
- Accessing "Demonstrated Reserve Base," coal that has been found or measured but not currently mined, would extend this while additional uses for coal, such as the Fischer-Tropsch synthesis would shorten this life
- Ten countries control 90.5% of the worldwide coal reserves: the United States, Russia, China, Australia, India, Germany, Ukraine, Kazakhstan, Indonesia, and Poland

Nuclear

- Worldwide reserves are 6,142,200 tonnes (one tonne = 1,000 kg)
- US reserves are 47,200 tonnes, 0.8% of world total reserves

- Worldwide uranium consumption for 2017 was 62,825 tonnes
- Assuming current consumption levels and no significant increase in reserves, there are 98 years remaining
- Reserves are defined as reasonably assured resources (RARs) and inferred resources, meaning uranium deposits that have been discovered by direct measurements. In addition, reserves are defined as uranium recoverable for a cost at or below $130/kg uranium. Relaxing this constraint would increase reserves, as uranium also exists in the ocean and granites, but is not recoverable at this cost
- Ten countries control 88.1% of the worldwide uranium reserves: Australia, Kazakhstan, Russia, Canada, Niger, Namibia, South Africa, Brazil, China, and Uzbekistan

Hydroelectric

- Worldwide potential for electricity is 10,000 TWh/year (terawatt-hour/year)
- US potential for electricity is 279 TWh/year, 2.8% of world potential
- As hydroelectric is renewable, it will always be available
- Ten countries control 59.5% of the worldwide potential: Russia, China, Canada, India, Brazil, Indonesia, Peru, Democratic Republic of the Congo, Tajikistan, and the United States

Wind Energy

- Worldwide potential for wind power ranges from 72,000 GW to 479,452 GW
- Assuming a worldwide potential of 94,500 GW and a capacity factor of 40%, the electricity generation would be 330,000 TWh
- Potential is difficult to determine, and involves many technical factors such as turbine rating, turbine height, the land and ocean area available for use, and wind speed available at a given height
- US potential for wind is 10,955 GW, 11.6% of world potential (based on a worldwide potential of 94,500 GW)
- Assuming a capacity factor of 40%, these 10,955 GW would generate 38,400 TWh
- As wind is renewable, it will always be available
- Based on a worldwide potential of 94,500 GW, ten countries control 81.5% of the worldwide potential: Russia, Canada, the United States, India, China, France, Sweden, Finland, the United Kingdom, and Germany

Solar PV and CSP

- Potential of top ten countries is 14,183 TW (terawatt)
- Potential is difficult to determine, and involves many technical factors such as the solar flux striking the earth, the amount of sunlight reflected back into space and absorbed by the atmosphere and clouds, the available land area, and the efficiency of converting solar energy into electricity. As this efficiency is higher for solar PV versus solar thermal, estimates were based on solar PV
- US potential for solar is 2,019 TW, 14.2% of top ten countries
- Assuming a capacity factor of 26%, these 2,019 TW would generate 4.6 million TWh

- As solar is renewable, it will always be available
- The top ten countries for solar potential are Russia, the United States, China, Brazil, Canada, Australia, India, Kazakhstan, Algeria, and Argentina

Bioethanol

- Worldwide potential for biomass to produce ethanol is around 2.9 trillion gallons per year
- Potential for the low end is with only rain used to grow the biomass while for the high end, irrigation is added along with a higher crop yield
- Current US potential is 702 million tons of biomass, which could produce around 65,000 million gallons of ethanol per year
- This potential could grow to 1,520 million tons by 2040, which would yield around 160,000 million gallons per year
- Biomass resources compete with alternate uses, such as electricity, food, and heating
- As bioethanol is renewable, it will always be available

Biomass to electricity

- Worldwide potential for biomass to generate electricity range is 232 EJ/year ($1 \text{ EJ} = 1 \times 10^{18}$ Joules) by 2050, which could produce 16,100 TWh/year
- Potential is based on many factors, such as the types plants and waste to include as biomass, land availability versus food crops, economic conditions, how worldwide food consumption competes with biomass for energy, increases in yields due to plant varieties and agricultural practices, loss of land due to soil degradation, and increased land utilization for nonagricultural purposes
- The US potential for biomass by 2030 is about 1,150 million tons per year, which could produce 1,200 TWh/year
- By 2040, the potential for biomass could grow to 1,520 million tons/ which could produce 1,600 TWh/year
- As biomass is renewable, it will always be available

Geothermal

- Worldwide potential is 210.9 GW
- Assuming a capacity factor of 75%, this 210.9 GW could generate 1,386 TWh
- Potentials are affected by how deep you are willing to drill, and at what cost
- US potential is 9.1 GW, 4.3% of world potential; this potential could grow to 16.5 GW if drilling to depths of 6.5 km (~4 miles) are considered
- As geothermal is renewable, it will always be available
- Assuming a capacity factor of 75%, this 9.1 GW could generate 60 TWh
- Ten countries control 63.3% of the worldwide potential: Indonesia, Japan, Chile, Pakistan, Mexico, India, Kenya, the United States, Taiwan, and China

Hydrogen

- Worldwide potential to make hydrogen is 44,300 million metric tonnes from natural gas, 136,000 million metric tonnes from coal, and 815 million metric tonnes per year from biomass

- US potential to make hydrogen is 2,010 million metric tonnes from natural gas, 33,000 million metric tonnes from coal, and 55 million metric tonnes per year from biomass
- It is noteworthy that the biomass is a yearly yield while the yields for natural gas and coal are based on total existing reserves
- Hydrogen produced from natural gas and coal is not renewable, but hydrogen produced from biomass, and other renewable technologies is renewable

Fischer-Tropsch Synthesis:

- GTL
  - Worldwide reserves for natural gas are 6,832 trillion cubic feet and US reserves are 308.5 trillion cubic feet in 2017
  - Using an estimate of 5,000 gallons/MM SCFD, if all these worldwide reserves were used for making FT transportation fuels, the total potential would be 34.2 trillion gallons, while if all the US reserves were used, the total potential would be 1.5 trillion gallons
- CTL
  - Worldwide reserves for coal in 2017 are 1,140,904 million tons and US reserves were reported as 276,587 million tons
  - Using a liquid product yield of 97 gallons per ton, if all these worldwide reserves were used for making FT transportation fuels, the total potential would be 110.7 trillion gallons, while if all the US reserves were used, the total potential would be 26.8 trillion gallons
- BTL
  - Worldwide biomass reserves are estimated with an energy content of 232 EJ/year, about 16,900 million tons per year
  - US potential is about 1,150 million tons per year
  - Using a liquid product yield of 49.3 gallons per ton, if all these worldwide reserves were used for making FT transportation fuels, the total potential would be 0.83 trillion gallons per year, while if all the US reserves were used, the total potential would be 0.057 trillion gallons per year

14.5. Footprint and Energy Density

Footprint is the amount of area used for a given energy type and energy density is the amount of energy derived from a given energy type per area per year. These cannot be calculated by simply using the land footprint of the refinery or electrical plant because other land area is used for the overall "Well-to-Wheels" analysis.

For example, the footprint and energy density for using petroleum crude oil to produce gasoline and diesel and deliver them to the customer includes land area (1) to drill and produce oil, (2) to transport the oil to the refinery, (3) for the refinery, and (4) to transport the refined products of gasoline and diesel to the fueling station. Similarly, for a coal power plant, there is land area (1) to mine coal, (2) to upgrade the coal before transport, (3) to transport the coal to the coal plant via rail, (4) to make electricity at the coal electrical plant, and (5) to transmit and distribute the electricity via the grid to commercial customers. As it turns out, some parts of the footprints dominate others, and this depends on the technology. A summary of the results, including which part of the footprint dominates, are shown in Table 14.11. Specific details for each technology may be found in the individual chapters. The energy densities are given in units of GWh/(year-square mile). Although the energy unit of gigawatt-hours (GWh) is normally used for electric power plants, this terminology was also used for technologies dedicated to making transportation fuels in order to compare all energy types on a common basis. For comparative purposes, the table also shows the land area needed to support a 1,000 MW plant, even though this is not directly applicable to those technologies used to make transportation fuels.

Of the traditional energy resources, including petroleum crude oil, natural gas, coal, and nuclear, the footprints for crude oil and natural gas are dictated by the land area needed for drilling and production while coal and nuclear are dictated by the land area of the power plant. This is not surprising for nuclear energy, since at a typical enrichment level of 4.3%, the energy release for the U-235 is 1,536,820,000 BTU/lb. Compared to 20,155 BTU/lb for crude oil, 22,792 BTU/lb for natural gas, and 12,712 BTU/lb for bituminous coal, nuclear energy contains from 67,000 to 120,000 times more energy than these fossil fuels on a mass basis.

However, it is surprising that the land needed to mine coal does not dominate the energy density as it does for petroleum and natural gas. The data reviewed in Chapters 1, 2, and 3 show that for a given land area, mining coal extracts about 700 times more energy than natural gas and nearly 900 times more energy than petroleum. Unfortunately, the data used for surface mining coal were based on new acres permitted, as specific data were not available (Chapter 3). Another possible factor for coal having a smaller footprint is that there is now less guesswork for finding coal than crude oil and natural gas. According to the World Coal Association,[25] "after centuries of mineral exploration, the location, size and characteristics of most countries' coal resources are quite well known." In contrast, finding petroleum crude oil and natural gas is not as exact, although the technology involved is quite varied and complicated. While a complete analysis is beyond the scope of this book, locating land that could produce oil and natural gas involves using geologic mapping, such that seismic data are collected to develop a picture below the surface followed by geologic mapping of the data to locate a potential drilling site. Then, since seismic imaging only provides locations with the potential to produce, a core sample is obtained by drilling, sometimes as deep as several thousand feet, to confirm the presence of oil and natural gas. Therefore if there is more certainty in mining coal than natural gas and crude oil, this would lower the footprint for coal relative to natural gas and crude oil.

For renewable fuels, the footprints for wind and solar were dictated by the plant area. For ethanol from corn, biomass to electricity, hydroelectric, and geothermal, the land used to harvest the energy dominated the footprint. In fact, for biomass to electricity, the 2,650 square miles needed for a 1,000 MW biomass power plant is a little

Table 14.11 "Well-to-Wheels" and Plant Energy Densities

Technology	Well-to-Wheels, GWh/(year-square mile)	Plant, GWh/(year-square mile)	Footprint to Support 1,000 MW Plant, Square Miles	Well-to-Wheels Dominating Factor
Fuel Technology				
Petroleum Crude Oil Refining	36.8	56,000	238	Drilling and production
Fischer-Tropsch Gas-to-Liquids	46.7	58,120	188	Gas extraction from ground
Fischer-Tropsch Coal-to-Liquids	25,000	25,000	0.4	CTL plant
Fischer-Tropsch Biomass-to-Liquids	3.3	18,700	2,650	Land to grow biomass
Ethanol (dry mill corn grain)	7.0	6,900	1,260	Land to grow corn
Ethanol (cellulosic corn stover)	1.9	5,700	4,600	Land to grow corn
Hydrogen (from natural gas)	46.7	5,800	187.7	Gas extraction from ground
Hydrogen (from onshore wind)	51.8	5,800	169	Wind farm
Power Plant Technology				
Natural Gas Combined Cycle (CC)	46.7	160,500	188	Gas extraction from ground
Supercritical Pulverized Coal	8,500	8,500	1.3	Power plant
Nuclear	5,900	5,900	1.5	Power plant
Hydroelectric (storage dam)	31.5	14,700	278	Reservoir
Wind Onshore	51.8 (2,470 direct impact)	51.8	169 (3.6 direct impact)	Wind farm
Wind Offshore	67.3	67.3	130	Wind farm
Solar Photovoltaic (PV)	204	204	43	Solar power plant
Concentrating Solar Power	277	277	32	Solar power plant
Biomass (Different types of Power Plants)	3.3	6,600	2,650	Land to grow biomass
Geothermal (Dry steam, flash, and binary)	524	11,900	16.7	Geothermal field

more than the land area for the state of Delaware, at 2,489 square miles. To put this into perspective, the International Energy Agency reports that the US electric power consumption per capita was 12.8 MWh in 2016. Assuming a capacity factor of 83%, a 1,000 MW plant will power about 570,000 homes. As the installed capacity for all US power plants was 1,203,092 MW in 2017,[19] it would take about 1,200 states the size of Delaware to support a United States where all electricity came from biomass.

The reservoir area for the hydroelectric dam dominated its footprint and leads to a calculated land area of 278 square miles to support a 1,000 MW plant. This is a little deceiving, however, as the reservoir has other uses than generating electricity, such as water control, providing water during dry seasons, and recreational activities such as boating and fishing. If we base the hydroelectric footprint on just the land area of the power station, the footprint for a 1,000 MW power station is only 0.60 square miles.

For geothermal energy, the footprint was dictated by the geothermal field, resulting in a footprint of 16.7 square miles for a 1,000 MW geothermal power plant. However, this footprint is still not that large and the geothermal field may not have a variety of other uses anyway. The power plant alone occupies only 0.7 square miles. If enhanced geothermal systems (EGS) become a reality, where drilling taps geothermal heat 3 to 5 km below the surface of the earth, this would decrease the footprint for the geothermal field. However, EGS currently suffers from a very high LCOE, $611.5/MWh versus $98.8/MWh for a geothermal flash plant.

For the two fastest growing renewable energies, wind and solar PV, the footprints for a 1,000 MW plant are less than that for an NGCC plant. The NGCC plant footprint, calculated to be 188 square miles, is dictated by the land area used to extract the natural gas from the ground. In comparison, the footprints for onshore and offshore wind farms are 169 and 130 square miles, respectively, while the footprint for a solar PV plant is calculated to be 43 square miles. In the case of an offshore wind farm, the land area is probably not that important, presuming it does not disrupt boat traffic. The problem for offshore wind is the high LCOE, $237.9/MWh versus $99.8/MWh for onshore wind. This is caused by the high capital cost for an offshore wind farm, about $4,800/kW versus $1,500/kW for an onshore farm. Laying the foundation for the turbine pads is more difficult, and must accommodate heavy storms and waves. In addition, it is expensive to connect the farm to the electric grid as the power cables must be installed under the ocean.

In fact, the calculated 169 square miles for the 1,000 MW onshore wind farm is a little deceiving, as there is a lot of unused area between turbines, with typical spacings from 200 m to 800 m. This space can still be used for farming and ranching. The components of the onshore wind farm only affect about 2% of the total area (see Chapter 6), and the footprint for this direct impact area is only 3.6 square miles for the 1,000 MW wind farm. This is different than solar power plants. For a solar PV plant, power plant components typically occupy 91% of the total area and for concentrating solar power (CSP), 77% of the total area. Even so, the 43 square miles calculated for the solar PV plant is still less than the 188 square miles impact of the NGCC plant.

References

1. "Statistical Review of World Energy—All Data, 1965–2018," 2018, https://www.bp.com/en/global/corporate/energy-economics/statistical-review-of-world-energy.html
2. "By All Energy Sources and by Producing Sector, All Sectors," Energy Information Administration, 2018, https://www.eia.gov/electricity/data.php
3. Direct Federal Financial Interventions and Subsidies in Energy in Fiscal Year 2016, https://www.eia.gov/analysis/requests/subsidy/pdf/subsidy.pdf
4. "Average US Wind Price Falls to $20 per Megawatt-Hour," Emma Foehringer Merchant, Green Tech Media, August 24, 2018, https://www.greentechmedia.com/articles/read/us-wind-prices-20-megawatt-hour#gs.a7q2sr
5. "Petroleum & Other Liquids," Energy Information Administration, 2018, https://www.eia.gov/dnav/pet/hist/LeafHandler.ashx?n=PET&s=MGFUPUS1&f=A
6. "Alternative Fuels Data Center," US Department of Energy, Energy Efficiency & Renewable Energy, 2018, https://www.afdc.energy.gov/data/categories/fuels-infrastructure
7. "U.S. Energy Consumption by the Transportation Sector," Tables 4.4 and 4.6, Bureau of Transportation Statistics, 2018, https://www.bts.gov/content/us-energy-consumption-transportation-sector
8. "DoE—U.S. Plug-in Electric Cars Consumed 2 TWh of Electricity in 2017," Inside EVs, 2017, https://insideevs.com/news/337946/doe-us-plug-in-electric-cars-consumed-2-twh-of-electricity-in-2017/
9. "U.S. Biodiesel Production, Exports, and Consumption," Alternative Fuels Data Center, Energy Efficiency & Renewable Energy, US Department of Energy, 2018, https://afdc.energy.gov/data/10325
10. "Next Generation Hydrogen Station Composite Data Products: Retail Stations Data through Quarter 4 of 2018," Sam Sprik et al., April 2019, https://www.nrel.gov/docs/fy19osti/73658.pdf
11. "Fact of the Month March 2019: There Are More Than 6,500 Fuel Cell Vehicles on the Road in the U.S.," Office of Energy Efficiency & Renewable Energy, March 2019, https://www.energy.gov/eere/fuelcells/fact-month-march-2019-there-are-more-6500-fuel-cell-vehicles-road-us
12. "The U.S. Has 1 Million Electric Vehicles, but Does It Matter?," Maxine Joselow, Scientific American E&E News, October 12, 2018, https://www.scientificamerican.com/article/the-u-s-has-1-million-electric-vehicles-but-does-it-matter/
13. "Natural Gas Vehicles," Alternative Fuels Data Center, Energy Efficiency & Renewable Energy, US Department of Energy, 2018, https://afdc.energy.gov/vehicles/natural_gas.html
14. "Renewable Fuel Standard (RFS): Overview and Issues," Randy Schnepf and Brent D. Yacobucci, Congressional Research Service, March 14, 2013, file:///C:/Users/paulf/Documents/Book/References_Figures_Tables_Revised/Conclusions/R40155.pdf
15. "U.S. EPA proposes higher 2019 biofuels mandate in line with expectations," June 26, 2018, Reuters Environment, https://www.reuters.com/article/us-usa-biofuels/u-s-epa-proposes-higher-2019-biofuels-mandate-in-line-with-expectations-idUSKBN1JM2B8
16. "Electric Power Monthly, Energy Information Administration, Data for June 2018, Table 5.6.A. Average Price of Electricity to Ultimate Customers by End-Use Sector, https://www.eia.gov/electricity/monthly/epm_table_grapher.php?t=epmt_5_6_a
17. "Compare Electric Cars and Plug-in Hybrids by Features, Price, Range," 2019, http://www.plugincars.com/cars

18. "What's the Average Miles Driven per Year? (Car Lease Guide)," Scott Oldham, November 1, 2018, https://www.autogravity.com/autogravitas/money/whats-average-miles-driven-per-year-car-lease-guide
19. "Table 3.1.A. Net Generation by Energy Source: Total (All Sectors), Energy Information Administration, 2017, https://www.eia.gov/electricity/annual/html/epa_03_01_a.html
20. "Ice Core Basics," Bethan Davies, Antarctic Glaciers, September 1, 2015, http://www.antarcticglaciers.org/glaciers-and-climate/ice-cores/ice-core-basics/
21. "Ice Cores and Climate Change," British Antarctic Survey, March 1, 2014, https://www.bas.ac.uk/data/our-data/publication/ice-cores-and-climate-change/
22. "Is the Sun Causing Global Warming?," NASA Global Climate Change, April 3, 2020, https://climate.nasa.gov/faq/14/is-the-sun-causing-global-warming/
23. "Electric Power Annual 2017," Energy Information Administration, October 2018, Revised December 2018, https://www.eia.gov/electricity/annual/pdf/epa.pdf
24. "Annual Energy Review," Energy Information Administration, Tables 4.3, 6.2, and 3.7a, 2017, https://www.eia.gov/totalenergy/data/annual/
25. "Where Is Coal Found," World Coal Association, 2018, https://www.worldcoal.org/coal/where-coal-found
26. "Electric Power Consumption (kWh per capita)," IEA Statistics for 2016, http://energyatlas.iea.org/#!/tellmap/-1118783123/1

APPENDIX A

Capital Cost, Operating Cost, Well-to-Wheels Levelized Cost of Electricity, and Well-to-Wheels Levelized Cost of Fuel

The first part of this appendix describes the equations and approach used for all levelized cost of energy calculations. Subsequent sections give calculation details for each energy type.

A.i.1. General Approach for All Calculations

Capital cost is defined simply as the initial, one-time, cost to build a plant, while operating costs are the continuous, ongoing cost of operating the plant. The Energy Information Administration (EIA) uses the term "overnight" capital costs, referring to the cost of the project as if no interest were incurred during its construction.[1]

The levelized cost of energy is calculated for each energy type, using a common approach, such that they can be compared and ranked on their economic viability. For this book, the "Levelized Cost of Electricity" (LCOE) and the "Levelized Cost of Fuel" (LCOF) are used. For each of these, the "Well-to-Wheels" levelized cost is calculated to take into account more than just production, but also include delivery, any marketing costs, and distribution. For electric power plants, we could also use terms such as "Well-to-Outlet" or "Mine-to-Outlet," to capture the concept of the overall cost to bring energy from a natural gas well or a coal mine to the business or home that uses the electricity. For example, in the case of making electricity from natural gas, the "Well-to-Outlet" concept is applicable, such that the cost of making electricity must consider the natural gas drilling and production costs, natural gas processing, transportation to the electric plant (e.g. by pipeline), operation of the power plant, and high-voltage transportation through the electric grid before coming to your home. For consistency however, the terminology of "Well-to-Wheels" is used throughout, even though the terminology is not specifically applicable to each technology.

The LCOE is the price, reported in \$/kilowatt-hour or \$/megawatt-hour, that is needed for an electric plant project to break even over the life of the project. That is, the LCOE is the flow of income, over the expected energy production life of a plant, which allows the owners to recover all costs, including capital costs, operating costs, the cost of financing, and an assumed return on investment.

There are many ways to calculate this value, and the method used for this book is shown in equation (1), adapted from two different sources.[2,3] In this equation, the term "8,760" is the number of hours in 1 year. The heat rate is the amount of energy used by the electric plant to generate 1 kilowatt-hour (kWh) of electricity. As such, it measures the efficiency of the electric plant to convert fuel into electricity, so the lower the heat rate the higher the efficiency.

$$LCOE = \left[\frac{Capital\ Cost * CRF * (1 - T * DPV)}{8760 * Capacity\ Factor * (1 - T)}\right] + \left[\frac{Fixed\ O\&M}{8760 * Capacity\ Factor}\right]$$
$$+ \left[\frac{Variable\ O\&M}{1{,}000}\right] + \left[\frac{Fuel\ Price * Heat\ Rate}{1{,}000{,}000}\right]$$
$$+ \left[Product\ \ Transportation + Marketing\right] \tag{1}$$

$$CRF = \left[\frac{D * (1 + D)^N}{((1 + D)^N - 1)}\right] \tag{2}$$

$$DPV = \sum_{i=1}^{N_{MACRS}} \left[\frac{MACRS\%(i)}{((1 + D)^i}\right] \tag{3}$$

As you can see, this equation contains many terms, each of which has, to a varying degree, an effect on the calculation. The different terms are defined in what follows. However, a few items deserve further explanation.

The Capital Recovery Factor, or CRF, is a ratio used to calculate the present value of an annuity for a given length of time. This is defined by equation (2).

To finance the capital cost to build the energy plant, companies may borrow the money or use existing capital. Even if the company has existing capital to build the plant, that money could, alternatively, have been invested at current interest rates to earn interest rather than investing in the energy plant. Therefore, the CRF is included in the equations for the levelized cost, and takes into account the current interest rate and the useful life of the plant. Essentially it is a method to see when that initial investment is recovered. The interest rate was assumed to be 7% for all calculations. The useful life or book life ("N") of different plants are taken from an Environmental Protection Agency (EPA) 2010 study.[4] As well, the levelized cost equations include a combined federal and state tax, set to 38%.

Depreciation, the reduction in value for an asset over time, is also a factor included in the levelized cost. Table A.i.1 shows the useful plant life and depreciation period used for each electric plant technology. The present value of depreciation, or DPV, was calculated using IRS Publication 946[5] for the Modified Accelerated Cost Recovery System, or MACRS schedule. Different types of power plants have different depreciation schedules, and the different years are also shown in Table A.i.1. Percentages for each year are taken from Table A.1 of the IRS publication, using the "half-year convention."

The percentages are shown in Table A.i.2. Using these percentages, the discount rate "D," and equation (3), the DPV can be calculated as a factor for different plant lives. These factors are summarized in Table A.i.1. Therefore, this LCOE equation contains capital cost, fixed costs, variables costs, fuel costs, product transportation to market costs, marketing costs, and adjustments made for the cost of money, depreciation, and the overall tax rate. Examples for fixed operating costs include maintenance, rent, insurance, wages, and utilities. Examples of variable operating costs, which are directly related to the volume of feed fuel processed, include chemicals, catalysts, fuel, and power generation. The difference between fixed and variable operating costs is that fixed costs do not change with the amount of fuel used for power generation, while variable costs are directly affected by the amount of fuel consumption.

The "Fuel Price" used in the calculations is assumed to include the cost of delivery to the plant. Best available fuel prices from 2018 were used in all the calculations. Note that the LCOE equation does not contain any government subsidies that may be available to a renewable energy technology, such as solar, geothermal, and wind. While these incentives are necessary to

Table A.i.1 Book Life, IRS MACRS Depreciation Schedule, and the DPV Calculated Using the MACRS Schedule

Energy Type	Book Life, years	US—MACRS Depreciation Schedule, year	US MACRS DPV
Advanced Pulverized Coal	40	20	0.54407
Advanced Pulverized Coal with CCS	40	20	0.54407
Integrated Gasification Combined Cycle Coal	40	20	0.54407
Integrated Gasification Combined Cycle Coal with CCS	40	20	0.54407
Natural Gas Conventional Combined Cycle	30	20	0.54407
Natural Gas Advanced Combined Cycle	30	20	0.54407
Natural Gas Advanced Combined Cycle with CCS	30	20	0.54407
Natural Gas Combustion Turbine	30	15	0.61548
Advanced Nuclear	40	15	0.61548
Wind—onshore	20	7	0.79019
Wind—offshore	20	7	0.79019
Solar Thermal	20	5	0.83155
Solar Photovoltaic	20	5	0.83155
Geothermal—Binary	20	5	0.83155
Biomass—BFB and BCC	40	7	0.79019
Hydroelectric	30	20	0.54407

Note: CCS stands for carbon capture and sequestration.

help renewable energies develop quickly and reduce some economic disadvantages, this book compares technologies on as equal a basis as possible.

It is worth noting that any economic evaluation is subject to change and even criticism, as improvements and maturation of different energy technologies will result in reductions in capital and operating costs. Projections of capital and operating costs presume some costs for materials and labor, and these can certainly change, and they can between different countries. Nevertheless, the best available economic data were used to evaluate each energy type.

Capital Cost	Cost of the plant in $/kW
CRF	Capital recovery factor in units of 1/year (defined by equation (2)
D	Interest rate is the annuity rate, which is a constant 0.07, or 7%, for these calculations
N	Useful lifetime of the new plant (taken from Table 8.4, reference 4), Table A.i.1
N_{MACRS}	MACRS depreciation schedule, years

T	Combined federal and state tax rate, assumed to be 38%
Capacity Factor	Yearly average percentage of a power as a fraction of capacity
DPV	Present value of depreciation, using the Modified Accelerated Cost Recovery System, or MACRS schedule, Table A.i.1
Fixed O&M	Fixed operations and maintenance cost of the plant per capacity in \$/kW
Variable O&M	Variable operations and maintenance cost of the plant per capacity in \$/kW
Fuel Price	Fuel cost for the plant (including delivery costs) in \$/MM BTU
Heat Rate	Efficiency of the power plant converting fuel into electricity in \$/MM BTU
Product Transportation and Marketing	Cost to market and deliver the product to a distribution center, \$/kW

Table A.i.2 Percentages for Different Years to Depreciate a Power Plant, Taken from Table A-1 of an IRS Publication Using a "Half-Year Convention"

Year	5-Year	7-Year	10-Year	15-Year	20-Year
1	20	14.29	10.0	5	3.75
2	32	24.49	18.0	9.5	7.219
3	19.2	17.49	14.4	8.55	6.677
4	11.52	12.49	11.52	7.7	6.177
5	11.52	8.93	9.22	6.93	5.713
6	5.76	8.92	7.37	6.23	5.285
7		8.93	6.55	5.9	4.888
8		4.46	6.55	5.9	4.522
9			6.56	5.91	4.462
10			6.55	5.9	4.461
11			3.28	5.91	4.462
12				5.9	4.461
13				5.91	4.462
14				5.9	4.461
15				5.91	4.462
16				2.95	4.461
17					4.462
18					4.461
19					4.462
20					4.461
21					2.231

Like LCOE, an LCOF can be calculated for transportation fuel-producing plants. This calculation includes the capital cost of the fuel production plant, distributed over the lifetime of the plant, plus production costs, transportation to market, and marketing costs.[3] The calculation result is the price of gasoline, in \$/gallon or gasoline equivalent (gge), that is needed for the plant investment to break even. The LCOF equation is shown in Equation (4) and the different terms are defined later. It is worth noting that, for some of the fuels, there is a "coproduct" production that will reduce the LCOF. For example, ethanol production from corn will also make "Distiller's dried grains with solubles (DDGS)," a nutrient rich coproduct. Likewise, in the process of making gasoline and diesel by the Fischer-Tropsch synthesis, some electricity will be made from excess heat, as will other useful products.

Another important point is that the LCOF is calculated in terms of a gallon of gasoline equivalent, or "gge," the equivalent energy in a gallon of gasoline. Therefore, some fuels need to be corrected to the equivalent energy of gasoline. For example, the energy content of gasoline is 1.64 times larger than ethanol, so the energy correction for ethanol is 1.64.

$$LCOF = \left[\frac{Total\ Capital\ Cost * CRF * (1 - T * DPV)}{Plant\ Scale * 1{,}000000 \dfrac{gal}{Mgal} * Capacity\ Factor} + O\&M + \frac{Feedstock\ Costs}{Total\ Fuel\ Yield} - \begin{matrix} Coproduct + Product \\ Transportation + Marketing \end{matrix} \right] * \frac{BTU\ /\ gallon\ gasoline}{BTU\ /\ gallon\ fuel} \quad (4)$$

Total Capital Cost:	Cost of the plant in \$/gge
CRF:	Capital recovery factor in units of 1/year (defined by equation (2)
D:	Interest rate is the annuity rate, which is a constant 0.07, or 7%, for these calculations
N:	Useful lifetime of the new plant, Table A.i.2
N_{MACRS}:	MACRS depreciation schedule, years
T:	Combined federal and state tax rate, assumed to be 38%
Capacity Factor:	Yearly average percentage of a power as a fraction of capacity
DPV:	Present value of depreciation, using the Modified Accelerated Cost Recovery System, or MACRS schedule, Table A.i.3
Plant Scale:	Million gallons of fuel produced in a year, M gallons/year
O&M:	Operations and maintenance cost of the plant, \$/gge
Feedstock Cost:	Cost of the feedstock (including delivery costs), \$/gge
Total Fuel Yield:	Fuel yield in gallons fuel/dry short ton
Coproduct	Coproduct credit, \$/gge

Product Transportation and Marketing:
Cost to market and deliver the product to a distribution center, \$/gge

$\left[\dfrac{BTU/gallon\ gasoline}{BTU/gallon\ Fuel} \right]$: Energy correction to a gallon of gasoline equivalent

As for electrical plants, the book life and MACRS values are given in Table A.i.3 for fuels.

Table A.i.3 Book Life, IRS MACRS Depreciation Schedule, and the DPV Calculated Using the MACRS Schedule

Energy Type	Book Life, years	US—MACRS Depreciation Schedule, year	US MACRS DPV
Refining	20	10	0.73443
Fischer-Tropsch Coal to Liquids (CTL)	20	7	0.79019
Fischer-Tropsch Gas to Liquids (GTL)	15	7	0.79019
Fischer-Tropsch Biofuels to Liquids (BTL)	20	7	0.79019
Dry Mill Ethanol	15	7	0.79019
Cellulosic Ethanol	15	7	0.79019
Hydrogen	20	7	0.79019

References

1. Monthly Energy Review, Energy Information Administration (EIA), DOE/EIA-0035, February 2019, https://www.eia.gov/totalenergy/data/monthly/pdf/mer.pdf
2. "2011 Cost of Wind Energy Review," S. Tegen et al., National Renewable Energy Laboratory, 2011, (http://www.nrel.gov/docs/fy13osti/56266.pdf)
3. "Levelized Cost Calculations," Open Energy Information (OpenEI), Transparent Cost Database, 2017, http://en.openei.org/apps/TCDB/levelized_cost_calculations.html
4. "Documentation for EPA Base Case v.4.10 Using the Integrated Planning Model," EPA #430R10010, August 2010, Chapter 8: Financial Assumptions, Table 8.4 (http://www.epa.gov/airmarkets/documents/ipm/Chapter_8.pdf)
5. "How to Depreciate Property: Section 179 Deduction—Special Depreciation Allowance—MACRS" IRS Publication 946, Cat. No. 13081F, 2019, https://www.irs.gov/publications/p946.

A.1. Petroleum Crude Oil: Levelized Cost of Fuel (LCOF)

For petroleum crude oil, the LCOF was calculated according to Equation (4) in the general approach described above. The book life used for the plant was 20 years and the number of years the IRS allows the plant depreciation used was 10 years, according to Table A.i.3. Likewise, the DPV needed in Equation (4) was also taken from Table A.i.3.

For the size of the plant, a refinery capacity of 200,000 barrels per day (BPD) was chosen to make the calculations, a size typical for US refineries. Unlike the other energy technologies, there are no complete economic studies available for a new refinery, complete with operating costs, feedstock costs, product transportation, and marketing. Therefore, each of these items was obtained from a variety of independent sources. For this 200,000 BPD refinery, the liquid product yield is assumed to be 204,000 BPD, a liquid volume expansion of 2%. The units for LCOF are in $/gallon of gasoline equivalent, or $/gge.

The capacity factor was set to 93.1%, based on 2018 data from the EIA. The capacity factor refers to the percent of the rated capacity used each year. According to the EIA,[1] this capacity factor has increased almost every year in the 9 years since 2010. In 2010, the capacity factor was 86.3%, increasing to 90.4% in 2014, and now 93.1%.

It then remains to establish the capital cost, operations and maintenance (O&M) cost, feedstock cost, the cost of transportation to the distribution center, and marketing costs. Transportation energy derived from petroleum crude oil is created over several steps including (1) crude oil production, (2) transportation to the refinery, (3) refining, (4) transportation of products to market, (5) taxes (both federal and state), and (6) marketing cost at the filling station. The cost for these different steps will yield the "Well-to-Wheels" LCOF in $/gge.

In general, some oil companies are fully integrated, meaning that they drill and produce crude oil and also refine the crude oil to make transportation fuels. On the other hand, some oil companies are exclusively refining companies, meaning that they must buy their crude oil at market prices. Even in a fully integrated company, however, there are internal costs that have the refining part of the company paying market prices, or near market prices, for a barrel of crude oil. Therefore, from the refinery point of view, crude oil production costs and the cost of transporting the crude oil to the refinery are included in the purchase price of the crude oil feed. In other words, the (1) crude oil production and (2) transportation to the refinery are captured in the cost of crude oil.

The best way to establish the true cost of crude oil is from global crude oil prices. West Texas Intermediate, or WTI, is used as an economic benchmark crude oil for many analyses and is used here. Unfortunately, there has been a lot of volatility in crude oil prices since 2005. Looking at 2-year intervals, the average annual price of WTI in 2005, 2007, 2009, 2011, 2013, 2015, and 2017 were $56.44, $72.26, $61.73, $94.88, $97.98, $48.66, and $50.88.[2] For 2018, the average price for WTI was $64.94, comparable to what it was in 2009 but far below 2011 and 2013 prices. The 2018 value is used for the LCOF calculations but, because of the volatility in crude oil prices, a sensitivity study was made using different prices for the cost of WTI to the refinery to see the impact of the crude oil cost. Based on the previous data, a low of $40/barrel ($40/BBL) and a high of $100/BBL were used for the sensitivity study.

For the capital cost, data were taken from new "greenfield" refinery and refinery expansion projects. A "greenfield" refinery is defined as completely new construction, also called a "grass-roots" refinery. For "greenfield" refinery construction in the United States, two new refineries did begin operation in 2015 including Petromax Refining in Houston (25,000 BPD) and Buckeye Partners LP (46,250 BPD). However, the last new refinery built with significant capacity was in 1977, the Marathon Garyville,[3] Louisiana, refinery. Initially the capacity was 200,000 BPD but it has since been upgraded to 539,000 BPD. There are a number of new refineries that have been built outside the United States, as well as upgrades to existing refineries. Several examples are shown in Table A.1.1.

Excepting the Petrochina Qinzhou refinery, the range of capital cost for "greenfield" refineries and refinery expansions is $20,000 to about $90,000/BPD capacity with about $40,000/BPD being the average. Therefore, $40,000/BPD is used for the base case and $20,000 and $90,000/BPD is included in the sensitivity study. This means that for a $200,000 BPD refinery, the base case capital cost is $8 billion dollars.

For fixed and variable O&M costs, several references were used to establish typical values. One reference[17] indicates European operating costs range from $3.67 to $4.86/BBL. An EIA 2009 study[18] shows O&M costs of $5.45 and $4.52 for 2008 and 2009, respectively. Another study[19] shows that, for 2012, operating costs range from $2.20 to $4.00/BBL with $3.30/BBL being typical, while a book by Fahim et al.[20] shows 2010 values of $2.71/BBL for the combination of fixed and variables costs. Finally, a study by Balfanz et al.[21] cites a value of $7/BBL, which includes personnel costs, maintenance, insurance, chemicals, and catalysts. Based on these studies, the base case fixed and variable O&M was set to $5/BBL and a range of $3 to $8/BBL was used for the sensitivity study.

The two remaining items we need to calculate the LCOF are (4) transportation of products to market and (6) marketing cost at the filling station. For the sum of these two steps, an April

Table A.1.1 New (Greenfield) Refinery and Expansion Projects

Owner	Location	Refinery	Feed Rate, BPD	Completion Data	Cost, $ billion	$/BPD Cost
Greenfield Refineries						
Sinopec[4]	Zhanjiang City, Guangdong, China	Zhanjiang	300,000	2016	9.0	30,000
Petrochina[5]	Qinzhou, Guangxi, China	Qinzhou	200,000	2010	2.2	11,000
Vietnam Oil & Gas[6]	Thanh Hóa Province, Vietnam	Nghi Son	200,000	2017	9.0	45,000
Petrobras[7]	Recife, Brazil	Abreu e Lima	230,000	2014	20.0	86,957
Saudi Aramco[8]	Jubail, Saudi Arabia	Jubail	400,000	2013	14.0	35,000
CNPC[9]	Zinder, Niger	Zinder	20,000	2011	0.6	30,000
Yasref Saudi Aramco[10]	Yanbu, Saudi Arabia	Yanbu	400,000	2014	10.0	25,000
MDU Resources[11]	Dickinson, ND, United States	Dakota Prairie	20,000	2015	0.43	21,500
Kitimat Clean Ltd[12]	Kitimat, BC, Canada	Kitimat	400,000	~2023	22.0	55,000
					Average	37,717
Refinery Expansion						
Reficar[13]	Cartagena, Columbia	Refineria de Cartagena	85,000 (added)	2015	3.89	45,765
Sonatrach[14]	Algeria	Skikda and Arzew	~100,000 (added)	2013	4.0	40,000
Ecopetrol[15]	Barrancabermeja, Columbia	Barrancabermeja	50,000 (added)	2017	3.39	67,800
Grupa Lotos[16]	Gdansk, Poland	Gdansk	90,000 (added)	2011	2.0	22,222
					Average	43,947
					Average of both	39,634

2019 update by the EIA[22] shows values of $0.32/gallon for gasoline and $0.60/gallon for diesel. Generally, the distribution cost for diesel is higher than gasoline since it has a higher density; therefore, a gallon of diesel weighs more than a gallon of gasoline.

With these data, the LCOF was calculated for the base case and sensitivity study cases, shown in Table A.1.2. For the base case, the price of crude oil is $64.94/BBL, the fixed and non-fuel variable O&M is $5/BBL, and the capital cost is $40,000/BPD installed capacity.

Table A.1.2 Levelized Cost of Fuel (LCOF) for Petroleum Crude Oil

	Levelized Capital Cost	Fixed and Non-Fuel Variable O&M	Feed Cost	Transportation and Marketing	Total System Levelized Cost	Total System Levelized Cost	Total System Levelized Cost
Units	$/gge	$/gge	$/gge	$/gge	$/gge	$/MWh	$/BOE
Base Case	$0.30	$0.12	$1.52	$0.32	$2.25	$67.41	$114.58
Capex is $20,000/BBL	$0.15	$0.12	$1.52	$0.32	$2.10	$62.90	$106.92
Capex is $90,000/BBL	$0.68	$0.12	$1.52	$0.32	$2.63	$78.68	$133.75
Opex is $3/BBL	$0.30	$0.07	$1.52	$0.32	$2.21	$66.01	$112.21
Opex is $8/BBL	$0.30	$0.19	$1.52	$0.32	$2.32	$69.50	$118.14
Crude is $30/BBL	$0.30	$0.12	$0.93	$0.32	$1.67	$50.00	$84.99
Crude is $100/BBL	$0.30	$0.12	$2.33	$0.32	$3.07	$91.88	$156.18

References

1. "US Percent Utilization of Refinery Operable Capacity," Energy Information Administration, 2018 annual average, https://www.eia.gov/dnav/pet/hist/LeafHandler.ashx?n=PET&s=MOPUEUS2&f=M
2. "Average Annual West Texas Intermediate (WTI) Crude Oil Price from 1976 to 2016 (in U.S. dollars per barrel)," The Statistics Portal, 2017, http://www.statista.com/statistics/266659/west-texas-intermediate-oil-prices/
3. "Frequently Asked Questions: When Was the Last Refinery Built in the United States," Energy Information Administration, January 1, 2019, http://www.eia.gov/tools/faqs/faq.cfm?id=29&t=6
4. "Total Joins KPC and Sinopec JV for the $9 Bn Zhanjiang Refinery & PC," 2b1st Consulting, March 27, 2012,http://www.2b1stconsulting.com/total-joins-kpc-and-sinopec-jv-for-the-9-bn-zhanjiang-refinery-pc/
5. "The Tanker Market's Battles to join the Recovery," Charles Weber Company Tanker Report, December 2010,
6. "Vietnam to Build Another Refinery with Kuwait, Japan Firms," Downstream Today, July 7, 2008.
7. "Exclusive: Petrobras May Delay Building Two Refineries due to Costs—Company Source," Reuters News, October 28, 2014, http://www.reuters.com/article/us-petrobras-refineries-delay-idUSKBN0IH2H720141028
8. "Saudi Aramco, Total set to begin exports at new Jubail refinery venture," Hydrocarbon Processing, September 12, 2013, http://www.hydrocarbonprocessing.com/Article/3254442/Saudi-Aramco-Total-set-to-begin-exports-at-new-Jubail-refinery-venture.html

9. "Niger Republic to export gasoline to Nigeria," Oil Price.com, August 11, 2011, http://oil-price.com/Latest-Energy-News/World-News/Niger-Republic-To-Export-Gasoline-To-Nigeria.html
10. "Saudi Aramco, Sinopec Sign Yanbu Refinery Deal," Market Watch, January 14, 2012, http://www.marketwatch.com/story/saudi-aramco-sinopec-sign-yanbu-refinery-deal-2012-01-14
11. "Dakota Prairie Refinery Commences Startup Operations," MDU Resources, May 4, 2015, http://www.mdu.com/news/2015/05/04/dakota-prairie-refinery-commences-startup-operations
12. "Kitimat Clean Fast Facts," 2019, http://www.kitimatclean.ca/fast-facts-2/
13. "Refineria de Cartagena (Reficar) Refinery Expansion Colombia," Hydrocarbon Technology.com, 2019, http://www.hydrocarbons-technology.com/projects/refineria-de-cartagena-reficar-refinery-expansion/
14. "Time to Build," Refining and Petrochemicals Middle East, p. 33, August 2010, https://issuu.com/itpbusiness/docs/pet_88_01082010
15. "Barrancabermeja Refinery Modernization," Oil & Gas Columbia, March 8, 2016, http://www.bnamericas.com/project-profile/en/barrancabermeja-refinery-modernization-modernizacion-refineria-barrancabermeja
16. "Key Lessons from Successful Hydrocracker Projects," Special Supplement to Hydrocarbon Processing, September 13, 2011, http://www.google.com/search?biw=1366&bih=624&no-j=1&q=gdansk+refinery+expansion+capital+costs&oq=gdansk+refinery+expansion+-capital+costs&gs_l=serp.3...4594.11627.0.11805.31.31.0.0.0.0.210.3644.2j20j4.26.0....0...1c.1.64.serp..11.5.709...30i10j33i21.sqt1cnDz5_g
17. "Oil Refining Fitness Check," Robert Marschinski, European Commission Joint Research Centre (JRC), slide 28, 2014,
18. "Performance Profiles of Major Energy Producers 2009," Energy Information Administration, Report Number DOE/EIA-0206 (2009), Table 14, February 25, 2011, http://www.eia.gov/finance/performanceprofiles/refining_marketing.cfm
19. "IEA Refinery Margins—Methodology Notes; Oil Industry and Market Division," September 2012.
20. "Fundamentals of Petroleum Refining," M.A. Fahim et al., Elsevier, 2010, Chapter 16: Refinery Economics, Table 16.3.
21. "The Estimated Forward Cost of EU Legislation for the EU Refining Industry," Ulrich Balfanz et al. Concawe Report Technical Report 11-2014, November, 2014.
22. "Gasoline and Diesel Fuel Update," April 1, 2019, Energy Information Administration, http://www.eia.gov/petroleum/gasdiesel/

A.2. Natural Gas: Levelized Cost of Electricity

For a natural gas power plant, the LCOE was calculated according to Equation (1) in the general approach described above. The book life used for all natural gas plants was 30 years. For plant depreciation, the IRS allows 20 years for combined cycle plants and 15 years for combustion turbine, according to Table A.i.1. Likewise, the DPV needed in Equation (1) was also taken from Table A.i.1.

The units for LCOE are in $/megawatt-hour, or $/MWh. Data for the different cases, including capacity factor, capital costs, O&M costs, and transmission capital costs were taken from two references.[1,2] For transmission capital costs, these are the cost for new transmission lines to tie into existing substations or existing transmission lines.

According to the EIA,[3] the cost of transmitting and distributing electricity in 2013 was 0.9 cents/kWh (or $9/MWh) for transmission and 2.6 cents/kWh (or $26/MWh) for distribution.

Table A.2.1 Economic Analysis for Natural Gas-Fired Electric Plants

Energy Type	Capacity Factor	Capital Cost	Levelized Capital Cost	Fixed O&M	Variable O&M (including fuel)	Transmission Investment	Total System Levelized Cost (LCOE) to Generate	LCOE Including Transmission and Distribution
		$/kW	$/MWh	$/MWh	$/MWh	$/MWh	$/MWh	$/MWh
Conventional CC[1]	0.87	$999	$13.52	$1.49	$24.5	$1.20	$40.73	$75.73
Conventional CC[2]	0.80	$1,000	$14.71	$0.82	$24.00	$1.20	$40.72	$75.72
Advanced CC[1]	0.87	$794	$10.74	$1.35	$22.0	$1.20	$35.32	$70.32
Average	0.85	$931	$12.99	$1.22	$23.52	$1.20	$38.93	$73.93
Advanced CC with CCS[1]	0.87	$2,205	$29.83	$4.52	$31.19	$1.20	$66.74	$101.74
Average	0.87	$2,205	$29.83	$4.52	$31.19	$1.20	$66.74	$101.74
Combustion Turbine[2]	0.10	$ 825	$93.78	$17.1	$34.3	$1.20	$146.42	$181.42
Combustion Turbine[1]	0.30	$1,126	$42.67	$1.4	$49.2	$3.50	$96.76	$131.76
Advanced Combustion Turbine[1]	0.30	$691	$26.18	$2.7	$42.1	$3.50	$74.44	$109.44
Average	0.23	$881	$54.21	$7.05	$41.88	$2.73	$105.87	$140.87

Here, transmission is defined as the high-voltage movement of electricity from the power station while distribution is the cost of electrical lines that take power from a substation to the customer. The cost of transmission and distribution is essential to get a true "Well-to-Wheels" analysis of the total cost. For a natural gas generating plant, the "Well-to-Wheels" costs will include (1) the cost of natural gas production and transportation to the power plant, (2) power plant generating costs, (3) cost of connecting the power plant to the electrical grid, (4) transmission and distribution costs, and (5) taxes (both federal and state).

Calculations were made using a natural gas fuel cost of $3.17MM BTU, based on Henry Hub spot price 2018 data.[4] This price captures the cost of natural gas production and transportation to the power plant.

Calculations for LCOE are shown in Table A.2.1. After calculating the LCOE to generate, $35/MWh is added to include the additional cost of transmission and distribution.

A sensitivity study was made for combined cycle generated electricity by varying the capital cost, the combined fixed and variable operating expenses (excluding the fuel cost), and the fuel cost. For the capital cost and operating expenses, the reference data were used as a guide.

The base case used the operating conditions for the 2019 EIA conventional combined cycle case,[1] so the capital cost was $999/kW, the combined fixed and variable O&M was $5.11/MWh excluding the cost of fuel, and the cost of natural gas was $3.17/MM BTU. For capital cost, the range for combined cycle without CCS was $700–$1,300/kW. Excluding the cost of fuel in the variable operating cost, the combined fixed and variable operating costs ranged from $2.80 to $5.10/MWh. Since the base case also represented the high cost for fixed and variable operating costs, the range was extended to a high value of $7.40/MWh.

Summarizing, the lows and highs for the sensitivity study for capital cost used prices of $700 and $1,300/kW for capital cost and $2.80 and $7.40/MWh for the combined fixed and variable operating costs.

Like crude oil, the price for natural gas has undergone great change over the last 14 years. Table A.2.2 shows historical Henry Hub natural gas prices,[4] indicating a range of $2.52 to

Table A.2.2 Henry Hub Historical Natural Gas Prices, $/MM BTU[4]

Year	$/MM BTU
2005	8.69
2006	6.73
2007	6.97
2008	8.86
2009	3.94
2010	4.37
2011	4.00
2012	2.75
2013	3.73
2014	4.37
2015	2.62
2016	2.52
2017	2.99
2018	3.17

Table A.2.3 Sensitivity Study using the EIA case for Natural Gas Combined Cycle[1]

	Natural Gas Price	Capital Cost	Fixed and Variable O&M (no fuel cost)	LCOE Including Transmission and Distribution
Units	**\$/MM BTU**	**\$/kW**	**\$/MWh**	**\$/MWh**
Base Case	\$3.17	\$999	\$5.10	\$75.73
Capex is \$700/kW	\$3.17	\$700	\$5.10	\$71.69
Capex is \$1,300/kW	\$3.17	\$1,300	\$5.10	\$79.81
Opex is \$2.8/MWh	\$3.17	\$999	\$2.80	\$73.44
Opex is \$7.4/MWh	\$3.17	\$999	\$7.40	\$78.04
N.G. price is \$2.50/MM BTU	\$2.50	\$999	\$5.10	\$71.31
N.G. price is \$6.00/MM BTU	\$6.00	\$999	\$5.10	\$94.41
N.G. price is \$9.00/MM BTU	\$9.00	\$999	\$5.10	\$114.21

\$8.86/MM BTU over this time period. Using these data as a guide, the sensitivity study will consider natural gas prices of \$2.50, \$6.00, and \$9.00/MM BTU.

Finally, all other data for the calculations were set using the EIA case for an advanced combined cycle natural gas plant.[1] The results are shown in Table 2.2.3.

References

1. "Cost and Performance Characteristics of New Generating Technologies," Annual Energy Outlook 2019, Table 2, Energy Information Administration, January 2019, https://www.eia.gov/outlooks/aeo/
2. "Lazard's Levelized Cost of Energy Analysis—Version 12.0," November 2018, https://www.lazard.com/media/450784/lazards-levelized-cost-of-energy-version-120-vfinal.pdf
3. "Annual Energy Outlook 2015 with projections to 2040," DOE/EIA-0383, Table A8, April 2015.
4. "Natural Gas: Henry Hub Natural Gas Spot Price," Energy Information Administration, 2018, https://www.eia.gov/dnav/ng/hist/rngwhhdm.htm

A.3. Coal: Levelized Cost of Electricity

For a coal power plant, the LCOE was calculated according to Equation (1) in the general approach described above.

Finding current data for capital costs, operating costs, and other coal plant parameters is a little less straightforward than for other technologies. For example, in 2013, the EIA provided data for pulverized coal with and without carbon capture and sequestration (CCS) as well as data for integrated gasification combined cycle (IGCC) with and without CCS.[1] However, for 2018 costs, the EIA only reported coal plant cases with 30% and 90% CCS.[2] The cases were limited because of the assumption that new coal plants without CCS cannot be built because of new plant emission standards.

In August 2015, the EPA released a final rule to limit greenhouse gas emission for new power plants.[3] This "Carbon Pollution Standard for New Power Plants" limits CO_2 emissions of no more than 1,000 lb of CO_2 per megawatt-hour (MWh) of electricity produced. Since coal plants typically generate more than 2,000 lb/MWh, CCS will be required to meet this standard.

Therefore, EIA data from both 2013 and 2018 as well as Lazard data from 2014,[4] 2017,[5] and 2018[6] were used to calculate LCOE for pulverized coal and IGCC plants, with and without CCS.

The units for LCOE are in $/megawatt-hour, or $/MWh. The discount rate and book life used for all types of coal technology to calculate the capital recovery factor (CRF) were set to 7% and 40 years, respectively. The MACRS depreciation schedule is 20 years for all types of coal technology, and the capacity factor was set according to either the EIA or Lazard data, as appropriate.

The capital cost, O&M cost, and transmission costs were taken from the authors. For transmission costs, the authors provide the capital cost for new transmission lines to tie into existing substations or existing transmission lines.

According to the EIA,[7] the cost of transmitting and distributing electricity in 2013 was 0.9 cents/kWh (or $9/MWh) for transmission and 2.6 cents/kWh (or $26/MWh) for distribution. Here, transmission is defined as the high-voltage movement of electricity from the power station while distribution is the cost of electrical lines that take power from a substation to the customer. The cost of transmission and distribution is essential to get a true "Well-to-Wheels" analysis of the total cost. For a coal generating plant, the "Well-to-Wheels" cost will include (1) the cost of coal production and transportation to the power plant, (2) power plant generating costs, (3) cost of connecting the power plant to the electrical grid, (4) transmission and distribution costs, and (5) taxes (both federal and state).

The cost of coal (based on Illinois No. 6 bituminous coal) was $2.06/MM BTU based on a 12-month average of 2018 EIA data.[8]

Results of the LCOE calculations are shown in Table A.3.1.

A sensitivity study was made for both pulverized coal combustion and the IGCC technology. The sensitivity study was made by varying the capital cost, the combined fixed and variable operating expenses (excluding the fuel cost), and the fuel cost. For the capital cost and operating expenses, all of the available reference data were used as a guide.

For capital cost, the cases with and without CCS were used to establish the maximum and minimum, since the "Carbon Pollution Standard for New Power Plants" will force new plants to be built with CCS. For pulverized coal, these were $3,000 and $8,400/kW, respectively while for IGCC these were $4,000 and $8,350/kW, respectively. Excluding the cost of fuel in the variable operating cost, the combined fixed and variable operating costs minimum and maximum were $6.9/MWh and $21.1/MWh for pulverized coal and $15.6/MWh and $19.6/MWh for IGCC.

Unlike crude oil and natural gas, the price for coal has been relatively stable over the last decade. Table A.3.2 shows historical prices for Illinois No. 6 bituminous coal, ranging from $2.06 to $2.39/MM BTU. Using a BTU content of 12,712 BTU/lb for the bituminous coal, these prices are equivalent to a range of $52.4 to $60.8 per ton. Although the range for the last decade is not large, the price of coal was as low as $1.20/MM BTU in 2000. In order to better understand the possible impact of coal prices on LCOE, the range was broadened to a low of $1.50/MM BTU to a high of $2.50/MM BTU.

The base cases were set using EIA data[1] for the pulverized coal and IGCC plants. The results are shown in Table A.3.3.

Table A.3.1 Economic Analysis for Coal-Fired Electric Plants

Energy Type	Capacity Factor	Capital Cost	Levelized Capital Cost	Fixed O&M	Variable O&M (including fuel)	Transmission Investment	Total System Levelized Cost (LCOE) to Generate	LCOE Including Transmission and Distribution
		$/kW	$/MWh	$/MWh	$/MWh	$/MWh	$/MWh	$/MWh
Pulverized coal[1]	0.85	$3,246	$41.84	$5.08	$22.60	$1.20	$70.71	$105.71
Pulverized coal[4,6]	0.93	$3,000	$35.34	$4.91	$20.03	$1.20	$61.48	$96.48
Average	0.89	$3,123	$38.59	$4.99	$21.31	$1.20	$66.09	$101.09
Pulverized coal with 30% CCS[2]	0.85	$5,169	$66.62	$9.69	$27.40	$1.20	$104.90	$139.90
Pulverized coal with 90% CCS[2]	0.85	$5,716	$73.67	$11.25	$33.89	$1.20	$120.01	$155.01
Pulverized with 90% CCS[1]	0.85	$5,227	$67.37	$10.82	$34.23	$1.20	$113.61	$148.61
Pulverized with 90% CCS[4,6]	0.93	$8,400	$98.95	$9.82	$29.72	$1.20	$139.69	$174.69
Average	0.88	$6,448	$80.00	$10.63	$32.61	$1.20	$124.44	$159.44
IGCC[1]	0.85	$4,400	$56.71	$8.36	$25.14	$1.20	$91.41	$126.41
IGCC[5]	0.75	$4,175	$60.99	$11.11	$32.60	$1.20	$105.90	$140.90
IGCC[4]	0.75	$4,000	$58.43	$9.47	$25.13	$1.20	$94.23	$129.23
Average	0.78	$4,192	$58.71	$9.65	$27.62	$1.20	$97.18	$132.18
IGCC with 90% CCS[1]	0.85	$6,599	$85.05	$9.78	$30.49	$1.20	$126.53	$161.53
IGCC with 90% CCS[5]	0.75	$8,350	$121.97	$11.11	$32.60	$1.20	$168.88	$201.88
IGCC with 90% CCS[4]	0.75	$8,000	$116.86	$11.11	$30.17	$1.20	$159.34	$194.34
Average	0.78	$7,650	$107.96	$10.67	$31.09	$1.20	$151.58	$185.92

Note: The pulverized coal data from Lazard had the same data for 2014 and 2018 (references 4 and 6).

Table A.3.2 Illinois No. 6 Bituminous Coal Historical Prices, $/MM BTU[8]

Year	$/MM BTU
2009	2.21
2010	2.27
2011	2.39
2012	2.38
2013	2.34
2014	2.37
2015	2.22
2016	2.11
2017	2.06
2018	2.06

Table A.3.3 Sensitivity Study Using the EIA Case for Integrated Gas Combined Cycle

	Coal Price	Capital Cost	Fixed and Variable O&M (no fuel cost)	LCOE Including Transmission and Distribution
Units	**$/MM BTU**	**$/kW**	**$/MWh**	**$/MWh**
Pulverized Coal				
Base Case	$2.06	$3,246	$9.55	$105.71
Capex is $3,000/kW	$2.06	$3,000	$9.55	$102.54
Capex is $8,400/kW	$2.06	$8,400	$9.55	$172.14
Opex is $6.9/MWh	$2.06	$3,246	$6.90	$103.06
Opex is $21.1/MWh	$2.06	$3,246	$21.10	$117.26
Coal price is $1.50/MM BTU	$1.50	$3,246	$9.55	$100.79
Coal price is $2.50/MM BTU	$2.50	$3,246	$9.55	$109.59
IGCC				
Base Case	$2.06	$4,400	$8.36	$126.41
Capex is $4,000/kW	$2.06	$4,000	$15.58	$121.26
Capex is $8,350/kW	$2.06	$8,350	$15.58	$177.32
Opex is $15.6/MWh	$2.06	$4,400	$15.60	$126.43
Opex is $19.6/MWh	$2.06	$4,400	$19.60	$130.43
Coal price is $1.50/MM BTU	$1.50	$4,400	$15.58	$121.54
Coal price is $2.50/MM BTU	$2.50	$4,400	$15.58	$130.24

References

1. "Updated Capital Cost Estimates for Utility Scale Electricity Generating Plants," Energy Information Administration, April 2013; and "Levelized Cost and Levelized Avoided Cost of New Generation Resources in the Annual Energy Outlook 2015," June 2015, Table 1.
2. "Cost and Performance Characteristics of New Generating Technologies," Annual Energy Outlook 2019, Energy Information Administration, Table 2, January 2019, https://www.eia.gov/outlooks/aeo/
3. "Regulating Power Sector Carbon Emissions," Center for Climate and Energy Solutions, 2019, https://www.c2es.org/content/regulating-power-sector-carbon-emissions/
4. "Lazard's Levelized Cost of Energy Analysis—Version 8," September 2014, https://www.lazard.com/media/1777/levelized_cost_of_energy_-_version_80.pdf
5. "Lazard's Levelized Cost of Energy Analysis—Version 11.0," November 2017, https://www.lazard.com/media/450337/lazard-levelized-cost-of-energy-version-110.pdf
6. "Lazard's Levelized Cost of Energy Analysis—Version 12.0," November 2018, https://www.lazard.com/media/450784/lazards-levelized-cost-of-energy-version-120-vfinal.pdf
7. "Annual Energy Outlook 2015 with projections to 2040," DOE/EIA-0383, Table A8, April 2015.
8. "Table 9.9 Cost of Fossil-Fuel Receipts at Electric Generating Plants," Energy Information Administration / Monthly Energy Review, p. 163, March 2019, http://www.eia.gov/totalenergy/data/monthly/pdf/sec9_13.pdf

A.4. Nuclear: Levelized Cost of Electricity

For a nuclear power plant, the LCOE was calculated according to Equation (1) in the general approach described above. In this case, the nuclear fuel is uranium with enriched levels of U-235.

The units for LCOE are in \$/megawatt-hour, or \$/MWh and data for nuclear power plants were taken from two sources.[1,2] Although the EIA reference[1] only specifies the reactor design as "advanced nuclear," an earlier EIA reference[3] further identifies it as a pressurized water reactor (PWR). The data of Lazard[2] do not specify reactor type. Regardless, Figure 1 from Mott MacDonald[4] shows that the capital cost for a PWR and boiling water reactor (BWR) are the same and Figure 2 from the same report shows that the LCOE for the two technologies are very nearly the same. Therefore, the various costs reviewed in this section are applied equally to either reactor design.

The tax rate for both federal and state is set to 38%. The discount rate and book life used to calculate the CRF were 7% and 40 years, respectively. The MACRS depreciation schedule is 15 years.

Capital costs, capacity factors, O&M costs, and transmission costs were taken from the EIA reports. For transmission costs, they provide the capital cost for new transmission lines to tie into existing substations or existing transmission lines.

According to the EIA,[5] the cost of transmitting and distributing electricity in 2013 was 0.9 cents/kWh (or \$9/MWh) for transmission and 2.6 cents/kWh (or \$26/MWh) for distribution. Here, transmission is defined as the high-voltage movement of electricity from the power station while distribution is the cost of electrical lines that take power from a substation to the customer. The cost of transmission and distribution is essential to get a true "Well-to-Wheels" analysis of the total cost. For a nuclear power plant, the "Well-to-Wheels" costs will include (1) the cost of uranium mining and production, preparation for the nuclear power plant through conversion to UF_6, enrichment, and manufacture of the fuel elements, (2) power plant generating costs, (3) transmission and distribution costs, and (4) taxes (both federal and state). The costs must also include (5) the cost of storing nuclear waste.

A price is needed for the uranium ore. According to one source,[6] uranium does not trade on an open market like other commodities but, rather, buyers and sellers negotiate private contracts. Spot prices are, however, given by independent consultants such as Ux Consulting and TradeTech. The 12-month average spot price for uranium in 2018 was $24.59/lb, or $54.21/kg, for U_3O_8.

Some more work has to be done to get the cost of nuclear fuel for the power plant, however. To obtain usable nuclear fuel, the U_3O_8 has to be converted to UF_6, enriched using a gas centrifuge, and fabricated into fuel elements. According to the World Nuclear Association,[7] the cost to obtain 1 kg of uranium as reactor fuel included 46% for the uranium ore, 6% to convert it to UF_6, 32% to enrich it, and 16% to fabricate the fuel elements.

Assuming a U-235 enrichment level of 4.3%, a starting concentration of 0.711% U-235 in the natural uranium, and a tails concentration of 0.24%, the feed requirements are 8.6 times the product requirements. Therefore, at a price of $54.21/kg, the different preparation costs are $466/kg to purchase the U_3O_8, $61/kg to convert it to UF_6, $324/kg to enrich it to 4.3% U-235, and $162/kg to fabricate the fuel elements. This gives a total cost of $1,013/kg or $459/lb for the uranium reactor fuel.

The nuclear fission of a single atom of U-235 releases 202.5 MeV, equivalent to 35.74 Giga BTU/lb (35.74×10^9 BTU/lb). At an enrichment level of 4.3%, the energy release on a mass basis is 1,536,820,000 BTU/lb. At a cost of $459/lb for the reactor fuel, this is equivalent to $0.30/MM BTU on an energy basis.

For the cost for nuclear waste disposal, one report available was published by the Organisation for Economic Co-Operation and Develop (OECD), of which the United States is a member.[8] The report discusses the economics of both direct disposal and partial recycling. Direct disposal is a process such that the fuel is used once and is then regarded as waste for disposal, and partial recycling is a process such that the spent fuel is reprocessed to recover unused uranium and plutonium, which are recycled back into reactors. According to Figure ES.1 in the report, the cost of direct disposal is about $1/MWh while the cost of direct disposal with partial recycling is about $7/MWh. In another study, Berry and Tolley[9] report direct disposal costs at $1.09/MWh and recycling costs of $2.40/MWh. The choice between direct disposal and recycling is an environmental issue, as well as the choice of temporary on-site storage versus permanent disposal. For the purpose of completing the economic analysis, a cost of $1/MWh was assumed for nuclear waste storage, based on direct disposal.

Results of the LCOE calculations are shown in Table A.4.1.

A sensitivity study was made by varying the capital cost, the combined fixed and variable operating expenses (excluding the fuel cost), and the fuel cost. For the capital cost and operating expenses, the values shown in Table A.4.1 were used as a guide.

For capital cost, the sensitivity study used the lower and upper values of $6,034/kW and $12,250/kW, based on the range of the reference data. Excluding the cost of fuel in the variable operating cost, the combined fixed and variable operating costs ranged from only $15.34 to $17.87/MWh. To broaden this range for the sensitivity study, a change of about 50% around the average was used, resulting in a low of $8/MWh and a high of $24/MWh.

Table A.4.2 shows annual averages for the spot price of uranium since 2012 and, as the table illustrates, the price of uranium ore has been steadily decreasing, losing nearly half its value since 2012. Using the 2018 average price, the total cost to make the uranium fuel was calculated earlier to be $1,013/kg or $0.30/MM BTU. For the sensitivity study, the lower and upper values of Table A.4.2 were used, or $21.66 and $48.40/lb, respectively. In terms of energy, these are $0.26/MM BTU and $0.59/MM BTU.

All other data for the calculations used the EIA case.[1] The results of the sensitivity study are shown in Table A.4.3.

Table A.4.1 Economic Analysis for Nuclear-Powered Electric Plants

Energy Type	Capacity Factor	Capital Cost	Levelized Capital Cost	Fixed O&M	Variable O&M (including fuel)	Transmission Investment	Total System Levelized Cost (LCOE) to Generate	LCOE Including Transmission and Distribution	LCOE Including Direct Disposal
EIA[1]	0.90	$6,034	$70.94	$13.10	$5.51	$1.10	$90.65	$125.65	$126.65
Lazard, low cost case[2]	0.90	$6,500	$76.42	$14.59	$3.89	$1.10	$95.99	$130.99	$131.99
Lazard, high cost case[2]	0.90	$12,250	$144.01	$17.12	$3.89	$1.10	$166.12	$201.12	$202.12
Average	0.90	$8,261	$97.12	$14.94	$4.43	$1.10	$117.59	$152.59	$153.59

Table A.4.2 Historical Spot Prices for Uranium, \$/lb of U_3O_8[33]

Year	\$/lb U_3O_8
2012	48.40
2013	38.17
2014	33.21
2015	36.55
2016	25.64
2017	21.66
2018	24.59

Table A.4.3 Sensitivity Study Using the EIA Nuclear Case[1]

	Nuclear Fuel Preparation Cost	Capital Cost	Fixed and Variable O&M (no fuel cost)	Total System Levelized Cost (LCOE) to Generate	Total LCOE Cost to Generate, Transmit, and Distribute
Units	\$/MM BTU	\$/kW	\$/MWh	\$/MWh	\$/MWh
Base Case	\$0.30	\$6,034	\$15.47	\$90.65	\$125.65
Capex is \$6,034/kW	\$0.30	\$6,034	\$15.47	\$90.65	\$125.65
Capex is \$12,250/kW	\$0.30	\$12,250	\$15.47	\$163.72	\$198.72
Opex is \$8/MWh	\$0.30	\$6,034	\$8.00	\$83.18	\$118.18
Opex is \$24/MWh	\$0.30	\$6,034	\$24.00	\$99.18	\$134.18
Nuclear price is \$0.26/MM BTU	\$0.26	\$6,034	\$15.47	\$90.23	\$125.23
Nuclear price is \$0.59/MM BTU	\$0.59	\$6,034	\$15.47	\$93.68	\$128.68

References

1. "Cost and Performance Characteristics of New Generating Technologies," Annual Energy Outlook 2019, January 2019, Energy Information Administration, Table 2, https://www.eia.gov/outlooks/aeo/
2. "Lazard's Levelized Cost of Energy Analysis—Version 12.0," November 2018, https://www.lazard.com/media/450784/lazards-levelized-cost-of-energy-version-120-vfinal.pdf
3. "Updated Capital Cost Estimates for Utility Scale Electricity Generating Plants," April 2013, Energy Information Administration.
4. "Costs of Low-Carbon Generation Technologies," Committee on Climate Change, May 2011, Mott MacDonald, https://www.theccc.org.uk/archive/aws/Renewables%20Review/MML%20final%20report%20for%20CCC%209%20may%202011.pdf
5. "Annual Energy Outlook 2015 with Projections to 2040," DOE/EIA-0383, Table A8, April 2015
6. "Uranium Price," Cameco, 2019, https://www.cameco.com/invest/markets/uranium-price

7. "The Economics of Nuclear Power," World Nuclear Association, July 2016, http://www.world-nuclear.org/information-library/economic-aspects/economics-of-nuclear-power.aspx
8. "The Economics of the Back End of the Nuclear Fuel Cycle," Organisation for Economic Co-Operation and Develop (OECD), NEA No. 7061, October 23, 2013, https://www.oecd-nea.org/ndd/pubs/2013/7061-ebenfc.pdf
9. "Nuclear Reprocessing: Technological, Economic, and Social Problems," R. Stephen Berry and George S. Tolley, final paper for BPRO 29000: Energy and Energy Policy, October 21, 2015, http://franke.uchicago.edu/bigproblems/BPRO29000-2015/Team25-NuclearReprocessing.pdf

A.5. Hydroelectric: Levelized Cost of Electricity

For a hydroelectric power plant, the LCOE was calculated according to Equation (1) in the general approach described above. Data for conventional hydroelectric and pumped storage plants were taken from two references.[1,2] For conventional hydroelectric, storage dams and run-of-river are treated as the same for the calculation of LCOE. The units for LCOE are in $/megawatt-hour, or $/MWh.

The tax rate for both federal and state is set to 38%. The discount rate and book life used to calculate the CRF were set to 7% and 30 years, respectively. The MACRS depreciation schedule is 20 years and the capacity factor was set according to the EIA reports.[1,2]

Capital costs, capacity factors, O&M costs, and transmission costs were taken from the authors. For transmission costs, they provide the capital cost for new transmission lines to tie into existing substations or existing transmission lines.

The capital costs include the cost for new transmission lines to tie into existing electrical substations or existing transmission lines. Therefore, the generating station supplies the transmission lines and steps up the voltage to tie into the "electrical grid," a network that delivers electricity to the customer. The grid has high-voltage transmission lines (sometimes distant) that carry power to demand centers, transformers to step down the voltage, and distribution lines that connect individual customers.

The costs of transmitting and distributing electricity were 0.9 cents/kWh (or $9/MWh) and 2.6 cents/kWh (or $26/MWh), taken from the EIA.[3] Here, transmission is defined as the high-voltage movement of electricity from the power station while distribution is the cost of electrical lines that take power from a substation to the customer. The cost of transmission and distribution is essential to get a true "Well-to-Wheels" analysis of the total cost. For the "Well-to-Wheels" cost, we need (1) the cost of constructing the hydroelectric plant, (2) power plant generating costs, (3) cost of connecting the power plant to the electrical grid, (4) transmission and distribution costs, and (5) taxes (both federal and state). Unlike nuclear, natural gas, and coal, there are no fuel costs for a hydroelectric plant so there are no drilling or mining costs.

Table A.5.1 shows the LCOE for the cost of generating electricity and the additional cost of transmission and distribution.

A sensitivity study was made for electricity generated from Storage Dam / Run-of-River and Pumped Storage by varying the capital cost, fixed O&M, and capacity factor.

A report by the International Renewable Energy Agency (IRENA) shows capital costs for large hydroelectric plants (>20 MW) with a range of $1,050/kW to $7,650/kW while small hydroelectric plants have a range of $1,300/kW to $8,000/kW.[4] Given these data, the sensitivity study used a range of $1,000/kW to $8,000/kW for the capital cost.

For O&M costs, the IRENA report indicates it is quoted as a percent of capital cost, with a typical value of 2 to 2.5%. However, for the EIA base case,[1] the fixed O&M is actually 1.4% of capital cost. A report by the International Finance Corporation[5] suggests that 1.5% of capital costs is more typical of projects they have analyzed. Also, for 19 projects analyzed from

Table A.5.1 Economic Analysis for Hydroelectric Plants

Energy Type	Capacity Factor	Capital Cost	Levelized Capital Cost; with MACRS depreciation	Fixed O&M	Variable O&M (including fuel)	Transmission Investment	Total System Levelized Cost (LCOE) to Generate	LCOE Including Transmission and Distribution
		$/kW	$/MWh	$/MWh	$/MWh	$/MWh	$/MWh	$/MWh
Storage/Run-of-River	0.375	2,948	92.53	12.44	1.36	2.00	108.32	143.32
Pumped Storage	0.118	5,288	527.46	17.41	0.00	2.00	546.87	581.87

countries around the world, fixed O&M ranged from \$2.3/kW to \$63.9/kW. They attribute this large spread as primarily due to labor costs, which vary widely among the different countries. Based on this information, the sensitivity study used a low of \$2.3/kW and high of \$63.9/kW for the fixed O&M costs.

For capacity factors, a 2017 hydroelectric report prepared by Oak Ridge National Laboratory (ORNL) indicates that for the US hydroelectric fleet, capacity factors as low as 25% and as high as 75% are not exceptional,[6] so these are used as the lower and upper range of the sensitivity study. Otherwise, the other data used in the sensitivity study were the same as used to prepare Table A.5.1.

For the capital cost of a pumped storage plant, Table A.5.1 shows a value of \$5,288/kW. However, this value given by the EIA is much higher than those reported by the US Army Corps of Engineers and the DOE. In the Corps of Engineers study, the capital cost range is given as \$617–\$1,752/kW,[7] while in the DOE study, the capital cost range is given as \$600–\$1,800/kW.[8] Given these data, the sensitivity study used \$600/kW as a minimum cost and \$5,288/kW as a maximum.

Likewise, for fixed O&M costs, the EIA data used to prepare Table A.5.1 had a fixed O&M cost of \$18.00/(kW-yr).[1] In contrast, the Corps of Engineers study presents a range of \$2.43–\$22.23/MWh[7] while the DOE study presents a range of \$2.51–\$22.98/(kW-yr).[8] Given these data, the sensitivity study used \$2.5/kW-yr as a minimum and \$23.0/kW-yr as a maximum.

Since the capacity factor is the percent of rated capacity used per year, this definition is not as easy to apply for a pumped storage plant, since these plants are really used as batteries to store energy from another type of electric plant, be it a fossil fuel plant or one that uses renewable

Table A.5.2 Sensitivity Study for Storage / Run-of-River and Pumped Storage

	Capacity Factor	Capital Cost	Fixed O&M	Total System Levelized Cost (LCOE) to Generate	Total LCOE Cost to Generate, Transmit, and Distribute
Units		**\$/kW**	**\$/(kW-yr)**	**\$/MWh**	**\$/MWh**
Base Case—Storage/ Run-of-River	0.375	\$2,948	\$40.85	\$108.32	\$143.32
Capex is \$1,000/kW	0.375	\$1,000	\$40.85	\$47.18	\$82.18
Capex is \$8,000/kW	0.375	\$8,000	\$40.85	\$266.89	\$301.89
Opex is \$2.3/kW-yr	0.375	\$2,948	\$2.30	\$96.59	\$131.59
Opex is \$63.9/kW-yr	0.375	\$2,948	\$63.90	\$115.34	\$150.34
Capacity Factor is 0.25	0.25	\$2,948	\$40.85	\$160.81	\$195.81
Capacity Factor is 0.75	0.75	\$2,948	\$40.85	\$55.84	\$90.84
Base Case—Pumped Storage	0.118	\$5,288	\$18.00	\$546.87	\$581.87
Capex is \$600/kW	0.118	\$600	\$18.00	\$79.26	\$114.26
Capex is \$3,000/kW	0.118	\$3,000	\$18.00	\$318.65	\$353.65
Capex is \$5,288/kW	0.118	\$5,288	\$18.00	\$546.87	\$581.87
Opex is \$2.5/kW-yr	0.118	\$5,288	\$2.50	\$531.88	\$566.88
Opex is \$23.0/kW-yr	0.118	\$5,288	\$23.00	\$551.71	\$586.71
Capacity Factor is 0.10	0.10	\$5,288	\$18.00	\$644.95	\$679.95
Capacity Factor is 0.40	0.40	\$5,288	\$18.00	\$162.74	\$197.74

energy. While the capacity factor for a pumped storage plant can certainly be affected by equipment failure or routine maintenance, the plant operation really has more to do with how many hours of low economic value are available to use electricity from the other plant.

Nevertheless, the capacity factor still has a significant impact on the LCOE, so it is included in the sensitivity study. The EIA data used to prepare Table A.5.1 used a capacity factor of 11.8%.[1] However, a study of several US pumped storage plants showed capacity factors ranging from around 10% to 40%.[9] Given these data, the sensitivity study used a minimum of 10% and a maximum of 40%. Results of the sensitivity study are shown in Table A.5.2.

References

1. "Cost and Performance Characteristics of New Generating Technologies," Annual Energy Outlook 2019, Energy Information Administration, Table 2, January 2019, https://www.eia.gov/outlooks/aeo/
2. "Updated Capital Cost Estimates for Utility Scale Electricity Generating Plants," Energy Information Administration, April 2013; and "Levelized Cost and Levelized Avoided Cost of New Generation Resources in the Annual Energy Outlook 2015," Table 1, June 2015.
3. "Annual Energy Outlook 2015 with projections to 2040," DOE/EIA-0383, Table A8, April 2015.
4. "Renewable Energy Technologies: Cost Analysis Series: Hydropower," International Renewable Energy Agency (IRENA) 1 (3/5), June 2012, https://www.irena.org/document-downloads/publications/re_technologies_cost_analysis-hydropower.pdf
5. "Hydroelectric Power: A Guide for Developers and Investors," International Finance Corporation, February 2015, http://www.ifc.org/wps/wcm/connect/topics_ext_content/ifc_external_corporate_site/ifc+sustainability/learning+and+adapting/knowledge+products/publications/hydroelectric_power_a_guide_for_developers_and_investors
6. "2017 Hydropower Market Report April 2018," Office of Energy Efficiency and Renewable Energy, Department of Energy, prepared by Oak Ridge National Laboratory, file:///C:/Users/paulf/Documents/Book/References_Figures_Tables_Revised/Hydroelectric/Hydropower%20Market%20Report.pdf
7. "Technical Analysis of Pumped Storage and Integration with Wind Power in the Pacific Northwest," Prepared for US Army Corps of Engineers, Northwest Division, Hydroelectric Design Center, August 2009, http://www.hydro.org/wp-content/uploads/2011/07/PS-Wind-Integration-Final-Report-without-Exhibits-MWH-3.pdf
8. "Pumped Storage and Potential Hydropower from Conduits," Report to Congress, February 2015, US Department of Energy, https://energy.gov/sites/prod/files/2015/06/f22/pumped-storage-potential-hydropower-from-conduits-final.pdf
9. "The Cost of Pumped Hydroelectric Storage," Oscar Galvan-Lopez, December 11, 2014, http://large.stanford.edu/courses/2014/ph240/galvan-lopez2/

A.6. Wind: Levelized Cost of Electricity

For both onshore and offshore wind farm power plants, the LCOE was calculated according to Equation (1) in the general approach described above. Input for the calculations came from 2019 EIA data[1] and 2018 data from Lazard.[2] The units for LCOE are in $/megawatt-hour, or $/MWh.

The tax rate for both federal and state is set to 38%. The discount rate and book life used to calculate the CRF were set to 7% and 20 years, respectively. The MACRS depreciation schedule is 7 years and the capacity factor was set according to the authors.

Table A.6.1 Economic Analysis for Onshore and Offshore Wind Farms

Energy Type	Capacity Factor	Capital Cost	Levelized Capital Cost; with MACRS depreciation	Fixed O&M	Transmission Investment	Total System Levelized Cost (LCOE) to Generate	LCOE Including Transmission and Distribution
		$/kW	$/MWh	$/MWh	$/MWh	$/MWh	$/MWh
Onshore							
EIA[1]	0.41	$1,624	$48.17	$13.48	$3.10	$64.75	$99.75
Lazard[2]	0.465	$1,350	$35.31	$7.92	$3.10	$46.32	$81.32
Average	0.44	$1,487	$41.74	$10.70	$3.10	$55.54	$90.54
Offshore							
EIA[1]	0.45	$6,542	$176.80	$20.33	$5:80	$202.93	$237.93
Lazard[2]	0.50	$3,025	$73.57	$21.69	$5.80	$101.06	$136.06
Average	0.48	$4,784	$125.19	$21.01	$5.80	$151.99	$186.99

The capital cost, O&M cost, and transmission costs were taken from the authors. For transmission costs, they provide the cost for new transmission lines to tie into existing electrical substations or existing transmission lines. In other words, the generating station provides transmission lines and steps up the voltage to tie into the "electrical grid," a network that delivers the electricity to the customer. The grid has high-voltage transmission lines (sometimes distant) that carry power to demand centers, transformers to step down the voltage, and distribution lines that connect individual customers.

The costs of transmitting and distributing electricity were 0.9 cents/kWh (or $9/MWh) and 2.6 cents/kWh (or $26/MWh), taken from the EIA.[3] Here, transmission is defined as the high-voltage movement of electricity from the power station while distribution is the cost of electrical lines that take power from a substation to the customer. The cost of transmission and distribution is essential to get a true "Well-to-Wheels" analysis of the total cost. For a wind farm, the "Well-to-Wheels" costs will include (1) the cost of constructing the wind farm, (2) power plant generating costs, (3) cost of connecting the power plant to the electrical grid, (4) transmission and distribution costs, and (5) taxes (both federal and state). Unlike nuclear, natural gas, and coal, there are no fuel costs for a wind farm so there are no drilling or mining costs.

Table A.6.1 shows the LCOE for the cost of generating electricity and the additional cost of transmission and distribution.

A sensitivity study was made for both onshore and offshore electricity generation by varying the capital cost, the fixed O&M expenses, and the capacity factor. Annual technology cost data from the National Renewable Energy Laboratory (NREL)[4] were used to set the ranges for these different variables. The base cases for both the onshore and offshore wind farms used the EIA data.[1]

Table A.6.2 Sensitivity Study for Onshore and Offshore Wind Farms

	Capacity Factor	Capital Cost	Fixed and Variable O&M (no fuel cost)	Total System Levelized Cost (LCOE) to Generate	Total LCOE Cost to Generate, Transmit, and Distribute
		$/kW	$/MWh	$/MWh	$/MWh
Base Case—Onshore Wind Farm (1)	0.41	$1,624	$13.48	$64.75	$99.75
Capex is $1,495/kW	0.41	$1,495	$13.48	$60.92	$95.92
Capex is $1,713/kW	0.41	$1,713	$13.48	$67.39	$102.39
Opex is $31/kW-yr	0.41	$1,624	$8.63	$59.90	$94.90
Opex is $51/kW-yr	0.41	$1,624	$14.20	$65.47	$100.47
Capacity Factor is 0.11	0.11	$1,624	$50.25	$232.89	$267.89
Capacity Factor is 0.48	0.48	$1,624	$11.52	$55.76	$90.76
Base Case—Offshore Wind Farm (1)	0.45	$6,542	$20.33	$202.93	$237.93
Capex is $3,439/kW	0.45	$3,439	$20.33	$119.07	$154.07
Capex is $6,458/kW	0.45	$6,458	$20.33	$200.66	$235.66
Opex is $106/kW-yr	0.45	$6,542	$26.89	$209.49	$244.49
Opex is $159/kW-yr	0.45	$6,542	$40.33	$222.93	$257.93
Capacity Factor is 0.31	0.31	$6,542	$29.51	$291.95	$326.95
Capacity Factor is 0.51	0.51	$6,542	$17.94	$179.73	$214.73

The capacity factor for onshore wind farms had a wide range, 11% to 48%. The NREL report used ten different areas of wind speed across the United States, so the wide range for capacity factor reflects wind speed differences. For offshore wind farms, the capacity factor had a smaller range, from 31% to 55%.

For an onshore wind farm, the capital cost range was $1,495 to $1,713/kW and the fixed O&M cost range was $8.63 to $14.20/MWh ($31 to $51/kW-yr). For an offshore wind farm, the capital cost range was $3,439 to $6,458/kW and the fixed O&M cost range was $26.89 to $40.33/MWh ($106 to $159/kW-yr). The results of the sensitivity study are shown in Table A.6.2.

References

1. "Cost and Performance Characteristics of New Generating Technologies," Annual Energy Outlook 2019, Table 2, Energy Information Administration, January 2019, https://www.eia.gov/outlooks/aeo/
2. "Lazard's Levelized Cost of Energy Analysis—Version 12.0," November 2018, https://www.lazard.com/media/450784/lazards-levelized-cost-of-energy-version-120-vfinal.pdf
3. "Annual Energy Outlook 2015 with Projections to 2040," DOE/EIA-0383, Table A8, April 2015.
4. "2018 Annual Technology Baseline Cost and Performance Summary," National Renewable Energy Laboratory, 2018, https://atb.nrel.gov/

A.7. Solar: Levelized Cost of Electricity

For both solar photovoltaic (PV) and concentrating solar power (CSP) electric plants, the LCOE was calculated according to Equation (1) in the general approach described above. Input for the calculations came from 2019 EIA data[1] and 2018 data from Lazard.[2] The units for LCOE are in $/megawatt-hour, or $/MWh. To show the improvement in LCOE over the last 4–5 years, the LCOE was also calculated using EIA data from 2015[3] and 2014 data from Lazard.[4]

The tax rate for both federal and state is set to 38%. The discount rate and book life used to calculate the CRF were set to 7% and 20 years, respectively. The MACRS depreciation schedule is 5 years and the capacity factor was set according to the authors.

The capital cost, O&M cost, and transmission costs were taken from the authors. For transmission costs, they provide the cost for new transmission lines to tie into existing electrical substations or existing transmission lines. In other words, the generating station provides transmission lines and steps up the voltage to tie into the "electrical grid," a network that delivers the electricity to the customer. The grid has high-voltage transmission lines (sometimes distant) that carry power to demand centers, transformers to step down the voltage, and distribution lines that connect individual customers.

The costs of transmitting and distributing electricity were 0.9 cents/kWh (or $9/MWh) and 2.6 cents/kWh (or $26/MWh), taken from the EIA.[5] Here, transmission is defined as the high-voltage movement of electricity from the power station while distribution is the cost of electrical lines that take power from a substation to the customer. The cost of transmission and distribution is essential to get a true "Well-to-Wheels" analysis of the total cost. For a solar energy plant, the "Well-to-Wheels" costs will include (1) the cost of constructing the solar plant, (2) power plant generating costs, (3) cost of connecting the power plant to the electrical grid, (4) transmission and distribution costs, and (5) taxes (both federal and state). Unlike nuclear, natural gas, and coal, there are no fuel costs for a solar plant, so there are no drilling or mining costs.

Table A.7.1 Economic Analysis for Solar PV and CSP

Energy Type	Capacity Factor	Capital Cost	Levelized Capital Cost; with MACRS depreciation	Fixed and Variable O&M	Transmission Investment	Total System Levelized Cost (LCOE) to Generate	LCOE Including Transmission and Distribution
Units		$/kW	$/MWh	$/MWh	$/MWh	$/MWh	$/MWh
Solar PV							
EIA[1]	0.25	$1,783	$84.78	$10.26	$4.1	$99.14	$134.14
Lazard[2]	0.27	$1,100	$49.35	$4.52	$4.1	$57.97	$92.97
Average	0.26	$1,442	$67.07	$7.39	$4.1	$78.55	$113.55
EIA[3]	0.25	$4,183	$198.91	$12.67	$4.1	$215.68	$250.68
Lazard[4]	0.26	$1,625	$75.76	$7.39	$4.1	$87.24	$122.24
Average	0.25	$2,904	$137.33	$10.03	$4.1	$151.46	$186.46
Solar CSP							
EIA[1]	0.29	$4,291	$175.90	$28.67	$6.0	$210.57	$245.57
Lazard[2]	0.48	$6,925	$173.31	$18.63	$6.0	$197.94	$232.94
Average	0.38	$5,608	$174.61	$23.65	$6.0	$204.26	$239.26
EIA[3]	0.20	$5,067	$301.18	$38.39	$6.0	$345.57	$380.57
Lazard[4]	0.52	$8,400	$192.04	$21.40	$6.0	$219.44	$254.44
Average	0.36	$6,734	$246.61	$29.90	$6.0	$282.51	$317.51

Table A.7.1 shows the LCOE for the cost of generating electricity and the additional cost of transmission and distribution.

A sensitivity study was made for both solar electricity generation technologies by varying the capital cost, the fixed O&M expenses, and the capacity factor. The base cases for both solar PV and CSP used the EIA data.[1]

For solar PV capital cost, a 2019 report from IRENA[6] shows that, over the last 5 years, the global weighted average capital cost for utility scale solar PV plants has dropped dramatically. Capital costs decreased from $2,323/kW in 2014 to $1,210/kW in 2018. Based on these data, the capital cost for solar PV is examined from a low of $1,200/kW to a high of $2,400/kW. For O&M costs, the data of Lazard[2] showed a range of about 25%, so this was used to establish the low and high. As the EIA base case O&M was about $10/MWh, the low was set to $7.5/MWh and the high to $12.5/MWh. Also, Lazard shows the capacity factor varying from 0.21 to 0.32, so these were used as the low and high relative to the base case value of 0.25.

For CSP, the 2018 data from Lazard[2] were used to set the ranges for all three variables. Capital cost was varied from a low of $3,850/kW to a high of $10,000/kW, O&M was varied from a low of $16.7/MWh to a high of $38.4/MWh, and the capacity factors of 0.43 and 0.52 were used in the sensitivity study. Both of these capacity factors were higher than the base case capacity factor of 0.29. For CSP technology, there is some relationship between capital cost and capacity factor. Thermal energy storage (TES), where heated fluids are stored for later use, can be built into the system to provide electricity generation after the solar conditions are no longer suitable for heating the fluids. Thus, if the system is built with 10-hour versus 4-hour storage, it will have a higher capacity factor but the capital costs will also be higher.

The results of the sensitivity study are shown in Table A.7.2.

Table A.7.2 Sensitivity Study for Solar PV and CSP

	Capacity Factor	Capital Cost	Fixed and Variable O&M (no fuel cost)	Total System Levelized Cost (LCOE) to Generate	Total LCOE Cost to Generate, Transmit, and Distribute
		$/kW	$/MWh	$/MWh	$/MWh
Solar PV Base Case (1)	0.25	$1,783	$10.26	$99.14	$134.14
Capex is $1,200/kW	0.25	$1,200	$10.26	$71.42	$106.42
Capex is $2,400/kW	0.25	$2,400	$10.26	$128.48	$163.48
Opex is $7.5/MWh	0.25	$1,783	$7.50	$96.39	$131.39
Opex is $12.5/MWh	0.25	$1,783	$12.50	$101.39	$136.39
Capacity Factor is 0.21	0.21	$1,783	$12.21	$117.24	$152.24
Capacity Factor is 0.32	0.32	$1,783	$8.01	$78.35	$113.35
Solar CSP Base Case (1)	0.29	$4,291	$28.67	$210.57	$245.57
Capex is $3,850/kW	0.29	$3,850	$28.67	$192.50	$227.50
Capex is $10,000/kW	0.29	$10,000	$28.67	$444.60	$479.60
Opex is $16.7/MWh	0.29	$4,291	$22.04	$203.94	$238.94
Opex is $38.4/MWh	0.29	$4,291	$36.06	$217.96	$252.96
Capacity Factor is 0.43	0.43	$4,291	$19.34	$143.97	$178.97
Capacity Factor is 0.52	0.52	$4,291	$15.99	$120.09	$155.09

References

1. "Cost and Performance Characteristics of New Generating Technologies," Annual Energy Outlook 2019, Table 2, Energy Information Administration, January 2019, https://www.eia.gov/outlooks/aeo/
2. "Lazard's Levelized Cost of Energy Analysis—Version 12.0," November 2018, https://www.lazard.com/media/450784/lazards-levelized-cost-of-energy-version-120-vfinal.pdf
3. "Updated Capital Cost Estimates for Utility Scale Electricity Generating Plants," Energy Information Administration, April 2013; and "Levelized Cost and Levelized Avoided Cost of New Generation Resources in the Annual Energy Outlook 2015," June 2015, Table 1
4. "Lazard's Levelized Cost of Energy Analysis—Version 8," September 2014, https://www.lazard.com/media/1777/levelized_cost_of_energy_-_version_80.pdf
5. "Annual Energy Outlook 2015 with Projections to 2040," DOE/EIA-0383, Table A8, April 2015.
6. "Renewable Power Generation Costs in 2018," International Renewable Energy Agency 2019, https://www.irena.org/publications/2019/May/Renewable-power-generation-costs-in-2018

A.8. Ethanol: Levelized Cost of Fuel

For the conversion of corn grain or corn stover to ethanol, the LCOF was calculated according to Equation (4) in the general approach described above. Calculations were based on an ethanol plant size of 100 million gallons per year. For corn grain, the dry milling process was used since 90% of the grain ethanol produced in the United States today comes from the dry milling process and only 10% from wet milling. The units for LCOF are in $/gallon of gasoline equivalent, or $/gge.

The book life used for ethanol the plant was 15 years and the number of years the IRS allows the plant depreciation used was 7 years, according to Table A.i.3. Likewise, the DPV needed in Equation (4) was also taken from Table A.i.3.

To determine the cost for a gge for ethanol, the cost of producing a gallon of ethanol has to be corrected to the cost of producing the same amount of energy in a gallon of gasoline. As the energy content of ethanol is lower than gasoline, 75,700 BTU versus 124,238 BTU for gasoline, a gge for ethanol is determined by adjusting the LCOF for ethanol using the energy ratio of gasoline to ethanol, or 1.64. In other words, the cost of making one gge of ethanol is 1.64 times that of making 1 gallon of ethanol.

Transportation fuel in the form of ethanol, derived from either corn grain or corn stover, is created over several steps including (1) planting, growing, and harvesting the corn or corn stover, (2) transporting the corn or stover to the ethanol plant, (3) the ethanol plant, (4) transporting the ethanol product to market, (5) taxes (both federal and state), and (6) marketing cost at the filling station. The costs for these different steps yield the "Well-to-Wheels" LCOF in $/gge.

For ethanol produced from corn grain, the cost of steps (1) and (2) are assumed to be the delivered cost of corn, shown in Table A.8.1 for the last 10 years.[1] As the table illustrates, the price of corn has had wide variation, from a low of $3.30/bushel in 2017 to a high of $6.95/bushel in 2012. Also shown in the table are the average sales price of ethanol as well as the value of distiller's dry grain solids (DDGS), a product used to feed farm animals that offsets part of the cost for operating the ethanol plant, step (3). For the base case study, the price of corn and DDGS are set to the annual average values for 2018.

The cost of operating the dry mill ethanol plant includes the capital cost of building the plant, the fixed and variable operating costs, the capacity factor, and the value of any coproducts.

Table A.8.1 Annual Average Prices for Ethanol, DDGS, and Corn[1]

Year Average	Ethanol per Gallon	DDGS per Ton	Corn per Bushel	Corn per Ton
2009	$1.63	$111.76	$3.56	$127.01
2010	$1.77	$116.96	$3.98	$142.15
2011	$2.56	$195.83	$6.61	$236.25
2012	$2.24	$238.20	$6.95	$248.39
2013	$2.34	$231.40	$6.16	$220.07
2014	$2.11	$160.78	$4.06	$144.95
2015	$1.43	$148.31	$3.62	$129.22
2016	$1.43	$120.97	$3.33	$118.93
2017	$1.43	$103.55	$3.30	$117.75
2018	$1.32	$142.30	$3.39	$120.90

For the capacity factor, according to the Renewable Fuels Association (RFA)[2] the existing US production capacity for 2019 was 16,501 million gallons per year (MGY) while the operating production was 15,975 MGY. This yields a capacity factor of 96.8%. The capacity factor is defined as the percent of the rated capacity used each year.

For the capital cost to build the dry mill ethanol plant, there are several studies available. A 2002 USDA (United States Department of Agriculture) study showed that, for 21 plants surveyed, the cost for new plants ranged from $1.05 to $3.00 per gallon of installed capacity.[3] Another study cited a cost of $1.30 to $1.40 per gallon of capacity for 2002, but increasing to $2.00 to $2.30 per gallon in 2007.[4] The increasing capital cost was attributed to limited worker availability, as well as rising costs of construction material and production equipment. Similar numbers were reported in 2010 by the US DOE, with the capital cost of a dry mill ethanol plant as $2.00 to $2.25 per gallon of capacity.[5] A 2016 report from the University of Illinois gave a capital cost of $2.11 per gallon of capacity,[6] as did a model from Iowa State University used to track ethanol profitability.[1] Both the University of Illinois and Iowa State studies assume a plant capacity of 100 million gallons of ethanol production per year and a yield of 2.80 gallons of ethanol per bushel of corn processed. Based on these two more recent studies, the capital cost is set to $2.11 per gallon of installed capacity, so a 100 million gallon plant would have a capital cost of $211 million.

For the fixed and variable operating costs, data from the Iowa State University study were used.[1] The fixed operating costs include labor and management as well as property taxes and were $4.4 million per year, or $0.05 per gallon of ethanol produced. Variable operating costs, not including the cost of corn grain, were $24.1 million per year, or $0.25 per gallon of ethanol produced. Variable operating costs include the enzyme, yeast, chemicals, the denaturant, repairs and maintenance, water, and electricity. The denaturant is added to the ethanol to make it undrinkable.

Step (5), federal and state taxes are set to a total of 38%. Step (4) involves the cost of transporting the ethanol product to market. One study states that rail accounted for 60% of ethanol transport in 2015.[8] However, normally ethanol is transported by both rail and truck, rail for long distance and truck for the final delivery. In this study, they used transport by rail from Illinois to California as a typical supply chain transfer of ethanol, with a cost of $0.24 per gallon.[7] A USDA study showed rail rates varying from $0.13 to $0.24 per gallon, with cost varying by distance.[8] As most ethanol plants are concentrated in the Midwest corn belt, ethanol

Table A.8.2 LCOF for Dry Mill Ethanol for a 100 Million Gallon per Year Plant and a Capital Cost of $211 Million.

Year	Corn Cost, $/bushel	DDGS Value, $/ton	Levelized Capital Cost, $/gallon	LCOF, $/gallon of Ethanol	LCOF, gge
2018	$3.39	$142.30	$0.27	$1.58	$2.60
2017	$3.30	$103.55	$0.27	$1.68	$2.76
2016	$3.33	$120.97	$0.27	$1.63	$2.68
2015	$3.62	$148.31	$0.27	$1.64	$2.69
2014	$4.06	$160.78	$0.27	$1.75	$2.87
2013	$6.16	$231.40	$0.27	$2.23	$3.66
2012	$6.95	$238.20	$0.27	$2.48	$4.07
2011	$6.61	$195.83	$0.27	$2.51	$4.12
2010	$3.98	$116.96	$0.27	$1.87	$3.07
2009	$3.56	$111.76	$0.27	$1.75	$2.86

typically has to be transported long distances to arrive at major metropolitan areas. Taking an average of the USDA rail rate gives a transportation cost of $0.185/gallon.

Step (6) is the markup cost at the filling station. Although there are many factors involved in the average markup of ethanol at the fueling stations, values of $0.10 to 0.20/gallon are typical. Combining the cost of transportation and the markup cost gives a value of $0.335/gallon. This is similar to the value of $0.32/gallon for transportation and marketing of gasoline from petroleum crude oil, taken from the EIA.

With these data, the LCOF was calculated for the base case and shown in Table A.8.2. As 10 years of data were available for the cost of corn and the DDGS, all 10 years are shown in the table.

Next, a sensitivity study was made for the dry milling ethanol plant by varying capital cost, O&M expenses, and the price of corn. The earlier discussion on capital cost gave a range of $1.05 to $3.00 per gallon of installed capacity with $2.11 used in the base case. Therefore, upper and lower values of $3.00 and $1.05 per gallon of capacity are used. For the cost of corn, Table A.8.1 showed a range of $3.30 to $6.95 per bushel over the last 10 years with $3.39 per bushel used in the base case. These upper and lower values are used in the sensitivity study. For O&M, the fixed and variable O&M (excluding the cost of corn) were a combined $0.30 per gallon of ethanol produced. Since there was not much literature data for O&M, a range of plus or minus 50% was arbitrarily chosen, resulting in a range of $0.15 to $0.45 per gallon of ethanol. Results of the sensitivity study are shown in Table A.8.3, where the base case was based on data for 2018.

We next turn to cellulosic ethanol by examining corn stover as a feedstock. The "Well-to-Wheels" steps are basically identical except that steps (1) and (2) involve the delivered cost of corn stover instead of corn grain and the ethanol plant for step (3) is different. Otherwise, the cost of transporting the ethanol product to market and the federal and state taxes are the same.

Results for five studies were used to establish the various costs for cellulosic ethanol from corn stover, two which are NREL studies and three which take data from actual commercial plants. Results for the LCOF calculations, and important input data, are shown in Table A.8.4. Although this is discussed in more detail in Chapter 8 for ethanol, it is interesting to note that all three commercial cellulosic ethanol plants used in the table have some issues. The DuPont plant in Nevada, Iowa, was sold to the German company Verbio Vereinigte, which will use

Table A.8.3 Sensitivity Study for a Dry Mill Ethanol Plant

Cases	Capital Cost, MM$	Corn Cost, $/bushel	Combined O&M, $/gallon	LCOF, $/gallon of Ethanol	LCOF, gge
Base Case	$211	$3.39	$0.30	$1.58	$2.60
Capex = $1.05/gallon capacity	$105	$3.39	$0.30	$1.45	$2.37
Capex = $1.05/gallon capacity	$300	$3.39	$0.30	$1.69	$2.78
O&M = $0.15/gallon	$211	$3.39	$0.15	$1.44	$2.36
O&M = $0.45/gallon	$211	$3.39	$0.45	$1.74	$2.85
Corn Cost = $3.30/bushel	$211	$3.30	$0.30	$1.55	$2.54
Corn Cost = $6.95/bushel	$211	$6.95	$0.30	$2.81	$4.62

Table A.8.4 LCOF for a Corn Stover Cellulosic Ethanol Plant

Data Source	Capital Cost	Corn Stover Cost	Liquid Product Yield	Capital Cost	Ethanol Yield	Combined O&M	LCOF, $/gallon of Ethanol	LCOF, gge
	$/gallon of capacity	$/ton	MM gallons/ year	$/BPD	Gallons/ dry ton	$/gge	$/gallon	$/gge
NREL[9]	$4.87	$65.30	56	$74,596	72.5	$0.97	$2.68	$4.08
NREL[10]	$6.93	$58.50	59	$106,108	75.8	$0.64	$2.50	$3.80
DuPont[11]	$6.67	$60.00	30	$102,200	80.0	$0.91	$2.82	$4.29
Poet[12]	$13.75	$70.00	20	$210,788	71.2	$0.91	$3.56	$5.42
Abengoa[13]	$9.18	$60.00	25	$140,688	68.5	$1.09	$3.15	$4.78
Average	$8.28	$62.76	38	$126,876	73.6	$0.90	$2.94	$4.47

the plant to make renewable natural gas.[14] The Abengoa plant in Hugoton, Kansas, was sold in 2016 to Synata Bio Inc. because the parent company went bankrupt, and it is not clear if or when production will resume.[15] Finally, the Poet Project Liberty plant in Emmetsburg, Iowa, has been slow to meet its 20 million gallon per year target for ethanol, reportedly due to difficulty in the chemical pretreatment step used to perform enzymatic saccharification of the corn stover.[16]

A sensitivity study was made for cellulosic ethanol using corn stover as a feed by varying capital cost, O&M expenses, and the price of corn stover. The average of the data in Table A.8.4 were used as the base case. Also, the range for capital cost was established using the lower and upper range of the data in Table A.8.4, rounded to $5 and $14/gallon of capacity. A lower and upper range for the corn stover is more difficult to establish as the price will depend on the costs associated with harvest, collection, and storage. On source provides details for these steps. For example, harvesting corn stover must be accompanied by adding fertilizer to replace nutrients removed with the stover.[17] And, harvesting the stover requires some capital intensive operations, such as mowing, raking, and baling which require expensive equipment. Based on their analysis, the breakeven prices for stover range from $45.50 to $74.70 per ton. This range is

Table A.8.5 Sensitivity Study for a Corn Stover Cellulosic Ethanol Plant

Cases	Capital Cost, MM$	Corn Stover Cost, $/ton	Combined O&M, $/gallon	LCOF, $/gallon of Ethanol	LCOF, gge
Base Case	$316	$62.76	$0.90	$2.94	$4.47
Capex $5/gallon of capacity	$190	$62.76	$0.90	$2.53	$3.84
Capex $14/gallon of capacity	$532	$62.76	$0.90	$3.64	$5.54
Combined O&M is $0.45/gallon	$316	$62.76	$0.45	$2.52	$3.83
Combined O&M is $1.35/gallon	$316	$62.76	$1.35	$3.42	$5.20
Price of corn stover is $60/ton	$316	$60.00	$0.90	$2.90	$4.41
Price of corn stover is $100/ton	$316	$100.00	$0.90	$3.44	$5.23

similar to that given in another study of $40 to $80/ton.[18] However, transportation costs are estimated to add an additional $20/ton, giving an overall delivered cost range of $60 to $100/ton. This is the range that is used for the sensitivity study. Since literature data were not widely available for O&M, a range of plus or minus 50% was arbitrarily chosen, resulting in a range of $0.45 to $1.35 per gallon of ethanol. Results of the sensitivity study are shown in Table A.8.5.

References

1. "Tracking Ethanol Profitability," Iowa State University Ag Decision Maker, D1–10, Ethanol Profitability, Don Hofstrand, Excel Model, 2019, https://www.extension.iastate.edu/agdm/energy/html/d1-10.html
2. "2019 Ethanol Industry Outlook," Renewable Fuels Association, 2019, https://ethanolrfa.org/wp-content/uploads/2019/02/RFA2019Outlook.pdf
3. "USDA's 2002 Ethanol Cost-of-Production Survey," Hosein Shapouri and Paul Gallagher, US Department of Agriculture, July 2005, https://lib.dr.iastate.edu/cgi/viewcontent.cgi?referer=https://www.google.com/&httpsredir=1&article=1021&context=econ_reportspapers
4. "Ethanol Plant Construct. Costs Are on the Rise," Flow Control, August 10, 2007, https://www.flowcontrolnetwork.com/ethanol-plant-construct-costs-are-on-the-rise/
5. "Current State of the U.S. Ethanol Industry," US Department of Energy Office of Biomass Programs, November 30, 2010, Fulfillment of Subcontract No. 02-5025, https://www.energy.gov/sites/prod/files/2014/04/f14/current_state_of_the_us_ethanol_industry.pdf
6. "The Profitability of Ethanol Production in 2016," Scott Irwin, Department of Agricultural and Consumer Economics, University of Illinois, February 1, 2017, https://farmdocdaily.illinois.edu/2017/02/the-profitability-of-ethanol-production-in-2016.html
7. "Biomass Feedstock Preprocessing and Long-Distance Transportation Logistics," Tao Lin et al., Global Change Biology Bioenergy 8 (2016), 160–170, November 21, 2014,

file:///C:/Users/paulf/Documents/Book/References_Figures_Tables_Revised/Ethanol/New%20folder/Biomass_Feedstock_Preprocessing.pdf
8. "U.S. Ethanol: An Examination of Policy, Production, Use, Distribution, and Market Interactions," J.A. Duffield et al., US Department of Agriculture Office of Energy Policy and New Uses, September 2015, https://www.usda.gov/oce/reports/energy/EthanolExamination102015.pdf
9. "Biochemical Production of Ethanol from Corn Stover: 2008 State of Technology Model," D. Humbird and A. Aden, Technical Report NREL/TP-510-46214, August 2009, https://www.nrel.gov/docs/fy09osti/46214.pdf
10. "Process Design and Economics for Biochemical Conversion of Lignocellulosic Biomass to Ethanol," D. Humbird et al., Technical Report NREL/TP-5100-47764, May 2011, https://www.nrel.gov/docs/fy11osti/47764.pdf
11. "DuPont Advances Commercialization of Cellulosic Ethanol with Iowa Biorefinery Groundbreaking," Nevada, IA, November 2012, http://biosciences.dupont.com/media/news-archive/news/2012/dupont-advances-commercialization-of-cellulosic-ethanol-with-iowa-biorefinery-groundbreaking/archive/news/2012/dupont-advances-commercialization-of-cellulosic-ethanol-with-iowa-biorefinery-groundbreaking/
12. "First Commercial-Scale Cellulosic Ethanol Plant in the U.S. Opens for Business," Emmetsburg, IA: Project Liberty, September 3, 2014, http://poet.com/pr/first-commercial-scale-cellulosic-plant
13. "Abengoa Celebrates Grand Opening of Cellulosic Ethanol Plant," Erin Voegele, Hugoton, KS, October 17, 2014, http://biomassmagazine.com/articles/11068/abengoa-celebrates-grand-opening-of-cellulosic-ethanol-plant
14. "DuPont Sells Iowa Ethanol Plant to German Company; It Will Soon Make Renewable Natural Gas," Donnelle Eller, Des Moines Register, November 9, 2018, https://www.desmoinesregister.com/story/money/agriculture/2018/11/08/dupont-cellulosic-ethanol-plant-nevada-sold-german-company-verbio-north-america-claus-sauter/1938321002/
15. "Is Cellulosic Ethanol Dead? Despite Setbacks, Signs of Progress," November 9, 2017, Environmental and Energy Study Institute (EESI), https://www.eesi.org/articles/view/is-cellulosic-ethanol-dead-despite-setbacks-signs-of-progress
16. "Zero to 10 Million in 5 Years," Susanne Retka Schill, Ethanol Producer Magazine, June 26, 2018, http://www.ethanolproducer.com/articles/15344/zero-to-10-million-in-5-years
17. "To Harvest Stover or Not: Is It Worth It?," Madhu Khanna and Nick Paulson, Department of Agricultural and Consumer Economics University of Illinois, Farmdoc Daily, February 10, 2016, https://farmdocdaily.illinois.edu/wp-content/uploads/2016/04/fdd180216.pdf
18. "Corn Stover for Bioenergy Production: Cost Estimates and Farmer Supply Response," Jena Thompson and Wallace E. Tyner, Department of Agricultural Economics Purdue University, September 2011, https://www.extension.purdue.edu/extmedia/EC/RE-3-W.pdf

A.9. Biomass: Levelized Cost of Electricity

For both biomass bubbling fluidized bed (BFB) and biomass combined cycle (BCC) power plants, the LCOE was calculated according to Equation (1) in the general approach described above. Input for the BFB calculations came from 2019 EIA data,[1] 2017 data from Lazard,[2] and NREL 2017 data.[3] The units for LCOE are in \$/megawatt-hour, or \$/MWh. To show the change in capital and operating costs over the previous 3–4 years, the LCOE was also calculated using EIA data from 2015[4] and 2014 data from Lazard.[5] This had further importance for the comparison to the LCOE calculation for BCC, as the most recent data for this technology was from

2013. This technology, also known as BIGCC (Biomass Integrated Combined Cycle Plants), is mostly in the research and demonstration plant phase, with little large-scale commercial success.

The tax rate for both federal and state is set to 38%. The discount rate and book life used to calculate the CRF were set to 7% and 40 years, respectively. The MACRS depreciation schedule is 7 years, and the capacity factor was set according to the authors.

The transmission costs were taken from the EIA reports. For transmission costs, they provide the cost for new transmission lines to tie into existing electrical substations or existing transmission lines. In other words, the generating station provides transmission lines and steps up the voltage to tie into the "electrical grid," a network that delivers the electricity to the customer. The grid has high-voltage transmission lines (sometimes distant) that carry power to demand centers, transformers to step down the voltage, and distribution lines that connect individual customers.

The costs of transmitting and distributing electricity were 0.9 cents/kWh (or $9/MWh) and 2.6 cents/kWh (or $26/MWh), taken from the EIA.[6] Here, transmission is defined as the high-voltage movement of electricity from the power station while distribution is the cost of electrical lines that take power from a substation to the customer. The cost of transmission and distribution is essential to get a true "Well-to-Wheels" analysis of the total cost. For a biomass power plant, the "Well-to-Wheels" costs will include (1) the cost of harvesting (and perhaps growing), preparing, and transporting the biomass to the power plant, (2) the cost of constructing the biomass plant, (3) power plant generating costs, (4) the cost of connecting the power plant to the electrical grid, (5) transmission and distribution costs, and (6) taxes (both federal and state).

Table A.9.1 shows input used to make the LCOE calculations.

The most notable difference between the three references is the capacity factor, which is high for the EIA and Lazard studies but low for the NREL study. This is likely because the EIA and Lazard studies assumed a performance similar to a coal-fired power plant, but the NREL study calculated a capacity factor from 2014 US operating data. For that year, 13.4 GW of capacity generated 64.3 TWh of electricity, giving a capacity factor of 55%. Of the 13.4 GW of biomass

Table A.9.1 Input Data for the LCOE Calculations

	Capacity Factor, %	Heat Rate, BTU/kWh	Capital Cost, $/kW	Fixed O&M, $/kW-yr	Variable O&M, $/MWh
Biomass Bubbling Fluidized Bed (BFB)					
EIA 2019 (50 MW plant)[1]	83	13,500	3,900	114.39	5.70
Lazard 2017[2]	80–85	14,500	1,700–4,000	50.00	10.00
NREL, 2017[3]	55	13,500	3,679	107.22	5.34
EIA 2015 (50 MW plant)[4]	83	13,500	4,114	105.63	5.26
Lazard 2014[5]	85–90	14,500	3,000–4,000	95.00	15.00
Biomass Combined Cycle (BCC)					
EIA[4]	0.83	12,350	8,180	356.07	17.49

capacity, 61.9% used wood and wood-derived fuels, 15.7% used landfill gas, 16.4% used municipal solid waste (MSW), and 6.0% used other waste biomass.

Since wood and wood-derived fuels had the highest percent use, that is the basis for the fuel used for the LCOE calculations. The 2013 EIA report[4] used a wood fuel cost of $2.40/MM BTU. And, while the report does not specify the wood type or the disposition of the wood (chips, pellets, briquettes, etc.), it does give the water and heat content as 17.27% and 6,853 BTU/lb, respectively. On this basis, the dry heat content is 8,284 BTU/lb and, at $2.40/MM BTU, this translates into a cost of $40/dry ton of biomass.

This value seems too low, as a value of $7.50/MM BTU at $121/dry ton was used in Chapter 8 for corn grain and a value of $3.75/MM BTU at $63/dry ton was used for corn stover biomass. Indeed, just the transportation cost for corn stover was about $20/ton.[7] And a California report which evaluated the delivery of wood biomass to the power plant found that the cost was dependent on distance, with $10 per dry ton for a distance of 20 miles, $20 for a distance of 41 miles, and $30 for a distance of 61 miles.[8] Cost depended on whether the road used was a US highway, paved double lane, gravel single lane, or dirt single lane, but worked out to about $0.05/mile.

Also, the cost of biomass can vary a lot, as some power plants use feedstock blending to control the overall average feedstock price. For example, a model feed blend used in a DOE report[9] contained 45% pine wood ($99.49/dry ton), 32% logging residues ($67.51/dry ton), 3% switchgrass ($66.68/dry ton), and 20% waste ($58.12/dry ton), giving an average delivered cost of $80/dry ton. So, for the LCOE calculations, a delivered cost of $80/dry ton was used. This distribution of the 2014 biomass feedstock data given earlier yields a blended cost close to $80/dry ton and, moreover, this value is in between the cost of corn gran and corn stover. Later, the sensitivity study will examine the impact of feedstock cost on the LCOE.

Table A.9.2 shows the LCOE for the cost of generating electricity and the additional cost of transmission and distribution. The average for the more recent data did not include the NREL study because of the low capacity factor. Although that capacity factor is based on actual operating data, it made for a better direct comparison with the BCC calculation to exclude it. Also, the current low capacity factor for commercial biomass plants may have more to do with current economics and demand for electricity, rather than down time due to normal maintenance issues.

A sensitivity study was made for both biomass technologies by varying the capital cost, the feedstock cost, and the capacity factor. The base case for the BFB technology used data from the 2019 EIA case while the base case for the BCC technology used the 2015 EIA case. This was done for consistency in comparing the two technologies.

For the BFB technology, the data in Table A.9.1. show a capital cost range of $1,700/kW to $4,000/kW. A 2017 IRENA report shows a much broader range, from $2,000 to $7,000/kW in Europe and North America and worldwide from $500 to $8,000/kW, where less expensive capital costs are found in Asia and South America.[10] Based on these data, the capital cost will examine a range of $500 to $8,000/kW. Since there are no commercial data available for the BCC technology, the spread around the mean for the BFB technology was used as a rough guide for the BCC technology, and the capital cost range will examine a range of $4,000 to $12,000/kW.

For capacity factor, the value of 55% obtained from 2014 US operating data is used for the lower limit and 90% is used for the upper limit. For the delivered cost of biomass feedstock, the DOE data for different feedstocks are used to set the upper and lower limits, namely about $100/dry ton based on pine wood and $60/dry ton based on waste. These ranges are applied to both the BFB and BCC technologies.

The results of the sensitivity study are shown in Table A.9.3.

Table A.9.2 Economic Analysis for Biomass BFB and BCC

Technology	Capacity Factor	Capital Cost	Levelized Capital Cost; with MACRS depreciation	Fixed O&M	Variable O&M (including fuel)	Transmission Investment	Total System Levelized Cost (LCOE) to Generate	LCOE Including Transmission and Distribution
		$/kW	$/MWh	$/MWh	$/MWh	$/MWh	$/MWh	$/MWh
BFB—EIA 2019[1]	0.83	$3,900	$37.3	$15.7	$73.2	$1.2	$135.5	$170.5
BFB—Lazard 2017[2]	0.85	$1,700	$19.3	$6.7	$82.5	$1.2	$109.7	$144.7
BFB—Lazard 2017[2]	0.80	$4,000	$48.3	$7.1	$82.5	$1.2	$139.2	$174.2
Average	0.83	$3,200	$35.0	$9.8	$79.4	$1.2	$128.1	$163.1
BFB—NREL 2017[3]	0.55	$3,679	$64.6	$22.3	$72.8	$1.2	$160.9	$195.9
BFB—EIA 2015[4]	0.83	$4,114	$47.9	$14.5	$72.8	$1.2	$136.4	$171.4
BFB—Lazard 2014[5]	0.90	$3,000	$32.2	$12.0	$87.5	$1.2	$133.0	$168.0
BFB—Lazard 2014[5]	0.85	$4,000	$45.5	$12.8	$87.5	$1.2	$146.9	$181.9
Average	0.86	$3,705	$41.9	$13.1	$82.6	$1.2	$138.8	$173.8
BCC—EIA 2015[4]	0.83	$8,180	$95.2	$49.0	$79.2	$1.2	$224.7	$259.7

Table A.9.3 Sensitivity Study for Biomass Bubbling Fluidized Bed (BFB) and Biomass Combined Cycle (BCC)

	Capacity Factor	Capital Cost	Fixed and Variable O&M (including fuel cost)	Total System Levelized Cost (LCOE) to Generate	Total LCOE Cost to Generate, Transmit, and Distribute
Units		**\$/kW**	**\$/MWh**	**\$/MWh**	**\$/MWh**
Base Case: BFB—EIA 2019	0.83	\$3,900	\$88.9	\$135.5	\$170.5
Capex = \$500/kW	0.83	\$500	\$88.9	\$96.0	\$131.0
Capex = \$8,000/kW	0.83	\$8,000	\$88.9	\$183.3	\$218.3
Capacity Factor = 0.55	0.55	\$3,900	\$96.9	\$166.7	\$201.7
Capacity Factor = 0.90	0.90	\$3,900	\$87.7	\$130.8	\$165.8
Feed Cost = \$60/ton	0.83	\$3,900	\$72.0	\$118.7	\$153.7
Feed Cost = \$100/ton	0.83	\$3,900	\$105.8	\$152.4	\$187.4
Base Case: BCC—EIA 2015	0.83	\$8,180	\$128.2	\$224.7	\$259.7
Capex = \$4,000/kW	0.83	\$4,000	\$128.2	\$176.0	\$211.0
Capex = \$12,000/kW	0.83	\$12,000	\$128.2	\$269.1	\$304.1
Capacity Factor = 0.55	0.55	\$8,180	\$153.1	\$298.1	\$333.1
Capacity Factor = 0.90	0.90	\$8,180	\$124.4	\$213.4	\$248.4
Feed Cost = \$60/ton	0.83	\$8,180	\$112.8	\$209.2	\$244.2
Feed Cost = \$100/ton	0.83	\$8,180	\$146.7	\$240.1	\$275.1

References

1. "Cost and Performance Characteristics of New Generating Technologies," Annual Energy Outlook 2019, Table 2, Energy Information Administration, January 2019, https://www.eia.gov/outlooks/aeo/
2. "Lazard's Levelized Cost of Energy Analysis—Version 11.0," November 2017, https://www.lazard.com/media/450337/lazard-levelized-cost-of-energy-version-110.pdf
3. "Electricity Generation Baseline Report," Jeffrey Logan et al., National Renewable Energy Laboratory, NREL Report 67645, January 2017, https://www.nrel.gov/docs/fy17osti/67645.pdf
4. "Updated Capital Cost Estimates for Utility Scale Electricity Generating Plants," Energy Information Administration, April 2013; and "Levelized Cost and Levelized Avoided Cost of New Generation Resources in the Annual Energy Outlook 2015," June 2015, Table 1
5. "Lazard's Levelized Cost of Energy Analysis—Version 8," September 2014, https://www.lazard.com/media/1777/levelized_cost_of_energy_-_version_80.pdf
6. "Annual Energy Outlook 2015 with projections to 2040," DOE/EIA-0383, Table A8, April 2015.
7. "Corn Stover for Bioenergy Production: Cost Estimates and Farmer Supply Response," Jena Thompson and Wallace E. Tyner, Department of Agricultural Economics Purdue University, September 2011, https://www.extension.purdue.edu/extmedia/EC/RE-3-W.pdf

8. "Biomass Power Plant Feedstock Procurement: Modeling Transportation Cost Zones and the Potential for Competition," Anil R. Kizha et al., California Agriculture 69(3):184–190, July 1, 2015, http://calag.ucanr.edu/Archive/?article=ca.v069n03p184
9. "Advancing Systems and Technologies to Produce Cleaner Fuels Technology Assessments," Quadrennial Technology Review 2015 Chapter 7, US Department of Energy, https://www.energy.gov/sites/prod/files/2016/01/f28/QTR2015-7B-Biomass-Feedstocks-and-Logistics.pdf
10. "Renewable Power Generation Costs in 2017," International Renewable Energy Agency (IRENA), file:///C:/Users/paulf/Documents/Book/References_Figures_Tables_Revised/Biomass/IRENA_2017_Power_Costs_2018.pdf

A.10. Geothermal: Levelized Cost of Electricity

For flash, binary, and enhanced geothermal systems (EGS) power plants, the LCOE was calculated according to Equation (1) in the general approach described above. For EGS, the LCOE was calculated when applied to both the flash and binary technologies. Input for the calculations came from 2019 EIA data,[1] NREL 2016 data,[2] and 2018 data from Lazard.[3] The units for LCOE are in $/megawatt-hour, or $/MWh.

The tax rate for both federal and state is set to 38%. The discount rate and book life used to calculate the CRF were set to 7% and 40 years, respectively. The MACRS depreciation schedule is 5 years and the capacity factor was set according to the authors.

The transmission costs were taken from the EIA report. For transmission costs, they provide the cost for new transmission lines to tie into existing electrical substations or existing transmission lines. In other words, the generating station provides transmission lines and steps up the voltage to tie into the "electrical grid," a network that delivers the electricity to the customer. The grid has high-voltage transmission lines (sometimes distant) that carry power to demand centers, transformers to step down the voltage, and distribution lines that connect individual customers.

The costs of transmitting and distributing electricity were 0.9 cents/kWh (or $9/MWh) and 2.6 cents/kWh (or $26/MWh), taken from the EIA.[4] Here, transmission is defined as the high-voltage movement of electricity from the power station while distribution is the cost of electrical lines that take power from a substation to the customer. The cost of transmission and distribution is essential to get a true "Well-to-Wheels" analysis of the total cost. For a geothermal power plant, the "Well-to-Wheels" costs will include (1) the cost of constructing the geothermal power plant, (2) power plant generating costs, (3) cost of connecting the power plant to the electrical grid, (4) transmission and distribution costs, and (5) taxes (both federal and state).

Table A.10.1 shows the LCOE for the cost of generating electricity and the additional cost of transmission and distribution.

Sensitivity studies for flash, binary, and EGS applied to binary technology were made by varying the capital cost, the fixed operating costs, and the capacity factor. The base case for each calculation was based on the 2016 NREL data.[2]

For the range in capital cost, a US DOE report gave possible capital cost reductions if larger geothermal plants are built and if improvements are made in exploration, drilling, and reservoir creation.[5] In the case of geothermal flash plants, this could result in a reduction from $4,229/kW to $3,319/kW while in the case of geothermal binary plants, this could result in a reduction from $5,445/kW to $4,273/kW. For the EGS technology applied to a binary geothermal plant, the capital cost could potentially be reduced from $32,268/kW to $5,509/kW.

For fixed O&M costs, the range of $10/MWh reported by Lazard[3] was applied to both the flash and binary technologies. So, for the flash plant, the fixed O&M costs ranged from the base case value of $16.6/MWh to $26.6/MWh while that for the binary plant ranged from the base case value of $24.5/MWh to $34.5/MWh. For the EGS plant, the range given in the NREL report[2] was used, namely from the base case value of $95.6/MWh to $23.1/MWh.

Table A.10.1 Economic Analysis for Geothermal Flash, Binary, and Enhanced Geothermal System (EGS) Technologies. For EGS, cases are shown for EGS applied to both the flash and binary technologies.

Energy Type	Capacity Factor	Capital Cost	Levelized Capital Cost; with MACRS depreciation	Fixed O&M	Variable O&M (including fuel)	Transmission Investment	Total System Levelized Cost (LCOE) to Generate	LCOE Including Transmission and Distribution
		$/kW	$/MWh	$/MWh	$/MWh	$/MWh	$/MWh	$/MWh
Flash								
EIA[1]	0.90	$2,787	$36.8	$15.5	$0.0	$1.4	$53.7	$88.7
NREL[2]	0.90	$4,229	$55.9	$16.6	$0.0	$1.4	$73.9	$108.9
Average	0.90	$3,508	$46.3	$16.1	$0.0	$1.4	$ 63.8	$98.8
Binary								
NREL[2]	0.80	$5,455	$81.1	$24.5	$0.0	$1.4	$107.0	$142.0
Lazard[3]	0.90	$4,000	$52.8	$25.0	$0.0	$1.4	$79.2	$114.2
Lazard[3]	0.85	$6,400	$89.5	$35.0	$0.0	$1.4	$125.9	$160.9
Average	0.85	$5,285	$74.5	$28.2	$0.0	$1.4	$104.0	$139.0
EGS								
EGS Flash—NREL[2]	0.90	$14,512	$191.7	$33.1	$0.0	$1.4	$226.2	$261.2
EGS Binary—NREL[2]	0.80	$32,268	$479.5	$95.6	$0.0	$1.4	$576.5	$611.5

For capacity factors, the values shown in Table A.10.1, ranging from 80% to 90%, are ideal numbers used for design and economic calculations. Commercial values are, however, generally lower. One of the problems with capacity factors for commercial geothermal plants is that the flow rate of the water or steam, as well as the temperature of the hot fluid, may decline over the life of the geothermal field. This will affect the plant's ability to meet their design capacity. Examining data from the EIA for commercial geothermal plants in the United States, the capacity factors were 78.1, 74.0, 74.3, 74.7, and 71.1% for the years of 2008 to 2012.[6] Breaking this down by technology, dry steam varied from 72.4 to 76.6%, flash systems varied from 76.9 to 89.3%, and binary systems varied from 49.3 to 69.7%. Although this report is a bit dated, more recent years for commercial US geothermal plants still have similar capacity factors. Although a breakdown by technology was not available, for the years 2013 to 2018 capacity factors were 73.6%, 74.0%, 74.3%, 73.9%, 74.0%, and 77.3%, respectively.[7]

Another reason these capacity factors are low is because of something called a "parasitic" load, the energy the power plant uses to operate and maintain the electrical plant. As this energy is not delivered to customers in the form of electricity, it lowers the ability of the plant to meet its rated capacity. Compared to nuclear and fossil fuel plants, in which the energy sources have higher temperature and thus higher thermal efficiencies, geothermal plants consume more parasitic energy because the fluid temperature from the geothermal source is lower. This is especially true for the binary plants, which, as reported in Chapter 10, have inherently lower fluid temperatures than dry steam and flash steam. So, for the sensitivity study, the capacity

Table A.10.2 Sensitivity Study for Geothermal Flash, Binary, and Enhanced Geothermal System (EGS) Technologies

	Capital Cost	Fixed O&M	Capacity Factor	Total System Levelized Cost (LCOE)	Total LCOE Cost to Generate, Transmit, and Distribute
Units	**$/kW**	**$/MWh**		**$/MWh**	**$/MWh**
Flash Sensitivity Case					
Base Case, NREL[2]	$4,229	$16.6	0.90	$73.9	$108.9
Capex is $3,319	$3,319	$16.6	0.90	$61.9	$96.9
Fixed O&M is $26.6/MWh	$4,229	$26.6	0.90	$83.9	$118.9
Capacity Factor is 0.769	$4,229	$19.4	0.769	$86.2	$121.2
Binary Sensitivity Case					
Base Case, NREL[2]	$5,455	$24.5	0.80	$107.0	$142.0
Capex is $4,273/kW	$4,273	$24.5	0.80	$89.4	$124.4
Fixed O&M is $34.5/MWh	$5,455	$34.5	0.80	$117.0	$152.0
Capacity Factor is 0.493	$5,455	$39.8	0.493	$172.8	$207.8
EGS Sensitivity Case					
Base Case, NREL[2]	$32,268	$95.6	0.80	$576.5	$611.5
Capex is $5,509/kW	$5,509	$95.6	0.80	$178.9	$213.9
Fixed O&M is $23.1/MWh	$32,268	$23.1	0.80	$504.0	$539.0
Capacity Factor is 0.493	$32,268	$155.1	0.493	$934.6	$969.6

factors will range from the high values shown in Table A.10.1 to the lowest observed commercial value for each technology, namely 76.9% for a geothermal flash plant and 49.3% for a binary plant. Since the EGS case is used combination with the binary geothermal plant, the same lower value of 49.3% for the binary system is also used for the EGS technology.

The results of the sensitivity study are shown in Table A.10.2.

References for Geothermal

1. "Cost and Performance Characteristics of New Generating Technologies," Annual Energy Outlook 2019, Table 2, Energy Information Administration, January 2019, https://www.eia.gov/outlooks/aeo/
2. "2018 Annual Technology Baseline ATB Cost and Performance Data for Electricity Generation Technologies—Interim Data without Geothermal Updates," National Renewable Energy Laboratory, 2018, https://data.nrel.gov/submissions/89
3. "Lazard's Levelized Cost of Energy Analysis—Version 12.0," November 2018, https://www.lazard.com/media/450784/lazards-levelized-cost-of-energy-version-120-vfinal.pdf
4. "Annual Energy Outlook 2015 with projections to 2040," DOE/EIA-0383, Table A8, April 2015.
5. "Geo Vision: Harnessing the Heat Beneath Our Feet," US Department of Energy, 218 pp, https://www.energy.gov/sites/prod/files/2019/05/f63/GeoVision-full-report.pdf
6. "Geothermal Plant Capacity Factors," Greg Mines et al., Proceedings, Fortieth Workshop on Geothermal Reservoir Engineering, Stanford University, Stanford, California, January 26–28, 2015, https://pangea.stanford.edu/ERE/db/GeoConf/papers/SGW/2015/Nathwani.pdf
7. "Table 6.7.B. Capacity Factors for Utility Scale Generators Not Primarily Using Fossil Fuels, January 2013–June 2019, Electric Power Monthly, Energy Information Administration, https://www.eia.gov/electricity/monthly/epm_table_grapher.php?t=epmt_6_07_b

A.11. Hydrogen: Levelized Cost of Fuel

For hydrogen, the LCOF was calculated according to Equation (4) in the general approach described above. The units for LCOF are in \$/gallon of gasoline equivalent, or \$/gge. As one kg of hydrogen is equivalent in energy content to one gallon of gasoline, or one gge, no correction is needed in Equation (4) to convert from \$/kg H_2 to \$/gge.

Calculations were made for four routes to produce hydrogen including (1) distributed natural gas reforming, (5) central biomass delivered by truck as a gas, (9) central electrolysis using wind electricity and delivered by pipeline, and (10) central coal gasification delivered by pipeline. "Distributed" means the hydrogen is produced at the refueling site while "central" means the hydrogen is produced offsite and then delivered to the refueling site (by pipeline or truck).

Table A.i.3 gives a book life of 20 years and a MACRS depreciation schedule of 7 years for hydrogen. This is consistent with the book life and MACRS depreciation for the different technologies used to produce that hydrogen, shown in Table A.i.1 and Table A.i.3. If the hydrogen is produced from natural gas, coal, or biomass, it essentially follows a Fischer-Tropsch type route of producing carbon monoxide and hydrogen by either reforming or gasification, followed by the water gas shift reaction to increase the amount of hydrogen. If the hydrogen is produced by electrolysis using electricity produced at an onshore wind farm, the book life and depreciation are also 20 and 7 years, respectively.

The combined tax rate for federal and state was 38%.

Input for the different cases were taken from an NREL report.[1] Compared to this report, however, different feed costs were used.

For Case (1) using natural gas, the cost used was the same as that used in Chapter 2 for natural gas to electricity, a value of \$3.17/MM BTU based on the Henry Hub average spot price for 2018.[2] For Case (5) using biomass, a DOE study[3] was used in Chapter 9 to calculated a blended cost of \$80/dry ton, or \$72.6/dry tonne. For Case (10) using coal, the cost used in Chapter 3 was based on Illinois No. 6 bituminous coal using EIA data[4] for 2018, a value of \$2.06/MM BTU or \$47.2/tonne based on an energy content of 12,712 BTU/lb.

For Case (9) using wind farm electricity to perform electrolysis and make hydrogen, the NREL report[1] used an electricity cost of \$0.0574/kWh. As the LCOE for an onshore wind farm was calculated to be \$0.0998/kWh in Chapter 6 and the average US price of electricity delivered to the customers for 2018 was \$0.1078/kWh,[5] this value seems too optimistic. Therefore, the cost of electricity from the wind farm for this case was set to the average US price of \$0.1078/kWh. It is true that the LCOE for wind farm electricity is dependent on the average wind speed for its location, so this variable is investigated in a sensitivity study. Input data for the four cases are shown in Table A.11.1.

With these data, the LCOF was calculated for the four different cases, shown in Table A.11.2.

A sensitivity analysis was made for the levelized cost of hydrogen. For the sensitivity analysis, only the highest and lowest cost cases were selected, namely Case (1), distributed natural

Table A.11.1 Input Data for the Four Cases. All data, Except Feed Cost, from an NREL Report.[1]

	Case 1: Distributed Natural Gas	Case 5: Central Biomass, Truck Delivery	Case 9: Central Wind, Pipeline Delivery	Case 10: Central Coal, Pipeline Delivery
Production				
Capacity Factor	0.89	0.90	0.90	0.90
Process Energy Efficiency	0.714	0.460	0.719	0.568
Energy Processing efficiency	0.972	1.000	1.000	1.000
Feed Delivered Cost	\$3.17/MM BTU	\$72.6/dry tonne	\$0.1078/kWh	\$47.2/tonne
H_2 Product Yield, kg/day	1,500	155,200	379,400	307,700
Capital Cost (per kg H2), \$/kg	\$1,152	\$1,296	\$1,320	\$2,497
Fixed O&M, \$/gge	\$0.19	\$0.23	\$0.23	\$0.30
Non-Fuel Variable O&M, \$/gge	\$0.12	\$0.40	\$0.06	\$0.18
Delivery				
Levelized Cost of Delivery, \$/gge		\$1.88	\$2.20	\$2.20
Compression, Storage, and Dispensing (CSD)				
Capital Cost (per kg H_2), \$/kg	\$3,806	\$2,189	\$2,648	\$2,648
Other O&M, \$/gge	\$0.54	\$0.46	\$0.41	\$0.41
Energy & Fuel, \$/gge	\$0.40	\$0.20	\$0.43	\$0.43

Note: O&M refers to operations and maintenance.

Table A.11.2 Levelized Cost of Fuel (LCOF) for Hydrogen. All values are in units of $/gge.

	Case 1: Distributed Natural Gas	Case 5: Central Biomass, Truck Delivery	Case 9: Central Wind, Pipeline Delivery	Case 10: Central Coal, Pipeline Delivery
Production				
Fixed O&M	$0.19	$0.23	$0.23	$0.30
Non-Fuel Variable O&M	$0.12	$0.40	$0.06	$0.18
Feed Cost	$0.67	$0.98	$6.25	$0.37
Levelized Capital Cost	$0.34	$0.38	$0.39	$0.73
Total Cost of Production	$1.32	$1.99	$6.93	$1.58
Delivery				
Levelized Cost of Delivery		$1.88	$2.20	$2.20
Compression, Storage, and Dispensing (CSD)				
Levelized Capital Cost	$1.11	$0.64	$0.77	$0.77
Other O&M	$0.54	$0.46	$0.41	$0.41
Energy & Fuel	$0.40	$0.20	$0.43	$0.43
Total Cost of CSD	$2.05	$1.30	$1.61	$1.61
LCOF	$3.37	$5.17	$10.74	$5.39

gas reforming, and Case (9), central electrolysis using wind electricity and delivered by pipeline. For both cases, the sensitivity analysis focuses on the cost of production.

For Case (1), variations in the production capital cost, the energy efficiency, and the price of fuel are considered. The energy efficiency is the percent of fuel energy that ends up as energy in the form of hydrogen. Using the NREL report as a guideline,[1] the capital costs cover a range from $800/kg H_2 to $1,500/kg H_2, with $1,152/kg H_2 from Table A.11.1 used as the base case. Also, the energy efficiency of 0.714 was varied from 0.6 to 0.8. The value of $3.17/MM BTU for natural gas was based on the Henry Hub average spot price for 2018.[2] Fourteen years of average Henry Hub data were shown in Chapter 2, taken from the same reference and shown again in Table A.11.3. As the data show, there has been great variation in the spot price for natural gas, as low as about $3/MM BTU and as high as $9/MM BTU. As the base case value is near the 14-year low, the two other values examined will be $6 and $9/MM BTU.

For Case (9), the variation in the production capital cost, energy efficiency, and the price of electricity are considered. Using the NREL report as a guideline,[1] the capital costs cover a range from $1,000/kg H_2 to $2,000/kg H_2, with $1,320/kg H_2 in Table A.11.1 used as the base case. Also, the base case energy efficiency of 0.719 is varied from 0.6 to 0.8. For the price of electricity from wind energy, the base case used a value of $0.1078/kWh based on the average price US customers paid for electricity in 2018.[5] A report from the DOE Office of Energy Efficiency[6] shows that the cost of onshore wind farm electricity is highly dependent on location, with the lowest cost projects in the so-called wind belt US interior, where wind speeds can top 8.5 meters per second, or 19 mph. The LCOE from plants in the US interior range from $0.02 to $0.06/kWh, compared to $0.06 to $0.12/kWh in other, less windy, parts of the United States. Based on this information, the cost of electricity used to produce hydrogen from electrolysis also looks at values of $0.04 and $0.08/kWh.

The results of the sensitivity study are shown in Table A.11.4.

Table A.11.3 Henry Hub Historical Natural Gas Prices, \$/MM BTU

Year	\$/MM BTU (million BTU)
2005	8.69
2006	6.73
2007	6.97
2008	8.86
2009	3.94
2010	4.37
2011	4.00
2012	2.75
2013	3.73
2014	4.37
2015	2.62
2016	2.52
2017	2.99
2018	3.17

Table A.11.4 Sensitivity Study for Distributed Natural Gas Reforming (Case (1)) and Central Electrolysis Using Wind Electricity and Delivered by Pipeline (Case (9))

	Energy Efficiency	Feed Cost	Capital Cost	Production Levelized Capital Cost	Total Cost of Production	LCOF
Units		\$/MM BTU	\$/kg H2	\$/gge	\$/gge	\$/gge
Case 1						
Base Case	0.714	\$3.17	\$1,152	\$0.34	\$1.32	\$3.37
Capex is \$800/kg H_2	0.714	\$3.17	\$800	\$0.23	\$1.22	\$3.27
Capex is \$1,500/kg H_2	0.714	\$3.17	\$1,500	\$0.44	\$1.42	\$3.47
Energy efficiency is 0.6	0.600	\$3.17	\$1,152	\$0.34	\$1.45	\$3.50
Energy efficiency is 0.8	0.800	\$3.17	\$1,152	\$0.34	\$1.25	\$3.30
Natural gas is \$6/MM BTU	0.714	\$6.00	\$1,152	\$0.34	\$1.92	\$3.97
Natural gas is \$9/MM BTU	0.714	\$9.00	\$1,152	\$0.34	\$2.56	\$4.61
Case 9		\$/kWh				
Base Case	0.719	\$0.1078	\$1,320	\$0.39	\$6.93	\$10.74
Capex is \$1,000/kg H_2	0.719	\$0.1078	\$1,000	\$0.29	\$6.83	\$10.65
Capex is \$2,000/kg H_2	0.719	\$0.1078	\$2,000	\$0.58	\$7.13	\$10.94
Energy efficiency is 0.6	0.600	\$0.1078	\$1,320	\$0.39	\$8.17	\$11.98
Energy efficiency is 0.8	0.800	\$0.1078	\$1,320	\$0.39	\$6.29	\$10.11
Electricity is \$0.04/kWh	0.719	\$0.0400	\$1,320	\$0.39	\$3.00	\$6.81
Electricity is \$0.08/kWh	0.719	\$0.0800	\$1,320	\$0.39	\$5.32	\$9.13

References

1. "Hydrogen Pathways Updated Cost, Well-to-Wheels Energy Use, and Emissions for the Current Technology Status of Ten Hydrogen Production, Delivery, and Distribution Scenarios," T. Ramsden et al., National Renewable Energy Laboratory, Technical Report NREL/TP-6A10-60528, March 2013, https://www.nrel.gov/docs/fy14osti/60528.pdf
2. "Natural Gas: Henry Hub Natural Gas Spot Price," Energy Information Administration, 2018, https://www.eia.gov/dnav/ng/hist/rngwhhdm.htm
3. "Advancing Systems and Technologies to Produce Cleaner Fuels Technology Assessments," Quadrennial Technology Review 2015 Chapter 7, US Department of Energy, https://www.energy.gov/sites/prod/files/2016/01/f28/QTR2015-7B-Biomass-Feedstocks-and-Logistics.pdf
4. "Table 9.9 Cost of Fossil-Fuel Receipts at Electric Generating Plants," Energy Information Administration / Monthly Energy Review, p. 163, March 2019, http://www.eia.gov/totalenergy/data/monthly/pdf/sec9_13.pdf
5. "Electric Power Monthly, Energy Information Administration, Data for June 2018, Table 5.6.A. Average Price of Electricity to Ultimate Customers by End-Use Sector, https://www.eia.gov/electricity/monthly/epm_table_grapher.php?t=epmt_5_6_a
6. "Average US Wind Price Falls to $20 per Megawatt-Hour," Emma Foehringer Merchant, Green Tech Media, August 24, 2018, https://www.greentechmedia.com/articles/read/us-wind-prices-20-megawatt-hour#gs.a7q2sr

A.12. Fischer-Tropsch Synthesis: Levelized Cost of Fuel

For the Fischer-Tropsch synthesis, the LCOF was calculated according to Equation (4) in the general approach described above. Three different Fischer-Tropsch approaches were evaluated, namely gas-to-liquids (GTL), coal-to-liquids (CTL), and biomass-to-liquids (BTL). In the case of GTL, the gas is natural gas. The LCOF was calculated for each of these three cases using literature data. However, to be consistent among these cases and with economic calculations made in other chapters of this book, the tax and discount rate will be the same as that used for other technologies, and therefore different than those used by the various authors. Also, the cost of fuel will be the same as used in other chapters. For GTL, the same cost used for natural gas in Chapter 2 is also used here. Likewise, the cost of coal used in Chapter 3 is used for CTL. For BTL, the three literature studies examined used different feed types including switchgrass, corn stover, and prairie grass. For these LCOF calculations, the fuel costs were kept the same as reported by the authors. For the sensitivity study, however, the same cost of biomass used in Chapter 9, the chapter for biomass to electricity, was used.

The units for LCOF are in $/gallon of gasoline equivalent, or $/gge. It should be noted that for most commercial Fischer-Tropsch plants, the main product is diesel, with gasoline secondary. We could therefore calculate the LCOF in terms of $/gallon of diesel. Since the energy contents of gasoline and diesel are 124,238 and 138,690 BTU/gallon (see Appendix B, Equations (28) and (29)), the LCOF for diesel would be about 90% of that for gasoline on an equivalent energy basis. However, to be consistent with other chapters, the LCOF is calculated based on gasoline.

Using Table A.i.3 given earlier in this chapter, the book life for CTL and BTL plants is set to 20 years while that for a GTL plant is set to 15 years. The MACRS depreciation schedule is 7 years for all Fischer-Tropsch cases.

As is the case with other technologies, the LCOF for the Fischer-Tropsch cases are analyzed in terms of "Well-to-Wheels" cost. For GTL, the "Well-to-Wheels" costs include (1) the cost of natural gas production and transportation to the Fischer-Tropsch plant, (2) the

Fischer-Tropsch plant including post-Fischer-Tropsch processing units, (3) transportation of products to market, (4) taxes (both federal and state), and (5) marketing cost at the filling station. For CTL and BTL, the "Well-to-Wheels" costs include (1) the cost of coal or biomass production, preparation, and transportation to the Fischer-Tropsch plant, (2) the Fischer-Tropsch plant including post-Fischer-Tropsch processing units, (3) transportation of products to market, (4) taxes (both federal and state), and (5) marketing cost at the filling station.

The cost for step (1) is included in the delivered cost of the fuel, whether natural gas, coal, or biomass. For GTL, the cost of natural gas was $3.17/MM BTU based on the Henry Hub spot price for 2018.[1] This same cost was also used in Chapter 2, the chapter for natural gas to electricity. For CTL, the price of coal was based on Illinois No. 6 bituminous coal. This value was $2.06/MM BTU, taken from a 12-month average of 2018 EIA data[2] and the same value used in Chapter 3 for coal to electricity. For BTL, the price of biomass was based on a DOE study[3] and the same value used in Chapter 9 for biomass to electricity. That value is $80/dry ton, or $72.6/dry tonne.

For steps (3) and (5), transportation of gasoline and diesel to the market plus the marketing cost at the filling station, an April 2019 update by the EIA[4] reports values of $0.32/gallon for gasoline and $0.60/gallon for diesel. Since LCOF is calculated for gasoline, $0.32/gallon is used. The taxes for step (4) are set to 38% for both federal and state.

Data for the other steps were taken from various published articles. For GTL, the data and calculated LCOF are shown in Table A.12.1. With the exception of the NETL report, the other data are based on existing commercial units.

For CTL, the data and calculated LCOF are shown in Table A.12.2. None of these data came from existing commercial units. Two of the studies add cost for CCS. Also, for most cases a power credit is given for electricity generated from unconverted syngas and light (C_1–C_4) gases. All of the cases were designed with liquid product yields of 50,000 BPD. Adding CCS increases the LCOF by $0.06 to $0.09, or about $2.5 to $4 per barrel (42 gallons in a barrel).

For BTL, the data and calculated LCOF are shown in Table A.12.3. None of these data came from existing commercial units. Two of the studies added the cost of CCS, increasing the LCOF by $0.10 to $0.17, or about $4 to $7 per barrel. For four of the cases, a power credit is given for electricity generated from unconverted syngas and light (C_1–C_4) gases. Although the feed costs were similar for the cases taken from these three studies, the biomass types were different and included switchgrass, corn stover, and prairie grass. Compared to Tables A.12.1 and A.12.2 for GTL and CTL, the BTL cases had small liquid product yields, from about 2,500 to 5,000 BPD.

Before discussing the sensitivity studies, it is instructive to look collectively at the capital costs for all three types of plants, shown in Table A.12.4. On average, GTL and CTL plants cost about $100,000 per barrel per day installed capacity and BTL plants cost about $180,000. These values are significantly different compared to average installed refinery costs of $40,000/BPD (see Table A.1.1). This is a large drawback for building new Fischer-Tropsch plants, because a typical GTL or CTL plant that produces 50,000 BPD of transportation fuels will cost five billion dollars at $100,000/BPD and a small 5,000 BPD BTL plant will cost $0.9 billion dollars at $180,000/BPD! When I last worked on Fischer-Tropsch R&D in the late 1990s and early 2000s, the capital cost goal was $25,000/BPD, so capital cost reduction has a long way to go to meet this target. There is, however, large variability in the commercial data for GTL plants, ranging from ~$35,000/BPD for the Oryx plant in Qatar to $180,000/BPD for the Escravos plant in Nigeria, suggesting that capital costs are affected by location and that the $25,000/BPD could be achieved under the right circumstances.

For the sensitivity study, capital cost, feed cost, and the combined fixed and non-fuel variable O&M costs were examined. For GTL, a plant with a 50,000 BPD production of transportation fuels was chosen. Using the previous table as a guide, the capital cost for commercial plants ranged from $35,000/BPD to $180,000/BPD with an average of about $100,000/BPD.

Table A.12.1 Economic Analysis for GTL Plants

Reference	Capacity Factor	Feed Cost	Liquid Product Yield	Levelized Capital Cost	Fixed O&M	Non-Fuel Variable O&M	Feed Cost	T&M (a)	LCOF
Units		$/MM BTU	BPD	$/gge	$/gge	$/gge	$/gge	$/gge	$/gge
NETL Report[5]	0.90	$3.17	50,000	$0.77	$0.37	$0.25	$0.64	$0.38	$2.42
Pearl, Qatar[6]	0.90	$3.17	140,000	$1.22	$0.14	$0.07	$0.89	$0.38	$2.70
Bintulu, Malaysia[7]	0.90	$3.17	14,700	$0.55	$0.14	$0.07	$0.58	$0.38	$1.72
Mossel Bay, South Africa[8]	0.90	$3.17	45,000	$0.80	$0.14	$0.07	$0.62	$0.38	$2.01
Oryx, Qatar[8]	0.90	$3.17	34,000	$0.32	$0.14	$0.07	$0.75	$0.38	$1.66
Escravos, Nigeria[9]	0.90	$3.17	34,000	$1.62	$0.14	$0.07	$0.75	$0.38	$2.96
Average	0.90	$3.17	52,951	$0.88	$0.18	$0.10	$0.71	$0.38	$2.25

a. Transportation and Marketing

Table A.12.2 Economic Analysis for CTL Plants

Reference	Capacity Factor	Feed Cost	Levelized Capital Cost	Fixed O&M	Non-Fuel Variable O&M	Feed Cost	T&M (a)	Power Credit	LCOF
Units		$/ton	$/gge	$/gge	$/gge	$/gge	$/gge	$/gge	$/gge
NETL Report (no CCS)[10]	0.90	$48.06	$0.68	$0.37	$0.08	$0.50	$0.38	–$0.04	$1.96
NETL Report (w/CCS)[10]	0.90	$48.06	$0.70	$0.38	$0.08	$0.49	$0.38	$0.00	$2.02
Princeton (no CCS)[11]	0.90	$54.03	$0.75	$0.28	$0.07	$0.61	$0.38	–$0.29	$1.81
Princeton (w/CCS)[11]	0.90	$54.03	$0.76	$0.29	$0.07	$0.61	$0.38	–$0.22	$1.90
NETL Report (no CCS)[12]	0.85	$54.08	$0.74	$0.31	$0.02	$0.63	$0.38	–$0.07	$2.01
Average (no CCS)	0.88	$52.06	$0.72	$0.32	$0.06	$0.58	$0.38	–$0.14	$1.92

a. Transportation and marketing.

Table A.12.3 Economic Analysis for BTL Plants

Reference	Capacity Factor	Feed Cost	Liquid Product Yield	Levelized Capital Cost	Fixed O&M	Non-Fuel Variable O&M	Feed Cost	T&M (a)	Power Credit	LCOF
Units		$/ton	BPD	$/gge	$/gge	$/gge	$/gge	$/gge	$/gge	$/gge
NETL Report (no CCS)[10] (b)	0.85	$85.60	5,000	$1.78	$0.89	$0.11	$1.66	$0.38	$0.00	$4.82
NETL Report (w/CCS)[10] (b)	0.85	$85.60	5,000	$1.88	$0.94	$0.11	$1.69	$0.38	$0.00	$4.99
NREL Report[13] (c)	0.85	$81.61	3,200	$1.55	$0.34	$0.29	$1.23	$0.38	–$0.13	$3.66
NREL Report[13] (d)	0.85	$81.61	2,479	$1.64	$0.38	$0.40	$1.59	$0.38	–$0.20	$4.19
Princeton (no CCS)[11] (e)	0.90	$85.04	4,410	$0.88	$0.33	$0.00	$1.54	$0.38	–$0.26	$2.87
Princeton (w/CCS)[11] (e)	0.90	$85.04	4,415	$0.90	$0.34	$0.00	$1.54	$0.38	–$0.19	$2.97
Average (no CCS)	0.86	$83.47	3,772	$1.46	$0.49	$0.20	$1.51	$0.38	–$0.15	$3.89

a. Transportation and marketing.

b. The biomass fuel is switchgrass.

c. Gasification is done at high temperature and the biomass fuel is corn stover.

d. Gasification is done at low temperature and the biomass fuel is corn stover.

e. The biomass fuel is prairie grass.

Table A.12.4 Total Capital Costs and Cost on a $/BPD basis for GTL, CTL, and BTL

	Total Project Investment	Liquid Product Yield	Capital Cost (per BBL product)
Units	**$ MM**	**BPD**	**$/BPD**
GTL Plants			
NETL Report[5]	$4,310	50,003	$86,189
Pearl Qatar[6]	$19,000	140,000	$135,714
Bintulu Malaysia[7]	$900	14,700	$61,225
Mossel Bay, South Africa[8]	$4,000	45,000	$88,889
Oryx, Qatar[8]	$1,200	34,000	$35,294
Escravos, Nigeria[9]	$6,120	34,000	$180,000
GTL Average			**$97,885**
CTL Plants			
NETL Report (no CCS)[10]	$5,125	50,000	$102,500
NETL Report (w/CCS)[10]	$5,310	50,000	$106,200
Princeton (no CCS)[11]	$4,878	50,000	$97,568
Princeton (w/CCS)[11]	$4,945	50,000	$98,908
NETL Report (no CCS)[12]	$4,528	49,992	$90,574
CTL Average			**$99,150**
BTL Plants			
NETL Report (no CCS) [10]	$1,086	5,000	$217,200
NETL Report (w/CCS) [10]	$1,148	5,000	$229,600
NREL Report[13]	$606	3,200	$189,350
NREL Report[13]	$498	2,479	$201,000
Princeton (no CCS)[11]	$503	4,410	$114,131
Princeton (w/CCS)[11]	$515	4,415	$116,726
BTL Average			**$178,001**

Based on this, the base case will use $100,000/BPD, with lower and upper limits of $35,000 and $180,000/BPD.

Table A.12.1 shows that, on average, the combined fixed and non-fuel variable O&M costs are $0.28/gge, or $11.76/BBL. In the same table, most of the calculations used a combined fixed and non-fuel variable O&M of $0.21/gge, or $8.82/BBL. This is similar to the value of $8.70/BBL reported by Maitlis and de Klerk.[14] According to Hobbs and Adair,[15] the combined costs can be as high as $12–$18/BBL. Based on these data, the combined fixed and non-fuel variable O&M also considers $12/BBL and $18/BBL, in addition to the base case value of $8.82/BBL.

Finally, over the last 14 years, Henry Hub average annual prices for natural gas[1] have varied from a low of $2.52/MM BTU to a high $8.86/MM BTU, as shown in Table A.12.5. Therefore, in addition to the base case value of $3.17/MM BTU, natural gas costs of $2.5, $6, and $9/MM BTU will also be evaluated. The results of the sensitivity study are shown in Table A.12.6.

Table A.12.5 Henry Hub Historical Natural Gas Prices, $/MM BTU

Year	$/MM BTU (million BTU)
2005	8.69
2006	6.73
2007	6.97
2008	8.86
2009	3.94
2010	4.37
2011	4.00
2012	2.75
2013	3.73
2014	4.37
2015	2.62
2016	2.52
2017	2.99
2018	3.17

Table A.12.6 Sensitivity Study for Gas-to-Liquids (GTL) Plant

Reference	Capacity Factor	Capital Cost	Feed Cost	Fixed and Non-Fuel O&M	LCOF
Units		$/BPD	$/MM BTU	$/gge	$/gge
Base Case	0.90	$100,000	$3.17	$0.21	$2.13
Low Capex	0.90	$35,000	$3.17	$0.21	$1.55
High Capex	0.90	$180,000	$3.17	$0.21	$2.85
Opex is $12/BBL	0.90	$100,000	$3.17	$0.29	$2.20
Opex is $18/BBL	0.90	$100,000	$3.17	$0.43	$2.35
Natural Gas is $2.5/MM BTU	0.90	$100,000	$2.50	$0.21	$1.99
Natural Gas is $6/MM BTU	0.90	$100,000	$6.00	$0.21	$2.70
Natural Gas is $9/MM BTU	0.90	$100,000	$9.00	$0.21	$3.31

Like GTL, the CTL plant was chosen to have a daily transportation fuels production of 50,000 BPD. For capital cost, Table A.12.4 shows an average of about $100,000/BPD for capital cost, but these model studies do not show much variation in capital cost. Therefore, the commercial data for GTL plants was used as a guide, so the same lower and upper limits of $35,000/BPD and $180,000/BPD used for GTL will also be used for CTL.

For the cases reported in Table A.12.2, the cost of coal was about $48/ton. For the low and high ranges, the cost of coal is reduced or increased by about 50%, resulting in coal prices of

Table A.12.7 Sensitivity Study for Coal-to-Liquids (CTL) Plant

	Capacity Factor	Capital Cost	Feed Cost	Fixed and Non-Fuel O&M	LCOF
Units		$/BPD	$/ton	$/gge	$/gge
Base Case	0.90	$100,000	$48.06	$0.43	$1.97
Capex is $35,000/BBL	0.90	$35,000	$48.06	$0.16	$1.26
Capex is $180,000/BBL	0.90	$180,000	$48.06	$0.78	$2.85
Opex is 3% fixed and 1% variable	0.90	$100,000	$48.06	$0.29	$1.83
Opex is 7% fixed and 1% variable	0.90	$100,000	$48.06	$0.58	$2.12
Coal is $20/ton	0.90	$100,000	$20.00	$0.43	$1.68
Coal is $80/ton	0.90	$100,000	$80.00	$0.43	$2.30

Table A.12.8 Sensitivity Study for Biomass-to-Liquids (BTL) Plant

	Capacity Factor	Capital Cost	Feed Cost	Fixed Operating Cost	Total LCOF
Units		$/BPD	$/ton	$/gge	$/gge
Base Case	0.90	$180,000	$80	$0.65	$4.11
Capex is $120,000/BBL	0.90	$120,000	$80	$0.43	$3.38
Capex is $240,000/BBL	0.90	$240,000	$80	$0.87	$4.83
Fixed Opex is 3% of Capex	0.90	$180,000	$80	$0.39	$3.85
Fixed Opex is 7% of Capex	0.90	$180,000	$80	$0.91	$4.37
Biomass is $40/ton	0.90	$180,000	$40	$0.65	$3.33
Biomass is $120/ton	0.90	$180,000	$120	$0.65	$4.89

$20/ton and $80/ton. Also, for the cases shown in Table A.12.2, the fixed operating costs were typically 5% of the capital cost while the variable operating costs, not including the cost of fuel, were 1% of the capital costs. Based on this, the fixed operating costs will use a range from 3% to 7% of the capital cost, keeping the variable operating costs constant. The results are shown in Table A.12.7.

A biomass plant is expected to have a smaller daily production of transportation fuels, and 5,000 BPD was chosen for the plant size. Based on Table A.12.4, the capital cost for the base case was set to $180,000/BPD, with low and high ranges of $120,000/BPD and $240,000/BPD. For the base case biomass feed price, the same price used in Chapter 9 for biomass to electricity, $80/day ton, was used. This was based on a DOE study.[3] For the variation in biomass cost, a $40/ton reduction and increase is applied, resulting in a low of $40/ton and a high of $120/ton. As was the case for CTL, the fixed operating costs were typically 5% of the capital cost while the variable operating costs, not including the cost of fuel, were 1% of the capital costs. Based on this, the fixed operating costs will use a range from 3% to 7% of the capital cost, keeping the variable operating costs constant. The results are shown in Table A.12.8.

References

1. "Natural Gas: Henry Hub Natural Gas Spot Price," Energy Information Administration, 2018, https://www.eia.gov/dnav/ng/hist/rngwhhdm.htm
2. "Table 9.9 Cost of Fossil-Fuel Receipts at Electric Generating Plants," Energy Information Administration / Monthly Energy Review, p. 163, March 2019, http://www.eia.gov/totalenergy/data/monthly/pdf/sec9_13.pdf
3. "Advancing Systems and Technologies to Produce Cleaner Fuels Technology Assessments," Quadrennial Technology Review 2015 Chapter 7, US Department of Energy, https://www.energy.gov/sites/prod/files/2016/01/f28/QTR2015-7B-Biomass-Feedstocks-and-Logistics.pdf
4. "Gasoline and Diesel Fuel Update," Energy Information Administration, April 1, 2019, http://www.eia.gov/petroleum/gasdiesel/
5. "Analysis of Natural Gas-to-Liquid Transportation Fuels via Fischer-Tropsch," Erik Shuster, National Energy Technology Laboratory, September 13, 2013, DOE/NETL-2013/1597, http://www.academia.edu/30467535/Analysis_of_Natural_Gas-to_Liquid_Transportation_Fuels_via_Fischer-Tropsch_OFFICE_OF_FOSSIL_ENERGY
6. "Pearl GTL Still Generating Work," MEED (Middle East Business Intelligence), November 24, 2013, http://www.meed.com/sectors/oil-and-gas/gas/pearl-gtl-still-generating-work/3186978.fullarticle
7. "The Shell GTL Process: Towards a World Scale Project in Qatar: The Pearl Project," Arend Hoek; Shell Global Solutions International, Amsterdam; DGMK-Conference "Synthesis Gas Chemistry"; October 4–6, 2006, Dresden, Bintulu, Malaysia GTL (1993), https://www.dgmk.de/petrochemistry/abstracts_content14/Hoek.pdf
8. "Gas to Liquids Plants: Turning Louisiana Natural Gas into Marketable Liquid Fuels," E. O'Brien, Louisiana Department of Natural Resources/Technology Assessment Division, November 2013, http://www.dnr.louisiana.gov/assets/TAD/newsletters/2013/2013-11_topic_1.pdf
9. "Gas to Liquids Conversion," A. de Klerk, U. of Alberta, Natural Gas Conversion Technologies Workshop of ARPA-E, US DOE, Houston, January 13, 2012, https://www.arpa-e.energy.gov/sites/default/files/documents/files/De_Klerk_NatGas_Pres.pdf
10. "Affordable, Low-Carbon Diesel Fuel from Domestic Coal and Biomass," January 14, 2009, DOE/NETL-2009/1349, Thomas J. Tarka, https://www.netl.doe.gov/energy-analyses/pubs/CBTL%20Final%20Report.pdf
11. "Fischer-Tropsch Fuels from Coal and Biomass," Thomas G. Kreutz et al., Princeton Environmental Institute, 25th Annual International Pittsburgh Coal Conference, August 21, 2008, https://acee.princeton.edu/wp-content/uploads/2016/10/Kreutz-et-al-PCC- 2008-10-7-08.pdf
12. "Baseline Technical and Economic Assessment of a Commercial Scale Fischer-Tropsch Facility," Lawrence Van Bibber et al., DOE/NETL-2007/1260; April 9, 2007, https://www.netl.doe.gov/energy-analyses/pubs/Small-Scale%20F-T%20Liquids%20Facilities_Final.pdf
13. "Techno-Economic Analysis of Biofuels Production Based on Gasification," R.M. Swanson et al., NREL Technical Report NREL/TP-6A20-46587, November 1, 2010, https://www.nrel.gov/docs/fy11osti/46587.pdf
14. "Greener Fischer-Tropsch Processes for Fuels and Feedstocks," P.M. Maitlis and A. de Klerk, Chapter 7: Fischer-Tropsch Process Economics, Table 7.4, Wiley-VCH, 2013.
15. "Analysis Shows GTL Viable Alternative for US Gas Producers," H.O. Hobbs Jr. and L.S. Adair, Oil & Gas Journal, August 6, 2012, file:///C:/Users/paulf/Documents/Book/References_Figures_Tables_Revised/Fischer_Tropsch/Analysis-shows-GTL-viable-alternative-for-US-gas-producers-PDF-File.pdf

APPENDIX B

Units, Unit Prefixes, and Basic Definitions

There is probably nothing more frustrating, and nothing more important, than the understanding of units. Units, and the conversions between units, are essential to understand the variety of information reported in the field of energy. Ideally, the entire world would use the metric system, greatly simplifying the comparative understanding of different energy types. The metric system is more formally defined as SI units, or *système international d'unités* in French. In the metric system, you use meter for length, kilogram for mass, seconds for time, and Kelvin for temperature. The greatest advantage is the easy conversion between greater and lesser units by powers of the ten. For example, 1,000 grams is one kilogram and one kilometer is 1,000 meters.

Unfortunately some countries, notably the United States, still use the English system of units, making it necessary to provide some basic information for the purpose of understanding this book. English units come from England and evolved from the Roman system. Compared to the SI unit system, conversion between units is not as simple as the metric system. For example, one pound is 16 ounces and one mile is 5,280 feet.

B.1. Units

B.1.1. Temperature

Polish scientist Daniel Fahrenheit lived most of his life in The Netherlands and invented the Fahrenheit temperature scale in 1724. However, all of Europe now uses the Celsius scale. In the Fahrenheit scale, water, which is the most abundant and arguably the most important liquid on earth, is defined to boil at 212°F and freeze at 32°F. In contrast, in the Celsius scale water is defined to boil at 100°C and freeze at 0°C. Conversion between the two scales is shown in equations (1) and (2). You may also be familiar with the term "Centigrade," which is equivalent to Celsius. The Celsius scale is named after Anders Celsius (1701–1744), a Swedish scientist.

$$°C = (°F - 32) * 5/9 \tag{1}$$

$$°F = (°C * 9/5) + 32 \tag{2}$$

Also, it is important to know that these are relative scales, as denoted by the degree sign. In most equations using temperature, such as the equation for the maximum efficiency of an engine, absolute temperature is used. Equations to convert either Celsius or Fahrenheit to their absolute temperature counterparts, Kelvin and Rankine, respectively, are shown below. Since these are absolute scales, no degree sign is used.

$$K = °C + 273.15 \tag{3}$$

$$R = °F + 459.67 \tag{4}$$

Incidentally, the concept and measurement of absolute zero is an interesting topic unto itself. It is a hypothetical temperature such that there is no molecular motion or heat. While some scientific studies suggest this may be an oversimplification, these are beyond the scope of this book.

B.1.2. Energy

For the purpose of this book, energy is defined as the ability to do work. Work is force acting through distance. In the SI or metric system, force is a Newton, with units of kg-m/s^2. Therefore, the SI unit for energy is the work done when a force of one Newton acts through a distance of one meter, or a Joule (J). A Joule has the units of kg-m^2/s^2. Because a unit like the Joule can be expressed in terms of the base units of kg, m, and s, it is often referred to as a "derived unit."

More commonly known units of energy are the calorie and the British Thermal Unit, or BTU. These two units are based on the energy to heat water. A calorie is the energy needed to raise the temperature of 1 gram of water by 1°C or 1K.

$$1 \text{ Calorie} = 4.184 \text{ J} \tag{5}$$

Similarly, the BTU is the energy needed to raise the temperature of 1 pound of water by 1°F or 1R.

$$1 \text{ BTU} = 1.0548 \times 10^3 \text{J} \tag{6}$$

Thus, for the three energy types, one is a "derived unit" while the other two are based on the energy needed to raise the temperature of water.

In order to compare technologies used for transportation fuels with those used for electricity generation, it is also useful to convert BTU units to Watt-hours, or Wh. A Watt is also a "derived unit" with units of kg-m^2/s^3. Given the units for one Joule, we can see that one Watt is one Joule per second. While the Watt is a unit of power, the amount of energy per unit time, the Wh is a unit of energy. The conversion from BTU to Wh is shown here:

$$1 \text{ BTU} = 0.293071 \text{ Wh} \tag{7}$$

In the case of nuclear energy, it is also useful to be familiar with the electron volt, or "eV." It is the amount of energy gained or lost by a single electron charge moving across an electric potential difference of one volt. It is a common unit used in physics and, by definition, is an empirical unit of measurement, such as the BTU and calorie. This unit is useful because of the small amounts of energy involved at an atomic or molecular level, making it more practical to use than other units of energy. Equation (8) shows the conversion of BTUs to eV.

$$1 \text{ BTU} = 6.585 \times 10^{21} \text{eV} \tag{8}$$

B.1.3. Mass

The SI unit of mass is the kilogram (kg), and the metric tonne (t) is 1,000 kg.

$$1 \text{ tonne} = 1{,}000 \text{ kg} \tag{9}$$

The English system of mass is the pound and the abbreviation, lb, comes from the Latin *libra pondo*. The English system ton, or short ton, is 2,000 lb.

$$1 \text{ ton} = 2{,}000 \text{ lb} \tag{10}$$

The following equation converts pounds to kilograms.

$$1 \text{ lb} = 0.45359 \text{ kg} \tag{11}$$

Using these conversion factors, 1 metric tonne is 2,204.6 lb and 1 English ton is 907.2 kg. Therefore, a metric tonne is about 10% larger than an English ton.

Causing even greater confusion, the British measurement system often uses the term “long ton,” which is 2,240 pounds or 160 stones, with a “stone” defined as 14 pounds. The US ton, defined as 2,000 pounds, is sometimes referred to as a “short ton.” In the absence of the preceding word “metric,” masses can also be differentiated by the different spelling of “ton” and “tonne.”

B.1.4. Length, Area, and Volume

The SI unit of length is meter (m), while the common English system units of length are feet (ft) and miles (mi). To convert between feet and meters, use the following equation.

$$1 \text{ ft} = 0.3048 \text{ m} \tag{12}$$

One kilometer (km) is 1,000 m. Since there are 5,280 ft per mile, Equation (12) can be used to calculate that one mile is 1.609 km.

$$1 \text{ mi} = 1.609 \text{ km} \tag{13}$$

For area and volume, square feet or cubic feet can be easily converted to square meters or cubic meters.

$$1 \text{ ft}^2 = 0.092903 \text{ m}^2 \tag{14}$$

$$1 \text{ ft}^3 = 0.028317 \text{ m}^3 \tag{15}$$

B.1.5. Power

Power is a unit of energy per unit time and the SI unit for power is the Watt (W), which is one Joule per second. Although there are many other units for power, such as BTU per second or horsepower, the Watt is the commonly used unit for power in the electrical power generation world. Like the SI unit for energy, the Joule, the Watt is a “derived unit.” Since the Joule has units of kg-m^2/s^2, the derived units for the Watt are shown here:

$$1 \text{ W} = 1 \text{ J/s} = 1 \text{ kg} \cdot \text{m}^2/\text{s}^3 \tag{16}$$

The common unit of measurement for an electrical plant is the kilowatt (kW) and kilowatt-hours (kWh). A kW is 1,000 Watts.

$$1\ \text{kW} = 1{,}000\ \text{W} \tag{17}$$

As there are 3,600 seconds in an hour and 1,000 W in a kW, multiplying 3,600 by 1,000 gives 3.6×10^6 Joules, the amount of energy in a kWh.

$$1\ \text{kWh} = 3.6 \times 10^6\ \text{J} \tag{18}$$

B.2. Unit Prefixes

Depending on the context, unit prefixes and scientific notation are used to simplify expressions for units of energy, length, mass, and the like. For example, if you are talking about how much electricity you consumed last month, your bill will likely show units of kilowatt hours (kWh). However, if your local electric plant has the capacity of one billion watts, it is cumbersome to describe this as 1,000,000 kilowatts or 1,000,000,000 Watts. And, this becomes even more unwieldy if you are calculating the amount of incoming solar radiation or the tons of coal in the world. Therefore, the science and engineering world commonly uses prefixes and scientific notation. The conversion between prefixes and scientific notation is shown here:

$$\text{Kilo (k)} = 1\ \text{thousand},\ 1 \times 10^3 \tag{19}$$

$$\text{Mega (M)} = 1\ \text{million},\ 1 \times 10^6 \tag{20}$$

$$\text{Giga (G)} = 1\ \text{billion},\ 1 \times 10^9 \tag{21}$$

$$\text{Tera (T)} = 1\ \text{trillion},\ 1 \times 10^{12} \tag{22}$$

$$\text{Peta (P)} = 1\ \text{quadrillion},\ 1 \times 10^{15} \tag{23}$$

If your local electric plant has the capacity of 1,000,000,000 Watts, it can be described a one GW plant, a one billion W plant, or a plant with a capacity of 1×10^9 W. As will become evident from the different chapters on energy, the terms used depend on the different energy fields. Thus, while the electricity industry typically describes energy using prefixes, such as MWh (megawatt-hour), the transportation industry typically uses terms like one billion BTU or 1×10^9 BTU.

B.3. Basic Definitions

There is no perfect way to compare or define different energy types. Certainly, comparing the energy content of a crude oil with a power plant poses some difficulties, as there are many different types and qualities of crude oil and since the energy produced is used for different end purposes, primarily transportation and electrical power generation, respectively. Nevertheless, standard values have been adopted for different gaseous, liquid, and solid fuels, and these are shown at the end of this section. While different types of crude oil do not have exactly 5.8×10^6 BTU of energy, the small differences are not enough to change the general comparisons made in this book.

Defining the energy of coal is a more difficult task than for crude oil or gasoline, since there are four different types of coal with different compositions, namely lignite, sub-bituminous, bituminous, and anthracite. In addition, petroleum coke, generated at refineries, is also used for electrical power generation.

$$1 \text{ Barrel Crude Oil (42 gallons)} = 5{,}800{,}000 \text{ BTU} \quad (24)$$

$$1 \text{ ft}^3 \text{natural gas} = 1{,}032 \text{ BTU or } 22{,}792 \text{ BTU/lb} \quad (25)$$

$$1 \text{ Gallon Crude Oil} = 138{,}095 \text{ BTU} \quad (26)$$

$$1 \text{ Gallon Propane} = 91{,}033 \text{ BTU} \quad (27)$$

$$1 \text{ Gallon Gasoline} = 124{,}238 \text{ BTU} \quad (28)$$

$$1 \text{ Gallon Diesel} = 138{,}690 \text{ BTU} \quad (29)$$

$$1 \text{ Gallon Residual Fuel Oil} = 149{,}691 \text{ BTU} \quad (30)$$

$$1 \text{ Gallon Ethanol} = 75{,}700 \text{ BTU} \quad (31)$$

$$1 \text{ ton Coal} = 16{,}000{,}000 \text{ to } 30{,}000{,}000 \text{ BTU} \quad (32)$$

$$1 \text{ ton Bituminous coal} = 25{,}424{,}000 \text{ BTU (12,712 BTU/lb)} \quad (33)$$

$$1 \text{ cubic foot (ft}^3\text{) Hydrogen} = 780 \text{ BTU} \quad (34)$$

$$1 \text{ atom U-235} = 202.5 \text{ MeV or } 83.14 \text{ TJ/kg } (83.14 \times 10^{12} \text{ J/kg}) \quad (35)$$

$$1 \text{ lb U-235} = 35.74 \text{ G BTU } (35.74 \times 10^{9} \text{ BTU/lb}) \quad (36)$$

Index